Atomic Masses of the Elements

Name	Symbol	Atomic Number	Atomic Mass[a]	Name	Symbol	Atomic Number	Atomic Mass[a]
Actinium	Ac	89	(227)	Molybdenum	Mo	42	95.94
Aluminum	Al	13	26.98	Neodymium	Nd	60	144.2
Americium	Am	95	(243)	Neon	Ne	10	20.18
Antimony	Sb	51	121.8	Neptunium	Np	93	(237)
Argon	Ar	18	39.95	Nickel	Ni	28	58.69
Arsenic	As	33	74.92	Niobium	Nb	41	92.91
Astatine	At	85	(210)	Nitrogen	N	7	14.01
Barium	Ba	56	137.3	Nobelium	No	102	(259)
Berkelium	Bk	97	(247)	Osmium	Os	76	190.2
Beryllium	Be	4	9.012	Oxygen	O	8	16.00
Bismuth	Bi	83	209.0	Palladium	Pd	46	106.4
Bohrium	Bh	107	(264)	Phosphorus	P	15	30.97
Boron	B	5	10.81	Platinum	Pt	78	195.1
Bromine	Br	35	79.90	Plutonium	Pu	94	(244)
Cadmium	Cd	48	112.4	Polonium	Po	84	(209)
Calcium	Ca	20	40.08	Potassium	K	19	39.10
Californium	Cf	98	(251)	Praseodymium	Pr	59	140.9
Carbon	C	6	12.01	Promethium	Pm	61	(145)
Cerium	Ce	58	140.1	Protactinium	Pa	91	231.0
Cesium	Cs	55	132.9	Radium	Ra	88	(226)
Chlorine	Cl	17	35.45	Radon	Rn	86	(222)
Chromium	Cr	24	52.00	Rhenium	Re	75	186.2
Cobalt	Co	27	58.93	Rhodium	Rh	45	102.9
Copper	Cu	29	63.55	Roentgenium	Rg	111	(272)
Curium	Cm	96	(247)	Rubidium	Rb	37	85.47
Darmstadtium	Ds	110	(271)	Ruthenium	Ru	44	101.1
Dubnium	Db	105	(262)	Rutherfordium	Rf	104	(261)
Dysprosium	Dy	66	162.5	Samarium	Sm	62	150.4
Einsteinium	Es	99	(252)	Scandium	Sc	21	44.96
Erbium	Er	68	167.3	Seaborgium	Sg	106	(266)
Europium	Eu	63	152.0	Selenium	Se	34	78.96
Fermium	Fm	100	(257)	Silicon	Si	14	28.09
Fluorine	F	9	19.00	Silver	Ag	47	107.9
Francium	Fr	87	(223)	Sodium	Na	11	22.99
Gadolinium	Gd	64	157.3	Strontium	Sr	38	87.62
Gallium	Ga	31	69.72	Sulfur	S	16	32.07
Germanium	Ge	32	72.64	Tantalum	Ta	73	180.9
Gold	Au	79	197.0	Technetium	Tc	43	(98)
Hafnium	Hf	72	178.5	Tellurium	Te	52	127.6
Hassium	Hs	108	(269)	Terbium	Tb	65	158.9
Helium	He	2	4.003	Thallium	Tl	81	204.4
Holmium	Ho	67	164.9	Thorium	Th	90	232.0
Hydrogen	H	1	1.008	Thulium	Tm	69	168.9
Indium	In	49	114.8	Tin	Sn	50	118.7
Iodine	I	53	126.9	Titanium	Ti	22	47.87
Iridium	Ir	77	192.2	Tungsten	W	74	183.8
Iron	Fe	26	55.85	Uranium	U	92	238.0
Krypton	Kr	36	83.80	Vanadium	V	23	50.94
Lanthanum	La	57	138.9	Xenon	Xe	54	131.3
Lawrencium	Lr	103	(260)	Ytterbium	Yb	70	173.0
Lead	Pb	82	207.2	Yttrium	Y	39	88.91
Lithium	Li	3	6.941	Zinc	Zn	30	65.41
Lutetium	Lu	71	175.0	Zirconium	Zr	40	91.22
Magnesium	Mg	12	24.31	—	—	112	(285)
Manganese	Mn	25	54.94	—	—	113	(284)
Meitnerium	Mt	109	(268)	—	—	114	(289)
Mendelevium	Md	101	(258)	—	—	115	(288)
Mercury	Hg	80	200.6				

[a] Values in parentheses are the mass number of the most stable isotope.

The Chemistry Place

Log on.

Tune in.

Succeed.

To help you succeed in Chemistry, your professor has arranged for you to enjoy access to great media resources, on an interactive CD-ROM and on The Chemistry Place website. You'll find that these resources that accompany your textbook will enhance your course materials.

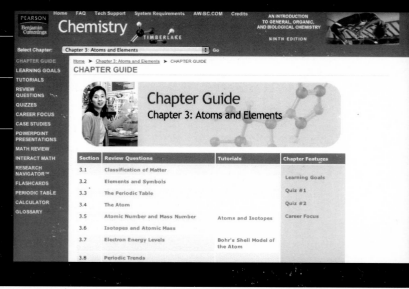

What your system needs to use these media resources:

WINDOWS™
- Windows 98, 2000, XP
- Pentium 366 MHz or higher
- Minimum 64 MB of RAM
- Screen resolution 800x600 or higher
- Minimum 64 MB RAM
- Internet Explorer 5.5, 6.0, Netscape 7.0

MACINTOSH™
- Mac 9.2, 10.2, 10.3
- G3 or higher Macintosh required
- Minimum 64 MB available RAM
- Screen resolution 800x600 or higher
- Internet Explorer 5.2, Netscape 7.0, Safari 1.2

SOFTWARE/PLUG-IN DOWNLOADS
- Macromedia Shockwave 8.0 and Flash 6.0 or later
- MDL Chime

Got technical questions?
Call 1-800-677-6337. Phone support is available 8:00am – 8:00pm Eastern, Monday-Friday and 5:00pm - 12:00am (midnight) Eastern, Sunday.

Visit http://247.aw.com. Email technical support is available 24/7.

Here's your personal ticket to success:

How to log on to www.aw-bc.com/chemplace

1. Go to www.aw-bc.com/chemplace
2. Click *Chemistry: An Introduction to General, Organic, and Biological Chemistry*, Ninth Edition.
3. Click "Register."
4. Scratch off the silver foil coating below to reveal your pre-assigned access code.
5. Enter your pre-assigned access code exactly as it appears below.
6. Complete the online registration form to create your own personal user Login Name and Password.
7. Once your personal Login Name and Password are confirmed by email, go back to www.aw-bc.com/chemplace, click *Chemistry: An Introduction to General, Organic, and Biological Chemistry*, Ninth Edition, type in your new Login Name and Password, and click "Log In."

Your Access Code is:

USTC-QESHM-BAYLE-COREO-PRATO-MOLES

If there is no silver foil covering the access code above, the code may no longer be valid. In that case, you need to purchase access online using a major credit card. Go to www.aw-bc.com/chemplace. Click *Chemistry: An Introduction to General, Organic, and Biological Chemistry*, Ninth Edition, and click Buy Now.

Chemistry

AN INTRODUCTION
TO GENERAL,
ORGANIC, AND
BIOLOGICAL
CHEMISTRY

NINTH EDITION

Chemistry

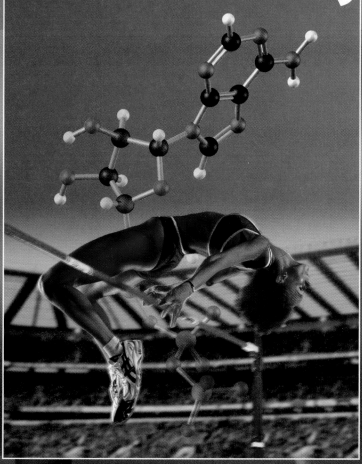

AN INTRODUCTION
TO GENERAL,
ORGANIC, AND
BIOLOGICAL
CHEMISTRY

NINTH EDITION

■

KAREN C.
TIMBERLAKE

PEARSON

Benjamin
Cummings

San Francisco   Boston   New York
Capetown   Hong Kong   London
Madrid   Mexico City   Montreal
Munich   Paris   Singapore
Sydney   Tokyo   Toronto

Publisher:	Jim Smith
Project Editor:	Lisa Leung
Senior Marketing Manager:	Scott Dustan
Senior Production Supervisor:	Corinne Benson
Production Editor:	Jean Lake
Compositor:	Progressive Information Technologies
Illustrator:	J.B. Woolsey Associates, LLC
Senior Manufacturing Buyer:	Michael Early
Text and Cover Designer:	Seventeenth Street Studios
Text Printer and Binder:	Courier Kendallville
Cover Printer:	Phoenix Color Corporation
Cover Photo Credit:	Alan Thornton/Stone/Getty Images
Cover Molecular Art:	J.B. Woolsey Associates, LLC

Library of Congress Cataloging-in-Publication Data

Timberlake, Karen.
  Chemistry: an introduction to general, organic, and biological chemistry / Karen C.
Timberlake.—9th ed.
    p. cm.
Includes index.
ISBN 0-8053-3015-1
  1. Chemistry—Textbooks. I. Title.

QD31.3.T55 2006
540—dc22

                                                                        2005047585

ISBN 0-8053-3015-1 (Student)
ISBN 0-8053-3028-3 (Instructor)

PEARSON

Benjamin
Cummings   **www.aw-bc.com/chemistry**   2 3 4 5 6 7 8 9 10-CRK-09 08 07 06 05

Brief Contents

Contents

9 Nuclear Radiation   299

10 Introduction to Organic Chemistry: Alkanes   331

11 Unsaturated Hydrocarbons 361

12 Organic Compounds with Oxygen and Sulfur 384

16 Amino Acids, Proteins, and Enzymes   532

17 Nucleic Acids and Protein Synthesis   573

18 Metabolic Pathways and Energy Production 607

Applications and Activities

CAREER FOCUS

ENVIRONMENTAL NOTE

About the Author

The whole art of teaching is only the art of awakening the natural curiosity of young minds.
—ANATOLE FRANCE

One must learn by doing the thing; though you think you know it, you have no certainty until you try.
—SOPHOCLES

Discovery consists of seeing what everybody has seen and thinking what nobody has thought.
—ALBERT SZENT-GYORGI

Karen Timberlake is professor emeritus of chemistry at Los Angeles Valley College, where she taught chemistry for allied health and preparatory chemistry for 36 years. She received her bachelor's degree in chemistry from the University of Washington and her Master's degree in biochemistry from the University of California at Los Angeles.

Professor Timberlake has been writing chemistry textbooks for 30 years. During that time, her name has become associated with the strategic use of pedagogical tools that promote student success in chemistry and the application of chemistry to real-life situations. More than one million students have learned chemistry using texts, laboratory manuals, and study guides written by Karen Timberlake. In addition to *Basic Chemistry*, she is also the author of *General, Organic, and Biological Chemistry, Structures of Life* and *Chemistry: An Introduction to General, Organic, and Biological Chemistry, Eighth Edition* with the accompanying *Study Guide with Solutions for Selected Problems, Laboratory Manual*, and *Essentials Laboratory Manual*.

Professor Timberlake belongs to numerous science and educational organizations including the American Chemical Society (ACS) and the National Science Teachers Association (NSTA). In 1987, she was the Western Regional Winner of Excellence in College Chemistry Teaching Award given by the Chemical Manufacturers Association. In 2004, she received the McGuffey Award in Physical Sciences by the Textbook Author Association, awarded for textbooks whose excellence has been demonstrated over time. She has participated in education grants for science teaching including the Los Angeles Collaborative for Teaching Excellence (LACTE) and a Title III grant at her college. She often speaks at conferences and educational meetings on the use of student-centered teaching methods in chemistry to promote the learning success of students.

Her husband, Bill, is also a professor emeritus of chemistry and contributes much to the preparation of the textbooks. When the Professors Timberlake are not writing textbooks, they relax by hiking, traveling, trying new restaurants, and playing lots of tennis. The Timberlakes' son, John, and daughter-in-law, Cindy, are involved in preparing materials for English language learners. Their grandson, Daniel, and granddaughter, Emily, don't know what they want to be yet.

TO THE STUDENT

I hope that this textbook helps you discover exciting new ideas and gives you a rewarding experience as you develop an understanding and appreciation of the role of chemistry in your life. If you would like to share your experience with chemistry, or have questions and comments about this text, I would appreciate hearing from you.

Karen Timberlake
Email: khemist@aol.com

Preface

Welcome to the ninth edition of *Chemistry: An Introduction to General, Organic, and Biological Chemistry.* This is a chemistry text designed to help students prepare for careers in health-related professions, such as nursing and respiratory therapy, and a text that assumes no prior knowledge of chemistry by the student. This new edition introduces a new problem-solving strategy and new conceptual and teamwork problems.

This text contains fundamental topics of general, organic, and biological chemistry. My goal is to provide a learning environment that makes the study of chemistry an engaging and positive experience. It is also my goal to help every student become a critical thinker by understanding scientific concepts that will form a basis for making important decisions about issues concerning health and the environment. Thus, I have utilized materials that

- Help students to develop problem-solving skills that lead to success in chemistry;
- Motivate students to learn and enjoy chemistry;
- Relate chemistry to careers that interest students; and
- Provide pedagogy that promotes learning.

Features of the Text

Students are often challenged by the study of chemistry and relevance of chemistry to their career paths. A common view of chemistry is that it is just a lot of facts to be memorized. To change this perception, I have included many features to give students an appreciation of chemistry in their lives and the skills to learn chemistry successfully. These pedagogical features include connections to real life, a visual guide to problem solving, and in-chapter problem sets that immediately reinforce the learning of a small group of new concepts. A successful learning program in each chapter provides many learning tools, which are discussed below.

Interviews with Health Care Professionals   Chapter openers begin with interviews with professionals in the health field who discuss the importance of chemistry in their careers.

Learning Goals Directed Learning   Learning Goals in every section preview the concepts the student is to learn.

Connections to Real Life   Health Notes and Environmental Notes throughout the text relate chemistry chapters to real-life topics in science and medicine that are interesting and motivating to students and support the role of chemistry in the real world. Explore Your World Features allow students to explore chemistry topics using everyday materials. The Career Focus interviews provide additional examples of professionals using chemistry.

Guide to Problem Solving Visualizations   As part of a comprehensive learning program, abundant Sample Problems and Study Checks in every chapter model successful problem-solving techniques for the student. A unique Guide to Problem Solving (GPS) strategy illustrates solutions to problems with color blocks that visually guide students through the solution pathway.

Question and Problem Sets in Every Section   Every section includes a comprehensive set of Questions and Problems that encourage students to apply critical thinking and problem solving to a small set of concepts. Within each problem set, the questions are in order of simple to more complex to enable students to build a knowledge base for problem solving. By solving a set of problems after studying each section, students immediately reinforce newly learned concepts rather than waiting until they get to the end of the chapter. Students are encouraged to be active learners by working problems as they progress through each chapter. Answers for all the Study Checks are located at the end of each chapter.

Matched Pairs of Problems   All the questions and problems in the text are written as pairs of matched problems. Thus, each question or problem is followed with a similar type of question or problem. Answers for the odd-numbered problems are located at the end of each chapter rather than at the end of the text, which gives immediate feedback to problem-solving efforts by students. The even-numbered problems do not have answers, allowing instructors to use them for homework and/or quiz questions.

End-of-Chapter Questions   Additional Questions and Problems also found at the end of the chapter provide more in-depth questions that integrate the topics from the entire chapter to promote further study and critical thinking. A new set of problems, Understanding the Concepts, encourages students to think about the concepts they have learned. Questions for Teamwork, a new feature, found at the end of each chapter can be utilized in cooperative learning environments for group work. The *Instructor's Manual* has the answers and solutions to all the problems in the text including Questions for Teamwork.

Macro-to-Micro Art Illustrations of Atomic Organization   Throughout the text, a robust and vibrant art program visually connects the real-life world of materials familiar to students with their atomic-level structures. Macro-to-micro illustrations show students that everyday things have an atomic level of organization and structure that determines their behavior and functions. Every figure in the text also contains a question that encourages the student to study that figure and relate the visual representation to the content in the text. The plentiful use of three-dimensional structures takes the descriptions of molecules to a visual level, which stimulates the imagination and aids the understanding of structures by the student.

End-of-Chapter Aids   A concept map connects and guides students through all the topics and concepts in the chapter. Chapter Reviews give a brief overview of the important concepts in each section of the chapter. Key Terms remind students of the new vocabulary presented in each chapter.

Chapter Organization of the Ninth Edition

Prologue
This new section, which precedes Chapter 1, introduces students to the concepts of chemicals and chemistry, discusses the scientific method, and asks students to develop a study plan for learning chemistry.

Chapter 1   Measurements
Chapter 1 looks at measurement and the need to understand numerical structures of the metric system in the sciences. An explanation of scientific notation and working with a

calculator has been incorporated into the chapter. The section on Problem Solving has been divided into two sections: *Writing Conversion Factors* and *Problem Solving*. New to this edition are conversion factors stated within a problems and percentage. A new problem feature, Guide to Problem Solving (GPS), uses color blocks that visually guide students through the solution pathway. Along with word maps of the problem-solving strategy, the GPS system provides students a visual format for understanding problem solving. Temperature has been moved to Chapter 2 with the energy discussion.

Chapter 2   Energy and Matter
Chapter 2 is now the discussion of energy and matter, moved from Chapter 6. The chapter looks at the topics of energy and heat, nutritional energy values, temperature conversions, states of matter, and energy involved in changes of state. More energy calculations now use the SI unit of the joule (J). Sample Problems were reworked to utilize the Guide to Problem Solving strategy. A Health Note on *Steam Burns* was added.

Chapter 3   Atoms and Elements
In Chapter 3, we look at elements, atoms, and subatomic particles. This chapter begins with a new discussion of the classification of matter including pure substances, mixtures, and types of mixtures. *The Periodic Table* emphasizes the numbering of groups from 1–18. The elements Darmstadtium 110 and Roentgenium 111 were added. *Energy-Level Changes* are now discussed. A new section *Periodic Trends* has been added that discusses periodic properties of elements including valence electrons, atomic size, and ionization energy. The Health Note *Chemical Sense of Taste and Smell* was deleted.

Chapter 4   Compounds and Their Bonds
With the discussion of valence electrons in Chapter 3, this chapter now begins with *Octet Rule and Ions*. Chapter 4 describes how the structures of atoms lead to ionic and covalent bonding. The use of electron-dot formulas illustrates the three-dimensional shapes of molecules and ions and provides a basis for the shape of organic and biochemical compounds. Electronegativity leads to a discussion of the polarity of bonds and molecules. The values for electronegativity are now included in *Electronegativity and Bond Polarity*.

Chapter 5   Chemical Reactions and Quantities
In Chapter 5, we look at the interaction of atoms and molecules in chemical reactions. Chemical equations are balanced and organized into combination, decomposition, and replacement reactions as well as oxidation and reduction. Students

learn patterns of chemical behavior as they organize and recognize the types of chemical reactions. With a basic understanding of equations, we can then look at some quantitative aspects of reactions including the mole and mass calculations. *The Mole* introduces the counting of particles, and new calculations of atoms and molecules in a mole have been added. In *Mole Relationships in Chemical Equations*, the concept of *Conservation of Mass* helps students calculate the mole and mass relationships among the reactants and products. The section on *Mass Calculations for Reactions* includes more quantitative problems with equations. This chapter now includes *Energy in Chemical Reactions*, which discusses activation energy and heat of reaction for exothermic and endothermic reactions.

Chapter 6   Gases

In Chapter 6 visual problem-solving strategies enhance the discussion and calculations with gas laws. New in this ninth edition are calculations for quantities of gases in chemical reactions as well as new problem sets Understanding the Concepts and Questions for Teamwork.

Chapter 7   Solutions

Chapter 7 introduces the student to important topics in healthcare: solutions, solubility, electrolytes, concentrations, and osmosis and dialysis. New problem-solving strategies clarify the use of concentrations to determine volume or mass of solute needed. A new section, *Solutions in Chemical Reactions,* has been added that gives problems using the volumes and molarities of solutions to calculate reactants and products in chemical reactions. The sections *Colloids and Suspensions* and *Osmosis and Dialysis* have been combined into a section titled *Properties of Solutions.*

Chapter 8   Acids and Bases

Chapter 8 discusses acids and bases and their strengths, conjugate acid–base pairs, the ionization of water, pH, and buffers. The section *Reactions of Acids and Bases* now includes acid–base titration using the neutralization reactions between acids and bases to calculate quantities of acid in a sample.

Chapter 9   Nuclear Radiation

This chapter, which was Chapter 3 in the eighth edition, studies the behavior of radioactive atoms and the type of radioactive particles that are emitted from the nuclei of radioactive atoms. Equations are written and balanced for both naturally occurring radioactivity and artificially produced radioactivity. The half-lives of radioisotopes are discussed and the amount of time for a sample to decay is calculated. Radioisotopes, important in the field of nuclear medicine, are described. The new Guide to Problem Solving illustrates the balancing of nuclear equations and new art clarifies the changes in the nucleus when radiation is emitted.

Chapter 10
Introduction to Organic Chemistry: Alkanes

This chapter introduces organic compounds beginning with the alkane family. The section *Bonding in Organic Compounds* has been combined with *Organic Compounds*. The section *Haloalkanes* has been shortened and is now part of *Alkanes with Substituents*. The visual GPS guides have been added to clarify the rules for nomenclature. The chapter concludes with the introduction of *Functional Groups* for each family of organic compounds, which forms a basis for understanding the biomolecules of living systems.

Chapter 11   Unsaturated Hydrocarbons

This chapter discusses alkenes and alkynes, cis-trans isomers, addition reactions, polymers of alkenes, and aromatic compounds. The Explore Your World feature *Modeling Cis-Trans Isomers* was rewritten. The reactions in *Additional Reactions* are hydrogenation and hydration, which are important in biological systems. Synthetic polymers that are used in everyday items and their recycling symbols are discussed. Nomenclature of aromatic compounds was simplified.

Chapter 12
Organic Compounds with Oxygen and Sulfur

This chapter discusses alcohols, ethers, thiols, aldehydes, and ketones. The section *Ethers* is now included in *Alcohols, Ethers, and Thiols*. The sections *Important Alcohols and Phenols* and *Some Important Aldehydes and Ketones* are now Health Notes. The final section, *Chiral Molecules,* uses simple compounds to introduce chiral molecules and chirality early in the text. I revisit chirality in carbohydrates and amino acids to show the concept's importance in the nature of biomolecules, chiral drugs, and their behavior in living systems.

Chapter 13
Carboxylic Acids, Esters, Amines, and Amides

This chapter completes the organic sequence. The families and reactions that are discussed are those that are most applicable to biochemical systems. The section *Acidity of Carboxylic Acids* was combined with *Properties of Carboxylic Acids*, and *Amines* now includes the section *Properties of Amines*. The nomenclature of amines and amides has been simplified.

Chapter 14   Carbohydrates

The study of biomolecules begins with carbohydrates, which contain some of the functional groups discussed in previous chapters. To the student, the application of organic chemistry to biochemistry is often the most exciting part of their study of chemistry because of its relationship to health and medicine. The section *Classification of Monosaccharides* was combined with *Carbohydrates*. The *D and L Notations from Fischer Projections* is now *Structures of Monosaccharides* and includes *Some Important Monosaccharides*.

Chapter 15   Lipids

This chapter continues our study of biomolecules in the lipid family including fatty acids, triacylglycerols, glycerophospholipids, and steroids. Structures for fatty acids now include both the formula and the line-bond formula. Health Notes of interest to students include olestra, trans fatty acids, and lipoproteins. We discuss the role of lipids in cell membranes as well as the lipids that function as steroid hormones. In the new Questions for Teamwork, students are asked to do calculations for the hydrogenation and saponification of a triacylglycerol.

Chapter 16   Amino Acids, Proteins, and Enzymes

This chapter discusses amino acids, formation of proteins, structural levels of proteins, enzymes, and enzyme action. Amino acids are written as their ionized forms in physiological solutions. *Levels of Protein Structure* describes the importance of protein structure at the primary, secondary, tertiary, and quaternary levels. Enzymes are discussed as biological catalysts and we see that proteins change shape and lose activity with changes in pH and temperature. Finally, we discuss the impact of substrate concentration, cofactors, and inhibitors on enzyme action.

Chapter 17   Nucleic Acids and Protein Structure

The discussion on nucleic acids is similar to the eighth edition. The components and structures of nucleic acids are described. The role of complementary base pairing is highlighted in both DNA replication and the formation of mRNA during protein synthesis. We look at the genetic code and its relationship to the order of amino acids in a protein and how mutations can occur when the nucleotide sequence is altered. In *Viruses*, we look at how DNA or RNA in viruses utilizes host cells to produce more viruses.

Chapter 18
Metabolic Pathways and Energy Production

In this chapter, we follow the metabolic pathways of biomolecules from the digestion of foodstuffs to the synthesis of ATP. Students look at the stages of metabolism and the digestion of carbohydrates along with the coenzymes required in metabolic pathways. The breakdown of glucose to pyruvate is described using the glycolytic pathway, which is followed by the decarboxylation of pyruvate to acetyl CoA. We look at the entry of acetyl CoA into the citric acid cycle and the production of reduced coenzymes. We describe electron transport, oxidative phosphorylation, and the synthesis of ATP. The oxidation of lipids and the degradation of amino acids are also discussed.

Instructional Package

Chemistry: An Introduction to General, Organic, and Biological Chemistry is the nucleus of an integrated teaching and learning pack of support material for both students and professors.

For Students

Student Study Guide and Selected Solutions for *Chemistry* by Karen Timberlake is keyed to the learning goals in the text and designed to promote active learning through a variety of exercises with answers as well as mastery exams. The *Study Guide* also contains complete solutions to odd-numbered problems. (ISBN 0-8053-3024-0)

Special Edition of the Chemistry Place™ for *Chemistry*
www.aw-bc.com/chemplace

This website, tailored specifically to *Chemistry*, offers comprehensive and interactive teaching and learning resources for chemistry. Features include **tutorials**, **review questions**, **quizzes**, **career resources**, a **glossary**, **flashcards**, an **interactive periodic table**, and **case studies** that show how chemical concepts apply to familiar situations. Over 2000 **PowerPoint® slides**—complete with art, photos, and tables from the text—teach key concepts from every section of the book. With **Research Navigator™** students can find topical news and journal articles in *The New York Times* **Search by Subject™ Archive** and **EBSCO ContentSelect™**. **InterAct Math** provides interactive exercises that allow the student to review and build the math skills needed in chemistry. Access to the Chemistry Place is packaged with every new copy of the book.

Laboratory Manual   by Karen Timberlake. This best-selling lab manual coordinates 42 experiments with the topics in *Chemistry*, uses new terms during the lab, and explores chemical concepts. Laboratory investigations develop skills of manipulating equipment, reporting data, solving problems, making calculations, and drawing conclusions. (0-8053-3025-9)

Essentials Laboratory Manual   by Karen Timberlake contains 25 experiments for the standard course sequence of topics. Features safety procedures, pre-lab questions, and repost pages. (0–8053–3023–2)

The Chemistry Tutor Center
www.aw-bc.com/tutorcenter

The Tutor Center provides one-to-one tutoring by internet, phone, fax, and email. Qualified instructors answer questions to help students understand concepts and solve problems in the textbook.

Media Grid, Walkthrough, and Media Icons

Located in the front of the book, the Media Grid correlates all the media available on the ChemistryPlace website with the content in *Chemistry*. The Walkthrough in the front of the book helps students and instructors find the book's features easily. Media Icons in the margins throughout the text direct students to tutorials and case studies on The Chemistry-Place™ website for *Chemistry*.

For Instructors

Instructor's Manual and Complete Solutions by Karen Timberlake, highlights chapter topics, includes answers and solutions for all problems in the text, and contains suggestions for the laboratory. (0-8053-3026-7)

Also visit the Benjamin-Cummings catalog page for Timberlake's *Chemistry*, ninth edition, at **www.aw-bc.com** to download available instructor supplements.

Online Instructor's Manual to Laboratory Manual contains answers to report pages for the *Laboratory Manual* and *Essential Laboratory Manual*. **www.aw-bc.com**.

Printed Test Bank (0-8053-3030-5)
Computerized Test Bank contains over 1500 questions in multiple-choice, matching, true/false, and short-answer format. (0-8053-3016-x)

The Chemistry Place™, Special Edition for *Chemistry*, includes many online resources for instructors, such as art from the textbook and downloadable PowerPoint® presentations. The online syllabus manager allows instructors to post class notes and assignments on the Web.

Transparency Acetates A set of 200 color transparencies are available. (0-8053-3029-1)

Image Presentation Library CD-ROM This CD-ROM includes every figure and table from the book in high resolution (150 dpi) format. Downloadable, customizable PowerPoint® lecture outlines, featuring over 2000 slides, are also available on the CD-ROM. (0-8053-3027-5)

Acknowledgments

The preparation of a new edition is a continuous effort of many people. As in my work on other textbooks, I am thankful for the support, encouragement, and dedication of many people who put in hours of tireless effort to produce a high quality book that provides an outstanding learning package. Once again the editorial team at Benjamin Cummings has done an exceptional job. I appreciate the work of my publisher, Jim Smith, who supported my vision of this ninth edition text with a new visual learning strategy and art program. My project editor, Lisa Leung, continually encouraged me at each step during the development of this edition, while skillfully coordinating reviews, art, website materials, and all the things it takes to make a book come together. Lisa also worked carefully and swiftly to complete all the supplements and media materials that accompany the text. Corinne Benson, production supervisor, and Jean Lake, production editor, brilliantly coordinated all phases of the manuscript to the final pages of a beautiful book. Cathy Cobb, copyeditor, precisely edited the manuscript to make sure the words were correct to help students learn chemistry.

I am especially proud of the art program in this text, which lends beauty and understanding to chemistry. I would like to thank Mark Ong, art director, and Randall Goodall, book designer, whose creative ideas provided beautiful art and an outstanding design for the cover and pages of the book. Laura Murray, photo researcher, was invaluable in researching and selecting vivid photos for the text so that students can see the beauty of chemistry. The macro-to-micro illustrations designed by J.B. Woolsey Associates give students visual impressions of the atomic and molecular organization of everyday things and are a fantastic learning tool. I want to thank Martha Ghent, proofreader, for the hours of proofreading all the pages. I also appreciate all the hard work in the field put in by the marketing team, Stacy Treco, director of marketing, and Scott Dustan, senior chemistry marketing manager. Without them, no one would know about this text.

This text also reflects the contributions of many professors who took the time to review and edit the manuscript and provide outstanding comments, help, and suggestions. I am extremely grateful to an incredible group of peers for their careful assessment of all the new ideas for the text; for their suggested additions, corrections, changes, and deletions; and for providing an incredible amount of feedback about the best direction for the text. In addition, I am grateful for the time scientists took to let us take photos and discuss their work with them. I admire and appreciate every one of you.

Reviewers

CAREER FOCUS AND REAL WORLD APPLICATIONS

THIS TEXT WAS DESIGNED TO HELP STUDENTS ATTAIN THEIR CAREER GOALS.

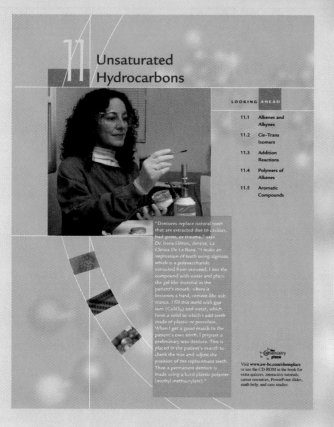

Chapter Opener

Each chapter begins with an interview with a professional in allied health or other relevant fields. These interviews illustrate how health professionals interact with chemistry in their careers.

Career Focus

Within the chapters are additional examples of professionals using chemistry.

■ Occupational Therapist

"Occupational therapists teach children and adults the skills they need for the job of living," says occupational therapist Leslie Wakasa. "When working with the pediatric population, we are crucial in educating children with disabilities, their families, caregivers, and school staff in ways to help them be as independent as they can be in all aspects of their daily lives. It's rewarding when you can show children how to feed themselves, which is a huge self-esteem issue for them. The opportunity to help people become more independent is very rewarding."

A combination of technology and occupational therapy helps children who are nonverbal to communicate and interact with their environment. By leaning on a red switch, Alex is learning to use a computer.

On the Web

The ChemistryPlace website features in-depth resources for each of the health professions featured in the book and takes students through interactive case studies.

Environmental Note

Environmental Notes delve into high interest issues such as global warming, radon, acid rain, pheromones, and pesticides.

ENVIRONMENTAL NOTE

■ Global Warming

The amount of carbon dioxide (CO_2) gas in our atmosphere is on the increase as we burn more gasoline, coal, and natural gas. The algae in the oceans and the plants and trees in the forests normally absorb carbon dioxide, but they cannot keep up with the continued increase. The cutting of trees in the rain forests (deforestation) reduces the amount of carbon dioxide removed from the atmosphere. Many of the trees are also burned as land is cleared. It has been estimated that deforestation may account for 15–30% of the carbon dioxide that remains in the atmosphere each year.

The carbon dioxide in the atmosphere acts like the glass in a greenhouse. When sunlight warms the Earth's surface, some of the heat is absorbed by carbon dioxide. It is not yet clear how severe the effects of global warming might be. Some scientists estimate that by around the year 2030, the atmospheric level of carbon dioxide could double

snow and ice could raise the ocean levels by as much as 2 m, which is enough to flood many cities located on the ocean shorelines.

Worldwide efforts are being made to reduce fossil fuel use and to slow or stop deforestation. It will require coopera-

HEALTH NOTE

■ Radiation and Food

Foodborne illnesses caused by pathogenic bacteria such as *Salmonella*, *Listeria*, and *Escherichia coli* have become a major health concern in the United States. The Centers for Disease Control and Prevention estimates that each year *E. coli* in contaminated foods infects 20,000 people in the United States, and that 500 people die. *E. coli* has been responsible for outbreaks of illness from contaminated ground beef, fruit juices, lettuce, and alfalfa sprouts.

The Food and Drug Administration (FDA) has approved the use of 0.3 kilogray (0.3 kGy) to 1 kGy of ionizing radiation produced by cobalt-60 or cesium-137 for the treatment of foods. The irradiation technology is much like that used to sterilize medical supplies. Cobalt pellets are placed in stainless steel tubes, which are arranged in racks. When food

(a)

(b)

Health Note

A rich array of **Health Notes** in each chapter apply chemical concepts to relevant topics of health and medicine. These topics include weight loss and weight gain, artificial fats, anabolic steroids, alcohol, genetic diseases, viruses, and cancer.

Explore Your World

Explore Your World includes hands-on activities that use everyday materials to encourage students to actively explore selected chemistry topics, either individually or in group-learning environments. Each activity is followed by questions to encourage critical thinking.

EXPLORE YOUR WORLD

■ Forming a Gas

Obtain baking soda and a jar or a plastic bottle. You will also need an elastic glove that fits over the mouth of the jar or a balloon that fits snugly over the top of the plastic bottle. Place a cup of vinegar in the jar or bottle. Sprinkle some baking soda into the fingertips of the glove or into the balloon. Carefully fit the glove or balloon over the top of the jar or bottle. Slowly lift the fingers of the glove or the balloon so that the baking soda falls into the vinegar. Watch what happens. Squeeze the glove or balloon.

QUESTION

1. Describe the properties of gas that you observe as the reaction takes place between vinegar and baking soda.
2. How do you know that a gas was formed?

KEEPING STUDENTS ENGAGED IS THE ULTIMATE GOAL

Learning Goals

At the beginning of each section, a **Learning Goal** clearly identifies the key concept of the section, providing a roadmap for studying. All information contained in that section relates back to the Learning Goal.

13.7 Draw the condensed structural formulas of each of the following carboxylic acids:
   a. propionic acid   **b.** benzoic acid
   c. chloroethanoic acid   **d.** 3-hydroxypropanoic acid
   e. butyric acid   **f.** heptanoic acid

13.8 Draw the condensed structural formulas of each of the following carboxylic acids:
   a. pentanoic acid   **b.** 3-ethylbenzoic acid
   c. 2-hydroxyacetic acid   **d.** 2,4-dibromobutanoic acid
   e. 3-methylbenzoic acid   **f.** hexanoic acid

■ 13.2 Properties of Carboxylic Acids

Carboxylic acids are among the most polar organic compounds because the functional group consists of two polar groups: a hydroxyl (—OH) group and a carbonyl (C=O) group. This C=O double bond is similar to that of the aldehydes and ketones.

Therefore, carboxylic acids form hydrogen bonds with other carboxylic acid molecules and water. This ability to form hydrogen bonds has a major influence on both their boiling points and solubility in water.

LEARNING GOAL

Describe the boiling points, solubility, and ionization of carboxylic acids in water.

WEB TUTORIAL
Carboxylic Acids

Writing Style

Karen Timberlake is known for her accessible writing style, based on a carefully paced and simple development of chemical ideas, suited to the background of allied health students. She precisely defines terms and sets clear goals for each section of the text. Her clear analogies help students to visualize and understand key chemical concepts.

Concept Maps

Each chapter ends with a **concept map** that reviews the key concepts of each chapter and how they fit together.

CONCEPT MAP

Art Program

The art program is not only beautifully rendered, but pedagogically effective as well.

FIGURE 15.9

A spherical lipoprotein particle surrounds nonpolar lipids with polar lipids and protein for transport to body cells.

Q **Why are the polar components on the surface of a lipoprotein particle and the nonpolar components at the center?**

Questions paired with figures challenge students to think critically about photos and illustrations.

Macro-to-Micro Art

Photographs and drawings illustrate the atomic structure of recognizable objects, putting chemistry in context and connecting the atomic world to the macroscopic world.

FIGURE 4.1

(a) The elements sodium and chlorine react to form the ionic compound sodium chloride, the compound that makes up table salt. (b) Crystals of NaCl under magnification. (c) A diagram of the arrangements of Na$^+$ and Cl$^-$ packed together in a NaCl crystal.

Q **What is the type of bonding between Na$^+$ and Cl$^-$ ions in salt?**

PROBLEM SOLVING

A Visual Guide to Problem Solving

The author understands the learning challenges facing students in this course, so she walks students through the problem-solving process step by step. For each type of problem, she uses a unique, color-coded flow chart that is coordinated with parallel worked examples to visually guide students through each problem-solving strategy.

Guide to Using Mole Factors

STEP 1
Write the given and needed moles.

STEP 2
Write a plan to convert the given to the needed moles.

STEP 3
Use coefficients to write relationships and mole–mole factors.

STEP 4
Set up problem using the mole factor that cancels given moles.

Sample Problems with Study Checks

Numerous **Sample Problems** appear throughout the text to immediately demonstrate the application of each new concept. The worked-out solution gives step-by-step explanations, provides a problem-solving model, and illustrates required calculations. Each Sample Problem is followed by a **Study Check** question that allows students to test their understanding of the problem-solving strategy.

SAMPLE PROBLEM 5.14

■ Mass of Product from Mass of Reactant

Acetylene (C_2H_2), used in welding, burns with oxygen.

$$2C_2H_2(g) + 5O_2(g) \longrightarrow 4CO_2(g) + 2H_2O(g)$$

How many grams of carbon dioxide are produced when 54.6 g C_2H_2 are burned?

SOLUTION

STEP 1 **Given** grams of C_2H_2 **Need** grams of CO_2

STEP 2 **Plan** Once we convert grams of C_2H_2 to moles of C_2H_2 using its molar mass, we can use a mole–mole factor to find the moles of CO_2. Then the molar mass of CO_2 will give us the grams of CO_2.

grams of C_2H_2 → Molar mass → moles of C_2H_2 → Mole–mole factor → moles of CO_2 → Molar mass → grams of CO_2

STEP 3 **Equalities/Conversion Factors** We need the molar mass of C_2H_2 and CO_2. The mole–mole factor that converts moles of C_2H_2 to moles of CO_2 is derived from the coefficients in the balanced equation.

1 mole C_2H_2 = 26.0 g		
$\dfrac{26.0 \text{ g } C_2H_2}{1 \text{ mole } C_2H_2}$	and	$\dfrac{1 \text{ mole } C_2H_2}{26.0 \text{ g } C_2H_2}$

1 mole C_2H_2 = 2 moles CO_2		
$\dfrac{2 \text{ moles } CO_2}{2 \text{ moles } C_2H_2}$	and	$\dfrac{1 \text{ mole } CO_2}{2 \text{ moles } C_2H_2}$

1 mole CO_2 = 44.0 g		
$\dfrac{44.0 \text{ g } CO_2}{1 \text{ mole } CO_2}$	and	$\dfrac{1 \text{ mole } CO_2}{44.0 \text{ g } CO_2}$

STEP 4 **Set Up Problem** Using our plan, we first convert grams of C_2H_2 to moles of C_2H_2.

$$54.6 \text{ g } C_2H_2 \times \frac{1 \text{ mole } C_2H_2}{26.0 \text{ g } C_2H_2} = 2.10 \text{ moles } C_2H_2$$

Then we change moles of C_2H_2 to moles of CO_2 by using the mole–mole factor.

$$2.10 \text{ moles } C_2H_2 \times \frac{4 \text{ moles } CO_2}{2 \text{ moles } C_2H_2} = 4.20 \text{ moles } CO_2$$

Finally, we can convert moles of CO_2 to grams of CO_2.

$$4.20 \text{ moles } CO_2 \times \frac{44.0 \text{ g } CO_2}{1 \text{ mole } CO_2} = 185 \text{ g } CO_2$$

The solution can be obtained using the conversion factors in sequence.

$$54.6 \text{ g } C_2H_2 \times \frac{1 \text{ mole } C_2H_2}{26.0 \text{ g } C_2H_2} \times \frac{4 \text{ moles } CO_2}{2 \text{ moles } C_2H_2} \times \frac{44.0 \text{ g } CO_2}{1 \text{ mole } CO_2} = 185 \text{ g C}$$

STUDY CHECK

Using the equation in Sample Problem 5.14, calculate the grams of CO_2 that can be produced when 25.0 g O_2 reacts.

Integrated Questions and Problems

Questions and Problems at the end of each section encourage students to apply concepts and begin problem solving after each section. Answers to odd-numbered problems are given at the end of each chapter.

Paired Problems

By pairing each even-numbered problem with a similar odd-numbered problem whose answer is at the end of the chapter, students are able to guide themselves through solving problems.

Additional Questions and Problems

Additional Questions and Problems at the end of each chapter combine topics from that chapter and previous chapters. These problems are more mathematically complex and require more critical thinking.

Understanding the Concepts questions encourage students to think about the concepts they have learned.

Questions for Teamwork are designed for group work in cooperative learning environments.

MEDIA AND ADDITIONAL SUPPLEMENTS

VALUABLE ANCILLARIES FOR BOTH STUDENTS AND INSTRUCTORS.

Interactivity

The ChemistryPlace website is packed with comprehensive sets of interactive **Tutorials** that bring chemical reactions and concepts to life. The website has a multitude of **Quizzes**, a **Glossary**, and **Flashcards** specifically tailored to the text. Over 2000 **PowerPoint slides**, complete with art, photos, and tables from the text, teach key concepts from every section of the book. **InterAct Math** provides exercises that allow students to review and build the math skills needed in chemistry.

Applications

The website, like the text, is committed to students. In-depth looks at health careers and **Case Studies** will provide a relevant context for learning.

THE CHEMISTRY PLACE

SUPPLEMENTS FOR THE STUDENTS

Study Guide and Selected Solutions
by Karen Timberlake
The *Study Guide* is keyed to the learning goals in the text and designed to promote active learning. It includes complete solutions to odd-numbered problems. (0-8053-3024-0)

Essential Laboratory Manual
by Karen Timberlake
This lab manual contains 25 experiments for the standard course sequence of topics. (0-8053-3023-2)

Laboratory Manual
by Karen Timberlake
This best-selling lab manual contains 42 experiments for the standard course sequence of topics. (0-8053-3025-9)

Tutor Center
www.aw-bc.com/tutorcenter
This center provides one-to-one tutoring by phone, fax, and e-mail. Qualified instructors answer questions and provide instructions regarding examples, exercises, and other content found in the text.

IMAGE PRESENTATIONS TOOLS

Image Presentation Library

This CD-ROM puts the outstanding art of Timberlake's *Chemistry* in the palm of your hand. Contained on the CD are jpeg files at 150 dpi to use in classroom presentations. Virtually all of the art and tables from the text are available on the CD. Downloadable, customizable PowerPoint® lecture outlines, featuring over 2000 slides, are also available on the CD-ROM. (0-8053-3027-5)

Transparency Acetates

This package includes 200 full-color transparency acetates. (0-8053-3029-1)

ADDITIONAL INSTRUCTOR SUPPLEMENTS

Instructor's Manual and Complete Solutions

By Karen Timberlake
Highlights chapter topics and includes suggestions for the laboratory.
Contains answers and solutions to all problems in the text.

Printed Test Bank
(0-8053-3030-5)

Computerized Test Bank

Over 1500 questions in multiple-choice, matching, true-false, and short-answer format. Available in print or on a dual-platform CD-ROM. (0-8053-3016-X)

Online Instructor's Manual to Laboratory Manual

Contains answers to report pages for the Laboratory Manual and Essential Laboratory Manual.

Also, visit the Benjamin Cummings catalog page for Timberlake's *Chemistry*, Ninth Edition, at www.aw-bc.com to download available instructor supplements.

PROLOGUE Chemistry in Our Lives

LOOKING AHEAD

A. Chemistry and Chemicals

B. Scientific Method: Thinking like a Scientist

C. A Study Plan for Learning Chemistry

"Chemistry plays an integral part in all aspects of medical legal death investigation," says Charles L. Cecil, forensic anthropologist and medical legal death investigator, San Francisco Medical Examiner's Office. "Crime-scene analysis of blood droplets determines whether they are human or nonhuman, the analyses of toxicological samples of blood and/or other fluids help determine the cause and time of death. Specialists in forensic anthropology can analyze trace-element ratios in bones to identify the number of individuals in mixed human bone situations. These conditions are quite often found during investigations of massive human-rights violations, such as the site of El Mozote in El Salvador."

Forensic anthropologists help police identify skeletal remains by determining the gender, approximate age, height, and cause of death. Analysis of bone can also provide information about illnesses or trauma a person may have experienced.

the
Chemistry
place

Visit **www.aw-bc.com/chemplace** or use the CD-ROM in the book for extra quizzes, interactive tutorials, career resources, PowerPoint slides, math help, and case studies.

What are some questions in science you have been curious about? How does car exhaust produce the smog that hangs over our cities? Why has the ozone been depleted in certain parts of the atmosphere? Why does an iron nail rust when exposed to air and rain? Why does aspirin or ibuprofen help a headache? Just like you, chemists are curious about the world we live in. They have also been interested in understanding smog, the causes of ozone depletion, formation of rust, and the effect of aspirin in the body.

Answers obtained by scientists may provide you with an understanding of the chemistry involved in your questions. One component of car exhaust is nitrogen monoxide (NO) formed in car engines where high temperatures convert nitrogen gas (N_2) and oxygen gas (O_2) to NO. In chemistry, these reactions are written in the form of equations such as $N_2(g) + O_2(g) \rightarrow 2NO(g)$. The reaction of NO with oxygen in the air produces NO_2, which gives the reddish brown color to smog. In the 1970s, scientists associated substances called chlorofluorocarbons (CFCs) with the depletion of ozone over Antarctica. As CFCs are broken down by UV light, chlorine (Cl) is released that causes the breakdown of ozone (O_3) molecules and destroys the ozone layer: $Cl + O_3 \rightarrow ClO + O_2$. When iron (Fe) in a nail reacts with oxygen (O_2) in the air, the oxidation of iron forms rust (Fe_2O_3): $4Fe(s) + 3O_2(g) \rightarrow 2Fe_2O_3(s)$. When a part of the body is injured, substances called prostaglandins are produced, which cause inflammation and pain. Aspirin acts to block the production of prostaglandins, thereby reducing inflammation, pain, and fever.

Other scientists design new fuels and more efficient ways to use them. Researchers in the medical field look for evidence that will help them understand and design new treatments for diabetes, genetic defects, cancer, AIDS, and aging. For the chemist in the laboratory, the physician in the dialysis unit, or the agricultural scientist, chemistry plays a central role in providing understanding, assessing solutions, and making important decisions.

■ A. Chemistry and Chemicals

Chemistry is the study of the composition, structure, properties, and reactions of matter. Matter is another word for all the substances that make up our world. Perhaps you imagine that chemistry is done only in a laboratory by a chemist wearing a lab coat and goggles. Actually, chemistry happens all around you every day and has a big impact on everything you use and do. You are doing chemistry when you cook food, add chlorine to a swimming pool, or start your car. A chemical reaction has taken place when you see a rusty nail or an antacid tablet dropped into water. Plants grow because chemical reactions convert carbon dioxide, water, and energy to carbohydrates. Chemical reactions take place when you digest food and break it down into substances that you need for energy and health.

TABLE P.1

Chemicals Commonly Used in Toothpaste

Chemical	Function
calcium carbonate	An abrasive used to remove plaque
sorbitol	Prevents loss of water and hardening of toothpaste
carrageenan (seaweed extract)	Keeps the toothpaste from hardening or separating
glycerin	Makes the toothpaste foam in the mouth
sodium lauryl sulfate	A detergent used to loosen plaque
titanium dioxide	Makes the toothpaste base white and opaque
triclosan	An antibacterial agent used to inhibit bacteria that cause plaque and gum disease
sodium monofluorophosphate	Prevents formation of cavities by strengthening tooth enamel with fluoride
methyl salicylate	Gives a pleasant flavor of wintergreen

All the things you see around you right now are composed of one or more chemicals. A *chemical* is any material used in or produced by a chemical process. Chemical processes take place in chemistry laboratories, manufacturing plants, and pharmaceutical labs as well as every day in nature and in the body. A *substance* is a chemical that consists of one type of matter and always has the same composition and properties wherever it is found. Often the terms chemical and substance are used interchangeably to describe a specific type of matter.

Each day you use products containing substances that were developed and prepared by chemists. When you shower in the morning, chemicals in soaps and shampoos combine with oils on your skin and scalp and are removed by rinsing with water. When you brush your teeth, the chemicals in toothpaste clean your teeth and prevent plaque and tooth decay. Toothpaste alone contains chemicals such as abrasives, antibacterial agents, enamel strengtheners, colorings, and flavorings. Some of the substances used to make toothpaste are listed in Table P.1. In cosmetics and lotions, chemicals are used to moisturize, prevent deterioration of the product, fight bacteria, and thicken the product. Your clothes may be made of natural polymers such as cotton or synthetic polymers such as nylon or polyester. Perhaps you wear a ring or watch made of gold, silver, or platinum. Your breakfast cereal is probably fortified with iron, calcium, and phosphorus, while the milk you drink is enriched with vitamins A and D. Antioxidants are chemicals added to your cereal to prevent it from spoiling. Some of the chemicals you may encounter when you cook in the kitchen are shown in Figure P.1.

Silicon dioxide (glass)

Caffeine (coffee)

Chemically treated water

Polymer material

Metal alloy

Baking carbohydrates, fats, and proteins

Natural gas

Fruits grown with fertilizers and pesticides

FIGURE P.1

Many of the items found in a kitchen are obtained using chemical reactions.

Q **What are some other chemicals found in a kitchen?**

■ **Everyday Chemicals**

Identify the chemical described by each of the following statements.

a. Aluminum used to make soda cans.
b. Salt (sodium chloride) used throughout history to preserve meat and fish.
c. Sugar (sucrose) used to sweeten coffee or tea.

SOLUTION

a. aluminum   **b.** salt (sodium chloride)   **c.** sugar (sucrose)

■ B. Scientific Method: Thinking like a Scientist

When you were very young, you explored the things around you by touching and tasting. When you got a little older, you asked questions about the world you live in. What is lightning? Where does a rainbow come from? Why is water blue? As an adult you may have wondered how antibiotics work. Each day you ask questions and seek answers as you organize and make sense of the world you live in.

When Nobel Laureate Linus Pauling described his student life in Oregon, he recalled that he read many books on chemistry, mineralogy, and physics. "I mulled over the properties of materials: why are some substances colored and others not, why are some minerals or inorganic compounds hard and others soft." He said, "I was building up this tremendous background of empirical knowledge and at the same time asking a great number of questions." Linus Pauling won two Nobel Prizes: the first, in 1954, was in chemistry and for his work on the structure of proteins, and the second, in 1962, was the Peace Prize.

■ Scientific Method

Although the process of trying to understand nature is unique to each scientist, there is a set of general principles, called the *scientific method*, that describes the thinking of a scientist.

1. **Observations** The first step in the scientific method is to observe, describe, and measure some event in nature. Observations based on measurements are called *data*.
2. **Hypothesis** After sufficient data are collected, a *hypothesis* is proposed that states a possible interpretation of the observations. The hypothesis must be stated in a way that it can be tested by experiments.
3. **Experiments** Experiments are tests that determine the validity of the hypothesis. Often many experiments are performed and a large amount of data collected. If the results of experiments provide different results than predicted by the hypothesis, a new or modified hypothesis is proposed and a new group of experiments are performed.
4. **Theory** When experiments can be repeated by many scientists with consistent results that confirm the hypothesis, the hypothesis becomes a *theory*. Each theory, however, continues to be tested and, based on new data, sometimes needs to be modified or even replaced. Then a new hypothesis is proposed, and the process of experimentation takes place once again.

Using the Scientific Method in Everyday Life

You may be surprised to realize that you use the scientific method in your everyday life. Let's suppose that you visit a friend in her home. Soon after you arrive, your eyes start to itch and you begin to sneeze. Then you observe that your friend has a new cat. Perhaps you ask yourself why you are sneezing and form the hypothesis that you are allergic to cats. To test your hypothesis, you leave your friend's home. If the sneezing stops, perhaps your hypothesis is correct. You test your hypothesis further by visiting another friend who also has a cat. If you start to sneeze again, your experimental results indicate that you are allergic to cats. However, if you continue sneezing after you leave your friend's home, your hypothesis is not supported. Now you need to form a new hypothesis, which could be that you have a cold.

THINK ABOUT IT!

■ Scientific Method

Identify each of the following statements as an observation or a hypothesis.

a. A silver tray turns a dull gray color when left uncovered.
b. Water freezes at 0°C.
c. Ice cubes have a greater volume than the liquid from which they were formed because the water molecules are further apart in the ice structure than in the liquid form.

SOLUTION

a. observation   **b.** observation   **c.** hypothesis

THINK ABOUT IT!

■ Designing an Experiment

The following statements are found in a student's notebook. Identify each of the following as **(1)** observation, **(2)** hypothesis, or **(3)** experiment.

a. "Today, I planted two tomato seedlings in the garden. Two more tomato seedlings are placed in a closet. I will give all the plants the same amount of water and fertilizer."

b. "After fifty days, the tomato plants in the garden are 3 feet high with green leaves. The plants in the closet are eight inches tall and yellow."

c. "Tomato plants need sunlight to grow."

SOLUTION

a. 3 **b.** 1 **c.** 2

Science and Technology

When scientific information is applied to industrial and commercial uses, it is called technology. Such uses have made the chemical industry one of the largest industries in the United States. Every year, technology provides new materials or procedures that produce more energy, cure diseases, improve crops, and produce new kinds of synthetic materials. Table P.2 lists some of the important scientific discoveries, laws, theories, and technological innovations that were made over the past 300 years.

TABLE P.2

Some Important Scientific and Technological Discoveries

Discovery	Date	Name	Country
Law of gravity	1687	Isaac Newton	England
Oxygen	1774	Joseph Priestley	England
Electric battery	1800	Alessandro Volta	Italy
Atomic theory	1803	John Dalton	England
Anesthesia, ether	1842	Crawford Long	U.S.
Nitroglycerine	1847	Ascanio Sobrero	Italy
Germ theory	1865	Louis Pasteur	France
Antiseptic surgery	1865	Joseph Lister	England
Discovery of nucleic acids	1869	Friedrich Miescher	Switzerland
Radioactivity	1896	Henri Becquerel	France
Discovery of radium	1898	Marie and Pierre Curie	France
Quantum theory	1900	Max Planck	Germany
Theory of relativity	1905	Albert Einstein	Germany
Identification of components of RNA and DNA	1909	Phoebus Theodore Levene	U.S.
Insulin	1922	Frederick Banting, Charles Best, John Macleod	Canada
Penicillin	1928	Alexander Fleming	England
Nylon	1937	Wallace Carothers	U.S.
Discovery of DNA as genetic material	1944	Oswald Avery	U.S.
Synthetic production of transuranium elements	1944	Glenn Seaborg, Arthur Wahl, Joseph Kennedy, Albert Ghiorso	U.S.
Determination of DNA structure	1953	Francis Crick, Rosalind Franklin, James Watson	England U.S.
Polio vaccine	1954 1957	Jonas Salk Albert Sabin	U.S. U.S.
Laser	1958 1960	Charles Townes Theodore Maiman	U.S. U.S.
MRI	1980	Paul Lauterbur	U.S
Cellular phones	1973	Martin Cooper	U.S.
Prozac	1988	Ray Fuller	U.S.
HIV protease inhibitor	1995	Joseph Martin, Sally Redshaw	U.S.

ENVIRONMENTAL NOTE

▪ DDT, Good Pesticide, Bad Pesticide

DDT (Dichlorodiphenyltrichloroethane) was once one of the most commonly used pesticides. While it was first made in 1874, it was not used as an insecticide until 1939. Before DDT was widely used, insect-borne diseases such as malaria and typhus were rampant in many parts of the world. Paul Mueller, who discovered that DDT was an effective pesticide, was recognized for saving many lives and received the Nobel Prize for medicine and physiology in 1948. DDT was considered the ideal pesticide because it was toxic to many insects, had a low toxicity to humans and animals, and was inexpensive to prepare.

In the United States, DDT was extensively used on home gardens as well as farm crops, particularly cotton and soybeans. Because of its stable chemical structure, DDT did not break down quickly in the environment, which meant that it did not have to be applied as often. At first, everyone was pleased with DDT as crop yields increased and diseases such as malaria and typhus were under control.

However, in the early 1950s, problems attributed to DDT began to arise. Insects were becoming more resistant to the pesticide. At the same time, there was increasing public concern about the long-term impact of a substance that could remain in the environment for many years. The metabolic systems of humans and animals cannot break down DDT, which is soluble in fats but not in water and stored in the fatty tissues of the body. Although the concentrations of DDT applied to crops were very low, the concentrations

of DDT found in fish and the birds that ate fish were as much as 10 million times greater. Although the birds did not directly perish, DDT was found to reduce the calcium in their eggshells. As a result, the incubating eggs cracked open early. Due to this difficulty with reproduction, the populations of birds such as the bald eagles and brown pelicans dropped significantly.

By 1972, DDT was banned in the United States. The EPA reported that the DDT levels were reduced by 90% in fish in Lake Michigan by 1978. Today new types of pesticides that are more water soluble and do not persist in the environment have replaced the long-lasting pesticides such as DDT. However, these new pesticides are much more toxic to humans.

However, there have also been improper applications of scientific research. The production of some substances has contributed to the development of hazardous conditions in our environment. We have become concerned about the energy requirements of new products and how some materials may cause changes in our oceans and atmosphere. We want to know if the new materials can be recycled, how they are broken down, and whether there are processes that are safer. The ways in which we continue to utilize scientific research will strongly impact our planet and its community in the future. These decisions can best be made if every citizen has an understanding of science.

C. A Study Plan for Learning Chemistry

Here you are taking chemistry, perhaps for the first time. Whatever your reasons are for choosing to study chemistry, you can look forward to learning many new and exciting ideas.

Features in This Text To Help You Study Chemistry

This text has been designed with a variety of study aids to complement different learning styles. On the inside of the front cover is a periodic table of the elements that provides information about the elements. On the inside of the back cover are tables that summarize useful information needed throughout the study of chemistry. Each chapter begins with *Looking Ahead*, which outlines the topics in the chapter. A *Learning Goal* at the beginning of each section previews the concepts you are to learn.

Before you begin reading the text, obtain an overview of the chapter by reviewing the *Looking Ahead* list of topics. As you prepare to read a section of the chapter, look at the section title and turn it into a question. When you are ready to read through that section, review the *Learning Goal*. A *Learning Goal* tells you what to expect in that section and what you need to accomplish. As you read, try to answer the questions you wrote for the section. The *Sample Problems* illustrate a step-by-step approach, enhanced, where appropriate, by a visual *Guide to Problem Solving (GPS)*. When you come to a sample problem, take the time to work it through, try each *Study Check*, and check your answer. If your answer matches, you most likely understand the topic. If your answer does not match, you need to study the section again. At the end of each section, you will find a set of paired *Questions and Problems* that allow you to apply problem solving immediately to the new concepts.

Items appear throughout each chapter to help you connect chemistry to real-life events. The figures and diagrams use macro-to-micro illustrations to depict the atomic level of organization of ordinary objects. These visual models illustrate the concepts described in the text and allow you to "see" the world in a microscopic way. Boxes called *Health or Environmental Notes* connect the chemical concepts you are studying to real-life situations.

At the end of each chapter, you will find several study aids that complete the chapter. The *Key Terms* are bold-faced in the text and listed again at the end of the chapter. *Chapter Reviews* and *Concept Maps* at the end of each chapter give a summary and show the connections of important concepts. *Additional Questions and Problems* at the end of each chapter provide more problems to test your understanding of the topics in the chapter. *Understanding the Concepts* is a set of questions that utilize art and structures to help you visualize concepts and *Questions for Teamwork* encourage you to study and learn with a group. *Answers to Study Checks* and *Answers to Selected Questions and Problems* are provided at the end of the chapter.

Using Active Learning to Learn Chemistry

A student who is an active learner continually interacts with the chemical ideas while reading the text and attending lecture. Let's see how this is done.

As you read and practice problem solving, you remain actively involved in studying, which enhances the learning process. In this way, you learn small bits of information at a time and establish the necessary foundation for understanding the next section. You may also note questions you have about the reading to discuss with your professor and laboratory instructor. Table P.3 summarizes these steps for active learning. The time you spend in lecture can also be useful as a learning time. By keeping track of the class schedule and reading the assigned material before lecture, you become aware of the new terms and concepts you need to learn. Some questions that occur during your reading may be answered during the lecture. If not, you can ask for further clarification from your professor.

Many studies now indicate that studying with a group can be beneficial to learning for many students. In a group, students motivate each other to study, fill in gaps, and correct misunderstandings by teaching and learning together. Studying alone

TABLE P.3

Steps in Active Learning

1. Read the set of *Looking Ahead* topics for an overview of the material.
2. Form a question from the title of the section you are going to read.
3. Read the section looking for answers to your question.
4. Self-test by working *Sample Problems* and *Study Checks* within each section.
5. Complete the *Questions and Problems* that follow each section and check the odd-numbered answers.
6. Proceed to the next section and repeat the above steps.

does not allow the process of peer correction that takes place when you work with a group of students in your class. In a group, you can cover the ideas more thoroughly as you discuss the reading and problem solving with other students. Waiting to study until the night before an exam does not give you time to understand concepts and practice problem solving. Ideas may be ignored or avoided that turn out to be important on test day.

Thinking Scientifically About Your Study Plan

As you embark on your journey into the world of chemistry, think about your approach to studying and learning chemistry. You might consider some of the ideas in the following list. Check those ideas that will help you learn chemistry successfully. Commit to them now. Your success depends on you.

My study of chemistry will include the following:

_____ Reviewing the *Learning Goals*.

_____ Keeping a problem notebook.

_____ Reading the text as an active learner.

_____ Self-testing by working the chapter problems and checking solutions in the text.

_____ Reading the chapter before lecture.

_____ Being an active learner in lecture.

_____ Going to lecture.

_____ Organizing a study group.

_____ Seeing the professor during office hours.

_____ Attending review sessions.

_____ Organizing my own review sessions.

_____ Studying a little bit as often as I can.

THINK ABOUT IT!

■ A Study Plan for Learning Chemistry

Which of the following activities would you include in a study plan for learning chemistry successfully?

a. skipping lecture

b. forming a study group

c. keeping a problem notebook

d. waiting to study the night before the exam

e. becoming an active learner

f. working assigned problems

g. attending the professor's office hours

h. staying up all night before an exam

SOLUTION

Your success in chemistry can be helped if you

b. form a study group

c. keep a problem notebook

e. become an active learner

f. work assigned problems

g. attend the professor's office hours

Measurements

LOO

"I use measurement in just about every part of my nursing practice," says registered nurse Vicki Miller. "When I receive a doctor's order for a medication, I have to verify that order. Then I draw a carefully measured volume from an IV or a vial to create that particular dose. Some dosage orders are specific to the size of the patient. I measure the patient's weight and calculate the dosage required for the weight of that patient."

Nurses use measurement each time they take a patient's temperature, height, weight, or blood pressure. Measurement is used to obtain the correct amounts for injections and medications and to determine the volumes of fluid intake and output. For each measurement, the amounts and units are recorded in the patient's records.

the Chemistry place

Visit **www.aw-bc.com/chemplace** or use the CD-ROM in the book for extra quizzes, interactive tutorials, career resources, PowerPoint slides, math help, and case studies.

Chemistry and measurement are an important part of our everyday life. Levels of toxic materials in our air, soil, and water are discussed in our newspaper. We read about radon in our homes, holes in the ozone layer, trans fatty acids, global warming, and DNA analysis. Understanding chemistry and measurement helps us make proper choices about our world.

Think about your day; you probably made some measurements. Perhaps you checked your weight by stepping on a scale. If you did not feel well, you may have taken your temperature. To make some soup, you added 2 cups of water to a package mix. If you stopped at the gas station, you watched the gas pump measure the number of gallons of gasoline you put in the car.

Measurement is an essential part of health careers such as nursing, dental hygiene, respiratory therapy, nutrition, and veterinary technology. The temperature, height, and weight of a patient are measured and recorded. Samples of blood and urine are collected and sent to a laboratory where glucose, pH, urea, and protein are measured by the lab technicians.

By learning about measurement, you will develop skills for solving problems and learn how to work with numbers in chemistry. If you intend to go into a health career, measurement and an understanding of measured values will be an important part of your evaluation of the health of a patient.

LEARNING GOAL

Write the names and abbreviations for the units used in measurements of length, volume, and mass; write a number in scientific notation.

1.1 Units of Measurement

The **metric system** is used by scientists and health professionals throughout the world. It is also the common measuring system in all but a few countries in the world. In 1960, a modification of the metric system called the *International System of Units*, Système International (**SI**), was adopted by scientists to provide additional uniformity for units used in the sciences. In this text, we will use metric units and introduce some of the SI units that are in use today, as listed in Table 1.1.

Length

The metric and SI unit of length is the **meter (m)**. It is 39.4 inches (in.), which makes a meter slightly longer than a yard (yd). A smaller unit of length, the **centimeter (cm)**, is more commonly used in chemistry and is about as wide as your little finger. For comparison, there is 2.54 cm in 1 in. (See Figure 1.1.)

TABLE 1.1

Units of Measurement		
Measurement	**Metric**	**SI**
Length	Meter (m)	Meter (m)
Volume	Liter (L)	Cubic meter (m^3)
Mass	Gram (g)	Kilogram (kg)
Time	Second (s)	Second (s)
Temperature	Celsius (°C)	Kelvin (K)

FIGURE 1.1

Length in the metric (SI) system is based on the meter, which is slightly longer than a yard.

Q How many centimeters are in a length of one inch?

$$1 \text{ m} = 100 \text{ cm}$$
$$1 \text{ m} = 39.4 \text{ in.}$$
$$2.54 \text{ cm} = 1 \text{ in.}$$

Volume

Volume is the amount of space a substance occupies. A **liter (L)**, which is slightly larger than the quart (qt), is commonly used to measure volume. The **milliliter (mL)** is more convenient for measuring smaller volumes of fluids in hospitals and laboratories. A comparison of metric and U.S. units for volume appears in Figure 1.2.

$$1 \text{ L} = 1000 \text{ mL}$$
$$1 \text{ L} = 1.06 \text{ qt}$$
$$946 \text{ mL} = 1 \text{ qt}$$

1 L

946 mL = 1 qt

FIGURE 1.2

Volume is the space occupied by a substance. In the metric system, volume is based on the liter, which is slightly larger than a quart.

Q How many milliliters are in 1 quart?

FIGURE 1.3

On an electronic balance, mass is shown in grams as a digital readout.

Q How many grams are in 1 pound of candy?

FIGURE 1.4

A thermometer is used to determine the temperature of a substance.

Q What kinds of temperature readings have you made today?

Mass

The **mass** of an object is a measure of the quantity of material it contains. You may be more familiar with the term *weight* than with mass. Weight is a measure of the gravitational pull on an object. On Earth, an astronaut with a mass of 75.0 kg has a weight of 165 lb. On the moon where the gravitational pull is one-sixth that of Earth, the astronaut has a weight of 27.5 lb. However, the mass of the astronaut is the same as on Earth, 75.0 kg. Scientists measure mass rather than weight because mass does not depend on gravity.

In the metric system, the unit for mass is the **gram (g)**. The SI unit of mass, the **kilogram (kg)**, is used for larger masses such as body weight. It takes 2.20 lb to make 1 kg, and 454 g is needed to equal 1 pound.

$$1 \text{ kg} = 1000 \text{ g}$$
$$1 \text{ kg} = 2.20 \text{ lb}$$
$$454 \text{ g} = 1 \text{ lb}$$

In a chemistry laboratory, a balance is used to measure the mass of a substance as shown in Figure 1.3.

Temperature

You probably use a thermometer to see how hot something is, or how cold it is outside, or perhaps to determine if you have a fever. (See Figure 1.4.) The **temperature** of an object tells us how hot or cold that object is. In the metric system, temperature is measured on the Celsius temperature scale. On the **Celsius (°C) scale**, water freezes at 0°C and boils at 100°C. On the Fahrenheit (°F) scale, water freezes at 32°F and boils at 212°F. In the SI system, temperature is based on the **Kelvin (K) temperature scale**, where the lowest temperature possible is assigned as 0 K. We will discuss the relationships between these temperature scales in Chapter 2.

SAMPLE PROBLEM 1.1

■ Units of Measurement

State the type of measurement (mass, length, volume, or temperature) indicated by the unit in each of the following.

a. 45.6 kg   **b.** 1.85 L   **c.** 45 m   **d.** 315 K

SOLUTION

a. mass   **b.** volume   **c.** length   **d.** temperature

STUDY CHECK

Give the SI unit and abbreviation that would be used to express the following measurements.

a. the length of a football field   **b.** the daytime temperature on Mars
c. the mass of an electric car

Answers to all of the Study Checks can be found at the end of this chapter in the Answer section. Checking your answers will let you know if you understand the material in this section.

Scientific Notation

In chemistry and science, we use numbers that are very, very small to measure things as tiny as the width of a human hair, which is 0.000 008 m, or things that are even smaller. Or perhaps we want to know the number of hairs on the average scalp, which is about 100 000 hairs. (See Figure 1.5.) In both measurements, it is convenient to write the numbers in scientific notation. (In this section we have added spaces to help make the places easier to count.)

Width of a human hair	0.000 008 m	8×10^{-6} m
Hairs on a human scalp	100 000 hairs	1×10^5 hairs

Writing a Number in Scientific Notation

When a number is written in **scientific notation**, there are two parts: a coefficient and a power of 10. For example, the number 2400 in scientific notation is 2.4×10^3. The coefficient is 2.4 and 10^3 shows the power of 10. The coefficient is determined by moving the decimal point three places to the left to give a number between 1 and 10. Because we moved the decimal point three places to the left, the power of ten is a positive 3, which is written as 10^3. For a number greater than one, the power of 10 is positive.

$$2\ 4\ 0\ 0. = 2.4 \times 1000 = 2.4 \times 10 \times 10 \times 10 = 2.4 \quad \times \quad 10^3$$

$\longleftarrow$ 3 places   Coefficient   Power of ten

When a number less than one is written in scientific notation, the power of ten is negative. For example, to write the number 0.000 86 in scientific notation the decimal point is moved to the right four places to give a coefficient of 8.6, which is greater than 1 but less than 10. By moving the decimal point four places to the right, the power of 10 is a negative 4 or 10^{-4}.

$$0.00086 = \frac{8.6}{10\ 000} = \frac{8.6}{10 \times 10 \times 10 \times 10} = 8.6 \quad \times \quad 10^{-4}$$

4 places $\longrightarrow$ Coefficient   Power of ten

Table 1.2 gives some examples of numbers written as positive and negative powers of ten. The powers of tens are really a way of keeping track of the decimal

8×10^{-6} m

FIGURE 1.5

Humans have an average of 1×10^5 hairs on their scalps. Each hair is about 8×10^{-6} m wide.

Q Why are large and small numbers written in scientific notation?

TABLE 1.2

Some Powers of Ten

Number	Multiples of Ten	Scientific Notation	
10 000	$10 \times 10 \times 10 \times 10$	1×10^4	
1 000	$10 \times 10 \times 10$	1×10^3	
100	10×10	1×10^2	Some positive powers of ten
10	10	1×10^1	
1	0	1×10^0	
0.1	$\dfrac{1}{10}$	1×10^{-1}	
0.01	$\dfrac{1}{10} \times \dfrac{1}{10} = \dfrac{1}{100}$	1×10^{-2}	Some negative powers of ten
0.001	$\dfrac{1}{10} \times \dfrac{1}{10} \times \dfrac{1}{10} = \dfrac{1}{1\,000}$	1×10^{-3}	
0.0001	$\dfrac{1}{10} \times \dfrac{1}{10} \times \dfrac{1}{10} \times \dfrac{1}{10} = \dfrac{1}{10\,000}$	1×10^{-4}	

point in the decimal number. Table 1.3 gives examples of writing measurements in scientific notation.

Scientific Notation and Calculators

You can enter numbers in scientific notation on many calculators using the EE or EXP key. After you enter the coefficient, push the EXP (or EE) key and enter only the power of ten because the EXP function key includes the **× 10** value. To enter a negative power of ten, push the plus/minus ($+/-$) key or the minus ($-$) key (depending on your calculator), but **not** the key that performs the subtraction operation "$-$". Some calculators require entering the sign before the power.

Number to Enter	Method	Display reads
4×10^6	4 EXP (EE) 6	4 06 or 4^{06}
2.5×10^{-4}	2.5 EXP (EE) +/−4	2.5 −04 or 2.5^{-04}

When a calculator answer appears in scientific notation, it is usually shown in the display as a number between 1 and 10 followed by a space and the power of ten. To express this display in scientific notation, write the number, insert $\times$ 10, and use the power of ten as an exponent.

TABLE 1.3

Some Measurements Written in Scientific Notation

Measured Quantity	Measurement	Scientific Notation
Diameter of Earth	12 800 000 m	1.28×10^7 m
Depth of Lake Tanganyika	1 460 m	1.46×10^3 m
Mass of a typical human	68 kg	6.8×10^1 kg
Mass of a hummingbird	0.002 kg	2×10^{-3} kg
Length of a pox virus	0.000 000 3 m	3×10^{-7} m
Mass of a bacterium (mycoplasma)	0.000 000 000 000 000 000 1 kg	1×10^{-19} kg

Calculator Display	Expressed in Scientific Notation
7.52^{04}	7.52×10^4
5.8^{-02}	5.8×10^{-2}

On many scientific calculators, a number can be converted into scientific notation using the appropriate keys. For example, the number 0.000 52 can be entered followed by hitting the 2nd or 3rd function key and the SCI key. The scientific notation appears in the calculator display as a coefficient and the power of ten.

$$0.000\ 52 \quad \boxed{2^{nd}\ or\ 3^{rd}\ function\ key} \quad \boxed{SCI} \quad = \quad 5.2^{-4} \text{ or } 5.2-04 \quad = 5.2 \times 10^{-4}$$
Key   Key   Display

SAMPLE PROBLEM 1.2

■ Scientific Notation

Write the following measurements in scientific notation.

a. 350 g　　　　**b.** 0.000 16 L　　　　**c.** 5 220 000 m

SOLUTION

a. 3.5×10^2 g　　　**b.** 1.6×10^{-4} L　　　**c.** 5.22×10^6 m

STUDY CHECK

Write the following measurements in scientific notation.

a. 425 000 m　　　　**b.** 0.000 000 8 g

QUESTIONS AND PROBLEMS

■ Units of Measurement

In every chapter, each magenta, odd-numbered exercise in Questions and Problems is paired with the next even-numbered exercise. The answers for all the Study Checks and the magenta, odd-numbered Questions and Problems are given at the end of this chapter. The complete solutions to the odd-numbered Questions and Problems are in the *Study Guide*.

1.1 Compare the units you would use and the units that a student in France would use to measure the following:
 a. mass
 b. height
 c. the amount of gasoline to fill the gas tank
 d. temperature

1.2 Suppose that a friend tells you the following information. Why are the statements confusing and how would you make them clear using a metric (SI) unit?
 a. I rode my bicycle for 15 today.
 b. My dog weighs 25.
 c. It is hot today. It is 30.
 d. I lost 1.5 last week.

1.3 What is the name of the unit and the type of measurement (mass, volume, length, or temperature) indicated for each of the following quantities?
 a. 4.8 m　　**b.** 325 g　　**c.** 1.5 mL　　**d.** 480 m　　**e.** 28°C

1.4 What is the name of the unit and the type of measurement (mass, volume, length, or temperature) indicated for each of the following quantities?
 a. 0.8 L　　**b.** 3.6 m　　**c.** 4 kg　　**d.** 35 g　　**e.** 373 K

1.5 Write the following measurements in scientific notation:
   a. 55 000 m   **b.** 480 g   **c.** 0.000 005 cm
   d. 0.000 14 s   **e.** 0.0072 L

1.6 Write the following measurements in scientific notation:
   a. 180 000 000 g   **b.** 0.000 06 m   **c.** 750 000 g
   d. 0.15 m   **e.** 0.024 kg

FIGURE 1.6

The lengths of the rectangular objects are measured as **(a)** 4.5 cm and **(b)** 4.55 cm.

Q **What is the length of the object in (c)?**

WEB TUTORIAL
Significant Figures

■ 1.2 Measured Numbers and Significant Figures

Whenever you make a measurement, you compare an object to a standard. For example, you may use a meterstick to measure your height, a scale to check your weight, and a thermometer to take your temperature. **Measured numbers** are the numbers you obtain when you use a measuring tool to determine your height, weight, or temperature.

■ Measured Numbers

Suppose you are going to measure the lengths of the objects in Figure 1.6. You would select a ruler with a scale marked on it. By observing the lines on the scale, you determine the measurement for each object. Perhaps the divisions on the scale are 1 cm. Another ruler might be marked in divisions of 0.1 cm. To report the length, you would first read the numerical value of the marked line. Finally, you *estimate* between the smallest marked lines. This estimated value is the final digit in a measured number.

For example, in Figure 1.6a, the end of the object falls between the lines marked 4 cm and 5 cm. That means that the length is 4 cm plus an estimated digit. If you estimate that the end is halfway between 4 cm and 5 cm, you would report its length as 4.5 cm. However, someone else might report the length as 4.4 cm. The last digit in a measured number may differ because people do not always estimate in the same way. The ruler shown in Figure 1.6b is marked with lines at 0.1 cm. With this ruler, you can estimate to the hundredths place (0.01 cm). Perhaps you would report the length of the object as 4.55 cm, while someone else may report its length as 4.56 cm. Both results are acceptable.

There is always some *uncertainty* in every measurement. When a measurement ends right on a marked line, a zero is written for the estimated digit. For example, in Figure 1.6c, the length measurement is written as 3.0 cm, not 3.

■ Significant Figures

In a measured number, the **significant figures** are all the digits including the estimated digit. All nonzero numbers are counted as significant figures. Zeros may or may not be significant depending on their position in a number. Table 1.4 gives the rules and examples of counting significant figures.

When one or more zeros in a large number are significant digits, they are shown more clearly by writing the number using scientific notation. For example, if the first zero in the measurement 500 m is significant, it can be shown by writing the measurement as 5.0×10^2 m. In this text, we will place a decimal

TABLE 1.4

Significant Figures in Measured Numbers

Rule	Measured Number	Number of Significant Figures
1. A number is a *significant figure* if it is		
a. not a zero	4.5 g	2
	122.35 m	5
b. a zero between nonzero digits	205 m	3
	5.082 kg	4
c. a zero at the end of a decimal number	50. L	2
	25.0°C	3
	16.00 g	4
d. any digit in the coefficient of a number written in scientific notation	4.0×10^5 m	2
	5.70×10^{-3} g	3
2. A zero is *not significant* if it is		
a. at the beginning of a decimal number	0.0004 lb	1
	0.075 m	2
b. used as a placeholder in a large number without a decimal point	850 000 m	2
	1 250 000 g	3

point after a significant zero at the end of a number. For example, a measurement written as 250. g has three significant figures, which includes the zero. It could also be written as 2.50×10^2 g. Unless noted otherwise, we will assume that zeros at the end of large standard numbers are not significant; that is, we would interpret 400 000 g as 4×10^5 g.

Exact Numbers

Exact numbers are numbers obtained by counting items or from a definition that compares two units in the same measuring system. Suppose a friend asks you to tell her the number of coats in your closet or the number of bicycles in your garage or the number of classes you are taking in school. Your answer would be given by counting the items. It was not necessary for you to use any type of measuring tool. Suppose someone asks you to state the number of seconds in one minute. Without using any measuring device, you would give the definition: 60 seconds in one minute. Exact numbers are not measured, do not have a limited number of significant figures, and do not affect the number of significant figures in a calculated answer. For more examples of exact numbers, see Table 1.5.

TABLE 1.5

Examples of Some Exact Numbers

Counted Numbers	Defined Equalities	
	U.S. System	Metric System
Eight doughnuts	1 ft = 12 in.	1 L = 1000 mL
Two baseballs	1 qt = 4 cups	1 m = 100 cm
Five capsules	1 lb = 16 ounces	1 kg = 1000 g

SAMPLE PROBLEM 1.3

■ Measured Numbers and Significant Figures

Identify each of the following numbers as measured or exact; give the number of significant figures in the measured numbers.

a. 42.2 g   **b.** 3 eggs   **c.** 0.0005 cm
d. 450 000 km   **e.** 9 planets

SOLUTION

a. measured; three   **b.** exact   **c.** measured; one
d. measured; two   **e.** exact

STUDY CHECK

State the number of significant figures in each of the following measured numbers:

a. 0.00035 g   **b.** 2000 m   **c.** 2.0045 L

QUESTIONS AND PROBLEMS

■ Measured Numbers and Significant Figures

1.7 Are the numbers in each of the following statements measured or exact?
   a. A patient weighs 155 lb.
   b. The basket holds 8 apples.
   c. In the metric system, 1 kg is equal to 1000 g.
   d. The distance from Denver, Colorado, to Houston, Texas, is 1720 km.

1.8 Are the numbers in each of the following statements measured or exact?
   a. There are 31 students in the laboratory.
   b. The oldest known flower lived 120 000 000 years ago.
   c. The largest gem ever found, an aquamarine, has a mass of 104 kg.
   d. A laboratory test shows a blood cholesterol level of 184 mg/dL.

1.9 For each measurement, indicate if the zeros are significant figures or not.
   a. 0.0038 m   **b.** 5.04 cm   **c.** 800. L
   d. 3.0×10^{-3} kg   **e.** 85 000 m

1.10 For each measurement, indicate if the zeros are significant figures or not.
   a. 20.05 g   **b.** 5.00 m   **c.** 0.000 02 L
   d. 120 000 years   **e.** 8.05×10^2 L

1.11 How many significant figures are in each of the following measured quantities?
   a. 11.005 g   **b.** 0.000 32 m   **c.** 36 000 000 km
   d. 1.80×10^4 g   **e.** 0.8250 L   **f.** 30.0°C

1.12 How many significant figures are in each of the following measured quantities?
   a. 20.60 mL   **b.** 1036.48 g   **c.** 4.00 m
   d. 20.8°C   **e.** 60 800 000 g   **f.** 5.0×10^{-3} L

LEARNING GOAL

Adjust calculated answers to the correct number of significant figures.

■ 1.3 Significant Figures in Calculations

In the sciences, we measure many things: the length of a bacterium, the volume of a gas sample, the temperature of a reaction mixture, or the mass of iron in a sample. The numbers obtained from these types of measurements are often used in calculations. The number of significant figures in the measured numbers limits the number of significant figures that can be given in the calculated answer.

Using a calculator will usually help do calculations faster. However, calculators cannot think for you. It is up to you to enter the numbers correctly, press the right function keys, and adjust the calculator display to give an answer with the correct number of significant figures.

Rounding Off

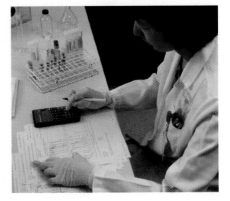

To calculate the area of a carpet that measures 5.5 m by 3.5 m, you multiply 5.5 times 3.5 to obtain the number 19.25 as the area in square meters. However, all four digits cannot be given in the answer because they are not all significant figures. Each measurement of length and width has only two significant figures. This means that the calculated result must be *rounded off* to give an answer that also has two significant figures, 19 m². When you obtain a calculator result, determine the number of significant figures for the answer and round off using the following rules.

Rules for Rounding Off

1. If the first digit to be dropped is *4 or less*, it and all following digits are simply dropped from the number.
2. If the first digit to be dropped is *5 or greater*, the last retained digit of the number is increased by 1.

		Three Significant Figures	Two Significant Figures
Example 1:	8.4234 rounds off to	8.42	8.4
Example 2:	14.780 rounds off to	14.8	15

SAMPLE PROBLEM 1.4

■ Rounding Off

Round off each of the following numbers to three significant figures:

a. 35.7823 m   **b.** 0.002627 L
c. 3826.8 g   **d.** 1.2836 kg

SOLUTION

a. 35.8 m   **b.** 0.00263 L
c. 3830 g   **d.** 1.28 kg

STUDY CHECK

Round off each of the numbers in Sample Problem 1.4 to two significant figures.

Multiplication and Division

In multiplication or division, the final answer is written so it has the same number of significant figures as the measurement with the *fewest significant figures* (SFs).

Example 1

Multiply the following measured numbers: 24.65×0.67.

24.65	$\times$	0.67	$=$	*16.5155*	$\longrightarrow$	17
Four SFs		Two SFs		Calculator display		Final answer, rounded to two SFs

The answer in the calculator display has more digits than the data allows. The measurement 0.67 has the least number of significant figures, two. Therefore, the calculator answer is rounded off to two significant figures.

Example 2

Solve the following:

$$\frac{2.85 \times 67.4}{4.39}$$

To do this problem on a calculator, enter the numbers and then press the operation keys. In this case, we might press the keys in the following order:

2.85	$\times$	67.4	$\div$	4.39	$=$	*43.756264*	$\longrightarrow$	43.8
Three SFs		Three SFs		Three SFs		Calculator display		Final answer, rounded to three SFs

All of the measurements in this problem have three significant figures. Therefore, the calculator result is rounded off to give an answer, 43.8, that has three significant figures.

Adding Significant Zeros

Sometimes, a calculator displays a small whole number. To give an answer with the correct number of significant figures, significant zeros may need to be written after the calculator result. For example, suppose the calculator display is 4, but you used measurements that have three significant numbers. The answer 4.00 is obtained by placing two significant zeros after the 4.

$\dfrac{8.00}{2.00} =$	*4*	$\longrightarrow$	4.00
Three SFs	Calculator display		Final answer, two zeros added to give three SFs

■ Significant Figures in Multiplication and Division

Perform the following calculations of measured numbers. Give the answers with the correct number of significant figures.

a. 56.8×0.37 **b.** $\dfrac{71.4}{11}$ **c.** $\dfrac{(2.075)(0.585)}{(8.42)(0.0045)}$ **d.** $\dfrac{25.0}{5.00}$

SOLUTION

a. 21   **b.** 6.5   **c.** 32   **d.** 5.00 (Add significant zeros)

STUDY CHECK

Solve:

a. 45.26 × 0.01088   **b.** 2.6 ÷ 324   **c.** $\dfrac{4.0 \times 8.00}{16}$

Addition and Subtraction

In addition or subtraction, the final answer is written so it has the same number of decimal places as the measurement with the *fewest decimal places*.

Example 3

Add:

2.045	Three decimal places
+ 34.1	One decimal place
36.145	Calculator display
36.1	Answer, rounded to one decimal place

When numbers are added or subtracted to give answers ending in zero, the zero does not appear after the decimal point in the calculator display. For example, 14.5 g − 2.5 g = 12.0 g. However, if you do the subtraction on your calculator, the display shows 12. To give the correct answer, a significant zero is written after the decimal point.

Example 4

14.5 g	One decimal place
− 2.5 g	One decimal place
12.	Calculator display
12.0 g	Answer, zero written after the decimal point

SAMPLE PROBLEM 1.6

■ Significant Figures in Addition and Subtraction

Perform the following calculations and give the answer with the correct number of decimal places:

a. 27.8 cm + 0.235 cm
b. 153.247 g − 14.82 g

SOLUTION

a. 28.0 cm   **b.** 138.43 g

STUDY CHECK

Solve:

a. 82.45 mg + 1.245 mg + 0.00056 mg   **b.** 4.259 L − 3.8 L

QUESTIONS AND PROBLEMS

▪ Significant Figures in Calculations

1.13 Why do we usually need to round off calculations that use measured numbers?

1.14 Why do we sometimes add a zero to a number in a calculator display?

1.15 Round off each of the following numbers to three significant figures.
   a. 1.854   **b.** 184.2038   **c.** 0.004738265
   d. 8807   **e.** 1.832149

1.16 Round off each of the numbers in problem 1.15 to two significant figures.

1.17 For the following problems, give answers with the correct number of significant figures:
   a. 45.7×0.034 **b.** 0.00278×5
   c. $\dfrac{34.56}{1.25}$ **d.** $\dfrac{(0.2465)(25)}{1.78}$

1.18 For the following problems, give answers with the correct number of significant figures:
   a. 400×185 **b.** $\dfrac{2.40}{(4)(125)}$
   c. $0.825 \times 3.6 \times 5.1$ **d.** $\dfrac{3.5 \times 0.261}{8.24 \times 20.0}$

1.19 For the following problems, give answers with the correct number of decimal places:
   a. 45.48 cm + 8.057 cm   **b.** 23.45 g + 104.1 g + 0.025 g
   c. 145.675 mL − 24.2 mL   **d.** 1.08 L − 0.585 L

1.20 For the following problems, give answers with the correct number of decimal places:
   a. 5.08 g + 25.1 g   **b.** 85.66 cm + 104.10 cm + 0.025 cm
   c. 24.568 mL − 14.25 mL   **d.** 0.2654 L − 0.2585 L

LEARNING GOAL

▪ Use the numerical values of prefixes to write a metric equality.

WEB TUTORIAL
Metric System

▪ 1.4 Prefixes and Equalities

The special feature of the metric system of units is that a **prefix** can be attached to any unit to increase or decrease its size by some factor of 10. For example, the prefixes *milli* and *micro* are used to make the smaller units, milligram (mg) and microgram (μg). Table 1.6 lists some of the metric prefixes, their symbols, and their decimal values.

TABLE 1.6

Metric and SI Prefixes				
Prefix	**Symbol**	**Meaning**	**Numerical Value**	**Scientific Notation**
Prefixes That Increase the Size of the Unit				
giga	G	one billion	1 000 000 000	10^9
mega	M	one million	1 000 000	10^6
kilo	k	one thousand	1 000	10^3
Prefixes That Decrease the Size of the Unit				
deci	d	one-tenth	0.1	10^{-1}
centi	c	one-hundredth	0.01	10^{-2}
milli	m	one-thousandth	0.001	10^{-3}
micro	μ	one-millionth	0.000 001	10^{-6}
nano	n	one-billionth	0.000 000 001	10^{-9}

The prefix *centi* is like cents in a dollar. One cent would be a centidollar or $\frac{1}{100}$ of a dollar. That also means that one dollar is the same as 100 cents. The prefix *deci* is like dimes in a dollar. One dime would be a decidollar or $\frac{1}{10}$ of a dollar. That also means that one dollar is the same as 10 dimes.

The U.S. Food and Drug Administration has determined the daily values (DV) of nutrients for adults and children age 4 or older. Some of these recommended daily values, which use prefixes, are listed in Table 1.7.

The relationship of a prefix to a unit can be expressed by replacing the prefix with its numerical value. For example, when the prefix *kilo* in *kilometer* is replaced with its value of 1000, we find that a kilometer is equal to 1000 meters. Other examples follow.

1 **kilo**meter (1 km) = **1000** meters (1000 m)

1 **kilo**liter (1 kL) = **1000** liters (1000 L)

1 **kilo**gram (1 kg) = **1000** grams (1000 g)

TABLE 1.7

Daily Values for Selected Nutrients

Nutrient	Amount Recommended
vitamin C	60 mg
vitamin B_{12}	6 μg
calcium	1 g
iron	18 mg
iodine	150 μg
sodium	2400 mg
zinc	15 mg

SAMPLE PROBLEM 1.7

■ **Prefixes**

Fill in the blanks with the correct numerical value:

a. kilogram = _____ grams
b. millisecond = _____ second
c. deciliter = _____ liter

SOLUTION

a. The numerical value of *kilo* is 1000; 1000 grams.
b. The numerical value of *milli* is 0.001; 0.001 second.
c. The numerical value of *deci* is 0.1; 0.1 liter.

STUDY CHECK

Write the correct prefix in the blanks:

a. 1 000 000 seconds = _____ seconds
b. 0.01 meter = _____ meter

Measuring Length

An ophthalmologist may measure the diameter of the retina of an eye in centimeters (cm), whereas a surgeon may need to know the length of a nerve in millimeters (mm). When the prefix *centi* is used with the unit *meter*, it indicates the unit *centimeter*, a length that is one-hundredth of a meter (0.01 m). A *millimeter* measures a length of 0.001 m. There are 1000 mm in a meter.

If we compare the lengths of a millimeter and a centimeter, we find that 1 mm is 0.1 cm; there are 10 mm in 1 cm. These comparisons are examples of **equalities**, which show the relationship between two units that measure the same quantity. For example, in the equality 1 m = 100 cm, each quantity describes the same length but in a different unit. Note that each quantity in the equality expression has both a number and a unit.

First Quantity		Second Quantity	
1	m =	100	cm
Number + unit		Number + unit	

FIGURE 1.7

The metric length of one meter is the same length as 10 dm, 100 cm, and 1000 mm.

Q **How many millimeters (mm) are in 1 centimeter (cm)?**

Some Length Equalities

$$1 \text{ m} = 100 \text{ cm}$$

$$1 \text{ m} = 1000 \text{ mm}$$

$$1 \text{ cm} = 10 \text{ mm}$$

Some metric units for length are compared in Figure 1.7.

TABLE 1.8

Some Typical Laboratory Test Values

Substance in Blood	Typical Range
albumin	3.5–5.0 g/dL
ammonia	20–150 μg/dL
calcium	8.5–10.5 mg/dL
cholesterol	105–250 mg/dL
iron (male)	80–160 μg/dL
protein (total)	6.0–8.0 g/dL

Measuring Volume

Volumes of 1 L or smaller are common in the health sciences. When a liter is divided into 10 equal portions, each portion is a deciliter (dL). There are 10 dL in 1 L. Laboratory results for blood work are often reported in mass per deciliter. Table 1.8 lists typical laboratory tests for some substances in the blood.

When a liter is divided into a thousand parts, each of the smaller volumes is called a milliliter. In a 1-L container of physiological saline, there are 1000 mL of solution (see Figure 1.8).

Some Volume Equalities

$$1 \text{ L} = 10 \text{ dL}$$

$$1 \text{ L} = 1000 \text{ mL}$$

$$1 \text{ dL} = 100 \text{ mL}$$

The **cubic centimeter (cm^3 or cc)** is the volume of a cube whose dimensions are 1 cm on each side. A cubic centimeter has the same volume as a milliliter, and the units are often used interchangeably.

$$1 \text{ cm}^3 = 1 \text{ cc} = 1 \text{ mL}$$

When you see *1 cm*, you are reading about length; when you see *1 cc* or *1 cm³* or *1 mL*, you are reading about volume. A comparison of units of volume is illustrated in Figure 1.9.

Measuring Mass

When you get a physical examination, your mass is recorded in kilograms, whereas the results of your laboratory tests are reported in grams, milligrams (mg), or micrograms (μg). A kilogram is equal to 1000 g. One gram represents the same mass as 1000 mg, and one mg equals 1000 μg.

Some Mass Equalities

1 kg = 1000 g

1 g = 1000 mg

1 mg = 1000 μg

FIGURE 1.8

A plastic intravenous fluid container contains 1000 mL.

Q **How many liters of solution are in the intravenous fluid container?**

SAMPLE PROBLEM 1.8

■ **Writing Metric Relationships**

Complete the following list of metric equalities:

a. 1 L = _____ dL   **b.** 1 km = _____ m
c. 1 m = _____ cm   **d.** 1 cm³ = _____ mL

SOLUTION

a. 10 dL   **b.** 1000 m
c. 100 cm   **d.** 1 mL

STUDY CHECK

Complete the following equalities:

a. 1 kg = _____ g   **b.** 1 mL = _____ L

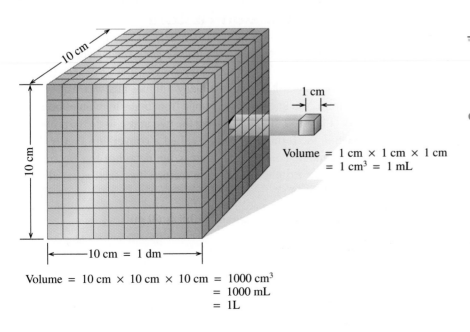

Volume = 1 cm × 1 cm × 1 cm
= 1 cm³ = 1 mL

Volume = 10 cm × 10 cm × 10 cm = 1000 cm³
= 1000 mL
= 1L

FIGURE 1.9

A cube measuring 10 cm on each side has a volume of 1000 cm³, or 1 L; a cube measuring 1 cm on each side has a volume of 1 cm³ (cc) or 1 mL.

Q **What is the relationship between a milliliter (mL) and a cubic centimeter (cm³)?**

QUESTIONS AND PROBLEMS

■Prefixes and Equalities

1.21 The speedometer in the margin is marked in both km/h and mi/h or mph. What is the meaning of each abbreviation?

1.22 In Canada, a highway sign gives a speed limit as 80 km/h. If you were going the maximum speed allowed in Canada, would you be exceeding the speed limit in the United States of 55 mph?

1.23 How does the prefix *kilo* affect the gram unit in *kilogram*?

1.24 How does the prefix *centi* affect the meter unit in *centimeter*?

1.25 Write the abbreviation for each of the following units:
   a. milligram   **b.** deciliter   **c.** kilometer
   d. kilogram   **e.** microliter

1.26 Write the complete name for each of the following units:
   a. cm   **b.** kg   **c.** dL   **d.** Gm   **e.** μg

1.27 Write the numerical values for each of the following prefixes:
   a. centi   **b.** kilo   **c.** milli   **d.** deci   **e.** mega

1.28 Write the complete name (prefix + unit) for each of the following numerical values:
   a. 0.10 g   **b.** 0.000 001 g   **c.** 1000 g
   d. $\frac{1}{100}$ g   **e.** 0.001 g

1.29 Complete the following metric relationships:
   a. 1 m = _____ cm   **b.** 1 km = _____ m
   c. 1 mm = _____ m   **d.** 1 L = _____ mL

1.30 Complete the following metric relationships:
   a. 1 kg = _____ g   **b.** 1 mL = _____ L
   c. 1 g = _____ kg   **d.** 1 g = _____ mg

1.31 For each of the following pairs, which is the larger unit?
   a. milligram or kilogram   **b.** milliliter or microliter
   c. cm or km   **d.** kL or dL

1.32 For each of the following pairs, which is the smaller unit?
   a. mg or g   **b.** centimeter or millimeter
   c. mm or μm   **d.** mL or dL

■ 1.5 Writing Conversion Factors

Many problems in chemistry and the health sciences require a change of units. You make changes in units every day. For example, suppose you spent 2.0 hours (hr) on your homework, and someone asked you how many minutes that was. You would answer 120 minutes (min). You knew how to change from hours to minutes because you knew an equality (1 hr = 60 min) that related the two units. To do the problem, the equality is written in the form of a fraction called a **conversion factor**. One of the quantities is the numerator, and the other is the denominator. Be sure to include the units when you write the conversion factors. Two factors are always possible from any equality.

Two Conversion Factors for the Equality 1 hr = 60 min

$$\frac{\text{Numerator} \longrightarrow}{\text{Denominator} \longrightarrow} \qquad \frac{60\ \text{min}}{1\ \text{hr}} \quad \text{and} \quad \frac{1\ \text{hr}}{60\ \text{min}}$$

TABLE 1.9

Some Common Equalities

Quantity	U.S.	Metric (SI)	Metric–U.S.
Length	1 ft = 12 in.	1 km = 1000 m	2.54 cm = 1 in.
	1 yard = 3 ft	1 m = 1000 mm	1 m = 39.4 in.
	1 mile = 5280 ft	1 cm = 10 mm	1 km = 0.621 mi
Volume	1 qt = 4 cups	1 L = 1000 mL	946 mL = 1 qt
	1 qt = 2 pints	1 dL = 100 mL	1 L = 1.06 qt
	1 gallon = 4 qts	1 mL = 1 cm^3	
Mass	1 lb = 16 oz	1 kg = 1000 g	1 kg = 2.20 lb
		1 g = 1000 mg	454 g = 1 lb
Time		1 hr = 60 min	
		1 min = 60 s	

These factors are read as "60 minutes per 1 hour," and "1 hour per 60 minutes." The term *per* means "divide." Some common relationships are given in Table 1.9. It is important that the equality you select to construct a conversion factor is a true relationship.

When an equality shows the relationship for two units from the same system (metric or U.S.), it is considered a definition and exact. It is not used to determine significant figures. When an equality shows the relationship of units from two different systems, the number is measured and counts toward the significant figures in a calculation. For example, in the equality 1 lb = 454 g, the measured number 454 has three significant figures. The number one in 1 lb is considered as exact. An exception is the relationship of 1 in. = 2.54 cm: the value 2.54 has been defined as exact.

Metric Conversion Factors

We can write metric conversion factors for the metric relationships we have studied. For example, from the equality for meters and centimeters, we can write the following factors:

Metric Equality	Conversion Factors		
1 m = 100 cm	$\dfrac{100 \text{ cm}}{1 \text{ m}}$	and	$\dfrac{1 \text{ m}}{100 \text{ cm}}$

Both are proper conversion factors for the relationship; one is just the inverse of the other. The usefulness of conversion factors is enhanced by the fact that we can turn a conversion factor over and use its inverse.

Metric–U.S. System Conversion Factors

Suppose you need to convert from pounds, a unit in the U.S. system, to kilograms in the metric (or SI) system. A relationship you could use is

1 kg = 2.20 lb

FIGURE 1.10

In the U.S., the contents of many packaged foods are listed in both U.S. and metric units.

Q **What are some advantages of using the metric system?**

The corresponding conversion factors would be

$$\frac{2.20 \text{ lb}}{1 \text{ kg}} \quad \text{and} \quad \frac{1 \text{ kg}}{2.20 \text{ lb}}$$

Figure 1.10 illustrates the contents of some packaged foods in both metric and U.S. units.

SAMPLE PROBLEM 1.9

■ Writing Conversion Factors from Equalities

Write conversion factors for the relationship for the following pairs of units:

a. Milligrams and grams
b. Minutes and hours
c. Quarts and milliliters

SOLUTION

Equality	Conversion Factors		
a. 1 g = 1000 mg	$\dfrac{1 \text{ g}}{1000 \text{ mg}}$	and	$\dfrac{1000 \text{ mg}}{1 \text{ g}}$
b. 1 hr = 60 min	$\dfrac{1 \text{ hr}}{60 \text{ min}}$	and	$\dfrac{60 \text{ min}}{1 \text{ hr}}$
c. 1 qt = 946 mL	$\dfrac{1 \text{ qt}}{946 \text{ mL}}$	and	$\dfrac{946 \text{ mL}}{1 \text{ qt}}$

STUDY CHECK

Write the equality and conversion factors for the relationship between inches and centimeters.

Conversion Factors Stated Within a Problem

Many times, an equality is specified within a problem that is true only for that problem. It might be the cost of one kilogram of oranges or the speed of a car in kilometers per hour. Such equalities are easy to miss when you first read a problem. Let's see how conversion factors are written from statements made within a problem.

EXPLORE YOUR WORLD

■ SI and Metric Equalities on Product Labels

Read the labels on some food products on your kitchen shelves and in your refrigerator or use the labels in Figure 1.10. List the amount of product given in different units. Write a relationship for two of the amounts for the same product and container. Look for measurements of grams and pounds or quarts and milliliters.

QUESTIONS

1. Use the stated measurement to derive a metric–U.S. conversion factor.
2. How do your results compare to the conversion factors we have described in this text?

1. The motorcycle was traveling at a speed of 85 kilometers per hour.

   Equality: 1 hr = 85 km

   Conversion factors: $\dfrac{85 \text{ km}}{1 \text{ hour}}$ and $\dfrac{1 \text{ hour}}{85 \text{ km}}$

2. One tablet contains 500 mg of vitamin C.

   Equality: 1 tablet = 500 mg vitamin C

   Conversion factors: $\dfrac{500 \text{ mg vitamin C}}{1 \text{ tablet}}$ and $\dfrac{1 \text{ tablet}}{500 \text{ mg vitamin C}}$

Conversion Factors for a Percentage

Sometimes a percentage is given in a problem. The term percent (%) means parts per 100 parts. To write a percentage as a conversion factor, we choose a unit and express the numerical relationship of the parts of this unit to 100 parts of the whole. For example, an athlete might have 18% (percent) body fat by mass. The percent quantity can be written as 18 mass units of body fat in every 100 mass units of body mass. Different mass units such as grams, kilograms (kg), or pounds (lb) can be used, but both units in the factor must be the same.

   Percent quantity: 18% body fat by mass

   Equality: 18 kg body fat = 100 kg body mass

   Conversion factors: $\dfrac{100 \text{ kg body mass}}{18 \text{ kg body fat}}$ and $\dfrac{18 \text{ kg body fat}}{100 \text{ kg body mass}}$

or

   Equality: 18 lb body fat = 100 lb body mass

   Conversion factors: $\dfrac{100 \text{ lb body mass}}{18 \text{ lb body fat}}$ and $\dfrac{18 \text{ lb body fat}}{100 \text{ lb body mass}}$

SAMPLE PROBLEM 1.10

Conversion Factors Stated in a Problem

Write the possible conversion factors for each of the following statements:

a. There are 325 mg of aspirin in 1 tablet.
b. One kilogram of bananas costs $1.25.
c. Wine is 12% alcohol by volume.

SOLUTION

a. $\dfrac{325 \text{ mg aspirin}}{1 \text{ tablet}}$ and $\dfrac{1 \text{ tablet}}{325 \text{ mg aspirin}}$

b. $\dfrac{\$1.25}{1 \text{ kg bananas}}$ and $\dfrac{1 \text{ kg bananas}}{\$1.25}$

c. A typical volume unit to use could be mL or L.

$\dfrac{12 \text{ mL alcohol}}{100 \text{ mL wine}}$ and $\dfrac{100 \text{ mL wine}}{12 \text{ mL alcohol}}$

or

$$\frac{12 \text{ L alcohol}}{100 \text{ L wine}} \quad \text{and} \quad \frac{100 \text{ L wine}}{12 \text{ L alcohol}}$$

STUDY CHECK

What conversion factors can be written for the following statements?

a. A cyclist in the Tour de France bicycle race rides at the average speed of 62.2 km/hr.

b. A 10.0 g sample of sterling silver has a volume of 94.8 cm^3.

QUESTIONS AND PROBLEMS

■ Writing Conversion Factors

1.33 Why can two conversion factors be written for an equality such as 1 m = 100 cm?

1.34 How can you check that you have written the correct conversion factors for an equality?

1.35 Write the numerical relationship and conversion factors for the following pairs of units:
  a. centimeters and meters   **b.** milligrams and grams
  c. liters and milliliters   **d.** deciliters and milliliters

1.36 Write the numerical relationship and corresponding conversion factors for the following pairs of units:
  a. centimeters and inches   **b.** pounds and kilograms
  c. pounds and grams   **d.** quarts and milliliters

■ 1.6 Problem Solving

The process of problem solving in chemistry often requires the conversion of an initial quantity given in one unit to the same quantity but in different units. By using one or more of the conversion factors we discussed in the previous section, the initial unit can be converted to the final unit.

Given quantity × One or more conversion factors = Desired quantity
(Initial unit) ————————————————→ (Final unit)

You may use a sequence similar to the steps in the following guide for problem solving (GPS):

Guide to Problem Solving (GPS) with Conversion Factors

STEP 1 **Given/Need**   State the initial unit given in the problem and the final unit needed.

STEP 2 **Plan**   Write out a sequence of units that starts with the initial unit and progresses to the final unit for the answer. Be sure you can supply the equality for each unit conversion.

STEP 3 **Equalities/Conversion Factors**   For each change of unit in your plan, state the equality and corresponding conversion factors. Recall that equalities are derived from the metric (SI) system, the U.S. system, and statements within a problem.

STEP 4 Set Up Problem Write the initial quantity and unit and set up conversion factors that connect the units. Be sure to arrange the units in each factor so the unit in the denominator cancels the preceding unit in the numerator. Check that the units cancel properly to give the final unit. Carry out the calculations, count the significant figures in each measured number, and give a final answer with the correct number of significant figures.

Suppose a problem requires the conversion of 164 lb to kilograms. One part of this statement (164 lb) is the given quantity (initial unit), while another part (kilograms) is the final unit needed for the answer. Once you identify these units, you can determine which equalities you need to convert the initial unit to the final unit.

STEP 1 Given 164 lb   **Need** kg

STEP 2 Plan It is helpful to decide on a plan of units. When we look at the initial units given and the final units needed, we see that one is a metric unit and the other is a unit in the U.S. system of measurement. Therefore, the connecting conversion factor must be one that includes a metric and a U.S. unit.

lb → Metric–U.S. factor → kg

STEP 3 Equalities/Conversion Factors From the discussion on U.S. and metric equalities, we can write the following equality and conversion factors:

$$1 \text{ kg} = 2.20 \text{ lb}$$

$$\frac{2.20 \text{ lb}}{1 \text{ kg}} \quad \text{and} \quad \frac{1 \text{ kg}}{2.20 \text{ lb}}$$

Guide to Problem Solving
Using Conversion Factors

STEP 1
State the given and needed units.

STEP 2
Write a unit plan to convert the given unit to the final unit.

STEP 3
State the equalities and conversion factors to cancel units.

STEP 4
Set up problem to cancel units and calculate answer.

STEP 4 Set Up Problem Now we can write the setup to solve the problem using the unit plan and a conversion factor. First, write down the initial unit, 164 lb. Then multiply by the conversion factor that has the unit lb in the denominator to cancel out the initial unit. The unit kg in the numerator (top number) gives the final unit for the answer.

Unit for answer goes here

$$164 \ \cancel{\text{lb}} \quad \times \quad \frac{1 \text{ kg}}{2.20 \ \cancel{\text{lb}}} \quad = \quad 74.5 \text{ kg}$$

Given   Conversion factor   Answer
(initial unit)   (cancels initial unit)   (desired unit)

Take a look at how the units cancel. The unit that you want in the answer is the one that remains after all the other units have cancelled out. This is a helpful way to check that a problem is set up properly.

$$\cancel{\text{lb}} \times \frac{\text{kg}}{\cancel{\text{lb}}} = \text{kg} \qquad \text{Unit needed for answer}$$

The calculation done on a calculator gives the numerical part of the answer. The calculator answer is adjusted to give a final answer with the proper number of significant figures (SFs).

$$164 \times \frac{1}{2.20} = 164 \div 2.20 = \boxed{74.54545455} = 74.5$$

3 SF   3 SF   Calculator answer   3 SF (rounded)

The value of 74.5 combined with the final unit, kg, gives the final answer of 74.5 kg. With few exceptions, answers to numerical problems contain a number and a unit.

SAMPLE PROBLEM 1.11

■ **Problem Solving Using Metric Factors**

The recommended amount of sodium in the diet per day is 2400 mg. How many grams of sodium is that?

SOLUTION

STEP 1 Given   2400 mg   **Need**   g

STEP 2 Plan   When we look at the initial units given and the final units needed, we see that both are metric units. Therefore, the connecting conversion factor must relate two metric units.

mg   Metric factor   g

STEP 3 Equalities/Conversion Factors   From the discussion on prefixes and metric equalities, we can write the following equality and conversion factors:

$$1 \text{ g} = 1000 \text{ mg}$$

$$\frac{1 \text{ g}}{1000 \text{ mg}} \quad \text{and} \quad \frac{1000 \text{ mg}}{1 \text{ g}}$$

STEP 4 Set Up Problem   We write the setup using the unit plan and a conversion factor starting with the initial unit, 2400 mg. The final answer (g) is obtained by using the conversion factor that cancels the unit mg.

Unit for answer goes here

$$2400 \text{ mg} \times \frac{1 \text{ g}}{1000 \text{ mg}} = 2.4 \text{ g}$$

Given   Metric factor   Answer

STUDY CHECK

If 1890 mL of orange juice is prepared from orange juice concentrate, how many liters of orange juice is that?

Using Two or More Conversion Factors

In many problems, two or more conversion factors are needed to complete the change of units. In setting up these problems, one factor follows the other. Each factor is arranged to cancel the preceding unit until the final unit is obtained. Up to this point, we have used the conversion factors one at a time and calculated an answer. You can work all problems in single steps, but if you do, be sure to keep one or two extra digits in the intermediate answers and only round off the final answer to the correct number of significant figures. A more efficient way to do these problems is to use a series of two or more conversion factors set up so that the unit in the denominator of each factor cancels the unit in the preceding numerator. Both these approaches are illustrated in the following Sample Problem.

SAMPLE PROBLEM 1.12

■ **Problem Solving Using Two Factors**

A recipe for salsa requires 3 cups of tomato sauce, which can be measured accurately as 3.0 cups. If only metric measures are available, how many liters of tomato sauce are needed? (There are 4 cups in 1 quart.)

SOLUTION

STEP 1 **Given**   3.0 cups tomato sauce   **Need**   liters (L)

STEP 2 **Plan**   We see that the initial unit, cups, needs to be changed to a final unit of liters, but we do not know an equality for cups and liters. However, we do know the U.S. equality for cups and quarts and the metric–U.S. equality for quarts and liters as shown in the following plan:

cups   U.S. factor   quarts   Metric–U.S. factor   liters

STEP 3 **Equalities/Conversion Factors**   From the discussion on U.S. and metric equalities, we can write the following equalities and conversion factors:

$$1 \text{ qt} = 4 \text{ cups} \qquad 1 \text{ L} = 1.06 \text{ qt}$$

$$\frac{1 \text{ qt}}{4 \text{ cups}} \quad \text{and} \quad \frac{4 \text{ cups}}{1 \text{ qt}} \qquad \frac{1 \text{ L}}{1.06 \text{ qt}} \quad \text{and} \quad \frac{1.06 \text{ qt}}{1 \text{ L}}$$

STEP 4 **Set Up Problem**   Working in single steps we can use the U.S. factor to convert from cups to quarts:

$$3.0 \text{ cups} \times \frac{1 \text{ qt}}{4 \text{ cups}} = 0.75 \text{ qt}$$

Then we use the metric–U.S. factor to cancel quarts and give liters as the unit.

$$0.75 \text{ qt} \times \frac{1 \text{ L}}{1.06 \text{ qt}} = 0.71 \text{ L}$$

When set up as a series, the first factor cancels cups, and the second factor cancels quarts, which gives liters as the final unit for the answer.

$$\text{cups} \times \frac{\text{qt}}{\text{cups}} \times \frac{\text{L}}{\text{qt}} = \text{L}$$

$$3.0 \text{ cups} \times \frac{1 \text{ qt}}{4 \text{ cups}} \times \frac{1 \text{ L}}{1.06 \text{ qt}} = 0.71 \text{ L}$$

Given   U.S.   Metric–U.S.   Answer
(initial unit) factor   factor   (desired unit)

The calculation is done in a sequence on a calculator to give the numerical part of the answer. The calculator answer is adjusted to give a final answer with the proper number of significant figures (SF).

$$3.0 \;\boxed{\div}\; 4 \;\boxed{\div}\; 1.06 = \boxed{0.707547} = 0.71$$

2 SF   Exact   3 SF   Calculator answer   2 SF (rounded)

STUDY CHECK

One medium bran muffin contains 4.2 g of fiber. How many ounces (oz) of fiber are obtained by eating three medium bran muffins if 1 lb = 16 oz? (*Hint*: number of muffins $\longrightarrow$ g of fiber $\longrightarrow$ lb $\longrightarrow$ oz.)

■

Using a sequence of two or more conversion factors is a very efficient way to set up and solve problems, especially if you are using a calculator. Once you have the problem set up, the calculations can be done without writing out the intermediate values. This process is worth practicing until you understand unit cancellation and the mathematical calculations.

Clinical Calculations Using Conversion Factors

Conversion factors are also useful for calculating medications. For example, if an antibiotic is available in 5-mg tablets, the dosage can be written as a conversion factor, 5 mg/1 tablet. In many hospitals, the apothecary unit of grains (gr) is still in use; there are 65 mg in 1 gr. When you do a medication problem, you often start with a doctor's order that contains the quantity to give the patient. The medication dosage is used as a conversion factor.

SAMPLE PROBLEM 1.13

■ **Clinical Calculations**

Synthroid is used as a replacement or supplemental therapy for diminished thyroid function. A dosage of 0.200 mg is prescribed with tablets that contain 50 μg Synthroid. How many tablets are required to provide the prescribed medication?

SOLUTION

STEP 1 Given   0.200 mg synthroid   **Need**   tablets

STEP 2 Plan   mg   [Metric factor] μg   [Clinical factor]   tablets

STEP 3 Equalities/Conversion Factors   In the problem, the information for the dosage is given as 50 μg per tablet. We will use this as one of the equalities as well as the metric equality for milligrams and micrograms and write conversion factors for each.

1 mg = 1000 μg		1 tablet = 50 μg	
$\dfrac{1 \text{ mg}}{1000 \text{ } \mu g}$ and $\dfrac{1000 \text{ } \mu g}{1 \text{ mg}}$		$\dfrac{1 \text{ tablet}}{50 \text{ } \mu g}$ and $\dfrac{50 \text{ } \mu g}{1 \text{ tablet}}$	

STEP 4 Set Up Problem   The problem can be set up using the metric factor to cancel "milligrams," and then the clinical factor to obtain "tablets" as the final unit.

$$0.200 \; \cancel{mg} \; \times \; \frac{1000 \; \cancel{\mu g}}{1 \; \cancel{mg}} \; \times \; \frac{1 \; \text{tablet}}{50 \; \cancel{\mu g}} \; = 4 \; \text{tablets}$$

STUDY CHECK

An antibiotic dosage of 500 mg is ordered. If the antibiotic is supplied in liquid form as 250 mg in 5.0 mL, how many mL would be given?

QUESTIONS AND PROBLEMS

■ Problem Solving

1.37 When you convert one unit to another, how do you know which unit to place in the denominator?

1.38 When you convert one unit to another, how do you know which unit to place in the numerator?

1.39 Use metric conversion factors to solve the following problems:
  a. The height of a student is 175 centimeters. How tall is the student in meters?
  b. A cooler has a volume of 5500 mL. What is the capacity of the cooler in liters?
  c. A hummingbird has a mass of 0.0055 kg. What is the mass of the hummingbird in grams?

1.40 Use metric conversion factors to solve the following problems:
  a. The daily value of phosphorus is 800 mg. How many grams of phosphorus are recommended?
  b. A glass of orange juice contains 0.85 dL of juice. How many milliliters of orange juice is that?
  c. A package of chocolate instant pudding contains 2840 mg of sodium. How many grams of sodium is that?

1.41 Solve the following problems using one or more conversion factors:
  a. A container holds 0.500 qt of liquid. How many milliliters of lemonade will it hold?
  b. What is the mass in kilograms of a person who weighs 145 lb?
  c. An athlete normally has 15% (by mass) body fat. How many lb of fat does a 74-kg athlete have?

1.42 Solve the following problems using one or more conversion factors:
  a. You need 4.0 ounces of a steroid ointment. If there are 16 oz in 1 lb, how many grams of ointment does the pharmacist need to prepare?
  b. During surgery, a patient receives 5.0 pints of plasma. How many milliliters of plasma were given? (1 quart = 2 pints)
  c. Wine is 12% by volume alcohol. How many mL of alcohol are in a 0.750 L bottle of wine?

1.43 Using conversion factors, solve the following clinical problems:
  a. You have used 250 L of distilled water for a dialysis patient. How many gallons of water is that?
  b. A patient needs 0.024 g of a sulfa drug. There are 8-mg tablets in stock. How many tablets should be given?
  c. The daily dose of ampicillin for the treatment of an ear infection is 115 mg/kg of body weight. What is the daily dose for a 34-lb child?

1.44 Using conversion factors, solve the following clinical problems:
  a. The physician has ordered 1.0 g of tetracycline to be given every 6 hours to a patient. If your stock on hand is 500-mg tablets, how many will you need for 1 day's treatment?
  b. An intramuscular medication is given at 5.00 mg/kg of body weight. If you give 425 mg of medication to a patient, what is the patient's weight in pounds?
  c. A physician has ordered 0.50 mg of atropine, intramuscularly. If atropine were available as 0.10 mg/mL of solution, how many milliliters would you need to give?

Cork (D = 0.26 g/mL)
Ice (D = 0.92 g/mL)
H_2O (D = 1.00 g/mL)
Aluminum (D = 2.70 g/mL)
Lead (D = 11.3 g/mL)

1.7 Density

Differences in density determine whether an object will sink or float. In Figure 1.11, the density of lead is greater than the density of water, and the lead object sinks. The cork floats because cork is less dense than water.

The mass and volume of any object can be measured. However, the separate measurements do not tell us how tightly packed the substance might be. If we compare the mass of the object to its volume, we obtain a relationship called **density**.

$$\text{Density} = \frac{\text{mass of substance}}{\text{volume of substance}}$$

In the metric system, the densities of solids and liquids are usually expressed as grams per cubic centimeter (g/cm^3) or grams per milliliter (g/mL). The density of gases is usually stated as grams per liter (g/L). Table 1.10 gives the densities of some common substances.

Density of Solids

The density of a solid is calculated from its mass and volume. When a solid is completely submerged, it displaces a volume of water that is *equal to its own volume.*

TABLE 1.10

Densities of Some Common Substances

Solids (at 25°C)	Density (g/mL)	Liquids (at 25°C)	Density (g/mL)	Gases (at 0°C)	Density (g/L)
Cork	0.26	Gasoline	0.66	Hydrogen	0.090
Wood (maple)	0.75	Ethanol	0.79	Helium	0.179
Ice	0.92	Olive oil	0.92	Methane	0.714
Sugar	1.59	Water (at 4°C)	1.00	Neon	0.90
Bone	1.80	Plasma (blood)	1.03	Nitrogen	1.25
Aluminum	2.70	Urine	1.003–1.030	Air (dry)	1.29
Cement	3.00	Milk	1.04	Oxygen	1.43
Diamond	3.52	Mercury	13.6	Carbon dioxide	1.96
Silver	10.5				
Lead	11.3				
Gold	19.3				

Mass of zinc object   Submerged zinc object

FIGURE 1.12

The density of a solid can be determined by volume displacement because a submerged object displaces a volume of water equal to its own volume.

Q How is the volume of zinc determined?

In Figure 1.12, the water level rises from 35.5 mL to 45.0 mL. This means that 9.5 mL of water is displaced and that the volume of the object is 9.5 mL. The density of the zinc is calculated as follows:

$$\text{Density} = \frac{68.60 \text{ g zinc}}{9.5 \text{ mL}} = 7.2 \text{ g/mL}$$

SAMPLE PROBLEM 1.14

■ Using Volume Displacement to Calculate Density

A lead weight used in the belt of a scuba diver has a mass of 226 g. When the weight is carefully placed in a graduated cylinder containing 200.0 mL of water, the water level rises to 220.0 mL. What is the density of the lead weight (g/mL)?

SOLUTION

STEP 1 Given mass = 226 g; water level before object submerged = 200.0 mL; water level after object submerged = 220.0 mL
Need density (g/mL)

STEP 2 Plan To calculate density, substitute the mass (g) and the volume (mL) of the lead weight into the expression for density.

STEP 3 Equality/Conversion Factor

$$\text{Density} = \frac{\text{mass of substance}}{\text{volume of substance}}$$

STEP 4 Set Up Problem The volume of the lead weight is equal to the volume of water displaced, which is calculated as follows:

Water level after object submerged	= 220.0 mL
− Water level before object submerged	= 200.0 mL
Water displaced (volume of lead weight) =	20.0 mL

The density is calculated by dividing the mass (g) by the volume (mL). Be sure to use the volume of water the object displaced and *not* the original volume of water.

$$\text{Density} = \underset{\text{3 SF}}{\frac{226 \text{ g}}{20.0 \text{ mL}}} = \underset{\text{3 SF}}{\frac{11.3 \text{ g}}{1 \text{ mL}}} = 11.3 \text{ g/mL}$$

STUDY CHECK

A total of 0.50 lb of glass marbles is added to 425 mL of water. The water level rises to a volume of 528 mL. What is the density (g/mL) of the glass marbles?

Problem Solving Using Density

Density can be used as a conversion factor. For example, if the volume and the density of a sample are known, the mass in grams of the sample can be calculated.

■ Problem Solving Using Density

If the density of milk is 1.04 g/mL, how many grams of milk are in 0.50 qt of milk?

SOLUTION

STEP 1 Given 0.50 qt   **Need** g

STEP 2 Plan

qt → U.S.–metric factor → L → Metric factor → mL → Density factor → g

STEP 3 Equalities/Conversion Factors

$$1 \text{ L} = 1.06 \text{ qt}$$
$$\frac{1 \text{ L}}{1.06 \text{ qt}} \text{ and } \frac{1.06 \text{ qt}}{1 \text{ L}}$$

$$1 \text{ L} = 1000 \text{ mL}$$
$$\frac{1 \text{ L}}{1000 \text{ mL}} \text{ and } \frac{1000 \text{ mL}}{1 \text{ L}}$$

$$1 \text{ mL} = 1.04 \text{ g}$$
$$\frac{1 \text{ mL}}{1.04 \text{ g}} \text{ and } \frac{1.04 \text{ g}}{1 \text{ mL}}$$

STEP 4 Set Up Problem

$$0.50 \text{ qt} \times \underset{\text{3 SF}}{\frac{1 \text{ L}}{1.06 \text{ qt}}} \times \underset{\text{Exact}}{\frac{1000 \text{ mL}}{1 \text{ L}}} \times \underset{\text{3 SF}}{\frac{1.04 \text{ g}}{1 \text{ mL}}} = \underset{\text{2 SF}}{490 \text{ g} \ (4.9 \times 10^2 \text{g})}$$

(2 SF)

STUDY CHECK

How many mL of mercury are in a thermometer that contains 20.4 g of mercury? (See Table 1.10 for the density of mercury.)

EXPLORE YOUR WORLD

■ Sink or Float?

1. Fill a large container or bucket with water. Place a can of diet and a can of nondiet soft drink in the water. What happens? Using information on the label, how might you account for your observations?
2. Design an experiment to determine the substance that is the most dense in each of the following:
   a. water and vegetable oil
   b. water and ice
   c. rubbing alcohol and ice
   d. vegetable oil, water, and ice

Specific Gravity

Specific gravity (sp gr) is a ratio between the density of a substance and the density of water. Specific gravity is calculated by dividing the density of a sample by the density of water. In this text, we will use the metric value of 1.00 g/mL for the density of water. A substance with a specific gravity of 1.00 has the same density as water. A substance with a specific gravity of 3.00 is three times as dense as water, whereas a substance with a specific gravity of 0.50 is just one-half as dense as water.

$$\text{Specific gravity} = \frac{\text{density of sample}}{\text{density of water}}$$

In the calculations for specific gravity, the units of density must match. Then all units cancel to leave only a number. *Specific gravity is one of the few unitless values you will encounter in chemistry.*

An instrument called a hydrometer is often used to measure the specific gravity of fluids such as battery fluid or a sample of urine. In Figure 1.13, a hydrometer is used to measure the specific gravity of a fluid.

SAMPLE PROBLEM 1.16

■ **Specific Gravity**

What is the specific gravity of coconut oil that has a density of 0.925 g/mL?

SOLUTION

$$\text{sp gr oil} = \frac{\text{density of oil}}{\text{density of water}} = \frac{0.925 \ \cancel{\text{g/mL}}}{1.00 \ \cancel{\text{g/mL}}} = 0.925 \quad \text{No units}$$

STUDY CHECK

What is the specific gravity of ice if 35.0 g of ice has a volume of 38.2 mL?

SAMPLE PROBLEM 1.17

■ **Problem Solving with Specific Gravity**

John took 2.0 teaspoons (tsp) of cough syrup (sp gr 1.20) for a persistent cough. If there is 5.0 mL in 1 tsp, what was the mass (in grams) of the cough syrup?

SOLUTION

STEP 1 Given 2.0 tsp   **Need** grams

STEP 2 Plan

tsp → [U.S.–metric factor] → mL → [Density factor] → g

STEP 3 Equalities/Conversion Factors For problem solving, it is convenient to convert the specific gravity value (1.20) to density

$$\text{Density} = (\text{sp gr}) \times 1.00 \text{ g/mL} = 1.20 \text{ g/mL}$$

1 tsp = 5.0 mL		1 mL = 1.20 g	
$\dfrac{5.0 \text{ mL}}{1 \text{ tsp}}$ and $\dfrac{1 \text{ tsp}}{5.0 \text{ mL}}$		$\dfrac{1 \text{ mL}}{1.20 \text{ g}}$ and $\dfrac{1.20 \text{ g}}{1 \text{ mL}}$	

FIGURE 1.13

When the specific gravity of beer measures 1.010 or less with a hydrometer, the fermentation process is complete.

Q **If the hydrometer reading is 1.006, what is the density of the liquid?**

▪ Determination of Percentage of Body Fat

Body mass is made up of protoplasm, extracellular fluid, bone, and adipose tissue. One way to determine the amount of adipose tissue is to measure the whole-body density. After the on-land mass of the body is determined, the underwater body mass is obtained by submerging the person in water. Because water helps support the body by giving it buoyancy, the underwater body mass is less. A higher percentage of body fat will make a person more buoyant, causing the underwater mass to be even lower. This occurs because fat has a lower density than the rest of the body.

The difference between the on-land mass and underwater mass, known as the buoyant force, is used to determine the body volume. Then the mass and volume of the person are used to calculate body density. For example, suppose a 70.0-kg person has a body volume of 66.7 L. The body density is calculated as

$$\frac{\text{Body mass}}{\text{Body volume}} = \frac{70.0 \text{ kg}}{66.7 \text{ L}}$$

$$= 1.05 \text{ kg/L} \quad \text{or} \quad 1.05 \text{ g/mL}$$

When the body density is determined, it is compared with a chart that correlates the percentage of adipose tissue with body density. A person with a body density of 1.05 g/mL has 21% body fat, according to such a chart. This procedure is used by athletes to determine exercise and diet programs.

STEP 4 Set Up Problem

$$2.0 \text{ tsp} \times \frac{5.0 \text{ mL}}{1 \text{ tsp}} \times \frac{1.20 \text{ g}}{1 \text{ mL}} = 12 \text{ g syrup}$$

STUDY CHECK

An ebony carving has a mass of 275 g. If ebony has a specific gravity of 1.33, what is the volume of the carving?

QUESTIONS AND PROBLEMS

▪ Density

1.45 What is the density (g/mL) of each of the following samples?
- **a.** A 20.0-mL sample of a salt solution that has a mass of 24.0 g.
- **b.** A solid object with a mass of 1.65 lb and a volume of 170 mL.
- **c.** A gem has a mass of 45.0 g. When the gem is placed in a graduated cylinder containing 20.0 mL of water, the water level rises to 34.5 mL.

1.46 What is the density of each of the following samples?
- **a.** A medication, if 3.00 mL has a mass of 3.85 g.
- **b.** The fluid in a car battery, if it has a volume of 125 mL and a mass of 155 g.
- **c.** A 5.00-mL urine sample from a patient suffering from symptoms resembling those of diabetes mellitus. The mass of the urine sample is 5.025 g.

1.47 Use the density value to solve the following problems:
- **a.** What is the mass, in grams, of 150 mL of a liquid with a density of 1.4 g/mL?
- **b.** What is the mass of a glucose solution that fills a 0.500-L intravenous bottle if the density of the glucose solution is 1.15 g/mL?
- **c.** A sculptor has prepared a mold for casting a bronze figure. The figure has a volume of 225 mL. If bronze has a density of 7.8 g/mL, how many ounces of bronze are needed in the preparation of the bronze figure?

1.48 Use the density value to solve the following problems:
- **a.** A graduated cylinder contains 18.0 mL of water. What is the new water level after 35.6 g of silver metal with a density of 10.5 g/mL is submerged in the water?
- **b.** A thermometer containing 8.3 g of mercury has broken. If mercury has a density of 13.6 g/mL, what volume spilled?
- **c.** A fish tank holds 35 gallons of water. Using the density of 1.0 g/mL for water, determine the number of pounds of water in the fish tank.

1.49 Solve the following specific gravity problems:
- **a.** A urine sample has a density of 1.030 g/mL. What is the specific gravity of the sample?
- **b.** A liquid has a volume of 40.0 mL and a mass of 45.0 g. What is the specific gravity of the liquid?
- **c.** The specific gravity of a vegetable oil is 0.85. What is its density?

1.50 Solve the following specific gravity problems:
- **a.** A 5.0% glucose solution has a specific gravity of 1.02. What is the mass of 500. mL of glucose solution?
- **b.** A bottle containing 325 g of cleaning solution is used for carpets. If the cleaning solution has a specific gravity of 0.850, what volume of solution was used?
- **c.** Butter has a specific gravity of 0.86. What is the mass, in grams, of 2.15 L of butter?

▪ Specific Gravity of Urine

The specific gravity of urine is often determined as part of a laboratory evaluation of the health of an individual. The specific gravity of urine is normally in the range 1.003–1.030. This is somewhat greater than the specific gravity of water because compounds such as salts and urea are dissolved in urine.

If the specific gravity of a person's urine is too low or too high, a physician might suspect a kidney problem. For example, if a urine sample shows a specific gravity of 1.001, significantly lower than normal, malfunctioning of the kidneys is a possibility. When water has been lost because of dehydration, the urine may have a specific gravity above normal.

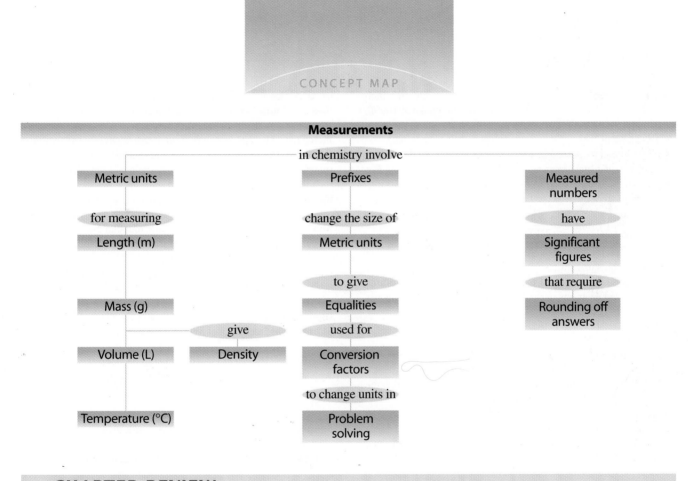

CONCEPT MAP

CHAPTER REVIEW

1.1 Units of Measurement

In science, physical quantities are described in units of the metric or International System (SI). Some important units are meter (m) for length, liter (L) for volume, gram (g) and kilogram (kg) for mass, and Celsius (°C) for temperature. Large and small numbers can be written using scientific notation in which the decimal point is moved to give a coefficient between 1 and 10 and the number of decimal places moved shown as a power of 10. A large number will have a positive power of 10, while a small number will have a negative power of 10.

1.2 Measured Numbers and Significant Figures

A measured number is any number obtained by using a measuring device. An exact number is obtained by counting items or from a definition; no measuring device is used. Significant figures are the numbers reported in a measurement including the last estimated digit. Zeros in front of a decimal number or at the end of a large number are not significant.

1.3 Significant Figures in Calculations

In multiplication or division, the final answer is written so it has the same number of significant figures as the measurement with the fewest significant figures. In addition or subtraction, the final answer is written so it has the same number of decimal places as the measurement with the fewest decimal places.

1.4 Prefixes and Equalities

Prefixes placed in front of a unit change the size of the unit by factors of 10. Prefixes such as *centi, milli,* and *micro* provide smaller units; prefixes such as *kilo* provide larger units. An equality relates two metric units that measure the same quantity of length, volume, or mass. Examples of metric equalities are 1 m = 100 cm; 1 L = 1000 mL; 1 kg = 1000 g.

1.5 Writing Conversion Factors

Conversion factors are used to express a relationship in the form of a fraction. Two factors can be written for any relationship in the metric or U.S. system.

1.6 Problem Solving

Conversion factors are useful when changing a quantity expressed in one unit to a quantity expressed in another unit. In the process, a given unit is multiplied by one or more conversion factors that cancel units until the desired answer is obtained.

1.7 Density

The density of a substance is a ratio of its mass to its volume, usually g/mL or g/cm³. The units of density can be used as a factor to convert between the mass and volume of a substance. Specific gravity (sp gr) compares the density of a substance to the density of water, 1.00 g/mL.

KEY TERMS

Celsius (°C) temperature scale A temperature scale on which water has a freezing point of 0°C and a boiling point of 100°C.

centimeter (cm) A unit of length in the metric system; there are 2.54 cm in 1 in.

conversion factor A ratio in which the numerator and denominator are quantities from an equality or given relationship. For example, the conversion factors for the relationship 1 kg = 2.20 lb are written as the following:

$$\frac{2.20 \text{ lb}}{1 \text{ kg}} \quad \text{and} \quad \frac{1 \text{ kg}}{2.20 \text{ lb}}$$

cubic centimeter (cm³, cc) The volume of a cube that has 1-cm sides, equal to 1 mL.

density The relationship of the mass of an object to its volume expressed as grams per cubic centimeter (g/cm³), grams per milliliter (g/mL), or grams per liter (g/L).

equality A relationship between two units that measure the same quantity.

exact number A number obtained by counting or definition.

gram (g) The metric unit used in measurements of mass.

Kelvin (K) temperature scale A temperature scale on which the lowest possible temperature is 0 K.

kilogram (kg) A metric mass of 1000 g, equal to 2.20 lb. The kilogram is the SI standard unit of mass.

liter (L) The metric unit for volume that is slightly larger than a quart.

mass A measure of the quantity of material in an object.

measured number A number obtained when a quantity is determined by using a measuring device.

meter (m) The metric unit for length that is slightly longer than a yard. The meter is the SI standard unit of length.

metric system A system of measurement used by scientists and in most countries of the world.

milliliter (mL) A metric unit of volume equal to one-thousandth of a L (0.001 L).

prefix The part of the name of a metric unit that precedes the base unit and specifies the size of the measurement. All prefixes are related on a decimal scale.

scientific notation A form of writing large and small numbers using a coefficient between 1 and 10, followed by a power of 10.

significant figures The numbers recorded in a measurement.

SI units The International System of units that modifies the metric system.

specific gravity (sp gr) A relationship between the density of a substance and the density of water:

$$\text{sp gr} = \frac{\text{density of sample}}{\text{density of water}}$$

temperature An indicator of the hotness or coldness of an object.

volume The amount of space occupied by a substance.

ADDITIONAL QUESTIONS AND PROBLEMS

This group of questions and problems is related to the topics in this chapter. However, they do not all follow the chapter order and they require you to combine concepts and skills from several sections. These problems will help you increase your critical thinking skills and help you prepare for your next exam.

1.51 During a workout at the gym, you set the treadmill at a pace of 55.0 meters per minute. How many minutes will you walk if you cover a distance of 7500 feet?

1.52 A fish company delivers 22 kg of salmon, 5.5 kg of crab, and 3.48 kg of oysters to your seafood restaurant.
   a. What is the total mass, in kilograms, of the seafood?
   b. What is the total number of pounds?

1.53 In France, grapes are 1.40 Euros per kilogram. What is the cost of grapes in dollars per pound, if the exchange rate is 1.30 dollars per Euro?

1.54 In Mexico, avocados are 125 pesos per kilogram. What is the cost in cents of an avocado that weighs 0.45 lb if the exchange rate is 11 pesos to the dollar?

1.55 Bill's recipe for onion soup calls for 4.0 lb of thinly sliced onions. If an onion has an average mass of 115 g, how many onions does Bill need?

1.56 The price of 1 pound (lb) of potatoes is $1.75. If all the potatoes sold today at the store bring in $1420, how many kilograms (kg) of potatoes did grocery shoppers buy?

1.57 The following nutrition information is listed on a box of crackers:
Serving size 0.50 oz (6 crackers)
Fat 4 g per serving   Sodium 140 mg per serving
  a. If the box has a net weight (contents only) of 8.0 oz, about how many crackers are in the box?
  b. If you ate 10 crackers, how many ounces of fat are you consuming?
  c. How many grams of sodium are used to prepare 50 boxes of crackers? (*Hint*: Refer to part a.)

1.58 A dialysis unit requires 75,000 mL of distilled water. How many gallons of water are needed? (1 gal = 4 qt)

1.59 To prevent bacterial infection, a doctor orders 4 tablets of amoxicillin per day for 10 days. If each tablet contains 250 mg of amoxicillin, how many ounces of the medication are given in 10 days?

1.60 Celeste's diet restricts her intake of protein to 24 g per day. If she eats 1.2 oz of protein, has she exceeded her protein limit for the day?

1.61 What is a cholesterol level of 1.85 g/L in the standard units of mg/dL?

1.62 An object has a mass of 3.15 oz. When it is submerged in a graduated cylinder initially containing 325.2 mL of water, the water level rises to 442.5 mL. What is the density (g/mL) of the object?

1.63 The density of lead is 11.3 g/mL. The water level in a graduated cylinder initially at 215 mL rises to 285 mL after a piece of lead is submerged. What is the mass in grams of the lead?

1.64 A graduated cylinder contains 155 mL of water. A 15.0-g piece of iron (density = 7.86 g/cm^3) and a 20.0-g piece of lead (density = 11.3 g/cm^3) are added. What is the new water level in the cylinder?

1.65 How many cubic centimeters (cm^3) of olive oil have the same mass as 1.00 L of gasoline? (See Table 1.10.)

1.66 Ethyl alcohol has a specific gravity of 0.79. What is the volume, in quarts, of 1.50 kg of alcohol?

1.67 In a process called liposuction, a doctor removes some adipose tissue from a person's body. If body fat has a density of 0.94 g/mL and 3.0 liters of fat are removed, how many pounds of fat were removed from the patient? (See the Health Note "Determination of Percentage of Body Fat.")

1.68 When a 50.0-kg person is immersed in a water tank for body-fat testing, she has an underwater mass of 2.0 kg. This difference in body mass is equal to the mass of water displaced by the body. The density of water is 1.00 g/mL. (See the Health Note "Determination of Percentage of Body Fat.")
  a. What volume of water is displaced?
  b. What is the volume, in L, of the person?
  c. What is the body density, in g/mL, of that person?

1.69 A urinometer is a hydrometer used to measure the specific gravity of a sample of urine.
  a. What is the density of a urine sample that has a specific gravity of 1.012?
  b. What is the mass of a 5.00-mL urine sample with a specific gravity of 1.022?

1.70 A normal range for specific gravity of urine is 1.003 to 1.030. A 10.0-mL sample of urine has a mass of 10.31 g. What is the specific gravity of the urine? Is the urine considered normal? Why or why not?

UNDERSTANDING THE CONCEPTS

1.71 Indicate if each of the following is answered with an exact number or a measured number.

  a. Number of legs   **b.** Height of table
  c. Number of chairs at the table   **d.** Area of table top

1.72 Measure the length of each of the objects in figure **(a)**, **(b)**, and **(c)** using the metric rule in the figure. Indicate the number of significant figures for each and the estimated digit for each.

(a)

(b)

(c)

1.73 Measure the length and width of the rectangle using a metric rule.

a. What is the length and width of this rectangle measured in centimeters?

b. What is the length and width of this rectangle measured in millimeters?

c. How many significant figures are in the length measurement?

d. How many significant figures are in the width measurement?

e. What is the area of the rectangle in cm^2?

f. How many significant figures are in the calculated answer for area?

1.74 Each of the following diagrams represents a container of water and a cube. Some cubes float while others sink. Match diagrams A, B, C, or D with one of the following descriptions and explain your choices.

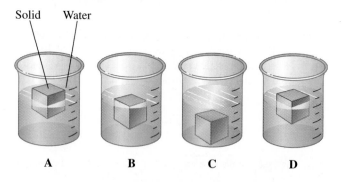

a. The cube has a greater density than water.

b. The cube has a density that is 0.60–0.80 g/mL.

c. The cube has a density that is $\frac{1}{2}$ the density of water.

d. The cube has the same density of water.

1.75 What is the density of the solid object that is weighed and submerged in water?

1.76 Consider the following solids.
The solids A, B, and C represent lead, copper, and aluminum. If each has a mass of 10.0 g, what is the identity of each solid?

Density aluminum = 2.70 g/mL

Density gold = 19.3 g/mL

Density silver = 10.5 g/mL

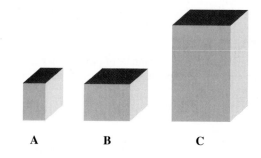

QUESTIONS FOR TEAMWORK

1.77 A balance measures mass to 0.001 g. If you determine the mass of an object that weighs about 30 g, would you record the mass as 30 g, 32.5 g, 31.25 g, 34.075 g, or 3000 g? Explain your choice by writing 2–3 complete sentences that describe your thinking.

1.78 When three students use the same meter stick to measure the length of a paper clip, they obtain results of 5.8 cm, 5.75 cm, and 5.76 cm. If the meter stick has millimeter markings, what are some reasons for the different values?

1.79 A car travels at 55 miles per hour and gets 11 kilometers per liter of gasoline. How many gallons of gasoline are needed for a 3.0-hour trip?

1.80 A 50.0-g silver object and a 50.0-g gold object are both added to 75.5 mL of water contained in a graduated cylinder. What is the new water level in the cylinder?

1.81 In the manufacturing of computer chips, cylinders of silicon are cut into thin wafers that are 3.00 inches in diameter and have a mass of 1.50 g of silicon. How thick (mm) is each wafer if silicon has a density of 2.33 g/cm^3? (The volume of a cylinder is $V = \pi r^2 h$.)

1.82 A sunscreen preparation contains 2.50% by mass benzyl salicylate. If a tube contains 4.0 ounces of sunscreen, how many kilograms of benzyl salicylate are needed to manufacture 325 tubes of sunscreen?

ANSWERS

Answers to Study Checks

1.1 **a.** meter (m)   **b.** kelvin (K)
     c. kilogram (kg)

1.2 **a.** 4.25×10^5 m   **b.** 8×10^{-7} g

1.3 **a.** two   **b.** one   **c.** five

1.4 **a.** 36 m   **b.** 0.0026 L
     c. 3800 g   **d.** 1.3 kg

1.5 **a.** 0.4924   **b.** 0.0080 or 8.0×10^{-3}
     c. 2.0

1.6 **a.** 83.70 mg   **b.** 0.5 L

1.7 **a.** mega   **b.** centi

1.8 **a.** 1000   **b.** 0.001 or $\frac{1}{1000}$

1.9 Equality: 1 in. = 2.54 cm

Conversion Factors: $\dfrac{1 \text{ in.}}{2.54 \text{ cm}}$ and $\dfrac{2.54 \text{ cm}}{1 \text{ in.}}$

1.10 **a.** $\dfrac{62.2 \text{ km}}{1 \text{ hr}}$; $\dfrac{1 \text{ hr}}{62.2 \text{ km}}$

     b. $\dfrac{94.8 \text{ cm}^3}{10.0 \text{ g}}$; $\dfrac{10.0 \text{ g}}{94.8 \text{ cm}^3}$

1.11 1.89 L

1.12 0.44 oz

1.13 10 mL

1.14 2.2 g/mL

1.15 1.50 mL

1.16 0.916

1.17 207 mL

Answers to Selected Questions and Problems

1.1 In the U.S., weight is measured in pounds (lb), height in feet and inches, gasoline in gallons, and temperature in Fahrenheit (°F). In France, weight is measured in kilograms, height in meters, gasoline in liters, and temperature in Celsius (°C).

1.3 **a.** meter; length
     b. gram; mass
     c. milliliter; volume
     d. meter; length
     e. Celsius; temperature

1.5 **a.** 5.5×10^4 m   **b.** 4.8×10^2 g   **c.** 5×10^{-6} cm
     d. 1.4×10^{-4} s   **e.** 7.2×10^{-3} L

1.7 **a.** measured   **b.** exact
     c. exact   **d.** measured

1.9 **a.** not significant   **b.** significant   **c.** significant
     d. significant   **e.** not significant

1.11 **a.** 5   **b.** 2
     c. 2   **d.** 3
     e. 4   **f.** 3

1.13 A calculator often gives more digits than the number of significant figures allowed in the answer.

1.15 **a.** 1.85   **b.** 184   **c.** 0.00474
     d. 8810   **e.** 1.83

1.17 **a.** 1.6   **b.** 0.01
     c. 27.6   **d.** 3.5

1.19 **a.** 53.54 cm   **b.** 127.6 g
     c. 121.5 mL   **d.** 0.50 L

1.21 km/h is kilometers per hour; mi/h is miles per hour

1.23 The prefix *kilo* means to multiply by 1000. One kg is the same mass as 1000 g.

1.25 **a.** mg   **b.** dL   **c.** km
     d. kg   **e.** μL

1.27 **a.** 0.01   **b.** 1000   **c.** 0.001
     d. 0.1   **e.** 1 000 000

1.29 **a.** 100 cm   **b.** 1000 m
     c. 0.001 m   **d.** 1000 mL

1.31 **a.** kilogram   **b.** milliliter
     c. km   **d.** kL

1.33 A conversion factor can be inverted to give a second conversion factor.

1.35 **a.** 100 cm = 1 m; $\dfrac{100 \text{ cm}}{1 \text{ m}}$ and $\dfrac{1 \text{ m}}{100 \text{ cm}}$

     b. 1000 mg = 1 g; $\dfrac{1000 \text{ mg}}{1 \text{ g}}$ and $\dfrac{1 \text{ g}}{1000 \text{ mg}}$

     c. 1 L = 1000 mL; $\dfrac{1000 \text{ mL}}{1 \text{ L}}$ and $\dfrac{1 \text{ L}}{1000 \text{ mL}}$

     d. 1 dL = 100 mL; $\dfrac{100 \text{ mL}}{1 \text{ dL}}$ and $\dfrac{1 \text{ dL}}{100 \text{ mL}}$

1.37 The unit in the denominator must cancel with the preceding unit.

1.39 **a.** 1.75 m   **b.** 5.5 L   **c.** 5.5 g

1.41 **a.** 472 mL   **b.** 65.9 kg   **c.** 24 lb

1.43 **a.** 66 gal   **b.** 3 tablets   **c.** 1800 mg

1.45 **a.** 1.20 g/mL   **b.** 4.4 g/mL   **c.** 3.10 g/mL

1.47 **a.** 210 g   **b.** 575 g   **c.** 62 oz

1.49 **a.** 1.030   **b.** 1.13   **c.** 0.85 g/mL

1.51 42 min

1.53 $0.83 per lb

1.55 16 onions

1.57 **a.** 96 crackers
    b. 0.2 oz of fat
    c. 110 g of sodium

1.59 0.35 oz

1.61 185 mg/dL

1.63 790 g

1.65 720 cm^3

1.67 6.2 lb

1.69 **a.** 1.012 g/mL   **b.** 5.11 g

1.71 **a.** exact   **b.** measured
    c. exact   **d.** measured

1.73 **a.** length = 6.96 cm; width = 4.75 cm
    b. length = 69.6 mm; width = 47.5 mm
    c. 3 significant figures
    d. 3 significant figures
    e. 33.1 cm^2
    f. 3 significant figures

1.75 1.8 g/mL

2 Energy and Matter

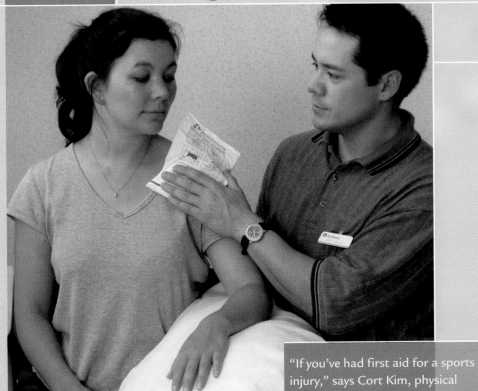

"If you've had first aid for a sports injury," says Cort Kim, physical therapist at the Sunrise Sports Medicine Clinic, "you've likely been treated with a cold pack or hot pack. We use them for several kinds of injury. Here, I'm showing how I can use a cold pack to reduce swelling in my patient's shoulder."

A hot or cold pack is just a packaged chemical reaction. When you hit or open the pack to activate it, your action mixes chemicals together and thus initiates the reaction. In a cold pack, the reaction is one that absorbs heat energy, chills the pack, and draws heat from the injury. Hot packs use reactions that release energy, thus warming the pack. In both cases, the reaction proceeds at a moderate pace, so that the pack stays active for a long time and doesn't get too cold or hot.

the **C**hemistry place

Visit **www.aw-bc.com/chemplace** or use the CD-ROM in the book for extra quizzes, interactive tutorials, career resources, PowerPoint slides, math help, and case studies.

ost of everything we do involves energy. We use energy when we walk, play tennis, study, and breathe. We also use energy when we warm water, cook food, turn on lights, use computers, use a washing machine, or drive our cars. Of course, that energy has to come from something. In our bodies, the food we eat provides us with energy. If we don't eat for a while, we run out of energy. In our homes, schools, and automobiles, burning fossil fuels such as oil, propane, or gasoline provides energy.

When we look around us, we see that matter takes the physical form of a solid, liquid, or gas. A familiar example is water, which is a compound that exists in all three states. In an ice cube or an ice rink, water is in the solid state. Water is a liquid when it comes out of a faucet or fills a pool. Water forms a gas when it evaporates from wet clothes or boils in a pan. Substances change states by losing or gaining energy. For example, energy is needed to melt ice cubes and to boil water in a teakettle. In contrast, energy is removed when water vapor (gas) condenses to liquid and liquid water in an ice cube tray freezes.

LEARNING GOAL

Identify energy as potential or kinetic.

2.1 Energy

When you are running, walking, dancing, or thinking, you are using energy to do **work**. In fact, **energy** is defined as the ability to do work. Suppose you are climbing a steep hill. Perhaps you become too tired to go on. We could say that you do not have sufficient energy to do any more work. Now suppose you sit down and have lunch. In a while you will have obtained some energy from the food, and you will be able to do more work and complete the climb. (See Figure 2.1.)

Potential and Kinetic Energy

All energy can be classified as potential energy or kinetic energy. **Potential energy** is stored energy, whereas **kinetic energy** is the energy of motion. Any object that is moving has kinetic energy. A boulder resting on top of a mountain has potential energy because of its location. If the boulder rolls down the mountain, the potential energy becomes kinetic energy. Water stored in a reservoir has potential energy. When the water goes over the dam, the potential energy is converted to kinetic energy. Even the food you eat has potential energy. When you digest food, you convert its potential energy to kinetic energy to do work in the body.

■ Forms of Energy

Identify the energy in each of the following as potential or kinetic:

a. gasoline
b. skating
c. a candy bar

SOLUTION

a. potential energy
b. kinetic energy
c. potential energy

STUDY CHECK

Would the energy in a stretched rubber band be kinetic or potential?

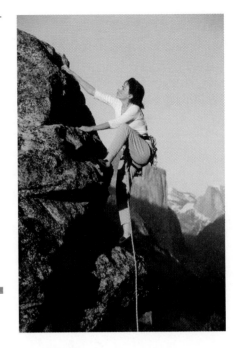

■ Heat and Units of Energy

Heat is the energy associated with the motion of particles in a substance. A frozen pizza feels cold because the particles in the pizza are moving very slowly. As heat is added, the motions of the particles increase, and the pizza becomes warm. Eventually the particles have enough energy to make the pizza hot and ready to eat.

FIGURE 2.1

Work is done as the rock climber moves up the cliff. At the top, the climber has more potential energy than when she started the climb.

Q **What happens to the potential energy of the climber as she descends?**

The SI unit of energy and work is the **joule (J)** pronounced, "jewel." The joule is a small amount of energy, so scientists often use the kilojoule (kJ), 1000 joules. When you heat water for one cup of tea, you use about 75,000 J or 75 kJ of heat.

Energy in joules

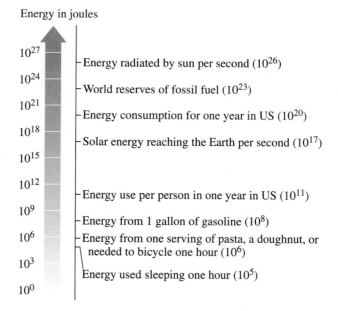

10^{27} — Energy radiated by sun per second (10^{26})

10^{24} — World reserves of fossil fuel (10^{23})

10^{21} — Energy consumption for one year in US (10^{20})

10^{18} — Solar energy reaching the Earth per second (10^{17})

10^{15}

10^{12} — Energy use per person in one year in US (10^{11})

10^{9} — Energy from 1 gallon of gasoline (10^{8})

10^{6} — Energy from one serving of pasta, a doughnut, or needed to bicycle one hour (10^{6})

10^{3} — Energy used sleeping one hour (10^{5})

10^{0}

You may be more familiar with the older unit **calorie (cal)**, from the Latin *caloric*, meaning "heat." The calorie was originally defined as the amount of energy (heat) needed to raise the temperature of one gram of water by 1°C (from 14.5°C to 15.5°C). Now one calorie is defined as exactly 4.184 J. This equality can also be written as a conversion factor.

1 cal = 4.184 J (exact)

$$\frac{4.184 \text{ J}}{1 \text{ cal}} \quad \text{and} \quad \frac{1 \text{ cal}}{4.184 \text{ J}}$$

When we talk about food and nutrition, we often use **Calorie (Cal)** with a capital *C*. The capital C indicates this kind of "Calorie" is actually a kilocalorie.

1 Cal = 1 kcal

1 Cal = 4.184 kJ

One **kilocalorie (kcal)** is equal to 1000 calories (small *c*). The international unit of nutritional energy is the kilojoule (kJ). For example, a baked potato has a nutritional value of 120 Calories, which is the same as 120 kcal or about 500 kJ of energy.

1 kcal = 1000 cal

1 kJ = 1000 J

SAMPLE PROBLEM 2.2

■ **Energy Units**

When 10. g octane in gasoline burns in an automobile engine, 8500 J are released. Convert this quantity of energy to each of the following units:

a. calories
b. kilojoules

SOLUTION

a. calories

STEP 1 **Given**   8500 J   **Need**   calories (cal)

STEP 2 **Plan**   J | Energy factor | cal

STEP 3 **Equalities/Conversion Factors**

$$1 \text{ J} = 4.184 \text{ cal}$$

$$\frac{1 \text{ J}}{4.184 \text{ cal}} \quad \text{and} \quad \frac{4.184 \text{ cal}}{1 \text{ J}}$$

STEP 4 **Set Up Problem**

$$8500 \cancel{J} \quad \times \quad \frac{1 \text{ cal}}{4.184 \cancel{J}} \quad = \quad 2000 \text{ or } 2.0 \times 10^3 \text{ cal}$$

b. kilojoules

STEP 1 **Given**   8500 J   **Need**   kilojoules

STEP 2 **Plan**   J | Energy factor | kJ

STEP 3 **Equalities/Conversion Factors**

$$1 \text{ kJ} = 1000 \text{ J}$$

$$\frac{1000 \text{ J}}{1 \text{ kJ}} \quad \text{and} \quad \frac{1 \text{ kJ}}{1000 \text{ J}}$$

STEP 4 **Set Up Problem** $8500 \cancel{J} \quad \times \quad \dfrac{1 \text{ kJ}}{1000 \cancel{J}} \quad = \quad 8.5 \text{ kJ}$

STUDY CHECK

How many cal are in a jelly doughnut if it provides 26 kcal of energy?

QUESTIONS AND PROBLEMS

■ **Energy**

2.1 Discuss the changes in the potential and kinetic energy of a roller-coaster ride as the roller-coaster climbs up a ramp and goes down the other side.

2.2 Discuss the changes in the potential and kinetic energy of a ski jumper taking the elevator to the top of the jump and going down the ramp.

2.3 Indicate whether each statement describes potential or kinetic energy:
a. water at the top of a waterfall   **b.** kicking a ball
c. the energy in a lump of coal   **d.** a skier at the top of a hill

2.4 Indicate whether each statement describes potential or kinetic energy:
a. the energy in your food   **b.** a tightly wound spring
c. an earthquake   **d.** a car speeding down the freeway

2.5 Convert each of the following energy units.
a. 3500 cal to kcal   **b.** 415 J to cal   **c.** 28 cal to J   **d.** 4.5 kJ to cal

■ Global Warming

The amount of carbon dioxide (CO_2) gas in our atmosphere is on the increase as we burn more gasoline, coal, and natural gas. The algae in the oceans and the plants and trees in the forests normally absorb carbon dioxide, but they cannot keep up with the continued increase. The cutting of trees in the rain forests (deforestation) reduces the amount of carbon dioxide removed from the atmosphere. Many of the trees are also burned as land is cleared. It has been estimated that deforestation may account for 15–30% of the carbon dioxide that remains in the atmosphere each year.

The carbon dioxide in the atmosphere acts like the glass in a greenhouse. When sunlight warms the Earth's surface, some of the heat is absorbed by carbon dioxide. It is not yet clear how severe the effects of global warming might be. Some scientists estimate that by around the year 2030, the atmospheric level of carbon dioxide could double and cause the temperature of Earth's atmosphere to rise by 2–5°C. If that should happen, it would have a profound impact on Earth's climate. For example, an increase in the melting of

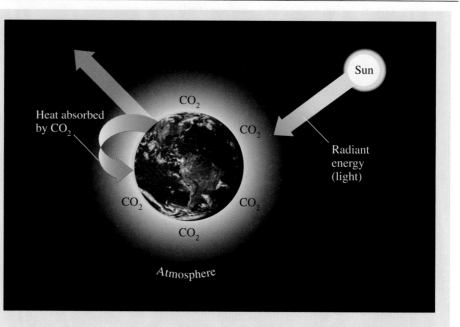

snow and ice could raise the ocean levels by as much as 2 m, which is enough to flood many cities located on the ocean shorelines.

Worldwide efforts are being made to reduce fossil fuel use and to slow or stop deforestation. It will require cooperation throughout the world to avoid the bleak future that some scientists predict should global warming continue unchecked.

2.6 Convert each of the following energy units.
    a. 8.1 kcal to cal   **b.** 325 J to kJ
    c. 2550 cal to kJ   **d.** 2.50 kcal to J

the **C**hemistry place

CASE STUDY
Calories from Hidden Sugar

■ 2.2 Energy and Nutrition

When you are watching your food intake, the Calories you are counting are actually kilocalories. In the field of nutrition, it is common to use the Calorie, Cal (with an uppercase C) to mean 1000 cal, or 1 kcal.

Nutritional Calories

    1 Cal = 1 kcal

    1 kcal = 1000 cal

The number of Calories in a food is determined by using an apparatus called a calorimeter, shown in Figure 2.2. A sample is placed in a steel container within the calorimeter and water is added to fill a surrounding chamber. When the food burns (combustion), the heat released causes an increase in the temperature of the surrounding water. By calculating the heat in kilocalories

Thermometer

Ignition wires

Insulated container

Oxygen

Steel combustion chamber

Food sample

Water

FIGURE 2.2

The caloric value of a food sample temperature change is determined by the combustion of a food sample in a calorimeter.

Q **What happens to the temperature of water in a calorimeter during the combustion of a food sample?**

(or Calories) absorbed by the water, we can determine the caloric content of the food.

Caloric Food Values

The **caloric values** are the kilocalories per gram of the three types of food: carbohydrates, fats, and proteins. These values are listed in Table 2.1.

If the composition of a food is known in terms of the mass of each food type, the caloric content (the total number of Calories) can be calculated.

$$\text{Kilocalories} = \underset{\substack{\text{Mass of carbohydrate,} \\ \text{fat, or protein}}}{\cancel{g}} \times \underset{\text{Caloric value}}{\frac{\text{kcal}}{\cancel{g}}}$$

The caloric content of a packaged food is listed in the nutritional information on the package, usually in terms of the number of Calories in a serving. The general composition and caloric content of some foods are given in Table 2.2.

TABLE 2.1

Caloric Values for the Three Food Types

Food Type	Carbohydrate	Fat (lipid)	Protein
Caloric value	$\dfrac{4 \text{ kcal}}{1 \text{ g}}$	$\dfrac{9 \text{ kcal}}{1 \text{ g}}$	$\dfrac{4 \text{ kcal}}{1 \text{ g}}$

SAMPLE PROBLEM 2.3

Caloric Content for a Food

How many kilocalories are in a piece of chocolate cake that contains 35 g of carbohydrate, 10 g of fat, and 5 g of protein? Round the final answer to the nearest tens place.

Snack Crackers

Nutrition Facts

Serving Size 14 crackers (31g)
Servings Per Container About 7

Amount Per Serving

Calories 120   Calories from Fat 35

	% Daily Value*
Total Fat 4g	**6%**
Saturated Fat 0.5g	**3%**
Polyunsaturated Fat 0.5%	
Monounsaturated Fat 1.5g	
Cholesterol 0mg	**0%**
Sodium 310mg	**13%**
Total Carbohydrate 19g	**6%**
Dietary Fiber Less than 1g	**4%**
Sugars 2g	
Proteins 2g	

Vitamin A 0%	•	Vitamin C 0%
Calcium 4%	•	Iron 6%

*Percent Daily Values are based on a 2,000 calorie diet. Your daily values may be higher or lower depending on your calorie needs.

		Calories: 2,000	2,500
Total Fat	Less than	65g	80g
Sat Fat	Less than	20g	25g
Cholesterol	Less than	300mg	300mg
Sodium	Less than	2,400mg	2,400mg
Total Carbohydrate		300g	375g
Dietary Fiber		25g	30g

Calories per gram:
Fat 9 • Carbohydrate 4 • Protein 4

EXPLORE YOUR WORLD

▪ Counting Calories

Obtain a food item that has a nutrition label. From the information on the label, determine the number of grams of carbohydrate, fat, and protein in one serving. Using the caloric values, calculate the total Calories for one serving. (Most products round off to the tens place.)

QUESTION

How does your total for the Calories in one serving compare to the Calories stated on the label for a single serving?

TABLE 2.2

General Composition and Caloric Content of Some Foods

Food	Protein (g)	Fat (g)	Carbohydrate (g)	Calories (kcal[a])
Banana, 1 medium	1	trace	26	110
Beans, red kidney, 1 cup	15	1	42	240
Beef, lean, 3 oz	22	5	trace	130
Carrots, raw, 1 cup	1	trace	11	50
Chicken, no skin, 3 oz	20	3	0	110
Egg, 1 large	6	6	trace	80
Milk, 4% fat, 1 cup	9	9	12	165
Milk, nonfat, 1 cup	9	trace	12	85
Oil, olive, 1 tbs	0	14	0	130
Potato, baked	3	trace	23	105
Salmon, 3 oz	17	5	0	115
Steak, 3 oz	20	27	0	320
Yogurt, lowfat, 1 cup	8	4	13	120

[a]Values rounded to nearest 5 kcal.

SOLUTION

Using the caloric values for carbohydrate, fat, and protein (Table 2.1), we can calculate the total number of kcal:

$$\text{Carbohydrate:} \quad 35 \text{ g} \times \frac{4 \text{ kcal}}{1 \text{ g}} = 140 \text{ kcal}$$

$$\text{Fat:} \quad 10 \text{ g} \times \frac{9 \text{ kcal}}{1 \text{ g}} = 90 \text{ kcal}$$

$$\text{Protein:} \quad 5 \text{ g} \times \frac{4 \text{ kcal}}{1 \text{ g}} = 20 \text{ kcal}$$

$$\text{Total caloric content:} \quad 250 \text{ kcal}$$

STUDY CHECK

A 1-oz (28 g) serving of oat-bran hot cereal with half a cup of whole milk contains 22 g of carbohydrate, 7 g of fat, and 10 g of protein. If you eat two servings of the oat bran for breakfast, how many kilocalories will you obtain?

QUESTIONS AND PROBLEMS

▪ Energy and Nutrition

2.7 Using the caloric values for foods, determine each of the following:
   a. The total Calories for 1 cup of orange juice that has 26 g of carbohydrate, no fat, and 2 g of protein.
   b. The grams of carbohydrate in 1 apple if the apple has no fat and no protein and provides 72 kcal of energy.
   c. The number of Calories in 1 tablespoon of vegetable oil, which contains 14 g of fat and no carbohydrate or protein.
   d. How many Calories are in 1 breakfast roll if it has 30. g of carbohydrate, 15 g of fat, and 5 g of protein?

HEALTH NOTE

▪ Losing and Gaining Weight

The number of kilocalories needed in the daily diet of an adult depends on sex and physical activity. Some general levels of energy needs are given in Table 2.3.

When food intake exceeds energy output, a person's body weight increases. Food intake is usually regulated by the hunger center in the hypothalamus, located in the brain. The regulation of food intake is normally proportional to the nutrient stores in the body. If these nutrient stores are low, you feel hungry; if they are high, you do not feel like eating.

Weight reduction occurs when food intake is less than energy output. Many diet products contain cellulose, which has no nutritive value but provides bulk and makes you feel full. Some diet drugs depress the hunger center and must be used with caution because they excite the nervous system and can elevate blood pressure. Because muscular exercise is an important way to expend energy, an increase in daily exercise aids weight loss. Table 2.4 lists some activities and the amount of energy they require.

TABLE 2.3

Typical Energy Requirements for 70-kg Adult	
Adult	**Energy (kcal)**
Female	2200
Male	3000

TABLE 2.4

Energy Expended by a 70-kg (155-lb) Person	
Activity	**Energy Expended (kcal/hr)**
Sleeping	60
Sitting	100
Walking	200
Swimming	500
Running	750

2.8 Using the caloric values for foods, determine each of the following:
   a. The total Calories for 2 tablespoons of crunchy peanut butter that contains 6 g of carbohydrate, 16 g of fat, and 7 g of protein.
   b. The grams of protein in 1 cup of soup that has 110 Cal with 7 g of fat and 9 g of carbohydrate.
   c. How many grams of sugar (carbohydrate) are in 1 can of cola if there are 140 Cal and no fat and no protein?
   d. How many grams of fat are in 1 avocado if there are 405 Calories, 13 g of carbohydrate, and 5 g of protein?

2.9 One cup of clam chowder contains 9 g of protein, 12 g of fat, and 16 g of carbohydrate. How many kilocalories are in the clam chowder?

2.10 A high-protein diet contains 70. g of carbohydrate, 150 g of protein, and 5.0 g of fat. How many kilocalories does this diet provide?

▪ 2.3 Temperature Conversions

LEARNING GOAL

▪ Given a temperature, calculate a corresponding temperature on another scale.

Temperature is a measure of how hot or cold a substance is compared to another substance. Heat always flows from a substance with a higher temperature to a substance with a lower temperature until the temperatures of both are the same. When you drink hot coffee or touch a hot pan, heat flows to your mouth or hand, which is at a lower temperature. When you touch an ice cube, it feels cold because heat flows from your hand to the colder ice cube.

Celsius and Fahrenheit Temperatures

Temperatures in science, and in most of the world, are measured and reported in *Celsius* (°C) units. In the United States, everyday temperatures are commonly reported in *Fahrenheit* (°F) units. A typical room temperature of 22°C would be the same as 72°F. A normal body temperature of 37.0°C is 98.6°F.

On the Celsius and Fahrenheit scales, the temperatures of melting ice and boiling water are used as reference points. On the Celsius scale, the freezing point of water is defined as 0°C, and the boiling point as 100°C. On a Fahrenheit scale, water freezes at 32°F and boils at 212°F. On each scale, the temperature difference between freezing and boiling is divided into smaller units, or degrees. On the Celsius scale, there are 100 units between the freezing and boiling of water compared with 180 units on the Fahrenheit scale. (See Figure 2.3.)

$$180 \text{ Fahrenheit degrees} \quad = \quad 100 \text{ Celsius degrees}$$

$$\frac{180 \text{ Fahrenheit degrees}}{100 \text{ Celsius degrees}} \quad = \quad \frac{1.8°F}{1°C}$$

To convert to a Fahrenheit temperature, the Celsius temperature is multiplied by 1.8, and then 32 degrees is added. The 32 degrees adjusts the freezing point of 0°C on the Celsius scale to 32°F on the Fahrenheit scale. Both values, 1.8 and 32, are exact numbers. The equation for this conversion follows:

$$T_F = \frac{1.8°F(T_C)}{1°C} \quad + \quad 32° \quad \text{or} \quad T_F = 1.8(T_C) + 32°$$

Changes   Adjusts freezing
°C to °F   point

FIGURE 2.3

A comparison of the Fahrenheit, Celsius, and Kelvin temperature scales between the freezing and boiling points of water.

Q What is the difference in the values for freezing on the Celsius and Fahrenheit temperature scales?

SAMPLE PROBLEM 2.4

■ Converting Celsius to Fahrenheit

While traveling in China, you discover that your temperature is 38.2°C. What is your temperature in Fahrenheit degrees?

SOLUTION

STEP 1 **Given**   38.2°C   **Need** T_F

STEP 2 **Plan**

$$T_C \boxed{\text{Temperature equation}} T_F$$

STEP 3 **Equality/Conversion Factor**

$$T_F = 1.8(T_C) + 32°$$

STEP 4 **Set Up Problem**   Substitute the Celsius temperature into the equation and solve.

$$T_F = 1.8(38.2) + 32°$$

$$T_F = 68.8° + 32° \qquad \text{1.8 is exact; 32 is exact}$$
$$= 100.8°F \qquad \text{Answer to the first decimal place}$$

In the equation, *the values of 1.8 and 32 are exact numbers.* Therefore, the answer is reported to the same decimal place as the initial temperature.

STUDY CHECK

When making ice cream, rock salt is used to chill the mixture. If the temperature drops to −11°C, what is it in °F? ■

To convert from Fahrenheit to Celsius, the temperature equation is rearranged for T_C.

$$T_C = \frac{T_F - 32°}{1.8}$$

SAMPLE PROBLEM 2.5

■ Converting Fahrenheit to Celsius

You are going to cook a pizza at 325°F. What is that temperature in Celsius degrees?

SOLUTION

STEP 1 **Given**   325°F   **Need** T_C

STEP 2 **Plan**

$$T_F \boxed{\text{Temperature equation}} T_C$$

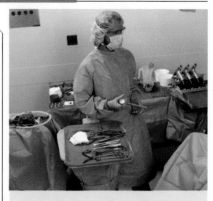

STEP 3 **Equality/Conversion Factor**

$$T_C = \frac{T_F - 32°}{1.8}$$

STEP 4 **Set Up Problem** To solve for T_C, substitute the Fahrenheit temperature into the equation and solve.

$$T_C = \frac{T_F - 32°}{1.8}$$

$$T_C = \frac{(325° - 32°)}{1.8} \quad \text{32 is exact; 1.8 is exact}$$

$$= \frac{293°}{1.8} = 163°C \quad \text{Answer to the one's place}$$

STUDY CHECK

A child has a temperature of 103.6°F. What is this temperature on a Celsius thermometer?

Kelvin Temperature Scale

Scientists tell us that the coldest temperature possible is −273°C (more precisely, −273.15°C). On the Kelvin scale, this temperature, called *absolute zero*, has the value of 0 Kelvin (0 K). Units on the Kelvin scale are called kelvins (K); no degree symbol is used. Because there are no lower temperatures, the Kelvin scale has no negative numbers. Between the freezing and boiling points of water, there are 100 kelvins, which makes a kelvin equal in size to a Celsius unit.

  1 K = 1°C

To calculate a Kelvin temperature, add 273 to the Celsius temperature:

  $T_K = T_C + 273$

Table 2.5 gives a comparison of some temperatures on the three scales.

TABLE 2.5

A Comparison of Temperatures

Example	Fahrenheit (°F)	Celsius (°C)	Kelvin (K)
Sun	9937	5503	5776
A hot oven	450	232	505
A desert	120	49	322
A high fever	104	40	313
Room temperature	72	22	295
Water freezes	32	0	273
A northern winter	−22	−30	243
Helium boils	−452	−269	4
Absolute zero	−459	−273	0

HEALTH NOTE

▪ Variation in Body Temperature

Normal body temperature is considered to be 37.0°C, although it varies throughout the day and from person to person. Oral temperatures of 36.1°C are common in the morning and climb to a high of 37.2°C between 6 P.M. and 10 P.M. Temperatures above 37.2°C for a person at rest are usually an indication of disease. Individuals involved in prolonged exercise may also experience elevated temperatures. Body temperatures of marathon runners can range from 39°C to 41°C as heat production during exercise exceeds the body's ability to lose heat.

Changes of more than 3.5°C from the normal body temperature begin to interfere with bodily functions. Temperatures above 41°C can lead to convulsions, particularly in children, which may cause permanent brain damage. Heatstroke occurs above 41.1°C. Sweat production stops and the skin becomes hot and dry. The pulse rate is elevated, and respiration becomes weak and rapid. The person can become lethargic and lapse into a coma. Damage to internal organs is a major concern, and treatment, which must be immediate, may include immersing the person in an ice-water bath.

In hypothermia, body temperature can drop as low as 28.5°C. The person may appear cold and pale and have an irregular heartbeat. Unconsciousness can occur if the body temperature drops below 26.7°C. Respiration becomes slow and shallow, and oxygenation of the tissues decreases. Treatment involves providing oxygen and increasing blood volume with glucose and saline fluids. Internal temperature may be restored by injecting warm fluids (37.0°C) into the peritoneal cavity.

SAMPLE PROBLEM 2.6

▪ Converting from Celsius to Kelvin Temperature

In a cryogenic laboratory, a technician freezes a sample using liquid nitrogen at −196°C. What is the temperature in K?

SOLUTION

To find the Kelvin temperature, we use the equation:

$$T_K = T_C + 273$$

$$T_K = -196°C + 273$$
$$= 77 \text{ K}$$

STUDY CHECK

On the planet Mercury, the average night temperature is 13 K, and the average day temperature is 683 K. What are these temperatures in Celsius degrees?

QUESTIONS AND PROBLEMS

▪ Temperature Conversions

2.11 Your friend who is visiting from France just took her temperature. When she reads 99.8, she becomes concerned that she is quite ill. How would you explain the temperature to your friend?

2.12 You have a friend who is using a recipe for flan from a Mexican cookbook. You notice that he set your oven temperature at 175°F. What would you advise him to do?

2.13 Solve the following temperature conversions:
    a. 37.0°C = _____ °F   **b.** 65.3°F = _____ °C
    c. −27°C = _____ K   **d.** 62°C = _____ K
    e. 114°F = _____ °C   **f.** 72°F   = _____ K

2.14 Solve the following temperature conversions:
    a. 25°C   = _____ °F   **b.** 155°C = _____ °F
    c. −25°F = _____ °C   **d.** 224 K = _____ °C
    e. 545 K = _____ °C   **f.** 875 K = _____ °F

2.15 a. A patient with heat stroke has a temperature of 106°F. What does this read on a Celsius thermometer?
    b. Because high fevers can cause convulsions in children, the doctor wants to be called if the child's temperature goes over 40.0°C. Should the doctor be called if a child has a temperature of 103°F?

2.16 a. Hot compresses are being prepared for the patient in room 32B. The water is heated to 145°F. What is the temperature of the hot water in °C?
    b. During extreme hypothermia, a young woman's temperature dropped to 20.6°C. What was her temperature on the Fahrenheit scale?

◼ **2.4 Specific Heat**

Every substance can absorb heat. When you want to bake a potato, you place it in a hot oven. If you are cooking pasta, you add the pasta to boiling water. You also know that the longer you heat something such as a potato or water, the higher the temperature becomes. Every substance has its own characteristic ability to absorb heat. Some substances must absorb more heat than others to reach a certain temperature. These energy requirements for different substances are described in terms of a property called specific heat. **Specific heat** (*SH*) is the amount of heat needed to raise the temperature of exactly 1 g of a substance by exactly 1°C. This temperature change is written as ΔT (*delta T*).

$$\text{Specific heat } (SH) = \frac{\text{heat}}{\text{grams} \times \Delta T} = \frac{\text{J (or cal)}}{1\text{ g} \times 1°C}$$

Now we can write the specific heat for water using our definition of the calorie and joule.

$$\text{Specific heat } (SH) \text{ of } H_2O(l) = 4.184\ \frac{\text{J}}{\text{g }°C} = 1.00\ \frac{\text{cal}}{\text{g }°C}$$

Note that the liquid state of water is indicated by an *l* in parentheses following the formula for water. The specific heat of a substance will depend on its state: solid (*s*), liquid (*l*), or gas (*g*). If we look at Table 2.6, we see that the specific heat of water is much larger than the specific heat of aluminum or copper. We see that 1 g of water requires 4.184 J to increase its temperature by 1°C. However, adding the same amount of heat (4.814 J) will raise the temperature of 1 g of aluminum by about 5°C and 1 g of copper by 10°C. Because of its high specific heat, the water in the body can absorb or release large amounts of heat to maintain an almost constant temperature.

TABLE 2.6

Specific Heats of Some Substances		
Substance	Specific Heat (cal/g °C)	Specific Heat (J/g °C)
Aluminum, Al	0.214	0.897
Copper, Cu	0.0920	0.385
Gold, Au	0.0308	0.129
Iron, Fe	0.108	0.450
Silver, Ag	0.0562	0.235
Ammonia, $NH_3(g)$	0.488	2.04
Ethanol, $C_2H_5OH(l)$	0.588	2.46
Sodium chloride, $NaCl(s)$	0.207	0.864
Water, $H_2O(l)$	1.00	4.184

Calculations Using Specific Heat

When we know the specific heat of a substance, we can calculate the heat lost or gained by measuring the mass of the substance and the initial and final temperature. We can substitute these measurements into the specific heat expression that is rearranged to solve for heat, which we call the *heat equation*.

Heat = mass × temperature change × specific heat

Heat = mass × ΔT × *SH*

cal = grams × °C × $\dfrac{\text{cal}}{\text{g °C}}$

J = grams × °C × $\dfrac{\text{J}}{\text{g °C}}$

SAMPLE PROBLEM 2.7

Calculating Heat with Temperature Increase

How many calories are absorbed by 45.2 g aluminum if its temperature rises from 12.5°C to 76.8°C? (See Table 2.6.)

SOLUTION

STEP 1 List given and needed data.

Given mass = 45.2 g

SH for aluminum = 0.214 cal/g °C

Initial temperature = 12.5°C
Final temperature = 76.8°C

Need heat in calories (cal)

STEP 2 Calculate the temperature change. The temperature change ΔT is the difference between the two temperatures.

$$\Delta T = T_{\text{final}} - T_{\text{initial}} = 76.8°C - 12.5°C = 64.3°C$$

Guide to Calculations
Using Specific Heat

STEP 1
List given and needed data.

STEP 2
Calculate temperature change (ΔT).

STEP 3
Write the equation for heat.
Heat = mass × ΔT × *SH*

STEP 4
Substitute given values and solve, making sure units cancel.

STEP 3 **Write the heat equation.**

$$\text{Heat} = \text{mass} \times \Delta T \times SH$$

STEP 4 **Substitute the given values into the equation and solve, making sure units cancel.**

$$\text{Heat} = 45.2 \text{ g} \times 64.3°\text{C} \times \frac{0.214 \text{ cal}}{\text{g} °\text{C}} = 622 \text{ cal}$$

STUDY CHECK

Some cooking pans have a layer of copper on the bottom. How many kilocalories are needed to raise the temperature of 125 g copper from 22°C to 325°C if the specific heat of copper is 0.0920 cal/g °C?

QUESTIONS AND PROBLEMS

■ Specific Heat

2.17 If the same amount of heat is supplied to samples of 10.0 g each of aluminum, iron, and copper all at 15°C, which sample would reach the highest temperature? (See Table 2.6.)

2.18 Substances A and B are the same mass and at the same initial temperature. When they are heated, the final temperature of A is 55°C higher than the temperature of B. What does this tell you about the specific heats of A and B?

2.19 What is the amount of heat required in each of the following?
 a. calories to heat 25 g of water from 15°C to 25°C
 b. calories to heat 150 g of water from 0°C to 75°C
 c. kilocalories to heat 150 g of water in a kettle from 15°C to 77°C

2.20 What is the amount of heat involved in each of the following?
 a. calories given off when 85 g of water cools from 45°C to 25°C
 b. calories given off when 25 g of water cools from 86°C to 61°C
 c. kilocalories absorbed when 5.0 kg of water warms from 22°C to 28°C

2.21 Calculate the energy in joules and calories
 a. required to heat 25.0 g water from 12.5°C to 25.7°C
 b. required to heat 38.0 g copper (Cu) from 122°C to 246°C
 c. lost when 15.0 g ethanol, C_2H_5OH, cools from 60.5°C to −42.0°C
 d. lost when 125 g iron, Fe, cools from 118°C to 55°C

2.22 Calculate the energy in joules and calories
 a. required to heat 5.25 g water, H_2O, from 5.5°C to 64.8°C
 b. lost when 75.0 g water, H_2O, cools from 86.4°C to 2.1°C
 c. required to heat 10.0 g silver (Ag) from 112. °C to 275°C
 d. lost when 18.0 g gold (Au) cools from 224°C to 118°C

2.23 Using the following data, determine the kilocalories for each food burned in a calorimeter:
 a. one stalk of celery that heats 505 g water from 25.2°C to 35.7°C
 b. a waffle that heats 4980 g water from 20.6°C to 62.4°C

2.24 Using the following data, determine the kilocalories for each food burned in a calorimeter:
 a. one cup of popcorn that changes the temperature of 1250 g water from 25.5°C to 50.8°C
 b. a sample of olive oil that increases the temperature of 357 g water from 22.7°C to 38.8°C

▪ **2.5 States of Matter**

LEARNING GOAL

▪ Identify the physical state of a substance as a solid, liquid, or gas.

Matter is anything that occupies space and has mass. This book, the food you eat, the water you drink, your cat or dog, and the air you breathe are just a few examples of matter. On Earth, matter exists in one of three *physical states*: solid, liquid, or gas. All matter is made up of tiny particles. In a **solid**, very strong attractive forces hold the particles close together. They are arranged in such a rigid pattern they can only vibrate slowly in their fixed positions. This gives a solid a definite shape and volume. For many solids, this rigid structure produces a crystal such as seen in quartz and amethyst (see Figure 2.4).

In a **liquid**, the particles have enough energy to move freely in random directions. They are still close to each other and have enough attractions to maintain a definite volume, but there is no rigid structure. Thus, when oil, water, or vinegar is poured from one container to another, the liquid maintains its own volume but takes the shape of the new container (see Figure 2.5).

The air you breathe is made of gases, mostly nitrogen and oxygen. In a **gas**, the molecules move at high speeds, which creates great distances between molecules. This behavior allows gases to fill their container. Gases have no definite shape or volume of their own; they take the shape and volume of their container, as shown in Figure 2.6. Table 2.7 compares some of the properties of the three states of matter.

SAMPLE PROBLEM 2.8

▪ States of Matter

Identify each of the following as a solid, a liquid, or a gas:

a. oxygen in the air
b. a substance with a fixed arrangement of particles
c. safflower oil
d. a substance that has great distances between its particles

(a)   **(b)**

FIGURE 2.4

The solid states of **(a)** quartz and **(b)** amethyst, a purple form of quartz.

Q Why do these crystals have a regular shape?

FIGURE 2.5

A liquid with a volume of 100 mL takes the shape of its container.

Q Why does a liquid have a definite volume, but not a definite shape?

FIGURE 2.6

A gas takes the shape and volume of its container.

Q Why does a gas fill the volume of a container?

TABLE 2.7

Some Properties of Solids, Liquids, and Gases

Ice:
H_2O (s)

Water:
H_2O (l)

Water vapor:
H_2O (g)

Property	Solid	Liquid	Gas
Shape	Has a definite shape	Takes the shape of the container	Takes the shape of its container
Volume	Has a definite volume	Has a definite volume	Fills the volume of the container
Arrangement of particles	Fixed, very close	Random, close	Random, far apart
Interaction between particles	Very strong	Strong	Essentially none
Movement of particles	Very slow	Moderate	Very fast
Examples	Ice, salt, iron	Water, oil, vinegar	Water vapor, helium, air

SOLUTION

a. gas   **b.** solid   **c.** liquid   **d.** gas

STUDY CHECK

What state has a definite volume but takes the shape of its container?

QUESTIONS AND PROBLEMS

■ States of Matter

2.25 Indicate whether each of the following describes a gas, a liquid, or a solid:
   a. This substance has no definite volume or shape.
   b. The particles in this substance do not interact strongly with each other.
   c. The particles of this substance are held in a definite structure.

2.26 Indicate whether each of the following describes a gas, a liquid, or a solid:
a. The substance has a definite volume but takes the shape of the container.
b. The particles of this substance are very far apart.
c. This substance occupies the entire volume of the container.

2.6 Changes of State

You are probably already familiar with most of these phase changes shown in Figure 2.7. Matter undergoes a **change of state** when it is converted from one state to another state.

When heat is added to a solid, the particles in the rigid structure begin to move faster. At a temperature called the **melting point (mp)**, the particles in the solid gain sufficient energy to overcome the attractive forces that hold them together. The particles in the solid separate and move about in random patterns. The substance is **melting**, changing from a solid to a liquid.

If the temperature of a liquid is lowered, the reverse process takes place. Kinetic energy is lost, the particles slow down, and attractive forces pull the particles close together. The substance is **freezing**. A liquid changes to a solid at the **freezing point (fp)**, which is the same temperature as the melting point. Every substance has its own freezing (melting) point: water freezes (melts) at 0°C; gold freezes (melts) at 1064°C; nitrogen freezes (melts) at −210°C. During any phase change, the temperature of a substance remains constant.

Heat of Fusion

During melting, energy called the **heat of fusion** is needed to separate the particles of a solid. For example, 80. calories of heat are needed to melt exactly 1 g of ice at its melting point.

LEARNING GOAL

Describe the changes of state between solids, liquids, and gases; calculate the energy involved.

Heat absorbed

Heat released

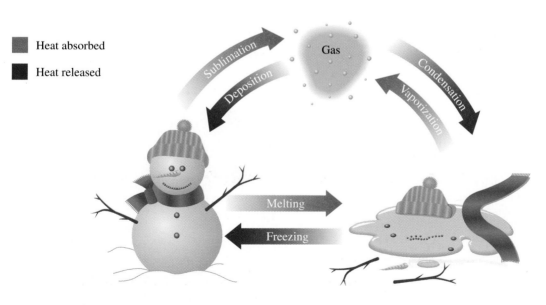

FIGURE 2.7

A summary of the changes of state.

Q Is heat added or released when liquid water freezes?

Heat of Fusion for Water

$$\frac{80.\ cal}{1\ g\ ice}$$

The heat of fusion (80. cal/g) is also the heat that must be removed to freeze 1 g of water at its freezing point. Water is sometimes sprayed in fruit orchards during very cold weather. If the air temperature drops to 0°C, the water begins to freeze. Heat is released as the water molecules bond together, which warms the air and protects the fruit.

To determine the heat needed to melt a sample of ice, multiply the mass of the ice by its heat of fusion. There is no temperature change in the calculation because temperature remains constant as long as the ice is melting.

Calculating Heat to Melt (or Freeze) Water

Heat = mass × heat of fusion

cal = g̶   × 80. cal/g̶

Guide to Calculations Using Heat of Fusion/Vaporization

STEP 1
List grams of substance and change of state.

STEP 2
Write the plan to convert grams to heat and desired unit.

STEP 3
Write the heat conversion factor and metric factor if needed.

STEP 4
Set up the problem with factors.

SAMPLE PROBLEM 2.9

■ **Heat of Fusion**

Ice cubes at 0°C with a mass of 26 g are added to your soft drink.

a. How much heat (cal) must be absorbed to melt all the ice at 0°C?
b. What happens to the temperature of your soft drink. Why?

SOLUTION

a. The heat in calories required to melt the ice is calculated as follows.

STEP 1 **List grams of substance and change of state.**

　　Given　26 g of $H_2O(s)$　　**Need**　calories to melt ice

STEP 2 **Write the plan to convert grams to heat and desired unit.**

　　　g ice　| Heat of fusion |　cal

■ **Histologist**

"While a patient is in surgery for skin cancer, some tissue around the cancer is sent to us," says Mary Ann Pipe, histology technician. "Using the Mohs surgical technique, we place the tissue block on a glass slide, chill it to −30°C in a machine called a cryostat, and freeze it for longer in another machine called a heat extractor. From this frozen block of tissue, we cut extremely thin slices—one-thousandth of an inch—from different depths. We prepare three separate slides from skin at three different depths up to the surface of the skin. We stain the cells pink and blue by placing the slides in hemotoxin, and then in eosin. The slices are a tissue map that the doctor can easily read to determine if the margins around the skin cancer are clear or whether more tissue must be removed."

STEP 3 Equalities/Conversion Factors

$$1 \text{ g } H_2O(s \longrightarrow l) = 80. \text{ cal}$$

$$\frac{80. \text{ cal}}{1 \text{ g } H_2O} \quad \text{and} \quad \frac{1 \text{ g } H_2O}{80. \text{ cal}}$$

STEP 4 Set Up Problem

$$26 \text{ g } H_2O \times \frac{80. \text{ cal}}{1 \text{ g } H_2O} = 2100 \text{ cal}$$

b. The soft drink will be colder because heat from the soft drink is providing the energy to melt the ice.

STUDY CHECK

In a freezer, 150 g water at 0°C is placed in an ice cube tray. How much heat in kilocalories must be removed to form ice cubes at 0°C?

Sublimation

In a process called **sublimation**, the particles on the surface of a solid absorb enough heat to change directly to a gas with no temperature change and without going through the liquid state. For example, dry ice, which is solid carbon dioxide, sublimes at −78°C. It is called "dry" because it does not form a liquid upon warming. In very cold areas, snow does not melt but sublimes directly to vapor. In a frostfree refrigerator, ice on the walls of the freezer sublimes when warm air is circulated through the compartment during the defrost cycle.

Freeze-dried foods prepared by sublimation are convenient for long-term storage and for camping and hiking. A food that has been frozen is placed in a vacuum chamber where it dries as the ice sublimes. The dried food retains all of its nutritional value and needs only water to be edible. A food that is freeze-dried does not need refrigeration because bacteria cannot grow without moisture.

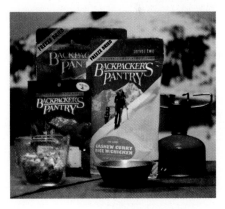

Boiling and Condensation

Water in a mud hole disappears, unwrapped food dries out, and clothes hung on a line dry. **Evaporation** is taking place as water molecules with sufficient energy escape from the liquid surface and enter the gas phase. (See Figure 2.8a.) The loss of the "hot" water molecules removes heat, which cools the remaining liquid water. With additional heat, more and more water molecules evaporate. At the **boiling point (bp)**, all the molecules of the liquid have the energy needed to change into a gas. The **boiling** of a liquid occurs as gas bubbles form throughout the liquid, then rise to the surface and escape. (See Figure 2.8b.)

In a reverse process, water molecules in the gas cool, lose kinetic energy, and slow down. They undergo the process of **condensation** as water vapor is converted back to liquid. You may have noticed that condensation occurs when you take a hot shower and the water vapor forms water droplets on a mirror. Because a substance loses heat as it condenses, its surroundings become warmer. That is why, when a rainstorm is approaching, we notice a warming of the air as gaseous water molecules condense to rain.

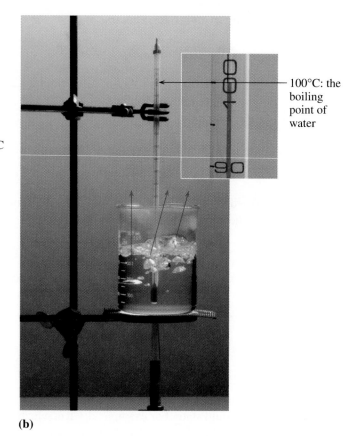

100°C: the boiling point of water

50°C

(a)   (b)

FIGURE 2.8

(a) Evaporation occurs at the surface of a liquid. (b) Boiling occurs as bubbles of gas form throughout the liquid.

Q **Why does water evaporate faster at 80°C than at 20°C?**

Heat of Vaporization

The energy needed to vaporize exactly 1 g of liquid to gas at its boiling point is called the **heat of vaporization**. For water, 540 calories are needed to convert 1 g of water to vapor at 100°C.

Heat of Vaporization for Water

$$\frac{540 \text{ cal}}{1 \text{ g water}}$$

When 1 g of water condenses, the heat of vaporization, 540 calories, is the amount of energy that must be removed. Therefore, 540 cal/g is also the *heat of condensation* of water. To calculate the amount of heat needed to vaporize (or condense) a sample of water, multiply the mass of the sample by the heat of vaporization. As before, no temperature change occurs during a change of state.

Calculating Heat to Vaporize (or Condense) Water

Heat = mass × heat of vaporization

cal = g̶   × 540 cal/g̶

▪ Steam Burns

Hot water at 100°C will cause burns and damage to the skin. However, getting steam on the skin is even more dangerous. Let us consider 25 g hot water at 100°C. If this water falls on a person's skin, the temperature of the water will drop to body temperature, 37°C. The heat released during cooling burns the skin. The amount of heat can be calculated from the temperature change, 63°C.

$$25 \text{ g} \times 63°\text{C} \times \frac{1.00 \text{ cal}}{\text{g }°\text{C}} = 1600 \text{ cal heat}$$

For comparison, we can calculate the amount of heat released when 25 g steam at 100°C hits the skin. First, the steam condenses to water (liquid) at 100°C:

$$25 \text{ g} \times \frac{540 \text{ cal}}{1 \text{ g}} = 14\,000 \text{ cal heat released}$$

Now the temperature of the 25 g water drops from 100°C to 37°C, releasing still more heat as we saw earlier. We can calculate the total amount of heat released from the condensation and cooling of the steam as follows:

Condensation (100°C) = 14 000 cal

Cooling (100°C to 37°C) = 1600 cal

Heat released = 16 000 cal (rounded)

The amount of heat released from steam is 10 times greater than the heat from the same amount of hot water.

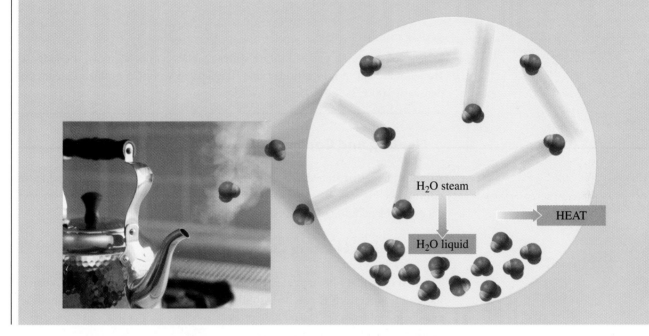

H₂O steam

H₂O liquid

HEAT

SAMPLE PROBLEM 2.10

▪ Using Heat of Vaporization

In a sauna, 150 g of water is converted to steam at 100°C. How many kilocalories of heat are needed?

SOLUTION

STEP 1 List grams of substance and change of state.

Given 150 g $H_2O(l)$ to $H_2O(g)$

Need kilocalories of heat to change state

STEP 2 **Write the plan to convert grams to heat and desired unit.**

Plan g H_2O → Heat of vaporization → cal → Metric factor → kcal

STEP 3 **Write the heat conversion factor and metric factor if needed.**

Equalities/Conversion Factors

1 g H_2O ($l \rightarrow g$) = 540 cal

$$\frac{540 \text{ cal}}{1 \text{ g } H_2O} \quad \text{and} \quad \frac{1 \text{ g } H_2O}{540 \text{ cal}}$$

1 kJ = 1000 J

$$\frac{1000 \text{ J}}{1 \text{ kJ}} \quad \text{and} \quad \frac{1 \text{ kJ}}{1000 \text{ J}}$$

STEP 4 **Set up the problem with factors.**

Problem Setup

$$150 \text{ g } H_2O \times \frac{540 \text{ cal}}{1 \text{ g } H_2O} \times \frac{1 \text{ kcal}}{1000 \text{ cal}} = 81 \text{ kcal}$$

STUDY CHECK

When steam from a pan of boiling water reaches a cool window, it condenses. How much heat in kilocalories (kcal) is released when 25 grams of steam condenses at 100°C?

Heating and Cooling Curves

All the changes of state can be illustrated visually as a diagram called a **heating curve**. (See Figure 2.9a.) In a heating curve, the temperature is shown on the vertical axis. The addition of heat is shown on the horizontal axis.

Steps on a Heating Curve

The first diagonal line indicates a warming of a solid as heat is added. When the melting temperature is reached, solid begins to change to a liquid without any change in temperature. The melting process appears as a flat line, or plateau, on the heating curve.

After all the particles are in the liquid state, the temperature of the liquid begins to rise again as shown by the next diagonal line. As more heat is added, the particles move about more vigorously until the boiling point is reached. Now the liquid begins to boil, indicated as another flat line on the heating curve. Again there is no temperature change as liquid changes into gas. After all the liquid becomes a gas, any additional heat will cause the temperature of the gas to rise again.

Steps on a Cooling Curve

A **cooling curve** is a diagram of the cooling process. (See Figure 2.9b.) When heat is removed from a substance such as water vapor, the gas cools and condenses. On the cooling curve, the temperature is plotted on the vertical axis and the removal of

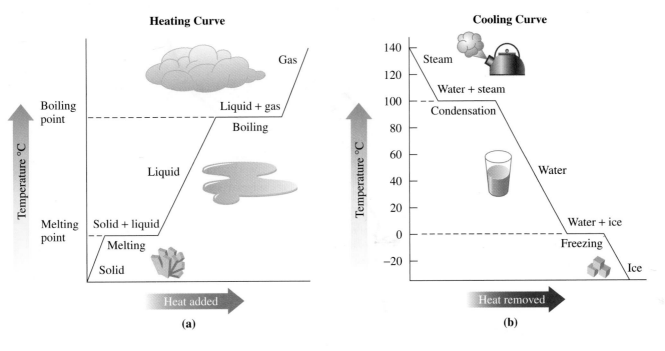

Heating Curve

Cooling Curve

FIGURE 2.9

(a) A heating curve diagrams the temperature increases and changes in state as heat is added. (b) A cooling curve for water.

Q **What does the plateau at 100°C represent on the cooling curve for water?**

heat on the horizontal axis. As a gas cools, heat is lost and the temperature drops. At the condensation point (same as the boiling point), the gas begins to condense, forming liquid. This process is indicated by a flat line (plateau) on the cooling curve at the condensation point.

After all of the gas has changed into liquid, the particles within the liquid cool as indicated by the downward sloping line that shows the temperature decrease. At the freezing point, a second flat line indicates the change of state from liquid to solid. Once all of the substance is frozen, the loss of more heat lowers the solid's temperature below its freezing point.

QUESTIONS AND PROBLEMS

■ Changes of State

2.27 Identify each of the following changes of state as melting, freezing, or sublimation.
   a. The solid structure of a substance breaks down as liquid forms.
   b. Coffee is freeze-dried.
   c. Water on the street turns to ice during a cold wintry night.

2.28 Identify each of the following changes of state as melting, freezing, or sublimation.
   a. Dry ice in an ice-cream cart disappears.
   b. Snow on the ground turns to liquid water.
   c. Heat is removed from 125 g of liquid water at 0°C.

2.29 Calculate the heat needed at 0°C to make each of the following changes of state. Indicate whether heat was absorbed or released.
   a. calories to melt 65 g of ice
   b. calories to melt 17 g of ice
   c. kilocalories to freeze 225 g of water

2.30 Calculate the heat needed at 0°C to make each of the following changes of state. Indicate whether heat was absorbed or released.
  a. calories to freeze 35 g of water
  b. calories to freeze 250 g of water
  c. kilocalories to melt 140 g of ice

2.31 Identify each of the following changes of state as evaporation, boiling, or condensation:
  a. The water vapor in the clouds changes to rain.
  b. Wet clothes dry on a clothesline.
  c. Lava flows into the ocean and steam forms.
  d. After a hot shower, your bathroom mirror is covered with water.

2.32 Identify each of the following changes of state as evaporation, boiling, or condensation:
  a. At 100°C, the water in a pan changes to steam.
  b. On a cool morning, the windows in your car fog up.
  c. A shallow pond dries up in the summer.
  d. Your teakettle whistles when the water is ready for tea.

2.33 a. How does perspiration during heavy exercise cool the body?
  b. Why do clothes dry more quickly on a hot summer day than on a cold winter day?

2.34 a. When a sports injury occurs during a game, a spray such as ethyl chloride may be used to numb an area of the skin. Explain how a substance like ethyl chloride that evaporates quickly can numb the skin.
  b. Why does water in a wide, flat, shallow dish evaporate more quickly than the same amount of water in a tall, narrow glass?

2.35 Calculate the heat change at 100°C in each of the following problems. Indicate whether heat was absorbed or released.
  a. calories to vaporize 10.0 g of water
  b. kilocalories to vaporize 50.0 g of water
  c. kilocalories to condense 8.0 kg of steam

2.36 Calculate the heat change at 100°C in each of the following problems. Indicate whether heat was absorbed or released.
  a. calories to condense 10.0 g of steam
  b. kilocalories to condense 75 g of steam
  c. kilocalories to vaporize 44 g of water

2.37 Draw a heating curve for a sample of ice that is heated from −20°C to 140°C. Indicate the segment of the graph that corresponds to each of the following:
  a. solid
  b. melting point
  c. liquid
  d. boiling point
  e. gas

2.38 Draw a cooling curve for a sample of steam that cools from 110°C to −10°C. Indicate the segment of the graph that corresponds to each of the following:
  a. solid
  b. freezing point
  c. liquid
  d. condensation point (boiling point)
  e. gas

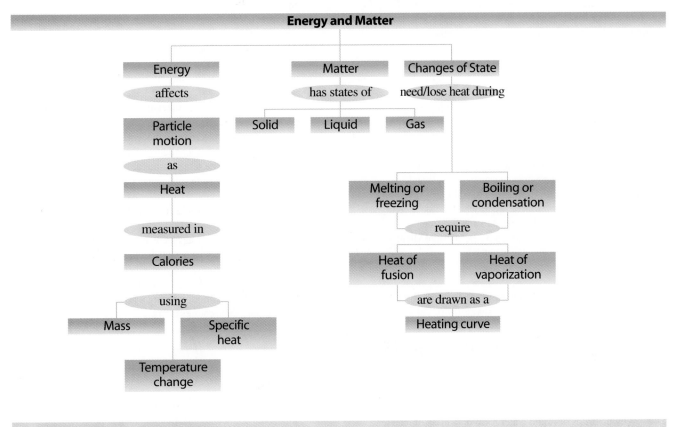

CHAPTER REVIEW

2.1 Energy
Energy is the ability to do work. Potential energy is stored energy; kinetic energy is the energy of motion. Heat, a common form of energy, is measured in calories or joules. One cal is equal to 4.184 J.

2.2 Energy and Nutrition
The nutritional Calorie is the same amount of energy as 1 kcal or 1000 calories. The caloric content of a food is the sum of kilocalories from carbohydrate, fat, and protein.

2.3 Temperature Conversions
On the Celsius scale, there are 100 units between the freezing point of water (0°C) and the boiling point (100°C). On the Fahrenheit scale, there are 180 units between the freezing point of water (32°F) and the boiling point (212°F). A Fahrenheit temperature is related to its Celsius temperature by the equation $T_F = 1.8\,T_C + 32°$.

2.4 Specific Heat
Specific heat is the amount of energy required to raise the temperature of exactly 1 g of a substance by exactly 1°C. The heat lost or gained by a substance is determined by multiplying its mass (g), the temperature change (ΔT), and its specific heat (cal/g °C).

2.5 States of Matter
Matter is anything that has mass and occupies space. The three states of matter are solid, liquid, and gas.

2.6 Changes of State
Melting occurs when the particles in a solid absorb enough energy to break apart and form a liquid. The amount of energy required to convert exactly 1 g of solid to liquid is called the heat of fusion. Boiling is the vaporization of a liquid at its boiling point. The heat of vaporization is the amount of heat needed to convert exactly 1 g of liquid to vapor. A heating or cooling curve illustrates the changes in temperature and state as heat is added to or removed from a substance. Plateaus on the graph indicate changes of state with no change in temperature.

KEY TERMS

boiling The formation of bubbles of gas throughout a liquid.

boiling point (bp) The temperature at which a liquid changes to gas (boils) and gas changes to liquid (condenses).

caloric value The kilocalories obtained per gram of the food types: carbohydrate, fat, and protein.

calorie (cal) The amount of heat energy that raises the temperature of exactly 1 g of water exactly 1°C.

Calorie (Cal) The dietary unit of energy, equal to 1000 cal, or 1 kcal.

change of state The transformation of one state of matter to another; for example, solid to liquid, liquid to solid, liquid to gas.

condensation The change of state of a gas to a liquid.

cooling curve A diagram that illustrates temperature changes and changes of states for a substance as heat is removed.

energy The ability to do work.

evaporation The formation of a gas (vapor) by the escape of high-energy molecules from the surface of a liquid.

freezing A change of state from liquid to solid.

freezing point (fp) The temperature at which a liquid changes to a solid (freezes), a solid changes to a liquid (melts).

gas A state of matter characterized by no definite shape or volume. Particles in a gas move rapidly.

heat The energy associated with the motion of particles in a substance.

heat of fusion The energy required to melt exactly 1 g of a substance at its melting point. For water, 80. cal are needed to melt 1 g of ice; 80. cal are released when 1 g of water freezes.

heat of vaporization The energy required to vaporize 1 g of a substance at its boiling point. For water, 540 calories are needed to vaporize exactly 1 g of liquid; 1 g of steam gives off 540 cal when it condenses.

heating curve A diagram that shows the temperature changes and changes of state of a substance as it is heated.

joule (J) The SI unit of heat energy; 4.184 J = 1 cal.

kinetic energy The energy of moving particles.

kilocalorie (kcal) An amount of heat energy equal to 1000 calories.

liquid A state of matter that takes the shape of its container but has a definite volume.

matter Anything that has mass and occupies space.

melting The conversion of a solid to a liquid.

melting point (mp) The temperature at which a solid becomes a liquid (melts). It is the same temperature as the freezing point.

potential energy An inactive type of energy that is stored for future use.

solid A state of matter that has its own shape and volume.

specific heat A quantity of heat that changes the temperature of exactly 1 g of a substance by exactly 1°C.

sublimation The change of state in which a solid is transformed directly to a gas without forming a liquid first.

work An activity that requires energy.

ADDITIONAL QUESTIONS AND PROBLEMS

2.39 On a hot day, the beach sand gets hot, but the water stays cool. Compare the specific heat of sand to that of water.

2.40 Why do drops of liquid water form on a glass of iced tea?

2.41 When it rains or snows, the air temperature seems warmer. Explain.

2.42 Water is sprayed on the ground of an orchard when temperatures are near freezing to keep the fruit from freezing. Explain.

2.43 A typical diet in the United States provides 15% of the calories from protein, 45% from carbohydrates, and the remainder from fats. Calculate the grams of protein, carbohydrate, and fat to be included each day in diets having the following caloric requirements:
a. 1200 kcal
b. 1900 kcal
c. 2600 kcal

2.44 Your friend has just eaten a pizza, cola soft drink, and ice cream. What is the total kilocalories your friend obtained from this meal? How many hours will your friend need to swim to "burn off" the kilocalories in this meal if your friend has a mass of 70.0-kg? (See Table 2.4.)

Item	Protein (g)	Fat (g)	Carbohydrate (g)
Pizza	13	10	29
Cola	0	0	51
Ice cream	8	28	44

2.45 If you want to lose 1 pound of "fat," which is 15% water, how many kilocalories do you need to expend?

2.46 Calculate the Cal (kcal) in 1 cup of whole milk: 12 g of carbohydrate, 9 g of fat, and 9 g of protein.

2.47 A hot-water bottle contains 725 g of water at 65°C. If the water cools to body temperature (37°C), how many kilocalories of heat could be transferred to sore muscles?

2.48 The melting point of benzene is 5.5°C and its boiling point is 80.1°C. Sketch a heating curve for benzene from 0°C to 100°C.
a. What is the state of benzene at 15°C?
b. What happens on the curve at 5.5°C?
c. What is the state of benzene at 63°C?
d. What is the state of benzene at 98°C?
e. At what temperature will both liquid and gas be present?

UNDERSTANDING THE CONCEPTS

2.49 Read the temperature on each of the Celsius thermometers.

A   B   C

2.50 A 70.0-kg person has just eaten a quarter-pound cheese-burger, French fries, and a chocolate shake. According to Table 2.4, determine each of the following:

a. The number of hours of sleep needed to "burn off" the kilocalories in this meal.
b. The number of hours of running needed to "burn off" the kilocalories in this meal.

Item	Protein (g)	Fat (g)	Carbohydrate (g)
Cheeseburger	31	29	34
French fries	3	11	26
Chocolate shake	11	9	60

2.51 Determine the energy to heat three cubes (gold, aluminum, and silver) each with a volume of 10.0 cm³ from 15°C to 25°C. Refer to Tables 1.10 and 2.6. What do you notice about the energy needed for each?

2.52 If you used the 8400 kilojoules you expend in energy in one day to heat 50 000 g of water at 20°C, what would be the rise in temperature? What would be the new temperature of the water?

2.53 The following is a heating curve for chloroform, a solvent for fats, oils, and waxes.

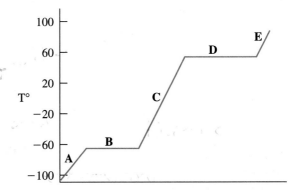

a. What is the melting point of chloroform?
b. What is the boiling point of chloroform?
c. On the heating curve, identify the segments A, B, C, D, and E as solid, liquid, gas, melting, or boiling.
d. At the following temperatures, is chloroform a solid, liquid, or gas?
    $-80°C$; $-40°C$; 25°C; 80°C

2.54 Associate the following diagrams with a segment on the heating curve for water.

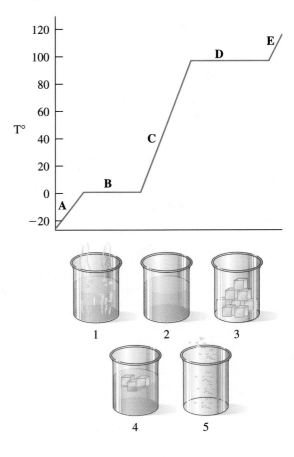

QUESTIONS FOR TEAMWORK

2.55 One liquid has a temperature of 140.°F and another liquid has a temperature of 60.0°C. Are the liquids at the same temperature or at different temperatures?

2.56 A 0.50 g sample of vegetable oil is placed in a calorimeter. When the sample is burned, 18.9 kJ are given off. What is the caloric value in kcal/g of the oil?

2.57 A 25-g sample of an alloy at 98°C is placed in 50. g water at 15°C. If the final temperature reached by the alloy sample and water is 27°C, what is the specific heat (cal/g °C) of the alloy?

2.58 Rearrange the heat equation to solve for each of the following:
  a. The mass in grams of water that absorbs 8250 J when its temperature rises from 18.3°C to 92.6°C.

  b. The mass in grams of a gold sample that absorbs 225 J when the temperature rises from 15.0°C to 47.0°C.
  c. The rise in temperature (ΔT) when a 20.0 g sample of iron absorbs 1580 J.
  d. The specific heat of a metal when 8.50 g of the metal absorbs 28 cal and the temperature rises from 12°C to 24°C.

2.59 How many kilocalories of heat are released when 75 g of steam at 100°C is converted to ice at 0°C? (*Hint*: Requires the total of several steps.)

2.60 A 45-g piece of ice at 0°C is added to a sample of water at 8°C. All of the ice melts and the temperature of the water decreases to 0°C. How many grams of water were in the sample?

ANSWERS

Answers to Study Checks

2.1 potential energy

2.2 26 000 cal

2.3 380 kcal

2.4 12°F

2.5 39.8°C

2.6 night −260.°C; day 410.°C

2.7 3.48 kcal

2.8 liquid

2.9 12 kcal

2.10 14 kcal

Answers to Selected Questions and Problems

2.1 When the car is at the top of the ramp, it has its maximum potential energy. As it descends, potential energy changes to kinetic energy. At the bottom, all the energy is kinetic.

2.3 **a.** potential **b.** kinetic
 c. potential **d.** potential

2.5 **a.** 3.5 kcal **b.** 99.2 cal
 c. 120 J **d.** 1100 cal

2.7 **a.** 110 Cal **b.** 18 g
 c. 130 Cal **d.** 280 Cal

2.9 210 kcal

2.11 In the U.S. we still use the Fahrenheit temperature scale. In °F, normal body temperature is 98.6. On the Celsius scale, her temperature would be 37.7°C, a mild fever.

2.13 **a.** 98.6°F **b.** 18.5°C **c.** 246 K
 d. 335 K **e.** 46°C **f.** 295 K

2.15 **a.** 41°C
 b. No. The temperature is equivalent to 39°C.

2.17 Copper has the lowest specific heat of the samples and will reach the highest temperature.

2.19 **a.** 250 cal **b.** 11 000 cal **c.** 9.3 kcal

2.21 **a.** 1380 J; 330. cal **b.** 1810 J; 434 cal
 c. 3780 J; 904 cal **d.** 3500 J; 850 cal

2.23 **a.** 5.30 kcal **b.** 208 kcal

2.25 **a.** gas **b.** gas **c.** solid

2.27 **a.** melting **b.** sublimation **c.** freezing

2.29 **a.** 5200 cal absorbed
 b. 1400 cal absorbed
 c. 18 kcal released

2.31 **a.** condensation **b.** evaporation
 c. boiling **d.** condensation

2.33 **a.** The heat from the skin is used to evaporate the water (perspiration). Therefore the skin is cooled.
 b. On a hot day, there are more molecules with sufficient energy to become water vapor.

2.35 **a.** 5400 cal absorbed
 b. 27 kcal absorbed
 c. 4300 kcal released

2.37

2.39 Sand must have a lower specific heat than water. When both substances absorb the same amount of heat, the final temperature of the sand will be higher than that of water.

2.41 When water vapor condenses or liquid water freezes, heat is released, which warms the air.

2.43 **a.** 45 g protein, 140 g carbohydrate, 53 g fat
b. 71 g protein, 210 g carbohydrate, 84 g fat
c. 98 g protein, 290 g carbohydrate, 120 g fat

2.45 3500 kcal

2.47 20. kcal

2.49 **a.** 61.4°C **b.** 53.80°C **c.** 5.8°C

2.51 gold 250 J or 59 cal; aluminum 240 J or 58 cal; silver 250 J or 59 cal. The heat needed for all the metal samples is almost the same.

2.53 **a.** −60°C **b.** 60°C
c. A represents the solid state. B represents the change from solid to liquid or melting of the substance. C represents the liquid state as temperature increases. D represents the change from liquid to gas or boiling of the liquid. E represents the gas state.
d. At −80°C, solid; at −40°C, liquid; at 25°C, liquid; at 80°C, gas

3 Atoms and Elements

"Many of my patients have diabetes, ulcers, hypertension, and cardiovascular problems," says Sylvia Lau, registered dietitian. "If a patient has diabetes, I discuss foods that raise blood sugar such as fruit, milk, and starches. I talk about how dietary fat contributes to weight gain and complications from diabetes. For stroke patients, I suggest diets low in fat and cholesterol because high blood pressure increases the risk of another stroke."

If a lab test shows low levels of iron, zinc, iodine, magnesium, or calcium, a dietitian discusses foods that provide those essential elements. For instance, she may recommend more beef for an iron deficiency, whole grain for zinc, leafy green vegetables for magnesium, dairy products for calcium, and iodized table salt and seafood for iodine.

the Chemistry place

Visit **www.aw-bc.com/chemplace** or use the CD-ROM in the book for extra quizzes, interactive tutorials, career resources, PowerPoint slides, math help, and case studies.

Every day, we see around us a variety of materials with many shapes and forms. To a scientist, all of this material is *matter*. There is matter everywhere around you; the orange juice you had for breakfast, the water you put in the coffee maker, the aluminum foil you wrap your sandwich in, your toothbrush and toothpaste, the oxygen you inhale, and the carbon dioxide you exhale are all forms of matter.

All matter is composed of *elements*, of which there are 115 different kinds. Of the 88 elements that occur in nature, one or more is used to make all the substances in our world. Many elements are already familiar to you. Perhaps you use aluminum in the form of foil or drink a soft drink from an aluminum can. You may have a ring or necklace made of gold, or silver, or perhaps platinum. If you play tennis or golf, your racket or clubs may be made from the elements titanium or carbon. In the body, calcium and phosphorus form the structure of bones and teeth, iron and copper are needed in the formation of red blood cells, and iodine is required for the proper functioning of the thyroid.

Deficiencies in certain of these elements can lead to improper functioning of enzymes and abnormal growth and bone formation. Low levels of iron can lead to anemia, while low levels of iodine can cause hypothyroidism and goiter. Someone working in health care is aware of the physical symptoms that indicate a deficiency of an element. Lab tests assess whether elements such as iron, copper, zinc, or iodine are within normal ranges in a patient's blood serum. A dietitian may recommend beef for iron and zinc, whole grains and leafy green vegetables for magnesium, dairy products for calcium, and iodized table salt and seafood for iodine.

3.1 Classification of Matter

LEARNING GOAL

Classify matter as pure substances or mixtures.

Matter is the material that makes up a substance. Thus matter always has mass and occupies space. The materials we use such as water, wood, plates, plastic bags, clothes, and shoes are all made of matter. Because there are so many kinds, we categorize matter by the types of components it contains. A pure substance has a definite composition, while a mixture is made up of two or more substances in varying amounts. (See Figure 3.1.)

Pure Substances

A **pure substance** is a type of matter that has a fixed or definite composition. There are two kinds of pure substances, elements and compounds. **Elements** are the simplest type because they are composed of only one type of atom. Perhaps you are aware of the elements silver, iron, and aluminum, which contain matter of one type. **Compounds** are pure substances too, but they consist of a combination of two or more elements always in the same ratio. For example, in all samples of water, H_2O, there is the same proportion of the elements hydrogen and oxygen. In another compound, hydrogen peroxide, H_2O_2, hydrogen and oxygen are also combined, but in a different ratio. Water, H_2O, and hydrogen peroxide, H_2O_2, are different compounds with different properties.

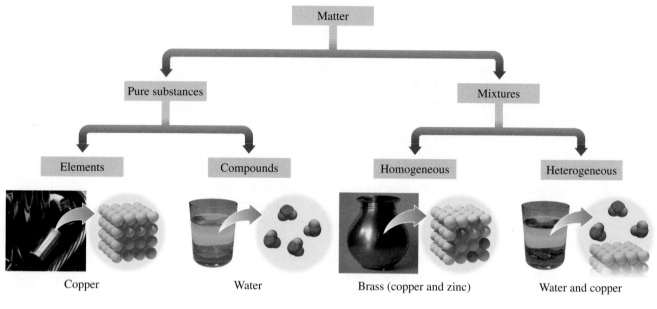

FIGURE 3.1

Matter is organized by its components: elements, compounds, and mixtures. An element or a compound has a fixed composition, whereas a mixture has a variable composition. A homogeneous mixture has a uniform composition, but a heterogeneous mixture does not.

Q **Why are copper and water pure substances, but brass is a mixture?**

An important difference between elements and compounds is that chemical processes can break down compounds into simpler substances such as elements. You may know that ordinary table salt consists of the compound NaCl, which can be broken down into sodium and chlorine as seen in Figure 3.2. Compounds are

FIGURE 3.2

The decomposition of salt, NaCl, produces the elements sodium and chlorine.

Q **How do elements and compounds differ?**

Physical method of separation

FIGURE 3.3

A mixture of spaghetti and water are separated using a strainer, a physical method of separation.

Q **Why can physical methods be used to separate mixtures but not compounds?**

not broken down through physical methods such as boiling or sifting. Elements cannot be broken down further by chemical or physical processes.

Mixtures

Much of the matter in our everyday lives consists of mixtures. In a **mixture**, two or more substances are physically mixed, but not chemically combined. The air we breathe is a mixture of mostly oxygen and nitrogen gases. The steel in buildings and railroad tracks is a mixture of iron, nickel, carbon, and chromium. The brass in knobs and fixtures is a mixture of zinc and copper. Solutions such as tea, coffee, and ocean water are mixtures too. In any mixture, the composition can vary. For example, two sugar–water mixtures would appear the same, but one would taste sweeter because it has a higher ratio of sugar to water. Different types of brass have different properties depending on the ratio of copper to zinc.

Physical processes can be used to separate mixtures because there are no chemical interactions between the components. For example, different coins such as nickels, dimes, and quarters can be separated by size; iron particles mixed with sand can be picked up with a magnet; and water is separated from cooked spaghetti using a strainer. (See Figure 3.3.)

SAMPLE PROBLEM 3.1

■ Pure Substances and Mixtures

Classify each of the following as a pure substance or a mixture.

a. sugar (sucrose)
c. coffee with milk and sugar

b. nickels and dimes in a box
d. aluminum

SOLUTION

a. Pure substance; one type of matter.
b. Mixture; the ratio of nickels and dimes can vary.
c. Mixture; the amount of coffee, milk, and sugar can vary.
d. Pure substance; one type of matter.

STUDY CHECK

Pure gold is labeled as 24-carat. Many jewelry items are made of 14-carat gold, which is 24-carat gold combined with another metal such as copper. Is 14-carat gold a pure substance or a mixture?

TABLE 3.1

Classification of Matter

Types of Mixtures

Mixtures can be classified further as homogeneous or heterogeneous. In a **homogeneous mixture**, also called a *solution*, the composition is uniform throughout the sample. Examples of familiar homogeneous mixtures are air, which contains oxygen and nitrogen gases; bronze, which is a mixture of copper and tin; and salt water, a solution of salt and water.

In a **heterogeneous mixture**, the components do not have a uniform composition throughout the sample. For example, a mixture of oil and water is heterogeneous because the oil floats on the surface of the water. Other examples of heterogeneous mixtures include the raisins in a cookie and the bubbles in a soda. Table 3.1 summarizes the classification of matter.

SAMPLE PROBLEM 3.2

▪ Classifying Mixtures

Classify each of the following mixtures as heterogeneous or homogeneous.

a. ice cream float   **b.** chocolate chip cookie

c. air in a scuba tank   **d.** vinegar (acetic acid and water)

SOLUTION

a. Heterogeneous; ice cream floats on soda, which is a nonuniform composition.

b. Heterogeneous; the chocolate chips are not uniformly distributed in a cookie.

c. Homogeneous; the gases oxygen and nitrogen have a uniform composition in air.

d. Homogeneous; acetic acid and water in vinegar have a uniform composition.

STUDY CHECK

A salad dressing is prepared with oil, vinegar, and chunks of blue cheese. Is this a homogeneous or heterogeneous mixture?

QUESTIONS AND PROBLEMS

▪ Classification of Matter

3.1 Classify each of the following as a pure substance or a mixture.
   a. baking soda   **b.** a blueberry muffin
   c. ice   **d.** aluminum foil

3.2 Classify each of the following as a pure substance or a mixture.
   a. a soft drink   **b.** vitamin C
   c. a cheese sandwich   **d.** an iron nail

3.3 Classify each of the following as a compound or element.
   a. a silicon chip   **b.** hydrogen peroxide H_2O_2
   c. oxygen   **d.** vitamin A

3.4 Classify each of the following as a compound or element.
   a. helium gas   **b.** mercury in a thermometer
   c. sugar   **d.** sulfur

3.5 Classify each of the following mixtures as homogeneous or heterogeneous.
   a. vegetable soup   **b.** salt water
   c. tea   **d.** tea with ice and lemon slice

3.6 Classify each of the following mixtures as homogeneous or heterogeneous.
   a. homogenized milk   **b.** chocolate chip ice cream
   c. gasoline   **d.** bubbly champagne

▪ 3.2 Elements and Symbols

LEARNING GOAL

▪ Given the name of an element, write its correct symbol; from the symbol, write the correct name.

Elements are primary substances from which all other things are built. They cannot be broken down into simpler substances. Many of the elements were named for planets, mythological figures, minerals, colors, geographic locations, and famous people. Some sources of names of elements are listed in Table 3.2.

TABLE 3.2

Some Elements and Their Names

Element	Source of Name
uranium	The planet Uranus
titanium	Titans (mythology)
chlorine	*Chloros*, "greenish yellow" (Greek)
iodine	*Ioeides*, "violet" (Greek)
magnesium	Magnesia, a mineral
californium	California
curium	Marie and Pierre Curie

▪ Chemical Symbols

Chemical symbols are one- or two-letter abbreviations for the names of the elements. Only the first letter of an element's symbol is capitalized; a second letter, if there is one, is lowercase. That way, we know when a different element is indicated. If both letters are capitalized, it represents the symbols of two different elements. For example, the element cobalt has the symbol Co. However, the two capital letters CO specify two elements, carbon (C) and oxygen (O).

One-Letter Symbols	Two-Letter Symbols
C carbon	Co cobalt
S sulfur	Si silicon
N nitrogen	Ne neon
I iodine	Ni nickel

Although most of the symbols use letters from the current names, some are derived from their ancient Latin or Greek names. For example, Na, the symbol for sodium, comes from the Latin word *natrium*. The symbol for iron, Fe, is derived from the Latin name *ferrum*. Table 3.3 lists the names and symbols of some common elements. Learning their names and symbols will greatly help your learning of chemistry. A complete list of all the elements and their symbols appears on the inside front cover of this text.

TABLE 3.3

Names and Symbols of Some Common Elements

Aluminum

Carbon

Sulphur

Name[a]	Symbol	Name[a]	Symbol	Name[a]	Symbol
aluminum	Al	gold (*aurum*)	Au	oxygen	O
argon	Ar	helium	He	phosphorus	P
arsenic	As	hydrogen	H	platinum	Pt
barium	Ba	iodine	I	potassium (*kalium*)	K
boron	B	iron (*ferrum*)	Fe	radium	Ra
bromine	Br	lead (*plumbum*)	Pb	silicon	Si
cadmium	Cd	lithium	Li	silver (*argentum*)	Ag
calcium	Ca	magnesium	Mg	sodium (*natrium*)	Na
carbon	C	manganese	Mn	strontium	Sr
chlorine	Cl	mercury (*hydrargyrum*)	Hg	sulfur	S
chromium	Cr	neon	Ne	tin (*stannum*)	Sn
cobalt	Co	nickel	Ni	titanium	Ti
copper (*cuprum*)	Cu	nitrogen	N	uranium	U
fluorine	F			zinc	Zn

[a]Names given in parentheses are ancient Latin or Greek words from which the symbols are derived.

■ Physical Properties of Elements

Suppose a friend asks you to look into a mirror and describe what you see. You might tell her the color of your hair, eyes, and skin, or the length and texture of your hair. Perhaps you have freckles or dimples. These are descriptions of some of your physical properties. Every element has **physical properties** too, which are those characteristics that can be observed or measured without affecting the identity of an element. The physical properties of a substance include its shape, color, odor, taste, density, hardness, melting point, and boiling point. For example, copper has the physical properties listed in Table 3.4.

SAMPLE PROBLEM 3.3

■ Writing Chemical Symbols

What are the chemical symbols for the following elements?

a. carbon   **b.** nitrogen   **c.** chlorine   **d.** copper

SOLUTION

a. C   **b.** N   **c.** Cl   **d.** Cu

STUDY CHECK

What are the chemical symbols for the elements silicon, sulfur, and silver?

SAMPLE PROBLEM 3.4

■ Naming Chemical Elements

Give the name of the element that corresponds to each of the following chemical symbols:

a. Zn   **b.** K   **c.** H   **d.** Fe

SOLUTION

a. zinc   **b.** potassium   **c.** hydrogen   **d.** iron

STUDY CHECK

What are the names of the elements with the chemical symbols Mg, Al, and F?

QUESTIONS AND PROBLEMS

■ Elements and Symbols

3.7 Write the symbols for the following elements:
   a. copper   **b.** silicon   **c.** potassium   **d.** nitrogen
   e. iron   **f.** barium   **g.** lead   **h.** strontium

3.8 Write the symbols for the following elements:
   a. oxygen   **b.** lithium   **c.** sulfur   **d.** aluminum
   e. hydrogen   **f.** neon   **g.** tin   **h.** gold

3.9 Write the name of the element for each symbol.
   a. C   **b.** Cl   **c.** I   **d.** Hg
   e. F   **f.** Ar   **g.** Zn   **h.** Ni

TABLE 3.4

Some Physical Properties of Copper

Color	Reddish-orange
Odor	Odorless
Melting point	1083°C
Boiling point	2567°C
State at 25°C	Solid
Luster	Very shiny
Conduction of electricity	Excellent
Conduction of heat	Excellent

3.10 Write the name of the element for each symbol.
  a. He   **b.** P   **c.** Na   **d.** Mg
  e. Ca   **f.** Br   **g.** Cd   **h.** Si

3.11 What elements are in the following substances?
  a. table salt, NaCl   **b.** plaster casts, $CaSO_4$
  c. Demerol, $C_{15}H_{22}ClNO_2$ **d.** antacid, $CaCO_3$

3.12 What elements are in the following substances?
  a. water, H_2O **b.** baking soda, $NaHCO_3$
  c. lye, NaOH   **d.** sugar, $C_{12}H_{22}O_{11}$

LEARNING GOAL

Use the periodic table to identify the group and the period of an element and decide whether it is a metal, nonmetal, or metalloid.

FIGURE 3.4

Groups and periods in the periodic table.

Q **What is the symbol of the alkali metal in Period 3?**

3.3 The Periodic Table

As more and more elements were discovered, it became necessary to organize them with some type of classification system. By the late 1800s, scientists recognized that certain elements looked alike and behaved much the same way. In 1872, a Russian chemist, Dmitri Mendeleev, arranged the 60 elements known at that

Periodic Table of Elements

Lanthanides

†Actinides

Metals   Metalloids   Nonmetals

time into groups with similar properties and placed them in order of increasing atomic masses. Today, this arrangement of 115 elements is known as the **periodic table**. (See Figure 3.4 on preceding page.)

Periods and Groups

Each horizontal row in the table is called a **period**. (See Figure 3.5.) The number of elements in the periods increase going down the periodic table. Each row is counted from the top of the table as Period 1 to Period 7. The first period contains only the elements hydrogen (H) and helium (He). The second period contains eight elements: lithium (Li), beryllium (Be), boron (B), carbon (C), nitrogen (N), oxygen (O), fluorine (F), and neon (Ne). The third period also contains eight elements beginning with sodium (Na) and ending with argon (Ar). The fourth period, which begins with potassium (K), and the fifth period, which begins with rubidium (Rb), have 18 elements each. The sixth period, which begins with cesium (Cs), has 32 elements. The seventh period as of today contains the 29 remaining elements although it could go up to 32.

Each vertical column on the periodic table contains a **group** (or family) of elements that have similar properties. At the top of each column is a number that is assigned to each group. The elements in the first two columns on the left of the periodic table and the last six columns on the right are called the *representative elements* or the *main group elements*. For many years, they have been given group numbers 1A–8A. On some periodic tables, the group numbers may be written with Roman numerals: IA–VIIIA. In the center of the periodic table is a block of elements known as the *transition elements* or *transition metals*, which are designated with the letter "B." A newer numbering system assigns group numbers of 1–18 going across the periodic table. Because both systems of group numbers are currently in use, they are both indicated on the periodic table in this text and are included in our discussions of elements and group numbers.

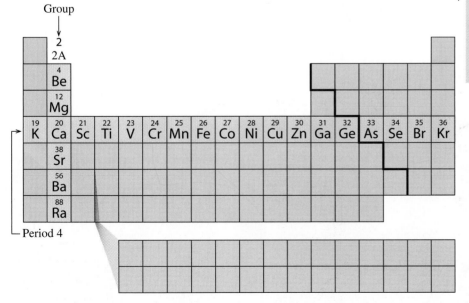

FIGURE 3.5

On the periodic table, each vertical column represents a group of elements and each horizontal row of elements represents a period.

Q **Are the elements Si, P, and S part of a group or a period?**

FIGURE 3.6

Certain groups on the periodic table have common names.

Q **What is the common name for the group of elements that includes helium and argon?**

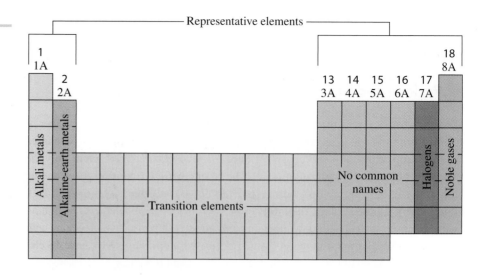

Classification of Groups

Several groups in the periodic table have special names. (See Figure 3.6.) Group 1A (1) elements, lithium (Li), sodium (Na), potassium (K), rubidium (Rb), cesium (Cs), and francium (Fr), are a family of elements known as the **alkali metals** (see Figure 3.7). The elements within this group are soft, shiny metals that are good conductors of heat and electricity and have relatively low melting points. Alkali

FIGURE 3.7

Lithium (Li), sodium (Na), and potassium (K) are some alkali metals from Group 1A (1).

Q **What physical properties do these alkali metals have in common?**

Lithium (Li)   Sodium (Na)   Potassium (K)

metals react vigorously with water and form white products when they combine with oxygen.

Although hydrogen (H) is at the top of Group 1A (1), hydrogen is not an alkali metal and has very different properties than the rest of the elements in this group. Thus hydrogen is not included in the classification of alkali metals. In some periodic tables, H is sometimes placed at the top of Group 7A (17).

Group 2A (2) elements, beryllium (Be), magnesium (Mg), calcium (Ca), strontium (Sr), barium (Ba), and radium (Ra), are called the **alkaline earth metals**. They are also shiny metals like those in Group 1A (1), but they are not as reactive.

The **halogens** are found on the right side of the periodic table in Group 7A (17). They include the elements fluorine (F), chlorine (Cl), bromine (Br), and iodine (I), as shown in Figure 3.8. The halogens, especially fluorine and chlorine, are strongly reactive and form compounds with most of the elements.

Group 8A (18) contains the **noble gases**, helium (He), neon (Ne), argon (Ar), krypton (Kr), xenon (Xe), and radon (Rn). They are quite unreactive and are seldom found in combination with other elements.

SAMPLE PROBLEM 3.5

■ Stating Group and Period Number

Identify the period number and the group number for the following elements:

a. calcium
b. tin

SOLUTION

The period is found by counting down the rows of the elements on the periodic table, and the group number is found at the top of the column that contains that element.

a. Calcium (Ca) is in Period 4 and Group 2A (2).
b. Tin (Sn) is in Period 5 and Group 4A (14).

STUDY CHECK

Give the symbols of the elements that are represented by the following period and group numbers:

a. Period 3, Group 5A (15)
b. Period 6, Group 8A (18)

Chlorine (Cl$_2$)   Bromine (Br$_2$)   Iodine (I$_2$)

FIGURE 3.8

Chlorine (Cl$_2$), bromine (Br$_2$), and iodine (I$_2$) are samples of halogens from Group 7A (17).

Q What elements are in the halogen group?

■ Metals, Nonmetals, and Metalloids

Another feature of the periodic table is the heavy zigzag line that separates the elements into the *metals* and the *nonmetals*. The metals are those elements on the left of the line *except for hydrogen*, and the nonmetals are the elements on the right. (See Figure 3.9.)

In general, most **metals** are shiny solids. They can be shaped into wires (ductile) or hammered into a flat sheet (malleable). Metals are good conductors of heat and electricity. They usually melt at higher temperatures than nonmetals. All of the metals are solids at room temperature, except for mercury (Hg), which is a liquid. Some typical metals are sodium (Na), magnesium (Mg), copper (Cu), gold (Au), silver (Ag), iron (Fe), and tin (Sn).

FIGURE 3.9

Along the heavy zigzag line on the periodic table that separates the metals and nonmetals are metalloids, which exhibit characteristics of both metals and nonmetals.

Q On which side of the heavy zigzag line are the nonmetals located?

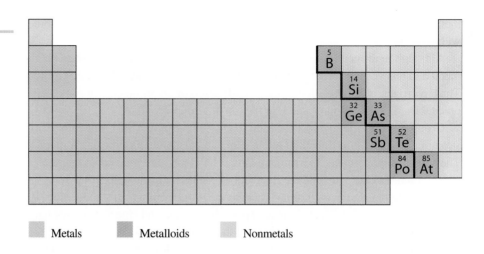

■ Metals   ■ Metalloids   ■ Nonmetals

Nonmetals are not very shiny, malleable, or ductile, and they are often poor conductors of heat and electricity. They typically have low melting points and low densities. You may have heard of nonmetals such as hydrogen (H), carbon (C), nitrogen (N), oxygen (O), chlorine (Cl), and sulfur (S).

The elements located along the heavy line are called **metalloids** and include B, Si, Ge, As, Sb, and Te. Metalloids are elements that exhibit some properties that are typical of the metals and other properties that are characteristic of the nonmetals. For example, they are better conductors of heat and electricity than the nonmetals, but not as good as the metals. The metalloids are semiconductors because they can act as conductors or insulators. Table 3.6 compares some characteristics of silver, a metal, with those of antimony, a metalloid, and sulfur, a nonmetal.

SAMPLE PROBLEM 3.6

■ Metals, Nonmetals, and Metalloids

Use a periodic table to classify each of the following elements as a metal, nonmetal, or metalloid.

a. Na   **b.** Si   **c.** I   **d.** Sn

SOLUTION

a. Sodium, which lies to the left of the heavy zigzag line, is a metal.
b. Silicon, which lies on the heavy zigzag line, is a metalloid.
c. Iodine, to the right of the zigzag line, is a nonmetal.
d. Tin, which lies on the left of the heavy line, is a metal.

STUDY CHECK

Identify each of the following elements as a metal or nonmetal.

a. A shiny, silver-colored element that melts at 962°C.
b. An element with a bright, lustrous reddish color that melts at 2570°C and is malleable, ductile, and an excellent conductor of heat and electricity.
c. A black, elemental substance that breaks into smaller chunks when hit with a hammer.

TABLE 3.6
Some Characteristics of a Metal, a Metalloid, and a Nonmetal

Silver (Ag)	Antimony (Sb)	Sulfur (S)
Metal	Metalloid	Nonmetal
Shiny	Blue-grey, shiny	Dull, yellow
Extremely ductile	Brittle	Brittle
Can be hammered into sheets (malleable)	Shatters when hammered	Shatters when hammered
Good conductor of heat and electricity	Poor conductor of heat and electricity	Poor conductor, good insulator
Used in coins, jewelry, tableware	Used to harden lead, color glass and plastics	Used in gunpowder, rubber, fungicides
Density 10.5 g/mL	Density 6.7 g/mL	Density 2.1 g/mL
Melting point 962°C	Melting point 630°C	Melting point 113°C

HEALTH NOTE

■ Some Important Trace Elements in the Body

Some metals and nonmetals known as trace elements are essential to the proper functioning of the body. Although they are required in very small amounts, their absence can disrupt major biological processes and cause illness. The trace elements listed in Table 3.7 are present in the body combined with other elements. The adult DV is the daily recommended amount for an adult.

TABLE 3.7
Some Important Trace Elements in the Body

Element	Adult DV	Biological Function	Deficiency Symptoms	Dietary Sources
iron (Fe)	10 mg (males) 18 mg (females)	Formation of hemoglobin; enzymes	Dry skin, spoon nails, decreased hemoglobin count, anemia	Liver and other organ meats, oysters, red or dark meat, green leafy vegetables, fortified breads and cereals, egg yolk
copper (Cu)	2.0–5.0 mg	Necessary in many enzyme systems; growth; aids formation of red blood cells and collagen	Uncommon; anemia, decreased white cell count; bone demineralization	Nuts, organ meats, whole grains, shellfish, eggs, poultry, leafy green vegetables
zinc (Zn)	15 mg	Amino acid metabolism; enzyme systems; energy production; collagen	Retarded growth and bone formation; skin inflammation; loss of taste and smell; poor healing	Oysters, crab, lamb, beef, organ meats, whole grains
manganese (Mn)	2.5–5.0 mg	Necessary for some enzyme systems; collagen formation; central nervous system; fat and carbohydrate metabolism; blood clotting	Abnormal skeletal growth; impairment of central nervous system	Whole grains, wheat germ, legumes, pineapple, figs
iodine (I)	150 μg	Necessary for activity of thyroid gland	Hypothyroidism; goiter; cretinism	Seafood, iodized salt
fluorine (F)	1.5–4.0 mg	Necessary for solid tooth formation and retention of calcium in bones with aging	Dental cavities	Tea, fish, water in some areas, supplementary drops, toothpaste

■ The Periodic Table

3.13 Identify the group or period number described by each of the following statements:
    a. contains the elements C, N, and O   **b.** begins with helium
    c. the alkali metals   **d.** ends with neon

3.14 Identify the group or period number described by each of the following statements:
    a. contains Na, K, and Rb   **b.** the row that begins with Li
    c. the noble gases   **d.** contains F, Cl, Br, and I

3.15 Classify the following as an alkali metal, alkaline earth metal, transition element, halogen, or noble gas:
    a. Ca   **b.** Fe   **c.** Xe   **d.** Na   **e.** Cl

3.16 Classify the following as an alkali metal, alkaline earth metal, transition element, halogen, or noble gas:
    a. Ne   **b.** Mg   **c.** Cu   **d.** Br   **e.** Ba

3.17 Give the symbol of the element described by the following:
    a. Group 4A, Period 2   **b.** a noble gas in Period 1
    c. an alkali metal in Period 3   **d.** Group 2, Period 4
    e. Group 13, Period 3

3.18 Give the symbol of the element described by the following:
    a. an alkaline earth metal in Period 2   **b.** Group 15, Period 3
    c. a noble gas in Period 4   **d.** a halogen in Period 5
    e. Group 4A, Period 4

3.19 Is each of the following elements a metal, nonmetal, or metalloid?
    a. calcium   **b.** sulfur   **c.** an element that is shiny
    d. does not conduct heat   **e.** located in Group 7A   **f.** phosphorus
    g. boron   **h.** silver

3.20 Is each of the following elements a metal, nonmetal, or metalloid?
    a. located in Group 2A   **b.** a good conductor of electricity   **c.** chlorine
    d. arsenic   **e.** an element that is not shiny   **f.** oxygen
    g. nitrogen   **h.** aluminum

LEARNING GOAL

Describe the electrical charge and location in an atom for a proton, a neutron, and an electron.

WEB TUTORIAL
Atoms and Isotopes

■ 3.4 The Atom

All the elements listed on the periodic table are made up of atoms. An **atom** is the smallest particle of an element that retains the characteristics of that element. You have probably seen the element aluminum. Imagine that you are tearing a piece of aluminum foil into smaller and smaller pieces. Now imagine that you have a piece so small that you can no longer break it down further. Then you would have an atom of aluminum, the smallest particle of an element that still retains the characteristics of that element.

Billions of atoms are packed together to build you and everything around you. The ink on this paper, even the dot over the letter *i*, contains huge numbers of atoms. There are as many atoms in that dot as there are seconds in 10 billion years.

The concept of the atom is relatively recent. Although the Greek philosophers in 500 BCE reasoned that everything must contain minute particles they called *atomos*, the idea of atoms did not become a scientific theory until 1808. Then John Dalton (1766–1844) developed an atomic theory that proposed that atoms were responsible for the combinations of elements found in compounds.

Dalton's Atomic Theory

1. All matter is made up of tiny particles called atoms.
2. All atoms of a given element are similar to one another and different from atoms of other elements.
3. Atoms of two or more different elements combine to form compounds. A particular compound is always made up of the same kinds of atoms and always has the same number of each kind of atom.
4. A chemical reaction involves the rearrangement, separation, or combination of atoms. Atoms are never created or destroyed during a chemical reaction.

Atoms are the building blocks of everything we see around us; yet, we cannot see an atom or even a billion atoms with the naked eye. However, when billions and billions of atoms are packed together, the characteristics of each atom are added to those of the next until we can see the characteristics we associate with the element. For example, a small piece of the shiny, copper-colored element we call copper consists of many, many copper atoms. Through a special kind of microscope called a scanning tunneling microscope, we can now see images of individual atoms, such as the atoms of carbon in graphite shown in Figure 3.10.

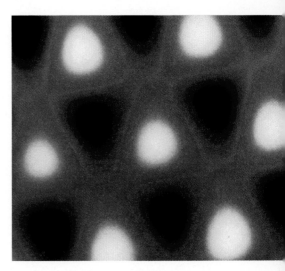

FIGURE 3.10

Graphite, a form of carbon, magnified millions of times by a scanning tunneling microscope. This instrument generates an image of the atomic structure. The round yellow objects are atoms.

Q Why is a microscope with extremely high magnification needed to see these atoms?

Parts of an Atom

By the early part of the twentieth century, growing evidence indicated that atoms were not solid spheres, as Dalton had imagined. New experiments showed that atoms were composed of even smaller bits of matter called **subatomic particles**. The chemistry of an element depends upon the subatomic particles that are the building blocks of the atoms.

There are three subatomic particles of interest to us: the proton, neutron, and electron. Two of these carry electrical charges. The **proton** has a positive charge (+), and the **electron** carries a negative charge (−). The **neutron** has no electrical charge; it is neutral.

Like charges repel; they push away from each other. When you brush your hair on a dry day, electrical charges that are alike build up on the brush and in your hair; as a result your hair flies away from the brush. Opposite or unlike charges attract. The crackle of clothes taken from the clothes dryer indicates the presence of electrical charges. The clinginess of the clothing is due to the attraction of opposite, unlike charges, as shown in Figure 3.11.

Mass

All of the subatomic particles are extremely small compared with the things you see around you. One proton has a mass of 1.7×10^{-24}g, and the neutron is about the same. The mass of the electron is even less. Because the mass of a subatomic particle is so minute, chemists use a small unit of mass called an **atomic mass unit (amu)**. An amu is defined as one-twelfth of the mass of the carbon atom with 6 protons and 6 neutrons, a standard with which the mass of every other atom is compared. In biology, the atomic mass unit is called a dalton in honor of John Dalton. On the amu scale, the proton and neutron each have a mass of about 1 amu. Because the electron mass is so small, it is usually ignored in atomic mass calculations. Table 3.8 summarizes some information about the subatomic particles in an atom.

Positive charges
repel

Negative charges
repel

Unlike charges
attract

FIGURE 3.11

Like charges repel and unlike charges attract.

Q Why are the electrons attracted to the protons in the nucleus of an atom?

TABLE 3.8

Particles in the Atom

Subatomic Particle	Symbol	Electrical Charge	Approximate Mass (amu)	Location in Atom
Proton	p or $p+$	1+	1	Nucleus
Neutron	n or n^0	0	1	Nucleus
Electron	$e-$	1−	0.0005 ($\frac{1}{2000}$)	Outside nucleus

Structure of the Atom

WEB TUTORIAL
Atoms and Isotopes

By the early 1900s, scientists knew that atoms contained subatomic particles, but they did not know how those particles were arranged. In 1911, an experiment led Ernest Rutherford to propose that most of the mass of the atom was contained in a small region at the center of the atom. In his gold-foil experiment, positively charged particles were aimed at a very thin sheet of gold foil as seen in Figure 3.12. He thought that these particles would travel in straight paths and go right on through the gold foil. But much to his surprise, some changed direction as they passed through the gold foil. A few particles were deflected so much that they reversed direction. According to Rutherford, it was as though he had shot a cannonball at a piece of tissue paper, and it bounced back at him.

From the gold-foil experiment, Rutherford concluded that the particles would change direction only if they came close to some type of dense, positively charged region in the atom. This dense core of an atom called the **nucleus**, where the protons and neutrons are located, has a positive charge and contains most of the mass of an atom. Most of an atom is empty space, which is occupied only by fast-moving electrons. (See Figure 3.13.) If an atom were the size of a football stadium, the nucleus would be about the size of a golf ball placed in the center of the field.

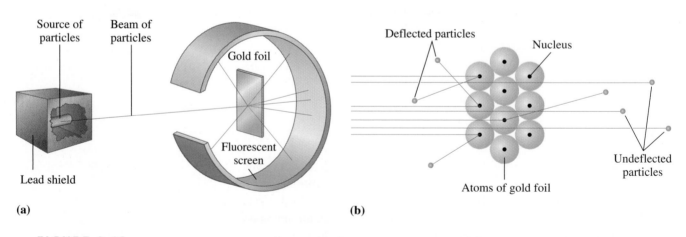

(a)

(b)

FIGURE 3.12

(a) Positive particles are aimed at a piece of gold foil. **(b)** Particles that come close to the atomic nuclei are deflected from their straight path.

Q **Why are some particles deflected while most pass through the gold foil undeflected?**

▪ Repulsion and Attraction

1. Obtain a tape dispenser with clear tape, a hairbrush, comb, and a piece of paper. Tear off a piece of the tape that is about 20 cm long (the length of your hand). Stick the tape to the edge of a table leaving the end hanging down. Tear off a second piece of tape and slowly bring it close to the first one. What happens? Is there an attraction or repulsion?

   Slide your thumb and finger along the tape you are holding. Bring it close to the piece that is hanging from the table. What happens? Is there an attraction or repulsion? Attach the second tape to the edge of the table. Brush your hair and bring the brush close to each piece of tape hanging from the table. What do you observe?

2. Tear a small piece of paper into bits. Brush your hair several times and place the brush just above the bits of paper. Use your knowledge of electrical charges to give an explanation for your observations. Try the same experiment with a comb.

QUESTIONS

1. What happens when objects with like charges are placed close together?
2. What happens when objects with unlike charges are placed close together?

SAMPLE PROBLEM 3.7

▪ Identifying Subatomic Particles

Complete the following table for subatomic particles:

Name	Symbol	Mass (amu)	Charge	Location in Atom
Electron	_____	_____	_____	_____
_____	_____	1	0	_____

SOLUTION

Name	Symbol	Mass (amu)	Charge	Location in Atom
Electron	e^-	0.0005 ($\frac{1}{2000}$)	1−	Outside nucleus
Neutron	n^0	1	0	Nucleus

STUDY CHECK

Give the symbol, mass, electrical charge, and location of a proton in the atom.

Electron cloud

Nucleus

● Proton
◐ Neutron

FIGURE 3.13

In an atom, the protons and neutrons are found in the nucleus; the electrons are located outside the nucleus.

Q **Why can we say that the atom is mostly empty space?**

QUESTIONS AND PROBLEMS

▪ The Atom

3.21 Is a proton, neutron, or electron described by each of the following?
   a. has the smallest mass   **b.** carries a positive charge
   c. is found outside the nucleus   **d.** is electrically neutral

3.22 Is a proton, neutron, or electron described by each of the following?
   a. has a mass about the same as a proton's
   b. is found in the nucleus
   c. is found in the larger part of the atom
   d. carries a negative charge

3.23 What did Rutherford's gold-foil experiment tell us about the organization of the subatomic particles in an atom?

3.24 Why does the nucleus in every atom have a positive charge?

3.25 Which of the following particles have opposite charges?
   a. two protons   **b.** a proton and an electron
   c. two electrons   **d.** a proton and a neutron

3.26 Which of the following pairs of particles have the same charges?
   a. two protons   **b.** a proton and an electron
   c. two electrons   **d.** an electron and a neutron

3.27 On a dry day, your hair flies away when you brush it. How would you explain this?

3.28 Sometimes clothes removed from the dryer cling together. What kinds of charges are on the clothes?

LEARNING GOAL

Given the atomic number and the mass number of an atom, state the number of protons, neutrons, and electrons.

■ 3.5 Atomic Number and Mass Number

All of the atoms of the same element always have the same number of protons. This feature distinguishes atoms of one element from atoms of all the other elements.

■ Atomic Number

An **atomic number**, which is equal to the number of protons in the nucleus of an atom, is used to identify each element.

Atomic number = number of protons in an atom

On the inside front cover of this text is a periodic table, which gives all of the elements in order of increasing atomic number. The atomic number is the whole number that appears above the symbol. For example, a hydrogen atom, with atomic number 1, has 1 proton; a lithium atom, with atomic number 3, has 3 protons; an atom of carbon, with atomic number 6, has 6 protons; and gold, with atomic number 79, has 79 protons.

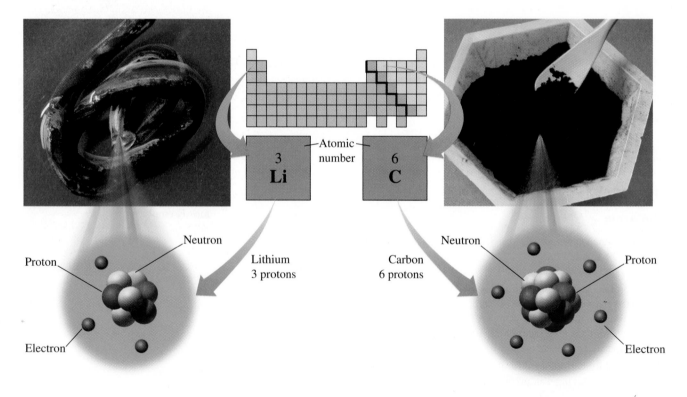

Lithium
3 protons

Carbon
6 protons

An atom is electrically neutral. That means that the number of protons in an atom is equal to the number of electrons. This electrical balance gives an atom an overall charge of zero. Thus, in every atom, the atomic number also gives the number of electrons.

■ Using Atomic Number to Find the Number of Protons and Electrons

Using the periodic table in Figure 3.4, state the atomic number, number of protons, and number of electrons for an atom of each of the following elements:

a. nitrogen   **b.** magnesium   **c.** bromine

SOLUTION

a. atomic number 7; 7 protons and 7 electrons
b. atomic number 12; 12 protons and 12 electrons
c. atomic number 35; 35 protons and 35 electrons

STUDY CHECK

Consider an atom that has 26 electrons.

a. How many protons are in its nucleus?
b. What is its atomic number?
c. What is its name, and what is its symbol?

■ Optician

"When a patient brings in a prescription, I help select the proper lenses, put them into a frame, and fit them properly on the patient's face," says Suranda Lara, optician, Kaiser Hospital. "If a prescription requires a thinner and lighter-weight lens, we formulate that lens. So we have to understand the different materials used to make lenses. Sometimes patients come in with their own glasses that they want to convert to sunglasses. We remove the lenses and put them into a tint bath, which turns them into sunglasses."

Opticians fit and adjust eyewear for patients who have had their eyesight tested by an ophthalmologist or optometrist. Optics and mathematics are used to select materials for frames and lenses that are compatible with patients' facial measurements and lifestyles.

■ Mass Number

We now know that the protons and neutrons determine the mass of the nucleus. For any atom, the **mass number** is the sum of the number of protons and neutrons in the nucleus.

Mass number = number of protons + number of neutrons

For example, an atom of oxygen that contains 8 protons and 8 neutrons has a mass number of 16. An atom of iron that contains 26 protons and 30 neutrons has a mass number of 56. Table 3.9 illustrates the relationship between atomic number, mass number, and the number of protons, neutrons, and electrons in some atoms of different elements.

TABLE 3.9

Composition of Some Atoms of Different Elements

Element	Symbol	Atomic Number	Mass Number	Number of Protons	Number of Neutrons	Number of Electrons
hydrogen	H	1	1	1	0	1
nitrogen	N	7	14	7	7	7
chlorine	Cl	17	37	17	20	17
iron	Fe	26	56	26	30	26
gold	Au	79	197	79	118	79

■ Calculating Numbers of Protons and Neutrons

For an atom of phosphorus that has a mass number of 31, determine the following:

a. the number of protons   **b.** the number of neutrons
c. the number of electrons

SOLUTION

a. On the periodic table, the atomic number of phosphorus is 15. A phosphorus atom has 15 protons.

b. The number of neutrons in this atom is found by subtracting the atomic number from the mass number. The number of neutrons is 16.

$$\text{Mass number} - \text{atomic number} = \text{number of neutrons}$$
$$31 \quad - \quad 15 \quad = \quad 16$$

c. Because an atom is neutral, there is an electrical balance of protons and electrons. Because the number of electrons is equal to the number of protons, the phosphorus atom has 15 electrons.

STUDY CHECK

How many neutrons are in the nucleus of a bromine atom that has a mass number of 80?

QUESTIONS AND PROBLEMS

■ Atomic Number and Mass Number

3.29 Would you use atomic number, mass number, or both to obtain the following?
  a. number of protons in an atom
  b. number of neutrons in an atom
  c. number of particles in the nucleus
  d. number of electrons in a neutral atom

3.30 What do you know about the subatomic particles from the following?
  a. atomic number
  b. mass number
  c. mass number − atomic number
  d. mass number + atomic number

3.31 Write the names and symbols of the elements with the following atomic numbers:
  a. 3   **b.** 9   **c.** 20   **d.** 30
  e. 10   **f.** 14   **g.** 53   **h.** 8

3.32 Write the names and symbols of the elements with the following atomic numbers:
  a. 1   **b.** 11   **c.** 19   **d.** 26
  e. 35   **f.** 47   **g.** 15   **h.** 2

3.33 How many protons and electrons are there in a neutral atom of the following?
  a. magnesium   **b.** zinc
  c. iodine   **d.** potassium

3.34 How many protons and electrons are there in a neutral atom of the following?
  a. carbon   **b.** fluorine
  c. calcium   **d.** sulfur

3.35 Complete the following table for neutral atoms.

Name of Element	Symbol	Atomic Number	Mass Number	Number of Protons	Number of Neutrons	Number of Electrons
	Al		27			
		12			12	
Potassium					20	
				16	15	
			56			26

3.36 Complete the following table for neutral atoms.

Name of Element	Symbol	Atomic Number	Mass Number	Number of Protons	Number of Neutrons	Number of Electrons
	N		15			
Calcium			42			
				38	50	
		14			16	
		56	138			

3.6 Isotopes and Atomic Mass

We have seen that all atoms of the same element have the same number of protons and electrons. However, the atoms of any one element are not completely identical because they can have different numbers of neutrons.

Isotopes

Isotopes are atoms of the same element that have different numbers of neutrons. For example, all atoms of the element magnesium (Mg) have 12 protons. However, some magnesium atoms have 12 neutrons, others have 13 neutrons, and still others have 14 neutrons. The differences in numbers of neutrons for these magnesium atoms cause their mass numbers to be different but not their chemical behavior. The three isotopes of magnesium have the same atomic number but different mass numbers.

To distinguish between the different isotopes of an element, we can write an **atomic symbol** that indicates the mass number and the atomic number of the atom.

Atomic Symbol for an Isotope of Magnesium

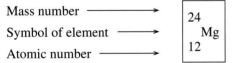

Mass number $\longrightarrow$
Symbol of element $\longrightarrow$ $^{24}_{12}\text{Mg}$
Atomic number $\longrightarrow$

An isotope may be referred to by its name or symbol followed by the mass number, such as magnesium-24 or Mg-24. Magnesium has three naturally occurring isotopes, as shown in Table 3.10.

TABLE 3.10

Isotopes of Magnesium			
Atomic symbol	$^{24}_{12}\text{Mg}$	$^{25}_{12}\text{Mg}$	$^{26}_{12}\text{Mg}$
Number of protons	12	12	12
Number of electrons	12	12	12
Mass number	**24**	**25**	**26**
% abundance	78.7%	10.1%	11.2%
Number of neutrons	**12**	**13**	**14**

LEARNING GOAL

Give the number of protons, electrons, and neutrons in the isotopes of an element.

WEB TUTORIAL
Atoms and Isotopes

Atomic structure of Mg

Isotopes of Mg

$^{24}_{12}\text{Mg}$ $^{25}_{12}\text{Mg}$ $^{26}_{12}\text{Mg}$

▪ Identifying Protons and Neutrons in Isotopes

State the number of protons and neutrons in the following isotopes of neon (Ne):

a. $^{20}_{10}Ne$ **b.** $^{21}_{10}Ne$ **c.** $^{22}_{10}Ne$

SOLUTION

The atomic number of Ne is 10; each isotope has 10 protons. The number of neutrons in each isotope is found by subtracting the atomic number (10) from each mass number.

a. 10 protons; 10 neutrons (20 − 10)   **b.** 10 protons; 11 neutrons (21 − 10)
c. 10 protons; 12 neutrons (22 − 10)

STUDY CHECK

Write a symbol for the following isotopes:

a. a nitrogen atom with 8 neutrons   **b.** an atom with 20 protons and 22 neutrons
c. an atom with mass number 27 and 14 neutrons

17 ——— 17 protons

Cl ——— Symbol for chlorine

35.45 ——— Atomic mass 35.45 amu

17p / 18n $^{35}_{17}Cl$ 17p / 20n $^{37}_{17}Cl$

75.8% of all Cl atoms   24.2%

Atomic Mass

In laboratory work, a scientist generally uses samples that contain many atoms of an element. Among those atoms are all of the various isotopes with their different masses. To obtain a convenient mass to work with, chemists use the mass of an "average atom" of each element. This average atom has an **atomic mass**, which is an average of the mass of all of the naturally occurring isotopes of that element.

To determine an atomic mass of an element, the masses of all the isotopes of that element must be determined experimentally. For example, in a sample of chlorine atoms, there are two isotopes, with ^{35}Cl being the more common isotope. In a large sample of chlorine atoms, there are three atoms of ^{35}Cl for every atom of ^{37}Cl. The atomic mass 35.45 of chlorine indicates a higher percentage of ^{35}Cl.

Most elements consists of several isotopes, and this is one reason atomic masses listed on the periodic table are seldom whole numbers.

Table 3.11 lists the isotopes of some selected elements and their atomic masses. On the periodic table, the atomic mass is given below the symbol of each element.

▪ Average Atomic Mass

Magnesium consists of three naturally occuring isotopes, ^{24}Mg, ^{25}Mg, and ^{26}Mg. Using the atomic mass on the periodic table, which isotope of magnesium is the most prevalent in a magnesium sample?

SOLUTION

The atomic mass for magnesium is 24.3, which means that ^{24}Mg must be the most prevalent isotope in a magnesium sample.

STUDY CHECK

If copper consists of two isotopes, ^{63}Cu and ^{65}Cu, are there more atoms of ^{63}Cu or ^{65}Cu in a sample of copper?

TABLE 3.11

The Atomic Mass of Some Elements

Element	Isotopes	Atomic Mass (weighted average)
lithium	^{6}Li, ^{7}Li	6.941 amu
carbon	^{12}C, ^{13}C, ^{14}C	12.01 amu
fluorine	^{19}F	19.00 amu
oxygen	^{16}O, ^{17}O, ^{18}O	16.00 amu
sulfur	^{32}S, ^{33}S, ^{34}S, ^{36}S	35.45 amu
copper	^{63}Cu, ^{65}Cu	63.55 amu

QUESTIONS AND PROBLEMS

■ Isotopes and Atomic Mass

3.37 What are the number of protons, neutrons, and electrons in the following isotopes?
    a. $^{27}_{13}Al$ **b.** $^{52}_{24}Cr$ **c.** $^{34}_{16}S$ **d.** $^{56}_{26}Fe$

3.38 What are the number of protons, neutrons, and electrons in the following isotopes?
    a. $^{2}_{1}H$ **b.** $^{14}_{7}N$ **c.** $^{26}_{14}Si$ **d.** $^{70}_{30}Zn$

3.39 Write the atomic symbols for isotopes with the following:
    a. 15 protons and 16 neutrons
    b. 35 protons and 45 neutrons
    c. 13 electrons and 14 neutrons
    d. a chlorine atom with 18 neutrons

3.40 Write the atomic symbols for isotopes with the following:
    a. an oxygen atom with 10 neutrons
    b. 4 protons and 5 neutrons
    c. 26 electrons and 30 neutrons
    d. a mass number of 24 and 13 neutrons

3.41 There are four isotopes of sulfur with mass numbers 32, 33, 34, and 36.
    a. Write the atomic symbol for each of these atoms.
    b. How are these isotopes alike?
    c. How are they different?
    d. Why is the atomic mass of sulfur listed on the periodic table not a whole number?

3.42 There are four isotopes of strontium with mass numbers 84, 86, 87, 88.
    a. Write the atomic symbol for each of these atoms.
    b. How are these isotopes alike?
    c. How are they different?
    d. Why is the atomic mass of strontium listed on the periodic table not a whole number?

■ 3.7 Electron Energy Levels

Electrons are constantly moving within the large space of an atom, which means they possess energy. However, they do not all have the same energy. We may think of the energy levels of an atom as similar to the rungs on a ladder. The lowest energy level would be the first rung of the ladder; the second energy level would be the second rung. As you climb up or down the ladder, you must step from one rung to the next. You cannot stop at a level between the rungs. In atoms, electrons exist only in the available energy levels. Generally, the energy levels closest to the nucleus contain electrons with the lowest energies, whereas energy levels farther away contain electrons with higher energies. Unlike the ladder, the lower energy levels are far apart compared to the higher energy levels that are closer together.

Electrons of similar energy are grouped in energy levels. The maximum number of electrons allowed in each energy level is given by the formula $2n^2$ where n is the number of the main energy level. As shown in Table 3.12, 1, the lowest energy level, can hold up to 2 electrons; level 2 can hold up to 8 electrons, level 3 can take 18 electrons, and level 4 has room for 32 electrons. In the atoms of the elements known today, electrons occupy 7 energy levels.

■ Orbitals

An **orbital** is a region in space around the nucleus in which an electron is most likely to be found. Each orbital can hold a maximum of two electrons. There are

LEARNING GOAL

■ Given the name or symbol of one of the first 18 elements in the periodic table, write the electron arrangement and use it to explain the periodic law.

Nucleus

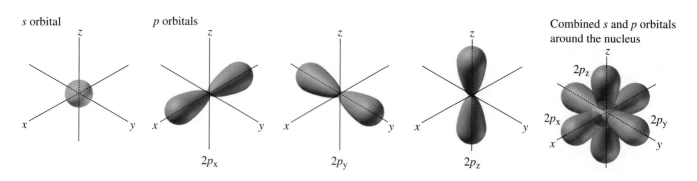

s orbital   *p* orbitals   Combined *s* and *p* orbitals around the nucleus

$2p_x$ $2p_y$ $2p_z$ $2p_z$ $2p_x$ $2p_y$

TABLE 3.12

Capacity of Some Energy Levels

Energy Level	Maximum Number of Electrons
1	2
2	8
3	18
4	32

WEB TUTORIAL
Bohr's Shell Model of the Atom

different types of orbitals. An *s* orbital is spherical, with the nucleus at the center. There is just one *s* orbital in every electron energy level.

A *p* orbital has two lobes. Starting with the second energy level, there is a set of three *p* orbitals arranged in three different directions, *x*, *y*, and *z*, around the nucleus. Starting with the third energy level, there are also *d* orbitals, and in energy level 4 and higher, there are *f* orbitals as listed in Table 3.13. However, their geometry is too complex for this text.

Electron Level Arrangements for the First 18 Elements

The electron level arrangement of an atom gives the number of electrons in each energy level. The electron level arrangements for the first 18 elements can be written by placing electrons in energy levels beginning with the lowest. The single electron of hydrogen and the 2 electrons of helium can be placed in energy level 1.

As shown in Table 3.14, the elements of the second period (lithium, Li, to neon, Ne) have enough electrons to fill the first energy level and part or all of the second energy level. For example, lithium has 3 electrons. Two of those electrons complete energy level 1. The remaining electron goes into the second energy level. As we go across Period 2, more and more electrons enter the second energy level. For example, an atom of carbon, with a total of 6 electrons, fills energy level 1 with 2 electrons, and 4 remaining electrons enter the second energy level. The last element in

TABLE 3.13

Orbitals in Energy Levels 1–4

Energy Level	Orbitals	Maximum Number of Electrons	Energy Level Capacity
1	1*s*	2	2
2	2*s*	2	8
	2*p* 2*p* 2*p*	6	
3	3*s*	2	18
	3*p* 3*p* 3*p*	6	
	3*d* 3*d* 3*d* 3*d* 3*d*	10	
4	4*s*	2	32
	4*p* 4*p* 4*p*	6	
	4*d* 4*d* 4*d* 4*d* 4*d*	10	
	4*f* 4*f* 4*f* 4*f* 4*f* 4*f* 4*f*	14	

TABLE 3.14

Electron Level Arrangements for the First 18 Elements

Element	Symbol	Atomic Number	Number of Electrons in Energy Level		
			1	2	3
hydrogen	H	1	1		
helium	He	2	2		
lithium	Li	3	2	1	
beryllium	Be	4	2	2	
boron	B	5	2	3	
carbon	C	6	2	4	
nitrogen	N	7	2	5	
oxygen	O	8	2	6	
fluorine	F	9	2	7	
neon	Ne	10	2	8	
sodium	Na	11	2	8	1
magnesium	Mg	12	2	8	2
aluminum	Al	13	2	8	3
silicon	Si	14	2	8	4
phosphorus	P	15	2	8	5
sulfur	S	16	2	8	6
chlorine	Cl	17	2	8	7
argon	Ar	18	2	8	8

Period 2 is neon. The 10 electrons in an atom of neon completely fill the first and second energy levels.

In an atom of sodium, atomic number 11, the first and second energy levels are filled and the last electron enters the third energy level. The rest of the elements in the third period continue to add to the third level. For example, a sulfur atom with 16 electrons has 2 electrons in the first level, 8 electrons in the second level, and 6 electrons in the third level. At the end of Period 3, we find that argon has 8 electrons in the third level. The filling of energy levels after argon is discussed in the next section.

SAMPLE PROBLEM 3.12

■ **Writing Electron Level Arrangements**

Write the electron level arrangement for each of the following:

a. oxygen   **b.** chlorine

SOLUTION

a. Oxygen has an atomic number of 8. Therefore, there are 8 electrons in the electron arrangement:

$2e^-$ $6e^-$

b. An atom of chlorine has 17 protons and 17 electrons. The electrons are arranged as follows:

$2e^-$ $8e^-$ $7e^-$

Ba   Sr

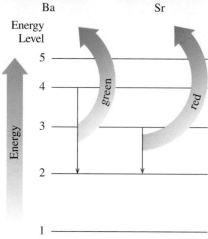

FIGURE 3.14

The heat of a flame provides energy for electrons in atoms of strontium and barium to jump to higher energy levels. When the electrons fall to a lower energy level, energy is emitted in the visible range of red light and green light.

Q **What causes electrons in barium and strontium to jump to higher energy levels?**

STUDY CHECK

What element has the following electron arrangement?

$2e^-$ $8e^-$ $2e^-$

Energy-Level Changes

Atoms can absorb energy from heat, light, or electricity. When some of the electrons in those atoms absorb a certain amount of energy, they jump to higher energy levels. When the electrons drop back to lower, more stable energy levels, energy is emitted. (See Figure 3.14.) If the energy emitted is in the visible range, we see color in the visible spectrum. Sodium streetlights and neon lights are examples of how electrons of atoms absorb energy and emit energy as yellow and red light.

QUESTIONS AND PROBLEMS

Electron Energy Levels

3.43 Electrons exist in specific energy levels. Explain.

3.44 In what order do electrons fill the energy levels 1–3 for the first 18 elements on the periodic table?

3.45 How many electrons are in energy level 2 of the following elements?
  a. sodium   **b.** nitrogen   **c.** sulfur
  d. helium   **e.** chlorine

3.46 How many electrons are in energy level 3 of the following elements?
  a. oxygen   **b.** sulfur   **c.** phosphorus
  d. argon   **e.** fluorine

3.47 Write the electron arrangement for each of the following elements:
  Example: sodium 2, 8, 1
  a. carbon   **b.** argon   **c.** sulfur   **d.** silicon
  e. an atom with 13 protons and 14 neutrons   **f.** nitrogen

3.48 Write the electron arrangement for each of the following atoms:
  Example: sodium 2, 8, 1
  a. phosphorus   **b.** neon   **c.** oxygen
  d. an atom with atomic number 18   **e.** aluminum   **f.** silicon

3.49 Identify the elements that have the following electron arrangements:

Energy Level:	1	2	3
a.	$2e^-$	$1e^-$	
b.	$2e^-$	$8e^-$	$2e^-$
c.	$1e^-$		
d.	$2e^-$	$8e^-$	$7e^-$
e.	$2e^-$	$6e^-$	

3.50 Identify the elements that have the following electron arrangements:

Energy Level:	1	2	3
a.	$2e^-$	$5e^-$	
b.	$2e^-$	$8e^-$	$6e^-$
c.	$2e^-$	$4e^-$	
d.	$2e^-$	$8e^-$	$8e^-$
e.	$2e^-$	$8e^-$	$3e^-$

3.51 Select the correct word. Electrons can jump to higher energy levels when they (absorb/emit) a photon.

3.52 Select the correct word. Electrons drop to lower energy levels when they (absorb/emit) a photon.

TABLE 3.16

Electron-Dot Symbols for Selected Elements in Periods 1–4

Group Number	1A (1)	2A (2)	3A (13)	4A (14)	5A (15)	6A (16)	7A (17)	8A (18)
Valence Electrons	1	2	3	4	5	6	7	8
Electron-Dot Symbols	H·							He:
	Li·	Be·	·B·	·C·	·N·	·O:	·F:	:Ne:
	Na·	Mg·	·Al·	·Si·	·P·	·S:	·Cl:	:Ar:
	K·	Ca·	·Ga·	·Ge·	·As·	·Se:	·Br:	:Kr:

SAMPLE PROBLEM 3.14

■ **Writing Electron-Dot Symbols**

Write the electron-dot symbol for each of the following elements:

a. bromine
b. aluminum

SOLUTION

a. Because the group number for bromine is 7A (17), bromine has 7 valence electrons.

   ·Br:

b. Aluminum, in Group 3A (3), has 3 valence electrons.

   ·Al·

STUDY CHECK

What is the electron-dot symbol for phosphorus?

Atomic Size

Although there are no fixed boundaries to atoms, scientists have a good idea of the typical volume of the electron clouds in atoms. This volume can be described in terms of its *atomic radius*, which is the distance from the nucleus to the valence (outermost) electrons. For representative elements, the atomic radius increases going down each group. We would expect an increase in radius because the outermost electrons in higher energy levels are farther from the nucleus. For example, in the alkali metals, Li has a valence electron in energy level 2; Na has a valence electron in energy level 3; K has a valence electron in energy level 4; and Rb has a valence electron in energy level 5. (See Figure 3.15.)

The atomic radii of representative elements typically decrease from left to right on the periodic table. Going across a period, there is an increase in the number of valence electrons with a corresponding increase in protons, which causes a stronger attraction for the outermost electrons. Going across a period, the valence electrons are pulled closer to the nucleus, which makes the atoms smaller.

FIGURE 3.15

The atomic radius increases going down a group, but decreases going from left to right across a period.

Q Why does the atomic radius increase going down a group?

Li atom

Na atom

K atom

Distance between the nucleus and valence electron

Ionization Energy Decreases

Ionization Energy

Electrons are held in atoms by their attraction to the nucleus. Therefore, energy is required to remove an electron from an atom. The **ionization energy** is the energy needed to remove the least tightly bound electron from an atom in the gaseous (g) state. When an electron is removed from a neutral atom, a particle called a cation, with a 1+ charge, is formed.

$$Na(g) + \text{energy (ionization)} \longrightarrow Na^+(g) + e^-$$

The ionization energy generally decreases going down a group. Less energy is needed to remove an electron as nuclear attraction for electrons decreases farther from the nucleus. *Going across a period from left to right, the ionization energy generally increases.* As the number of protons in the nucleus increases across a period, more energy is required to remove an electron. In general, the ionization energy is low for the metals and high for the nonmetals.

In Period 1, the valence electrons are close to the nucleus and strongly held. H and He have high ionization energies because a large amount of energy is required to remove an electron. The ionization energy for He is the highest of any element because He has a full, stable, energy level which is disrupted by removing an electron. The high ionization energies of the noble gases indicate that their electron arrangements are especially stable. (See Figure 3.16.)

SAMPLE PROBLEM 3.15

■ **Ionization Energy**

Indicate the element in each that has the higher ionization energy and explain your choice.

a. K or Na   **b.** Mg or Cl   **c.** F or N

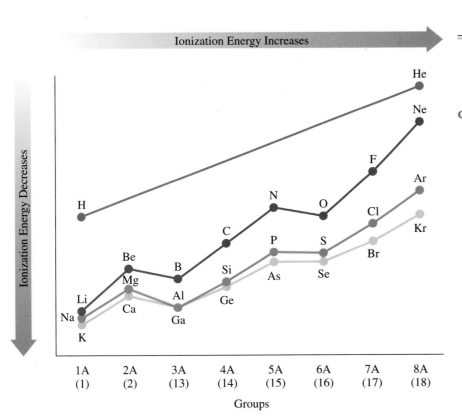

Ionization Energy Increases

Ionization Energy Decreases

| 1A (1) | 2A (2) | 3A (13) | 4A (14) | 5A (15) | 6A (16) | 7A (17) | 8A (18) |

Groups

FIGURE 3.16

Ionization energies for the representative elements tend to decrease going down a group and increase going across a period.

Q **Why is the ionization energy for F greater than for Cl?**

SOLUTION

a. Na. In Na, the valence electron is closer to the nucleus.

b. Cl. Attraction for the valence electrons increases across a period, going left to right.

c. F. With more protons, the F nucleus has a stronger attraction for valence electrons in the second energy level than N.

STUDY CHECK

Arrange Sn, Sr, and I in order of increasing ionization energy.

QUESTIONS AND PROBLEMS

■ Periodic Trends

3.53 The elements boron and aluminum are in the same group on the periodic table.
   a. Write the electron arrangements for B and Al.
   b. How many valence electrons are in the outer energy level of each atom?
   c. What is their group number?

3.54 The elements fluorine and chlorine are in the same group on the periodic table.
   a. Write the electron arrangements for F and Cl.
   b. How many valence electrons are in each of their outer energy levels?
   c. What is their group number?

3.55 What is the number of electrons in the outer energy level and the group number for each of the following elements?
   Example: fluorine $7e^-$; Group 7A (17)
   a. magnesium   **b.** chlorine
   c. oxygen   **d.** nitrogen
   e. barium   **f.** bromine

3.56 What is the number of electrons in the outer energy level and the group number for each of the following elements?
Example: fluorine $7e^-$; Group 7A (17)
a. lithium   **b.** silicon   **c.** neon
d. argon   **e.** tin   **f.** cesium

3.57 Why do Mg, Ca, and Sr have similar properties?

3.58 Name two elements that would exhibit physical and chemical behavior similar to chlorine.

3.59 Write the group number and electron-dot symbol for each element:
a. sulfur   **b.** nitrogen   **c.** calcium
d. sodium   **e.** potassium

3.60 Write the group number and electron-dot symbol for each element:
a. carbon   **b.** oxygen   **c.** fluorine
d. lithium   **e.** chlorine

3.61 Using the symbol M for a metal atom, draw the electron-dot symbol for an atom of a metal in the following groups:
a. Group 1A (1)
b. Group 2A (2)

3.62 Using the symbol Nm for a nonmetal atom, draw the electron-dot symbol for an atom of a nonmetal in the following groups:
a. Group 5A (15)
b. Group 7A (17)

3.63 The alkali metals are in the same family on the periodic table. What is their group number and how many valence electrons does each have?

3.64 The halogens are in the same family on the periodic table. What is their group number and how many valence electrons does each have?

3.65 Place the elements in each set in order of decreasing atomic radius.
a. Mg, Al, Si
b. Cl, Br, I
c. I, Sb, Sr

3.66 Place the elements in each set in order of decreasing atomic radius.
a. Cl, S, P
b. Ge, Si, C
c. Ba, Ca, Sr

3.67 Select the larger atom in each pair.
a. Na or Cl
b. Na or Rb
c. Na or Mg

3.68 Select the larger atom in each pair.
a. S or Cl
b. S or O
c. S or Se

3.69 Arrange each set of elements in order of increasing ionization energy.
a. F, Cl, Br
b. Na, Cl, Al
c. Na, K, Cs

3.70 Arrange each set of elements in order of increasing ionization energy.
a. C, N, O
b. P, S, Cl
c. As, P, N

3.71 Select the element in each pair with the higher ionization energy.
a. Br or I
b. Mg or Al
c. Si or P

3.72 Select the element in each pair with the higher ionization energy.
a. O or Ne   **b.** K or Br   **c.** Ca or Ba

CONCEPT MAP

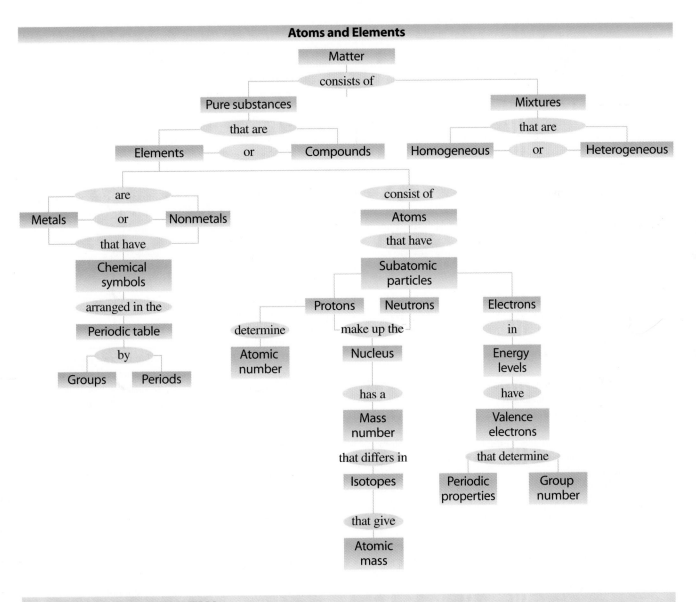

Atoms and Elements

Matter — consists of

Pure substances — that are — Elements — or — Compounds

Mixtures — that are — Homogeneous — or — Heterogeneous

Elements — are — Metals — or — Nonmetals — that have — Chemical symbols — arranged in the — Periodic table — by — Groups / Periods

consist of — Atoms — that have — Subatomic particles

Protons — determine — Atomic number

Protons / Neutrons — make up the — Nucleus — has a — Mass number — that differs in — Isotopes — that give — Atomic mass

Electrons — in — Energy levels — have — Valence electrons — that determine — Periodic properties / Group number

CHAPTER REVIEW

3.1 Classification of Matter
Matter is everything that occupies space and has mass. Matter is classified as pure substances or mixtures. Pure substances, which are elements or compounds, have fixed compositions, and mixtures have variable compositions. The substances in mixtures can be separated using physical methods.

3.2 Elements and Symbols
Elements are the primary substances of matter. Chemical symbols are one- or two-letter abbreviations of the names of the elements.

3.3 The Periodic Table
The periodic table is an arrangement of the elements by increasing atomic number. A vertical column on the periodic table containing elements with similar properties is called a group. A horizontal row is called a period. Elements in Group 1A (1) are called the alkali metals; Group 2A (2), alkaline earth metals; Group 7A (17), the halogens; and Group 8A (18), the noble gases. On the periodic table, metals are located on the left of the heavy zigzag line, and nonmetals are to the right of the heavy zigzag line. Elements located on the heavy line are called metalloids.

3.4 The Atom

An atom is the smallest particle that retains the characteristics of an element. Atoms are composed of three subatomic particles. Protons have a positive charge (+), electrons carry a negative charge (−), and neutrons are electrically neutral. The protons and neutrons are found in the tiny, dense nucleus. Electrons are located outside the nucleus.

3.5 Atomic Number and Mass Number

The atomic number gives the number of protons in all the atoms of the same element. In a neutral atom, there is an equal number of protons and electrons. The mass number is the total number of protons and neutrons in an atom.

3.6 Isotopes and Atomic Mass

Atoms that have the same number of protons but different numbers of neutrons are called isotopes. The atomic mass of an element is the weighted average mass of all the isotopes in a naturally occurring sample of that element.

3.7 Electron Energy Levels

Every electron has a specific amount of energy. In an atom, the electrons of similar energy are grouped in specific energy levels. The first level nearest the nucleus can hold 2 electrons, the second level can hold 8 electrons, the third level will take up to 18 electrons. Each level consists of orbitals, which represent the space where electrons of that energy are likely to be found. An *s* orbital has a spherical shape. A *p* orbital has two lobes like a dumbbell and there are three *p* orbitals in each level from $n = 2$.

3.8 Periodic Trends

The properties of elements are related to the valence electrons of the atoms. The electron arrangement is written by placing the number of electrons in that atom in order from the lowest energy levels and filling to higher levels. The similarity of behavior for the elements in a group is related to having the same number of electrons in their outermost, valence, level. The group number for an element gives the number of electrons in its outer most energy level. With only a few minor exceptions, each group of elements has the same arrangement of valence electrons differing only in the energy level. The radius of an atom increases going down a group and decreases going across a period. The energy required to remove a valence electron is the ionization energy, which generally decreases going down a group and generally increases going across a period.

KEY TERMS

alkali metals Elements of Group 1A (1) except hydrogen; these are soft, shiny metals with one outer shell electron.

alkaline earth metals Group 2A (2) elements, which have 2 electrons in their outer shells.

atom The smallest particle of an element that retains the characteristics of the element.

atomic mass The weighted average mass of all the naturally occurring isotopes of an element.

atomic mass unit (amu) A small mass unit used to describe the mass of very small particles such as atoms and subatomic particles; 1 amu is equal to one-twelfth the mass of a ^{12}C atom.

atomic number A number that is equal to the number of protons in an atom.

atomic symbol An abbreviation used to indicate the mass number and atomic number of an isotope.

chemical symbol An abbreviation that represents the name of an element.

compound A pure substance consisting of two or more elements, with a definite composition, that can be broken down into simpler substances by chemical methods.

electron A negatively charged subatomic particle having a very small mass that is usually ignored in calculations; its symbol is e^-.

electron-dot symbol The representation of an atom that shows valence electrons as dots around the symbol of the element.

element A pure substance that cannot be separated into any simpler substances by chemical methods.

group A vertical column in the periodic table that contains elements having similar physical and chemical properties.

group number A number that appears at the top of each vertical column (group) in the periodic table and indicates the number of electrons in the outermost energy level.

halogen Group 7A (17) elements of fluorine, chlorine, bromine, iodine, and astatine.

heterogeneous mixture A mixture of two or more substances that are not mixed uniformly.

homogeneous mixture A mixture of two or more substances that are mixed uniformly.

ionization energy The energy needed to remove the least tightly bound electron from the outermost energy level of an atom.

isotope An atom that differs only in mass number from another atom of the same element. Isotopes have the same atomic number (number of protons) but different numbers of neutrons.

mass number The total number of neutrons and protons in the nucleus of an atom.

matter Anything that has mass and occupies space.

metal An element that is shiny, malleable, ductile, and a good conductor of heat and electricity. The metals are located to the left of the zigzag line in the periodic table.

metalloid Elements with properties of both metals and nonmetals located along the heavy zigzag line on the periodic table.

mixture The physical combination of two or more substances that does not change the identities of the mixed substances.

neutron A neutral subatomic particle having a mass of 1 amu and found in the nucleus of an atom; its symbol is n or n^0.

noble gas An element in Group 8A (18) of the periodic table, generally unreactive and seldom found in combination with other elements.

nonmetal An element with little or no luster that is a poor conductor of heat and electricity. The nonmetals are located to the right of the zigzag line in the periodic table.

nucleus The compact, very dense center of an atom, containing the protons and neutrons of the atom.

orbital The region around the nucleus where electrons of a certain energy are more likely to be found. The *s* orbitals are spherical; the *p* orbitals have two lobes.

period A horizontal row of elements in the periodic table.

periodic table An arrangement of elements by increasing atomic number such that elements having similar chemical behavior are grouped in vertical columns.

physical property A characteristic that can be observed or measured without affecting the identity of an element, including shape, color, odor, taste, density, hardness, melting point, and boiling point.

proton A positively charged subatomic particle having a mass of 1 amu and found in the nucleus of an atom; its symbol is p or p^+.

pure substance A type of matter with a fixed composition: elements and compounds.

subatomic particle A particle within an atom; protons, neutrons, and electrons are subatomic particles.

valence electrons Electrons in the outermost energy level of an atom.

ADDITIONAL QUESTIONS AND PROBLEMS

3.73 Classify each of the following as an element, compound, or mixture.
   a. carbon in pencils
   b. carbon dioxide (CO_2) we exhale
   c. orange juice
   d. neon gas in lights
   e. a salad dressing of oil and vinegar

3.74 Classify each of the following as a homogenous or heterogeneous mixture.
   a. hot fudge sundae   **b.** herbal tea
   c. vegetable oil   **d.** water and sand
   e. mustard

3.75 Give the symbol and name of the element found in the following group and period on the periodic table.
   a. Group 2A, Period 3   **b.** Group 7A, Period 4
   c. Group 13, Period 3   **d.** Group 16, Period 2

3.76 Give the group and period number for the following elements:
   a. potassium   **b.** phosphorus
   c. carbon   **d.** neon

3.77 Write the names of two elements that are in the following groups:
   a. halogens   **b.** noble gases
   c. alkali metals   **d.** alkaline earth metals

3.78 The following are trace elements that have been found to be crucial to the biochemical and physiological processes in the body. Indicate whether each is a metal or nonmetal.
   a. zinc   **b.** cobalt
   c. manganese (Mn)   **d.** iodine
   e. copper   **f.** selenium (Se)

3.79 Indicate if each of the following statements is *true* or *false*.
   a. The proton is a negatively charged particle.
   b. The neutron is 2000 times as heavy as a proton.
   c. The atomic mass unit is based on a carbon atom with 6 protons and 6 neutrons.
   d. The nucleus is the largest part of the atom.
   e. The electrons are located outside the nucleus.

3.80 Indicate if each of the following statements is *true* or *false*.
   a. The neutron is electrically neutral.
   b. Most of the mass of an atom is due to the protons and neutrons.
   c. The charge of an electron is equal, but opposite, to the charge of a neutron.

   d. The proton and the electron have about the same mass.
   e. The mass number is the number of protons.

3.81 Complete the following statements:
   a. The atomic number gives the number of _____ in the nucleus.
   b. In an atom, the number of electrons is equal to the number of _____.
   c. Sodium and potassium are examples of elements called _____.

3.82 Complete the following statements:
   a. The number of protons and neutrons in an atom is also the _____ number.
   b. The elements in Group 7A (17) are called the _____.
   c. Elements that are shiny and conduct heat are called _____.

3.83 Write the names and symbols of the elements with the following atomic numbers:
   a. 3   **b.** 9   **c.** 20   **d.** 33
   e. 50   **f.** 55   **g.** 79   **h.** 8

3.84 Write the names and symbols of the elements with the following atomic numbers:
   a. 1   **b.** 11   **c.** 20   **d.** 26
   e. 35   **f.** 47   **g.** 83   **h.** 92

3.85 For the following atoms, determine the number of protons, neutrons, and electrons.
   a. $^{27}_{13}$Al   **b.** $^{52}_{24}$Cr   **c.** $^{34}_{16}$S
   d. $^{56}_{26}$Fe   **e.** $^{136}_{54}$Xe

3.86 For the following atoms, give the number of protons, neutrons, and electrons.
   a. $^{22}_{10}$Ne   **b.** $^{127}_{53}$I   **c.** $^{75}_{35}$Br
   d. $^{133}_{55}$Cs   **e.** $^{195}_{78}$Pt

3.87 Complete the following table:

Name	Nuclear Symbol	Number of Protons	Number of Neutrons	Number of Electrons
	$^{34}_{16}$S			
		30	40	
magnesium			14	
	$^{220}_{86}$Rn			

3.88 Complete the following table:

Name	Nuclear Symbol	Number of Protons	Number of Neutrons	Number of Electrons
potassium			22	
	$^{51}_{23}V$			
		48	64	
barium			82	

3.89 Write the symbol and mass number for each of the following:
  a. an atom with 4 protons and 5 neutrons
  b. an atom with 12 protons and 14 neutrons
  c. a calcium atom with a mass number of 46
  d. an atom with 30 electrons and 40 neutrons
  e. a copper atom with 34 neutrons

3.90 Write a symbol and mass number for each of the following:
  a. an aluminum atom with 14 neutrons
  b. an atom with atomic number 26 and 32 neutrons
  c. a strontium atom with 50 neutrons
  d. An atom with a mass number of 72 and atomic number 33

3.91 The most abundant isotope of lead is ^{208}Pb.
  a. How many protons, neutrons, and electrons are in the atom?
  b. What is the symbol of an isotope of lead with 132 neutrons?
  c. What is the name and symbol of an isotope with the same mass number as in part b and 131 neutrons?

3.92 The most abundant isotope of silver is ^{107}Ag.
  a. How many protons, neutrons, and electrons are in the atom?
  b. What is the symbol of an isotope of silver with 62 neutrons?
  c. What is the name and symbol of an isotope with the same mass number as in part b and 49 protons?

3.93 Write an electron arrangement and the group number for an atom of each of the following:
  a. nitrogen **b.** sodium **c.** sulfur **d.** boron

3.94 Write a electron arrangement and the group number for an atom of each of the following:
  a. carbon **b.** silicon **c.** phosphorus **d.** argon

3.95 Why is the ionization energy of Ca higher than K, but lower than Mg?

3.96 Why is the ionization energy of Cl lower than F, but higher than S?

3.97 Of the elements Na, P, Cl, and F, which
  a. is a metal?
  b. has the largest atomic radius?
  c. has the highest ionization energy?
  d. loses an electron most easily?
  e. is found in Group 7A (17), Period 3?

3.98 Of the elements: Mg, Ca, Br, Kr, which
  a. is a noble gas?
  b. has the smallest atomic radius?
  c. has the lowest ionization energy?
  d. requires the most energy to remove an electron?
  e. is found in Group 2A (2), Period 4?

UNDERSTANDING THE CONCEPTS

3.99 Identify the following as an element, compound, or mixture:

  a.

  b.

  c.

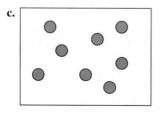

Consider the mixtures in the following diagrams.

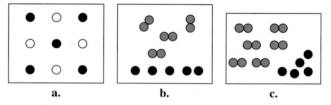

a. b. c.

3.100 Which diagram(s) illustrate a homogeneous mixture? Explain your choice.

3.101 Which diagram(s) illustrate a heterogeneous mixture? Explain your choice.

3.102 Cadmium, atomic number 48, consists of eight naturally occurring isotopes. Do you expect any of the isotopes to have the atomic mass listed on the periodic table for cadmium? Explain.

3.103 Consider the following atoms in which the chemical symbol of the element is represented by X.

$$^{16}_{8}X \quad ^{16}_{9}X \quad ^{18}_{10}X \quad ^{17}_{8}X \quad ^{18}_{8}X$$

a. What atoms have the same number of protons?
b. Which atoms are isotopes? Of what element?
c. Which atoms have the same mass number?
d. What atoms have the same number of neutrons?

3.104 The red light in a neon sign is given off because
a. electrons are removed.
b. electrons jump to higher energy levels.
c. electrons fall to lower energy levels.

QUESTIONS FOR TEAMWORK

3.105 Classify each of the following as a homogeneous mixture or heterogeneous mixture:
a. lemon-flavored water   **b.** stuffed mushrooms
c. chicken noodle soup   **d.** ketchup
e. hard-boiled egg   **f.** eye drops

3.106 According to Dalton's atomic theory, which of the following are true?
a. Atoms of an element are identical to atoms of other elements.
b. Every element is made of atoms.
c. Atoms of two different elements combine to form compounds.
d. In a chemical reaction, some atoms disappear and new atoms appear.

3.107 For each of the following, write the symbol and name for X and the number of protons and neutrons. Which are isotopes of each other?

a. $^{37}_{17}X$ **b.** $^{56}_{26}X$ **c.** $^{116}_{50}X$

d. $^{124}_{50}X$ **e.** $^{116}_{48}X$

3.108 The most abundant isotope of iron is Fe-56.
a. How many protons, neutrons, and electrons are in this isotope?
b. What is the symbol of another isotope of iron with 25 neutrons?
c. What is the symbol of a different atom with the same mass number and 27 neutrons?

3.109 Give the symbol of the element that has the
a. smallest atomic radius in Group 6A
b. smallest atomic radius in Period 3
c. highest ionization energy in Group 15
d. lowest ionization energy in Period 3

3.110 If the diameter of a sodium atom is 3.14×10^{-8} cm, how many sodium atoms would fit along a line exactly 1 inch long?

3.111 A lead atom has a mass of 3.4×10^{-22} g. How many lead atoms are in a cube of lead that has a volume of 2.00 cm³ if the density of lead is 11.3 g/cm³?

ANSWERS

Answers to Study Checks

3.1 24-carat gold is a pure substance; 14-carat gold is a mixture.

3.2 This is a heterogeneous mixture because it does not have a uniform composition.

3.3 Si, S, and Ag

3.4 magnesium, aluminum, and fluorine

3.5 **a.** P   **b.** Rn

3.6 **a.** metal   **b.** metal   **c.** nonmetal

3.7 A proton, symbol p or p^+ with a mass of 1 amu and a charge of 1^+, is found in the nucleus of an atom.

3.8 **a.** 26   **b.** 26   **c.** iron, Fe

3.9 45 neutrons

3.10 **a.** $^{15}_{7}N$ **b.** $^{42}_{20}Ca$ **c.** $^{27}_{13}Al$

3.11 There are more atoms of ^{63}Cu because the atomic mass 63.55 is closer to ^{63}Cu than ^{65}Cu.

3.12 magnesium

3.13 5A (15); phosphorus

3.14 $\cdot \ddot{P} \cdot$

3.15 Ionization energy increases going across a period: Sr is lowest, Sn is higher, and I is the highest of this set.

Answers to Selected Questions and Problems

3.1 **a.** pure substance   **b.** mixture
c. pure substance   **d.** pure substance

3.3 **a.** element   **b.** compound
c. element   **d.** compound

3.5 **a.** heterogeneous   **b.** homogeneous
c. homogeneous   **d.** heterogeneous

3.7 **a.** Cu   **b.** Si   **c.** K   **d.** N
e. Fe   **f.** Ba   **g.** Pb   **h.** Sr

3.9 **a.** carbon   **b.** chlorine   **c.** iodine
d. mercury   **e.** fluorine   **f.** argon
g. zinc   **h.** nickel

3.11 **a.** sodium, chlorine
b. calcium, sulfur, oxygen
c. carbon, hydrogen, chlorine, nitrogen, oxygen
d. calcium, carbon, oxygen

3.13 **a.** Period 2   **b.** Group 8A (18)
c. Group 1A (1)   **d.** Period 2

3.15 **a.** alkaline earth metal
b. transition element
c. noble gas
d. alkali metal
e. halogen

3.17 **a.** C   **b.** He   **c.** Na   **d.** Ca   **e.** Al

3.19 **a.** metal   **b.** nonmetal   **c.** metal
d. nonmetal   **e.** nonmetal   **f.** nonmetal
g. metalloid   **h.** metal

3.21 **a.** electron   **b.** proton
c. electron   **d.** neutron

3.23 Rutherford determined that an atom contains a small, compact nucleus that is positively charged.

3.25 **b.** a proton and an electron

3.27 In the process of brushing hair, strands of hair become charged with like charges that repel each other.

3.29 **a.** atomic number   **b.** both
c. mass number   **d.** atomic number

3.31 **a.** lithium, Li   **b.** fluorine, F   **c.** calcium, Ca
d. zinc, Zn   **e.** neon, Ne   **f.** silicon, Si
g. iodine, I   **h.** oxygen, O

3.33 **a.** 12   **b.** 30   **c.** 53   **d.** 19

3.35 *See Table 3.17*

3.37 **a.** 13 protons, 14 neutrons, 13 electrons
b. 24 protons, 28 neutrons, 24 electrons
c. 16 protons, 18 neutrons, 16 electrons
d. 26 protons, 30 neutrons, 26 electrons

3.39 **a.** $^{31}_{15}P$ **b.** $^{80}_{35}Br$ **c.** $^{27}_{13}Al$ **d.** $^{35}_{17}Cl$

3.41 **a.** $^{32}_{16}S$ $^{33}_{16}S$ $^{34}_{16}S$ $^{36}_{16}S$
b. They all have the same number of protons and electrons.
c. They have different numbers of neutrons, which gives them different mass numbers.
d. The atomic mass of S listed on the periodic table is the average atomic mass of all the isotopes.

3.43 The energy of electrons is of a specific quantity for each energy level. Electrons cannot have energies that are between the energy levels.

3.45 **a.** 8   **b.** 5   **c.** 8   **d.** 0   **e.** 8

3.47 **a.** 2, 4   **b.** 2, 8, 8   **c.** 2, 8, 6
d. 2, 8, 4   **e.** 2, 8, 3   **f.** 2, 5

3.49 **a.** Li   **b.** Mg   **c.** H   **d.** Cl   **e.** O

3.51 absorb

3.53 **a.** B 2,3 Al 2, 8, 3   **b.** 3   **c.** Group 3A (13)

3.55 **a.** $2e^-$, Group 2A (2)
b. $7e^-$, Group 7A (17)
c. $6e^-$, Group 6A (16)
d. $5e^-$, Group 5A (15)
e. $2e^-$, Group 2A (2)
f. $7e^-$, Group 7A (17)

3.57 They each have two electrons in their outer shells.

3.59 **a.** Group 6A (16) $\cdot\ddot{\underset{..}{S}}\cdot$
b. Group 5A (15) $\cdot\ddot{N}\cdot$
c. Group 2A (2) $\cdot Ca\cdot$
d. Group 1A (1) Na·
e. Group 1A (1) K·

3.61 **a.** M·   **b.** ·M·

3.63 They are all in Group 1A; each has 1 valence electron.

3.65 **a.** Mg, Al, Si   **b.** I, Br, Cl   **c.** Sr, Sb, I

3.67 **a.** Na   **b.** Rb   **c.** Na

3.69 **a.** Br, Cl, F   **b.** Na, Al, Cl   **c.** Cs, K, Na

3.71 **a.** Br   **b.** Mg   **c.** P

3.73 **a.** element   **b.** compound   **c.** mixture
d. element   **e.** mixture

3.75 **a.** Mg, magnesium   **b.** Br, bromine
c. Al, aluminum   **d.** O, oxygen

3.77 **a.** any two elements in Group 7A (17) such as fluorine, chlorine, bromine, or iodine
b. any two elements in Group 8A (18) such as helium, neon, argon, krypton, xenon, or radon
c. any two elements in Group 1A (1) such as lithium, sodium, and potassium, except hydrogen
d. any two elements in Group 2A (2) such as magnesium, calcium, and barium

3.79 **a.** false   **b.** false   **c.** true
d. false   **e.** true

3.81 **a.** protons   **b.** protons   **c.** alkali metals

3.83 **a.** lithium, Li   **b.** fluorine, F   **c.** calcium, Ca
d. arsenic, As   **e.** tin, Sn   **f.** cesium, Cs
g. gold, Au   **h.** oxygen, O

3.85 **a.** 13 protons, 14 neutrons, 13 electrons
b. 24 protons, 28 neutrons, 24 electrons
c. 16 protons, 18 neutrons, 16 electrons

TABLE 3.17

Name of Element	Symbol	Atomic Number	Mass Number	Number of Protons	Number of Neutrons	Number of Electrons
aluminum	Al	13	27	13	14	13
magnesium	Mg	12	24	12	12	12
potassium	K	19	39	19	20	19
sulfur	S	16	31	16	15	16
iron	Fe	26	56	26	30	26

d. 26 protons, 30 neutrons, 26 electrons
e. 54 protons, 82 neutrons, 54 electrons

3.87

Name	Nuclear Symbol	Number of Protons	Number of Neutrons	Number of Electrons
sulfur	$^{34}_{16}S$	16	18	16
zinc	$^{70}_{30}Zn$	30	40	30
magnesium	$^{26}_{12}Mg$	12	14	12
radon	$^{220}_{86}Rn$	86	134	86

3.89 **a.** ^9Be **b.** ^{26}Mg **c.** ^{46}Ca
d. ^{70}Zn **e.** ^{63}Cu

3.91 **a.** 82 protons, 126 neutrons, 82 electrons
b. ^{214}Pb **c.** ^{214}Bi, bismuth

3.93 **a.** N Group 5A (15)   2, 5
b. Na Group 2A (2)   2, 8, 1

c. S Group 6A (16)   2, 8, 6
d. B Group 3A (13)   2, 3

3.95 Calcium has a greater number of protons than K. The least tightly bound electron in Ca is further from the nucleus than in Mg and needs less energy to remove.

3.97 **a.** Na   **b.** Na   **c.** F   **d.** Na   **e.** Cl

3.99 **a.** compound   **b.** mixture   **c.** element

3.101 Diagrams b and c illustrate heterogeneous mixtures; they do not have uniform composition throughout the samples.

3.103 **a.** $^{16}_8X$, $^{17}_8X$, $^{18}_8X$ all have 8 protons

b. $^{16}_8X$, $^{17}_8X$, $^{18}_8X$ all isotopes of O

c. $^{16}_8X$ and $^{16}_9X$ have mass number 16, and $^{18}_{10}X$ and $^{18}_8X$ have a mass number of 18.

d. $^{16}_8X$ and $^{18}_{10}X$ have 8 neutrons each.

4 Compounds and Their Bonds

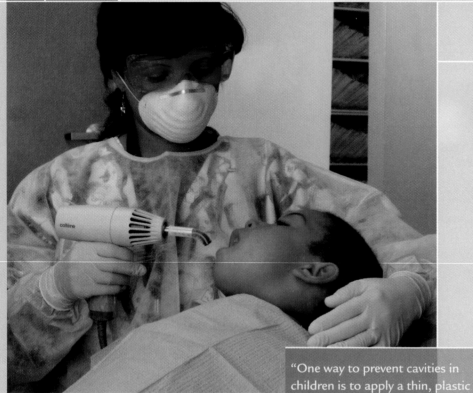

"One way to prevent cavities in children is to apply a thin, plastic coating called a sealant to their teeth," says Dr. Pam Alston, a dentist in private practice. "We look for teeth with deep grooves and pits that trap food. We clean the teeth and apply an etching agent, which helps the sealant bind to the teeth. Then we apply the liquid sealant, which fills in the grooves and pits, and use ultraviolet light to solidify the coating."

The use of fluoride compounds, such as SnF_2 in toothpaste and NaF in water, and mouth rinses have greatly reduced tooth decay. The fluoride ion replaces the hydroxide ion to form $Ca_{10}(PO_4)_6F_2$, which strengthens the enamel and makes it less susceptible to decay. Other compounds used in dentistry are the anesthetic known as laughing gas, N_2O, and Novocaine, $C_{13}H_{20}N_2O_2$.

the Chemistry place

Visit **www.aw-bc.com/chemplace** or use the CD-ROM in the book for extra quizzes, interactive tutorials, career resources, PowerPoint slides, math help, and case studies.

n nature, atoms of almost all the elements on the periodic table are found in combination with other atoms. Only the atoms of the noble gases — He, Ne, Ar, Kr, Xe, and Rn — are found as individual atoms. As discussed in Chapter 3, a compound is a pure substance, composed of two or more elements, with a definite composition. In a typical ionic compound, one or more electrons are transferred from the atoms of metals to atoms of nonmetals. The attraction that results is called an ionic bond.

We use many ionic compounds every day. When we cook or bake we use ionic compounds such as salt, NaCl, and baking soda, $NaHCO_3$. Epsom salts, $MgSO_4$, may be used to soak sore feet. Milk of magnesia, $Mg(OH)_2$, or calcium carbonate, $CaCO_3$, may be taken to settle an upset stomach. In a mineral supplement, iron is present as iron(II) sulfate, $FeSO_4$. Certain sunscreens contain zinc oxide, ZnO, and the tin(II) fluoride, SnF_2, in toothpaste provides fluoride to help prevent tooth decay.

The structures of ionic crystals result in the beautiful facets seen in gems. Sapphires and rubies are made of aluminum oxide, Al_2O_3. Impurities of chromium make rubies red, and iron and titanium make sapphires blue.

In compounds of nonmetals, covalent bonding occurs by atoms sharing one or more valence electrons. There are many more covalent compounds than there are ionic ones and many simple covalent compounds are present in our everyday lives. For example, water (H_2O), oxygen (O_2), and carbon dioxide (CO_2) are all covalent compounds.

Covalent compounds consist of molecules, which are discrete groups of atoms. A molecule of water (H_2O) consists of two atoms of hydrogen and one atom of oxygen. When you have iced tea, perhaps you add molecules of sugar (sucrose), which is a covalent compound ($C_{12}H_{22}O_{11}$). Other covalent compounds include propane C_3H_8, alcohol C_2H_6O, the antibiotic amoxicillin ($C_{16}H_{19}N_3O_5S$), and the antidepressant Prozac ($C_{17}H_{18}F_3NO$).

4.1 Octet Rule and Ions

LEARNING GOAL

Using the octet rule, write the symbols of the simple ions for the representative elements.

Most of the elements on the periodic table combine to form compounds except for the atoms of the noble gases. One explanation is that noble gases have a particularly stable **octet** of 8 valence electrons except for helium, which is stable with 2 electrons filling its first energy level.

Compounds are the result of the formation of chemical bonds between two or more different elements. Ionic bonds occur when atoms of one element lose valence electrons and the atoms of another element gain valence electrons. Ionic compounds typically occur between metals and nonmetals, which are far apart on the periodic table. For example, atoms of sodium and chlorine form NaCl, an ionic compound. However, atoms of elements that are close together on the periodic table share valence electrons and form covalent compounds. For example, atoms of nitrogen and chlorine form NCl_3, a covalent compound.

Loss and gain
of electrons

Sharing electrons

Ionic bond

Covalent bond

M is a metal
Nm is a nonmetal

In the formation of either an ionic bond or covalent bond, atoms adjust their valence electrons such that they each have the electron configuration of a noble gas. Known as the **octet rule**, this tendency for atoms to acquire a noble gas arrangement provides a key to our understanding of the ways in which atoms bond and form compounds.

Positive Ions

In ionic bonding, the valence electrons of a metal are transferred to a nonmetal. Because the ionization energies of metals of Groups 1A (1), 2A (2), and 3A (3) are low, these metal atoms readily lose their valence electrons to nonmetals. In doing so, they acquire the electron configuration of a noble gas (usually 8 valence electrons) and form **ions** with positive charges. For example, when a sodium atom loses its single valence electron, the remaining electrons have the noble gas configuration of neon. By losing an electron sodium has 10 electrons instead of 11. Because there are still 11 protons in its nucleus, the atom is no longer neutral. It has become a sodium ion and has an electrical charge, called an **ionic charge**, of $1+$. In the symbol for the sodium ion, the ionic charge is written in the upper right-hand corner, as Na^+.

Sodium atom
$Na^\bullet$

Sodium ion
Na^+

$11p^+$

e^-

Loss of
valence
electron

$11p^+$

$11e^-$

$10e^-$

2, 8, 1

2, 8

Metals in ionic compounds have lost their valence electrons to form positively charged ions. Positive ions are also called **cations** (pronounced *cat'-ions*). Magnesium, a metal in Group 2A (2), attains a noble gas arrangement like neon by losing 2 valence electrons to form a positive ion with a $2+$ ionic charge. A metal ion is named by its element name. Thus, Mg^{2+} is named the magnesium ion.

Magnesium atom
$Mg^\bullet$

Magnesium ion
Mg^{2+}

$12p^+$

e^-
e^-

Loss of
two valence
electrons

$12p^+$

$12e^-$

$10e^-$

2, 8, 2

2, 8

Negative Ions

Nonmetals form negative ions when they gain one or more electrons to complete their valence energy levels. For example, when an atom of chlorine with 7 valence electrons obtains one more electron, it attains the noble gas arrangement of argon. The resulting particle is a chloride ion having a negative ionic charge (Cl^-). Ions with negative charges are called **anions** (pronounced *an'-ions*). Nonmetal ions are named by using *ide* at the end of their element names.

Chlorine atom
$:\ddot{\underset{\cdot\cdot}{Cl}}\cdot$

e^-

Gain of one valence electron

$17p^+$

$17e^-$

2, 8, 7

Chloride ion
$[:\ddot{\underset{\cdot\cdot}{Cl}}:]^-$

$17p^+$

$18e^-$

2, 8, 8

HEALTH NOTE

■ Some Uses for Noble Gases

Noble gases may be used when it is necessary to have a substance that is unreactive. Scuba divers normally use a pressurized mixture of nitrogen and oxygen gases for breathing under water. However, when the air mixture is used at depths where pressure is high, the nitrogen gas is absorbed into the blood, where it can cause mental disorientation. To avoid this problem, a breathing mixture of oxygen and helium may be substituted. The diver still obtains the necessary oxygen, but the unreactive helium that dissolves in the blood does not cause mental disorientation. However, its lower density does change the vibrations of the vocal cords, and the diver will sound like Donald Duck.

Helium is also used to fill blimps and balloons. When dirigibles were first designed, they were filled with hydrogen, a very light gas. However, when they came in contact with any type of spark or heating source, they exploded violently because of the extreme reactivity of hydrogen gas with oxygen present in the air. Today blimps are filled with unreactive helium gas, which presents no danger of explosion.

Lighting tubes are generally filled with a noble gas such as neon or argon. While the electrically heated filaments that produce the light get very hot, the surrounding noble gases do not react with the hot filament. If heated in air, the elements that constitute the filament would soon burn up.

SAMPLE PROBLEM 4.1

■ Calculating Ionic Charge

Write the symbol and the ionic charge for each of the following ions:

a. a nitride ion that has 7 protons and 10 electrons
b. a calcium ion that has 20 protons and 18 electrons
c. a lithium ion that has 3 protons and 2 electrons

SOLUTION

a. $7p^+$ and $10e^- = N^{3-}$
b. $20p^+$ and $18e^- = Ca^{2+}$
c. $3p^+$ and $2e^- = Li^+$

STUDY CHECK

How many protons and electrons are in a bromide ion, Br^-?

Ionic Charges from Group Numbers

Group numbers can be used to determine the ionic charges for most ions of the representative elements. We have seen that metals lose electrons to form positive ions. The elements in Groups 1A (1), 2A (2), and 3A (13) lose 1, 2, and 3 electrons, respectively. Group 1A (1) metals form ions with 1+ charges, Group 2A (2) metals form ions with 2+ charges, and Group 3A (13) metals form ions with 3+ charges.

The nonmetals from Groups 5A (15), 6A (16), and 7A (17) form negative ions. Group 5A (15) nonmetals form ions with 3− charges, Group 6A (16) nonmetals form ions with 2− charges, and Group 7A (17) nonmetals form ions with 1− charges. The nonmetals of Group 4A (14) do not typically form ions. Table 4.1 lists the ionic charges for some common ions of representative elements.

TABLE 4.1

Ionic Charges for Representative Elements

Group Number	Number of Valence Electrons	Electron Change to Give an Octet	Ionic Charge	Examples
Metals				
1A (1)	1	Lose 1	1+	Li^+, Na^+, K^+
2A (2)	2	Lose 2	2+	Mg^{2+}, Ca^{2+}
3A (13)	3	Lose 3	3+	Al^{3+}
Nonmetals				
5A (15)	5	Gain 3	3−	N^{3-}, P^{3-}
6A (16)	6	Gain 2	2−	O^{2-}, S^{2-}
7A (17)	7	Gain 1	1−	F^-, Cl^-, Br^-, I^-

SAMPLE PROBLEM 4.2

■ Writing Ions

Consider the elements aluminum and oxygen.

a. Identify each as a metal or a nonmetal.
b. State the number of valence electrons for each.
c. State the number of electrons that must be lost or gained for each to acquire an octet.
d. Write the symbol of each resulting ion, including its ionic charge.

SOLUTION

Aluminum	**Oxygen**
a. metal	nonmetal
b. 3 valence electrons	6 valence electrons
c. loses $3e^-$	gains $2e^-$
d. Al^{3+}	O^{2-}

STUDY CHECK

What are the symbols for the ions formed by potassium and sulfur?

QUESTIONS AND PROBLEMS

■ Octet Rule and Ions

4.1 **a.** How does the octet rule explain the formation of a sodium ion?
   b. What noble gas has the same electron arrangement as the sodium ion?
   c. Why are Group 1A (1) and Group 2A (2) elements found in many compounds, but not Group 8A (18) elements?

4.2 **a.** How does the octet rule explain the formation of a chloride ion, Cl^-?
   b. What noble gas has the same electron arrangement as the chloride ion, Cl^-?
   c. Why are Group 7A (17) elements found in many compounds, but not Group 8A (18) elements?

4.3 State the number of electrons that must be lost by atoms of each of the following elements to acquire a noble gas electron arrangement:
   a. Li   **b.** Mg   **c.** Al   **d.** Cs   **e.** Ba

HEALTH NOTE

■ Some Important Ions in the Body

A number of ions in body fluids have important physiological and metabolic functions. Some of them are listed in Table 4.2.

TABLE 4.2

Ions in the Body

Ion	Occurrence	Function	Source	Result of Too Little	Result of Too Much
Na^+	Principal cation outside the cell	Regulation and control of body fluids	Salt, cheese, pickles	Hyponatremia, anxiety, diarrhea, circulatory failure, decrease in body fluid	Hypernatremia, little urine, thirst, edema
K^+	Principal cation inside the cell	Regulation of body fluids and cellular functions	Bananas, orange juice, milk, prunes, potatoes	Hypokalemia (hypopotassemia), lethargy, muscle weakness, failure of neurological impulses	Hyperkalemia (hyperpotassemia), irritability, nausea, little urine, cardiac arrest
Ca^{2+}	Cation outside the cell; 90% of calcium in the body in bone as $Ca_3(PO_4)_2$ or $CaCO_3$	Major cation of bone; muscle smoothant	Milk, yogurt, cheese, greens, and spinach	Hypocalcemia, tingling fingertips, muscle cramps, osteoporosis	Hypercalcemia, relaxed muscles, kidney stones, deep bone pain
Mg^{2+}	Cation outside the cell; 70% of magnesium in the body in bone structure	Essential for certain enzymes, muscles, and nerve control	Widely distributed (part of chlorophyll of all green plants), nuts, whole grains	Disorientation, hypertension, tremors, slow pulse	Drowsiness
Cl^-	Principal anion outside the cell	Gastric juice, regulation of body fluids	Salt	Same as for Na^+	Same as for Na^+

4.4 State the number of electrons that must be gained by atoms of each of the following elements to acquire a noble gas electron arrangement:
a. Cl   **b.** O   **c.** N   **d.** I   **e.** P

4.5 Write the symbols of the ions with the following number of protons and electrons:
a. 3 protons, 2 electrons   **b.** 9 protons, 10 electrons
c. 12 protons, 10 electrons   **d.** 26 protons, 23 electrons
e. 30 protons, 28 electrons

4.6 How many protons and electrons are in the following ions?
a. O^{2-} **b.** K^+ **c.** Br^- **d.** S^{2-} **e.** Sr^{2+}

4.7 Write the symbol for the ion of each of the following:
    a. chlorine   **b.** potassium   **c.** oxygen   **d.** aluminum

4.8 Write the symbol for the ion of each of the following:
    a. fluorine   **b.** calcium   **c.** sodium   **d.** lithium

LEARNING GOAL

Using charge balance, write the correct formula for an ionic compound.

4.2 Ionic Compounds

Ionic compounds consist of positive and negative ions. The ions are held together by strong attractions between the oppositely charged ions, called ionic bonds.

Properties of Ionic Compounds

The physical and chemical properties of an ionic compound such as NaCl are very different from those of the original elements. For example, the original elements of NaCl were sodium, a soft, shiny metal, and chlorine, a yellow-green poisonous gas. Yet, as positive and negative ions, they form table salt, NaCl, a white, crystalline substance that is common in our diet. In ionic compounds, the attraction between the ions is very strong, which makes the melting points of ionic compounds high, often more than 300°C. For example, the melting point of NaCl is 800°C. At room temperature, ionic compounds are solids.

The structure of an ionic solid depends on the arrangement of the ions. In a crystal of NaCl, which has a cubic shape, the larger Cl^- ions (green) are packed close together in a lattice structure as shown in Figure 4.1. The smaller Na^+ ions (silver) occupy the holes between the Cl^- ions.

(a)

(b)

(c)

    —Na^+
    —Cl^-

FIGURE 4.1

(a) The elements sodium and chlorine react to form the ionic compound sodium chloride, the compound that makes up table salt.
(b) Crystals of NaCl under magnification. **(c)** A diagram of the arrangements of Na^+ and Cl^- packed together in a NaCl crystal.

Q **What is the type of bonding between Na^+ and Cl^- ions in salt?**

Charge Balance in Ionic Compounds

The **formula** of an ionic compound indicates the number and kinds of ions that make up the ionic compound. The sum of the ionic charges in the formula is always zero. That means that the total amount of positive charge is equal to the total amount of negative charge. For example, the NaCl formula indicates that there is one sodium ion, Na^+, for every chloride ion, Cl^-, in the compound. Note that the ionic charges of the ions do not appear in the formula of the ionic compound.

Subscripts in Formulas

Consider a compound of magnesium and chlorine. To achieve an octet, a Mg atom loses its 2 valence electrons to form Mg^{2+}. Each Cl atom gains 1 electron to form Cl^-, which has a complete valence energy level. In this example, two Cl^- ions are needed to balance the positive charge of Mg^{2+}. This gives the formula, $MgCl_2$, magnesium chloride, in which a subscript of 2 shows that two Cl^- were needed for charge balance.

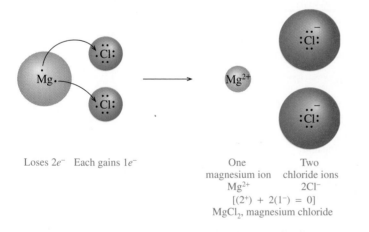

Writing Ionic Formulas from Ionic Charges

The subscripts in the formula of an ionic compound represent the number of positive and negative ions that give an overall charge of zero. Thus, we can now write a formula directly from the ionic charges of the positive and negative ions. Suppose we wish to write the formula of the ionic compound containing Na^+ and S^{2-} ions. To balance the ionic charge of the S^{2-} ion, we will need to place two Na^+ ions in the formula. This gives the formula Na_2S, which has an overall charge of zero. In the formula of an ionic compound, the cation is written first followed by the anion. Appropriate subscripts are used to show the number of ions.

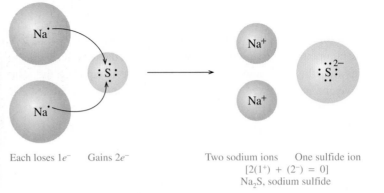

Each loses $1e^-$ Gains $2e^-$ Two sodium ions   One sulfide ion
$$[2(1^+) + (2^-) = 0]$$
Na_2S, sodium sulfide

SAMPLE PROBLEM 4.3

■ Writing Formulas from Ionic Charges

Use ionic charge balance to write the formula for the ionic compound containing K^+ and N^{3-}.

SOLUTION

Determine the number of each ion needed for charge balance. The charge for nitrogen $(3-)$ is balanced by three K^+ ions. Writing the positive ion first gives the formula K_3N.

STUDY CHECK

Use ionic charges to determine the formula of the compound that would form when calcium and oxygen react.

QUESTIONS AND PROBLEMS

■ Ionic Compounds

4.9 Which of the following pairs of elements are likely to form an ionic compound?
   a. lithium and chlorine   **b.** oxygen and chlorine
   c. potassium and oxygen   **d.** sodium and neon
   e. sodium and magnesium

4.10 Which of the following pairs of elements are likely to form an ionic compound?
   a. helium and oxygen   **b.** magnesium and chlorine
   c. chlorine and bromine   **d.** potassium and sulfur
   e. sodium and potassium

4.11 Write the correct ionic formula for compounds formed between the following ions:
   a. Na^+ and O^{2-} **b.** Al^{3+} and Br^- **c.** Ba^{2+} and O^{2-}
   d. Mg^{2+} and Cl^- **e.** Al^{3+} and S^{2-}

4.12 Write the correct ionic formula for compounds formed between the following ions:
   a. Al^{3+} and Cl^- **b.** Ca^{2+} and S^{2-} **c.** Li^+ and S^{2-}
   d. K^+ and N^{3-} **e.** K^+ and I^-

4.13 Write the correct formula for ionic compounds formed by the following metals and nonmetals:
   a. sodium and sulfur   **b.** potassium and nitrogen
   c. aluminum and iodine   **d.** lithium and oxygen

4.14 Write the correct formula for ionic compounds formed by the following metals and nonmetals:
   a. calcium and chlorine   **b.** barium and bromine
   c. sodium and phosphorus   **d.** magnesium and oxygen

TABLE 4.3

TABLE 4.3

Formulas and Names of Some Common Ions

Group Number	Formula of Ion	Name of Ion	Group Number	Formula of Ion	Name of Ion
	Metals			**Nonmetals**	
1A (1)	Li^+	lithium	5A (15)	N^{3-}	nitride
	Na^+	sodium		P^{3-}	phosphide
	K^+	potassium	6A (16)	O^{2-}	oxide
2A (2)	Mg^{2+}	magnesium		S^{2-}	sulfide
	Ca^{2+}	calcium	7A (17)	F^-	fluoride
	Ba^{2+}	barium		Cl^-	chloride
3A (3)	Al^{3+}	aluminum		Br^-	bromide
				I^-	iodide

4.3 Naming and Writing Ionic Formulas

LEARNING GOAL

Given the formula of an ionic compound, write the correct name; given the name of an ionic compound, write the correct formula.

As we mentioned in section 4.2, the name of a metal ion is the same as its elemental name. The name of a nonmetal ion is obtained by replacing the end of its elemental name with *ide*. Table 4.3 lists the names of some important metal and nonmetal ions.

Naming Ionic Compounds Containing Two Elements

In the name of an ionic compound made up of two elements, the metal ion is named followed by the name of the nonmetal ion. Subscripts are never mentioned; they are understood as a result of the charge balance of the ions in the compound. (See Table 4.4.)

Guide to Naming Ionic Compounds with Metals That Form a Single Ion

STEP 1
Identify the cation and anion.

STEP 2
Name the cation by its element name.

STEP 3
Name the anion by changing the last part of its element name to *ide*.

STEP 4
Write the name of the cation first and the name of the anion second.

TABLE 4.4

Names of Some Ionic Compounds

Compound	Metal Ion	Nonmetal Ion	Name
NaF	Na^+ sodium	F^- fluoride	sodium fluoride
$MgBr_2$	Mg^{2+} magnesium	Br^- bromide	magnesium bromide
Al_2O_3	Al^{3+} aluminum	O^{2-} oxide	aluminum oxide

SAMPLE PROBLEM 4.4

Naming Ionic Compounds

Write the name of each of the following ionic compounds:

a. Na_2O **b.** Mg_3N_2

SOLUTION

Compound	Ions and Names		Name of Compound
a. Na_2O	Na^+	O^{2-}	
	sodium	oxide	sodium oxide
b. Mg_3N_2	Mg^{2+}	N^{3-}	
	magnesium	nitride	magnesium nitride

STUDY CHECK

Name the compound $CaCl_2$.

TABLE 4.5

Some Metals That Form More Than One Positive Ion

Element	Possible Ions	Name of Ion
chromium	Cr^{2+}	chromium(II)
	Cr^{3+}	chromium(III)
copper	Cu^+	copper(I)
	Cu^{2+}	copper(II)
gold	Au^+	gold(I)
	Au^{3+}	gold(III)
iron	Fe^{2+}	iron(II)
	Fe^{3+}	iron(III)
lead	Pb^{2+}	lead(II)
	Pb^{4+}	lead(IV)
tin	Sn^{2+}	tin(II)
	Sn^{4+}	tin(IV)

Metals with Variable Charge

The transition metals typically form two or more kinds of positive ions because they lose their outer electrons as well as electrons from a lower energy level. For example, in some ionic compounds, iron is in the Fe^{2+} form, but in other compounds, it takes the Fe^{3+} form. Copper also forms two different ions: Cu^+ is present in some compounds and Cu^{2+} in others. When a metal can form two or more ions, it is not possible to predict the ionic charge from the group number. We say that it has a variable valence.

When different ions are possible for a metal, a naming system is needed to identify the particular cation in a compound. To do this, a Roman numeral that matches the ionic charge is placed in parentheses after the elemental name of the metal. For the cations of iron, Fe^{2+} is named iron(II), and Fe^{3+} is named iron(III). Table 4.5 lists the ions of some common metals that produce two or more ions.

Figure 4.2 shows some common ions and their location on the periodic table. Typically, the transition metals form more than one positive ion. However, zinc, cadmium, and silver form only one ion. The ionic charges of silver, cadmium, and

1 Group 1A	2 Group 2A	3 3B	4 4B	5 5B	6 6B	7 7B	8	9 8B	10	11 1B	12 2B	13 Group 3A	14 Group 4A	15 Group 5A	16 Group 6A	17 Group 7A	18 Group 8A
H^+																	
Li^+														N^{3-}	O^{2-}	F^-	
Na^+	Mg^{2+}											Al^{3+}		P^{3-}	S^{2-}	Cl^-	
K^+	Ca^{2+}				Cr^{2+} Cr^{3+}		Fe^{2+} Fe^{3+}			Cu^+ Cu^{2+}	Zn^{2+}					Br^-	
Rb^+	Sr^{2+}									Ag^+	Cd^{2+}	Sn^{2+} Sn^{4+}				I^-	
Cs^+	Ba^{2+}									Au^+ Au^{3+}		Pb^{2+} Pb^{4+}					

 Metals   Metalloids   Nonmetals

FIGURE 4.2

On the periodic table, positive ions are produced from metals and negative ions are produced from nonmetals.

Q **What are the typical ions produced by calcium, copper and oxygen?**

zinc are fixed like Group 1A (1), 2A (2), and 3A (13) metals, so their elemental names are sufficient when naming their ionic compounds.

The selection of the correct Roman numeral depends upon the calculation of the ionic charge of the transition metal in the formula. For example, we know that in the formula $CuCl_2$ the positive charge of the copper ion must balance the negative charge of two chloride ions. Because we know that chloride ions each have a $1-$ charge, there must be a total negative charge of $2-$. Balancing the $2-$ by the positive charge gives a charge of $2+$ for Cu (a Cu^{2+} ion):

$CuCl_2$

Cu charge $+$ Cl^- charge $= 0$

(?) $+ 2(1-)$ $= 0$

(2+) $+ 2-$ $= 0$

To indicate the copper ion Cu^{2+}, we place (II) after *copper* when naming the compound:

copper(II) chloride

Table 4.6 lists names of some ionic compounds in which the metals form two types of positive ion.

TABLE 4.6

Some Ionic Compounds of Metals That Form Two Kinds of Positive Ions

Compound	Systematic Name
$FeCl_2$	iron(II) chloride
$FeCl_3$	iron(III) chloride
Cu_2O	copper(I) oxide
$CuCl_2$	copper(II) chloride
$SnCl_2$	tin(II) chloride
$PbBr_4$	lead(IV) bromide

SAMPLE PROBLEM 4.5

■ **Naming Ionic Compounds with Variable Charge Metal Ions**

Write the name for Cu_2S.

SOLUTION

STEP 1   Determine the charge of the cation from the anion.   The nonmetal S in Group 6A (16) forms the S^{2-} ion. Because there are two Cu ions to balance the S^{2-}, the charge of each Cu ion is $1+$.

	Metal	**Nonmetal**
Elements	copper	sulfur
Groups	transition	6A (16)
Ions	Cu?	S^{2-}
Charge balance	$2(+1)$ +	$(2-) = 0$
Ions	Cu^+	S^{2-}

STEP 2   Name the cation by its element name and use a Roman numeral in parentheses for the charge.

copper(I)

STEP 3   Name the anion by changing the last part of its element names to *ide*.

sulfide

STEP 4   Write the name of the cation first and the name of the anion second.

copper(I) sulfide

Guide to Naming Ionic Compounds with Variable Charge Metals

STEP 1
Determine the charge of the cation from the anion.

STEP 2
Name the cation by its element name and a Roman numeral in parentheses for the charge.

STEP 3
Name the anion by changing the last part of its element name to *ide*.

STEP 4
Write the name of the cation first and the name of the anion second.

STUDY CHECK

Write the name of the compound whose formula is $AuCl_3$.

SAMPLE PROBLEM 4.6

■ **Writing Formulas of Ionic Compounds**

Write the formula for iron(III) chloride.

SOLUTION

Guide to Writing Formulas from the Name of an Ionic Compound

STEP 1
Identify the cation and anion.

STEP 2
Balance the charges.

STEP 3
Write the formula, cation first, using subscripts from charge balance.

STEP 1 **Identify the cation and anion.** The Roman numeral (III) indicates that the charge of the iron ion is 3+, Fe^{3+}.

	Metal	**Nonmetal**
Elements	iron(III)	chlorine
Groups	transition	7A (17)
Ions	Fe^{3+}	Cl^-

STEP 2 **Balance the charges.**

$$Fe^{3+} \quad Cl^-$$
$$Cl^-$$
$$Cl^-$$
$$\overline{1(3+) + 3(1-) = 0}$$

Becomes a subscript in the formula

STEP 3 **Write the formula, cation first, using the subscripts from the charge balance.**

$FeCl_3$

STUDY CHECK

Write the correct formula for chromium(III) oxide.

QUESTIONS AND PROBLEMS

■ **Naming and Writing Ionic Formulas**

4.15 Write names for the following ionic compounds:
   a. Al_2O_3 **b.** $CaCl_2$ **c.** Na_2O
   d. Mg_3N_2 **e.** KI

4.16 Write names for the following ionic compounds:
   a. $MgCl_2$ **b.** K_3P **c.** Li_2S
   d. LiBr   **e.** MgO

4.17 Why is a Roman numeral placed after the name of most transition metal ions?

4.18 The compound $CaCl_2$ is named calcium chloride; the compound $CuCl_2$ is named copper(II) chloride. Explain why a Roman numeral is used in one name but not the other.

4.19 Write the names of the following Group 4A (4) and transition metal ions (include the Roman numeral when necessary):
   a. Fe^{2+} **b.** Cu^{2+} **c.** Zn^{2+} **d.** Pb^{4+} **e.** Cr^{3+}

4.20 Write the names of the following Group 4A (4) and transition metal ions (include the Roman numeral when necessary):
   a. Ag^+ **b.** Cu^+ **c.** Fe^{3+} **d.** Sn^{2+} **e.** Au^{3+}

4.21 Write names for the following ionic compounds:
   a. $SnCl_2$ **b.** FeO   **c.** Cu_2S **d.** CuS   **e.** $CrBr_3$

4.22 Write names for the following ionic compounds:
   a. Ag_3P **b.** PbS   **c.** SnO_2 **d.** $AuCl_3$ **e.** ZnS

4.23 Indicate the charge on the metal ion in each of the following formulas:
   a. $AuCl_3$ **b.** Fe_2O_3 **c.** PbI_4 **d.** $SnCl_2$

4.24 Indicate the charge on the metal ion in each of the following formulas:
    a. $FeCl_2$ **b.** CuO   **c.** Fe_2S_3 **d.** $CrCl_3$

4.25 Write formulas for the following ionic compounds:
    a. magnesium chloride   **b.** sodium sulfide
    c. copper(I) oxide   **d.** zinc phosphide
    e. gold(III) nitride   **f.** chromium(II) chloride

4.26 Write formulas for the following ionic compounds:
    a. iron(III) oxide   **b.** barium fluoride
    c. tin(IV) chloride   **d.** silver sulfide
    e. copper(II) chloride   **f.** lithium nitride

◾ 4.4 Polyatomic Ions

LEARNING GOAL

Write the name and formula of a compound containing a polyatomic ion.

In a **polyatomic ion**, a group of atoms has acquired an electrical charge. Most polyatomic ions consist of a nonmetal such as phosphorus, sulfur, carbon, or nitrogen bonded to oxygen atoms. These oxygen-containing polyatomic ions have an ionic charge of $1-$, $2-$, or $3-$ because 1, 2, or 3 electrons were added to the atoms in the group to complete their octets. Only one of the common polyatomic ions, NH_4^+, is positively charged. Some models of polyatomic ions are shown in Figure 4.3.

◾ Naming Polyatomic Ions

The names of the most common polyatomic ions end in *ate*. The *ite* ending is used for the names of related ions that have one less oxygen atom. Recognizing these endings will help you identify polyatomic ions in the name of a compound. The

Window cleaner
NH_4OH

Plaster molding
$CaSO_4$

Fertilizer
$NaNO_3$

NH_4^+ OH^-
Ammonium ion   Hydroxide ion

Ca^{2+} SO_4^{2-}
Sulfate ion

Na^+ NO_3^-
Nitrate ion

FIGURE 4.3

Many products contain polyatomic ions, which are groups of atoms that carry an ionic charge.

Q Why does the sulfate ion have a 2− charge?

TABLE 4.7

Names and Formulas of Some Common Polyatomic Ions

Nonmetal	Formula of Ion[a]	Name of Ion
hydrogen	OH^-	hydroxide
nitrogen	NH_4^+	ammonium
	$\boxed{NO_3^-}$	**nitrate**
	NO_2^-	nitrite
chlorine	$\boxed{ClO_3^-}$	**chlorate**
	ClO_2^-	chlorite
carbon	$\boxed{CO_3^{2-}}$	**carbonate**
	HCO_3^-	hydrogen carbonate (or bicarbonate)
	CN^-	cyanide
	$C_2H_3O_2^-\ (CH_3COO^-)$	acetate
sulfur	$\boxed{SO_4^{2-}}$	**sulfate**
	HSO_4^-	hydrogen sulfate (or bisulfate)
	SO_3^{2-}	sulfite
	HSO_3^-	hydrogen sulfite (or bisulfite)
phosphorus	$\boxed{PO_4^{3-}}$	**phosphate**
	HPO_4^{2-}	hydrogen phosphate
	$H_2PO_4^-$	dihydrogen phosphate
	PO_3^{3-}	phosphite

[a]Boxed formulas are the most common polyatomic ion for that element.

hydroxide ion (OH^-) and cyanide ion (CN^-) are exceptions to this naming pattern. There is no easy way to learn polyatomic ions. You will need to memorize the number of oxygen atoms and the charge associated with each ion, as shown in Table 4.7. By memorizing the formulas and the names of the ions shown in the boxes, you can derive the related ions. For example, the sulfate ion is SO_4^{2-}. We write the formula of the sulfite ion, which has one less oxygen atom, as SO_3^{2-}. The formula of hydrogen carbonate, or *bicarbonate*, can be written by placing a hydrogen cation (H^+) in front of the formula for carbonate (CO_3^{2-}) and decreasing the charge from $2-$ to $1-$ to give HCO_3^-.

Writing Formulas for Compounds Containing Polyatomic Ions

No polyatomic ion exists by itself. Like any ion, a polyatomic ion must be associated with ions of opposite charge. The bonding between polyatomic ions and other ions is one of electrical attraction. For example, the compound sodium sulfate consists of sodium ions (Na^+) and sulfate ions (SO_4^{2-}) held together by ionic bonds.

To write correct formulas for compounds containing polyatomic ions we follow the same rules of charge balance that we used for writing the formulas of simple ionic compounds. The total negative and positive charges must equal zero. For example, consider the formula for a compound containing calcium ions and carbonate ions. The ions are written as

$$Ca^{2+} \qquad CO_3^{2-}$$

calcium ion   carbonate ion

Ionic charge: $(2+)$ + $(2-) = 0$

Polyatomic Ions in Bones and Teeth

Bone structure consists of two parts: a solid mineral material and a second phase made up primarily of collagen proteins. The mineral substance is a compound called hydroxyapatite, a solid formed from calcium ions, phosphate ions, and hydroxide ions. This material is deposited in the web of collagen to form a very durable bone material.

$$Ca_{10}(PO_4)_6(OH)_2$$
Calcium hydroxyapatite

In most individuals, bone material is continuously being absorbed and reformed. After age 40, more bone material may be lost than formed, a condition called osteoporosis. Bone mass reduction occurs at a faster rate in women than in men and at different rates in different parts of the skeleton. The reduction in bone mass can be as much as 50% over a period of 30 to

(a)　　　　　　　　　　　**(b)**

40 years. Scanning electron micrographs (SEMs) show **(a)** normal bone and **(b)** bone in osteoporosis due to calcium loss. It is recommended that persons over 35, especially women, include a daily calcium supplement in their diet.

Because one ion of each balances the charge, the formula is written as

$$CaCO_3$$
calcium carbonate

When more than one polyatomic ion is needed for charge balance, parentheses are used to enclose the formula of the ion. A subscript is written outside the closing parenthesis to indicate the number of polyatomic ions. Consider the formula for magnesium nitrate. The ions in this compound are the magnesium ion and the nitrate ion, a polyatomic ion.

$$Mg^{2+} \qquad NO_3^{-}$$
magnesium ion　　nitrate ion

To balance the positive charge of 2+, two nitrate ions are needed. The formula, including the parentheses around the nitrate ion, is as follows:

$$NO_3^{-}$$
$$Mg^{2+}$$
$$NO_3^{-}$$
$$(2+) + 2(1-) = 0$$

magnesium nitrate

$$Mg(NO_3)_2$$

Parentheses enclose the formula of the nitrate ions

Subscript outside the parentheses indicates the use of two nitrate ions

■Writing Formulas for Ionic Compounds with Polyatomic Ions

Write the formula of aluminum bicarbonate.

SOLUTION

STEP 1 **Identify the cation and anion.**   The cation is the aluminum ion, Al^{3+}, and the cation is bicarbonate, which is a polyatomic anion, HCO_3^-.

	Cation	Anion
Ions	Al^{3+}	HCO_3^-

STEP 2 **Balance the charges.**

$$Al^{3+} \qquad \begin{array}{c} HCO_3^- \\ HCO_3^- \\ HCO_3^- \\ \hline \end{array}$$

$$\mathbf{1}(3+) \quad + \quad \mathbf{3}(1-) = 0$$

Becomes a subscript in the formula

STEP 3 **Write the formula, cation first, using the subscripts from the charge balance.**   The formula for the compound is written by enclosing the formula of the bicarbonate ion, HCO_3^-, in parentheses and writing the subscript 3 outside the last parenthesis.

$$Al(HCO_3)_3$$

STUDY CHECK

Write the formula for a compound containing ammonium ion(s) and phosphate ion(s).

Naming Compounds Containing Polyatomic Ions

When naming ionic compounds containing polyatomic ions, we write the positive ion, usually a metal, first, and then we write the name of the polyatomic ion. It is important that you learn to recognize the polyatomic ion in the formula and name it correctly. As with other ionic compounds, no prefixes are used.

Na_2SO_4	$FePO_4$	$Al_2(CO_3)_3$
$Na_2\boxed{SO_4}$	$Fe\boxed{PO_4}$	$Al_2(\boxed{CO_3})_3$
sodium sulfate	iron(III) sulfate	aluminum carbonate

Table 4.8 lists the formulas and names of some ionic compounds that include polyatomic ions and also gives their uses in medicine and industry.

■Naming Compounds Containing Polyatomic Ions

Name the following ionic compounds:

a. $CaSO_4$ **b.** $Cu(NO_2)_2$

TABLE 4.8

Some Compounds That Contain Polyatomic Ions

Formula	Name	Use
$BaSO_4$	barium sulfate	Radiopaque medium
$CaCO_3$	calcium carbonate	Antacid, calcium supplement
$Ca_3(PO_4)_2$	calcium phosphate	Calcium replenisher
$CaSO_3$	calcium sulfite	Preservative in cider and fruit juices
$CaSO_4$	calcium sulfate	Plaster casts
$AgNO_3$	silver nitrate	Topical anti-infective
$NaHCO_3$	sodium bicarbonate *or* sodium hydrogen carbonate	Antacid
$Zn_3(PO_4)_2$	zinc phosphate	Dental cements
$FePO_4$	iron(III) phosphate	Food and bread enrichment
K_2CO_3	potassium carbonate	Alkalizer, diuretic
$Al_2(SO_4)_3$	aluminum sulfate	Antiperspirant, anti-infective
$AlPO_4$	aluminum phosphate	Antacid
$MgSO_4$	magnesium sulfate	Cathartic, Epsom salts

SOLUTION

Compound Formula	Ions Present	Name of Compound
a. $CaSO_4$	Ca^{2+}, calcium SO_4^{2-}, sulfate	calcium sulfate
b. $Cu(NO_2)_2$	Cu^{2+}, copper(II) NO_2^{-}, nitrite	copper(II) nitrite

One less oxygen than nitrate, NO_3^{-}

STUDY CHECK

What is the name of $Ca_3(PO_4)_2$?

QUESTIONS AND PROBLEMS

■ Polyatomic Ions

4.27 Write the formulas including the charge for the following polyatomic ions:
   a. hydrogen carbonate   **b.** ammonium
   c. phosphate   **d.** hydrogen sulfate

4.28 Write the formulas including the charge for the following polyatomic ions:
   a. nitrite   **b.** sulfite   **c.** hydroxide   **d.** phosphite

4.29 Name the following polyatomic ions:
   a. SO_4^{2-} **b.** CO_3^{2-} **c.** PO_4^{3-} **d.** NO_3^{-}

4.30 Name the following polyatomic ions:
   a. OH^{-} **b.** HSO_3^{-} **c.** CN^{-} **d.** NO_2^{-}

4.31 Complete the following table with the formula of the compound:

	OH^{-}	NO_2^{-}	CO_3^{2-}	HSO_4^{-}	PO_4^{3-}
Li^{+}					
Cu^{2+}					
Ba^{2+}					

4.32 Complete the following table with the formula of the compound:

	OH$^-$	NO$_3^-$	HCO$_3^-$	SO$_3^{2-}$	PO$_4^{3-}$
NH$_4^+$					
Al^{3+}					
Pb^{4+}					

4.33 Write the formula for the polyatomic ion in each of the following and name each compound:

a. Na$_2$CO$_3$ **b.** NH$_4$Cl   **c.** Li$_3$PO$_4$ **d.** Cu(NO$_2$)$_2$ **e.** FeSO$_3$

4.34 Write the formula for the polyatomic ion in each of the following and name each compound:

a. KOH   **b.** NaNO$_3$ **c.** CuCO$_3$ **d.** NaHCO$_3$ **e.** BaSO$_4$

4.35 Write the correct formula for the following compounds:

a. barium hydroxide   **b.** sodium sulfate   **c.** iron(II) nitrate
d. zinc phosphate   **e.** iron(III) carbonate

4.36 Write the correct formula for the following compounds:

a. aluminum chlorate   **b.** ammonium oxide   **c.** magnesium bicarbonate
d. sodium nitrite   **e.** copper(I) sulfate

LEARNING GOAL

Given the formula of a covalent compound, write its correct name; given the name of a covalent compound, write its formula.

the Chemistry place

WEB TUTORIAL
Covalent Bonds

4.5 Covalent Compounds

Earlier, we learned that atoms of metals and nonmetals become more stable by forming ionic compounds. However, atoms of nonmetals have high ionization energies and do not lose electrons easily. Thus, in *covalent compounds*, electrons are not transferred from one atom to another, but are shared between atoms of nonmetals to achieve stability. When atoms share electrons, they form **molecules**.

Formation of a Hydrogen Molecule

The simplest covalent molecule is hydrogen gas, H$_2$. When two hydrogen atoms are far apart, they are not attracted to each other. As the atoms move closer, the positive charge of each nucleus attracts the electron of the other atom. This attraction pulls the atoms closer until they share a pair of valence electrons and form a *covalent bond*. In the covalent bond in H$_2$, the shared electrons give the noble gas configuration of He to each of the H atoms. Thus the atoms bonded in H$_2$ are more stable than two individual H atoms.

In covalent compounds, the valence electrons and the shared pairs of electrons can be shown using electron-dot formulas. In an electron-dot formula, a shared pair of electrons is written as two dots or a single line between the atomic symbols. This notation is shown in the formation of the covalent bond in the electron-dot formula for the H$_2$ molecule.

$$\text{H}\cdot \; + \; \cdot\text{H} \longrightarrow \text{H}:\text{H} \longrightarrow \text{H}-\text{H} \; = \; \text{H}_2$$

Electrons to share   A shared pair of electrons   A covalent bond   A hydrogen molecule

Formation of Octets in Covalent Molecules

In most covalent compounds, atoms share electrons to achieve octets for the valence electrons. For example, a fluorine molecule, F$_2$, consists of two fluorine atoms. Each fluorine atom has 7 valence electrons. By sharing their unpaired valence electrons, each F atom achieves an octet. In the resulting F_2 molecule, each F atom has the noble gas configuration of neon. In the electron-dot formula, the shared electrons, or *bonding pair*, are written between atoms with the nonbonding pairs of electrons, or *lone pairs*, on the outside. This is shown in the formation of the covalent bond for the F_2 molecule.

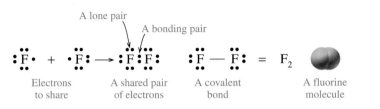

| Electrons to share | A shared pair of electrons | A covalent bond | A fluorine molecule |

Hydrogen (H_2) and fluorine (F_2) are examples of nonmetal elements whose natural state is diatomic; that is, they contain two like atoms. The elements that exist as diatomic molecules are listed in Table 4.9.

Sharing Electrons Between Atoms of Different Elements

In Period 2, the number of electrons that an atom shares and the number of covalent bonds it forms are usually equal to the number of electrons needed to acquire a noble gas arrangement. For example, carbon has 4 valence electrons. Because carbon needs to acquire 4 more electrons for an octet, it forms 4 covalent bonds by sharing its 4 valence electrons.

Methane, a component of natural gas, is a compound made of carbon and hydrogen. To attain an octet, each carbon shares 4 electrons, and each hydrogen shares 1 electron. In this molecule, a carbon atom forms four covalent bonds with four hydrogen atoms. The electron-dot formula for the molecule is written with the carbon atom in the center and the hydrogen atoms on the sides. Table 4.10 gives the formulas of some covalent molecules for Period 2 elements.

While the octet rule is useful, there are exceptions. We have already seen that a hydrogen (H_2) molecule requires just two electrons or a single bond to achieve the

TABLE 4.9

Elements That Exist as Diatomic, Covalent Molecules

Element	Diatomic Molecule	Name
H	H_2	hydrogen
N	N_2	nitrogen
O	O_2	oxygen
F	F_2	fluorine
Cl	Cl_2	chlorine
Br	Br_2	bromine
I	I_2	iodine

Methane, CH_4

the Chemistry place

WEB TUTORIAL Covalent Bonds

TABLE 4.10

Electron-Dot Formulas for Some Covalent Compounds

| CH_4 | NH_3 | H_2O |

Formulas Using Electron Dots Only

Formulas Using Bonds and Electron Dots

Molecular Models

| Methane molecule | Ammonia molecule | Water molecule |

TABLE 4.11

Typical Bonding Patterns of Some Nonmetals in Covalent Compounds

1A (1)	3A (13)	4A (14)	5A (15)	6A (16)	7A (17)
^{a}H 1 bond					
	^{a}B 3 bonds	C 4 bonds	N 3 bonds	O 2 bonds	F 1 bond
		Si 4 bonds	P 3 bonds	S 2 bonds	Cl, Br, I 1 bond

^{a}H and B do not form eight-electron octets. H atoms share one electron pair; B atoms share three electron pairs for a set of 6 electrons.

stability of the nearest noble gas, helium. Although the nonmetals typically form octets, atoms such as P, S, Cl, Br, and I can share more valence electrons and form stable valence shells of 10 or 12 electrons. For example, in PCl_3, the P atom has an octet, but in PCl_5 the P atom has 10 valence electrons or 5 bonds. In H_2S, the S atom has an octet, but in SF_6, there are 12 valence electrons or 6 bonds to the sulfur atom.

Table 4.11 gives the bonding patterns for some nonmetals.

SAMPLE PROBLEM 4.9

▪ Writing Electron-Dot Formulas for Covalent Compounds

Draw an electron-dot formula for H_2S.

SOLUTION

Sulfur, which has 6 valence electrons, shares 2 electrons to form two covalent bonds. Two atoms of hydrogen, each having 1 valence electron, will form two covalent bonds, one for each atom:

$$H:\overset{..}{\underset{..}{S}}:$$
$$H$$

Electron-dot structure for H_2S

STUDY CHECK

Write the electron-dot structure for PH_3.

▪ Multiple Covalent Bonds

Up to now, we have looked at covalent bonding in molecules having only single bonds. In many covalent compounds, atoms share two or three pairs of electrons to complete their octets. A **double bond** occurs when two pairs of electrons are shared and in a **triple bond**, three pairs of electrons are shared. Atoms of carbon, oxygen, nitrogen, and sulfur are most likely to form multiple bonds. For example, there are double bonds in CO_2 because two pairs of electrons are shared between the carbon atom and each oxygen atom to give octets. Atoms of hydrogen and the halogens do not form double or triple bonds.

$:\overset{..}{\underset{..}{O}}(::)C(::)\overset{..}{\underset{..}{O}}:$ $16e^-$ $:\overset{..}{O}=C=\overset{..}{O}:$ CO_2

Octets   Molecule of
                                    carbon dioxide

In the electron-dot structure for the covalent compound N_2, an octet is achieved when each nitrogen atom shares 3 electrons. Thus, three covalent bonds, or a triple bond, is formed.

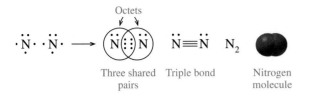

Naming Covalent Compounds

When naming a covalent compound, the first nonmetal in the formula is named by its elemental name; the second nonmetal is named by its elemental name with the ending changed to *ide*. Subscripts indicating two or more atoms of an element are expressed as prefixes placed in front of each name. Table 4.12 lists prefixes used in naming covalent compounds. The names of covalent compounds need prefixes because several different compounds can be formed from the same two nonmetals. For example, carbon and oxygen can form two different compounds, carbon monoxide, CO, and carbon dioxide, CO_2.

When the vowels *o* and *o* or *a* and *o* appear together, the first vowel is omitted as in carbon monoxide. In the name of a covalent compound, the prefix *mono* is usually omitted, as in NO, nitrogen oxide. Traditionally, however, CO is named carbon monoxide. The prefix *mono* is not used for the first nonmetal. Table 4.13 lists the formulas, names, and commercial uses of some covalent compounds.

TABLE 4.12

Prefixes Used in Naming Covalent Compounds

1 mono	6 hexa
2 di	7 hepta
3 tri	8 octa
4 tetra	9 nona
5 penta	10 deca

TABLE 4.13

Some Common Covalent Compounds

Formula	Name	Commercial Uses
CS_2	carbon disulfide	Manufacture of rayon
CO_2	carbon dioxide	Carbonation of beverages, fire extinguishers, propellant in aerosols, dry ice
NO	nitrogen oxide	Stabilizer
N_2O	dinitrogen oxide	Inhalation anesthetic, "laughing gas"
SiO_2	silicon dioxide	Manufacture of glass
SO_2	sulfur dioxide	Preserving fruits, vegetables; disinfectant in breweries; bleaching textiles
SF_6	sulfur hexafluoride	Electrical circuits

SAMPLE PROBLEM 4.10

Naming Covalent Compounds

Name the covalent compound NCl_3.

SOLUTION

We can use the guide to naming covalent compounds to name these covalent compounds.

STEP 1 Name the first nonmetal by its element name. In NCl_3, the first nonmetal (N) is nitrogen.

STEP 2 Name the second nonmetal by changing the last part of its element name to *ide*. The second nonmetal (Cl) is named chloride.

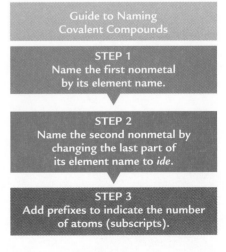

Guide to Naming Covalent Compounds

STEP 1
Name the first nonmetal by its element name.

STEP 2
Name the second nonmetal by changing the last part of its element name to *ide*.

STEP 3
Add prefixes to indicate the number of atoms (subscripts).

STEP 3 **Add prefixes to indicate the number of atoms of each nonmetal.**
Because nitrogen is the first nonmetal, the prefix *mono* is understood
and not used. The subscript indicating three Cl atoms is shown as the
prefix *tri*.

NCl_3 nitrogen trichloride

STUDY CHECK

Write the name of each of the following compounds:

a. $SiBr_4$ **b.** Br_2O

Writing Formulas from the Names of Covalent Compounds

In the name of a covalent compound, the names of two nonmetals are given along
with prefixes for the number of atoms of each. To obtain a formula, we write the
symbol for each element and a subscript if a prefix indicates two or more atoms.

SAMPLE PROBLEM 4.11

■ **Writing Formulas for Covalent Compounds**

Write the formula for each of the following covalent compounds:

a. sulfur dichloride   **b.** diboron trioxide

SOLUTION

a. sulfur dichloride

STEP 1 **Write the symbols in order of the elements in the names.**   The first
nonmetal is sulfur and the second nonmetal is chlorine.

S   Cl

STEP 2 **Write prefixes as subscripts.**   Since there is no prefix for sulfur, we
know that there is one atom of sulfur. The prefix *di* in *dichloride* indicates that there are two atoms of chlorine shown as a subscript 2 in the
formula.

SCl_2

b. diboron trioxide

STEP 1 **Write the symbols in order of the elements in the names.**   The first
nonmetal is boron and the second nonmetal is oxygen.

B   O

STEP 2 **Write prefixes as subscripts.**   The prefix *di* in *diboron* indicates that
there are two atoms of boron shown as a subscript 2 in the formula.
The prefix *tri* in *trioxide* indicates that there are three atoms of oxygen
shown as a subscript 3 in the formula.

B_2O_3

STUDY CHECK

What is the formula of iodine pentafluoride?

Guide to Writing Formulas for
Covalent Compounds

STEP 1
Write the symbols in the order
of the elements in the name.

STEP 2
Write any prefixes as subscripts.

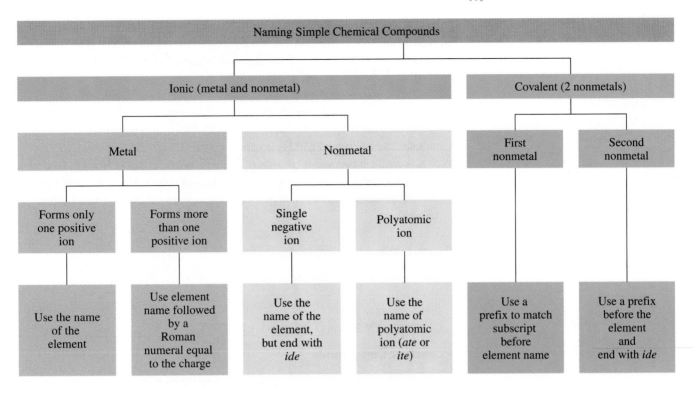

FIGURE 4.4

A flowchart of how ionic and covalent compounds are named.

Q **Why are the names of some metal ions followed by a Roman numeral in the name of a compound?**

Summary of Naming Compounds

Throughout this chapter we have examined strategies for naming ionic and covalent compounds. Now we can summarize the rules, as illustrated in Figure 4.4. In general, compounds having two elements are named by stating the first element, followed by the second element with an *ide* ending. If the first element is a metal, the compound is ionic; if the first element is a nonmetal, the compound is covalent. For ionic compounds, it is necessary to determine whether the metal can form more than one type of positive ion; if so, a Roman numeral following the name of the metal indicates the particular ionic charge. One exception is the ammonium ion NH_4^+, which is also written first as a positively charged polyatomic ion. In naming covalent compounds having two elements, prefixes are necessary to indicate the number of atoms of each nonmetal as shown in that particular formula. Ionic compounds having three or more elements include some type of polyatomic ion. They are named by ionic rules, but have an *ate* or *ite* ending when the polyatomic ion has a negative charge.

SAMPLE PROBLEM 4.12

■ **Naming Ionic and Covalent Compounds**

Name the following compounds:

a. Na_3P **b.** $CuSO_4$ **c.** SO_3

SOLUTION

a. Na_3P is an ionic compound. Na is a metal that forms a single ion, Na^+, which is named sodium. P^{3-}, a single negative ion, is named phosphide. The compound Na_3P is sodium phosphide.

b. $CuSO_4$ is an ionic compound. Cu is a transition metal that forms more than one ion. Cu^{2+} is the positive ion that balances the charge on SO_4^{2-}. Roman numeral II specifies the charge in the name copper(II). SO_4^{2-} is a polyatomic ion named sulfate. The compound $CuSO_4$ is named copper(II) sulfate.

c. SO_3 is covalent because it consists of two nonmetals. The first element S is named sulfur (no prefix is needed). The subscript three calls for the prefix *tri* and the element oxygen becomes oxide. The compound SO_3 is named sulfur trioxide.

STUDY CHECK

What is the name of $Fe(NO_3)_2$?

QUESTIONS AND PROBLEMS

■ Covalent Compounds

4.37 Write the electron-dot formula for each of the following covalent molecules:
  a. Br_2 **b.** H_2 **c.** HF   **d.** OF_2

4.38 Write the electron-dot formula for each of the following covalent molecules:
  a. NCl_3 **b.** CCl_4 **c.** Cl_2 **d.** SiF_4

4.39 Name the following covalent compounds:
  a. PBr_3 **b.** CBr_4 **c.** SiO_2 **d.** HF

4.40 Name the following covalent compounds:
  a. CS_2 **b.** P_2O_5 **c.** Cl_2O **d.** PCl_3

4.41 Name the following covalent compounds:
  a. N_2O_3 **b.** NCl_3 **c.** $SiBr_4$ **d.** PCl_5

4.42 Name the following covalent compounds:
  a. SiF_4 **b.** IBr_3 **c.** CO_2 **d.** SO_3

4.43 Write the formulas of the following covalent compounds:
  a. carbon tetrachloride   **b.** carbon monoxide
  c. phosphorus trichloride   **d.** dinitrogen tetroxide

4.44 Write the formulas of the following covalent compounds:
  a. sulfur dioxide   **b.** silicon tetrachloride
  c. iodine pentafluoride   **d.** dinitrogen oxide

4.45 Write the formulas of the following covalent compounds:
  a. oxygen difluoride   **b.** boron trifluoride
  c. dinitrogen trioxide   **d.** sulfur hexafluoride

4.46 Write the formulas of the following covalent compounds:
  a. sulfur dibromide   **b.** carbon disulfide
  c. tetraphosphorus hexoxide   **d.** dinitrogen pentoxide

4.47 Name the compounds that are found in the following sources:
  a. $Al_2(SO_4)_3$ antiperspirant
  b. $CaCO_3$ antacid
  c. N_2O "laughing gas," inhaled anesthetic
  d. Na_3PO_4 cathartic
  e. $(NH_4)_2SO_4$ fertilizer
  f. Fe_2O_3 pigment

4.48 Name the compounds that are found in the following sources:
 a. N_2 Earth's atmosphere
 b. $Mg_3(PO_4)_2$ antacid
 c. $FeSO_4$ iron supplement in vitamins
 d. $MgSO_4$ Epsom salts
 e. Cu_2O fungicide
 f. SnF_2 dental caries prophylactic

4.6 Electronegativity and Bond Polarity

In this chapter, we have seen that atoms form chemical bonds using their valence electrons. In bonds between identical nonmetal atoms, the bonding electrons are shared equally. However, in most compounds, bonds form between atoms of different elements. Then the bonding electrons are attracted to one atom more than the other.

Electronegativity

The ability of an atom to attract bonding electrons to itself is called its **electronegativity**. The electronegativity values assigned to the representative elements are shown in Figure 4.5. Nonmetals have high electronegativity values compared to metals because nonmetals have a greater attraction for electrons than metals. The nonmetals with the highest electronegativity values are fluorine (4.0) at the top of Group 7A (17) and oxygen (3.5) at the top of Group 6A (16). The metal cesium at the bottom of Group 1A (1) has the lowest electronegativity value of 0.7. Smaller atoms tend to have higher electronegativity values because the valence electrons they share are closer to their nuclei. Thus, the values of electronegativity increase going from left to right across each period of the periodic table and increase going up within each group. The values of electronegativity for the transition metals are

the
Chemistry
place

WEB TUTORIAL
Electronegativity

FIGURE 4.5

The electronegativity of representative elements indicates the ability of atoms to attract shared electrons. Electronegativity values increase across a period and going up a group.

Q **What element on the periodic chart has the strongest attraction for shared electrons?**

Electronegativity increases →

H	
2.1	

				18
				Group 8A

1 Group 1A	2 Group 2A		13 Group 3A	14 Group 4A	15 Group 5A	16 Group 6A	17 Group 7A
Li 1.0	Be 1.5		B 2.0	C 2.5	N 3.0	O 3.5	F 4.0
Na 0.9	Mg 1.2		Al 1.5	Si 1.8	P 2.1	S 2.5	Cl 3.0
K 0.8	Ca 1.0		Ga 1.6	Ge 1.8	As 2.0	Se 2.4	Br 2.8
Rb 0.8	Sr 1.0		In 1.7	Sn 1.8	Sb 1.9	Te 2.1	I 2.5
Cs 0.7	Ba 0.9		Tl 1.8	Pb 1.9	Bi 1.9	Po 2.0	At 2.1

Electronegativity increases ↑

also low, but we will not include them in our discussion. Note that there are no electronegativity values for the noble gases because they do not typically form bonds.

WEB TUTORIAL
Bonds and Bond Polarity

Types of Bonding

Earlier we discussed bonding as either *ionic*, in which electrons are transferred, or *covalent*, in which electrons were equally shared. Actually most bonds are somewhere between these types of bonds. The difference in the electronegativity values of two atoms gives an indication of the type of bond that forms. In H—H, the electronegativity difference is zero $(2.1 - 2.1 = 0)$, which means the bonding electrons are shared equally between the two hydrogen atoms. A covalent bond between atoms with identical or very similar electronegativities is a **nonpolar covalent bond**. However, most covalent bonds are between different atoms with different electronegativity values. (For example, in H—Cl, there is an electronegativity difference of $3.0 - 2.1 = 0.9$.) (See Figure 4.6.) When the electrons are shared unequally in a covalent bond, it is called a **polar covalent bond**.

In a polar covalent bond, the shared electrons are attracted to the more electronegative atom, which makes it partially negative. At the other end of the polar bond, the atom with the lower electronegativity becomes partially positive. Because a polar covalent bond has a separation of positive and negative charges or two poles, it is called a **dipole**. The positive and negative ends of a polar covalent bond are indicated by the lowercase Greek letter delta with a positive or negative sign δ^+ and δ^-. Sometimes an arrow pointing from the positive charge to the negative charge ($\longleftarrow$) is used to indicate the dipole.

Examples of Dipoles in Polar Covalent Bonds

$$\overset{\delta+}{C}-\overset{\delta-}{O} \qquad \overset{\delta+}{N}-\overset{\delta-}{O} \qquad \overset{\delta+}{Cl}-\overset{\delta-}{F}$$

Variations in Bonding

The variations in bonding are continuous; there is no definite point at which one type of bond stops and the next starts. However, for purposes of discussion, we can use some general ranges for predicting the type of bond between atoms. When electronegativity differences are from 0.0 to 0.4, the electrons are shared about equally in a nonpolar covalent bond. For example, H—H $(2.1 - 2.1 = 0)$ and

FIGURE 4.6

In the nonpolar covalent bond of H_2, electrons are shared equally. In the polar covalent bond of HCl, electrons are shared unequally.

Q **H_2 has a nonpolar covalent bond, but HCl has a polar covalent bond. Explain.**

H—H
Equal sharing of electrons in a nonpolar covalent bond

$H\overset{\delta+}{—}Cl^{\delta-}$
Unequal sharing of electrons in a polar covalent bond (dipole)

TABLE 4.14

Electronegativity Difference and Types of Bonds

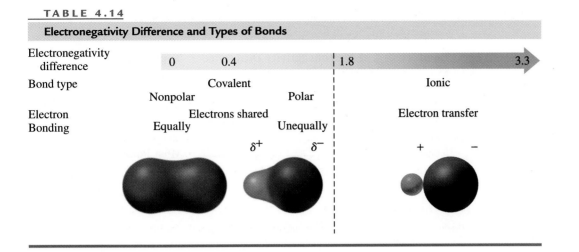

C—H $(2.5 - 2.1 = 0.4)$ are classified as nonpolar covalent bonds. As electronegativity differences increase, there is also an increase in the polarity of the covalent bond. For example, an O—H bond with a difference of 1.4 is much more polar than an O—F bond with a difference of only 0.5. Differences in electronegativity of about 1.8 or greater generally indicate a bond that is mostly ionic. For example, we would classify K—Cl with an electronegativity difference of $3.0 - 0.8 = 2.2$ as an ionic bond resulting from a transfer of electrons. (See Table 4.14.)

As the electronegativity difference increases, the shared electrons are attracted more strongly to the more electronegative atom. The *polarity*, which depends on the separation of charges, also increases. Eventually, the difference in electronegativity is great enough that the electrons are transferred from one atom to another, which results in an ionic bond. For example, the electronegativity difference for the ionic compound NaCl is $3.0 - 0.9 = 2.1$. Thus for large differences in electronegativity, we would predict an ionic bond. (See Table 4.15.)

SAMPLE PROBLEM 4.13

■ Bond Polarity

Using electronegativity values, place the following bonds in order of increasing polarity. Classify each bond as nonpolar covalent, polar covalent, or ionic.

N—N   O—H   Cl—As   O—K

TABLE 4.15

Predicting Bond Type from Electronegativity Differences

Molecule		Type of Electron Sharing	Electronegativity Difference[a]	Bond Type
H_2	H—H	Shared equally	$2.1 - 2.1 = 0$	Nonpolar covalent
Cl_2	Cl—Cl	Shared equally	$3.0 - 3.0 = 0$	Nonpolar covalent
HBr	$\overset{\delta^+}{H}—\overset{\delta^-}{Br}$	Shared unequally	$2.8 - 2.1 = 0.7$	Polar covalent
HCl	$\overset{\delta^+}{H}—\overset{\delta^-}{Cl}$	Shared unequally	$3.0 - 2.1 = 0.9$	Polar covalent
NaCl	Na^+Cl^-	Electron transfer	$3.0 - 0.9 = 2.1$	Ionic
MgO	$Mg^{2+}O^{2-}$	Electron transfer	$3.5 - 1.2 = 2.3$	Ionic

[a]Values are taken from Figure 4.5.

SOLUTION

For each bond, we obtain the electronegativity values and calculate the difference.

Bond	Electronegativity Difference	Type of Bond
N—N	$3.0 - 3.0 = 0.0$	Nonpolar covalent
O—H	$3.5 - 2.1 = 1.4$	Polar covalent
Cl—As	$3.0 - 2.0 = 1.0$	Polar covalent
O—K	$3.5 - 0.8 = 2.7$	Ionic

Least Polar	$\longrightarrow$		**Most Polar (ionic)**
N—N	Cl—As	O—H	O—K

STUDY CHECK

For each of the following pairs, identify the more polar bond.

a. Si—S or Si—N   **b.** O—P or N—P

QUESTIONS AND PROBLEMS

■ **Electronegativity and Bond Polarity**

4.49 Describe the trend in electronegativity going across a period.

4.50 Describe the trend in electronegativity going down a group.

4.51 Approximately what electronegativity difference would you expect for a nonpolar covalent bond?

4.52 Approximately what electronegativity difference would you expect for a polar covalent bond?

4.53 Predict whether each of the following bonds is ionic, polar covalent, or nonpolar covalent:
   a. Si—Br   **b.** Li—F   **c.** Br—F
   d. Br—Br   **e.** N—P   **f.** C—P

4.54 Predict whether each of the following bonds is ionic, polar covalent, or nonpolar covalent:
   a. Si—O   **b.** K—Cl   **c.** S—F
   d. P—Br   **e.** Li—O   **f.** N—P

4.55 For each of the following bonds, indicate the positive end with δ^+ and the negative end with δ^-. Write an arrow to show the dipole for each.
   a. N—F   **b.** Si—Br   **c.** C—O
   d. P—Br   **e.** B—Cl

4.56 For each of the following bonds, indicate the positive end with δ^+ and the negative end with δ^-. Write an arrow to show the dipole for each.
   a. Si—Br   **b.** Se—F   **c.** Br—F
   d. N—H   **e.** N—P

LEARNING GOAL

Predict the three-dimensional structure of a molecule and classify it as polar or nonpolar.

the Chemistry place

WEB TUTORIAL
The Shape of Molecules

■ 4.7 Shapes and Polarity of Molecules

We can now look at the three-dimensional shapes of molecules. The shape is important in our understanding of how molecules interact with other molecules, enzymes, or certain antibiotics.

To determine shape, we use the **valence-shell electron-pair repulsion (VSEPR) theory**, which uses the geometry of the electron groups around a central atom. In the VSEPR theory, the electron groups stay as far apart as possible to reduce the repulsion between the electrons. To determine the shape of a molecule, we first write its electron-dot formula. Then we determine the number of electron pairs bonded to atoms and the number of nonbonding (lone) pairs of electrons around the central atom. An electron group can be the electron pairs bonded to atoms or a lone pair. We will limit our discussion to central atoms with four pairs of electrons.

In a molecule of CH_4, four hydrogen atoms are bonded to the central carbon atom. The four bonds have a maximum separation when they are at an angle of about 109°, which forms a **tetrahedral** shape.

Tetrahedral

Tetrahedral
arrangement

Tetrahedral
shape

Shapes of Molecules with Lone Pairs Around the Central Atom

In many molecules, there are one or more lone (unshared) pairs of electrons around the central atom. Then we predict the shape of the molecule from the number of bonded atoms. For example, in NH_3, the four electron pairs around the central nitrogen atom have a tetrahedral arrangement. However, the shape of the molecule is based on the central nitrogen atom and three hydrogen atoms and is called a **pyramidal** shape.

Pyramidal

Tetrahedral
arrangement

Pyramidal
shape

In H_2O, the four electron groups have the geometry of a tetrahedron at angles of 109°. However, the shape determined by the central oxygen atom and two hydrogen atoms is called **bent**.

Bent

Tetrahedral
arrangement

Bent
shape

Some examples of shapes of molecules are shown in Table 4.16.

TABLE 4.16

Examples of Shapes of Molecules

Molecule	Electron-Dot Formula	Bonded Atoms	Molecular Shape (angle)	
CH_4	H H:C:H H	4	tetrahedral (109°)	
NH_3	H:N:H H	3	pyramidal (109°)	
H_2O	:O:H H	2	bent (109°)	

SAMPLE PROBLEM 4.14

■ **Shapes of Molecules**

Predict the shape of the following molecules:

a. PH_3 **b.** $SiCl_4$

SOLUTION

Guide to Predicting Molecular Shape (VSEPR Theory)
STEP 1 Write the electron-dot formula for the molecule.
STEP 2 Arrange the electron groups around the central atom to minimize repulsion.
STEP 3 Use the atoms bonded to the central atom to determine the molecular shape.

a. STEP 1 H:P:H
         H

STEP 2 The electron-dot formula for PH_3 indicates four electron groups around the central P atom, which arrange themselves in a tetrahedron to minimize repulsion.

STEP 3 The three bonded atoms would have a pyramidal shape.

b. STEP 1
         :Cl:
   :Cl:Si:Cl:
         :Cl:

STEP 2 The electron-dot formula for $SiCl_4$ has four bonded atoms.

STEP 3 With four bonded atoms and no lone pairs, the $SiCl_4$ molecule would have a tetrahedral shape.

STUDY CHECK

Predict the shape of CBr_4.

Polarity of Molecules

We have seen that covalent bonds in molecules can be polar or nonpolar. Molecules can also be polar or nonpolar depending on their shape. Diatomic molecules such as H_2 or Cl_2 are nonpolar because they contain one nonpolar covalent bond.

H—H   Cl—Cl
Nonpolar

Molecules with two or more polar bonds can also be **nonpolar** if the polar bonds have a symmetrical arrangement in the molecule.

In a **polar molecule**, one end of the molecule is more negatively charged than another end. Polarity in a molecule occurs when the polar bonds do not cancel each other. This cancellation depends on the type of atoms, the electron pairs around the central atom, and the shape of the molecule. For example, the HCl molecule is polar because electrons are shared unequally in a polar covalent bond.

Positive end   Negative end

In polar molecules with three or more atoms, the shape of the molecule determines whether the dipoles cancel or not. Often there are lone pairs around the central atom. In H_2O, the dipoles do not cancel, which makes the molecule positive at one end and negative at the other end. This gives the molecule a dipole.

Negative end

Net dipole

Positive end

In the molecule NH_3, there are three dipoles, but they do not cancel.

Net dipole

When the polar bonds or dipoles in a molecule cancel each other, the molecule is nonpolar. For example, CO_2 and CCl_4 contain polar bonds. However, the symmetry of the polar bonds cancels the dipoles, which makes CO_2 and CCl_4 molecules nonpolar.

Examples of Nonpolar Molecules with Polar Bonds

Net dipole = 0

The four individual bond polarities add up to zero (they cancel)

Net dipole = 0

SAMPLE PROBLEM 4.15

■**Polarity of Molecules**

Determine whether each of the following molecules is polar or nonpolar:

a. CBr_4

b. OF_2

SOLUTION

a. The electron-dot formula for CBr_4 has four bonded atoms.

$$
\begin{array}{c}
:\ddot{Br}: \\
:\ddot{Br}:\ddot{C}:\ddot{Br}: \\
:\ddot{Br}:
\end{array}
$$

The molecule would have a tetrahedral shape. With four identical atoms bonded to the central and no lone pairs, CBr_4 would be nonpolar.

b. The electron dot formula for OF_2 shows four electron groups with two bonded atoms.

$$
\begin{array}{c}
:\ddot{O}:\ddot{F}: \\
:\ddot{F}:
\end{array}
$$

The molecule would have a bent shape. With two lone pairs, the molecule is polar.

STUDY CHECK

Would PCl_3 be a polar or nonpolar molecule?

QUESTIONS AND PROBLEMS

■ Shapes and Polarity of Molecules

4.57 What is the shape of a molecule with four bonded atoms?

4.58 What is the shape of a molecule with two bonded atoms and two lone pairs?

4.59 In the molecule PCl_3, the four electron groups around the phosphorus atom are arranged in a tetrahedral geometry. However, the shape of the molecule is called pyramidal. Why does the shape of the molecule have a different name from the name of the electron group geometry?

4.60 In the molecule H_2S, the four electron groups around the sulfur atom are arranged in a tetrahedral geometry. However, the shape of the molecule is called bent. Why does the shape of the molecule have a different name from the name of the electron group geometry?

4.61 Compare the electron-dot formulas of PH_3 and NH_3. Why do these molecules have the same shape?

4.62 Compare the electron-dot formulas CH_4 and H_2O. Why do these molecules have approximately the same angles but different shapes?

4.63 Use the VSEPR theory to predict the shape of each molecule:
  a. OF_2 **b.** CCl_4

4.64 Use the VSEPR theory to predict the shape of each molecule:
  a. NCl_3 **b.** SCl_2

4.65 The molecule Cl_2 is nonpolar, but HCl is polar. Explain.

4.66 The molecules CH_4 and CH_3Cl both contain four bonds. Why is CH_4 nonpolar whereas CH_3Cl is polar?

4.67 Identify the following molecules as polar or nonpolar:
  a. HBr   **b.** NF_3 **c.** CBr_4

4.68 Identify the following molecules as polar or nonpolar:
  a. H_2S **b.** PBr_3 **c.** $SiCl_4$

CONCEPT MAP

Compounds and Their Bonds

- **Ionic compounds**
  - contain
  - **Ionic bonds**
  - between
    - **Metals**
      - that form
      - **Positive ions**
    - **Nonmetals**
      - that form
      - **Negative ions**
  - use
  - **Charge balance**
  - to write subscripts for the
  - **Chemical formula**

- **Covalent compounds**
  - contain
  - **Covalent bonds**
  - between
    - **Nonmetals**
      - can be
        - **Polar**
        - **Nonpolar**
  - use
  - **VSEPR theory**
  - to determine
    - **Shape**
    - **Polarity**

CHAPTER REVIEW

4.1 Octet Rule and Ions
The nonreactivity of the noble gases is associated with the stable arrangement of 8 electrons, an octet, in their valence shells; helium needs 2 electrons for stability. Atoms of elements other than the noble gases achieve stability by losing, gaining, or sharing their valence electrons with other atoms in the formation of compounds. Metals of the representative elements form an octet by losing their valence electrons to form positively charged cations: Group 1A (1), 1+, Group 2A (2), 2+, and Group 3A (13), 3+. When they react with metals, nonmetals gain electrons to form octets in their valence shells. As anions, they have negative charges: Group 5A (15), 3−, Group 6A (16), 2−, Group 7A (17), 1−.

4.2 Ionic Compounds
The total positive and negative ionic charge must balance in the formula of an ionic compound. Charge balance in a formula is achieved by using subscripts after each symbol so that the overall charge is zero.

4.3 Naming and Writing Ionic Formulas
In naming ionic compounds, the positive ion is given first, followed by the name of the negative ion. Ionic compounds containing two elements end with *ide*. When the metal can form more than one positive ion, its ionic charge is determined from the total negative charge in the formula. Typically transition metals form cations with two or more ionic charges. The charge is given as a Roman numeral in the name, such as iron(II) and iron(III) for the cations of iron with 2+ and 3+ ionic charges.

4.4 Polyatomic Ions
A polyatomic ion is a group of nonmetal atoms that carries an electrical charge; for example, the carbonate ion has the formula CO_3^{2-}. Most polyatomic ions have names that end with *ate* or *ite*.

4.5 Covalent Compounds
In a covalent bond, electrons are shared by atoms of nonmetals. By sharing, each of the atoms achieves a noble gas arrangement. In a nonpolar covalent bond, electrons are shared equally by atoms. In a polar covalent bond, electrons are unequally shared because they are attracted by the more electronegative atom. In some covalent compounds, double or triple bonds are needed to provide an octet. Two nonmetals can form two or more different covalent compounds. In their names, prefixes are used to indicate the subscript in the formula. The ending of the second nonmetal is changed to *ide*.

4.6 Electronegativity and Bond Polarity
Electronegativity is the ability of an atom to attract the electrons it shares with another atom. In general, the electronegativities of

metals are low, while nonmetals have high electronegativities. In a nonpolar covalent bond, atoms share the bonding pair of electrons equally. In a polar covalent bond, the bonding electrons are unequally shared because they are attracted to the more electronegative atom. As a result, the atom with the lower electronegativity is partially positive (δ^+) and the atom with the higher electronegativity is partially negative (δ^-). Atoms that form ionic bonds have large differences in electronegativity.

4.7 Shapes and Polarity of Molecules

The shape of a molecule is determined from the electron-dot formula and the number of bonded atoms and lone pairs. The electron arrangement of four electron groups around a central atom is tetrahedral. When all the electron groups are bonded to atoms, the shape has the same name as the electron arrangement. A central atom with two bonded atoms and two lone pairs has a bent shape. A central atom with three bonded atoms and one lone pair has a pyramidal shape.

Nonpolar molecules contain nonpolar covalent bonds or have an arrangement of bonded atoms that causes the dipoles to cancel out. In polar molecules, the dipoles do not cancel because there are nonidentical bonded atoms or lone pairs.

KEY TERMS

anion A negatively charged ion such as Cl^-, O^{2-}, or SO_4^{2-}.

bent The shape of a molecule with two bonded atoms and two lone pairs.

cation A positively charged ion such as Na^+, Mg^{2+}, Al^{3+}, and NH_4^+.

dipole The separation of positive and negative charge in a polar bond indicated by an arrow that is drawn from the more positive atom to the more negative atom.

double bond A sharing of two pairs of electrons by two atoms.

electronegativity The relative ability of an element to attract electrons.

formula The group of symbols and subscripts that represent the atoms or ions in a compound.

ion An atom or group of atoms having an electrical charge because of a loss or gain of electrons.

ionic charge The difference between the number of protons (positive) and the number of electrons (negative) written in the upper right corner of the symbol for the element or polyatomic ion.

molecule The smallest unit of two or more atoms held together by covalent bonds.

nonpolar covalent bond A covalent bond in which the electrons are shared equally between atoms.

nonpolar molecule A molecule that has only nonpolar bonds or in which the bond dipoles cancel.

octet 8 valence electrons.

octet rule Elements in Groups 1–7A (1, 2, 13–17) react with other elements by forming ionic or covalent bonds to produce a noble gas arrangement, usually 8 electrons in the outer shell.

polar covalent bond A covalent bond in which the electrons are shared unequally between atoms.

polar molecule A molecule containing bond dipoles that do not cancel.

polyatomic ion A group of covalently bonded nonmetal atoms that has an overall electrical charge.

pyramidal The shape of a molecule that has three bonded atoms and one lone pair around a central atom.

tetrahedral The shape of a molecule with four bonded atoms.

triple bond A sharing of three pairs of electrons by two atoms.

valence-shell electron-pair repulsion (VSEPR) theory A theory that predicts the shape of a molecule by placing the electron pairs on a central atom as far apart as possible to minimize the mutual repulsion of the electrons.

ADDITIONAL QUESTIONS AND PROBLEMS

4.69 Write the electron arrangement of the following:
a. N^{3-} b. Mg^{2+} c. F^-
d. Al^{3+} e. Li^+

4.70 Write the electron arrangement of the following:
a. K^+ b. Na^+ c. S^{2-}
d. Cl^- e. Ca^{2+}

4.71 Write the symbol for the ion of each of the following:
a. chloride   b. potassium
c. oxide   d. aluminum

4.72 Write the symbol for the ion of each of the following:
a. fluoride   b. calcium
c. sodium   d. lithium

4.73 What is the name of each of the following ions?
a. K^+ b. S^{2-} c. Ca^{2+} d. N^{2-}

4.74 What is the name of each of the following ions?
a. Mg^{2+} b. Ba^{2+} c. I^- d. Cl^-

4.75 Write the formula of the following ionic compounds:
a. gold(III) chloride   b. lead(IV) oxide
c. silver chloride   d. calcium nitride
e. copper(I) phosphide   f. chromium(II) chloride

4.76 Write the formula of the following ionic compounds:
a. tin(IV) oxide   b. iron(III) sulfide
c. lead(IV) sulfide   d. chromium(III) iodide
e. lithium nitride   f. gold(I) oxide

4.77 Name each of the following covalent compounds:
a. NCl_3 b. SCl_2
c. N_2O d. F_2
e. PCl_5 f. P_2O_5

4.78 Name each of the following covalent compounds:
  a. CBr_4 **b.** SF_6 **c.** Br_2
  d. N_2O_4 **e.** SO_2 **f.** CS_2

4.79 Give the formula for each of the following:
  a. carbon monoxide   **b.** diphosphorus pentoxide
  c. dihydrogen sulfide   **d.** sulfur dichloride

4.80 Give the formula for each of the following:
  a. silicon dioxide   **b.** carbon tetrabromide
  c. sulfur trioxide   **d.** dinitrogen monoxide

4.81 Classify each of the following compounds as ionic or covalent, and give its name.
  a. $FeCl_3$ **b.** Na_2SO_4 **c.** N_2O
  d. F_2 **e.** PCl_5 **f.** CF_4

4.82 Classify each of the following compounds as ionic or covalent, and give its name.
  a. $Al_2(CO_3)_3$ **b.** SF_6 **c.** Br_2
  d. Mg_3N_2 **e.** SO_2 **f.** $CrPO_4$

4.83 Write the formulas for the following:
  a. tin(II) carbonate   _____
  b. lithium phosphide   _____
  c. silicon tetrachloride _____
  d. iron(III) sulfide   _____
  e. carbon dioxide   _____
  f. calcium bromide   _____

4.84 Write the formulas for the following:
  a. sodium carbonate   _____
  b. nitrogen dioxide   _____
  c. aluminum nitrate   _____
  d. copper(I) nitride   _____
  e. potassium phosphate _____
  f. lead(IV) oxide   _____

4.85 Select the more polar bond in each of the following pairs:
  a. C—N or C—O   **b.** N—F or N—Br
  c. Br—Cl or S—Cl   **d.** Br—Cl or Br—I
  e. N—F or N—O

4.86 Select the more polar bond in each of the following pairs:
  a. C—C or C—O   **b.** P—Cl or P—Br

  c. Si—S or Si—Cl   **d.** F—Cl or F—Br
  e. P—O or P—S

4.87 Show the dipole arrow for each of the following bonds:
  a. Si—Cl   **b.** C—N   **c.** F—Cl
  d. C—F   **e.** N—O

4.88 Show the dipole arrow for each of the following bonds:
  a. C—O   **b.** N—F   **c.** O—Cl
  d. S—Cl   **e.** P—F

4.89 Classify each of the following bonds as nonpolar covalent, polar covalent, or ionic.
  a. Si—Cl   **b.** C—C   **c.** Na—Cl
  d. C—H   **e.** F—F

4.90 Classify each of the following bonds as nonpolar covalent, polar covalent, or ionic.
  a. C—N   **b.** Cl—Cl   **c.** K—Br
  d. H—H   **e.** N—F

4.91 Classify the following molecules as polar or nonpolar:
  a. NCl_3 **b.** CH_3Cl **c.** SiF_4

4.92 Classify the following molecules as polar or nonpolar:
  a. GeH_4 **b.** PCl_3 **c.** SCl_2

4.93 Predict the shape and polarity of each of the following molecules:
  a. A central atom with three identical bonded atoms and one lone pair.
  b. A central atom with two bonded atoms and two lone pairs.

4.94 Predict the shape and polarity of each of the following molecules:
  a. A central atom with four identical bonded atoms.
  b. A central atom with four bonded atoms that are not identical.

4.95 Predict the shape and polarity of each of the following molecules:
  a. H_2S **b.** NF_3

4.96 Predict the shape and polarity of each of the following molecules:
  a. H_2O **b.** CF_4

UNDERSTANDING THE CONCEPTS

4.97 Identify each of the following atoms or ions:

$15p^+$ $18e^-$ $16n$	$8p^+$ $8e^-$ $8n$	$30p^+$ $28e^-$ $35n$	$26p^+$ $23e^-$ $28n$
A	**B**	**C**	**D**

4.98 Consider the following bonds

Ca—O   C—O   K—O   O—O   N—O

  a. Which bonds are polar covalent?
  b. Which bonds are nonpolar covalent?
  c. Which bonds are ionic?
  d. Arrange the covalent bonds in order of decreasing polarity.

4.99 In the following electron-dot formulas, assume X and Y are atoms of nonmetals and all bonds are polar covalent.

a. X—Ÿ—X   **b.** X—Ÿ—X   **c.** X—Y—X (with X above and below central Y)

Match each of the molecules with the correct diagram of its shape and name the shape; indicate if each molecule is polar or nonpolar.

1.   2.   3.

4.100 As discussed in the Health Note "Polyatomic Ions in Bone and Teeth," the mineral component of bone and teeth is composed of calcium hydroxyapatite, $Ca_{10}(PO_4)_6(OH)_2$. Name the ions in the formula.

4.101 Identify the following as an atom or ion of
1. H
2. Li
3. Li^+
4. H^+
5. N^{3-}

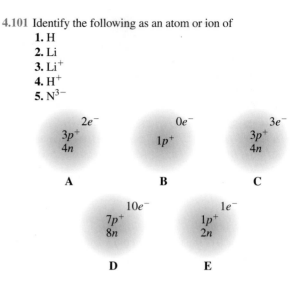

QUESTIONS FOR TEAMWORK

4.102 Consider the following electron-dot formulas for elements X and Y.

$$X \cdot \quad \cdot \ddot{\underset{..}{Y}} \cdot$$

a. What are the group numbers of X and Y?
b. Will a compound of X and Y be ionic or covalent?
c. What ions would be formed by X and Y?
d. What would be the formula of a compound of X and Y?
e. What would be the formula of a compound of X and chlorine?
f. What would be the formula of a compound of Y and chlorine?

4.103 Complete the following table for atoms or ions:

Atom or Ion	Number of Protons	Number of Electrons	Electrons Lost/Gained
K^+			
	$12p^+$	$10e^-$	
	$8p^+$		$2e^-$ gained
		$10e^-$	$3e^-$ lost

4.104 One of the ions of tin is tin(IV).
a. What is the symbol for this ion?
b. How many protons and electrons are in the ion?
c. What is the formula of tin(IV) oxide?
d. What is the formula of tin(IV) phosphate?

4.105 Classify the following compounds as ionic or covalent and name each.
a. Li_2O **b.** N_2O
c. CF_4 **d.** Cl_2O
e. MgF_2 **f.** CO
g. $CaCl_2$ **h.** K_3PO_4

4.106 Name the following compounds:
a. $FeCl_2$ **b.** Cl_2O_7
c. N_2 **d.** $Ca_3(PO_4)_2$
e. PCl_3 **f.** $Al(NO_3)_2$
g. $PbCl_4$ **h.** $MgCO_3$
i. NO_2 **j.** $SnSO_4$
k. $Ba(NO_3)_2$ **l.** CuS

ANSWERS

Answers to Study Checks

4.1 $35p^+$ and $36e^-$

4.2 K^+ and S^{2-}

4.3 CaO

4.4 calcium chloride

4.5 gold(III) chloride

4.6 Cr_2O_3

4.7 $(NH_4)_3PO_4$

4.8 calcium phosphate

4.9 $H : \ddot{P} : H$
$\quad\quad \overset{..}{H}$

4.10 a. silicon tetrabromide   **b.** dibromine oxide

4.11 IF_5

4.12 iron(II) nitrate

4.13 a. Si—N   **b.** O—P

4.14 With four bonded atoms, the shape would be tetrahedral.

4.15 polar

Answers to Selected Questions and Problems

4.1 **a.** By losing 1 valence electron from the third energy level, sodium achieves an octet in the second energy level.
 b. The sodium ion Na^+ has the same electron arrangement as Ne (2, 8).
 c. Group 1A (1) and 2A (2) elements become stable by losing electrons to form compounds. Group 8A (18) elements are stable.

4.3 **a.** 1   **b.** 2   **c.** 3   **d.** 1   **e.** 2

4.5 **a.** Li^+ **b.** F^- **c.** Mg^{2+}
 d. Fe^{3+} **e.** Zn^{2+}

4.7 **a.** Cl^- **b.** K^+ **c.** O^{2-} **d.** Al^{3+}

4.9 a, c

4.11 **a.** Na_2O **b.** $AlBr_3$ **c.** BaO
 d. $MgCl_2$ **e.** Al_2S_3

4.13 **a.** Na_2S **b.** K_3N **c.** AlI_3 **d.** Li_2O

4.15 **a.** aluminum oxide   **b.** calcium chloride
 c. sodium oxide   **d.** magnesium nitride
 e. potassium iodide

4.17 Most of the transition metals form more than one positive ion. The specific ion is indicated in a name by writing a Roman numeral that is the same as the ionic charge. For example, iron forms Fe^{2+} and Fe^{3+} ions, which are named iron(II) and iron(III).

4.19 **a.** iron(II)   **b.** copper(II)   **c.** zinc
 d. lead (IV)   **e.** chromium (III)

4.21 **a.** tin(II) chloride   **b.** iron(II) oxide
 c. copper(I) sulfide   **d.** copper(II) sulfide
 e. chromium(III) bromide

4.23 **a.** Au^{3+} **b.** Fe^{3+} **c.** Pb^{4+} **d.** Sn^{2+}

4.25 **a.** $MgCl_2$ **b.** Na_2S **c.** Cu_2O
 d. Zn_3P_2 **e.** AuN   **f.** $CrCl_2$

4.27 **a.** HCO_3^- **b.** NH_4^+ **c.** PO_4^{3-} **d.** HSO_4^-

4.29 **a.** sulfate   **b.** carbonate
 c. phosphate   **d.** nitrate

4.31

	OH^-	NO_2^-	CO_3^{2-}	HSO_4^-	PO_4^{3-}
Li^+	LiOH	$LiNO_2$	Li_2CO_3	$LiHSO_4$	Li_3PO_4
Cu^{2+}	$Cu(OH)_2$	$Cu(NO_2)_2$	$CuCO_3$	$Cu(HSO_4)_2$	$Cu_3(PO_4)_2$
Ba^{2+}	$Ba(OH)_2$	$Ba(NO_2)_2$	$BaCO_3$	$Ba(HSO_4)_2$	$Ba_3(PO_4)_2$

4.33 **a.** CO_3^{2-}, sodium carbonate
 b. NH_4^+, ammonium chloride
 c. PO_4^{3-}, lithium phosphate
 d. NO_2^-, copper(II) nitrite
 e. SO_3^{2-}, iron(II) sulfite

4.35 **a.** $Ba(OH)_2$
 b. Na_2SO_4
 c. $Fe(NO_3)_2$
 d. $Zn_3(PO_4)_2$
 e. $Fe_2(CO_3)_3$

4.37 **a.** $:\ddot{B}r:\ddot{B}r:$ **b.** H:H
 c. $H:\ddot{F}:$ **d.** $:\ddot{O}:\ddot{F}:$
 $\phantom{d.\ :\ddot{O}:}\ddot{F}:$

4.39 **a.** phosphorus tribromide
 b. carbon tetrabromide
 c. silicon dioxide
 d. hydrogen fluoride

4.41 **a.** dinitrogen trioxide
 b. nitrogen trichloride
 c. silicon tetrabromide
 d. phosphorus pentachloride

4.43 **a.** CCl_4 **b.** CO   **c.** PCl_3 **d.** N_2O_4

4.45 **a.** OF_2 **b.** BF_3 **c.** N_2O_3 **d.** SF_6

4.47 **a.** aluminum sulfate
 b. calcium carbonate
 c. dinitrogen oxide
 d. sodium phosphate
 e. ammonium sulfate
 f. iron(III) oxide

4.49 The electronegativity increases going across a period.

4.51 A nonpolar covalent bond would have an electronegativity difference of 0.0 to 0.4.

4.53 **a.** polar covalent
 b. ionic
 c. polar covalent
 d. nonpolar covalent
 e. polar covalent
 f. nonpolar covalent

4.55 **a.** $\overset{\delta+}{N}—\overset{\delta-}{F}$ **b.** $\overset{\delta+}{Si}—\overset{\delta-}{Br}$ **c.** $\overset{\delta+}{C}—\overset{\delta-}{O}$
 d. $\overset{\delta+}{P}—\overset{\delta-}{Br}$ **e.** $\overset{\delta+}{B}—\overset{\delta-}{Cl}$

4.57 tetrahedral

4.59 The four electron groups in PCl_3 have a tetrahedral arrangement, but three bonded atoms around a central atom give a pyramidal shape.

4.61 In both PH_3 and NH_3, there are four electron pairs of which three are bonded to atoms and one is a lone pair. The shapes of both are pyramidal.

4.63 **a.** bent   **b.** tetrahedral

4.65 Cl_2 is a nonpolar molecule because there is a nonpolar covalent bond between Cl atoms, which have identical electronegativity values. In HCl, the bond is a polar bond, which makes HCl a polar molecule.

4.67 **a.** polar   **b.** polar   **c.** nonpolar

4.69 **a.** 2, 8   **b.** 2, 8   **c.** 2, 8   **d.** 2, 8   **e.** 2

4.71 **a.** Cl^- **b.** K^+ **c.** O^{2-} **d.** Al^{3+}

4.73 **a.** potassium   **b.** sulfide
 c. calcium   **d.** nitride

4.75 **a.** $AuCl_3$ **b.** PbO_2 **c.** AgCl
     d. Ca_3N_2 **e.** Cu_3P **f.** $CrCl_2$

4.77 **a.** nitrogen trichloride
     b. sulfur dichloride
     c. dinitrogen monoxide
     d. fluorine
     e. phosphorus pentachloride
     f. diphosphorus pentoxide

4.79 **a.** CO   **b.** P_2O_5 **c.** H_2S **d.** SCl_2

4.81 **a.** ionic, iron(III) chloride
     b. ionic, sodium sulfate
     c. covalent, dinitrogen oxide
     d. covalent, fluorine
     e. covalent, phosphorus pentachloride
     f. covalent, carbon tetrafluoride

4.83 **a.** $SnCO_3$ **b.** Li_3P **c.** $SiCl_4$
     d. Fe_2S_3 **e.** CO_2 **f.** $CaBr_2$

4.85 **a.** C—O   **b.** N—F   **c.** S—Cl
     d. Br—I   **e.** N—F

4.87 **a.** Si—Cl   **b.** C—N   **c.** F—Cl

d. C—F   **e.** N—O

4.89 **a.** polar covalent
     b. nonpolar covalent
     c. ionic
     d. nonpolar covalent
     e. nonpolar covalent

4.91 **a.** polar   **b.** polar   **c.** nonpolar

4.93 **a.** pyramidal, polar   **b.** bent, polar

4.95 **a.** bent, polar   **b.** pyramidal, polar

4.97 **a.** P^{3-} ion   **b.** O atom
     c. Zn^{2+} ion   **d.** Fe^{3+} ion

4.99 **a.** 2 – pyramidal, polar
     b. 1 – bent, polar
     c. 3 – tetrahedral, nonpolar

4.101 **1.** H(E)   **2.** Li (C)   **3.** Li^+ (A)
     4. H^+ (B)   **5.** N^{3-} (D)

5 Chemical Reactions and Quantities

"In our food science laboratory I develop a variety of food products, from cake donuts to energy beverages," says Anne Cristofano, senior food technologist at Mattson & Company. "When I started the donut project, I researched the ingredients, then weighed them out in the lab. I added water to make a batter and cooked the donuts in a fryer. The batter and the oil temperature make a big difference. If I don't get the right taste or texture, I adjust the ingredients, such as sugar and flour, or adjust the temperature."

A food technologist studies the physical and chemical properties of food and develops scientific ways to process and preserve it for extended shelf life. The food products are tested for texture, color, and flavor. The results of these tests help improve the quality and safety of food.

the **C**hemistry place

Visit **www.aw-bc.com/chemplace** or use the CD-ROM in the book for extra quizzes, interactive tutorials, career resources, PowerPoint slides, math help, and case studies.

Chemical reactions occur everywhere. In the engines of our cars, gasoline burns to provide energy to move the car, play the radio, and run the air conditioner. Chemical reactions in our bodies provide energy to move our muscles. In nature, chemical reactions occur in the leaves of plants to convert carbon dioxide and water into carbohydrates.

Chemists use chemical equations to describe the reactants and products of chemical reactions. Some chemical reactions are simple, while others are quite complex. However, they can all be written with chemical equations. In every chemical reaction, the atoms in the reacting substances, called reactants, are rearranged to give new substances called products. When we know the chemical equation for a reaction, we can determine the amount of product that can be produced. We do much the same thing at home when we use a recipe to make cookies or add the right quantity of water to make soup. At the automotive repair shop, a mechanic adjusts the carburetor or fuel injection system of an engine to allow for the correct amounts of fuel and oxygen so the engine will run properly. In the hospital, a respiratory therapist evaluates the level of CO_2 and O_2 in the blood. A certain amount of oxygen must reach the tissues for efficient metabolic reactions. If the oxygenation of the blood is low, the therapist will oxygenate the patient and recheck the blood levels.

In this chapter, we will describe the quantity of a compound as a mole and calculate the mass related to the formula of a compound. From a balanced equation, we can determine the mass and number of moles of a reactant and calculate the amount of product. Knowing how to determine the quantitative results of a chemical reaction is important to both chemists and medical personnel such as pharmacists and respiratory therapists.

Identify a change in a substance as a chemical or a physical change.

TABLE 5.1

Comparison of Some Chemical and Physical Changes

Chemical Changes	Physical Changes
Rusting nail	Melting ice
Bleaching a stain	Boiling water
Burning a log	Sawing a log in half
Tarnishing silver	Tearing paper
Fermenting grapes	Breaking a glass
Souring of milk	Pouring milk

5.1 Chemical Changes

In a **physical change**, the appearance of a substance is altered, but not its composition. When liquid water becomes a gas, or freezes to a solid, it is still water (see Figure 5.1). If we smash a rock or tear a piece of paper, only the size of the material changes. The smaller pieces are still rock or paper because there was no change in the composition of the substances.

In a **chemical change**, the reacting substances change into new substances that have different formulas and different properties. New properties may involve a change in color or the formation of bubbles or a solid. For instance, when silver tarnishes, the bright silver metal (Ag) reacts with sulfur (S) to become the dull, blackish substance we call tarnish (Ag_2S) (see Figure 5.1). Table 5.1 gives examples of some typical physical and chemical changes.

A **chemical reaction** always involves chemical change because atoms of the reacting substances form new combinations with new properties. A chemical reaction takes place when rust forms on a nail or piece of iron: the iron (Fe) is reacting with oxygen (O_2) to produce the new substance, rust (Fe_2O_3). Perhaps you have placed an antacid tablet in a glass of water and noticed it fizzing and bubbling as sodium hydrogen carbonate ($NaHCO_3$) reacts to form carbon dioxide (CO_2) gas (see Figure 5.2).

A physical change:
the melting of ice

A chemical change:
the tarnishing of silver

H₂O (liquid)   H₂O (ice)

Water and ice are both made
of H₂O molecules

Ag   Ag₂S

Silver and tarnish are different substances

FIGURE 5.1

A chemical change produces new substances; a physical change does not.

Q Why is the formation of tarnish a chemical change?

SAMPLE PROBLEM 5.1

■ **Classifying Chemical and Physical Change**

Classify each of the following changes as physical or chemical:

a. water freezing into an icicle   **b.** burning a match
c. breaking a chocolate bar   **d.** digesting a chocolate bar

SOLUTION

a. Physical. Freezing water involves only a change from liquid water to ice. No change has occurred in the substance water.
b. Chemical. Burning a match causes the formation of ash that was not present before striking the match.
c. Physical. Breaking a chocolate bar does not affect its composition.
d. Chemical. The digestion of the chocolate bar converts its components into new substances.

STUDY CHECK

Classify the following changes as physical or chemical:

a. chopping a carrot   **b.** burning a candle   **c.** inflating a balloon

the
Chemistry
place

WEB TUTORIAL
What is Chemistry?

FIGURE 5.2

Examples of chemical reactions involve chemical change: iron (Fe) reacts with oxygen (O_2) to form rust (Fe_2O_3) and an antacid ($NaHCO_3$) tablet in water forms bubbles of carbon dioxide (CO_2).

Q **What is the evidence for chemical change in these chemical reactions?**

QUESTIONS AND PROBLEMS

■ Chemical Changes

5.1 Classify each of the following changes as chemical or physical:
  a. chewing a gum drop
  b. ignition of fuel in the space shuttle
  c. drying clothes
  d. neutralizing stomach acid with an antacid tablet
  e. formation of snowflakes
  f. an exploding dynamite stick

5.2 Classify each of the following changes as chemical or physical:
  a. fogging the mirror during a shower
  b. tarnishing of a silver bracelet
  c. breaking a bone
  d. mending a broken bone
  e. burning paper
  f. picking up the leaves in the yard

Write a balanced chemical equation from the formulas of the reactants and products for a reaction.

■ 5.2 Chemical Equations

When you build a model airplane, prepare a new recipe, mix a medication, or clean a patient's teeth, you follow a set of directions. These directions tell you what materials to use and the products you will obtain. In chemistry, a **chemical equation** tells us the materials we need and the products that will form in a chemical reaction.

■ Writing a Chemical Equation

Suppose you work in a bicycle shop, assembling wheels and bodies into bicycles. You could represent this process by a simple equation.

Equation: Wheels + Body ⟶ Bicycle

When you burn charcoal in a grill, the carbon in the charcoal combines with oxygen to form carbon dioxide. We can represent this reaction by a chemical equation that is much like the one for the bicycle:

Reactants   Product

Equation: $C(s)$ + $O_2(g)$ $\xrightarrow{\Delta}$ $CO_2(g)$

In an equation, the formulas of the **reactants** are written on the left of the arrow and the formulas of the **products** on the right. When there are two or more formulas on the same side, they are separated by plus (+) signs. The delta sign (Δ) indicates that heat was used to start the reaction.

Sometimes, as in the charcoal case, the formulas in an equation may include letters, in parentheses, that give the physical state of the substances as solid (s), liquid (l), or gas (g). If a substance is dissolved in water, it is an aqueous (aq) solution. Table 5.2 summarizes some of the symbols used in equations.

When a reaction takes place, the bonds between the atoms of the reactants are broken and new bonds are formed to give the products. Atoms cannot be gained, lost, or changed into other types of atoms during a chemical reaction. Therefore, a reaction must be written as a **balanced equation**, which shows the same number of atoms for each element on both sides of the arrow. Let's see if the equation we wrote above for burning carbon is balanced:

$C(s) + O_2(g) \longrightarrow CO_2(g)$

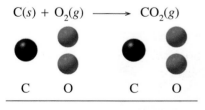

C   O   C   O

Reactant atoms   =   Product atoms

The answer is yes: this equation is *balanced* because there is one carbon atom and two oxygen atoms on each side of the arrow.

Now consider the reaction in which hydrogen reacts with oxygen to form water. First we write the formulas of the reactants and products:

$H_2(g) + O_2(g) \longrightarrow H_2O(g)$

TABLE 5.2

Some Symbols Used in Writing Equations

Symbol	Meaning
+	Separates two or more formulas
$\longrightarrow$	Reacts to form products
Δ	The reactants are heated
(s)	Solid
(l)	Liquid
(g)	Gas
(aq)	Aqueous

WEB TUTORIAL
Chemical Reactions and Equations

Is the equation balanced? To find out, we add up the atoms of each element on each side of the arrow. No; the equation is *not balanced*. The number of atoms on the left side does not match the number of atoms on the right side. To balance this equation, we place whole numbers called **coefficients** in front of some of the formulas. First we write a coefficient of 2 in front of the H_2O formula to represent the formation of 2 molecules of water. Now the product has four hydrogen atoms. That means we must also write a coefficient of 2 in front of the formula H_2 in the reactants to give 4 H atoms on the reactant side. Coefficients are used to balance an equation, but the subscripts in the formulas are never changed. Now the number of hydrogen atoms and oxygen atoms are the same in the reactants as in the products. The equation is *balanced*.

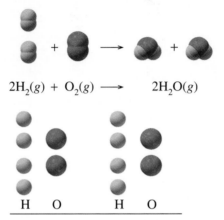

$$2H_2(g) + O_2(g) \longrightarrow 2H_2O(g)$$

H	O	H	O

Reactant atoms = Product atoms

Balancing a Chemical Equation

We have seen that a chemical equation must be balanced. In many cases, we can use a method of trial and error to balance an equation.

To demonstrate the process, let us balance the equation for the reaction of the gas methane, CH_4, with oxygen to produce carbon dioxide and water. This is the principal reaction that occurs in the flame of a gas burner you use in the laboratory and in the flame of a gas stove.

STEP 1 **Write an equation, using the correct formulas.** As a first step, we write the equation using the correct formulas for the reactants and products.

$$CH_4(g) + O_2(g) \longrightarrow CO_2(g) + H_2O(g)$$

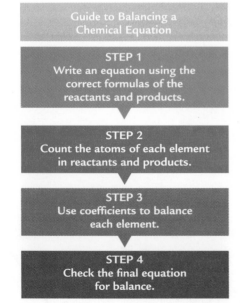

Guide to Balancing a Chemical Equation

STEP 1
Write an equation using the correct formulas of the reactants and products.

STEP 2
Count the atoms of each element in reactants and products.

STEP 3
Use coefficients to balance each element.

STEP 4
Check the final equation for balance.

CH_4 O_2 CO_2 H_2O

STEP 2 **Determine if the equation is balanced.** When we compare the atoms on the reactant side with the atoms on the product side, we see that there are more hydrogen atoms on the left side and more oxygen atoms on the right.

$$CH_4(g) + O_2(g) \longrightarrow CO_2(g) + H_2O(g)$$

1 C	1 C	
4 H	2 H	Not balanced
2 O	3 O	Not balanced

STEP 3 **Balance the equation one element at a time.** First we balance the hydrogen atoms by placing a coefficient of 2 in front of the formula for water. Then we balance the oxygen atoms by placing a coefficient of 2 in front of the formula for oxygen. There are now four oxygen atoms and four hydrogen atoms in both the reactants and products.

$$CH_4(g) + 2O_2(g) \longrightarrow CO_2(g) + 2H_2O(g)$$

STEP 4 **Check to see if the equation is balanced.** Rechecking the balanced equation shows that the numbers of atoms of carbon, hydrogen, and oxygen are the same for both the reactants and the products. The equation is balanced using the lowest possible whole numbers as coefficients.

$$CH_4(g) + 2O_2(g) \longrightarrow CO_2(g) + 2H_2O(g) \quad \text{Balanced}$$

Reactants		**Products**
1 C atom	=	1 C atom
4 H atoms	=	4 H atoms
4 O atoms	=	4 O atoms

Suppose you had written the equation as follows:

$$2CH_4(g) + 4O_2(g) \longrightarrow 2CO_2(g) + 4H_2O(g) \quad \text{Incorrect}$$

Although there are equal numbers of atoms on both sides of the equation, this is not written correctly. All the coefficients must be divided by 2 to obtain the lowest possible whole numbers.

SAMPLE PROBLEM 5.2

■ **Balancing Equations**

Balance the following equations:

a. $Al(s) + Cl_2(g) \longrightarrow AlCl_3(s)$
b. $Na_3PO_4(aq) + MgCl_2(aq) \longrightarrow Mg_3(PO_4)_2(s) + NaCl(aq)$

SOLUTION

a. STEP 1 The correct formulas are written in the equation.

$$Al(s) + Cl_2(g) \longrightarrow AlCl_3(s)$$

STEP 2 When we compare the number of atoms on the reactant and product sides, we find that the chlorine atoms are not balanced.

Reactants   **Products**

$$Al(s) + Cl_2(g) \longrightarrow AlCl_3(s)$$

1 Al   1 Al

2 Cl   3 Cl   Not balanced

STEP 3 In the formulas with chlorine, there is an odd-even relationship. A 2 placed in front of $AlCl_3$ will give an even number of 6 Cl atoms. Any number we use to multiply the Cl_2 will double the number of Cl atoms. To balance the 6 Cl in the product, we place a 3 in front of Cl_2. Finally, we must rebalance the 2 Al atoms that are now in the product by placing a 2 in front of the Al on the left side.

STEP 4 When we recheck the balance of atoms, we see that the equation is balanced.

Reactants   **Products**

$$2Al(s) + 3Cl_2(g) \longrightarrow 2AlCl_3(s) \quad \text{Balanced}$$

$$2 \text{ Al} = 2 \text{ Al}$$

$$6 \text{ Cl} = 6 \text{ Cl}$$

b. STEP 1 In the equation, the correct formulas are written.

$$Na_3PO_4(aq) + MgCl_2(aq) \longrightarrow Mg_3(PO_4)_2(s) + NaCl(aq)$$

STEP 2 When we compare the number of ions on the reactant and product sides, we find that the equation is not balanced. In this equation, it is more convenient to balance the phosphate polyatomic ion as a group instead of its individual atoms.

$$3MgCl_2(aq) + 2Na_3PO_4(aq) \longrightarrow 6NaCl(aq) + Mg_3(PO_4)_2(s)$$

Reactants **Products**

$Na_3PO_4(aq) + MgCl_2(aq) \longrightarrow Mg_3(PO_4)_2(s) + NaCl(aq)$

$3\,Na^+$	$1\,Na^+$	Not balanced
$1\,PO_4^{3-}$	$2\,PO_4^{3-}$	Not balanced
$1\,Mg^{2+}$	$3\,Mg^{2+}$	Not balanced
$2\,Cl^-$	$1\,Cl^-$	Not balanced

STEP 3 We begin with the formula of $Mg_3(PO_4)_2$, which is the most complex. A 3 in front of $MgCl_2$ balances magnesium and a 2 in front of Na_3PO_4 balances the phosphate ion. Looking again at each of the ions in the reactants and products, we see that the sodium and chloride ions are not yet equal. A 6 in front of the NaCl balances the equation.

STEP 4 A check of the atoms indicates the equation is balanced.

Reactants **Products**

$2Na_3PO_4(aq) + 3MgCl_2(aq) \longrightarrow Mg_3(PO_4)_2(s) + 6NaCl(aq)$

Balanced

$6Na^+$	$=$	$6Na^+$
$2PO_4^{3-}$	$=$	$2PO_4^{3-}$
$3Mg^{2+}$	$=$	$3Mg^{2+}$
$6\,Cl^-$	$=$	$6\,Cl^-$

STUDY CHECK

Balance the following equation:

$Fe(s) + O_2(g) \longrightarrow Fe_3O_4(s)$

QUESTIONS AND PROBLEMS

■ Chemical Equations

5.3 Determine whether each of the following equations is balanced or not balanced:
  a. $S(s) + O_2(g) \longrightarrow SO_3(g)$ **b.** $2Al(s) + 3Cl_2(g) \longrightarrow 2AlCl_3(s)$
  c. $H_2(g) + O_2(g) \longrightarrow H_2O(g)$ **d.** $C_3H_8(g) + 5O_2(g) \longrightarrow 3CO_2(g) + 4H_2O(g)$

5.4 Determine whether each of the following equations are balanced or not balanced:
  a. $PCl_3(s) + Cl_2(g) \longrightarrow PCl_5(s)$ **b.** $CO(g) + 2H_2(g) \longrightarrow CH_3OH(g)$
  c. $2KClO_3(s) \longrightarrow 2KCl(s) + O_2(g)$ **d.** $Mg(s) + N_2(g) \longrightarrow Mg_3N_2(s)$

5.5 Balance the following equations:
  a. $N_2(g) + O_2(g) \longrightarrow NO(g)$ **b.** $HgO(s) \longrightarrow Hg(l) + O_2(g)$
  c. $Fe(s) + O_2(g) \longrightarrow Fe_2O_3(s)$ **d.** $Na(s) + Cl_2(g) \longrightarrow NaCl(s)$

5.6 Balance the following equations:
  a. $Ca(s) + Br_2(l) \longrightarrow CaBr_2(s)$
  b. $P_4(s) + O_2(g) \longrightarrow P_4O_{10}(s)$
  c. $Sb_2S_3(s) + HCl(aq) \longrightarrow SbCl_3(s) + H_2S(g)$
  d. $Fe_2O_3(s) + C(s) \longrightarrow Fe(s) + CO(g)$

5.7 Balance the following equations:
  a. $Mg(s) + AgNO_3(aq) \longrightarrow Mg(NO_3)_2(aq) + Ag(s)$
  b. $Al(s) + CuSO_4(aq) \longrightarrow Cu(s) + Al_2(SO_4)_3(aq)$
  c. $Pb(NO_3)_2(aq) + NaCl(aq) \longrightarrow PbCl_2(s) + NaNO_3(aq)$
  d. $Al(s) + HCl(aq) \longrightarrow AlCl_3(aq) + H_2(g)$

5.8 Balance the following equations:
  a. $Zn(s) + H_2SO_4(aq) \longrightarrow ZnSO_4(aq) + H_2(g)$
  b. $Al(s) + H_2SO_4(aq) \longrightarrow Al_2(SO_4)_3(aq) + H_2(g)$

c. $K_2SO_4(aq) + BaCl_2(aq) \longrightarrow BaSO_4(s) + KCl(aq)$
d. $CaCO_3(s) \longrightarrow CaO(s) + CO_2(g)$

5.3 Types of Reactions

A great number of reactions occur in nature, in biological systems, and in the laboratory. However, there are some general patterns among all reactions that help us classify reactions. Most reactions fit into four general reaction types.

Combination Reactions

In a **combination reaction**, two or more elements or simple compounds bond together to form one product. For example, sulfur and oxygen combine to form the product sulfur dioxide.

Combination reaction

$$S(s) + O_2(g) \longrightarrow SO_2(g)$$

In Figure 5.3, the elements magnesium and oxygen combine to form a single product, magnesium oxide.

$$2Mg(s) + O_2(g) \longrightarrow 2MgO(s)$$

Two or more   combine   a single
reactants   to yield   product

A + B ⟶ A B

LEARNING GOAL

Identify a reaction as a combination, decomposition, or replacement.

the
Chemistry
place

WEB TUTORIAL
Chemical Reactions and Equations

FIGURE 5.3

This combination reaction is
$$2Mg(s) + O_2(g) \xrightarrow{\Delta} 2MgO(s)$$

Q **What happens to the atoms in the reactants in a combination reaction?**

Decomposition Reactions

In a **decomposition reaction**, a reactant splits into two or more simpler products. For example, calcium carbonate breaks apart into simpler compounds of calcium oxide and carbon dioxide when heated.

$$CaCO_3(s) \xrightarrow{\Delta} CaO(s) + CO_2(g)$$

Replacement Reactions

In a replacement reaction, elements in a compound are replaced by other elements. In a **single replacement reaction**, a reacting element switches place with an element in the other reacting compound.

Single replacement

One element   replaces   another element

In the single replacement reaction shown in Figure 5.4, zinc replaces hydrogen in hydrochloric acid, HCl(*aq*).

$$Zn(s) + 2HCl(aq) \longrightarrow ZnCl_2(aq) + H_2(g)$$

| Zn(*s*) | + | 2HCl(*aq*) | $\longrightarrow$ | ZnCl$_2$(*aq*) | + | H$_2$(*g*) |
| Zinc | | Hydrochloric acid | | Zinc chloride | | Hydrogen |

FIGURE 5.4

In a single replacement reaction, an atom or ion replaces an atom or ion in a compound.

Q **What changes in the formulas of the reactants identify this equation as a single replacement?**

Na$_2$SO$_4$(aq)   +   BaCl$_2$(aq) ————————→ BaSO$_4$(s)   +   2NaCl(aq)
Sodium sulfate   Barium chloride   Barium sulfate   Sodium chloride

FIGURE 5.5

In a double replacement reaction the positive ions in the reactants replace each other.

Q **How do the changes in the formulas of the reactants identify this equation as a double replacement reaction?**

In a **double replacement reaction**, the positive ions in the reacting compounds switch places.

Double replacement

For example, in the reaction shown in Figure 5.5, barium ions change places with sodium ions in the reactants to form sodium chloride and a white solid precipitate of barium sulfate. The formulas of the products depend on the charges of the ions.

$$BaCl_2(aq) + Na_2SO_4(aq) \longrightarrow BaSO_4(s) + 2NaCl(aq)$$

SAMPLE PROBLEM 5.3

■**Identifying Reactions and Predicting Products**

Classify the following reactions as combination, decomposition, or single or double replacement:

a. $2Fe_2O_3(s) + 3C(s) \longrightarrow 3CO_2(g) + 4Fe(s)$
b. $Fe_2S_3(s) \longrightarrow 2Fe(s) + 3S(s)$
c. $BaCl_2(aq) + K_2SO_4(aq) \longrightarrow BaSO_4(s) + 2KCl(aq)$

SOLUTION

a. In this *single replacement* reaction, a C atom replaces Fe in Fe_2O_3 to form the compound CO_2 and Fe atoms.

b. When one reactant breaks down into two products, the reaction is *decomposition*.

c. There are two reactants and two products, but the positive ions have exchanged places, which makes this a *double replacement* reaction.

STUDY CHECK

Nitrogen monoxide gas and oxygen gas react to form nitrogen dioxide gas. Write the balanced equation and identify the reaction type.

QUESTIONS AND PROBLEMS

▪ Types of Reactions

5.9 **a.** Why is the following reaction called a decomposition reaction?

$$2Al_2O_3(s) \xrightarrow{\Delta} 4Al(s) + 3O_2(g)$$

HEALTH NOTE

▪ Smog and Health Concerns

There are two types of smog. One, photochemical smog, requires sunlight to initiate reactions that produce pollutants such as nitrogen oxides and ozone. The other type of smog, industrial or London smog, occurs in areas where coal containing sulfur is burned and the unwanted product sulfur dioxide is emitted.

Photochemical smog is most prevalent in cities where people are dependent on cars for transportation. On a typical day in Los Angeles, for example, nitrogen monoxide (NO) emissions from car exhausts increase as traffic increases on the roads. The nitrogen monoxide is formed when N_2 and O_2 react at high temperatures in car and truck engines.

$$N_2(g) + O_2(g) \xrightarrow{Heat} 2NO(g)$$

Then NO reacts with oxygen in the air to produce NO_2, a reddish brown gas that is irritating to the eyes and damaging to the respiratory tract.

$$2NO(g) + O_2(g) \longrightarrow 2NO_2(g)$$

When NO_2 is exposed to sunlight, it is converted into NO and oxygen atoms.

$$NO_2(g) + \xrightarrow{Sunlight} NO(g) + O(g)$$
oxygen atoms

Oxygen atoms are so reactive that they combine with oxygen molecules in the atmosphere, forming ozone.

$$O(g) + O_2(g) \longrightarrow O_3(g)$$
ozone

In the upper atmosphere (the stratosphere), ozone is beneficial because it protects us from harmful ultraviolet radiation that comes from the sun. However, in the lower atmosphere, ozone irritates the eyes and respiratory tract, where it causes coughing, decreased lung function, and fatigue. It also causes deterioration of fabrics, cracks rubber, and damages trees and crops.

Industrial smog is prevalent in areas where coal with a high sulfur content is burned to produce electricity. During combustion, the sulfur is converted to sulfur dioxide:

$$S(s) + O_2(g) \longrightarrow SO_2(g)$$

The SO_2 is damaging to plants, suppresses growth, and it is corrosive to metals such as steel. SO_2 is also damaging to humans and can cause lung impairment and respiratory difficulties. The SO_2 in the air reacts with more oxygen to form SO_3. SO_3 combines with water in the air to form sulfuric acid, which makes the rain acidic.

$$2SO_2(g) + O_2(g) \longrightarrow 2SO_3(g)$$

$$SO_3(g) + H_2O(l) \longrightarrow H_2SO_4(aq)$$
sulfuric acid

The presence of sulfuric acid in rivers and lakes causes an increase in the acidity of the water, reducing the ability of animals and plants to survive.

b. Why is the following reaction called a single replacement reaction?

$$Br_2(g) + BaI_2(s) \longrightarrow BaBr_2(s) + I_2(g)$$

5.10 a. Why is the following reaction called a combination reaction?

$$H_2(g) + Br_2(g) \longrightarrow 2HBr(g)$$

b. Why is the following reaction called a double replacement reaction?

$$AgNO_3(aq) + NaCl(aq) \longrightarrow AgCl(s) + NaNO_3(aq)$$

5.11 Classify each of the following reactions as a combination, decomposition, single replacement, or double replacement.
a. $4Fe(s) + 3O_2(g) \longrightarrow 2Fe_2O_3(s)$
b. $Mg(s) + 2AgNO_3(aq) \longrightarrow Mg(NO_3)_2(aq) + 2Ag(s)$
c. $CuCO_3(s) \longrightarrow CuO(s) + CO_2(g)$
d. $NaOH(aq) + HCl(aq) \longrightarrow NaCl(aq) + H_2O(l)$
e. $ZnCO_3(s) \longrightarrow CO_2(g) + ZnO(s)$
f. $Al_2(SO_4)_3(aq) + 6KOH(aq) \longrightarrow 2Al(OH)_3(s) + 3K_2SO_4(aq)$
g. $Pb(s) + O_2(g) \longrightarrow PbO_2(s)$

5.12 Classify each of the following reactions as a combination, decomposition, single replacement, or double replacement.
a. $CuO(s) + 2HCl(aq) \longrightarrow CuCl_2(aq) + H_2O(l)$
b. $2Al(s) + 3Br_2(g) \longrightarrow 2AlBr_3(s)$
c. $Pb(NO_3)_2(aq) + 2NaCl(aq) \longrightarrow PbCl_2(s) + 2NaNO_3(aq)$
d. $2Mg(s) + O_2(g) \longrightarrow 2MgO(s)$
e. $Fe_2O_3(s) + 3C(s) \longrightarrow 2Fe(s) + 3CO(g)$
f. $C_6H_{12}O_6(aq) \longrightarrow 2C_2H_6O(aq) + 2CO_2(g)$
g. $BaCl_2(aq) + K_2CO_3(aq) \longrightarrow BaCO_3(s) + 2KCl(aq)$

5.4 Oxidation–Reduction Reactions

Perhaps you have never heard of oxidation and reduction reactions. Yet this type of reaction has many important applications in our everyday lives. When you see a rusty nail, tarnish on a silver spoon, or corrosion on metal, you are observing oxidation.

$$4Fe(s) + 3O_2(g) \longrightarrow 2Fe_2O_3(s) \qquad \text{Fe is oxidized}$$
$$\text{rust}$$

When you turn the lights on in your car, an oxidation–reduction reaction within your car battery provides the electricity. On a cold, wintry day, you might build a wood fire. Burning wood is an oxidation–reduction reaction. When you eat foods with starches in them, you digest the starches to give glucose, which is oxidized in your cells to give you energy along with carbon dioxide and water. Every breath you take provides oxygen to carry out oxidation in your cells.

$$C_6H_{12}O_6(aq) + 6O_2(g) \longrightarrow 6CO_2(g) + 6H_2O(l) + \text{energy}$$

Oxidation–Reduction

Oxidation–reduction reactions are defined as reactions with a loss and a gain of electrons among the reactants. In the oxidation part of the reaction, one substance loses electrons. Simultaneously, a reduction also occurs as another substance gains these electrons. You may like to use the shorthand phrase "LEO the lion goes GER" to remind yourself that a loss of electrons is oxidation (LEO), and a gain of electrons is a reduction (GER).

Oxidation–Reduction Involving Ions

When we looked at the formation of ionic compounds, we also saw that metals lost electrons and nonmetals gained electrons. Now we can say that metals are oxidized and nonmetals are reduced.

Consider the elements and ions in the reaction where calcium loses electrons to form calcium ions and where sulfur gains electrons to form sulfide ions. The elemental forms of calcium and sulfur have no ionic charge; they are neutral.

$$Ca + S \longrightarrow Ca^{2+} + S^{2-} = CaS$$

If we write the formation of each ion of the compound as half reactions, we find that one is an oxidation and the other is a reduction. In every oxidation–reduction reaction, the number of electrons exchanged among the reactants must be equal.

$$Ca \longrightarrow Ca^{2+} + 2e^- \quad \text{Oxidation}$$
$$S + 2e^- \longrightarrow S^{2-} \quad \text{Reduction}$$

The loss of electrons is an **oxidation**. We say that calcium was oxidized. At the same time, the sulfur atom gained electrons. This gain of electrons is called **reduction**. We say that the sulfur was reduced. The two parts must occur simultaneously and with the same total number of electrons. We can write the two parts as follows.

If we look at a single replacement reaction, we also find that it has two parts: an oxidation and a reduction. Consider the reaction of zinc and copper(II) sulfate. (See Figure 5.6.)

$$Zn(s) + CuSO_4(aq) \longrightarrow ZnSO_4(aq) + Cu(s)$$

We can rewrite the equation to show the ions of the compounds.

$$Zn(s) + Cu^{2+}(aq) + SO_4^{2-}(aq) \longrightarrow Zn^{2+}(aq) + SO_4^{2-}(aq) + Cu(s)$$

Now we can show that zinc loses 2 electrons to form Zn^{2+}, and Cu^{2+} gains 2 electrons to form Cu.

$$Zn(s) \longrightarrow Zn^{2+}(aq) + 2e^- \quad \text{Oxidation}$$
$$Cu^{2+}(aq) + 2e^- \longrightarrow Cu(s) \quad \text{Reduction}$$

FIGURE 5.6

In this single replacement reaction $Zn(s)$ is oxidized to Zn^{2+} when it provides two electrons to reduce Cu^{2+} to $Cu(s)$:

$$Zn(s) + Cu^{2+}(aq) \longrightarrow Cu(s) + Zn^{2+}(aq)$$

Q In the oxidation, does Zn(s) lose or gain electrons?

In this single replacement reaction, zinc was oxidized and copper(II) ion was reduced. No change occurred for the sulfate ion.

SAMPLE PROBLEM 5.4

▪ Oxidation–Reduction Reactions

In photographic film, the following decomposition reaction occurs in the presence of light. What is oxidized and what is reduced?

$$2AgBr(s) \xrightarrow{\text{Light}} 2Ag(s) + Br_2(g)$$

SOLUTION

In the AgBr compound, silver ion has an ionic charge of $1+$ and the bromide ion has an ionic charge of $1-$. We can write the reaction as follows.

$$2Ag^+(aq) + 2Br^-(aq) \longrightarrow 2Ag(s) + Br_2(g)$$

We can determine the oxidation reaction and the reduction reaction by writing the changes for each ion separately.

$$2Br^-(aq) \longrightarrow Br_2(g) + 2e^- \quad \text{Oxidation}$$
$$2Ag^+(aq) + 2e^- \longrightarrow 2Ag(s) \quad \text{Reduction}$$

In the reaction, bromide ion is oxidized, and silver ion is reduced.

STUDY CHECK

In the following combination reaction, what is oxidized and what is reduced?

$$2Li(s) + F_2(g) \longrightarrow 2LiF(s)$$

▪ Oxidation of Fruits and Vegetables

Freshly cut surfaces of fruits and vegetables discolor when exposed to oxygen in the air. Cut three slices of a fruit or vegetable such as apple, potato, avocado, or banana. Leave one piece on the kitchen counter (uncovered). Wrap one piece in plastic wrap and leave on the kitchen counter. Dip one piece in lemon juice and leave uncovered.

QUESTIONS

1. What changes take place in each sample after 1–2 hours?
2. Why would wrapping fruits and vegetables slow the rate of discoloration?
3. If lemon juice contains vitamin C (an antioxidant), why would dipping a fruit or vegetable in lemon juice affect the oxidation reaction on the surface of the fruit or vegetable?
4. Other kinds of antioxidants are vitamin E, citric acid, and BHT. Look for these antioxidants on the labels of cereals, potato chips, and other foods in your kitchen. Why are antioxidants added to food products that will be stored on our kitchen shelves?

Oxidation and Reduction in Biological Systems

In the cells of the body, oxidation of organic (carbon) compounds involves the transfer of hydrogen atoms (H), which are composed of electrons and protons. For example, the oxidation of a typical biochemical molecule can involve the transfer of two hydrogen atoms (or $2H^+$ and $2e^-$) to a proton acceptor such as the coenzyme FAD (flavin adenine dinucleotide). The coenzyme is reduced to $FADH_2$. In many biochemical oxidation–reduction reactions, the transfer of hydrogen atoms is necessary

Oxidation (loss of 2H)

Reduction (gain of 2H)

for the production of energy in the cells. For example, methyl alcohol (CH_3OH), a poisonous substance, is metabolized in the body by the following reactions.

$$CH_3OH \longrightarrow H_2CO + 2H \quad \text{Oxidation: loss of H atoms}$$
methyl alcohol   formaldehyde

The formaldehyde can be oxidized further, this time by the addition of oxygen, to produce formic acid.

$$2H_2CO + O_2 \longrightarrow 2H_2CO_2 \quad \text{Oxidation: addition of O atoms}$$
formaldehyde   formic acid

Finally, formic acid is oxidized to carbon dioxide and water.

$$2H_2CO_2 + O_2 \longrightarrow 2CO_2 + 2H_2O \quad \text{Oxidation: addition of O atoms}$$
formic acid

The intermediate products of the oxidation of methyl alcohol are quite toxic, causing blindness and possibly death as they interfere with key reactions in the cells of the body.

In summary, we find that the particular definition of oxidation and reduction we use depends on the process that occurs in the reaction. All of these definitions are summarized in Table 5.3. Oxidation always involves a loss of electrons, but it may also be seen as an addition of oxygen, or the loss of hydrogen atoms. A reduction always involves a gain of electrons, and may also be seen as the loss of oxygen, or the gain of hydrogen.

TABLE 5.3

Characteristics of Oxidation and Reduction

Oxidation	
Always Involves	**May Involve**
Loss of electrons	Addition of oxygen
	Loss of hydrogen

Reduction	
Always Involves	**May Involve**
Gain of electrons	Loss of oxygen
	Gain of hydrogen

QUESTIONS AND PROBLEMS

■ Oxidation–Reduction Reactions

5.13 Indicate whether each of the following describes an oxidation or a reduction:
  a. loss of hydrogen atoms   **b.** gain of electrons
  c. addition of oxygen atoms   **d.** always accompanies oxidation

5.14 Indicate whether each of the following describes an oxidation or a reduction:
  a. loss of electrons   **b.** gain of hydrogen atoms
  c. loss of oxygen atoms   **d.** always accompanies reduction

5.15 In the following reactions, identify what is oxidized, and what is reduced.
  a. $Zn(s) + Cl_2(g) \longrightarrow ZnCl_2(s)$
  b. $Cl_2(g) + 2NaBr(aq) \longrightarrow 2NaCl(aq) + Br_2(g)$
  c. $2PbO(s) \longrightarrow 2Pb(s) + O_2(g)$
  d. $2Fe^{3+}(aq) + Sn^{2+}(aq) \longrightarrow 2Fe^{2+}(aq) + Sn^{4+}(aq)$

5.16 In the following reactions, identify what is oxidized, and what is reduced.
  a. $2Li(s) + F_2(g) \longrightarrow 2LiF(s)$
  b. $Cl_2(g) + 2KI(aq) \longrightarrow 2KCl(aq) + I_2(g)$
  c. $Zn(g) + Cu^{2+}(aq) \longrightarrow Zn^{2+}(aq) + Cu(s)$
  d. $Fe(s) + CuSO_4(aq) \longrightarrow FeSO_4(aq) + Cu(s)$

5.17 In the mitochondria of human cells, energy for the production of ATP is provided by the oxidation and reduction reactions of the iron ions in the cytochromes of the electron transport chain. Identify each of the following reactions as oxidation or reduction.
a. $Fe^{3+} + e^- \longrightarrow Fe^{2+}$ $\qquad$ **b.** $Fe^{2+} \longrightarrow Fe^{3+} + e^-$

5.18 Chlorine (Cl_2) is a strong germicide used to disinfect drinking water and to kill microbes in swimming pools. If the product is Cl^-, was the elemental chlorine oxidized or reduced?

5.19 When linoleic acid, an unsaturated fatty acid, reacts with hydrogen, it forms a saturated fatty acid.

$$C_{18}H_{32}O_2 + 2H_2 \longrightarrow C_{18}H_{36}O_2$$

Is linoleic acid oxidized or reduced in the hydrogenation reaction?

5.20 In one of the reactions in the citric acid cycle, which provides energy for ATP synthesis, succinic acid is converted to fumaric acid.

$$C_4H_6O_4 \longrightarrow C_4H_4O_4 + 2H$$

$\qquad$ succinic acid $\qquad$ fumaric acid

The reaction is accompanied by a coenzyme, flavin adenine dinucleotide (FAD).

$$FAD + 2H \longrightarrow FADH_2$$

a. Is succinic acid oxidized or reduced?
b. Is FAD oxidized or reduced?
c. Why would the two reactions occur together?

LEARNING GOAL

Use Avogadro's number to determine the number of particles in a given number of moles.

5.5 The Mole

At the store, you buy eggs by the dozen. In an office, pencils are ordered by the gross and paper by the ream. In a restaurant, soda is ordered by the case. In each of these examples, the terms *dozen, gross, ream,* and *case* count the number of items present. For example, when you buy a dozen eggs, you know you will get 12 eggs in the carton.

$\qquad$ 24 cans = 1 case

$\qquad$ 144 pencils = 1 gross

$\qquad$ 500 sheets = 1 ream

$\qquad$ 12 eggs = 1 dozen

Avogadro's Number

In chemistry, particles such as atoms, molecules, and ions are counted by the **mole**, a unit that contains 6.02×10^{23} items. This very large number, called **Avogadro's number** after Amedeo Avogadro, an Italian physicist, looks like this when written with 3 significant figures:

Avogadro's Number

602 000 000 000 000 000 000 000 = 6.02×10^{23}

TABLE 5.4

Number of Particles in One-Mole Samples	
Substance	Number and Type of Particles
1 mole aluminum	6.02×10^{23} aluminum atoms
1 mole sulfur	6.02×10^{23} sulfur atoms
1 mole water (H_2O)	6.02×10^{23} H_2O molecules
1 mole NaCl	6.02×10^{23} NaCl formula units
1 mole vitamin C ($C_6H_8O_6$)	6.02×10^{23} vitamin C molecules

One mole of any element always contains Avogadro's number of atoms. For example, 1 mole of carbon contains 6.02×10^{23} carbon atoms; 1 mole of aluminum contains 6.02×10^{23} aluminum atoms; 1 mole of sulfur contains 6.02×10^{23} sulfur atoms.

$$1 \text{ mole of an element} = 6.02 \times 10^{23} \text{ atoms of that element}$$

One mole of a compound contains Avogadro's number of molecules or formula units. Molecules are the particles of covalent compounds; **formula units** are the group of ions given by the formula of an ionic compound. For example, 1 mole of CO_2, a covalent compound, contains 6.02×10^{23} molecules of CO_2. One mole of NaCl, an ionic compound, contains 6.02×10^{23} formula units of NaCl (Na^+, Cl^-). Table 5.4 gives examples of the number of particles in some 1-mole quantities.

Avogadro's number tells us that one mole of any element or compound contains 6.02×10^{23} of the particular type of particles that make up a substance. We can use Avogadro's number as a conversion factor to convert between the moles of a substance and number of particles it contains.

$$\frac{6.02 \times 10^{23} \text{ particles}}{1 \text{ mole}} \quad \text{and} \quad \frac{1 \text{ mole}}{6.02 \times 10^{23} \text{ particles}}$$

For example, we use Avogadro's number to convert 4.00 moles iron to atoms of iron.

$$4.00 \text{ moles Fe atoms} \times \frac{6.02 \times 10^{23} \text{ Fe atoms}}{1 \text{ mole Fe atoms}} = 2.41 \times 10^{24} \text{ Fe atoms}$$

Avogadro's number as a conversion factor

We can also use Avogadro's number to convert 3.01×10^{24} molecules CO_2 to moles of CO_2.

$$3.01 \times 10^{24} \text{ CO}_2 \text{ molecules} \times \frac{1 \text{ mole CO}_2 \text{ molecules}}{6.02 \times 10^{23} \text{ CO}_2 \text{ molecules}} = 5.00 \text{ moles CO}_2 \text{ molecules}$$

Avogadro's number as a conversion factor

Calculating Particles and Moles

Moles of element or compound

Avogadro's number

Particles: atoms, ions, molecules, or formula units

SAMPLE PROBLEM 5.5

■ **Calculating the Number of Molecules in a Mole**

How many molecules of ammonia, NH_3, are present in 1.75 moles ammonia?

SOLUTION

STEP 1 **Given** 1.75 mole NH_3 **Need** molecules of NH_3

STEP 2 **Plan** moles of NH_3 Avogadro's number   molecules of NH_3

STEP 3 Equalities/Conversion Factors

$$1 \text{ mole NH}_3 = 6.02 \times 10^{23} \text{ molecules NH}_3$$

$$\frac{6.02 \times 10^{23} \text{ molecules NH}_3}{1 \text{ mole NH}_3} \quad \text{and} \quad \frac{1 \text{ mole NH}_3}{6.02 \times 10^{23} \text{ molecules NH}_3}$$

STEP 4 Set Up Problem Calculate the number of NH_3 molecules:

$$1.75 \text{ moles NH}_3 \times \frac{6.02 \times 10^{23} \text{ molecules NH}_3}{1 \text{ mole NH}_3} = 1.05 \times 10^{24} \text{ molecules NH}_3$$

STUDY CHECK

How many moles of water, H_2O, contain 2.60×10^{23} molecules of water?

Moles of Elements in a Formula

We have seen that the subscripts in a chemical formula of a compound indicate the number of atoms of each type of element. For example, in a molecule of aspirin, chemical formula $C_9H_8O_4$, there are 9 carbon atoms, 8 hydrogen atoms, and 4 oxygen atoms. The subscripts also state the number of moles of each element in one mole of aspirin: 9 moles of carbon atoms, 8 moles of hydrogen atoms, and 4 moles of oxygen atoms.

$$C_9H_8O_4$$

	Carbon	**Hydrogen**	**Oxygen**
Atoms in 1 molecule	9 atoms C	8 atoms H	4 atoms O
Moles of atoms in 1 mole	9 moles C	8 moles H	4 moles O

Using the subscripts from the aspirin formula, $C_9H_8O_4$, we can write the following conversion factors for each of the elements in 1 mole of aspirin.

$$\frac{9 \text{ moles C}}{1 \text{ mole C}_9\text{H}_8\text{O}_4} \qquad \frac{8 \text{ moles H}}{1 \text{ mole C}_9\text{H}_8\text{O}_4} \qquad \frac{4 \text{ moles O}}{1 \text{ mole C}_9\text{H}_8\text{O}_4}$$

$$\frac{1 \text{ mole C}_9\text{H}_8\text{O}_4}{9 \text{ moles C}} \qquad \frac{1 \text{ mole C}_9\text{H}_8\text{O}_4}{8 \text{ moles H}} \qquad \frac{1 \text{ mole C}_9\text{H}_8\text{O}_4}{4 \text{ moles O}}$$

Aspirin $C_9H_8O_4$

Number of atoms in 1 molecule
Carbon (C)   Hydrogen (H)   Oxygen (O)

SAMPLE PROBLEM 5.6

■ **Calculating the Moles of an Element in a Compound**

How many moles of carbon atoms are present in 1.50 moles aspirin, $C_9H_8O_4$?

SOLUTION

   STEP 1 **Given**   1.50 moles $C_9H_8O_4$ **Need**   moles of C atoms

   STEP 2 **Plan**   Moles of $C_9H_8O_4$ Subscript   moles of C atoms

   STEP 3 **Equalities/Conversion Factors**

$$1 \text{ mole } C_9H_8O_4 = 9 \text{ moles C atoms}$$

$$\frac{9 \text{ moles C}}{1 \text{ mole } C_9H_8O_4} \quad \text{and} \quad \frac{1 \text{ mole } C_9H_8O_4}{9 \text{ moles C}}$$

   STEP 4 **Set Up Problem**

$$1.50 \text{ moles } C_9H_8O_4 \times \frac{9 \text{ moles C atoms}}{1 \text{ mole } C_9H_8O_4} = 13.5 \text{ moles C atoms}$$

STUDY CHECK

How many moles of aspirin $C_9H_8O_4$ contain 0.480 mole O atoms?

QUESTIONS AND PROBLEMS

■ **The Mole**

5.21 What is a mole?

5.22 What is Avogadro's number?

5.23 Consider the formula for quinine $C_{20}H_{24}N_2O_2$.
   a. How many moles of hydrogen are in 1.0 mole of quinine?
   b. How many moles of carbon are in 5.0 moles of quinine?
   c. How many moles of nitrogen are in 0.020 mole of quinine?

5.24 Consider the formula for $Al_2(SO_4)_3$, which is used in antiperspirants.
   a. How many moles of sulfur are present in 3.0 moles of $Al_2(SO_4)_3$?
   b. How many moles of aluminum ions are present in 0.40 mole $Al_2(SO_4)_3$?
   c. How many moles of sulfate ions (SO_4^{2-}) are present in 1.5 moles $Al_2(SO_4)_3$?

5.25 Calculate each of the following:
   a. number of C atoms in 0.500 mole C
   b. number of SO_2 molecules in 1.28 moles SO_2
   c. moles of Fe in 5.22×10^{22} atoms Fe
   d. moles of C_2H_5OH in 8.50×10^{24} molecules C_2H_5OH

5.26 Calculate each of the following:
   a. number of Li atoms in 4.5 moles Li
   b. number of CO_2 molecules in 0.0180 mole CO_2
   c. moles of Cu in 7.8×10^{21} atoms Cu
   d. moles of C_2H_6 in 3.754×10^{23} molecules C_2H_6

5.27 Calculate each of the following quantities in 2.00 moles H_3PO_4:
   a. moles of H   **b.** moles of O   **c.** atoms of P   **d.** atoms of O

5.28 Calculate each of the following quantities in 0.185 mole $(C_3H_5)_2O$:
   a. moles of C   **b.** moles of O
   c. atoms of H   **d.** atoms of C

WEB TUTORIAL
Stoichiometry

6.02 × 10²³ atoms of C

↕

1 mole of C atoms

↕

12.01 g of C atoms

5.6 Molar Mass

A single atom or molecule is much too small to weigh, even on the most accurate balance. In fact, it takes a huge number of atoms or molecules to make a piece of a substance that you can see. An amount of water that contains Avogadro's number of water molecules is only a few sips of water. In the laboratory, we can use a balance to weigh out Avogadro's number of particles or 1 mole of a substance.

For any element, the quantity called its **molar mass** is the number of grams equal to the atomic mass of that element. We obtain 6.02×10^{23} atoms of an element when we weigh out the number of grams equal to its molar mass. For example, carbon has an atomic mass of 12.01 on the periodic table. To obtain one mole of carbon atoms, I would weigh out 12.01 g of carbon. Thus, we can use the periodic table to determine the molar mass of an element.

47	6	16
Ag	**C**	**S**
107.9	12.01	32.06

1 mole silver atoms has a mass of 107.9 g

1 mole carbon atoms has a mass of 12.01 g

1 mole sulfur atoms has a mass of 32.06 g

Molar Mass of a Compound

To determine the molar mass of a compound, multiply the molar mass of each element by its subscript in the formula, and add the results. For example, the molar mass of sulfur trioxide, SO_3, is obtained by adding the molar masses of 1 mole of sulfur and 3 moles of oxygen. In our calculations, we round molar mass to the tenths (0.1 g) place.

Number of Moles in the Formula		Molar Mass		Total Mass for Each
1 mole S	×	$\dfrac{32.1 \text{ g S}}{1 \text{ mole S}}$	=	32.1 g S
3 moles O	×	$\dfrac{16.0 \text{ g O}}{1 \text{ mole O}}$	=	48.0 g O
		Molar mass of SO_3	=	80.1 g

One–Mole Quantities

| S | Fe | NaCl | $K_2Cr_2O_7$ | $C_{12}H_{22}O_{11}$ |

FIGURE 5.7

One-mole samples: sulfur, S (32.1 g); iron, Fe (55.9 g); salt, NaCl (58.5 g); potassium dichromate, $K_2Cr_2O_7$ (294.2 g); and sugar, sucrose, $C_{12}H_{22}O_{11}$ (342.0 g).

Q How is the molar mass for $K_2Cr_2O_7$ obtained?

Figure 5.7 shows some 1-mole quantities of substances. Table 5.5 lists the molar mass for several 1-mole samples.

TABLE 5.5

The Molar Mass of Selected Elements and Compounds

Substance	Molar Mass
1 mole carbon (C)	12.0 g
1 mole sodium (Na)	23.0 g
1 mole iron (Fe)	55.9 g
1 mole NaF (preventative for dental caries)	42.0 g
1 mole $CaCO_3$ (antacid)	100.1 g
1 mole $C_6H_{12}O_6$ (glucose)	180.0 g
1 mole $C_8H_{10}N_4O_2$ (caffeine)	194.0 g

SAMPLE PROBLEM 5.7

■ **Calculating Molar Mass of Compounds**

Find the molar mass of each compound.

a. $CaCO_3$, a compound used as an antacid and to replenish calcium in the diet
b. $Mg(NO_3)_2$ used in fireworks

SOLUTION

a. *Number of Moles in the Formula* — *Molar Mass* — *Total Mass for Each Element*

$$1 \text{ mole Ca} \times \frac{40.1 \text{ g Ca}}{1 \text{ mole Ca}} = 40.1 \text{ g Ca}$$

$$1 \text{ mole C} \times \frac{12.0 \text{ g C}}{1 \text{ mole C}} = 12.0 \text{ g C}$$

$$3 \text{ moles O} \times \frac{16.0 \text{ g O}}{1 \text{ mole O}} = 48.0 \text{ g O}$$

$$\text{Molar mass of } CaCO_3 = 100.1 \text{ g}$$

b. *Number of Moles in the Formula* — *Molar Mass* — *Total Mass for Each Element*

$$1 \text{ mole Mg} \times \frac{24.3 \text{ g Mg}}{1 \text{ mole Mg}} = 24.3 \text{ g Mg}$$

$$2 \text{ moles N} \times \frac{14.0 \text{ g N}}{1 \text{ mole N}} = 28.0 \text{ g N}$$

$$6 \text{ moles O} \times \frac{16.0 \text{ g O}}{1 \text{ mole O}} = 96.0 \text{ g O}$$

$$\text{Molar mass of } Mg(NO_3)_2 = 148.3 \text{ g}$$

EXPLORE YOUR WORLD

■ Calculating Moles in the Kitchen

The labels on food products list the components in grams and milligrams. It is possible to convert those values to moles, which is also interesting. Read the labels of some products in the kitchen and convert the amounts given in grams or milligrams to moles using molar mass. Here are some examples.

QUESTIONS

1. How many moles of NaCl are in a box of salt containing 746 g NaCl?
2. How many moles of sugar are contained in a 5-lb bag of sugar if sugar has the formula $C_{12}H_{22}O_{11}$?
3. A serving of cereal contains 90 mg potassium. If there are 11 servings of cereal in the box, how many moles of K^+ are present in the cereal in the box?

STUDY CHECK

Calculate the molar mass of salicylic acid, $C_7H_6O_3$.

■ Calculations Using Molar Mass

The molar mass of an element or a compound is one of the most useful conversion factors in chemistry. Molar mass is used to change from moles of a substance to grams, or from grams to moles. To do these calculations, we use the molar mass as a conversion factor. For example, 1 mole of magnesium has a mass of 24.3 g. To express molar mass as an equality, we can write

$$1 \text{ mole Mg} = 24.3 \text{ g Mg}$$

From this equality, two conversion factors can be written.

$$\frac{24.3 \text{ g Mg}}{1 \text{ mole Mg}} \quad \text{and} \quad \frac{1 \text{ mole Mg}}{24.3 \text{ g Mg}}$$

Conversion factors are written for compounds in the same way. For example, the molar mass of the compound H_2O is 18.0 g.

$$1 \text{ mole } H_2O = 18.0 \text{ g } H_2O$$

The conversion factors from the molar mass of H_2O are written as

$$\frac{18.0 \text{ g } H_2O}{1 \text{ mole } H_2O} \quad \text{and} \quad \frac{1 \text{ mole } H_2O}{18.0 \text{ g } H_2O}$$

We can now change from moles to grams, or grams to moles, using the conversion factors derived from the molar mass. (Remember, you must determine the molar mass of the substance first.)

SAMPLE PROBLEM 5.8

■ Converting Moles of an Element to Grams

Silver metal is used in the manufacture of tableware, mirrors, jewelry, and dental alloys. If the design for a piece of jewelry requires 0.750 mole silver, how many grams of silver are needed?

SOLUTION

Calculating Mass and Moles

Mass (grams)

⇅

Molar mass

⇅

Moles

STEP 1 **Given** 0.750 mole Ag   **Need** grams of Ag

STEP 2 **Plan** moles of Ag Molar mass factor grams of Ag

STEP 3 **Equalities/Conversion Factors**

$$1 \text{ mole Ag} = 107.9 \text{ g Ag}$$
$$\frac{107.9 \text{ g Ag}}{1 \text{ mole Ag}} \quad \text{and} \quad \frac{1 \text{ mole Ag}}{107.9 \text{ g Ag}}$$

STEP 4 **Set Up Problem** Calculate the grams of silver using the molar mass.

$$0.750 \text{ mole Ag} \times \frac{107.9 \text{ g Ag}}{1 \text{ mole Ag}} = 80.9 \text{ g Ag}$$

STUDY CHECK

Calculate the number of grams of gold (Au) present in 0.124 mole gold. ∎

SAMPLE PROBLEM 5.9

∎ Converting Mass of a Compound to Moles

A box of salt contains 737 g NaCl. How many moles of NaCl are present in the box?

SOLUTION

STEP 1 **Given** 737 g NaCl   **Need** moles of NaCl

STEP 2 **Plan** grams of NaCl [Molar mass factor] moles of NaCl

STEP 3 **Equalities/Conversion Factors** The molar mass of NaCl is the sum of the masses of one mole Na^+ and one mole Cl^-:

$$(1 \times 23.0 \text{ g/mole}) + (1 \times 35.5 \text{ g/mole}) = 58.5 \text{ g/mole}$$

$$1 \text{ mole NaCl} = 58.5 \text{ g}$$

$$\frac{58.5 \text{ g NaCl}}{1 \text{ mole NaCl}} \quad \text{and} \quad \frac{1 \text{ mole NaCl}}{58.5 \text{ g NaCl}}$$

STEP 4 **Set Up Problem** We calculate the moles of NaCl using the molar mass.

$$737 \text{ g NaCl} \times \frac{1 \text{ mole NaCl}}{58.5 \text{ g NaCl}} = 12.6 \text{ moles NaCl}$$

STUDY CHECK

One gel cap of an antacid contains 311 mg $CaCO_3$ and 232 mg $MgCO_3$. In a recommended dosage of two gel caps, how many moles each of $CaCO_3$ and $MgCO_3$ are present? ∎

QUESTIONS AND PROBLEMS

∎ Molar Mass

5.29 Calculate the molar mass for each of the following compounds:
   a. NaCl (table salt)
   b. Fe_2O_3 (rust)
   c. Li_2CO_3 (antidepressant)
   d. $Al_2(SO_4)_3$ (antiperspirant)
   e. $Mg(OH)_2$ (antacid)
   f. $C_{16}H_{19}N_3O_5S$ (amoxicillin, an antibiotic)

5.30 Calculate the molar mass for each of the following compounds:
   a. $FeSO_4$ (iron supplement)
   b. Al_2O_3 (absorbent and abrasive)
   c. $C_7H_5NO_3S$ (saccharin)
   d. C_3H_8O (rubbing alcohol)
   e. $(NH_4)_2CO_3$ (baking powder)
   f. $Zn(C_2H_3O_2)_2$ (zinc dietary supplement)

5.31 Calculate the number of grams in each of the following:
   a. 2.00 moles Na   **b.** 2.80 moles Ca   **c.** 0.125 mole Sn

5.32 Calculate the number of grams in each of the following:
   a. 1.50 moles K   **b.** 2.5 moles C   **c.** 0.25 mole P

5.33 Calculate the number of grams in each of the following:
   a. 0.500 mole NaCl   **b.** 1.75 moles Na_2O **c.** 0.225 mole H_2O

5.34 Calculate the number of grams in each of the following:
   a. 2.0 moles $MgCl_2$ **b.** 3.5 moles C_3H_8 **c.** 5.00 moles C_2H_6O

5.35 a. The compound $MgSO_4$ is called Epsom salts. How many grams will you need to prepare a bath containing 5.00 moles of Epsom salts?

b. In a bottle of soda, there is 0.25 mole CO_2. How many grams of CO_2 are in the bottle?

5.36 a. Cyclopropane, C_3H_6, is an anesthetic given by inhalation. How many grams are in 0.25 mole of cyclopropane?

b. The sedative Demerol hydrochloride has the formula $C_{15}H_{22}ClNO_2$. How many grams are in 0.025 mole of Demerol hydrochloride?

5.37 How many moles are contained in each of the following?

   a. 50.0 g Ag   **b.** 0.200 g C

   c. 15.0 g NH_3 **d.** 75.0 g SO_2

5.38 How many moles are contained in each of the following?

   a. 25.0 g Ca   **b.** 5.00 g S

   c. 40.0 g H_2O **d.** 100.0 g O_2

5.39 A can of Drãno contains 480 g of NaOH. How many moles of NaOH are in the can of Drãno?

5.40 A gold nugget weighs 35.0 g. How many moles of gold are in the nugget?

5.41 How many moles of S are in each of the following quantities?

   a. 25 g S   **b.** 125 g SO_2 **c.** 2.0 moles Al_2S_3

5.42 How many moles of C are in each of the following quantities?

   a. 75 g C   **b.** 0.25 mole C_2H_6 **c.** 88 g CO_2

LEARNING GOAL

Given a quantity in moles of reactant or product, use a mole–mole factor from the balanced equation to calculate the moles of another substance in the reaction.

5.7 Mole Relationships in Chemical Equations

In an earlier section, we saw that equations are balanced in terms of the numbers of each type of atom in the reactants and products. However, when experiments are done in the laboratory or medications prepared in the pharmacy, samples contain billions of atoms and molecules so it is impossible to count them individually. What can be measured conveniently is mass, using a balance. Because mass is related to the number of particles through the molar mass, measuring the mass is equivalent to counting the number of particles or moles.

Conservation of Mass

In any chemical reaction, the total amount of matter in the reactants is equal to the total amount of matter in the products. If all of the reactants were weighed, they would have a total mass equal to the total mass of the products. This is known as the *law of conservation of mass*, which says that there is no change in the total mass of the substances reacting in a chemical reaction. Thus, no material is lost or gained as original substances are changed to new substances.

For example, tarnish forms when silver reacts with sulfur to form silver sulfide.

$$2Ag(s) + S(s) \longrightarrow Ag_2S(s)$$

In this reaction, the number of silver atoms that react is two times the number of sulfur atoms. When 200 silver atoms react, 100 sulfur atoms are required. Normally, however, many more atoms would actually be present in this reaction. If we are dealing with molar amounts, then the coefficients in the equation can be interpreted in terms of moles. Thus, 2 moles silver react with each 1 mole sulfur. Since the molar mass of each can be determined, the quantities of silver and sulfur can

$$2Ag(s) \qquad + \qquad S(s) \qquad \longrightarrow \qquad Ag_2S(s)$$

Mass of reactants   =   Mass of products

also be stated in terms of mass in grams of each. Therefore, an equation for a chemical equation can be interpreted several ways, as seen in Table 5.6.

TABLE 5.6
Information Available from a Balanced Equation

	Reactants		Products
Equation	**2Ag(s)**	+ S(s)	⟶ **Ag₂S(s)**
Atoms	2 Ag atoms	+ 1 S atom	⟶ 1 Ag₂S formula unit
	200 Ag atoms	+ 100 S atoms	⟶ 100 Ag₂S formula units
Avogadro's number of atoms	$2(6.02 \times 10^{23})$ Ag atoms	+ $1(6.02 \times 10^{23})$ S atoms	⟶ $1(6.02 \times 10^{23})$ Ag₂S formula units
Moles	2 moles Ag	+ 1 mole S	⟶ 1 mole Ag₂S
Mass (g)	2(107.9 g) Ag	+ 1(32.1 g) S	⟶ 1(247.9 g) Ag₂S
Total mass (g)	247.9 g		⟶ 247.9 g

Mole–Mole Factors from an Equation

When iron reacts with sulfur, the product is iron(III) sulfide.

$$2Fe(s) + 3S(s) \longrightarrow Fe_2S_3(s)$$

Iron (Fe)	Sulfur (S)	Iron(III) sulfide (Fe₂S₃)
$2Fe(s)$ +	$3S(s)$ ⟶	$Fe_2S_3(s)$

Because the equation is balanced, we know the proportions of iron and sulfur in the reaction. For this reaction, we see that 2 moles iron reacts with 3 moles sulfur to form 1 mole iron(III) sulfide. From the coefficients, we can write **mole–mole factors** between reactants and between reactants and products.

Fe and S: $\dfrac{2 \text{ moles Fe}}{3 \text{ moles S}}$ and $\dfrac{3 \text{ moles S}}{2 \text{ moles Fe}}$

Fe and Fe_2S_3: $\dfrac{2 \text{ moles Fe}}{1 \text{ mole } Fe_2S_3}$ and $\dfrac{1 \text{ mole } Fe_2S_3}{2 \text{ moles Fe}}$

S and Fe_2S_3: $\dfrac{3 \text{ moles S}}{1 \text{ mole } Fe_2S_3}$ and $\dfrac{1 \text{ mole } Fe_2S_3}{3 \text{ moles S}}$

SAMPLE PROBLEM 5.10

■ **Writing Mole–Mole Factors**

Consider the following balanced equation:

$$4Na(s) + O_2(s) \longrightarrow 2Na_2O(s)$$

Write the mole–mole factors for

a. Na and O_2 **b.** Na and Na_2O

SOLUTION

a. $\dfrac{4 \text{ moles Na}}{1 \text{ mole } O_2}$ and $\dfrac{1 \text{ mole } O_2}{4 \text{ moles Na}}$

b. $\dfrac{4 \text{ moles Na}}{2 \text{ moles } Na_2O}$ and $\dfrac{2 \text{ moles } Na_2O}{4 \text{ moles Na}}$

STUDY CHECK

Using the equation in Sample Problem 5.10, write the mole–mole factors for O_2 and Na_2O.

Using Mole–Mole Factors in Calculations

Whenever you prepare a recipe, adjust an engine for the proper mixture of fuel and air, or prepare medicines in a pharmaceutical laboratory, you need to know the proper amounts of reactants to use and how much of the product will form. Earlier we wrote all the possible conversion factors that can be obtained from the balanced equation: $2Fe(s) + 3S(s) \longrightarrow Fe_2S_3(s)$. Now we will show how mole–mole factors are used in chemical calculations.

SAMPLE PROBLEM 5.11

■ **Calculating Moles of a Reactant**

In the reaction of iron and sulfur, how many moles of sulfur are needed to react with 6.0 moles iron?

$$2Fe(s) + 3S(s) \longrightarrow Fe_2S_3(s)$$

SOLUTION

STEP 1 **Write the given and needed number of moles.** In this problem, we need to find the number of moles of S that react with 6.0 moles Fe.

 Given moles of Fe   **Need** moles of S

STEP 2 **Write the plan to convert the given to the needed.**

moles of Fe │ Mole–mole factor │ moles of S

STEP 3 **Use coefficients to write relationships and mole–mole factors.** Use coefficients to write the mole–mole factors for the given and needed substances.

$$2 \text{ moles Fe} = 3 \text{ moles S}$$

$$\frac{2 \text{ moles Fe}}{3 \text{ moles S}} \quad \text{and} \quad \frac{3 \text{ moles S}}{2 \text{ moles Fe}}$$

STEP 4 **Set up problem using the mole–mole factor that cancels given moles.** Use a mole–mole factor to cancel the given moles and provide needed moles.

$$6.0 \text{ moles Fe} \times \frac{3 \text{ moles S}}{2 \text{ moles Fe}} = 9.0 \text{ moles S}$$

The answer is given with 2 significant figures because the given quantity 6.0 moles Fe has 2 SFs. The values in the mole–mole factor are exact.

Guide to Using Mole Factors

STEP 1
Write the given and needed moles.

STEP 2
Write a plan to convert the given to the needed moles.

STEP 3
Use coefficients to write relationships and mole–mole factors.

STEP 4
Set up problem using the mole factor that cancels given moles.

STUDY CHECK

Using the equation in Sample Problem 5.11, calculate the number of moles of iron needed to completely react with 2.7 moles sulfur.

■

SAMPLE PROBLEM 5.12

■ **Calculating Moles of a Product**

In a combustion reaction, propane (C_3H_8) reacts with oxygen. How many moles of CO_2 can be produced when 2.25 moles C_3H_8 reacts?

$$C_3H_8(g) + 5O_2(s) \longrightarrow 3CO_2(g) + 4H_2O(g)$$
Propane

SOLUTION

STEP 1 **Write the given and needed moles.** In this problem, we need to find the number of moles of CO_2 that can be produced when 2.25 moles C_3H_8 react.

 Given moles of C_3H_8 **Need** moles of CO_2

STEP 2 **Write the plan to convert the given to the needed.**

moles of C_3H_8 │ Mole–mole factor │ moles of CO_2

STEP 3 **Use coefficients to write relationships and mole–mole factors.**
Use coefficients to write the mole–mole factors for the given and needed substances.

$$1 \text{ mole } C_3H_8 = 3 \text{ moles } CO_2$$

$$\frac{1 \text{ mole } C_3H_8}{3 \text{ moles } CO_2} \quad \text{and} \quad \frac{3 \text{ moles } CO_2}{1 \text{ mole } C_3H_8}$$

STEP 4 **Set up problem using the mole–mole factor that cancels given moles.** Use a mole–mole factor to cancel the given moles and provide needed moles.

$$2.25 \text{ moles } C_3H_8 \times \frac{3 \text{ moles } CO_2}{1 \text{ mole } C_3H_8} = 6.75 \text{ moles } CO_2$$

The answer is given with 3 significant figures because the given quantity 2.25 moles C_3H_8 has 3 SFs. The values in the mole–mole factor are exact.

STUDY CHECK

Using the equation in Sample Problem 5.12, calculate the number of moles of oxygen that must react to produce 0.756 mole water.

QUESTIONS AND PROBLEMS

■ Mole Relationships in Chemical Equations

5.43 Give an interpretation of the following equations in terms of (1) number of particles and (2) number of moles:
a. $2SO_2(g) + O_2(g) \longrightarrow 2SO_3(g)$
b. $4P(s) + 5O_2(s) \longrightarrow 2P_2O_5(s)$

5.44 Give an interpretation of the following equations in terms of (1) number of particles and (2) number of moles:
a. $2Al(s) + 3Cl_2(g) \longrightarrow 2AlCl_3(s)$
b. $4HCl(g) + O_2(g) \longrightarrow 2Cl_2(g) + 2H_2O(g)$

5.45 Write all of the mole–mole factors for the equations listed in problem 5.43.

5.46 Write all of the mole–mole factors for the equations listed in problem 5.44.

5.47 The reaction of hydrogen with oxygen produces water.

$$2H_2(g) + O_2(g) \longrightarrow 2H_2O(g)$$

a. How many moles of O_2 are required to react with 2.0 moles H_2?
b. If you have 5.0 moles O_2, how many moles of H_2 are needed for the reaction?
c. How many moles of H_2O form when 2.5 moles O_2 reacts?

5.48 Ammonia is produced by the reaction of hydrogen and nitrogen.

$$N_2(g) + 3H_2(g) \longrightarrow 2NH_3(g)$$
$$\text{ammonia}$$

a. How many moles of H_2 are needed to react with 1.0 mole N_2?
b. How many moles of N_2 reacted if 0.60 mole NH_3 is produced?
c. How many moles of NH_3 are produced when 1.4 moles H_2 reacts?

5.49 Carbon disulfide and carbon monoxide are produced when carbon is heated with sulfur dioxide.

$$5C(s) + 2SO_2(g) \longrightarrow CS_2(l) + 4CO(g)$$

a. How many moles of C are needed to react with 0.500 mole SO_2?
b. How many moles of CO are produced when 1.2 moles C reacts?
c. How many moles of SO_2 are required to produce 0.50 mole CS_2?
d. How many moles of CS_2 are produced when 2.5 moles C reacts?

5.50 In the acetylene torch, acetylene gas (C_2H_2) burns in oxygen to produce carbon dioxide and water.

$$2C_2H_2(g) + 5O_2(g) \longrightarrow 4CO_2(g) + 2H_2O(g)$$

a. How many moles of O_2 are needed to react with 2.00 moles C_2H_2?
b. How many moles of CO_2 are produced when 3.5 moles C_2H_2 reacts?
c. How many moles of C_2H_2 are required to produce 0.50 mole H_2O?
d. How many moles of CO_2 are produced from 0.100 mole O_2?

■ 5.8 Mass Calculations for Reactions

LEARNING GOAL
■
Given the mass in grams of a substance in a reaction, calculate the mass in grams of another substance in the reaction.

When you perform a chemistry experiment in the laboratory, you use a laboratory balance to obtain a certain mass of reactant. From the mass in grams, you can determine the number of moles of reactant. By using mole–mole factors, you can predict the moles of product that can be produced. Then the molar mass of the product is used to convert the moles back into mass in grams. The following procedure can be used to set up and solve problems that involve calculations of quantities for substances in a chemical reaction.

Finding the Moles or Masses of Substances in a Chemical Reaction

1. **If necessary, convert the mass of substance A to moles using its molar mass (g/mole).**

$$\text{grams of A} \xrightarrow[\text{of A}]{\text{Molar mass}} \text{moles of A}$$

2. **Convert the moles of A to moles of substance B using the mole–mole factor.**

$$\text{moles of A} \xrightarrow[\text{(moles B/moles A)}]{\text{Mole–mole factor}} \text{moles of B}$$

3. **If necessary, convert the moles of substance B to grams using its molar mass (g/mole).**

$$\text{moles of B} \xrightarrow[\text{of B}]{\text{Molar mass}} \text{grams of B}$$

SAMPLE PROBLEM 5.13

■ **Mass of Products from Moles of Reactant**

In the formation of smog, nitrogen reacts with oxygen to produce nitrogen oxide. Calculate the grams of NO produced when 1.50 moles O_2 react.

$$N_2(g) + O_2(g) \longrightarrow 2NO(g)$$

SOLUTION

STEP 1 Given 1.50 moles of O_2 **Need** grams of NO

STEP 2 Plan

moles of O_2 [Mole–mole factor] moles of NO [Molar mass] grams of NO

STEP 3 Equalities/Conversion Factors The mole–mole factor that converts moles of O_2 to moles of NO is derived from the coefficients in the balanced equation.

$$1 \text{ mole } O_2 = 2 \text{ moles NO}$$
$$\frac{2 \text{ moles NO}}{1 \text{ mole } O_2} \quad \text{and} \quad \frac{1 \text{ mole } O_2}{2 \text{ moles NO}}$$

$$1 \text{ mole NO} = 30.0 \text{ g}$$
$$\frac{30.0 \text{ g NO}}{1 \text{ mole NO}} \quad \text{and} \quad \frac{1 \text{ mole NO}}{30.0 \text{ g NO}}$$

STEP 4 Set Up Problem First, we can change the given, 1.50 moles O_2, to moles of NO.

$$1.50 \text{ moles } O_2 \times \frac{2 \text{ moles NO}}{1 \text{ mole } O_2} = 3.00 \text{ moles NO}$$

Now, the moles of NO can be converted to grams of NO using its molar mass.

$$3.00 \text{ moles NO} \times \frac{30.0 \text{ g NO}}{1 \text{ mole NO}} = 90.0 \text{ g NO}$$

These two steps can also be written as a sequence of conversion factors that lead to the mass in grams of NO.

$$1.50 \text{ moles } O_2 \times \frac{2 \text{ moles NO}}{1 \text{ mole } O_2} \times \frac{30.0 \text{ g NO}}{1 \text{ mole NO}} = 90.0 \text{ g NO}$$

STUDY CHECK

Using the equation in Sample Problem 5.13, calculate the grams of NO that can be produced when 0.734 mole N_2 reacts.

■

SAMPLE PROBLEM 5.14

■ **Mass of Product from Mass of Reactant**

Acetylene (C_2H_2), used in welding, burns with oxygen.

$$2C_2H_2(g) + 5O_2(g) \longrightarrow 4CO_2(g) + 2H_2O(g)$$

How many grams of carbon dioxide are produced when 54.6 g C_2H_2 are burned?

SOLUTION

STEP 1 Given grams of C_2H_2 **Need** grams of CO_2

STEP 2 Plan Once we convert grams of C_2H_2 to moles of C_2H_2 using its molar mass, we can use a mole–mole factor to find the moles of CO_2. Then the molar mass of CO_2 will give us the grams of CO_2.

grams of C_2H_2 [Molar mass] moles of C_2H_2 [Mole–mole factor] moles of CO_2 [Molar mass] grams of CO_2

STEP 3 **Equalities/Conversion Factors**   We need the molar mass of C_2H_2 and CO_2. The mole–mole factor that converts moles of C_2H_2 to moles of CO_2 is derived from the coefficients in the balanced equation.

$$1 \text{ mole } C_2H_2 = 26.0 \text{ g}$$
$$\frac{26.0 \text{ g } C_2H_2}{1 \text{ mole } C_2H_2} \quad \text{and} \quad \frac{1 \text{ mole } C_2H_2}{26.0 \text{ g } C_2H_2}$$

$$1 \text{ mole } C_2H_2 = 2 \text{ moles } CO_2$$
$$\frac{2 \text{ moles } C_2H_2}{1 \text{ mole } CO_2} \quad \text{and} \quad \frac{1 \text{ mole } CO_2}{2 \text{ moles } C_2H_2}$$

$$1 \text{ mole } CO_2 = 44.0 \text{ g}$$
$$\frac{44.0 \text{ g } CO_2}{1 \text{ mole } CO_2} \quad \text{and} \quad \frac{1 \text{ mole } CO_2}{44.0 \text{ g } CO_2}$$

STEP 4 **Set Up Problem**   Using our plan, we first convert grams of C_2H_2 to moles of C_2H_2.

$$54.6 \text{ g } C_2H_2 \times \frac{1 \text{ mole } C_2H_2}{26.0 \text{ g } C_2H_2} = 2.10 \text{ moles } C_2H_2$$

Then we change moles of C_2H_2 to moles of CO_2 by using the mole–mole factor.

$$2.10 \text{ moles } C_2H_2 \times \frac{4 \text{ moles } CO_2}{2 \text{ moles } C_2H_2} = 4.20 \text{ moles } CO_2$$

Finally, we can convert moles of CO_2 to grams of CO_2.

$$4.20 \text{ moles } CO_2 \times \frac{44.0 \text{ g } CO_2}{1 \text{ mole } CO_2} = 185 \text{ g } CO_2$$

The solution can be obtained using the conversion factors in sequence.

$$54.6 \text{ g } C_2H_2 \times \frac{1 \text{ mole } C_2H_2}{26.0 \text{ g } C_2H_2} \times \frac{4 \text{ moles } CO_2}{2 \text{ moles } C_2H_2} \times \frac{44.0 \text{ g } CO_2}{1 \text{ mole } CO_2} = 185 \text{ g } C$$

STUDY CHECK

Using the equation in Sample Problem 5.14, calculate the grams of CO_2 that can be produced when 25.0 g O_2 reacts.

■

SAMPLE PROBLEM 5.15

■ **Calculating the Mass of Reactant**

Propane gas, C_3H_8, a fuel for camp stoves and some specially equipped automobiles, reacts with oxygen to produce carbon dioxide and water. How many grams of O_2 are required to react with 22.0 g C_3H_8?

$$C_3H_8(g) + 5O_2(g) \longrightarrow 3CO_2(g) + 4H_2O(g)$$

SOLUTION

STEP 1 **Given**   grams of C_3H_8 **Need**   grams of O_2

STEP 2 **Plan**   The 22.0 g C_3H_8 are first changed to moles of C_3H_8 using its molar mass. Then the moles of C_3H_8 are changed to moles of O_2, which can be converted to grams of O_2.

grams C_3H_8 | Molar mass | moles of C_3H_8 | Mole–mole factor | moles of O_2 | Molar mass | grams of O_2

STEP 3 Equalities/Conversion Factors

$$1 \text{ mole } C_3H_8 = 44.1 \text{ g}$$
$$\frac{44.1 \text{ g } C_3H_8}{1 \text{ mole } C_3H_8} \quad \text{and} \quad \frac{1 \text{ mole } C_3H_8}{44.1 \text{ g } C_3H_8}$$

$$1 \text{ mole } C_3H_8 = 5 \text{ moles } O_2$$
$$\frac{1 \text{ mole } C_3H_8}{5 \text{ moles } O_2} \quad \text{and} \quad \frac{5 \text{ moles } O_2}{1 \text{ mole } C_3H_8}$$

$$1 \text{ mole } O_2 = 32.0 \text{ g}$$
$$\frac{32.0 \text{ g } O_2}{1 \text{ mole } O_2} \quad \text{and} \quad \frac{1 \text{ mole } O_2}{32.0 \text{ g } O_2}$$

STEP 4 Set Up Problem   Using a sequence of three factors converts grams of C_3H_8 to grams of O_2.

$$22.0 \text{ g } C_3H_8 \times \frac{1 \text{ mole } C_3H_8}{44.1 \text{ g } C_3H_8} \times \frac{5 \text{ moles } O_2}{1 \text{ mole } C_3H_8} \times \frac{32.0 \text{ g } O_2}{1 \text{ mole } O_2} = 79.8 \text{ g } O_2$$

STUDY CHECK

Using the equation in Sample Problem 5.15, calculate the grams of C_3H_8 that are needed to produce 15.0 g H_2O.

QUESTIONS AND PROBLEMS

■ Mass Calculations for Reactions

5.51 Sodium reacts with oxygen to produce sodium oxide.

$$4Na(s) + O_2(g) \longrightarrow 2Na_2O(s)$$

a. How many grams of Na_2O are produced when 2.50 moles Na react?
b. If you have 18.0 g Na, how many grams of O_2 are required for reaction?
c. How many grams of O_2 are needed in a reaction that produces 75.0 g Na_2O?

5.52 Nitrogen gas reacts with hydrogen gas to produce ammonia by the following equation:

$$N_2(g) + 3H_2(g) \longrightarrow 2NH_3(g)$$

a. If you have 1.80 moles H_2, how many grams of NH_3 can be produced?
b. How many grams of H_2 are needed to react with 2.80 g N_2?
c. How many grams of NH_3 can be produced from 12.0 g H_2?

5.53 Ammonia and oxygen react to form nitrogen and water.

$$4NH_3(g) + 3O_2(g) \longrightarrow 2N_2(g) + 6H_2O(g)$$
Ammonia

a. How many grams of O_2 are needed to react with 8.00 moles NH_3?
b. How many grams of N_2 can be produced when 6.50 g O_2 reacts?
c. How many grams of water are formed from the reaction of 34.0 g NH_3?

5.54 Iron(III) oxide reacts with carbon to give iron and carbon monoxide.

$$Fe_2O_3(s) + 3C(s) \longrightarrow 2Fe(s) + 3CO(g)$$

a. How many grams of C are required to react with 2.50 moles Fe_2O_3?
b. How many grams of CO are produced when 36.0 g C reacts?
c. How many grams of Fe can be produced when 6.00 g Fe_2O_3 reacts?

5.55 Nitrogen dioxide and water react to produce nitric acid HNO_3 and nitrogen monoxide.

$$3NO_2(g) + H_2O(l) \longrightarrow 2HNO_3(aq) + NO(g)$$

a. How many grams of H_2O are required to react with 28.0 g NO_2?
b. How many grams of NO are obtained from 15.8 g NO_2?
c. How many grams of HNO_3 are produced from 8.25 g NO_2?

5.56 Calcium cyanamide reacts with water to form calcium carbonate and ammonia.

$$CaCN_2(s) + 3H_2O(l) \longrightarrow CaCO_3(s) + 2NH_3(g)$$

a. How many grams of water are needed to react with 75.0 g $CaCN_2$?
b. How many grams of NH_3 are produced from 5.24 g $CaCN_2$?
c. How many grams of $CaCO_3$ form if 155 g water react?

5.57 When the ore lead(II) sulfide burns in oxygen, the products are lead(II) oxide and sulfur dioxide.
a. Write the balanced equation for the reaction.
b. How many grams of oxygen are required to react with 0.125 mole lead(II) sulfide?
c. How many grams of sulfur dioxide can be produced when 65.0 g lead(II) sulfide react?
d. How many grams of lead(II) sulfide are used to produce 128 g lead(II) oxide?

5.58 When the gases dihydrogen sulfide and oxygen react, they form the gases sulfur dioxide and water vapor.
a. Write the balanced equation for the reaction.
b. How many grams of oxygen are required to react with 2.50 g dihydrogen sulfide?
c. How many grams of sulfur dioxide can be produced when 38.5 g oxygen react?
d. How many grams of oxygen are required to produce 55.8 g water vapor?

■ 5.9 Energy in Chemical Reactions

LEARNING GOAL

■ Describe exothermic and endothermic reactions and factors that affect the rate of a reaction.

For a chemical reaction to take place, the molecules of the reactants must come in contact with each other. If there are no collisions between the molecules (or atoms), no reaction is possible. When the molecules collide, bonds between atoms are broken and new bonds can form. The energy needed to break apart those bonds is called the **activation energy**. If the energy of a collision is less than the activation energy, the molecules bounce apart without reacting.

The concept of activation energy is analogous to climbing over a hill. To reach a destination on the other side, we must expend energy to climb to the top of the hill. Once we are at the top, we can easily roll down the other side. The energy needed to get us from our starting point to the top of the hill would be our activation energy.

Exothermic reaction — Energy of reactants, Activation energy, Heat of reaction (released), Energy of products, Energy, Progress of reaction

Endothermic reaction — Activation energy, Energy of products, Heat of reaction (absorbed), Energy of reactants, Energy, Progress of reaction

Exothermic and Endothermic Reactions

The *heat of reaction* is the energy difference between the reactants and the products. In **exothermic reactions**, the energy of the products is lower than that of the reactants and heat is given off. In **endothermic reactions**, the energy of the products is higher than the energy of the reactants and heat must be absorbed for products to form.

For example, the combustion of methane (CH_4) is an exothermic reaction. Heat (213 kcal per mole of CH_4 burned) is released along with the products CO_2 and H_2O.

Exothermic Reaction

$$CH_4(g) + 2O_2(g) \longrightarrow CO_2(g) + 2H_2O(g) + 213 \text{ kcal}$$

methane   heat released

The formation of hydrogen iodide is an endothermic reaction. In the overall energy change, heat is absorbed when the hydrogen and iodine undergo reaction.

Endothermic Reaction

$$H_2(g) + I_2(g) + 12 \text{ kcal of heat} \longrightarrow 2HI(g)$$

                          heat absorbed

Reaction	Energy Change	Heat in the Equation
Exothermic	Heat released	Product side
Endothermic	Heat absorbed	Reactant side

SAMPLE PROBLEM 5.16

▪ Exothermic and Endothermic Reactions

In the reaction of one mole of solid carbon with oxygen gas, the energy of the carbon dioxide product is 94 kcal lower than the energy of the reactants.

a. Is the reaction exothermic or endothermic?
b. What is the equation for the reaction?

HEALTH NOTE

▪ Hot Packs and Cold Packs

In a hospital, at a first-aid station, or at an athletic event, a *cold pack* may be used to reduce swelling from an injury, remove heat from inflammation, or decrease capillary size to lessen the effect of hemorrhaging. Inside the plastic container of a cold pack, there is a compartment containing solid ammonium nitrate (NH_4NO_3) that is separated from a compartment containing water. The pack is activated when it is hit or squeezed hard enough to break the walls between the compartments and cause the ammonium nitrate to mix with the water (shown as H_2O over the reaction arrow). In an endothermic process, one mole of NH_4NO_3 that dissolves

absorbs 6.3 kcal of heat from the water. The temperature drops and the pack becomes cold and ready to use:

Endothermic Reaction in a Cold Pack

$$6.3 \text{ kcal} + NH_4NO_3(s) \xrightarrow{H_2O} NH_4NO_3(aq)$$

Hot packs are used to relax muscles, lessen aches and cramps, and increase circulation by expanding capillary size. Constructed in the same way as cold packs, a hot pack may contain the salt $CaCl_2$. The dissolving of the salt in water is exothermic and releases 18 kcal per mole of salt. The temperature rises and the pack becomes hot and ready to use:

Exothermic Reaction in a Hot Pack

$$CaCl_2(s) \xrightarrow{H_2O} CaCl_2(aq) + 18 \text{ kcal}$$

SOLUTION

a. exothermic   **b.** $C(s) + O_2(g) \longrightarrow CO_2(g) + 94$ kcal

STUDY CHECK

When two moles of liquid ethanol (C_2H_5OH) react with oxygen gas (O_2), the products are carbon dioxide, water, and 326 kcal of heat. Write a balanced equation including the heat of reaction.

Rate of Reaction

The *rate* (or speed) *of reaction* is measured by the amount of reactant used up, or the amount of product formed, in a certain period of time. Reactions with low activation energies go faster than reactions with high activation energies. The rate of a reaction can be affected by changes in the temperature, the amounts of reactants in the container, and the addition of catalysts.

At higher temperatures, the increase in kinetic energy of the reactants makes them move faster and collide more often, and provides more collisions with the required energy of activation. Reactions almost always go faster at higher temperatures. For every 10°C increase in temperature, most reaction rates approximately double. If we want food to cook faster, we raise the temperature. When body temperature rises, there is an increase in the pulse rate, rate of breathing, and metabolic rate. On the other hand, we slow down a reaction by lowering the temperature. In some cardiac surgeries, body temperature is lowered to 28°C so the heart can be stopped and less oxygen is required by the brain. This is also the reason that some people have survived submersion in icy lakes for long periods of time. We refrigerate perishable foods to make them last longer.

The rate of a reaction also increases when reactants are added. Then there are more collisions between the reactants and the reaction goes faster. For example, a patient having difficulty breathing may be given a breathing mixture with a higher oxygen content than the atmosphere. The increase in the number of oxygen molecules in the lungs increases the rate at which oxygen combines with hemoglobin. The increased rate of oxygenation of the blood means that the patient can breathe more easily.

$$Hb \; + \; O_2 \; \longrightarrow \; HbO_2$$
Hemoglobin   Oxygen   Oxyhemoglobin

TABLE 5.7

Factors That Increase Reaction Rate

Factor	Reason
More reactants	More collisions
Higher temperature	More collisions with energy of activation
Adding a catalyst	Lowers energy of activation

Another way to speed up a reaction is to lower the energy of activation. This can be done by adding a **catalyst**. Earlier, we discussed the energy required to climb a hill. If instead we found a tunnel through the hill, we won't need as much energy to get to the other side. A catalyst acts by providing an alternate pathway with a lower energy requirement. As a result, more collisions form product successfully. Catalysts have found many uses in industry. In the production of margarine, the reaction of hydrogen with vegetable oils is normally very slow. However, when finely divided platinum is present as a catalyst, the reaction occurs rapidly. In the body, biocatalysts called enzymes make most metabolic reactions go at the rates necessary for proper cellular activity.

A summary of the factors affecting reaction rate is given in Table 5.7.

SAMPLE PROBLEM 5.17

■**Factors That Affect the Rate of Reaction**

Indicate whether the following changes will increase, decrease, or have no effect upon the rate of reaction:

a. increase in temperature
b. decrease in the number of reactants
c. addition of catalyst

SOLUTION

a. increase
b. decrease
c. increase

STUDY CHECK

How does lowering the temperature affect the rate of reaction?

QUESTIONS AND PROBLEMS

■**Energy in Chemical Reactions**

5.59 **a.** Why do chemical reactions require an energy of activation?
  b. What is the function of a catalyst?
  c. In an exothermic reaction, is the energy of the products higher or lower?
  d. Draw an energy diagram for an exothermic reaction.

5.60 **a.** What is measured by the heat of reaction?
  b. How does the heat of reaction differ in exothermic and endothermic reactions?

 c. In an endothermic reaction, is the energy of the products higher or lower than the reactants?

 d. Draw an energy diagram for an endothermic reaction.

5.61 Classify the following as exothermic or endothermic reactions:

 a. 125 kcal are released.

 b. In the energy diagram, the energy level of the products is higher than the reactants.

 c. The metabolism of glucose in the body provides energy.

5.62 Classify the following as exothermic or endothermic reactions:

 a. In the energy diagram, the energy level of the products is lower than the reactants.

 b. In the body, the synthesis of proteins requires energy.

 c. 30 kcal are absorbed.

5.63 Classify the following as exothermic or endothermic reactions:

 a. gas burning in a Bunsen burner:

$$CH_4(g) + 2O_2(g) \longrightarrow CO_2(g) + 2H_2O(g) + 213 \text{ kcal}$$

 b. dehydrating limestone:

$$Ca(OH)_2(s) + 15.6 \text{ kcal} \longrightarrow CaO(s) + H_2O(g)$$

 c. formation of aluminum oxide and iron from aluminum and iron(III) oxide:

$$2Al(s) + Fe_2O_3(s) \longrightarrow Al_2O_3(s) + 2Fe(s) + 204 \text{ kcal}$$

5.64 Classify the following as exothermic or endothermic reactions:

 a. the combustion of propane:

$$C_3H_8(g) + 5O_2(g) \longrightarrow 3CO_2(g) + 4H_2O(g) + 531 \text{ kcal}$$

 b. the formation of "table" salt:

$$2Na(s) + Cl_2(g) \longrightarrow 2NaCl(s) + 196 \text{ kcal}$$

 c. decomposition of phosphorus pentachloride:

$$PCl_5(s) + 16 \text{ kcal} \longrightarrow PCl_3(s) + Cl_2(g)$$

5.65 a. What is meant by the rate of a reaction?

 b. Why does bread grow mold more quickly at room temperature than in the refrigerator?

5.66 a. How does a catalyst affect the activation energy?

 b. Why is pure oxygen used in respiratory distress?

5.67 How would each of the following change the rate of the reaction shown here?

$$2SO_2(g) + O_2(g) \longrightarrow 2SO_3(g)$$

 a. adding SO_2

 b. raising the temperature

 c. adding a catalyst

 d. removing some SO_2

5.68 How would each of the following change the rate of the reaction shown here?

$$2NO(g) + 2H_2(g) \longrightarrow N_2(g) + 2H_2O(g)$$

 a. adding more NO

 b. lowering the temperature

 c. removing some H_2

 d. adding a catalyst

CONCEPT MAP

CHAPTER REVIEW

5.1 Chemical Changes
A chemical change occurs when the atoms of the initial substances rearrange to form new substances. When new substances form, a chemical reaction has taken place.

5.2 Chemical Equations
A chemical equation shows the formulas of the substances that react on the left side of a reaction arrow and the products that form on the right side of the reaction arrow. An equation is balanced by writing the smallest whole numbers (coefficients) in front of formulas to equalize the atoms of each element in the reactants and the products.

5.3 Types of Reactions
Many chemical reactions can be organized by reaction type: combination, decomposition, single replacement, or double replacement.

5.4 Oxidation–Reduction Reactions
When electrons are transferred in a reaction, it is an oxidation–reduction reaction. One reactant loses electrons, and another reactant gains electrons. Overall, the number of electrons lost and gained is equal.

5.5 The Mole
One mole of an element contains 6.02×10^{23} atoms; a mole of a compound contains 6.02×10^{23} molecules or formula units.

5.6 Molar Mass
The molar mass (g/mole) of any substance is the mass in grams equal numerically to its atomic mass, or the sum of the atomic masses, which have been multiplied by their subscripts in a formula. It becomes a conversion factor when it is used to change a quantity in grams to moles, or to change a given number of moles to grams.

5.7 Mole Relationships in Chemical Equations
In a balanced equation, the total mass of the reactants is equal to the total mass of the products. The coefficients in an equation describing the relationship between the moles of any two components are used to write mole–mole factors. When the number of moles for one substance is known, a mole–mole factor is used to find the moles of a different substance in the reaction.

5.8 Mass Calculations for Reactions
In calculations using equations, the molar masses of the substances and their mole factors are used to change the number of

grams of one substance to the corresponding grams of a different substance.

5.9 Energy in Chemical Reactions

In a reaction, the reacting particles must collide with energy equal to or greater than the energy of activation. The heat of reaction is the energy difference between the initial energy of the reactants and the final energy of the products. In exothermic reactions, heat is released. In endothermic reactions, heat is absorbed. The rate of a reaction, which is the speed at which the reactants are converted to products, can be increased by adding more reactants, raising the temperature, or adding a catalyst.

KEY TERMS

activation energy The energy needed upon collision to break apart the bonds of the reacting molecules.

Avogadro's number The number of items in a mole, equal to 6.02×10^{23}.

balanced equation The final form of a chemical equation that shows the same number of atoms of each element in the reactants and products.

catalyst A substance that increases the rate of reaction by lowering the activation energy.

chemical change The formation of a new substance with a different composition and properties than the initial substance.

chemical equation A shorthand way to represent a chemical reaction using chemical formulas to indicate the reactants and products and coefficients to show reacting ratios.

chemical reaction The process by which a chemical change takes place.

coefficients Whole numbers placed in front of the formulas to balance the number of atoms or moles of atoms of each element on both sides of an equation.

combination reaction A reaction in which reactants combine to form a single product.

decomposition reaction A reaction in which a single reactant splits into two or more simpler substances.

double replacement reaction A reaction in which parts of two different reactants exchange places.

endothermic reaction A reaction that requires heat; the energy of the products is higher than the energy of the reactants.

exothermic reaction A reaction that releases heat; the energy of the products is lower than the energy of the reactants.

formula unit The group of ions represented by the formula of an ionic compound.

molar mass The mass in grams of 1 mole of an element equal numerically to its atomic mass. The molar mass of a compound is equal to the sum of the masses of the elements in the formula.

mole A group of atoms, molecules, or formula units that contains 6.02×10^{23} of these items.

mole–mole factor A conversion factor that relates the number of moles of two compounds derived from the coefficients in an equation.

oxidation The loss of electrons by a substance. Biological oxidation may involve the addition of oxygen, or the loss of hydrogen.

oxidation–reduction reaction A reaction in which the oxidation of one reactant is always accompanied by the reduction of another reactant.

physical change A change in which the physical appearance of a substance changes but the chemical composition stays the same.

products The substances formed as a result of a chemical reaction.

reactants The initial substances that undergo change in a chemical reaction.

reduction The gain of electrons by a substance. Biological reduction may involve the loss of oxygen or the gain of hydrogen.

single replacement reaction An element replaces a different element in a compound.

ADDITIONAL QUESTIONS AND PROBLEMS

5.69 Balance each of the following unbalanced equations and identify the type of reaction.
   a. $NH_3(g) + HCl(g) \longrightarrow NH_4Cl(s)$
   b. $Fe_3O_4(s) + H_2(g) \longrightarrow Fe(s) + H_2O(g)$
   c. $Sb(s) + Cl_2(g) \longrightarrow SbCl_3(s)$
   d. $NI_3(s) \longrightarrow N_2(g) + I_2(g)$
   e. $KBr(aq) + Cl_2(aq) \longrightarrow KCl(aq) + Br_2(l)$
   f. $Fe(s) + H_2SO_4(aq) \longrightarrow Fe_2(SO_4)_3(aq) + H_2(g)$
   g. $Al_2(SO_4)_3(aq) + NaOH(aq) \longrightarrow$
   $$Na_2SO_4(aq) + Al(OH)_3(s)$$

5.70 Balance each of the following unbalanced equations and identify the type of reaction.
   a. $Li_3N(s) \longrightarrow Li(s) + N_2(g)$
   b. $Mg(s) + N_2(g) \longrightarrow Mg_3N_2(s)$
   c. $Al(s) + HCl(aq) \longrightarrow AlCl_3(aq) + H_2(g)$
   d. $Mg(s) + H_3PO_4(aq) \longrightarrow Mg_3(PO_4)_2(s) + H_2(g)$
   e. $Cr_2O_3(s) + H_2(g) \longrightarrow Cr(s) + H_2O(g)$

   f. $Al(s) + Cl_2(g) \longrightarrow AlCl_3(s)$
   g. $MgCl_2(aq) + AgNO_3(aq) \longrightarrow$
   $$Mg(NO_3)_2(aq) + AgCl(s)$$

5.71 Identify each of the following as an oxidation or a reduction reaction.
   a. $Zn^{2+} + 2e^- \longrightarrow Zn$ **b.** $Al \longrightarrow Al^{3+} + 3e^-$
   c. $Pb \longrightarrow Pb^{2+} + 2e^-$ **d.** $Cl_2 + 2e^- \longrightarrow 2Cl^-$

5.72 Write a balanced chemical equation for each of the following oxidation–reduction reactions.
   a. Sulfur reacts with molecular chlorine to form sulfur dichloride.
   b. Molecular chlorine and sodium bromide react to form molecular bromine and sodium chloride.
   c. Aluminum metal and iron(III) oxide react to produce aluminum oxide and elemental iron.
   d. Copper(II) oxide reacts with elemental C to form elemental copper and carbon dioxide.

5.73 During heavy exercise and workouts, lactic acid, $C_3H_6O_3$, accumulates in the muscles where it can cause pain and soreness.
a. How many molecules are in 0.500 mole lactic acid?
b. How many C atoms are in 1.50 moles lactic acid?
c. How many moles of lactic acid contain 4.5×10^{24} atoms of O?
d. What is the molar mass of lactic acid?

5.74 Ibuprofen, the anti-inflammatory ingredient in Advil, has the formula $C_{13}H_{18}O_2$.
a. How many molecules are in 0.200 mole ibuprofen?
b. How many H atoms are in 0.100 mole ibuprofen?
c. How many moles of ibuprofen contain 7.4×10^{25} atoms of C?
d. What is the molar mass of ibuprofen?

5.75 Calculate the molar mass of each of the following:
a. $FeSO_4$, ferrous sulfate, iron supplement
b. $Ca(IO_3)_2$, calcium iodate, iodine source in table salt
c. $C_5H_8NNaO_4$, monosodium glutamate, flavor enhancer
d. $C_6H_{12}O_2$, isoamyl formate used to make artificial fruit syrups

5.76 Calculate the molar mass of each of the following:
a. $Mg(HCO_3)_2$, magnesium hydrogen carbonate
b. $Au(OH)_3$, gold(III) hydroxide, used in gold plating
c. $C_{18}H_{34}O_2$, oleic acid from olive oil
d. $C_{21}H_{26}O_5$, prednisone, anti-inflammatory

5.77 How many grams are in 0.150 mole of each of the following?
a. K   **b.** Cl_2 **c.** Na_2CO_3

5.78 How many grams are in 2.25 moles of each of the following?
a. N_2 **b.** NaBr   **c.** C_6H_{14}

5.79 How many moles are in 25.0 g of each of the following compounds?
a. CO_2 **b.** $Al(OH)_3$ **c.** $MgCl_2$

5.80 How many moles are in 4.00 g of each of the following compounds?
a. NH_3 **b.** $Ca(NO_3)_2$ **c.** SO_3

5.81 At a winery, glucose ($C_6H_{12}O_6$) in grapes undergoes fermentation to produce ethanol (C_2H_6O) and carbon dioxide.

$$C_6H_{12}O_6 \longrightarrow 2C_2H_6O + 2CO_2$$
glucose   ethanol

a. How many moles of glucose are required to form 124 g of ethanol?
b. How many grams of ethanol would be formed from the reaction of 0.240 kg of glucose?

5.82 Gasohol is a fuel containing ethanol (C_2H_6O) that burns in oxygen (O_2) to give carbon dioxide and water.
a. State the reactants and products for this reaction in the form of a balanced equation.
b. How many moles of O_2 are needed to completely react with 4.0 moles of C_2H_6O?

c. If a car produces 88 g of CO_2, how many grams of O_2 are used up in the reaction?
d. If you add 125 g of C_2H_6O to your gas, how many grams of CO_2 and H_2O can be produced?

5.83 Balance the following equation:

$$NH_3(g) + F_2(g) \longrightarrow N_2F_4(g) + HF(g)$$

a. How many moles of each reactant are needed to produce 4.00 moles of HF?
b. How many grams of F_2 are required to react with 1.50 moles of NH_3?
c. How many grams of N_2F_4 can be produced when 3.40 g of NH_3 reacts?

5.84 When nitrogen dioxide (NO_2) from car exhaust combines with water in the air it forms nitric acid (HNO_3), which causes acid rain, and nitrogen monoxide.

$$3NO_2(g) + H_2O(l) \longrightarrow 2HNO_3(aq) + NO(g)$$

a. How many molecules of NO_2 are needed to react with 0.250 mole H_2O?
b. How many grams of HNO_3 are produced when 60.0 g NO_2 completely reacts?

5.85 Pentane gas, C_5H_{12}, reacts with oxygen to produce carbon dioxide and water.

$$C_5H_{12}(g) + 8O_2(g) \longrightarrow 5CO_2(g) + 6H_2O(g)$$

a. How many grams of pentane must react to produce 4.0 moles water?
b. How many grams of CO_2 are produced from 32.0 g oxygen and excess pentane?

5.86 Propane gas, C_3H_8, reacts with oxygen to produce water and carbon dioxide. Propane has a density of 2.02 g/L at room temperature.

$$C_3H_8(g) + 5O_2(g) \longrightarrow 3CO_2(g) + 4H_2O(l)$$
propane

a. How many moles of water form when 5.00 L propane gas (C_3H_8) completely react?
b. How many grams of CO_2 are produced from 18.5 g oxygen gas and excess propane?
c. How many grams of H_2O can be produced from the reaction of 8.50×10^{22} molecules propane gas, C_3H_8?

5.87 The equation for the formation of silicon tetrachloride from silicon and chlorine is

$$Si(s) + 2Cl_2(g) \longrightarrow SiCl_4(g) + 157 \text{ kcal}$$

a. Is the formation of $SiCl_4$ an endothermic or exothermic reaction?
b. Is the energy of the products higher or lower than the energy of the reactants?

5.88 The equation for the formation of nitrogen monoxide is

$$N_2(g) + O_2(g) + 90.2 \text{ kJ} \longrightarrow 2NO(g)$$

a. Is the formation of NO an endothermic or exothermic reaction?
b. Is the energy of the products higher or lower than the energy of the reactants?

UNDERSTANDING THE CONCEPTS

5.89 Identify some of the physical and chemical changes of a burning wax candle.

5.90 In each of the following, identify the change as physical or chemical.

a.

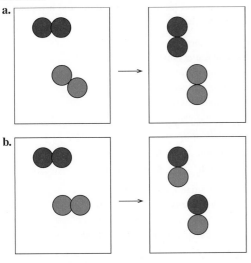

b.

5.91 Balance each of the following by adding coefficients and identify the type of reaction for each.

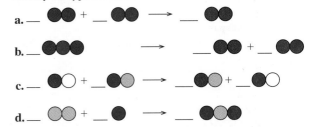

a. __

b. __

c. __

d. __

5.92 In an experiment, a piece of copper is weighed and allowed to react with oxygen.

$$2Cu(s) + O_2(g) \longrightarrow 2CuO(s)$$

a. How many grams of CuO could be produced according to the equation?
b. How many moles of O_2 are required to completely react the Cu?

QUESTIONS FOR TEAMWORK

5.93 Write a balanced equation for each of the following reaction descriptions and identify each type of reaction.
  a. An aqueous solution of lead(II) nitrate is mixed with aqueous sodium phosphate to produce solid lead(II) phosphate and aqueous sodium nitrate.
  b. Gallium metal heated in oxygen gas forms solid gallium(III) oxide.
  c. When solid sodium nitrate is heated, solid sodium nitrite and oxygen gas are produced.
  d. Solid bismuth(III) oxide and solid carbon react to form bismuth metal and carbon monoxide gas.

5.94 The gaseous hydrocarbon acetylene, C_2H_2, used in welders' torches, releases a large amount of heat when it burns according to the following equation:

$$2C_2H_2(g) + 5O_2(g) \longrightarrow 4CO_2(g) + 2H_2O(g)$$

  a. How many *moles of water* are produced from the complete reaction of 64.0 g oxygen?

  b. How many *moles of oxygen* are needed to react completely with 2.25×10^{24} molecules of acetylene?
  c. How many *grams of carbon dioxide* are produced from the complete reaction of 78.0 g acetylene?

5.95 Consider the following *unbalanced* equation:

$$Al(s) + O_2(g) \longrightarrow Al_2O_3(s)$$

  a. Balance the equation.
  b. Identify the type of reaction.
  c. How many moles of oxygen must react with 4.50 moles Al?
  d. How many grams of aluminum oxide are produced when 50.2 g aluminum react?
  e. When aluminum is reacted in a closed container with 8.00 g oxygen, how many grams of aluminum oxide can form?

5.96 A toothpaste contains 0.24% by mass sodium fluoride used to prevent dental caries and 0.30% by mass triclosan $C_{12}H_7Cl_3O_2$, a preservative and antigingivitis agent. One tube contains 119 g toothpaste.
  a. How many moles of NaF are in the tube of toothpaste?
  b. How many fluoride atoms F^- are in the tube of toothpaste?
  c. How many grams of sodium ion Na^+ are in 1.50 g toothpaste?
  d. How many molecules of triclosan are in the tube of toothpaste?

5.97 A gold bar is 2.31 cm long, 1.48 cm wide, and 0.0758 cm thick.
  a. If gold has a density of 19.3 g/mL, what is the mass of the gold bar?
  b. How many atoms of gold are in the bar?
  c. When the same mass of gold combines with oxygen, the oxide product has a mass of 5.61 g. How many moles of oxygen atoms are combined with the gold?
  d. What is the formula of the oxide product?

ANSWERS

Answers to Study Checks

5.1 **a.** physical   **b.** chemical   **c.** physical

5.2 $3Fe(s) + 2O_2(g) \longrightarrow Fe_3O_4(s)$

5.3 $2NO(g) + O_2(g) \longrightarrow 2NO_2(g)$ Combination reaction

5.4 Lithium is oxidized: $2Li \longrightarrow 2Li^+ + 2e^-$
       Fluorine is reduced: $F_2 + 2e^- \longrightarrow 2F^-$

5.5 0.432 mole H_2O

5.6 0.120 mole aspirin

5.7 138.0 g

5.8 24.4 g Au

5.9 0.00621 mole of $CaCO_3$, 0.00550 mole of $MgCO_3$

5.10 $\dfrac{1 \text{ mole } O_2}{2 \text{ moles } Na_2O}$ and $\dfrac{2 \text{ moles } Na_2O}{1 \text{ mole } O_2}$

5.11 1.8 moles of Fe

5.12 0.945 mole O_2

5.13 44.0 g NO

5.14 27.5 g CO_2

5.15 9.17 g C_3H_8

5.16 $C_2H_5OH(l) + 3O_2(g) \longrightarrow 2CO_2(g) + 3H_2O(g) + 163 \text{ kcal}$

5.17 The rate of reaction will decrease because the number of collisions between reacting particles will be less, and a smaller number of the collisions that do occur will have sufficient activation energy.

Answers to Selected Questions and Problems

5.1 **a.** physical   **b.** chemical   **c.** physical
       d. chemical   **e.** physical   **f.** chemical

5.3 **a.** not balanced   **b.** balanced
       c. not balanced   **d.** balanced

5.5 **a.** $N_2(g) + O_2(g) \longrightarrow 2NO(g)$
       b. $2HgO(s) \longrightarrow 2Hg(l) + O_2(g)$
       c. $4Fe(s) + 3O_2(g) \longrightarrow 2Fe_2O_3(s)$
       d. $2Na(s) + Cl_2(g) \longrightarrow 2NaCl(s)$

5.7 **a.** $Mg(s) + 2AgNO_3(aq) \longrightarrow Mg(NO_3)_2(aq) + 2Ag(s)$
       b. $2Al(s) + 3CuSO_4(aq) \longrightarrow 3Cu(s) + Al_2(SO_4)_3(aq)$
       c. $Pb(NO_3)_2(aq) + 2NaCl(aq) \longrightarrow$
               $PbCl_2(s) + 2NaNO_3(aq)$
       d. $2Al(s) + 6HCl(aq) \longrightarrow 2AlCl_3(aq) + 3H_2(g)$

5.9 **a.** A single reactant splits into two simpler substances (elements).
       b. One element in the reacting compound is replaced by the other reactant.

5.11 **a.** combination
        b. single replacement
        c. decomposition
        d. double replacement
        e. decomposition
        f. double replacement
        g. combination

5.13 **a.** oxidation   **b.** reduction
        c. oxidation   **d.** reduction

5.15 **a.** Zn is oxidized; Cl_2 is reduced.
        b. Br^- in NaBr is oxidized; Cl_2 is reduced.
        c. The O^{2-} in PbO is oxidized; the Pb^{2+} is reduced.
        d. Sn^{2+} is oxidized; Fe^{3+} is reduced.

5.17 **a.** reduction   **b.** oxidation

5.19 Linoleic acid gains hydrogen atoms and is reduced.

5.21 1.00 mole contains 6.02×10^{23} atoms of an element, molecules of a covalent substance, or formula units of an ionic substance.

5.23 **a.** 24 moles H
        b. 100
        c. 0.040 mole N

5.25 **a.** 3.01×10^{23} C atoms
        b. 7.71×10^{23} SO_2 molecules
        c. 0.0867 mole Fe
        d. 14.1 moles C_2H_5OH

5.27 **a.** 6.00 moles H   **b.** 8.00 moles O
        c. 1.20×10^{24} P atoms   **d.** 4.82×10^{24} O atoms

5.29 **a.** 58.5 g   **b.** 159.8 g   **c.** 73.8 g
        d. 342.3 g   **e.** 58.3 g   **f.** 365.1 g

5.31 **a.** 46.0 g **b.** 112 g **c.** 14.8 g

5.33 **a.** 29.3 g **b.** 109 g **c.** 4.05 g

5.35 **a.** 602 g **b.** 11 g

5.37 **a.** 0.463 mole **b.** 0.0167 mole
c. 0.882 mole **d.** 1.17 moles

5.39 12 moles

5.41 **a.** 0.78 mole S **b.** 1.95 moles S
c. 6.0 moles S

5.43 **a. (1)** Two molecules of sulfur dioxide react with 1 molecule of oxygen to produce 2 molecules of sulfur trioxide.
a. (2) Two moles of sulfur dioxide react with 1 mole of oxygen to produce 2 moles of sulfur trioxide.
b. (1) Four atoms of phosphorus react with 5 molecules of oxygen to produce 2 molecules of diphosphorus pentoxide.
b. (2) Four moles of phosphorus react with 5 moles of oxygen to produce 2 moles of diphosphorus pentoxide.

5.45 **a.** $\dfrac{2 \text{ moles } SO_2}{1 \text{ mole } O_2}$ and $\dfrac{1 \text{ mole } O_2}{2 \text{ moles } SO_2}$

$\dfrac{2 \text{ moles } SO_2}{2 \text{ moles } SO_3}$ and $\dfrac{2 \text{ moles } SO_3}{2 \text{ moles } SO_2}$

$\dfrac{2 \text{ moles } SO_3}{1 \text{ mole } O_2}$ and $\dfrac{1 \text{ mole } O_2}{2 \text{ moles } SO_3}$

b. $\dfrac{4 \text{ moles } P}{5 \text{ moles } O_2}$ and $\dfrac{5 \text{ moles } O_2}{4 \text{ moles } P}$

$\dfrac{4 \text{ moles } P}{2 \text{ moles } P_2O_5}$ and $\dfrac{2 \text{ moles } P_2O_5}{4 \text{ moles } P}$

$\dfrac{5 \text{ moles } O_2}{2 \text{ moles } P_2O_5}$ and $\dfrac{2 \text{ moles } P_2O_5}{5 \text{ moles } O_2}$

5.47 **a.** 1.0 mole O_2 **b.** 10. moles H_2
c. 5.0 moles H_2O

5.49 **a.** 1.25 moles C **b.** 0.96 mole CO
c. 1.0 mole SO_2 **d.** 0.50 mole CS_2

5.51 **a.** 77.5 g Na_2O **b.** 6.26 g O_2 **c.** 19.4 g O_2

5.53 **a.** 192 g O_2 **b.** 3.79 g N_2 **c.** 54.0 g H_2O

5.55 **a.** 3.65 g H_2O **b.** 3.43 g NO **c.** 7.53 g HNO_3

5.57 **a.** $2PbS(s) + 3O_2(g) \longrightarrow 2PbO(s) + 2SO_2(g)$
b. 6.00 g O_2 **c.** 17.4 g SO_2 **d.** 137 g PbS

5.59 **a.** The energy of activation is the energy required to break the bonds of the reacting molecules.
b. A catalyst provides a pathway that lowers the activation energy and speeds up a reaction.
c. In exothermic reactions, the energy of the products is lower than the reactants.
d.

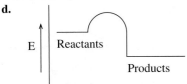

5.61 **a.** exothermic **b.** endothermic
c. exothermic

5.63 **a.** exothermic **b.** endothermic
c. exothermic

5.65 **a.** The rate of a reaction tells how fast the products are formed.
b. Reactions go faster at higher temperatures.

5.67 **a.** increase **b.** increase
c. increase **d.** decrease

5.69 **a.** $NH_3(g) + HCl(g) \longrightarrow NH_4Cl(s)$ combination
b. $Fe_3O_4(s) + 4H_2(g) \longrightarrow 3Fe(s) + 4H_2O(g)$
single replacement
c. $2Sb(s) + 3Cl_2(g) \longrightarrow 2SbCl_3(s)$ combination
d. $2NI_3(s) \longrightarrow N_2(g) + 3I_2(g)$ decomposition
e. $2KBr(aq) + Cl_2(aq) \longrightarrow 2KCl(aq) + Br_2(l)$
single replacement
f. $2Fe(s) + 3H_2SO_4(aq) \longrightarrow Fe_2(SO_4)_3(aq) + 3H_2(g)$
single replacement
g. $Al_2(SO_4)_3(aq) + 6NaOH(aq) \longrightarrow$
$3Na_2SO_4(aq) + 2Al(OH)_3(s)$
double replacement

5.71 **a.** reduction **b.** oxidation
c. oxidation **d.** reduction

5.73 **a.** 3.01×10^{23} molecules
b. 2.71×10^{24} C atoms
c. 2.5 moles lactic acid
d. 90.1 g

5.75 **a.** 152.0 g
b. 389.9 g
c. 169.1 g
d. 116.1 g

5.77 **a.** 5.87 g **b.** 10.7 g **c.** 15.9 g

5.79 **a.** 0.568 mole **b.** 0.321 mole **c.** 0.262 mole

5.81 **a.** 1.35 moles glucose **b.** 123 g ethanol

5.83 $2NH_3 + 5F_2 \longrightarrow N_2F_4 + 6HF$
a. 1.33 moles NH_3 and 3.33 moles F_2
b. 143 g F_2
c. 10.4 g N_2F_4

5.85 **a.** 48 g pentane **b.** 27.5 g CO_2

5.87 **a.** exothermic **b.** lower

5.89 Physical changes include the following: solid wax melts to form liquid, the candle becomes shorter, the liquid wax changes to solid, the shape of the wax changes, the wick becomes shorter.
Chemical changes include the following: heat and light are emitted, wax and the wick burn in the presence of oxygen.

5.91 **a.** 1, 1, 2 combination reaction
b. 2, 2, 1 decombination reaction
c. 1, 1, 1, 1 double replacement reaction
d. 1, 4, 2 combination reaction

6 Gases

"When oxygen levels in the blood are low, the cells in the body don't get enough oxygen," says Sunanda Tripathi, registered nurse, Santa Clara Valley Medical Center. "We use a nasal cannula to give supplemental oxygen to a patient. At a flow rate of 2 liters per minute, a patient breathes in a gaseous mixture that is about 28% oxygen compared to 21% in ambient air."

When a patient has a breathing disorder, the flow and volume of oxygen into and out of the lungs are measured. A ventilator may be used if a patient has difficulty breathing. When pressure is increased, the lungs expand. When the pressure of the incoming gas is reduced, the lung volume contracts to expel carbon dioxide. These relationships—known as gas laws—are an important part of ventilation and breathing.

the Chemistry place

Visit **www.aw-bc.com/chemplace** or use the CD-ROM in the book for extra quizzes, interactive tutorials, career resources, PowerPoint slides, math help, and case studies.

We all live at the bottom of a sea of gases called the atmosphere. The most important of these gases is oxygen, which constitutes about 21% of the atmosphere. Without oxygen, life on this planet would be impossible—oxygen is vital to all life processes of plants and animals. Ozone (O_3), formed in the upper atmosphere by the interaction of oxygen with ultraviolet light, absorbs some of the harmful radiation before it can strike Earth's surface. The other gases in the atmosphere include nitrogen (78% of the atmosphere), argon, carbon dioxide (CO_2), and water vapor. Carbon dioxide gas, a product of combustion and metabolism, is used by plants in photosynthesis, a process that produces the oxygen that is essential for humans and animals.

The atmosphere has become a dumping ground for other gases, such as methane, chlorofluorohydrocarbons (CFCs), sulfur dioxide, and nitrogen oxides. The chemical reactions of these gases with sunlight and oxygen in the air are contributing to air pollution, ozone depletion, global warming, and acid rain. Such chemical changes can seriously affect our health and our lifestyle. A knowledge of gases and some of the laws that govern gas behavior can help us understand the nature of matter and allow us to make decisions concerning important environmental and health issues.

6.1 Properties of Gases

The behavior of gases is quite different from that of liquids and solids. Gas particles are far apart, whereas particles of both liquids and solids are held close together as strong attractive forces become more important at lower temperatures. This means that a gas has no definite shape or volume and will completely fill any container. Because there are great distances between its particles, a gas is less dense than a solid or liquid and can be compressed. A model for the behavior of a gas, called the **kinetic molecular theory of gases**, helps us understand gas behavior.

Kinetic Molecular Theory of Gases

1. **A gas consists of small particles (atoms or molecules) that move randomly with rapid velocities.** Gas molecules moving in all directions at high speeds cause a gas to fill the entire volume of a container.
2. **The attractive forces between the particles of a gas can be neglected.** Gas particles move far apart and fill a container of any size and shape.
3. **The actual volume occupied by gas molecules is extremely small compared to the volume that the gas occupies.** The volume of the container is considered equal to the volume of the gas. Most of the volume of a gas is empty space, which allows gases to be easily compressed.
4. **The average kinetic energy of gas molecules is proportional to the Kelvin temperature.** Gas particles move faster as the temperature increases. At higher temperatures, gas particles hit the walls of the container with more force, which produces higher pressures.

LEARNING GOAL

Describe the kinetic theory of gases and the properties of gases.

the
Chemistry
place

WEB TUTORIAL
Properties of Gases

5. Gas particles are in constant motion, moving rapidly in straight paths. When gas particles collide, they rebound and travel in new directions. When they collide with the walls of the container, they exert gas pressure. An increase in the number or force of collisions against the walls of the container causes an increase in the pressure of the gas.

The kinetic theory helps explain some of the characteristics of gases. For example, we can quickly smell perfume from a bottle that is opened on the other side of a room, because its particles move rapidly in all directions. They move faster at higher temperatures, and more slowly at lower temperatures. Sometimes tires and gas-filled containers explode when temperatures are too high. From the kinetic theory, we know that gas particles move faster when heated, hit the walls of a container with more force, and cause a buildup of pressure inside a container.

SAMPLE PROBLEM 6.1

■ **Kinetic Theory of Gases**

How would you explain the following characteristics of a gas using the kinetic theory of gases?

a. The particles of gas are not attracted to each other.
b. Gas particles move faster when the temperature is increased.

SOLUTION

a. The distance between the particles of a gas is so great that the particles are not attracted to each other.
b. Temperature increases the kinetic energy of the gas particles, which makes them move faster.

STUDY CHECK

Why can you smell food that is cooking in the kitchen when you are in a different part of the house?

■

When we talk about a gas, we describe it in terms of four properties: pressure, volume, temperature, and the amount of gas.

Pressure (*P*)
Gas particles are extremely small. However, when billions and billions of gas particles hit against the walls of a container, they exert a force known as the pressure of the gas. The more molecules that strike the wall, the greater the pressure. (See Figure 6.1.) If we heat the container to make the molecules move faster, they smash into the walls of the container more often and with increased force, which increases the pressure. The gas molecules of oxygen and nitrogen in the air around us are exerting pressure on us all the time. We call the pressure exerted by the air **atmospheric pressure**. (See Figure 6.2.) As you go to higher altitudes the atmospheric pressure is less because there are fewer molecules of oxygen and nitrogen in the air. The most common units used for gas measurement are the atmosphere (atm) and millimeters of mercury (mm Hg). On the TV weather report, you may hear or see the atmospheric pressure given in inches of mercury or kilopascals in countries other than the United States. In a chemistry lab the unit torr may be used.

FIGURE 6.1

Gas particles move in straight lines within a container. The gas particles exert pressure when they collide with the walls of the container.

Q Why does heating the container increase the pressure of the gas within it?

Volume (*V*)

The volume of gas equals the size of the container in which the gas is placed. When you inflate a tire or a basketball, you are adding more gas particles, which increases the number of particles hitting the walls of the tire or basketball, and its volume increases. Sometimes, on a cool morning, a tire looks flat. The volume of the tire has decreased because a lower temperature decreases the speed of the molecules, which reduces the force of their impacts on the walls of the tire. The most common units for volume measurement are liters (L) and milliliters (mL).

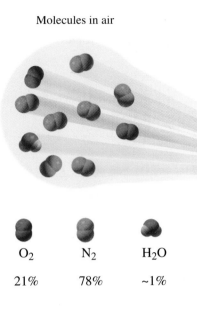

Molecules in air

O_2 N_2 H_2O

21%   78%   ~1%

Temperature (*T*)

The temperature of a gas is related to the kinetic energy of its particles. For example, if we have a gas at 200 K in a rigid container and heat it to a temperature of 400 K, the gas particles will have twice the kinetic energy that they did at 200 K. That also means that the gas at 400 K exerts twice the pressure of the gas at 200 K. Although you measure gas temperature using a Celsius thermometer, all comparisons of gas behavior and all calculations related to temperature must use the Kelvin temperature. No one has quite created the conditions for absolute zero (K), but we predict that the particles will have zero kinetic energy, and the gas will exert zero pressure at this temperature.

Amount of Gas (*n*)

When you add air to a bicycle tire, you are increasing the amount of gas, which results in a higher pressure in the tire. Usually we measure the amount of gas by its mass (grams). In gas law calculations, we need to change the grams of gas to moles.

A summary of the four properties of a gas are given in Table 6.1.

TABLE 6.1

Properties That Describe a Gas		
Property	**Description**	**Unit(s) of Measurement**
Pressure (*P*)	The force exerted by gas against the walls of the container	atmosphere (atm); mm Hg; torr; kPa
Volume (*V*)	The space occupied by the gas	liter (L); milliliter (mL)
Temperature (*T*)	Determines the kinetic energy and rate of motion of the gas particles	Celsius (°C); Kelvin (K) *required in calculations*
Amount (*n*)	The quantity of gas present in a container	grams (g); moles (mole) *required in calculations*

FIGURE 6.2

A column of air extending from the upper atmosphere to the surface of the Earth produces a pressure on each of us of about 1 atmosphere. While there is a lot of pressure on the body, it is balanced by the pressure inside the body.

Q **Why is there less pressure at higher altitudes?**

Atmospheric pressure

SAMPLE PROBLEM 6.2

■ Properties of Gases

Identify the property of a gas that is described by each of the following:

a. increases the kinetic energy of gas particles
b. the force of the gas particles hitting the walls of the container
c. the space that is occupied by a gas

■ Forming a Gas

Obtain baking soda and a jar or a plastic bottle. You will also need an elastic glove that fits over the mouth of the jar or a balloon that fits snugly over the top of the plastic bottle. Place a cup of vinegar in the jar or bottle. Sprinkle some baking soda into the fingertips of the glove or into the balloon. Carefully fit the glove or balloon over the top of the jar or bottle. Slowly lift the fingers of the glove or the balloon so that the baking soda falls into the vinegar. Watch what happens. Squeeze the glove or balloon.

QUESTION

1. Describe the properties of gas that you observe as the reaction takes place between vinegar and baking soda.
2. How do you know that a gas was formed?

LEARNING GOAL

Describe the units of measurement used for pressure and change from one unit to another.

FIGURE 6.3

A barometer: The pressure exerted by the gases in the atmosphere is equal to the downward pressure of a mercury column in a closed glass tube. The height of the mercury column measured in mm Hg is called atmospheric pressure.

Q Why does the height of the mercury column change from day to day?

SOLUTION

a. temperature   **b.** pressure   **c.** volume

STUDY CHECK

When helium is added to a balloon, the number of grams of gas increases. What property of a gas is described?

■

QUESTIONS AND PROBLEMS

■ Properties of Gases

6.1 Use the kinetic molecular theory of gases to explain each of the following:
   a. Gases move faster at higher temperatures.
   b. Gases can be compressed much more than liquids or solids.

6.2 Use the kinetic molecular theory of gases to explain each of the following:
   a. A container of non-stick cooking spray explodes when thrown into a fire.
   b. The air in a hot-air balloon is heated to make the balloon rise.

6.3 Identify the property of a gas that is measured in each of the following:
   a. 350 K   **b.** space occupied by a gas
   c. 2.00 g of O_2 **d.** force of gas particles striking the walls of the container

6.4 Identify the property of a gas that is measured in each of the following measurements:
   a. 425 K   **b.** 1.0 atm   **c.** 10.0 L   **d.** 0.50 mole of He

■ 6.2 Gas Pressure

When billions and billions of gas particles hit against the walls of a container, they exert **pressure**, which is defined as a force acting on a certain area.

$$\text{Pressure}(P) = \frac{\text{force}}{\text{area}}$$

The air that covers the surface of Earth, the atmosphere, contains vast numbers of gas particles. Because the air particles have mass, when they collide with the surface of Earth they exert an *atmospheric pressure*.

The atmospheric pressure can be measured using a barometer, as shown in Figure 6.3. At a pressure of exactly 1 atmosphere (atm), the mercury column would be exactly 760 mm high. We say that the atmospheric pressure is 760 mm Hg (millimeters of mercury), which is an **atmosphere (atm)**. One atmosphere of pressure may also be expressed as 760 *torr*, a pressure unit named to honor Evangelista Torricelli, the inventor of the barometer. Because they are equal, units of torr and mm Hg are used interchangeably.

   1 atm = 760 mm Hg = 760 torr

   1 mm Hg = 1 torr

In SI units, pressure is measured in pascals (Pa); 1 atm is equal to 101,325 Pa. Because a pascal is a very small unit, it is likely that pressures would be reported in kilopascals.

The American equivalent of 1 atm is 14.7 pounds per square inch (lb/in.[2]). When you use a pressure gauge to check the air pressure in the tires of a car it may read 30–35 psi. Table 6.2 summarizes the various units used in the measurement of pressure.

TABLE 6.2

Units for Measuring Pressure		
Unit	Abbreviation	Unit Equivalent to 1 atm
Atmosphere	atm	1 atm (exact)
Millimeters of Hg	mm Hg	760 mm Hg
Torr	torr	760 torr
Pounds per square inch	lb/in.^{2} (psi)	14.7 lb/in.^{2}
Kilopascal	kPa	101.325 kPa

Atmospheric pressure changes with variations in weather and altitude. On a hot, sunny day, the air is more dense, which causes more pressure on the mercury surface. The mercury column rises indicating a higher atmospheric pressure. On a rainy day, the atmosphere exerts less pressure, which causes the mercury column to go lower. In the weather report, this type of weather is called a low-pressure system. Atmosphere pressure is greatest at sea level. Above sea level, the density of the gases in the air decreases, which causes lower atmospheric pressures; the atmospheric pressure is greater in Death Valley because it is below sea level. (See Table 6.3.)

Divers must be concerned about increasing pressures on their ears and lungs when they dive below the surface of the ocean. Because water is denser than air, the pressure on a diver increases rapidly as the diver descends. At a depth of 33 ft below the surface of the ocean, an additional 1 atmosphere of pressure is exerted by the water on a diver, for a total of 2 atm. At 100 ft down, there is a total pressure of 4 atm on a diver. The air tanks a diver carries continuously adjust the pressure of the breathing mixture to match the increase in pressure.

CASE STUDY
Scuba Diving and Blood Gases

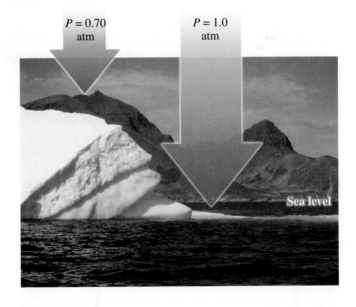

$P = 0.70$ atm

$P = 1.0$ atm

Sea level

SAMPLE PROBLEM 6.3

■ **Units of Pressure**

A sample of neon gas has a pressure of 0.50 atm. Give the pressure of the neon in millimeters of Hg.

SOLUTION

The equality 1 atm = 760 mm Hg can be written as conversion factors:

$$\frac{760 \text{ mm Hg}}{1 \text{ atm}} \quad \text{or} \quad \frac{1 \text{ atm}}{760 \text{ mm Hg}}$$

Using the appropriate conversion factor, the problem is set up as

$$0.50 \text{ atm} \times \frac{760 \text{ mm Hg}}{1 \text{ atm}} = 380 \text{ mm Hg}$$

TABLE 6.3

Altitude and Atmospheric Pressure		
Location	Altitude (km)	Atmospheric Pressure (mm Hg)
Sea level	0	760
Los Angeles	0.09	752
Las Vegas	0.70	700
Denver	1.60	630
Mount Whitney	4.50	440
Mount Everest	8.90	253

STUDY CHECK

What is the pressure in atmospheres for a gas that has a pressure of 655 torr?

■ Measuring Blood Pressure

The measurement of your blood pressure is one of the important measurements a doctor or nurse makes during a physical examination. Acting like a pump, the heart contracts to create the pressure that pushes blood through the circulatory system. During contraction, the blood pressure is called systolic and is at its highest. When the heart muscles relax, the blood pressure is called diastolic and falls. Normal range for systolic pressure is 100–120 mm Hg, and for diastolic pressure, 60–80 mm Hg, usually expressed as a ratio such as 100/80. These values are somewhat higher in older people. When blood pressures are elevated, such as 140/90, there is a greater risk of stroke, heart attack, or kidney damage. Low blood pressure prevents the brain from receiving adequate oxygen, causing dizziness and fainting.

The blood pressures are measured by a sphygmomanometer, an instrument consisting of a stethoscope and an inflatable cuff connected to a tube of mercury called a manometer. After the cuff is wrapped around the upper arm, it is pumped up with air until it cuts off the flow of blood through the arm. With the stethoscope over the artery, the air is slowly released from the cuff. When the pressure equals the systolic pressure, blood starts to flow again, and the noise it makes is heard through the stethoscope. As air

continues to be released, the cuff deflates until no sound in the artery is heard. That second pressure reading is noted as the diastolic pressure, the pressure when the heart is not contracting.

QUESTIONS AND PROBLEMS

■ Gas Pressure

6.5 What units are used to measure the pressure of a gas?

6.6 Which of the following statement(s) describes the pressure of a gas?
   a. the force of the gas particles on the walls of the container
   b. the number of gas particles in a container
   c. the volume of the container
   d. 3.00 atm   **e.** 750 torr

6.7 An oxygen tank contains oxygen (O_2) at a pressure of 2.00 atm. What is the pressure in the tank in terms of the following units?
   a. torr   **b.** mm Hg

6.8 On a climb up Mt. Whitney, the atmospheric pressure is 467 mm Hg. What is the pressure in terms of the following units?
   a. atm   **b.** torr

LEARNING GOAL

■ Use the pressure–volume relationship (Boyle's law) to determine the new pressure or volume of a certain amount of gas at a constant temperature.

■ 6.3 Pressure and Volume (Boyle's Law)

Imagine that you can see air particles hitting the walls inside a bicycle tire pump. What happens to the pressure inside the pump as we push down on the handle? As the air is compressed, the air particles are crowded together. In the smaller volume, more collisions occur, and the air pressure increases.

When a change in one property (in this case, volume) causes a change in the opposite direction of another property (in this case, pressure), the properties have an **inverse relationship**. The relationship between the pressure and volume of a gas is known as **Boyle's law**. The law states that the volume (V) of a sample of gas changes inversely with the pressure (P) of the gas as long as there has been no change in the temperature (T) or amount of gas (n), as illustrated in Figure 6.4.

If we change the volume or pressure of a gas sample without any change in the temperature or in the amount of the gas, the new pressure and volume will give the same PV product as the initial pressure and volume.

Boyle's Law

$$P_1 V_1 = P_2 V_2 \qquad \text{No change in number of moles and temperature}$$

SAMPLE PROBLEM 6.4

■ Calculating Pressure When Volume Changes

A sample of hydrogen gas (H_2) has a volume of 5.0 L and a pressure of 1.0 atm. What is the new pressure if the volume is decreased to 2.0 L at constant temperature?

SOLUTION

STEP 1 Organize the data in a table. In this problem, we want to know the final pressure (P_2) for the change in volume. In calculations with gas laws, it is helpful to organize the data in a table.

Conditions 1	Conditions 2
$V_1 = 5.0$ L	$V_2 = 2.0$ L
$P_1 = 1.0$ atm	$P_2 = ?$

STEP 2 Rearrange the gas law for the unknown. For a PV relationship, we use Boyle's law and solve for P_2 by dividing both sides by V_2.

$$P_1 V_1 = P_2 V_2$$

$$\frac{P_1 V_1}{V_2} = \frac{P_2 \cancel{V_2}}{\cancel{V_2}}$$

$$P_2 = P_1 \times \frac{V_1}{V_2}$$

STEP 3 Substitute values into the gas law to solve for the unknown. From the table, we see that the volume has decreased. Because pressure and volume are inversely related, the pressure must increase. When we substitute in the values, we see the ratio of the volumes (volume factor) is greater than 1, which increases the pressure.

$$P_2 = 1.0 \text{ atm} \times \frac{5.0 \cancel{L}}{2.0 \cancel{L}} = 2.5 \text{ atm}$$

Volume factor increases pressure

Note that the units of volume cancel and the final pressure is in atmospheres.

STUDY CHECK

A sample of helium gas has a volume of 150 mL at 750 torr. If the volume expands to 450 mL at constant temperature, what is the new pressure in torr?

Piston

$V = 4$ L
$P = 1$ atm

$V = 2$ L
$P = 2$ atm

FIGURE 6.4

Boyle's law: As volume decreases, gas molecules become more crowded, which causes the pressure to increase. Pressure and volume are inversely related.

Q **If the volume of a gas increases, what will happen to its pressure?**

Guide to Using the Gas Laws

STEP 1
Organize the data in a table of initial and final conditions.

STEP 2
Rearrange the gas law to solve for unknown quantity.

STEP 3
Substitute values into the gas law equation to solve for the unknown.

■ Calculating Volume When Pressure Changes

The gauge on a 12-L tank of compressed oxygen reads 4500 mm Hg. How many liters would this same gas occupy at a pressure of 750 mm Hg at constant temperature?

SOLUTION

STEP 1 **Organize the data in a table.** Placing our information in a table gives the following:

Conditions 1	Conditions 2
P_1 = 4500 mm Hg	P_2 = 750 mm Hg
V_1 = 12 L	V_2 = ?

STEP 2 **Rearrange the gas law for the unknown.** Using Boyle's law, we solve for V_2. According to Boyle's law, a decrease in the pressure will cause an increase in the volume.

$$V_2 = V_1 \times \frac{P_1}{P_2}$$

HEALTH NOTE

■ Pressure–Volume Relationship in Breathing

The importance of Boyle's law becomes more apparent when you consider the mechanics of breathing. Our lungs are elastic, balloon-like structures contained within an airtight chamber called the thoracic cavity. The diaphragm, a muscle, forms the flexible floor of the cavity.

INSPIRATION

The process of taking a breath of air begins when the diaphragm contracts, and the rib cage expands, causing an increase in the volume of the thoracic cavity. The elasticity of the lungs allows them to expand when the thoracic cavity expands. According to Boyle's law, the pressure inside the lungs will decrease when their volume increases. This causes the pressure inside the lungs to fall below the pressure of the atmosphere. This difference in pressures produces a *pressure gradient* between the lungs and the atmosphere. In a pressure gradient, molecules flow from an area of greater pressure to an area of lower pressure. Thus, we inhale as air flows into the lungs (*inspiration*), until the pressure within the lungs becomes equal to the pressure of the atmosphere.

Air inhaled

Rib cage expands as rib muscles contract

Lung

Diaphragm

Inhalation
Diaphragm contracts
(moves down)

Air exhaled

Rib cage gets smaller as rib muscles relax

Exhalation
Diaphragm relaxes
(moves up)

EXPIRATION

Expiration, or the exhalation phase of breathing, occurs when the diaphragm relaxes and moves back up into the thoracic cavity to its resting position. This reduces the volume of the thoracic cavity, which squeezes the lungs and decreases their volume. Now the pressure in the lungs is greater than the pressure of the atmosphere, so air flows out of the lungs. Thus, breathing is a process in which pressure gradients are continuously created between the lungs and the environment as a result of the changes in the volume and pressure.

STEP 3 **Substitute values into the gas law to solve for the unknown.**
When we substitute in the values, the ratio of pressures (pressure factor) is greater than 1, which increases the volume.

$$V_2 = 12 \text{ L} \times \frac{4500 \text{ mm Hg}}{750 \text{ mm Hg}} = 72 \text{ L}$$

Pressure factor increases volume

STUDY CHECK

A sample of methane gas (CH_4) has a volume of 125 mL at 0.600 atm pressure. How many milliliters will it occupy at a pressure of 1.50 atm at constant temperature?

QUESTIONS AND PROBLEMS

■ Pressure and Volume (Boyle's Law)

6.9 Why do scuba divers need to exhale air when they ascend to the surface of the water?

6.10 Why does a sealed bag of chips expand when you take it to a higher altitude?

6.11 The air in a cylinder with a piston has a volume of 220 mL and a pressure of 650 mm Hg.
 a. To obtain a higher pressure inside the cylinder at constant temperature, should the cylinder change as shown in A or B? Explain your choice.

Initial ⟶ A   or   B

 b. If the pressure inside the cylinder increases to 1.2 atm, what is the final volume of the cylinder? Complete the following data table:

Property	Initial	Final
Pressure (P)		
Volume (V)		

6.12 A balloon is filled with helium gas. When the following changes are made at constant temperature, which of these diagrams (A, B, or C) shows the new volume of the balloon?

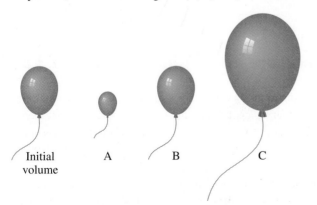

Initial volume   A   B   C

 a. The balloon floats to a higher altitude where the outside pressure is lower.
 b. The balloon is taken inside the house, but the atmospheric pressure remains the same.
 c. The balloon is put in a hyperbaric chamber in which the pressure is increased.

6.13 A gas with a volume of 4.0 L is contained in a closed container. Indicate what changes in pressure must have occurred if the volume undergoes the following changes at constant temperature.
a. The volume is compressed to 2 L.
b. The volume is allowed to expand to 12 L.
c. The volume is compressed to 0.40 L.

6.14 A gas at a pressure of 2.0 atm is contained in a closed container. Indicate the changes in its volume when the pressure undergoes the following changes at constant temperature.
a. The pressure increases to 6.0 atm.
b. The pressure drops to 1.0 atm.
c. The pressure drops to 0.40 atm.

6.15 A 10.0-L balloon contains He gas at a pressure of 655 mm Hg. What is the new pressure of the He gas at each of the following volumes if there is no change in temperature?
a. 20.0 L   **b.** 2.50 L   **c.** 1500 mL

6.16 The air in a 5.00-L tank has a pressure of 1.20 atm. What is the new pressure of the air when the air is placed in tanks that have the following volumes, if there is no change in temperature?
a. 1.00 L   **b.** 2500 mL   **c.** 750 mL

6.17 A sample of nitrogen (N_2) has a volume of 50.0 L at a pressure of 760. mm Hg. What is the volume of the gas at each of the following pressures if there is no change in temperature?
a. 1500 mm Hg   **b.** 2.0 atm   **c.** 0.500 atm

6.18 A sample of methane (CH_4) has a volume of 25 mL at a pressure of 0.80 atm. What is the volume of the gas at each of the following pressures if there is no change in temperature?
a. 0.40 atm   **b.** 2.00 atm   **c.** 2500 mm Hg

6.19 Cyclopropane, C_3H_6, is a general anesthetic. A 5.0-L sample has a pressure of 5.0 atm. What is the volume of the anesthetic given to a patient at a pressure of 1.0 atm?

6.20 The volume of air in a person's lungs is 615 mL at a pressure of 760 mm Hg. Inhalation occurs as the pressure in the lungs drops to 752 mm Hg. To what volume did the lungs expand?

6.21 Use the words *inspiration* and *expiration* to describe the part of the breathing cycle that occurs as a result of each of the following:
a. The diaphragm contracts (flattens out).
b. The volume of the lungs decreases.
c. The pressure within the lungs is less than the atmosphere.

6.22 Use the words *inspiration* and *expiration* to describe the part of the breathing cycle that occurs as a result of each of the following:
a. The diaphragm relaxes, moving up into the thoracic cavity.
b. The volume of the lungs expands.
c. The pressure within the lungs is greater than the atmosphere.

LEARNING GOAL

Use the temperature–volume relationship (Charles' law) to determine the new temperature or volume of a certain amount of gas at a constant pressure.

6.4 Temperature and Volume (Charles' Law)

Suppose that you are going to take a ride in a hot-air balloon. The captain turns on a propane burner to heat the air inside the balloon. As the temperature rises, the air particles move faster and spread out, causing the volume of the balloon to increase. When the air in the balloon becomes less dense than the air outside, the balloon and its passengers rise. In fact, it was in 1787 that Jacques Charles, a balloonist as well as a physicist, proposed that the volume of a gas is related to the temperature. This became **Charles' law**, which states that the volume (V) of a gas is directly related to the temperature (T) when there is no change in the pressure (P) or amount (n) of gas. (See Figure 6.5.) A **direct relationship** is one in which the related properties increase or decrease together. For two conditions, we can write Charles' law as follows.

Charles' Law

$$\frac{V_1}{T_1} = \frac{V_2}{T_2}$$ No change in number of moles and pressure

---

SAMPLE PROBLEM 6.6

■ **Calculating Volume When Temperature Changes**

A sample of neon gas has a volume of 5.40 L and a temperature of 15°C. Find the new volume of the gas after the temperature has been increased to 42°C at constant pressure.

SOLUTION

STEP 1 **Organize the data in a table.**   When the temperatures are given in degrees Celsius, they must be changed to Kelvins.

$$T_1 = 15°C + 273 = 288 \text{ K}$$
$$T_2 = 42°C + 273 = 315 \text{ K}$$

Conditions 1	Conditions 2
$T_1 = 288$ K	$T_2 = 315$ K
$V_1 = 5.40$ L	$V_2 = ?$

STEP 2 **Rearrange the gas law for the unknown.**   In this problem, we want to know the final volume (V_2) when the temperature increases. Using Charles' law, we solve for V_2 by multiplying both sides by T_2.

$$\frac{V_1}{T_1} = \frac{V_2}{T_2}$$

$$\frac{V_1}{T_1} \times T_2 = \frac{V_2}{\cancel{T_2}} \times \cancel{T_2}$$

$$V_2 = V_1 \times \frac{T_2}{T_1}$$

STEP 3 **Substitute values into the gas law to solve for the unknown.**   From the table, we see that the temperature has increased. Because temperature is directly related to volume, the volume must increase. When we substitute in the values, we see that the ratio of the temperatures (temperature factor) is greater than 1, which increases the volume.

$$V_2 = 5.40 \text{ L} \times \underbrace{\frac{315 \text{ K}}{288 \text{ K}}}_{\substack{\text{Temperature factor} \\ \text{increases volume}}} = 5.91 \text{ L}$$

$T = 200$ K $T = 400$ K
$V = 1$ L $V = 2$ L

FIGURE 6.5

Charles' law: The Kelvin temperature of a gas is directly related to the volume of the gas when there is no change in the pressure. When the temperature increases, making the molecules move faster, the volume must increase to maintain constant pressure.

Q **If the temperature of a gas decreases at constant pressure, how will the volume change?**

STUDY CHECK

A mountain climber inhales 486 mL of air at a temperature of −8°C. What volume will the air occupy in the lungs if the climber's body temperature is 37°C?

Remember that all temperatures used in gas law calculations must be converted to their corresponding Kelvin (K) temperatures.

QUESTIONS AND PROBLEMS

■ Temperature and Volume (Charles' Law)

6.23 Select the diagram that shows the new volume of a balloon when the following changes are made at constant pressure.

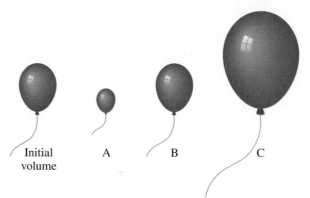

Initial volume   A   B   C

   a. The temperature is changed from 150 K to 300 K.
   b. The balloon is placed in a freezer.
   c. The balloon is first warmed, and then returned to its starting temperature.

6.24 Indicate whether the final volume of gas in each of the following is the same, larger, or smaller than the initial volume:
   a. A volume of 500 mL of air on a cold winter day at 5°C is breathed into the lungs, where body temperature is 37°C.
   b. The heater used to heat 1400 L of air in a hot-air balloon is turned off.
   c. A balloon filled with helium at the amusement park is left in a car on a hot day.

6.25 What change in volume occurs when gases for hot-air balloons are heated prior to their ascent?

6.26 On a cold, wintry morning, the tires on a car appear flat. How has their volume changed overnight?

6.27 A balloon contains 2500 mL of helium gas at 75°C. What is the new volume of the gas when the temperature changes to the following, if n and P are not changed?
   a. 55°C   **b.** 680 K   **c.** −25°C   **d.** 240 K

6.28 A gas has a volume of 4.00 L at 0°C. What final temperature in degrees Celsius is needed to cause the volume of the gas to change to the following, if n and P are not changed?
   a. 10.0 L   **b.** 1200 mL   **c.** 2.50 L   **d.** 50.0 mL

$T = 200$ K $T = 400$ K
$P = 1$ atm $P = 2$ atm

FIGURE 6.6

Gay-Lussac's law: The pressure of a gas is directly related to the temperature of the gas. When the Kelvin temperature of a gas is doubled, the pressure is doubled at constant volume.

Q How does a decrease in the temperature of a gas affect its pressure at constant volume?

LEARNING GOAL

■ Use the temperature–pressure relationship (Gay-Lussac's law) to determine the new temperature or pressure of a certain amount of gas at a constant volume.

■ 6.5 Temperature and Pressure (Gay-Lussac's Law)

If we could watch the molecules of a gas as the temperature rises, we would notice that they move faster and hit the sides of the container more often and with greater force. If we keep the volume of the container the same, we would observe an increase in the pressure. A temperature–pressure relationship, also known as **Gay-Lussac's law**, states that the pressure of a gas is directly related to its Kelvin temperature. This means that an increase in temperature increases the pressure of a gas, and a decrease in temperature decreases the pressure of the gas, provided the volume and number of moles of the gas remain the same. (See Figure 6.6.) The ratio of pressure (P) to temperature (T) is the same under all conditions as long as volume (V) and amount of gas (n) do not change.

Gay-Lussac's Law

$$\frac{P_1}{T_1} = \frac{P_2}{T_2}$$ No change in number of moles and volume

SAMPLE PROBLEM 6.7

■ Calculating Pressure When Temperature Changes

Aerosol containers can be dangerous if they are heated because they can explode. Suppose a container of hair spray with a pressure of 4.0 atm at a room temperature of 25°C is thrown into a fire. If the temperature of the gas inside the aerosol can reaches 402°C, what will be its pressure? The aerosol container may explode if the pressure inside exceeds 8.0 atm. Would you expect it to explode?

SOLUTION

STEP 1 Organize the data in a table. We must first change the temperatures to kelvins.

$$T_1 = 25°C + 273 = 298 \text{ K}$$

$$T_2 = 402°C + 273 = 675 \text{ K}$$

Conditions 1	Conditions 2
$P_1 = 4.0$ atm	$P_2 = ?$
$T_1 = 298$ K	$T_2 = 675$ K

STEP 2 Rearrange the gas law for the unknown. Using Gay-Lussac's law, we can solve for P_2.

$$\frac{P_1}{T_1} = \frac{P_2}{T_2}$$

$$\frac{P_1}{T_1} \times T_2 = \frac{P_2}{\cancel{T_2}} \times \cancel{T_2}$$

$$P_2 = P_1 \times \frac{T_2}{T_1}$$

STEP 3 Substitute values into the gas law to solve for the unknown. From the table, we see that the temperature has increased. Because pressure and temperature are directly related, the pressure must increase. When we substitute in the values, we see the ratio of the temperatures (temperature factor) is greater than 1, which increases pressure.

$$P_2 = 4.0 \text{ atm} \times \frac{675 \text{ K}}{298 \text{ K}} = 9.1 \text{ atm}$$

Temperature factor increases volume

Because the calculated pressure exceeds 8.0 atm, we expect the can to explode.

STUDY CHECK

In a storage area where the temperature has reached 55°C, the pressure of oxygen gas in a 15.0-L steel cylinder is 965 torr. To what temperature would the gas have to be cooled to reduce the pressure to 850. torr?

CAREER FOCUS

■ Nurse Anesthetist

"During surgery, I work with the surgeon to provide a safe level of anesthetics that renders the patient free from pain," says Mark Noguchi, nurse anesthetist (CRNA), Kaiser Hospital. "We do spinal and epidural blocks as well as general anesthetics, which means the patient is totally asleep. We use a variety of pharmaceutical agents including halothane ($C_2HBrClF_3$), and bupivacain ($C_{18}H_{28}N_2O$), as well as muscle relaxants such as midazolam ($C_{18}H_{13}ClFN_3$) to achieve the results we want for the surgical situation. We also assess the patient's overall hemodynamic status. If blood is lost, we replace components such as plasma, platelets, and coagulation factors. We also monitor the heart rate and run EKGs to determine cardiac function."

QUESTIONS AND PROBLEMS

■ Temperature and Pressure (Gay-Lussac's Law)

6.29 Why do aerosol cans explode if heated?

6.30 How can the tires on a car have a blowout when the car is driven on hot pavement in the desert?

6.31 In a fire at an oxygen-tank factory, some of the tanks filled with oxygen explode. Explain.

6.32 In Gay-Lussac's law, what properties of gases are considered as directly related and which properties are held constant?

6.33 Solve for the new pressure when each of the following temperature changes occurs, with n and V constant.
   a. A gas sample has a pressure of 1200 torr at 155°C. What is the final pressure of the gas after the temperature has dropped to 0°C?
   b. An aerosol can has a pressure of 1.40 atm at 12°C. What is the final pressure in the aerosol can if it is used in a room where the temperature is 35°C?

6.34 Solve for the new temperature in degrees Celsius when pressure is changed.
   a. A 10.0-L container of helium gas with a pressure of 250 torr at 0°C is heated to give a pressure of 1500 torr.
   b. A sample of air at 40.°C and 740 mm Hg is cooled to give a pressure of 680 mm Hg.

LEARNING GOAL

Use the combined gas law to find the new pressure, volume, or temperature of a gas when changes in two of these properties are given.

■ 6.6 The Combined Gas Law

All pressure–volume–temperature relationships for gases that we have studied may be combined into a single relationship called the **combined gas law**. This expression is useful for studying the effect of changes in two of these variables on the third as long as the amount of gas (number of moles) remains constant.

Combined Gas Law

$$\frac{P_1 V_1}{T_1} = \frac{P_2 V_2}{T_2} \qquad \text{No change in moles of gas}$$

SAMPLE PROBLEM 6.8

■ Using the Combined Gas Law

A 25.0-mL bubble is released from a diver's air tank at a pressure of 4.00 atm and a temperature of 11°C. What is the volume of the bubble when it reaches the ocean surface, where the pressure is 1.00 atm and the temperature is 18°C?

SOLUTION

STEP 1 Organize the data in a table. We must first change the temperature to kelvins.

$$T_1 = 11°C + 273 = 284 \text{ K}$$
$$T_2 = 18°C + 273 = 291 \text{ K}$$

Conditions 1	Conditions 2
$P_1 = 4.00$ atm	$P_2 = 1.00$ atm
$V_1 = 25.0$ mL	$V_2 = ?$
$T_1 = 284$ K	$T_2 = 291$ K

STEP 2 **Rearrange the gas law for the unknown.**   Because the pressure and
temperature are both changing, we must use the combined gas law to
solve for V_2.

$$\frac{P_1 V_1}{T_1} = \frac{P_2 V_2}{T_2}$$

$$\frac{P_1 V_1}{T_1} \times \frac{T_2}{P_2} = \frac{P_2 V_2 \times T_2}{T_2 \times P_2}$$

$$V_2 = V_1 \times \frac{P_1}{P_2} \times \frac{T_2}{T_1}$$

STEP 3 **Substitute the values into the gas law to solve for the unknown.**
From the data table, we determine that the pressure decrease and the
temperature increase will both increase the volume.

$$V_2 = 25.0 \text{ mL} \times \frac{4.00 \text{ atm}}{1.00 \text{ atm}} \times \frac{291 \text{ K}}{284 \text{ K}} = 102 \text{ mL}$$

Pressure   Temperature
factor   factor
increases   increases
volume   volume

STUDY CHECK

A weather balloon is filled with 15.0 L of helium at a temperature of 25°C and
a pressure of 685 mm Hg. What is the pressure of the helium in the balloon in
the upper atmosphere when the temperature is −35°C and the volume
becomes 34.0 L?

By remembering the combined gas law, we can derive any of the gas laws by
omitting those properties that do not change.

QUESTIONS AND PROBLEMS

■ The Combined Gas Law

6.35 A sample of helium gas has a volume of 6.50 L at a pressure of 845 mm Hg and a temperature of 25°C. What is the pressure of the gas in atm when the volume and temperature of the gas sample are changed to the following?
    a. 1850 mL and 325 K   **b.** 2.25 L and 12°C   **c.** 12.8 L and 47°C

6.36 A sample of argon gas has a volume of 735 mL at a pressure of 1.20 atm and a temperature of 112°C. What is the volume of the gas in milliliters when the pressure and temperature of the gas sample are changed to the following?
    a. 658 mm Hg and 281 K   **b.** 0.55 atm and 75°C   **c.** 15.4 atm and −15°C

6.37 A 100.0-mL bubble of hot gases at 225°C and 1.80 atm escapes from an active volcano. What is the new volume of the bubble outside the volcano where the temperature
is −25°C and the pressure is 0.80 atm?

6.38 A scuba diver 40 ft below the ocean surface inhales 50.0 mL of compressed air
in a scuba tank at a pressure of 3.00 atm and a temperature of 8°C. What is the
pressure of air in the lungs if the gas expands to 150.0 mL at a body temperature
of 37°C?

Describe the relationship between the amount of a gas and its volume and use this relationship in calculations.

$n = 1$ mole
$V = 1$ L

$n = 2$ moles
$V = 2$ L

FIGURE 6.7

Avogadro's law: The volume of a gas is directly related to the number of moles of the gas. If the number of moles is doubled, the volume must double at constant temperature and pressure.

Q **If a balloon has a leak, what happens to its volume?**

6.7 Volume and Moles (Avogadro's Law)

In our study of the gas laws, we have looked at changes in properties for a specified amount (n) of gas. Now we will consider how the properties of a gas change when there is a change in number of moles or grams.

Avogadro's Law

When you blow up a balloon, its volume increases because you add more air molecules. If a basketball gets a hole in it, and some of the air leaks out, its volume decreases. **Avogadro's law** states that the volume of a gas is directly related to the number of moles of a gas when temperature and pressure are not changed. If the number of moles of a gas are doubled, then the volume will double as long as we do not change the pressure or the temperature. (See Figure 6.7.) For two conditions, we can write Avogadro's law as follows.

Avogadro's Law

$$\frac{V_1}{n_1} = \frac{V_2}{n_2}$$ No change in pressure or temperature

SAMPLE PROBLEM 6.9

Calculating Volume for a Change in Moles

A balloon with a volume of 220 mL is filled with 2.0 moles helium. To what volume will the balloon expand if 3.0 moles helium are added, to give a total of 5.0 moles helium (the pressure and temperature do not change)?

SOLUTION

STEP 1 **Organize the data in a table.**   A data table for our given information can be set up as follows:

Conditions 1	Conditions 2
$V_1 = 220$ mL	$V_2 = ?$
$n_1 = 2.0$ moles	$n_2 = 5.0$ moles

STEP 2 **Rearrange the gas law for the unknown.**   Using Avogadro's law, we can solve for V_2.

$$\frac{V_1}{n_1} = \frac{V_2}{n_2}$$

$$n_2 \times \frac{V_1}{n_1} = \frac{V_2}{n_2} \times n_2$$

$$V_2 = V_1 \times \frac{n_2}{n_1}$$

STEP 3 **Substitute the values into the gas law to solve for the unknown.** From the table, we see that the number of moles has increased. Because the number of moles and volume are directly related, the volume must increase at constant pressure and temperature. When we substitute in the values, we see the ratio of the moles (mole factor) is greater than 1, which increases volume.

$$V_2 = 220 \text{ mL} \times \frac{5.0 \text{ moles}}{2.0 \text{ moles}} = 550 \text{ mL}$$

New volume   Initial volume   Mole factor that increases volume

STUDY CHECK

At a certain temperature and pressure, 8.00 g oxygen gas has a volume of 5.00 L. What is the volume after 4.00 g oxygen gas is added to the balloon?

STP and Molar Volume

Using Avogadro's law, we can say that any two gases will have equal volumes if they contain the same number of moles of gas at the same temperature and pressure. To help us make comparisons between different gases, arbitrary conditions called standard temperature (273 K) and pressure (1 atm) (abbreviated **STP**) were selected by scientists:

STP Conditions

Standard temperature is 0°C (273 K)
Standard pressure is 1 atm (760 mm Hg)

At STP, it was observed that 1 mole of any gas has a volume of 22.4 L. (See Figure 6.8.) When a gas is at STP, it has a volume of 22.4 L, which is called its **molar volume**.

Molar Volume

The volume of 1 mole of gas at STP = 22.4 L

As long as a gas is at STP conditions (0°C and 1 atm), its molar volume can be used as a conversion factor to convert between the number of moles of gas and its volume.

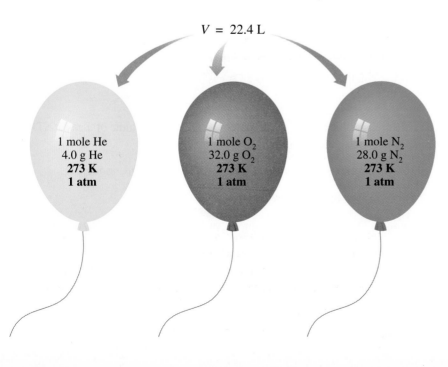

$V = 22.4 \text{ L}$

1 mole He	1 mole O₂	1 mole N₂
4.0 g He	32.0 g O₂	28.0 g N₂
273 K	**273 K**	**273 K**
1 atm	**1 atm**	**1 atm**

FIGURE 6.8

Avogadro's law indicates that one mole of any gas at STP has a molar volume of 22.4L.

Q **What volume of gas is occupied by 16.0 g methane gas, CH₄, at STP?**

Molar Volume Conversion Factors

$$\frac{1 \text{ mole gas (STP)}}{22.4 \text{ L}} \quad \text{and} \quad \frac{22.4 \text{ L}}{1 \text{ mole gas (STP)}}$$

Moles of gas	Molar volume 22.4 L/mole	Volume (L) of gas

SAMPLE PROBLEM 6.10

▪ Using Molar Volume

How many moles of helium are present in 5.25 L He at STP?

SOLUTION

The molar volume of a gas at STP can be used to calculate moles of He.

STEP 1 Given 5.25 L He at STP **Need** moles of He

STEP 2 Write a plan. liters of He | Molar volume | moles of He

STEP 3 Write conversion factors.

$$1 \text{ mole He (STP)} = 22.4 \text{ L}$$
$$\frac{22.4 \text{ L He}}{1 \text{ mole He}} \quad \text{and} \quad \frac{1 \text{ mole He}}{22.4 \text{ L He}}$$

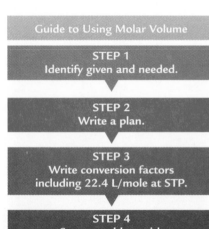

Guide to Using Molar Volume

STEP 1
Identify given and needed.

STEP 2
Write a plan.

STEP 3
Write conversion factors
including 22.4 L/mole at STP.

STEP 4
Set up problem with
factors to cancel units.

STEP 4 Set up problem with factors to cancel units.

$$5.25 \text{ L He} \times \frac{1 \text{ mole He}}{22.4 \text{ L He}} = 0.234 \text{ mole He (at STP)}$$

STUDY CHECK

How many moles of nitrogen (N_2) are present in 5.6 L of the gas at STP?

SAMPLE PROBLEM 6.11

▪ Using Molar Volume to Find Volume at STP

What is the volume in liters of 64.0 g O_2 gas at STP?

SOLUTION

Once we convert the mass of O_2 to moles O_2, the molar volume of a gas at STP can be used to calculate the volume (L) of O_2.

STEP 1 Given 64.0 g $O_2(g)$ at STP **Need** volume in liters (L)

STEP 2 Write a plan for solution.

grams of O_2 | Molar mass | moles of O_2 | Molar volume | = liters of O_2

STEP 3 Write conversion factors.

$$1 \text{ mole } O_2 = 32.0 \text{ g}$$
$$\frac{32.0 \text{ g } O_2}{1 \text{ mole } O_2} \quad \text{and} \quad \frac{1 \text{ mole } O_2}{32.0 \text{ g } O_2}$$

$$1 \text{ mole } O_2 \text{ (STP)} = 22.4 \text{ L}$$
$$\frac{22.4 \text{ L } O_2}{1 \text{ mole } O_2} \quad \text{and} \quad \frac{1 \text{ mole } O_2}{22.4 \text{ L } O_2}$$

STEP 4 Set up problem with factors to cancel units.

$$64.0 \text{ g } \cancel{O_2} \times \frac{1 \text{ mole } \cancel{O_2}}{32.0 \text{ g } \cancel{O_2}} \times \frac{22.4 \text{ L } O_2}{1 \text{ mole } \cancel{O_2}} = 44.8 \text{ L } O_2 \text{ (STP)}$$

STUDY CHECK

How many grams of $N_2(g)$ are in 5.6 L $N_2(g)$ at STP?

Gases in Reactions at STP

We can use the molar volume at STP to determine the moles of a gas in a reaction. Once we know the moles of gas in a reaction, we can use a mole factor to determine the moles of any other substance as we have done before.

SAMPLE PROBLEM 6.12

Gases in Chemical Reactions at STP

When potassium metal reacts with chlorine gas, the product is solid potassium chloride.

$$2K(s) + Cl_2(g) \longrightarrow 2KCl(s)$$

How many grams of potassium chloride are produced when 7.25 L chlorine gas at STP reacts with potassium?

SOLUTION

Guide to Using Molar Volume
for Reactions

STEP 1
Find moles of gas A using molar
volume or ideal gas law.

STEP 2
Determine moles of substance B
using mole–mole factor.

STEP 3
Convert moles of substance B
to grams or volume.

STEP 1 Find moles of gas A using molar volume.   At STP, we can use molar volume (22.4 L/mole) to determine moles of Cl_2 gas.

$$7.25 \text{ L } \cancel{Cl_2} \times \frac{1 \text{ mole } Cl_2}{22.4 \text{ L } \cancel{Cl_2}} = 0.324 \text{ mole } Cl_2$$

STEP 2 Determine moles of substance B using mole–mole factor from the balanced equation.

$$1 \text{ mole } Cl_2 = 2 \text{ moles KCl}$$

$$\frac{2 \text{ moles KCl}}{1 \text{ mole } Cl_2} \quad \text{and} \quad \frac{1 \text{ mole } Cl_2}{2 \text{ moles KCl}}$$

$$0.324 \text{ mole } \cancel{Cl_2} \times \frac{2 \text{ moles KCl}}{1 \text{ mole } \cancel{Cl_2}} = 0.648 \text{ mole KCl}$$

STEP 3 Convert moles of substance B to grams or volume.   Using the molar mass of KCl, we can determine the grams of KCl.

$$1 \text{ mole KCl} = 74.6 \text{ g KCl}$$

$$\frac{1 \text{ mole KCl}}{74.6 \text{ g KCl}} \quad \text{and} \quad \frac{74.6 \text{ g KCl}}{1 \text{ mole KCl}}$$

$$0.648 \text{ mole } \cancel{KCl} \times \frac{74.6 \text{ g KCl}}{1 \text{ mole } \cancel{KCl}} = 48.3 \text{ g KCl}$$

These steps can also be set up as a continuous solution.

$$7.25 \; \text{L Cl}_2 \times \frac{1 \; \text{mole Cl}_2}{22.4 \; \text{L Cl}_2} \times \frac{2 \; \text{moles KCL}}{1 \; \text{mole Cl}_2} \times \frac{74.6 \; \text{g KCl}}{1 \; \text{mole KCl}} = 48.3 \; \text{g KCl}$$

STUDY CHECK

Hydrogen gas forms when zinc metal reacts with aqueous HCl.

$$Zn(s) + 2HCl(aq) \longrightarrow ZnCl_2(aq) + H_2(g)$$

How many liters of hydrogen gas at STP are produced when 15.8 g zinc reacts?

QUESTIONS AND PROBLEMS

■ Volume and Moles (Avogadro's Law)

6.39 What happens to the volume of a bicycle tire or a basketball when you use an air pump to add air?

6.40 Sometimes when you blow up a balloon and release it, it flies around the room. What is happening to the air that was in the balloon and its volume?

6.41 A sample containing 1.50 moles neon gas has a volume of 8.00 L. What is the new volume of the gas in liters when the following changes occur in the quantity of the gas at constant pressure and temperature?
   a. A leak allows one-half of the neon atoms to escape.
   b. A sample of 25.0 g neon is added to the neon gas already in the container.
   c. A sample of 3.50 mole O_2 is added to the neon gas already in the container.

6.42 A sample containing 4.80 g O_2 gas has a volume of 15.0 L. Pressure and temperature remain constant.
   a. What is the new volume if 0.500 mole O_2 gas is added?
   b. Oxygen is released until the volume is 10.0 L. How many moles of O_2 are removed?
   c. What is the volume after 4.00 g He is added to the O_2 gas already in the container?

6.43 Use the molar volume of a gas to solve the following at STP:
   a. the number of moles of O_2 in 44.8 L O_2 gas
   b. the number of moles of CO_2 in 4.00 L CO_2 gas
   c. the volume (L) of 6.40 g O_2
   d. the volume (mL) occupied by 50.0 g neon

6.44 Use molar volume to solve the following problems at STP:
   a. the volume (L) occupied by 2.50 moles N_2
   b. the volume (mL) occupied by 0.420 mole He
   c. the number of grams of neon contained in 11.2 L Ne gas
   d. the number of moles of H_2 in 1620 mL H_2 gas

6.45 Mg metal reacts with HCl to produce hydrogen gas:

$$Mg(s) + 2HCl(aq) \longrightarrow MgCl_2(aq) + H_2(g)$$

What volume of hydrogen at STP is released when 8.25 g Mg reacts?

6.46 Aluminum oxide is formed from its elements.

$$4Al(s) + 3O_2(g) \longrightarrow 2Al_2O_3(s)$$

How many grams of Al will react with 12.0 L O_2 at STP?

6.8 Partial Pressures (Dalton's Law)

LEARNING GOAL

Use partial pressures to calculate the total pressure of a mixture of gases.

Many gas samples are a mixture of gases. For example, the air you breathe is a mixture of mostly oxygen and nitrogen gases. In gas mixtures, scientists observed that all gas particles behave in the same way. Therefore, the total pressure of the gases in a mixture is a result of the collisions of the gas particles regardless of what type of gas they are.

Dalton's Law

In a gas mixture, each gas exerts its **partial pressure**, which is the pressure it would exert if it were the only gas in the container. **Dalton's law** states that the total pressure of a gas mixture is the sum of the partial pressures of the gases in the mixture.

Dalton's Law

$$P_{total} = P_1 + P_2 + P_3 + \cdots$$

Total pressure   = Sum of the partial pressures
of a gas mixture   of the gases in the mixture

Suppose we have two separate tanks, one filled with helium at 2.0 atm and the other filled with argon at 4.0 atm. When the gases are combined in a single tank with the same volume and temperature, the number of gas molecules, not the type of gas, determines the pressure in a container. There the pressure of the gas mixture would be 6.0 atm, which is the sum of their individual or partial pressures.

$$
\begin{aligned}
P_{total} &= P_{He} + P_{Ar} \\
&= 2.0\ \text{atm} + 4.0\ \text{atm} \\
&= 6.0\ \text{atm}
\end{aligned}
$$

$P_{He} = 2.0\ \text{atm}$ $P_{Ar} = 4.0\ \text{atm}$

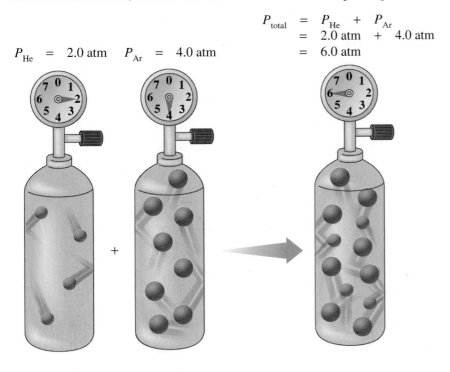

+

SAMPLE PROBLEM 6.13

Calculating the Total Pressure of a Gas Mixture

A 10-L gas tank contains propane (C_3H_8) gas at a pressure of 300. torr. Another 10-L gas tank contains methane (CH_4) gas at a pressure of 500. torr. In preparing a gas fuel mixture, the gases from both tanks are combined in a 10-L container at the same temperature. What is the pressure of the gas mixture?

SOLUTION

Using Dalton's law of partial pressure, we find that the total pressure of the gas mixture is the sum of the partial pressures of the gases in the mixture.

$$P_{total} = P_{propane} + P_{methane}$$
$$= 300.\ torr + 500.\ torr$$
$$= 800.\ torr$$

Therefore, when both propane and methane are placed in the same container, the total pressure of the mixture is 800. torr.

STUDY CHECK

A gas mixture consists of helium with a partial pressure of 315 mm Hg, nitrogen with a partial pressure of 204 mm Hg, and argon with a partial pressure of 422 mm Hg. What is the total pressure in atmospheres?

Air Is a Gas Mixture

The air you breathe is a mixture of gases. What we call the atmospheric pressure is actually the sum of the partial pressures of the gases in the air. Table 6.4 lists partial pressures for the gases in air on a typical day.

TABLE 6.4

Typical Composition of Air

Gas	Partial Pressure (mm Hg)	Percentage (%)
Nitrogen, N_2	594.0	78
Oxygen, O_2	160.0	21
Carbon dioxide, CO_2	0.3	1
Water vapor, H_2O	5.7	
Total air	760.0	100

SAMPLE PROBLEM 6.14

■ Partial Pressure of a Gas in a Mixture

A mixture of oxygen and helium is prepared for a scuba diver who is going to descend 200 ft below the ocean surface. At that depth, the diver breathes a gas mixture that has a total pressure of 7.0 atm. If the partial pressure of the oxygen in the tank at that depth is 1.5 atm, what is the partial pressure of the helium?

SOLUTION

From Dalton's law of partial pressures, we know that the total pressure is equal to the sum of the partial pressures:

$$P_{total} = P_{O_2} + P_{He}$$

To solve for the partial pressure of helium (P_{He}), we rearrange the expression to give the following:

$$P_{He} = P_{total} - P_{O_2}$$
$$P_{He} = 7.0\ atm - 1.5\ atm$$
$$= 5.5\ atm$$

Thus, in the gas mixture that the diver breathes, the partial pressure of the helium is 5.5 atm.

STUDY CHECK

An anesthetic consists of a mixture of cyclopropane gas, C_3H_6, and oxygen gas, O_2. If the mixture has a total pressure of 825 torr, and the partial pressure of the cyclopropane is 73 torr, what is the partial pressure of the oxygen in the anesthetic?

■ **Blood Gases**

Our cells continuously use oxygen and produce carbon dioxide. Both gases move in and out of the lungs through the membranes of the alveoli, the tiny air sacs at the ends of the airways in the lungs. An exchange of gases occurs in which oxygen from the air diffuses into the lungs and into the blood, while carbon dioxide produced in the cells is carried to the lungs to be exhaled. In Table 6.5, partial pressures are given for the gases in air that we inhale (inspired air), air in the alveoli, and the air that we exhale (expired air).

At sea level, oxygen normally has a partial pressure of 100 mm Hg in the alveoli of the lungs. Because the partial pressure of oxygen in venous blood is 40 mm Hg, oxygen diffuses

from the alveoli into the bloodstream. The oxygen combines with hemoglobin, which carries it to the tissues of the body where the partial pressure of oxygen can be very low, less than 30 mm Hg. Oxygen diffuses from the blood where the partial pressure of O_2 is high into the tissues where O_2 pressure is low.

As oxygen is used in the cells of the body during metabolic processes, carbon dioxide is produced, so the partial pressure of CO_2 may be as high as 50 mm Hg or more. Carbon dioxide diffuses from the tissues into the bloodstream and is carried to the lungs. There it diffuses out of the blood, where CO_2 has a partial pressure of 46 mm Hg, into the alveoli, where the CO_2 is at 40 mm Hg and is exhaled. Table 6.6 gives the partial pressures of blood gases in the tissues and in oxygenated and deoxygenated blood.

TABLE 6.5

Partial Pressures of Gases During Breathing

Gas	Partial Pressure (mm Hg)		
	Inspired Air	Alveolar Air	Expired Air
Nitrogen, N_2	594.0	573	569
Oxygen, O_2	160.0	100	116
Carbon dioxide, CO_2	0.3	40	28
Water vapor, H_2O	5.7	47	47
Total	760.0	760	760

TABLE 6.6

Partial Pressures of Oxygen and Carbon Dioxide in Blood and Tissues

Gas	Partial Pressure (mm Hg)		
	Oxygenated Blood	Deoxygenated Blood	Tissues
O_2	100	40	30 or less
CO_2	40	46	50 or greater

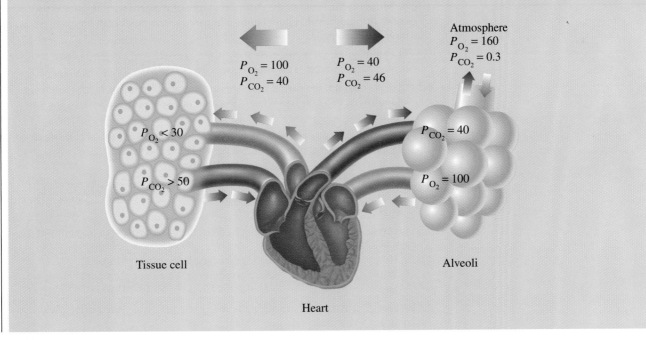

$P_{O_2} = 100$
$P_{CO_2} = 40$

$P_{O_2} = 40$
$P_{CO_2} = 46$

Atmosphere
$P_{O_2} = 160$
$P_{CO_2} = 0.3$

$P_{O_2} < 30$

$P_{CO_2} > 50$

$P_{CO_2} = 40$

$P_{O_2} = 100$

Tissue cell

Alveoli

Heart

QUESTIONS AND PROBLEMS

■ **Partial Pressures (Dalton's Law)**

6.47 A typical air sample in the lungs contains oxygen at 100 mm Hg, nitrogen at 573 mm Hg, carbon dioxide at 40 mm Hg, and water vapor at 47 mm Hg. Why are these pressures called partial pressures?

■ Hyperbaric Chambers

A burn patient may undergo treatment for burns and infections in a hyperbaric chamber, a device in which pressures can be obtained that are two to three times greater than atmospheric pressure. A greater oxygen pressure increases the level of dissolved oxygen in the blood and tissues, where it fights bacterial infections. High levels of oxygen are toxic to many strains of bacteria. The hyperbaric chamber may also be used during surgery to help counteract carbon monoxide (CO) poisoning, and to treat some cancers.

The blood is normally capable of dissolving up to 95% of the oxygen. Thus, if the partial pressure of the oxygen is 2280 mm Hg (3 atm), 95% of that or 2160 mm Hg of oxygen can dissolve in the blood where it saturates the tissues. In the case of carbon monoxide poisoning, this oxygen can replace the carbon monoxide that has attached to the hemoglobin.

A patient undergoing treatment in a hyperbaric chamber must also undergo decompression (reduction of pressure) at a rate that slowly reduces the concentration of dissolved oxygen in the blood. If decompression is too rapid, the oxygen dissolved in the blood may form gas bubbles in the circulatory system.

If divers do not decompress slowly, they suffer a similar condition called the bends. While below the surface of the ocean, divers breathe air at higher pressures. At such high pressures, nitrogen gas will dissolve in their blood. If they ascend to the surface too quickly, the dissolved nitrogen forms bubbles in the blood that can produce life-threatening blood clots. The gas bubbles can also appear in the joints and tissues of the body and be quite painful. A diver suffering from the bends is placed immediately in a decompression chamber where pressure is first increased and then slowly decreased. The dissolved nitrogen can then diffuse through the lungs until atmospheric pressure is reached.

6.48 Suppose a mixture contains helium and oxygen gases. If the partial pressure of helium is the same as the partial pressure of oxygen, what do you know about the number of helium atoms compared to the number of oxygen molecules? Explain.

6.49 In a gas mixture, the partial pressures are nitrogen 425 torr, oxygen 115 torr, and helium 225 torr. What is the total pressure (torr) exerted by the gas mixture?

6.50 In a gas mixture, the partial pressures are argon 415 mm Hg, neon 75 mm Hg, and nitrogen 125 mm Hg. What is the total pressure (atm) exerted by the gas mixture?

6.51 A gas mixture containing oxygen, nitrogen, and helium exerts a total pressure of 925 torr. If the partial pressures are oxygen 425 torr and helium 75 torr, what is the partial pressure (torr) of the nitrogen in the mixture?

6.52 A gas mixture containing oxygen, nitrogen, and neon exerts a total pressure of 1.20 atm. If helium added to the mixture increases the pressure to 1.50 atm, what is the partial pressure (atm) of the helium?

6.53 In certain lung ailments such as emphysema, there is a decrease in the ability of oxygen to diffuse into the blood.
a. How would the partial pressure of oxygen in the blood change?
b. Why does a person with severe emphysema sometimes use a portable oxygen tank?

6.54 An accident to the head can affect the ability of a person to ventilate (breathe in and out), and so can certain drugs.
a. What would happen to the partial pressures of oxygen and carbon dioxide in the blood if a person cannot properly ventilate?
b. When a person with hypoventilation is placed on a ventilator, an air mixture is delivered at pressures that are alternately above the air pressure in the person's lung, and then below. How will this move oxygen gas into the lungs, and carbon dioxide out?

CONCEPT MAP

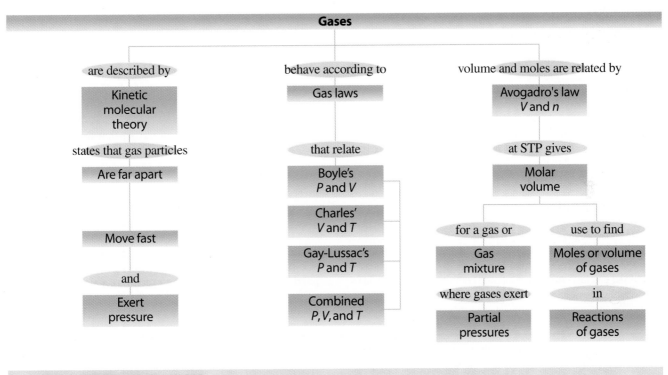

CHAPTER REVIEW

6.1 Properties of Gases
In a gas, particles are so far apart and moving so fast that their attractions are unimportant. A gas is described by the physical properties of pressure (P), volume (V), temperature (T), and amount in moles (n).

6.2 Gas Pressure
A gas exerts pressure, the force of the gas particles striking the surface of a container. Gas pressure is measured in units of torr, mm Hg, and atm.

6.3 Pressure and Volume (Boyle's Law)
The volume (V) of a gas changes inversely with the pressure (P) of the gas if there is no change in the amount and temperature: $P_1V_1 = P_2V_2$. This means that the pressure increases if volume decreases; pressure decreases if volume increases.

6.4 Temperature and Volume (Charles' Law)
The volume (V) of a gas is directly related to its Kelvin temperature (T) when there is no change in the amount and pressure of the gas:

$$\frac{V_1}{T_1} = \frac{V_2}{T_2}$$

Therefore, if temperature increases, the volume of the gas increases; if temperature decreases, volume decreases.

6.5 Temperature and Pressure (Gay-Lussac's Law)
The pressure (P) of a gas is directly related to its Kelvin temperature (T).

$$\frac{P_1}{T_1} = \frac{P_2}{T_2}$$

This means that an increase in temperature increases the pressure of a gas, or a decrease in temperature decreases the pressure, as long as the amount and volume stay constant.

6.6 The Combined Gas Law
Gas laws combine into a relationship of pressure (P), volume (V), and temperature (T).

$$\frac{P_1 V_1}{T_1} = \frac{P_2 V_2}{T_2}$$

This expression is used to determine the effect of changes in two of the variables on the third.

6.7 Volume and Moles (Avogadro's Law)

The volume (V) of a gas is directly related to the number of moles (n) of the gas when the pressure and temperature of the gas do not change.

$$\frac{V_1}{n_1} = \frac{V_2}{n_2}$$

If the moles of gas are increased, the volume must increase; or if the moles of gas are decreased, the volume decreases. At standard temperature (273 K) and pressure (1 atm) abbreviated STP, 1 mole of any gas has a volume of 22.4 L.

6.8 Partial Pressures (Dalton's Law)

In a mixture of two or more gases, the total pressure is the sum of the partial pressures of the individual gases.

$$P_{total} = P_1 + P_2 + P_3 + \cdots$$

The partial pressure of a gas in a mixture is the pressure it would exert if it were the only gas in the container.

KEY TERMS

atmosphere (atm) The pressure exerted by a column of mercury 760 mm high.

atmospheric pressure The pressure exerted by the atmosphere.

Avogadro's law A gas law that states that the volume of gas is directly related to the number of moles of gas in the sample when pressure and temperature do not change.

Boyle's law A gas law stating that the pressure of a gas is inversely related to the volume when temperature and moles of the gas do not change.

Charles' law A gas law stating that the volume of a gas changes directly with a change in Kelvin temperature when pressure and moles of the gas do not change.

combined gas law A relationship that combines several gas laws relating pressure, volume, and temperature.

$$\frac{P_1 V_1}{T_1} = \frac{P_2 V_2}{T_2}$$

Dalton's law A gas law stating that the total pressure exerted by a mixture of gases in a container is the sum of the partial pressures that each gas would exert alone.

direct relationship A relationship in which two properties increase or decrease together.

Gay-Lussac's law A gas law stating that the pressure of a gas changes directly with a change in temperature when the number of moles of a gas and its volume are held constant.

inverse relationship A relationship in which two properties change in opposite directions.

kinetic molecular theory of gases A model used to explain the behavior of gases.

molar volume A volume of 22.4 L occupied by 1 mole of a gas at STP conditions of 0°C (273 K) and 1 atm.

partial pressure The pressure exerted by a single gas in a gas mixture.

pressure The force exerted by gas particles that hit the walls of a container.

STP Standard conditions of 0°C (273 K) temperature and 1 atm pressure used for the comparison of gases.

ADDITIONAL QUESTIONS AND PROBLEMS

6.55 At a restaurant, a customer chokes on a piece of food. You put your arms around the person's waist and use your fists to push up on the person's abdomen, an action called the Heimlich maneuver.
a. How would this action change the volume of the chest and lungs?
b. Why does it cause the person to expel the food item from the airway?

6.56 An airplane is pressurized to 650 mm Hg, which is the atmospheric pressure at a ski resort at 13,000 ft altitude.
a. If air is 21% oxygen, what is the partial pressure of oxygen on the plane?
b. If the partial pressure of oxygen drops below 100 mm Hg, passengers become drowsy. If this happens, oxygen masks are released. What is the total cabin pressure at which oxygen masks are dropped?

6.57 A fire extinguisher has a pressure of 10. atm at 25°C. What is the pressure in atmospheres if the fire extinguisher is used at a temperature of 75°C?

6.58 A weather balloon has a volume of 750 L when filled with helium at 8°C at a pressure of 380 torr. What is the new volume of the balloon, where the pressure is 0.20 atm and the temperature is −45°C?

6.59 A sample of hydrogen (H_2) gas at 127°C has a pressure of 2.00 atm. At what temperature (°C) will the pressure of the H_2 decrease to 0.25 atm?

6.60 A sample of nitrogen (N_2) and helium has a volume of 250 mL at 30°C and a total pressure of 745 mm Hg.
a. If the pressure of helium is 32 mm Hg, what is the partial pressure of the nitrogen?
b. What is the volume of the nitrogen at STP?

6.61 A gas mixture with a total pressure of 2400 torr is used by a scuba diver. If the mixture contains 2.0 moles of helium and 6.0 moles of oxygen, what is the partial pressure of each gas in the sample?

6.62 What is the total pressure in mm Hg of a gas mixture containing argon gas at 0.25 atm, helium gas at 350 mm Hg, and nitrogen gas at 360 torr?

6.63 A gas mixture contains oxygen and argon at partial pressures of 0.60 atm and 425 mm Hg. If nitrogen gas added to the sample increases the total pressure to 1250 torr, what is the partial pressure in torr of the nitrogen added?

6.64 A gas mixture contains helium and oxygen at partial pressures of 255 torr and 0.450 atm. What is the total pressure in mm Hg of the mixture after it is placed in a container one-half the volume of the original container?

6.65 What is the density (g/L) of oxygen gas at STP?

6.66 At a party, a helium-filled balloon floats close to some hot lights and bursts. What might be a reason for this?

6.67 Compare the partial pressures of the following respiratory gases:
   a. oxygen in the lungs and in the blood coming to the alveoli
   b. oxygen in arterial blood and venous blood
   c. carbon dioxide in the tissues and arterial blood
   d. carbon dioxide in the venous blood and the lungs

6.68 For each of the comparisons in problem 6.67, describe the direction of diffusion for each gas.

6.69 When heated, calcium carbonate decomposes to give calcium oxide and carbon dioxide gas.

$$CaCO_3(s) \longrightarrow CaO(s) + CO_2(g)$$

If 2.00 moles of $CaCO_3$ react, how many liters of CO_2 gas are produced at STP?

6.70 Magnesium reacts with oxygen to form magnesium oxide. How many liters of oxygen gas at STP are needed to react completely with 8.0 g of magnesium?

$$2Mg(s) + O_2(g) \longrightarrow 2MgO(s)$$

6.71 Your space ship has docked at a space station above Mars. The temperature inside the space station is a carefully controlled 24°C at a pressure of 745 mm Hg. A balloon with a volume of 425 mL drifts into the airlock where the temperature is −95°C and the pressure is 0.115 atm. What is the new volume of the balloon? Assume that the balloon is very elastic.

6.72 How many liters of H_2 gas can be produced at STP from 25.0 g of Zn?

$$Zn(s) + 2HCl(aq) \longrightarrow ZnCl_2(aq) + H_2(g)$$

6.73 Aluminum oxide can be formed from its elements.

$$4Al(s) + 3O_2(g) \longrightarrow 2Al_2O_3(s)$$

What volume of oxygen is needed at STP to completely react 5.4 g of aluminum?

6.74 Glucose $C_6H_{12}O_6$ is metabolized in living systems according to the reaction

$$C_6H_{12}O_6(s) + 6O_2(g) \longrightarrow 6CO_2(g) + 6H_2O(l)$$

How many grams of water can be produced when 12.5 L O_2 reacts at STP?

UNDERSTANDING THE CONCEPTS

6.75 At 100°C, which of the following gases exerts
   a. the lowest pressure?
   b. the highest pressure?

1.   2.   3.

6.76 Indicate which diagram represents the volume of the gas sample in a flexible container when each of the following changes take place.

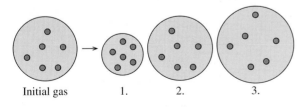

Initial gas   1.   2.   3.

   a. Temperature increases at constant pressure
   b. Temperature decreases at constant pressure
   c. Pressure increases at constant temperature

   d. Pressure decreases at constant temperature
   e. Doubling the pressure and doubling the Kelvin temperature

6.77 A balloon is filled with helium gas with a pressure of 1.00 atm and neon gas with a pressure of 0.50 atm. For each of the following changes of the initial balloon, select the diagram (A, B, or C) that shows the final (new) volume of the balloon.

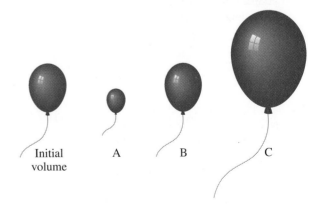

Initial volume   A   B   C

a. The balloon is put in a cold storage unit (*P* and *n* constant).

b. The balloon floats to a higher altitude where the pressure is less (*n, T* constant).

c. All of the neon gas is removed (*T* constant).

d. The Kelvin temperature doubles and 1/2 of the gas atoms leak out (*P* constant).

e. 2.0 mol O_2 gas is added at constant *T* and *P*.

6.78 Indicate if pressure increases, decreases, or stays the same in each of the following:

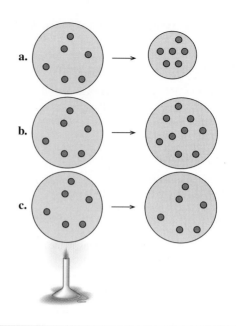

QUESTIONS FOR TEAMWORK

6.79 Two flasks of equal volume and at the same temperature contain different gases. One flask contains 10.0 g Ne and the other flask contains 10.0 g He. Which of the following statements are correct? If not, explain.

a. Both flasks contain the same number of atoms.

b. The pressures in the flasks are the same.

c. The flask that contains helium has a higher pressure than the flask that contains neon.

d. The densities of the gases are the same.

6.80 In the fermentation of glucose (wine making), a volume of 780 mL CO_2 gas was produced at 37°C and 1.00 atm. What is the volume (L) of the gas when measured at 22°C and 675 mm Hg?

6.81 A gas sample has a volume of 4250 mL at 15°C and 745 mm Hg. What is the new temperature (°C) after the sample is transferred to a new container with a volume of 2.50 L and a pressure of 1.20 atm?

6.82 A weather balloon has a volume of 750 L when filled with helium at 8°C at a pressure of 380 torr. What is the new volume of the balloon, where the pressure is 0.20 atm and the temperature is −45°C?

6.83 A liquid is placed in a 25.0-L flask. At 140°C, the liquid evaporates completely to give a pressure of 0.900 atm. If the

flask can withstand pressures up to 1.30 atm, calculate the maximum temperature that the gas can be heated to without breaking.

6.84 You are doing research on planet X, and you take your favorite balloon with you. The temperature inside the space station is a carefully controlled 24°C and the pressure is 755 mm Hg. Your balloon drifts into the airlock, which cycles, sending your balloon, which has a volume of 850 mL, into the atmosphere of planet X. If the night temperature on planet X is −103°C and the pressure is 0.150 atm, what volume will your balloon occupy assuming that it is very elastic and will not burst?

6.85 When sensors in a car detect a collision, they cause the reaction of sodium azide, NaN_3, which generates nitrogen gas to fill the air bags within 0.03 second.

$$2NaN_3(s) \longrightarrow 2Na(s) + 3N_2(g)$$

How many liters of N_2 are produced at STP if the air bag contains 132 g NaN_3?

6.86 Nitrogen dioxide reacts with water to produce oxygen and ammonia.

$$4NO_2(g) + 6H_2O(g) \longrightarrow 7O_2(g) + 4NH_3(g)$$

How many liters of O_2 at STP are produced when 2.5 × 10^{23} molecules NO_2 react?

ANSWERS

Answers to Study Checks

6.1 The gas molecules that carry the odor of the food move throughout the house until they reach the room that you are in.

6.2 The mass in grams gives the amount of gas.

6.3 0.862 atm

6.4 250 torr

6.5 50.0 mL

6.6 569 mL

6.7 16°C

6.8 241 mm Hg

6.9 7.50 L

6.10 0.25 mole N_2

6.11 7.0 g N_2

6.12 5.41 L H_2

6.13 1.24 atm

6.14 752 torr

Answers to Selected Questions and Problems

6.1 **a.** At a higher temperature, gas particles have greater kinetic energy, which makes them move faster.
b. Because there are great distances between the particles of a gas, they can be pushed closer together and still remain a gas.

6.3 **a.** temperature
b. volume
c. amount
d. pressure

6.5 atmospheres (atm), mm Hg, torr, lb/in.^{2}, kPa

6.7 **a.** 1520 torr
b. 1520 mm Hg

6.9 As a diver ascends to the surface, external pressure decreases. If the air in the lungs were not exhaled, its volume would expand and severely damage the lungs. The pressure in the lungs must adjust to changes in the external pressure.

6.11 **a.** The pressure is greater in cylinder A. According to Boyle's law, a decrease in volume pushes the gas particles closer together, which will cause an increase in the pressure.
b.

Property	Initial	Final
Pressure (P)	650 mm Hg	1.2 atm
Volume (V)	220 mL	160 mL

6.13 **a.** The pressure doubles.
b. The pressure falls to one-third the initial pressure.
c. The pressure increases to ten times the original pressure.

6.15 **a.** 328 mm Hg   **b.** 2620 mm Hg
c. 4400 mm Hg

6.17 **a.** 25 L   **b.** 25 L   **c.** 100. L

6.19 25 L

6.21 **a.** inspiration   **b.** expiration   **c.** inspiration

6.23 **a.** C   **b.** A   **c.** B

6.25 When a gas is heated at constant pressure, the volume of the gas increases to fill the hot-air balloon.

6.27 **a.** 2400 mL   **b.** 4900 mL
c. 1800 mL   **d.** 1700 mL

6.29 An increase in temperature increases the pressure inside the can. When the pressure exceeds the pressure limit of the can, it explodes.

6.31 When the temperature increases, the pressure of the oxygen inside the filled tanks increases causing the tanks to explode.

6.33 **a.** 770 torr   **b.** 1.51 atm

6.35 **a.** 4.25 atm   **b.** 3.07 atm   **c.** 0.605 atm

6.37 110 mL

6.39 The volume increases because the amount of gas particles is increased.

6.41 **a.** 4.00 L   **b.** 14.6 L   **c.** 26.7 L

6.43 **a.** 2.00 moles of O_2 **b.** 0.179 mole of CO_2
c. 4.48 L   **d.** 55 400 mL

6.45 7.60 L H_2

6.47 In a gas mixture, the pressure that each gas exerts as part of the total pressure is called the partial pressure of that gas. Because the air sample is a mixture of gases, the total pressure is the sum of the partial pressures of each gas in the sample.

6.49 765 torr

6.51 425 torr

6.53 **a.** The partial pressure of oxygen will be lower than normal.
b. Breathing a higher concentration of oxygen will help to increase the supply of oxygen in the lungs and blood and raise the partial pressure of oxygen in the blood.

6.55 **a.** The volume of the chest and lungs is decreased.
b. The decrease in volume increases the pressure, which can dislodge the food in the trachea.

6.57 12 atm

6.59 $-223°C$

6.61 He 600 torr, O_2 1800 torr

6.63 370 torr

6.65 1.43 g/L

6.67 **a.** The P_{O_2} is high in the lungs and low in the blood coming to the lungs.
b. The P_{O_2} is high in arterial blood and low in venous blood.
c. The P_{CO_2} is high in the tissues and low in arterial blood.
d. The P_{CO_2} is high in the venous blood and low in the lungs.

6.69 44.8 L CO_2

6.71 2170 mL

6.73 3.4 L of O_2 (g)

6.75 **a.** 2   **b.** 1

6.77 **a.** A   **b.** C   **c.** A
d. B   **e.** C

7 Solutions

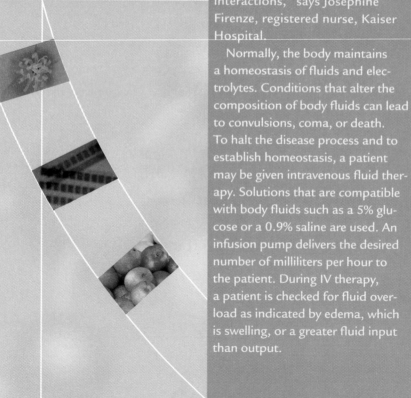

"There is a lot of chemistry going on in the body, including drug interactions," says Josephine Firenze, registered nurse, Kaiser Hospital.

Normally, the body maintains a homeostasis of fluids and electrolytes. Conditions that alter the composition of body fluids can lead to convulsions, coma, or death. To halt the disease process and to establish homeostasis, a patient may be given intravenous fluid therapy. Solutions that are compatible with body fluids such as a 5% glucose or a 0.9% saline are used. An infusion pump delivers the desired number of milliliters per hour to the patient. During IV therapy, a patient is checked for fluid overload as indicated by edema, which is swelling, or a greater fluid input than output.

the Chemistry place

Visit **www.aw-bc.com/chemplace** or use the CD-ROM in the book for extra quizzes, interactive tutorials, career resources, PowerPoint slides, math help, and case studies.

Solutions are everywhere around us. Most consist of one substance dissolved in another. The air we breathe is a solution of oxygen and nitrogen gases. Carbon dioxide gas dissolved in water makes carbonated drinks. When we make solutions of coffee or tea, we use hot water to dissolve substances from coffee beans or tea leaves. The ocean is also a solution, consisting of many salts such as sodium chloride dissolved in water. In a hospital, the antiseptic tincture of iodine is a solution of iodine dissolved in alcohol.

Our body fluids contain water and dissolved substances such as glucose and urea and electrolytes such as K^+, Na^+, Cl^-, Mg^{2+}, HCO_3^-, and HPO_4^{2-}. Proper amounts of each of these dissolved substances and water must be maintained in the body fluids. Small changes in electrolyte levels can seriously disrupt cellular process and endanger our health. Therefore, the measurement of their concentrations is a valuable diagnostic tool.

In the processes of osmosis and dialysis, water, essential nutrients, and waste products enter and leave the cells of the body. In osmosis, water flows in and out of the cells of the body. In dialysis, small particles in solution as well as water diffuse through semipermeable membranes. The kidneys utilize osmosis and dialysis to regulate the amount of water and electrolytes that are excreted.

■ 7.1 Solutions

A **solution** is a mixture in which one substance called the **solute** is uniformly dispersed in another substance called the **solvent**. Because the solute and the solvent do not react with each other, they can be mixed in varying proportions. A little salt dissolved in water tastes slightly salty. When more salt dissolves, the water tastes very salty. Usually, the solute (in this case, salt) is the substance present in the smaller amount, whereas the solvent (in this case, water) is present in the larger amount. In a solution, the particles of the solute are evenly dispersed among the molecules of the solvent. (See Figure 7.1.)

LEARNING GOAL

■ Identify the solute and solvent in a solution. Describe the formation of a solution.

SAMPLE PROBLEM 7.1

■ Identifying a Solute and a Solvent

Identify the solute and the solvent in each of the following solutions:

a. 15 g of sugar dissolved in 100 g of water
b. 75 mL of water mixed with 25 mL of isopropyl alcohol (rubbing alcohol)

SOLUTION

a. Sugar, the smaller quantity, is the solute; water is the solvent.
b. Isopropyl alcohol with the smaller volume is the solute. Water is the solvent.

STUDY CHECK

A tincture of iodine is prepared with 0.10 g of I_2 and 10.0 mL of ethyl alcohol. What is the solute, and what is the solvent?

Solute: The substance present in lesser amount

Salt

Water

Solvent: The substance present in greater amount

A solution of copper(II) sulfate ($CuSO_4$) forms as particles of solute dissolve, move away from the crystals, and become evenly dispersed among the solvent (water) molecules.

Q **What does the uniform blue color indicate about the $CuSO_4$ solution?**

H₂O

$CuSO_4$

Types of Solutes and Solvents

Solutes and solvents may be solids, liquids, or gases. When sugar is dissolved in a glass of water, a liquid sugar solution forms. Sugar is the solute, and water is the solvent. Soda water and soft drinks are prepared by dissolving CO_2 gas in water. The CO_2 gas is the solute, and water is the solvent. Table 7.1 lists some solutes and solvents and their solutions.

TABLE 7.1

Some Examples of Solutions

Type	Example	Solute	Solvent
Gas Solutions			
Gas in a gas	Air	Oxygen (gas)	Nitrogen (gas)
Liquid Solutions			
Gas in a liquid	Soda water	Carbon dioxide (gas)	Water (liquid)
	Household ammonia	Ammonia (gas)	Water (liquid)
Liquid in a liquid	Vinegar	Acetic acid (liquid)	Water (liquid)
Solid in a liquid	Seawater	Sodium chloride (solid)	Water (liquid)
	Tincture of iodine	Iodine (solid)	Alcohol (liquid)
Solid Solutions			
Liquid in a solid	Dental amalgam	Mercury (liquid)	Silver (solid)
Solid in a solid	Brass	Zinc (solid)	Copper (solid)
	Steel	Carbon (solid)	Iron (solid)

Water as a Solvent

Water is one of the most common solvents in nature. In the H_2O molecule, an oxygen atom shares electrons with two hydrogen atoms. Because the oxygen atom is much more electronegative, the O—H bonds are polar. In each of the polar bonds, the oxygen atom has a partial negative (δ^-) charge, and the hydrogen atom has a partial positive (δ^+) charge. Because of the arrangement of the polar bonds, water is a *polar substance.*

Hydrogen bonds occur between molecules where a partially positive hydrogen is attracted to the strongly electronegative atoms of O, N, or F in other molecules. In water, hydrogen bonds are formed by the attraction between the oxygen atom of one water molecule and a hydrogen atom in another water molecule. In the diagram, hydrogen bonds are shown as dots between the water molecules. Although hydrogen bonds are much weaker than covalent or ionic bonds, there are many of them linking molecules together. As a result hydrogen bonding plays an important role in the properties of water and biological compounds such as proteins, carbohydrates, and DNA.

WEB TUTORIAL
Hydrogen Bonding

Formation of Solutions

An ionic compound such as sodium chloride, NaCl, is held together by ionic bonds between positive Na^+ ions and negative Cl^- ions. It dissolves in water because water is a polar solvent. When solid NaCl crystals are placed in water, the process of dissolution begins as the ions on the surface of the crystal come in contact with water molecules. (See Figure 7.2.) The negatively charged oxygen atom at one end of a water molecule attracts the positive Na^+ ions. The positively charged hydrogen atoms at the other end of a water molecule attract the negative Cl^- ions. The attractive forces of several water molecules provide the energy to break the ionic bonds between the Na^+ and Cl^- ions in the NaCl crystal. As the water molecules pull the ions into solution, a new surface of the NaCl crystal is exposed to the solvent. During a process called **hydration**, the dissolved Na^+ and Cl^- ions are surrounded by water molecules, which diminishes their attraction to other ions and helps keep them in solution.

Like Dissolves Like

Gases form solutions easily because their particles are moving so rapidly that they are far apart and attractions to the other gas particles are not important. When solids or liquids form solutions, there must be an attraction between the solute particles

FIGURE 7.2

Ions on the surface of a crystal of NaCl dissolve in water as they are attracted to the polar water molecules that pull the ions into solution and surround them.

Q What helps keep the Na$^+$ and Cl$^-$ ions in solution?

H$_2$O

NaCl

Hydrated ions

Na$^+$

Cl$^-$

(a)   (b)   (c)

FIGURE 7.3

Like dissolves like. **(a)** The test tubes contain an upper layer of water (polar) and a lower layer of CH$_2$Cl$_2$ (nonpolar). **(b)** The nonpolar solute I$_2$ dissolves in the nonpolar layer. **(c)** The ionic solute Ni(NO$_3$)$_2$ dissolves in the water.

Q Which layer would dissolve polar molecules of sugar?

and the solvent particles. Then the particles of the solute and solvent will mix together. If there is no attraction between a solute and a solvent, their particles do not mix and no solution forms.

A salt such as NaCl will form a solution with water because the Na$^+$ and Cl$^-$ ions in the salt are attracted to the positive and negative parts of the individual water molecules. You can dissolve table sugar, C$_{12}$H$_{22}$O$_{11}$, in water because the molecule has many polar —OH groups that attract water molecules.

However, compounds containing nonpolar molecules such as iodine (I$_2$), oil, or grease do not dissolve in water because water is polar. Nonpolar solutes require nonpolar solvents for a solution to form. The expression "like dissolves like" is a way of saying that the polarities of a solute and a solvent must be similar in order to form a solution. Figure 7.3 illustrates the formation of some polar and nonpolar solutions.

◼ Water in the Body

The average adult contains about 60% water by weight, and the average infant about 75%. About 60% of the body's water is contained within the cells as intracellular fluids; the other 40% makes up extracellular fluids, which include the interstitial fluid in tissue and the plasma in the blood. These external fluids carry nutrients and waste materials between the cells and the circulatory system.

Every day you lose between 1500 and 3000 mL of water from the kidneys as urine, from the skin as perspiration, from the lungs as you exhale, and from the gastrointestinal tract. Serious dehydration can occur in an adult if there is a 10% net loss in total body fluid, and a 20% loss of fluid can be fatal. An infant suffers severe dehydration with a 5–10% loss in body fluid.

Water loss is continually replaced by the liquids and foods in the diet and from metabolic processes that produce water in the cells of the body. Table 7.2 lists the % by mass of water contained in some foods.

24 Hours

Water gain		Water loss	
		Urine	1500 mL
Liquid	1000 mL	Perspiration	300 mL
Food	1200 mL	Breath	600 mL
Metabolism	300 mL	Feces	100 mL
Total	2500 mL	Total	2500 mL

TABLE 7.2

Percentage of Water in Some Foods

Food	Water (% by mass)	Food	Water (% by mass)
Vegetables		**Meats/Fish**	
Carrot	88	Chicken, cooked	71
Celery	94	Hamburger, broiled	60
Cucumber	96	Salmon	71
Tomato	94	**Grains**	
Fruits		Cake	34
Apple	85	French bread	31
Banana	76	Noodles, cooked	70
Cantaloupe	91	**Milk Products**	
Orange	86	Cottage cheese	78
Strawberry	90	Milk, whole	87
Watermelon	93	Yogurt	88

SAMPLE PROBLEM 7.2

◼ Polar and Nonpolar Solutes

Indicate whether each of the following substances will dissolve in water. Explain.

a. KCl
b. octane, C_8H_{18}, a compound in gasoline
c. ethanol, C_2H_5OH, a substance in mouthwash

SOLUTION

a. Yes. KCl is an ionic compound.
b. No. C_8H_{18} is a nonpolar substance.
c. Yes. C_2H_5OH is a polar substance.

STUDY CHECK

Will oil, a nonpolar substance, dissolve in hexane, a nonpolar solvent?

Like Dissolves Like

Mix together small amounts of the following substances:
a. oil and water
b. water and vinegar
c. salt and water
d. sugar and water
e. salt and oil

QUESTIONS

1. Which of the mixtures formed a solution? Which did not?
2. Why do some mixtures form solutions, but others do not?

QUESTIONS AND PROBLEMS

Solutions

7.1 Identify the solute and the solvent in each solution composed of the following:
  a. 10.0 g of NaCl and 100.0 g of H_2O
  b. 50.0 mL of ethanol, $C_2H_5OH(l)$, and 10.0 mL of H_2O
  c. 0.20 L of O_2 and 0.80 L of N_2

7.2 Identify the solute and the solvent in each solution composed of the following:
  a. 50.0 g of silver and 4.0 g of mercury
  b. 100.0 mL of water and 5.0 g of sugar
  c. 1.0 g of I_2 and 50.0 mL of alcohol

7.3 Describe the formation of an aqueous KI solution.

7.4 Describe the formation of an aqueous LiBr solution.

7.5 Water is a polar solvent; CCl_4 is a nonpolar solvent. In which solvent is each of the following more likely to be soluble?
  a. KCl, ionic   **b.** I_2, nonpolar
  c. sugar, polar   **d.** gasoline, nonpolar

7.6 Water is a polar solvent; hexane is a nonpolar solvent. In which solvent is each of the following more likely to be soluble?
  a. vegetable oil, nonpolar   **b.** benzene, nonpolar
  c. $LiNO_3$, ionic   **d.** Na_2SO_4, ionic

7.2 Electrolytes and Nonelectrolytes

Solutes can be classified by their ability to conduct an electrical current. When solutes called **electrolytes** dissolve in water, they separate into ions, which are able to conduct electricity. When solutes called **nonelectrolytes** dissolve in water, they do not separate into ions and their solutions do not conduct electricity.

To test solutions for ions, we can use an apparatus that consists of a battery and a pair of electrodes connected by wires to a light bulb. The light bulb glows when electricity can flow, which only happens when electrolytes provide ions to complete the circuit.

Strong Electrolytes

A **strong electrolyte** is a solute that dissociates completely into ions when it dissolves in water. For example, when sodium chloride (NaCl) dissolves in water, the sodium and chloride ions are attracted to water molecules. In a process called *dissociation*, the ions separate from the solid. As ions form, they are hydrated by surrounding water molecules. In the equation for the dissociation of NaCl in water, the H_2O over the arrow indicates that water is needed for the dissociation process, but is not a reactant.

$$NaCl(s) \xrightarrow{H_2O} Na^+(aq) + Cl^-(aq)$$

When we write the equation for the dissociation, the electrical charges must balance. For example, magnesium nitrate dissolves in water to give one magnesium ion for every two nitrate ions. However, only the ionic bonds between Mg^{2+} and NO_3^- are broken, not the covalent bonds within the polyatomic ion. The dissociation for $Mg(NO_3)_2$ is written as follows:

$$Mg(NO_3)_2(s) \xrightarrow{H_2O} Mg^{2+}(aq) + 2NO_3^-(aq)$$

Strong electrolyte

Weak Electrolytes

A **weak electrolyte** is a solute that dissolves in water mostly as whole molecules. Only a few of the dissolved molecules separate, which produces a small number of ions in solution. Thus solutions of weak electrolytes do not conduct electrical current as well as solutions of strong electrolytes. For example, an aqueous solution of HF, which is a weak electrolyte, consists of mostly HF molecules, and a few H^+ and F^- ions. Within the solution, a few HF molecules dissociate into ions. As more H^+ and F^- ions form, some recombine to give HF molecules, which is indicated by a backwards arrow. Eventually the rate of formation of ions is equal to the rate at which they recombine. The use of two arrows indicates that the forward and reverse reactions are taking place at the same time.

$$HF(aq) \underset{\text{recombination}}{\overset{\text{dissociation}}{\rightleftharpoons}} H^+(aq) + F^-(aq)$$

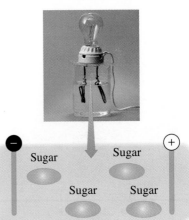

Weak electrolyte

Nonelectrolytes

Solutes that are nonelectrolytes dissolve in water as molecules and do not separate into ions; thus, solutions of nonelectrolytes do not conduct electricity. For example, sucrose (sugar) is a nonelectrolyte that dissolves in water as whole molecules only.

$$C_{12}H_{22}O_{11}(s) \xrightarrow{H_2O} C_{12}H_{22}O_{11}(aq)$$

Sucrose   Solution of sucrose molecules

Nonelectrolyte

Water

Sugar (sucrose)

Sugar (sucrose) molecules in solution

TABLE 7.3

Classification of Solutes in Aqueous Solutions

Types of Solute	Dissociation	Contained in Solution	Conducts Electricity	Examples
Strong electrolyte	Completely	Ions only	Yes	Ionic compounds such as NaCl, KBr, $MgCl_2$, $NaNO_3$; bases such as NaOH, KOH; acids such as HCl, HBr, HNO_3, $HClO_4$
Weak electrolyte	Partially	Mostly molecules and a few ions	Yes, but poorly	HF, H_2O, NH_3, CH_3COOH (acetic acid)
Nonelectrolyte	None	Molecules only	No	Carbon compounds such as CH_3OH, C_2H_5OH, $C_{12}H_{22}O_{11}$

Table 7.3 summarizes the classification of solutes in aqueous solutions.

SAMPLE PROBLEM 7.3

▪ Solutions of Electrolytes and Nonelectrolytes

Indicate whether solutions of each of the following contain ions, molecules, or both:

a. Na_2SO_4, a strong electrolyte
b. CH_3OH, a nonelectrolyte

SOLUTION

a. A solution of Na_2SO_4 contains the ions of the salt, Na^+ and SO_4^{2-}.
b. A nonelectrolyte such as CH_3OH dissolves in water as molecules.

STUDY CHECK

Boric acid, H_3BO_3, is a weak electrolyte. Would you expect a boric acid solution to contain ions, molecules, or both?

Equivalents

Body fluids typically contain a mixture of several electrolytes, such as Na^+, Cl^-, K^+, and Ca^{2+}. We measure each individual ion in terms of an **equivalent (Eq)**, which is the amount of that ion equal to 1 mole of positive or negative electrical charge. For example, 1 mole of Na^+ ions and 1 mole of Cl^- ions are each 1 equivalent because they each contain 1 mole of charge. For an ion with a charge of $2+$ or $2-$, there are 2 equivalents for each mole. Some examples of ions and equivalents are shown in Table 7.4.

TABLE 7.4

Equivalents of Electrolytes

Ion	Electrical Charge	Number of Equivalents in 1 Mole
Na^+	$1+$	1 Eq
Ca^{2+}	$2+$	2 Eq
Fe^{3+}	$3+$	3 Eq
Cl^-	$1-$	1 Eq
SO_4^{2-}	$2-$	2 Eq

SAMPLE PROBLEM 7.4

■**Electrolyte Concentration**

In body fluids, concentrations of electrolytes are often expressed as milliequivalent (mEq) per liter. A typical concentration for Na^+ in the blood is 138 mEq/L. How many moles of sodium ion are in 1.00 L of blood?

SOLUTION

Using the volume and the electrolyte concentration (in mEq/L) we can find the number of equivalents in 1.00 L of blood.

$$1.00 \; \cancel{L} \times \frac{138 \; \cancel{mEq}}{1 \cancel{L}} \times \frac{1 \; Eq}{1000 \; \cancel{mEq}} = 0.138 \; Eq \; Na^+$$

We can then convert equivalents to moles (for Na^+ there is 1 Eq per mole).

$$0.138 \; \cancel{Eq \; Na^+} \times \frac{1 \; mole \; Na^+}{1 \; \cancel{Eq \; Na^+}} = 0.138 \; mole$$

STUDY CHECK

A Ringer's solution for intravenous fluid replacement contains 155 mEq Cl^- per liter of solution. If a patient receives 1250 mL of Ringer's solution, how many moles of chloride were given?

QUESTIONS AND PROBLEMS

■**Electrolytes and Nonelectrolytes**

7.7 KF is a strong electrolyte, and HF is a weak electrolyte. How are they different?

7.8 NaOH is a strong electrolyte, and CH_3OH is a nonelectrolyte. How are they different?

7.9 The following soluble salts are strong electrolytes. Write a balanced equation for their dissociation in water:
a. KCl   **b.** $CaCl_2$ **c.** K_3PO_4 **d.** $Fe(NO_3)_3$

7.10 The following salts are strong electrolytes. Write a balanced equation for their dissociation in water:
a. LiBr   **b.** $NaNO_3$ **c.** $FeCl_3$ **d.** $Mg(NO_3)_2$

7.11 Indicate whether aqueous solutions of the following will contain ions only, molecules only, or molecules and some ions:
a. acetic acid ($HC_2H_3O_2$), found in vinegar, a weak electrolyte
b. NaBr, a strong electrolyte
c. fructose ($C_6H_{12}O_6$), a nonelectrolyte

7.12 Indicate whether aqueous solutions of the following will contain ions only, molecules only, or molecules and some ions:
a. Na_2SO_4, a strong electrolyte
b. ethanol, C_2H_5OH, a nonelectrolyte
c. HCN, hydrocyanic acid, a weak electrolyte

7.13 Indicate the type of electrolyte represented in the following equations:

a. $K_2SO_4(s) \xrightarrow{H_2O} 2K^+(aq) + SO_4{}^{2-}(aq)$

b. $NH_4OH(aq) \underset{}{\overset{H_2O}{\rightleftharpoons}} NH_4{}^+(aq) + OH^-(aq)$

c. $C_6H_{12}O_6(s) \xrightarrow{H_2O} C_6H_{12}O_6(aq)$

▪ Electrolytes in Body Fluids

The concentrations of electrolytes present in body fluids and in intravenous fluids given to a patient are often expressed in milliequivalents per liter (mEq/L) of solution.

$$1 \text{ Eq} = 1000 \text{ mEq}$$

Table 7.5 gives the concentrations of some typical electrolytes in blood plasma. There is a charge balance because the total number of positive charges is equal to the total number of negative charges. The use of a specific intravenous solution depends on the nutritional, electrolyte, and fluid needs of the individual patient. Examples of various types of solutions are given in Table 7.6.

TABLE 7.5

Some Typical Concentrations of Electrolytes in Blood Plasma

Electrolyte	Concentration (mEq/L)
Cations	
Na^+	138
K^+	5
Mg^{2+}	3
Ca^{2+}	4
Total	150
Anions	
Cl^-	110
HCO_3^-	30
HPO_4^{2-}	4
Proteins	6
Total	150

TABLE 7.6

Electrolyte Concentrations in Intravenous Replacement Solutions

Solution	Electrolytes (mEq/L)	Use
Sodium chloride (0.9%)	Na^+ 154, Cl^- 154	Replacement of fluid loss
Potassium chloride with 5.0% dextrose	K^+ 40, Cl^- 40	Treatment of malnutrition (low potassium levels)
Ringer's solution	Na^+ 147, K^+ 4, Ca^{2+} 4, Cl^- 155	Replacement of fluids and electrolytes lost through dehydration
Maintenance solution with 5.0% dextrose	Na^+ 40, K^+ 35, Cl^- 40, lactate$^-$ 20, HPO_4^{2-} 15	Maintenance of fluid and electrolyte levels
Replacement solution (extracellular)	Na^+ 140, K^+ 10, Ca^{2+} 5, Mg^{2+} 3, Cl^- 103, acetate$^-$ 47, citrate^{3-} 8	Replacement of electrolytes in extracellular fluids

7.14 Indicate the type of electrolyte represented in the following equations:

a. $CH_3OH(l) \xrightarrow{\text{H}_2\text{O}} CH_3OH(aq)$

b. $MgCl_2(s) \xrightarrow{\text{H}_2\text{O}} Mg^{2+}(aq) + 2Cl^-(aq)$

c. $HClO(aq) \xrightleftharpoons{\text{H}_2\text{O}} H^+(aq) + ClO^-(aq)$

7.15 Indicate the number of equivalents in each of the following:
 a. 1 mole K^+ **b.** 2 moles OH^-
 c. 1 mole Ca^{2+} **d.** 3 moles CO_3^{2-}

7.16 Indicate the number of equivalents in each of the following:
 a. 1 mole Mg^{2+} **b.** 0.5 mole H^+
 c. 4 moles Cl^- **d.** 2 moles Fe^{3+}

7.17 A physiological saline solution contains 154 mEq/L each of Na^+ and Cl^-. How many moles each of Na^+ and Cl^- are in 1.00 L of the saline solution?

7.18 A solution to replace potassium loss contains 40. mEq/L each of K^+ and Cl^-. How many moles each of K^+ and Cl^- are in 1.5 L of the solution?

7.19 A solution contains 40. mEq/L of Cl^- and 15 mEq/L of HPO_4^{2-}. If Na^+ is the only cation in the solution, what is the Na^+ concentration in milliequivalents per liter?

7.20 A sample of Ringer's solution contains the following concentrations (mEq/L) of cations: Na^+ 147, K^+ 4, and Ca^{2+} 4. If Cl^- is the only anion in the solution, what is the Cl^- concentration in milliequivalents per liter?

■ 7.3 Solubility

LEARNING GOAL

■ Define *solubility*; distinguish between an unsaturated and a saturated solution.

The term **solubility** is used to describe the amount of a solute that can dissolve in a given amount of solvent. Many factors such as the type of solute, the type of solvent, and temperature affect a solute's solubility. Solubility, usually expressed in grams of solute in 100 grams of solvent, is the maximum amount of solute that can be dissolved at a certain temperature. If a solute readily dissolves when added to the solvent, the solution does not contain the maximum amount of solute. We call the solution an **unsaturated solution**. When a solution contains all the solute that can dissolve, it is a **saturated solution**. If we try to add more solute, undissolved solute will remain on the bottom of the container.

Unsaturated solution

Saturated solution

A solution becomes saturated when the rates of the forward reaction that dissolves the solute and the reverse reaction of recrystallization become equal. Then there is no further change in the amount of solid solute. The rate at which

Gout and Kidney Stones: A Problem of Saturation in Body Fluids

The conditions of gout and kidney stones involve compounds in the body that exceed their solubility levels and form solid products. Gout affects adults, primarily men, over the age of 40. Attacks of gout may occur when the concentration of uric acid in blood plasma exceeds its solubility, which is 7 mg/100 mL of plasma at 37°C. Insoluble deposits of needle-like crystals of uric acid can form in the cartilage, tendons, and soft tissues where they cause painful gout attacks. They may also form in the tissues of the kidneys, where they can cause renal damage. High levels of uric acid in the body can be caused by an increase in uric acid production, failure of the kidneys to remove uric acid, or by a diet with an overabundance of foods containing purines, which are metabolized to uric acid in the body. Foods in the diet that contribute to high levels of uric acid include certain meats, sardines, mushrooms, asparagus, and beans. Drinking alcoholic beverages may also significantly increase uric acid levels and bring about gout attacks.

Treatment for gout involves diet changes and drugs. Depending on the levels of uric acid, a medication, such as probenecid, can be used to help the kidneys eliminate uric acid, or allopurinol, which blocks the production of uric acid by the body.

Kidney stones are solid materials that form in the urinary tract. Most kidney stones are composed of calcium phosphate and calcium oxalate, although they can be solid uric acid. The excessive ingestion of minerals and insufficient water intake can cause the concentration of mineral salts to exceed the solubility of the mineral salts and lead to the formation of kidney stones. When a kidney stone passes through the urinary tract, it causes considerable pain and discomfort, necessitating the use of painkillers and surgery. Sometimes ultrasound is used to break up kidney stones. Persons prone to kidney stones are advised to drink six to eight glasses of water every day to prevent saturation levels of minerals in the urine.

the Chemistry place

CASE STUDY
Kidney Stones and Saturated Solutions

solute particles are removed from the surface of the solid is equal to the rate at which particles in solution crystallize out as solid.

$$\text{Solid solute} \underset{\text{crystallizes}}{\overset{\text{dissolves}}{\rightleftharpoons}} \text{saturated solution}$$

SAMPLE PROBLEM 7.5

Saturated Solutions

At 20°C, the solubility of KCl is 34 g/100 g of water. In the laboratory, a student mixes 45 g of KCl with 100 g of water at a temperature of 20°C. Is the solution saturated or unsaturated?

SOLUTION

Only 34 g of the KCl will dissolve because that is its solubility at 20°C. The solution is saturated.

▪ Preparing Solutions

In your kitchen, place $\frac{1}{4}$ or $\frac{1}{2}$ cup of cold water in a glass. Begin adding 1 tablespoon of sugar at a time and stir thoroughly. Take a sip of the liquid in the glass as you proceed. As the sugar solution becomes more concentrated, you may need to stir for a few minutes until all the sugar dissolves. Each time observe the solution after several minutes to determine when it is saturated.

Repeat the above activity with $\frac{1}{4}$ or $\frac{1}{2}$ cup of warm water. Count the number of tablespoons of sugar you need to form a saturated solution.

QUESTIONS

1. What do you notice about how sweet each sugar solution tastes?
2. How did you know when you obtained a saturated solution?
3. How much sugar dissolved in the warm water compared to the cold water?

STUDY CHECK

At 50°C, the solubility of $NaNO_3$ is 114 g/100 g of water. How many grams of $NaNO_3$ are needed to make a saturated $NaNO_3$ solution with 50. g of water at 50°C?

▪ Effect of Temperature on Solubility

For most solids, solubilities increase as temperature increases. A few substances show little change in solubility at higher temperatures, and a few are less soluble. (See Figure 7.4.) For example, when you add sugar to iced tea, a layer of undissolved sugar may form on the bottom of the glass. Hot tea can dissolve more sugar than cold tea, which is why sugar is added before the tea is cooled.

The solubility of a gas in water decreases as the temperature increases. At higher temperatures, more gas molecules have the energy to escape from the solution. Perhaps you have observed the bubbles escaping from a cold carbonated soft drink as it warms. At high temperatures, bottles containing carbonated solutions may burst as more gas molecules leave the solution and increase the gas pressure inside the bottle. Biologists have found that increased temperatures in rivers and lakes cause the amount of dissolved oxygen to decrease until the warm water can no longer support a biological community. Electricity-generating plants are required to have their own ponds to use with their cooling towers to lessen the threat of thermal pollution.

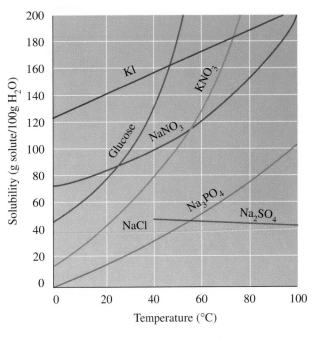

▪ Henry's Law

Henry's law states that the solubility of gas in a liquid is directly related to the pressure of that gas above the liquid. At higher pressures, there are more gas molecules available to enter and dissolve in the liquid. A can of soda is carbonated by using CO_2 gas at high pressure to increase the solubility of the CO_2 in the beverage. When you open the can at atmospheric pressure, the pressure on the CO_2 drops, which decreases the solubility of CO_2. As a result, bubbles of CO_2 rapidly escape from the solution. The burst of bubbles is even more noticeable when you open a warm can of soda.

FIGURE 7.4

In water, most common solids are more soluble as the temperature increases.

Q **Compare the solubility of $NaNO_3$ at 20°C and 60°C.**

CO₂ under high pressure

Lots of CO₂ dissolved in soda

Gas molecule

Cola

More gas molecules dissolve

Pressure released

CO₂ bubbles out of solution

Gas at low pressure

Fewer gas molecules dissolve

Cola

SAMPLE PROBLEM 7.6

■ Factors Affecting Solubility

Indicate whether the solubility of the solute will increase or decrease in each of the following situations:

a. using 80°C water instead of 25°C water to dissolve sugar
b. a lake, which contains dissolved O_2, warms

SOLUTION

a. An increase in the temperature increases the solubility of the sugar.
b. An increase in the temperature decreases the solubility of O_2 gas.

STUDY CHECK

At 10°C, the solubility of KNO_3 is 20 g/100 g H_2O. Would the value of 5 g/100 g H_2O or 65 g/100 g H_2O be the more likely solubility at 40°C? Explain.

QUESTIONS AND PROBLEMS

■ Solubility

7.21 State whether each of the following refers to a saturated or unsaturated solution:
    a. A crystal added to a solution does not change in size.
    b. A sugar cube completely dissolves when added to a cup of coffee.

7.22 State whether each of the following refers to a saturated or unsaturated solution:
    a. A spoonful of salt added to boiling water dissolves.
    b. A layer of sugar forms on the bottom of a glass of tea as ice is added.

Use this table for problems 7.23–7.26.

Solubility (g/100 g H_2O)		
Substance	20°C	50°C
KCl	34.0	42.6
$NaNO_3$	88.0	114.0
$C_{12}H_{22}O_{11}$ (sugar)	203.9	260.4

7.23 Using the above table, determine whether each of the following solutions will be saturated or unsaturated at 20°C:
  a. Adding 25.0 g KCl to 100 g H_2O
  b. Adding 11.0 g $NaNO_3$ to 25 g H_2O
  c. Adding 400.0 g sugar to 125 g H_2O

7.24 Using the above table, determine whether each of the following solutions will be saturated or unsaturated at 50°C:
  a. Adding 25.0 g KCl to 50 g H_2O
  b. Adding 150.0 g $NaNO_3$ to 75 g H_2O
  c. Adding 80.0 g sugar to 25 g H_2O

7.25 A solution containing 80.0 g KCl in 200 g H_2O at 50°C is cooled to 20°C.
  a. How many grams of KCl remain in solution at 20°C?
  b. How many grams of solid KCl came out of solution after cooling?

7.26 A solution containing 80.0 g of $NaNO_3$ in 75.0 g H_2O at 50°C is cooled to 20°C.
  a. How many grams of $NaNO_3$ remain in solution at 20°C?
  b. How many grams of solid $NaNO_3$ came out of solution after cooling?

7.27 Explain the following observations:
  a. More sugar dissolves in hot tea than in ice tea.
  b. Champagne in a warm room goes flat.
  c. A warm can of soda has more spray when opened than a cold one.

7.28 Explain the following observations:
  a. An open can of soda loses its "fizz" quicker at room temperature than in the refrigerator.
  b. Chlorine gas in tap water escapes as the sample warms to room temperature.
  c. Less sugar dissolves in iced coffee than in hot coffee.

■ 7.4 Percent Concentration

LEARNING GOAL

■ Calculate the percent concentration of a solute in a solution; use percent concentration to calculate the amount of solute or solution.

The amount of solute dissolved in a certain amount of solution is called the **concentration** of the solution. Although there are many ways to express a concentration, they all specify a certain amount of solute in a given amount of solution.

$$\text{Concentration of a solution} = \frac{\text{amount of solute}}{\text{amount of solution}}$$

■ Mass Percent

The **mass percent** (% m/m) concentration of a solution describes the mass of the solute in every 100 grams of solution. The mass in grams of the solution is the sum of the mass of the solute and the mass of the solvent.

$$\text{mass percent (\% m/m)} = \frac{\text{mass of solute (g)}}{\text{mass of solute (g)} + \text{mass of solvent (g)}} \times 100$$

$$= \frac{\text{mass of solute (g)}}{\text{mass of solution (g)}} \times 100$$

Add 8.00 g of KCl

Add water until the
solution weighs
50.00 g

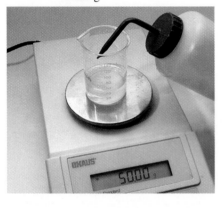

Suppose we prepared a solution by mixing 8.00 g of KCl (solute) with 42.00 g of water (solvent). Together the mass of the solute and mass of solvent give the mass of the solution (8.00 g + 42.00 g = 50.00 g). The mass % is calculated by substituting in the values into the mass percent expression.

$$\frac{8.00 \text{ g KCl}}{50.00 \text{ g solution}} \times 100\% = 16.0\% \text{ (m/m)}$$

$$\underbrace{8.00 \text{ g KCl} + 42.00 \text{ g H}_2\text{O}}$$
(solute   +   solvent)

SAMPLE PROBLEM 7.7

■ Percent Concentration (m/m)

What is the mass percent of a solution prepared by dissolving 30.0 g NaOH in 120.0 g of H_2O?

SOLUTION

STEP 1 Given   30.0 g NaOH and 120.0 g H_2O
Need   mass percent (% m/m) of NaOH

STEP 2 Plan   The mass percent is calculated by using the mass in grams of the solute and solution in the definition of mass percent.

STEP 3 Equalities/Conversion Factors

$$\text{mass percent (\% m/m)} = \frac{\text{mass of solute}}{\text{mass of solute + mass of solvent}} \times 100$$

$$\text{mass percent (\% m/m)} = \frac{\text{grams of solute}}{\text{grams of solution}} \times 100$$

STEP 4 Set Up Problem   The mass of the solute and the solution are obtained from the data.

$$\begin{array}{rl}
\text{mass of solute} & = \quad\; 30.0 \text{ g NaOH} \\
\text{mass of solvent} & = +120.0 \text{ g H}_2\text{O} \\
\hline
\text{mass of solution} & = \quad 150.0 \text{ g solution}
\end{array}$$

$$\text{mass percent (\% m/m)} = \frac{30.0 \text{ g NaOH}}{150.0 \text{ g solution}} \times 100$$

$$= 20.0\% \text{ (m/m) NaOH}$$

STUDY CHECK

What is the mass percent of NaCl in a solution made by dissolving 2.0 g of NaCl in 56.0 g of H_2O?

Guide to Calculating
Solution Concentrations

STEP 1
State the given and
needed concentration.

STEP 2
Write a plan to calculate
needed concentration.

STEP 3
Write equalities and conversion factors.

STEP 4
Set up problem to calculate answer.

Volume Percent

Because the volumes of liquids or gases are easily measured, the concentrations of their solutions are often expressed as **volume percent** (% v/v). The units of volume used in the ratio must be the same, for example, both in milliliters or both in liters.

$$\text{volume percent (\% v/v)} = \frac{\text{volume of solute}}{\text{volume of solution}} \times 100\%$$

We interpret a volume/volume percent as the volume of solute in 100 mL of solution. In the wine industry, a label that reads 12% (v/v) means 12 mL of alcohol in 100 mL of wine.

Mass/Volume Percent

A **mass/volume percent** (% m/v), or weight/volume percent (% w/v), is calculated by dividing the grams of the solute by the volume (mL) of solution and multiplying by 100. Widely used in hospitals and pharmacies, the preparation of intravenous solutions and medicines involves the mass/volume percent.

$$\text{mass/volume \%} = \frac{\text{grams of solute}}{\text{milliliters of solution}} \times 100\%$$

The mass/volume percent indicates the grams of a substance that are contained in 100 mL of a solution. For example, a 5% (m/v) glucose solution contains 5 g of glucose in 100 mL of solution. The volume of solution represents the combination of the glucose and H_2O.

SAMPLE PROBLEM 7.8

■ Calculating Percent Concentration

A student prepared a solution by dissolving 5.0 g KI in enough water to give a final volume of 250 mL. What is the mass/volume percent of the KI solution?

SOLUTION

$$\text{mass/volume \%} = \frac{\overset{\text{Mass of solute}}{5.0 \text{ g KI}}}{\underset{\text{Volume of solution}}{250 \text{ mL solution}}} \times 100\% = 2.0\% \text{ (m/v) KI}$$

STUDY CHECK

What is the mass/volume percent (% m/v), of Br_2 in a solution prepared by dissolving 12 g of bromine (Br_2) in enough carbon tetrachloride to make 250 mL of solution?

Water added to make a solution — 250 mL

5.0 g KI   2.0% (m/v) KI solution

Percent Concentrations as Conversion Factors

In the preparation of solutions, we often need to calculate the amount of solute or solution. Then the percent concentration is useful as a conversion factor. Some examples of percent concentrations, their meanings, and possible conversion factors are given in Table 7.7.

TABLE 7.7

Conversion Factors from Percent Concentrations		
Percent Concentration	**Meaning**	**Conversion Factors**
10% (m/m) KCl	There are 10 g of KCl in 100 g of solution.	$\dfrac{10 \text{ g KCl}}{100 \text{ g solution}}$ and $\dfrac{100 \text{ g solution}}{10 \text{ g KCl}}$
5% (m/v) glucose	There are 5 g of glucose in 100 mL of solution.	$\dfrac{5 \text{ g glucose}}{100 \text{ mL solution}}$ and $\dfrac{100 \text{ mL solution}}{5 \text{ g glucose}}$
12% (v/v) ethanol	There are 12 mL of ethanol in 100 mL of solution.	$\dfrac{12 \text{ mL ethanol}}{100 \text{ mL solution}}$ and $\dfrac{100 \text{ mL solution}}{12 \text{ mL ethanol}}$

▪ Using Mass/Volume Percent To Find Mass of Solute

A topical antibiotic is 1.0% (m/v) Clindamycin. How many grams of Clindamycin are in 60. mL of the 1.0% (m/v) solution?

SOLUTION

STEP 1 **Given** 1.0% (m/v) Clindamycin   **Need** grams of Clindamycin

STEP 2 **Plan** milliliters of solution → % (m/v) factor → grams of Clindamycin

STEP 3 **Equalities/Conversion Factors** The percent (m/v) indicates the grams of a solute in every 100 mL of a solution. The 1.0% (m/v) can be written as two conversion factors.

$$100 \text{ mL solution} = 1.0 \text{ g Clindamycin}$$

$$\frac{1.0 \text{ g Clindamycin}}{100 \text{ mL solution}} \quad \text{and} \quad \frac{100 \text{ mL solution}}{1.0 \text{ g Clindamycin}}$$

STEP 4 **Set Up Problem** The volume of the solution is converted to mass of solute using the conversion factor.

$$60. \text{ mL solution} \times \frac{1.0 \text{ g Clindamycin}}{100 \text{ mL solution}} = 0.60 \text{ g Clindamycin}$$

STUDY CHECK

Calculate the grams of KCl and grams of water in 225 g of an 8.00% (m/m) KCl solution.

QUESTIONS AND PROBLEMS

▪ Percent Concentration

7.29 What is the difference between a 5% (m/m) glucose solution and a 5% (m/v) glucose solution?

7.30 What is the difference between a 10% (v/v) methyl alcohol (CH_3OH) solution and a 10% (m/m) methyl alcohol solution?

7.31 Calculate the mass percent, % (m/m), for the solute in each of the following solutions:
a. 25 g of KCl and 125 g of H_2O
b. 12 g of sugar in 225 g of tea with sugar (solution)

7.32 Calculate the mass percent, % (m/m), for the solute in each of the following solutions:
a. 75 g of NaOH in 325 g of NaOH solution
b. 2.0 g of KOH in 20.0 g of H_2O

7.33 Calculate the mass/volume percent, % (m/v) for the solute in each of the following solutions:
a. 75 g of Na_2SO_4 in 250 mL of Na_2SO_4 solution
b. 39 g of sucrose in 355 mL of a carbonated drink

7.34 Calculate the mass/volume percent, % (m/v), for the solute in each of the following solutions:
a. 2.50 g KCl in 50.0 mL of solution
b. 7.5 g of casein in 120 mL of low-fat milk

7.35 Calculate the amount of solute needed to prepare the following solutions:
a. 50.0 mL of a 5.0% (m/v) KCl solution
b. 1250 mL of a 4.0% (m/v) NH_4Cl solution

7.36 Calculate the amount of solute needed to prepare the following solutions:
   a. 150 mL of a 40.0% (m/v) $LiNO_3$ solution
   b. 450 mL of a 2.0% (m/v) KCl solution

7.37 A mouthwash contains 22.5% alcohol by volume. If the bottle of mouthwash contains 355 mL, what is the volume in milliliters of the alcohol?

7.38 A bottle of champagne is 11% alcohol by volume. If there are 750 mL of champagne in the bottle, how many milliliters of alcohol are present?

7.39 A patient receives 100. mL of 20.% (m/v) mannitol solution every hour.
   a. How many grams of mannitol are given in 1 hour?
   b. How many grams of mannitol does the patient receive in 15 hours?

7.40 A patient receives 250 mL of a 4.0% (m/v) amino acid solution twice a day.
   a. How many grams of amino acids are in 250 mL of solution?
   b. How many grams of amino acids does the patient receive in 1 day?

7.41 A patient needs 100. g of glucose in the next 12 hours. How many liters of a 5% (m/v) glucose solution must be given?

7.42 A patient received 2.0 g of NaCl in 8 hours. How many milliliters of a 0.90% (m/v) NaCl (saline) solution were delivered?

7.5 Molarity and Dilution

LEARNING GOAL

Calculate the molarity of a solution; use molarity to calculate the moles of solute or the volume needed to prepare a solution. Describe the dilution of a solution.

When the solutes of solutions take part in reactions, chemists are interested in the number of reacting particles. For this purpose, chemists use **molarity (M)**, a concentration that states the number of moles of solute in exactly 1 liter of solution. The molarity of a solution can be calculated knowing the moles of solute and the volume of solution.

$$\text{Molarity (M)} = \frac{\text{moles of solute}}{\text{liters of solution}}$$

For example, if 1.0 mole of NaCl were dissolved in enough water to prepare 1.0 L of solution, the resulting NaCl solution has a molarity of 1.0 M. The abbreviation M indicates the units of moles per liter (moles/L).

$$M = \frac{\text{moles of solute}}{\text{liters of solution}} = \frac{1.0 \text{ mole NaCl}}{1 \text{ L}} = 1.0 \text{ M NaCl}$$

SAMPLE PROBLEM 7.10

■ **Calculating Molarity**

What is the molarity (M) of 60.0 g NaOH in 0.250 L of solution?

SOLUTION

STEP 1 **Given** 60.0 g NaOH in 0.250 L of solution
   Need molarity (moles/L)

STEP 2 **Plan** The calculation of molarity requires the moles of NaOH and the volume of the solution in liters.

$$\text{Molarity (M)} = \frac{\text{moles of solute}}{\text{liters of solution}}$$

g NaOH $\boxed{\text{Molar mass}}$ $\dfrac{\text{moles NaOH}}{\text{volume (L)}}$ = M NaOH solution

Volumetric flask

1.0 mole NaCl (58.5 g)

Add water until 1.0 liter mark is reached.

Mix

A 1.0 molar NaCl solution

STEP 3 **Equalities/Conversion Factors**

$$1 \text{ mole NaOH} = 40.0 \text{ g NaOH}$$

$$\frac{1 \text{ mole NaOH}}{40.0 \text{ g NaOH}} \quad \text{and} \quad \frac{40.0 \text{ g NaOH}}{1 \text{ mole NaOH}}$$

STEP 4 **Set Up Problem**

$$\text{moles NaOH} = 60.0 \text{ g NaOH} \times \frac{1 \text{ mole NaOH}}{40.0 \text{ g NaOH}} = 1.50 \text{ moles NaOH}$$

The molarity is calculated by dividing the moles of NaOH by the volume in liters.

$$\frac{1.50 \text{ moles NaOH}}{0.250 \text{ L}} = \frac{6.00 \text{ moles NaOH}}{1 \text{ L}} = 6.00 \text{ M NaOH}$$

STUDY CHECK

What is the molarity of a solution that contains 75.0 g KNO_3 dissolved in 0.350 L solution?

Molarity as a Conversion Factor

When we need to calculate the moles of solute or the volume of solution, the molarity is used as a conversion factor. Examples of conversion factors from molarity are given in Table 7.8.

TABLE 7.8

Some Examples of Molar Solutions

Molarity	Meaning	Conversion Factors		
6.0 M HCl	6.0 moles HCl in 1 liter of solution	$\dfrac{6.0 \text{ moles HCl}}{1 \text{ L}}$	and	$\dfrac{1 \text{ L}}{6.0 \text{ moles HCl}}$
0.20 M NaOH	0.20 mole NaOH in 1 liter of solution	$\dfrac{0.20 \text{ mole NaOH}}{1 \text{ L}}$	and	$\dfrac{1 \text{ L}}{0.20 \text{ mole NaOH}}$

To prepare a solution, we must convert the number of moles of solute needed into grams. Using the volume and the molarity of the solution with the molar mass of the solute, we can calculate the number of grams of solute necessary. This type of calculation is illustrated in Sample Problem 7.11.

SAMPLE PROBLEM 7.11

Using Molarity to Find Mass of Solute

How many grams of KCl would you need to weigh out to prepare 0.250 L of a 2.00 M KCl solution?

SOLUTION

STEP 1 **Given** 0.250 L of 2.00 M KCl solution   **Need** moles of KCl

STEP 2 **Plan** The grams of KCl are calculated by finding moles KCl using the volume and molarity of the KCl solution.

liters of KCl   Molarity   moles of KCl   Molar mass   grams of KCl

STEP 3 **Equalities/Conversion Factors**

1 L KCl = 2.00 moles KCl		1 mole KCl = 74.6 g KCl	
$\dfrac{1 \text{ L KCl}}{2.00 \text{ moles KCl}}$ and $\dfrac{2.00 \text{ moles KCl}}{1 \text{ L KCl}}$		$\dfrac{1 \text{ mole KCl}}{74.6 \text{ g KCl}}$ and $\dfrac{74.6 \text{ g KCl}}{1 \text{ mole KCl}}$	

STEP 4 **Set Up Problem**

$$\text{moles KCl} = 0.250 \text{ L solution} \times \frac{2.00 \text{ moles KCl}}{1 \text{ L KCl}} = 0.500 \text{ mole KCl}$$

The grams of KCl is calculated by multiplying the moles of KCl by the molar mass.

$$\text{gram KCl} = 0.500 \text{ mole KCl} \times \frac{74.6 \text{ g KCl}}{1 \text{ mole KCl}} = 37.3 \text{ g KCl}$$

Combining the steps, we can write the problem setup as follows:

$$0.250 \text{ L solution} \times \frac{2.00 \text{ moles KCl}}{1 \text{ L KCl}} \times \frac{74.6 \text{ g KCl}}{1 \text{ mole KCl}} = 37.3 \text{ g KCl}$$

STUDY CHECK

How many grams of $NaHCO_3$ are in 325 mL of a 4.50 M $NaHCO_3$ solution?

SAMPLE PROBLEM 7.12

■ **Using Molarity to Find Volume**

How many liters of a 2.00 M NaCl solution are needed to provide 67.3 g NaCl?

SOLUTION

STEP 1 **Given** 67.3 g NaCl from a 2.00 M NaCl solution
Need liters NaCl

STEP 2 **Plan** The volume of NaCl is calculated using the moles of NaCl and molarity of the NaCl solution.

grams NaCl Molar mass moles NaCl Molarity liters NaCl

STEP 3 **Equalities/Conversion Factors**

1 mole NaCl = 58.5 g NaCl		
$\dfrac{1 \text{ mole NaCl}}{58.5 \text{ g NaCl}}$ and $\dfrac{58.5 \text{ g NaCl}}{1 \text{ mole NaCl}}$		

The molarity of any solution can be written as two conversion factors.

1 L NaCl = 2.00 moles NaCl		
$\dfrac{1 \text{ L NaCl}}{2.00 \text{ moles NaCl}}$ and $\dfrac{2.00 \text{ moles NaCl}}{1 \text{ L NaCl}}$		

STEP 4 Set Up Problem

$$\text{liters of NaCl} = 67.3 \text{ g NaCl} \times \frac{1 \text{ mole NaCl}}{58.5 \text{ g NaCl}} \times \frac{1 \text{ L NaCl}}{2.00 \text{ moles NaCl}}$$

$$= 0.575 \text{ L NaCl}$$

STUDY CHECK

How many moles of HCl are present in 750 mL of a 6.0 M HCl solution?

Dilution

In chemistry, we often need to prepare a dilute solution from a more concentrated solution. In a process called **dilution**, we add water to a solution to make a larger volume. For example, you might prepare some orange juice by adding three cans of water to the original can of concentrated orange juice.

1 can orange   +   3 cans water   =   4 cans of orange juice
juice concentrate

When more water is added, the solution volume increases, causing a decrease in the concentration. However, *the amount of solute does not change.*

Amount of solute in the concentrated solution = amount of solute in the diluted solution

Therefore the initial amount of solute before dilution is equal to the final amount of solute in the diluted solution. (See Figure 7.5.) By diluting the solution, we decrease the concentration. The amount of the solute depends on the concentration, C, and the volume, V.

$$C_1V_1 = C_2V_2$$

We have seen that the concentration of a solution can be expressed in % or molarity, M.

$$\%_1V_1 = \%_2V_2$$

or

$$M_1V_1 = M_2V_2$$

FIGURE 7.5

When water is added to a concentrated solution, there is no change in the number of particles, but the solute particles can spread out as the volume of the diluted solution increases.

Q **What is the concentration of the diluted solution after an equal volume of water is added to a sample of 6 M HCl?**

SAMPLE PROBLEM 7.13

■ **Volume of a Diluted Solution**

What volume (mL) of a 2.5% (m/v) KOH solution can be prepared by diluting 50.0 mL of a 12% (m/v) KOH solution?

SOLUTION

STEP 1 **Give Data in a Table**   We make a table of the concentrations and volumes of the initial and diluted solutions.

Initial: $C_1 = 12\%$ (m/v) $V_1 = 50.0$ mL
Diluted: $C_2 = 2.5\%$ (m/v) $V_2 = ?$ mL

STEP 2 **Plan**   The volume of the dilute solution can be calculated by solving the dilution expression for V_2.

$$C_1V_1 = C_2\ V_2$$
$$\frac{C_1V_1}{C_2} = \frac{\mathcal{C}_2}{\mathcal{C}_2}\ V_2$$
$$V_2 = \frac{C_1V_1}{C_2}$$

STEP 3 **Set Up Problem**   The values from the table are in now used in the dilution expression to solve for V_2.

$$V_2 = \frac{C_1V_1}{C_2}$$
$$V_2 = \frac{12\%\ (\text{m/v}) \times 50.0\ \text{mL}}{2.5\%\ (\text{m/v})} = 240\ \text{mL (diluted KOH solution)}$$

Guide to Calculating Dilution Quantities
STEP 1 Prepare a table of the initial and diluted volumes and concentrations.
STEP 2 Write a plan that solves the dilution expression for the unknown quantity.
STEP 3 Set up problem by placing known quantities in dilution expression.

STUDY CHECK

What is the percent (% m/v) of the dilute solution when 25.0 mL of 15% (m/v) HCl is diluted to 125 mL?

SAMPLE PROBLEM 7.14

■ Molarity of a Diluted Solution

What is the molarity of a solution prepared when 75.0 mL of a 4.00 M KCl solution is diluted to a volume of 0.500 L?

SOLUTION

STEP 1 Give Data in a Table   We make a table of the molar concentrations and volumes of the initial and diluted solutions.

Initial: M_1 = 4.00 M KCl V_1 = 75.0 mL = 0.0750 L

Diluted: M_2 = ? M KCl V_2 = 0.500 L

STEP 2 Plan   The unknown molarity can be calculated by solving the dilution expression for M_2.

$$M_1 V_1 = M_2 V_2$$

$$\frac{M_1 V_1}{V_2} = M_2 \frac{\cancel{V_2}}{\cancel{V_2}}$$

$$M_2 = M_1 \times \frac{V_1}{V_2}$$

STEP 3 Set Up Problem   The diluted concentration is calculated by placing the values from the table into the dilution expression.

$$= 4.00 \text{ M} \times \frac{0.075 \cancel{L}}{0.500 \cancel{L}} = 0.600 \text{ M KCl (diluted solution)}$$

STUDY CHECK

You need to prepare 600. mL of 2.00 M NaOH solution from a 10.0 M NaOH solution. What volume of the 10.0 M NaOH solution do you use?

QUESTIONS AND PROBLEMS

■ Molarity and Dilution

7.43 Calculate the molarity (M) of the following solutions:
   a. 2.0 moles of glucose in 4.0 L of solution
   b. 4.0 g of KOH in 2.0 L of solution

7.44 Calculate the molarity (M) of the following solutions:
   a. 0.50 mole glucose in 0.200 L of solution
   b. 36.5 g of HCl in 1.0 L of solution

7.45 Calculate the moles of solute needed to prepare each of the following:
   a. 1.0 L of a 3.0 M NaCl solution
   b. 0.40 L of a 1.0 M KBr solution

7.46 Calculate the moles of solute needed to prepare each of the following:
   a. 5.0 L of a 2.0 M $CaCl_2$ solution
   b. 4.0 L of a 0.10 M NaOH solution

7.47 Calculate the grams of solute needed to prepare each of the following solutions:
   a. 2.0 L of a 1.5 M NaOH solution
   b. 4.0 L of a 0.20 M KCl solution

7.48 Calculate the grams of solute needed to prepare each of the following solutions:
   a. 2.0 L of a 6.0 M NaOH solution
   b. 5.0 L of a 0.10 M $CaCl_2$ solution

7.49 What volume in liters provides the following amounts of solute?
    a. 3.0 moles of NaOH from a 2.0 M NaOH solution
    b. 15 moles of NaCl from a 1.5 M NaCl solution

7.50 What volume in liters provides the following amounts of solute?
    a. 0.100 mole of KCl from a 4.0 M KCl solution
    b. 5.0 moles of HCl from a 6.0 M HCl solution

7.51 To make tomato soup, you add one can of water to the condensed soup. Why is this a dilution?

7.52 A can of frozen lemonade calls for the addition of three cans of water to make a pitcher of the beverage. Why is this a dilution?

7.53 Calculate the final concentration of each of the following diluted solutions:
    a. 2.0 L of a 6.0 M HCl solution is added to water so that the final volume is 6.0 L.
    b. Water is added to 0.50 L of a 12 M NaOH solution to make 3.0 L of a diluted NaOH solution.
    c. A 10.0-mL sample of 25% (m/v) KOH solution is diluted with water so that the final volume is 100.0 mL.
    d. A 50.0-mL sample of 15% (m/v) H_2SO_4 solution is added to water to give a final volume of 250 mL.

7.54 Calculate the final concentration of each of the following diluted solutions:
    a. 1.0 L of a 4.0 M HNO_3 solution is added to water so that the final volume is 8.0 L.
    b. Water is added to 0.25 L of a 6.0 M KOH solution to make 2.0 L of a diluted KOH solution.
    c. A 50.0-mL sample of an 8.0% (m/v) NaOH is diluted with water so that the final volume is 200.0 mL.
    d. A 5.0-mL sample of 50.0% (m/v) acetic acid ($HC_2H_3O_2$) solution is added to water to give a final volume of 25 mL.

7.55 What is the final volume in liters of each of the following diluted solutions?
    a. a 0.20 M HCl solution prepared from 20.0 mL of a 6.0 M HCl solution
    b. a 2.0% (m/v) NaOH solution prepared from 50.0 mL of a 10.0% (m/v) NaOH solution
    c. a 0.50 M H_3PO_4 solution prepared from 0.500 L of a 6.0 M H_3PO_4 solution

7.56 What is the final volume (L) of each of the following diluted solutions?
    a. a 1.0% (m/v) KOH solution prepared from 10.0 mL of a 20.0% KOH solution
    b. a 0.10 M HCl solution prepared from 25 mL of a 6.0 M HCl solution
    c. a 1.0 M NaOH solution prepared from 50.0 mL of a 12 M NaOH solution

---

▪ 7.6 Solutions in Chemical Reactions

LEARNING GOAL

Given the volume and molarity of a solution, calculate the amount of another reactant or product in the reaction.

In Section 7.5, we learned that the moles of a solute can be determined if we know the molarity and volume of a solution. We can also determine the volume of a solution if we know the number of moles of a solute and its molarity. These are the types of calculations we need to do for the substances involved in a chemical reaction that takes place in aqueous solution. If we have a balanced equation, we can use the molarity and volume of a solution to determine the moles of a substance required or produced in a chemical reaction. We can also use molarity and the number of moles of a solute to determine the volume of a solution as seen in Sample Problem 7.15.

SAMPLE PROBLEM 7.15

▪ Volume of a Solution in a Reaction

Zinc reacts with HCl to produce $ZnCl_2$ and hydrogen gas H_2.

$$Zn(s) + 2HCl(aq) \longrightarrow ZnCl_2(aq) + H_2(g)$$

How many liters of a 1.50 M HCl solution completely react with 5.32 g of zinc?

SOLUTION

STEP 1 **Given** 5.32 g Zn and a 1.50 M HCl solution
Need L HCl solution

STEP 2 **Plan** We can use the molar mass of Zn to find the moles of Zn and the mole–mole factor from the balanced equation to convert moles of Zn to moles of HCl. Since the concentration of the HCl solution is given, the molarity can be used to convert moles to volume in liters.

g Zn | Molar mass | moles Zn | Mole–mole factor | moles HCl | Molarity | L HCl

Guide to Calculations Involving Solutions in Chemical Reactions

STEP 1
State the given and needed quantities.

STEP 2
Write a plan to calculate needed quantity or concentration.

STEP 3
Write equalities and conversion factors including mole–mole and concentration factors.

STEP 4
Set up problem to calculate needed quantity or concentration.

STEP 3 **Equalities/Conversion Factors**

Molar mass of Zn

1 mole Zn = 65.4 g Zn

$$\frac{1 \text{ mole Zn}}{65.4 \text{ g Zn}} \quad \text{and} \quad \frac{65.4 \text{ g Zn}}{1 \text{ mole Zn}}$$

Mole–mole factor

1 mole Zn = 2 moles HCl

$$\frac{1 \text{ mole Zn}}{2 \text{ moles HCl}} \quad \text{and} \quad \frac{2 \text{ moles HCl}}{1 \text{ mole Zn}}$$

Molarity of HCl solution

1 L HCl = 1.50 moles HCl

$$\frac{1 \text{ L HCl}}{1.50 \text{ moles HCl}} \quad \text{and} \quad \frac{1.50 \text{ moles HCl}}{1 \text{ L HCl}}$$

STEP 4 **Set Up Problem** We can write the problem setup as seen in our plan.

$$5.32 \text{ g Zn} \times \frac{1 \text{ mole Zn}}{65.4 \text{ g Zn}} \times \frac{2 \text{ moles HCl}}{1 \text{ mole Zn}} \times \frac{1 \text{ L HCl}}{1.50 \text{ moles HCl}} = 0.108 \text{ L HCl}$$

STUDY CHECK

Using the reaction in Sample Problem 7.15, how many grams of zinc can react with 225 mL of 0.200 M HCl?

SAMPLE PROBLEM 7.16

■ Volume of a Reactant

How many L 0.250 M $BaCl_2$ is needed to react with 0.0325 L 0.160 M Na_2SO_4 solution?

$$Na_2SO_4(aq) + BaCl_2(aq) \longrightarrow BaSO_4(s) + 2NaCl(aq)$$

SOLUTION

We list the given volumes and molarities for the components in the chemical equation:

STEP 1 **Given** 0.0325 L of 0.160 M Na_2SO_4 and 0.250 M $BaCl_2$
Need $BaCl_2$

STEP 2 **Plan** We use the volume and molarity of the Na_2SO_4 solution to determine the moles of Na_2SO_4. The moles of Na_2SO_4 are converted to moles of $BaCl_2$ using the mole–mole factor from the equation. Finally, we use the molarity of $BaCl_2$ to calculate the volume of $BaCl_2$ in liters.

L Na_2SO_4 ▸ Molarity ▸ moles Na_2SO_4 ▸ Mole–mole factor ▸ mole $BaCl_2$ ▸ Molarity ▸ L $BaCl_2$

STEP 3 Equalities/Conversion Factors

Molarity of Na_2SO_4

1 L Na_2SO_4 = 0.160 mole Na_2SO_4

$$\frac{1\ L\ Na_2SO_4}{0.160\ mole\ Na_2SO_4} \quad and \quad \frac{0.160\ mole\ Na_2SO_4}{1\ L\ Na_2SO_4}$$

Mole–mole factor

1 mole Na_2SO_4 = 1 mole $BaCl_2$

$$\frac{1\ mole\ Na_2SO_4}{1\ mole\ BaCl_2} \quad and \quad \frac{1\ mole\ BaCl_2}{1\ mole\ Na_2SO_4}$$

Molarity of $BaCl_2$

1 L $BaCl_2$ = 0.250 mole $BaCl_2$

$$\frac{1\ L\ BaCl_2}{0.250\ mole\ BaCl_2} \quad and \quad \frac{0.250\ mole\ BaCl_2}{1\ L\ BaCl_2}$$

STEP 4 Set Up Problem

$$0.0325\ L\ \cancel{Na_2SO_4} \times \frac{0.160\ \cancel{moles\ Na_2SO_4}}{1\ L\ \cancel{Na_2SO_4}} \times \frac{1\ \cancel{mole\ BaCl_2}}{1\ \cancel{mole\ Na_2SO_4}} \times \frac{1\ L\ \cancel{BaCl_2}}{0.250\ \cancel{mole\ BaCl_2}} = 0.0208\ L$$

STUDY CHECK

For the reaction in Sample Problem 7.16, how many milliliters of 0.330 M Na_2SO_4 are needed to react with 26.8 mL of a 0.216 M $BaCl_2$ solution?

QUESTIONS AND PROBLEMS

■ Solutions in Chemical Reactions

7.57 Given the reaction

$$Pb(NO_3)_2(aq) + 2KCl(aq) \longrightarrow PbCl_2(s) + 2KNO_3(aq)$$

a. How many grams of $PbCl_2$ will be formed from 50.0 mL of 1.50 M KCl and excess $Pb(NO_3)_2$?

b. How many milliliters of 2.00 M $Pb(NO_3)_2$ will react with 50.0 mL of 1.50 M KCl?

7.58 In the reaction

$$NiCl_2(aq) + 2NaOH(aq) \longrightarrow Ni(OH)_2(s) + 2NaCl(aq)$$

a. How many milliliters of 0.200 M NaOH are needed to react with 18.0 mL of 0.500 M $NiCl_2$?

b. How many grams of $Ni(OH)_2$ are produced from the reaction of 35.0 mL of 1.75 M NaOH?

7.59 In the reaction

$$Mg(s) + 2HCl(aq) \longrightarrow MgCl_2(aq) + H_2(g)$$

a. How many milliliters of a 6.00 M HCl solution are required to react with 15.0 g magnesium?

b. How many moles of hydrogen gas form when 0.500 L of 2.00 M HCl reacts?

7.60 The calcium carbonate in limestone reacts with HCl to produce a calcium chloride solution and carbon dioxide gas.

$$CaCO_3(s) + 2HCl(aq) \longrightarrow CaCl_2(aq) + H_2O(l) + CO_2(g)$$

a. How many milliliters of 0.200 M HCl can react with 8.25 g $CaCO_3$?
b. How many moles of CO_2 form when 15.5 mL of 3.00 M HCl react with excess $CaCO_3$?

LEARNING GOAL

Identify a mixture as a solution, a colloid, or a suspension. Describe osmosis and dialysis.

■ 7.7 Properties of Solutions

The solute particles in a solution play an important role in determining the properties of that solution. In most of the solutions discussed so far, the solute is dissolved as small particles that are uniformly dispersed throughout the solvent to give a homogeneous solution. When you observe a solution, such as salt water, you cannot visually distinguish the solute from the solvent. The solution appears transparent. The particles are so small that they go through filters and through semipermeable membranes. A **semipermeable membrane** allows solvent molecules such as water and very small solute particles to pass through, but not large solute molecules.

HEALTH NOTE

■ Colloids and Solutions in the Body

Colloids in the body are isolated by semipermeable membranes. For example, the intestinal lining allows solution particles to pass into the blood and lymph circulatory systems. However, the colloids from foods are too large to pass through the membrane, and they remain in the intestinal tract. Digestion breaks down large colloidal particles, such as starch and protein, into smaller particles, such as glucose and amino acids, that can pass through the intestinal membrane and enter the circulatory system. Certain foods, such as bran, a fiber, cannot be broken down by human digestive processes, and they move through the intestine intact.

Because large proteins, such as enzymes, are colloids, they remain inside cells. However, many of the substances that must be obtained by cells, such as oxygen, amino acids, electrolytes, glucose, and minerals, can pass through cellular membranes. Waste products, such as urea and carbon dioxide, pass out of the cell to be excreted.

TABLE 7.9

Examples of Colloids

	Substance Dispersed	Dispersing Medium
Fog, clouds, sprays	Liquid	Gas
Dust, smoke	Solid	Gas
Shaving cream, whipped cream, soapsuds	Gas	Liquid
Styrofoam, marshmallows	Gas	Solid
Mayonnaise, butter, homogenized milk, hand lotions	Liquid	Liquid
Cheese, butter	Liquid	Solid
Blood plasma, paints (latex), gelatin	Solid	Liquid
Cement, pearls	Solid	Solid

Colloids

The particles in colloidal dispersions, or **colloids**, are much larger than solute particles in a solution. Colloidal particles are large molecules, such as proteins, or groups of molecules or ions. Colloids are homogeneous mixtures that do not separate or settle out. Colloidal particles are small enough to pass through filters but too large to pass through semipermeable membranes. Table 7.9 lists several examples of colloids.

Suspensions

Suspensions are heterogeneous, nonuniform mixtures that are very different from solutions or colloids. The particles of a suspension are so large that they can often be seen with the naked eye. They are trapped by filters and semipermeable membranes.

The weight of the suspended solute particles causes them to settle out soon after mixing. If you stir muddy water, it mixes but then quickly separates as the suspended

TABLE 7.10

Comparison of Solutions, Colloids, and Suspensions

Type of Mixture	Type of Particle	Settling	Separation
Solution	Small particles such as atoms, ions, or small molecules	Particles do not settle	Particles cannot be separated by filters or semipermeable membranes
Colloid	Larger molecules or groups of molecules or ions	Particles do not settle	Particles can be separated by semipermeable membranes but not by filters
Suspension	Very large particles that may be visible	Particles settle rapidly	Particles can be separated by filters

particles settle to the bottom and leave clear liquid at the top. You can find suspensions among the medications in a hospital or in your medicine cabinet. These include Kaopectate, calamine lotion, antacid mixtures, and liquid penicillin. It is important to "shake well before using" to suspend all the particles before giving a medication that is a suspension.

Water-treatment plants make use of the properties of suspensions to purify water. When coagulants such as aluminum sulfate or ferric sulfate are added to untreated water, they react with impurities to form large suspended particles called floc. In the water-treatment plant, a system of filters traps the suspended particles but clean water passes through.

Table 7.10 compares the different types of mixtures and Figure 7.6 illustrates some properties of solutions, colloids, and suspensions.

● Solution
▲ Colloid
■ Suspension

Filter

Semipermeable membrane

Settling

(a)   (b)   (c)

FIGURE 7.6

Properties of different types of mixtures: **(a)** suspensions settle out; **(b)** suspensions are separated by a filter; **(c)** solution particles go through a semipermeable membrane, but colloids and suspensions do not.

Q **A filter paper can be used to separate suspension particles from a solution, but a semipermeable membrane is needed to separate colloids from a solution. Explain.**

■ Classifying Types of Mixtures

Classify each of the following as a solution, colloid, or suspension:

a. a mixture that settles rapidly upon standing
b. a mixture whose solute particles pass through both filters and membranes

SOLUTION

a. suspension   **b.** solution

STUDY CHECK

Enzymes are large protein molecules that catalyze chemical reactions inside the cells of the body. If an aqueous mixture of enzymes cannot pass through the cell membrane, are they solutions or colloids?

◾ Osmosis and Dialysis

WEB TUTORIAL
Diffusion
Osmosis

The movement of water into and out of the cells of plants as well as our own bodies is an important biological process. In a process called **osmosis**, the solvent water moves through a semipermeable membrane from a solution that has a lower concentration of solute into a solution where the solute concentration is higher. This movement, or osmosis, of water happens in a direction that equalizes (or attempts to equalize) the concentrations on both sides of the membrane. For example, when water is separated from sucrose (sugar) solution by a semipermeable membrane, water molecules, but not the larger sucrose molecules, move through the membrane to dilute the sucrose solution. As more water flows across the membrane, the level of the sugar solution rises as the level of water on the other side decreases.

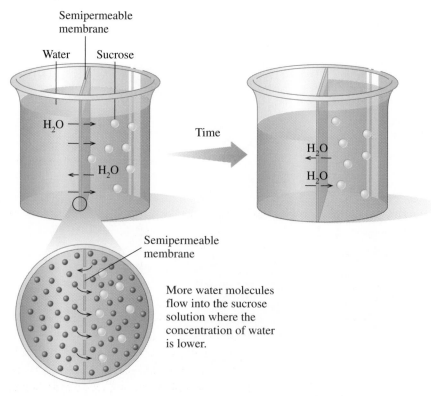

More water molecules flow into the sucrose solution where the concentration of water is lower.

Osmotic pressure is the pressure that prevents the flow of additional water into the more concentrated solution. The water molecules flow back and forth between the two compartments as the system comes to equilibrium, but the levels remain fixed. The osmotic pressure of a solution depends on the number of solute particles in the solution. Pure water has no osmotic pressure. The greater the number of particles dissolved in a solution, the higher its osmotic pressure.

If a pressure greater than the osmotic pressure is applied to a solution, osmosis is reversed and solvent flows out of the solution and the level in the solvent compartment increases. This process, known as reverse osmosis, is used in desalinization plants in some parts of the world to produce drinking water from sea (salt) water.

SAMPLE PROBLEM 7.18

▪ Osmotic Pressure

A 2% (m/v) sucrose solution and an 8% (m/v) sucrose solution are separated by a semipermeable membrane.

a. Which sucrose solution exerts the greater osmotic pressure?
b. In what direction does water flow initially?
c. Which solution will have the higher level of liquid at equilibrium?

SOLUTION

a. The 8% (m/v) sucrose solution has the higher solute concentration, more solute particles, and the greater osmotic pressure.
b. Initially, water will flow out of the 2% (m/v) solution into the more concentrated 8% (m/v) solution.
c. The level of the 8% (m/v) solution will be higher.

STUDY CHECK

If a 10% glucose solution is separated from a 5% (m/v) glucose solution by a semipermeable membrane, which solution will decrease in volume?

Isotonic Solutions

Because the cell membranes in biological systems are semipermeable, osmosis is an ongoing process. The solutes in body solutions such as blood, tissue fluids, lymph, and plasma all exert osmotic pressure. Most intravenous solutions are **isotonic solutions**, which exert the same osmotic pressure as body fluids. *Iso* means "equal to," and *tonic* refers to the osmotic pressure of the solution in the cell. In the hospital, isotonic solutions or **physiological solutions** include 0.90% (m/v) NaCl solution and 5.0% (m/v) glucose solution. Although they do not contain the same particles as the body fluids, they exert the same osmotic pressure.

Hypotonic and Hypertonic Solutions

A red blood cell placed in an isotonic solution retains its normal volume because there is an equal flow of water into and out of the cell. (See Figure 7.7a.) However, if a red blood cell is placed in a solution that is not isotonic, the differences in osmotic pressure inside and outside the cell can drastically alter the volume of the cell. When a red blood cell is placed in pure water, a **hypotonic solution** (*hypo* means "lower

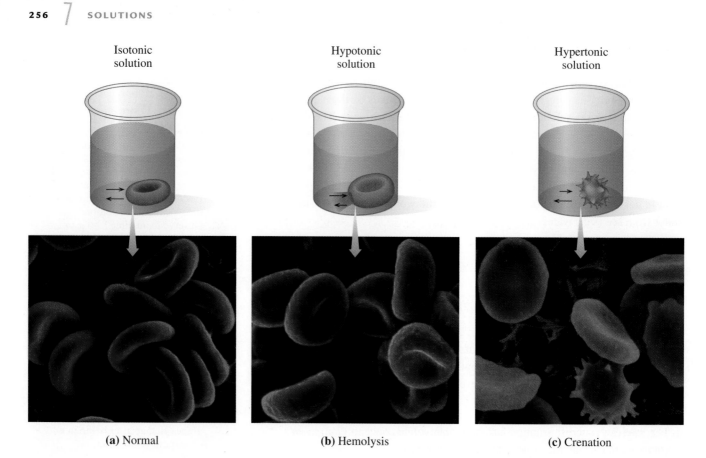

Isotonic
solution

Hypotonic
solution

Hypertonic
solution

(a) Normal

(b) Hemolysis

(c) Crenation

FIGURE 7.7

(a) In an isotonic solution, a red blood cell retains its normal volume. **(b)** Hemolysis: In a hypotonic solution, water flows into a red blood cell, causing it to swell and burst. **(c)** Crenation: In a hypertonic solution, water leaves the red blood cell, causing it to shrink.

Q   What happens to a red blood cell placed in a 4% NaCl solution?

than"), water flows into the cell by osmosis. (See Figure 7.7b.) The increase in fluid causes the cell to swell, and possibly burst, a process called **hemolysis**. A similar process occurs when you place dehydrated food, such as raisins or dried fruit, in water. The water enters the cells and the food becomes plump and smooth.

If a red blood cell is placed in a **hypertonic solution**, which has a higher solute concentration (*hyper* means "greater than"), water leaves the cell by osmosis. Suppose a red blood cell is placed in a 10% (m/v) NaCl solution. Because the osmotic pressure in the red blood cell is equal to that of a 0.90% (m/v) NaCl solution, the 10% (m/v) NaCl solution has a much greater osmotic pressure. As water is lost, the cell shrinks, a process called **crenation**. (See Figure 7.6c.) A similar process occurs when making pickles, which uses a hypertonic salt solution that causes the cucumbers to shrivel as they lose water.

SAMPLE PROBLEM   7.19

■ **Isotonic, Hypotonic, and Hypertonic Solutions**

Describe each of the following solutions as isotonic, hypotonic, or hypertonic. Indicate whether a red blood cell placed in each solution will undergo hemolysis, crenation, or no change.

a. a 5.0% (m/v) glucose solution

b. a 0.2% (m/v) NaCl solution

SOLUTION

a. A 5.0% (m/v) glucose solution is isotonic. A red blood cell will not undergo any change.

b. A 0.2% (m/v) NaCl solution is hypotonic. A red blood cell will undergo hemolysis.

STUDY CHECK

What is the effect of a 10% (m/v) glucose solution on a red blood cell?

Dialysis

Dialysis is a process that is similar to osmosis. In dialysis, a semipermeable membrane, called a dialyzing membrane, permits small solute molecules and ions as well as solvent water molecules to pass through, but it retains large particles, such as colloids. Dialysis is a way to separate solution particles from colloids.

Suppose we fill a cellophane bag with a solution containing NaCl, glucose, starch, and protein and place it in pure water. Cellophane is a dialyzing membrane, and the sodium ions, chloride ions, and glucose molecules will pass through it into the surrounding water. However, starch and protein remain inside because they are colloids. Water molecules will flow by osmosis into the cellophane bag. Eventually the concentrations of sodium ions, chloride ions, and glucose molecules inside and outside the dialysis bag become equal. To remove more NaCl or glucose, the cellophane bag must be placed in a fresh sample of pure water.

Initial   Final

○ Solution particles such as Na⁺, Cl⁻, glucose
● Colloidal particles such as protein, starch

■ Dialysis by the Kidneys and the Artificial Kidney

The fluids of the body undergo dialysis by the membranes of the kidneys, which remove waste materials, excess salts, and water. In an adult, each kidney contains about 2 million nephrons. At the top of each nephron, there is a network of arterial capillaries called the glomerulus.

As blood flows into the glomerulus, small particles, such as amino acids, glucose, urea, water, and certain ions, will move through the capillary membranes into the nephron. As this solution moves through the nephron, substances still of value to the body (such as amino acids, glucose, certain ions, and 99% of the water) are reabsorbed. The major waste product, urea, is excreted in the urine.

HEMODIALYSIS

If the kidneys fail to dialyze waste products, increased levels of urea can become life-threatening in a relatively short time. A person with kidney failure must use an artificial kidney, which cleanses the blood by **hemodialysis**.

A typical artificial kidney machine contains a large tank filled with about 100 L of water containing selected electrolytes. In the center of this dialyzing bath (dialysate), there is a dialyzing coil or membrane made of cellulose tubing. As the patient's blood flows through the dialyzing coil, the highly concentrated waste products dialyze out of the blood. No blood is lost because the membrane is not permeable to large particles such as red blood cells.

Dialysis patients do not produce much urine. As a result, they retain large amounts of water between dialysis treatments, which produces a strain on the heart. The intake of fluids for a dialysis patient may be restricted to as little as a few teaspoons of water a day. In the dialysis procedure, the pressure of the blood is increased as it circulates through the dialyzing coil so water can be squeezed out of the blood. For some dialysis patients, 2–10 L of water may be removed during one treatment. Dialysis patients have from two to three treatments a week, each treatment requiring about 5–7 hr. Some of the newer treatments require less time. For many patients, dialysis is done at home with a home dialysis unit.

Dialyzed blood

Blood

Pump

Dialysate

Dialyzing coil

○ = Urea and other waste products

QUESTIONS AND PROBLEMS

■ Properties of Solutions

7.61 Identify the following as characteristic of a solution, colloid, or suspension:
   a. a mixture that cannot be separated by a semipermeable membrane
   b. a mixture that settles out upon standing

7.62 Identify the following as characteristic of a solution, colloid, or suspension:
   a. Particles of this mixture remain inside a semipermeable membrane, but pass through filters.
   b. The particles of solute in this solution are very large and visible.

7.63 a. How do plants obtain water from the ground?
   b. How does a pickle get all shriveled up?

7.64 a. Why should you not drink large quantities of seawater?
   b. Why does salt preserve foods?

7.65 A 10% (m/v) starch solution is separated from pure water by an osmotic membrane.
   a. Which solution has the higher osmotic pressure?
   b. In which direction will water flow initially?
   c. In which compartment will the volume level rise?

7.66 Two solutions, a 0.1% (m/v) albumin solution and a 2% (m/v) albumin solution, are separated by a semipermeable membrane. (Albumins are colloidal proteins.)
a. Which compartment has the higher osmotic pressure?
b. In which direction will water flow initially?
c. In which compartment will the volume level rise?

7.67 Indicate the compartment (A or B) that will increase in volume for each of the following pairs of solutions separated by semipermeable membranes:

A	B
a. 5.0% (m/v) glucose	10% (m/v) glucose
b. 4% (m/v) albumin	8% (m/v) albumin
c. 0.1% (m/v) NaCl	10% (m/v) NaCl

7.68 Indicate the compartment (A or B) that will increase in volume for each of the following pairs of solutions separated by semipermeable membranes:

A	B
a. 20% (m/v) glucose	10% (m/v) glucose
b. 10% (m/v) albumin	2% (m/v) albumin
c. 0.5% (m/v) NaCl	5% (m/v) NaCl

7.69 Are the following solutions isotonic, hypotonic, or hypertonic compared with a red blood cell?
a. distilled H_2O
b. 1% (m/v) glucose
c. 0.90% (m/v) NaCl
d. 5.0% (m/v) glucose

7.70 Will a red blood cell undergo crenation, hemolysis, or no change in each of the following solutions?
a. 1% (m/v) glucose
b. 2% (m/v) NaCl
c. 5% (m/v) NaCl
d. 0.1% (m/v) NaCl

7.71 Each of the following mixtures is placed in a dialyzing bag and immersed in distilled water. Which substances will be found outside the bag in the distilled water?
a. NaCl solution
b. starch (colloid) and alanine (amino acid solution)
c. NaCl solution and starch (colloid)
d. urea solution

7.72 Each of the following mixtures is placed in a dialyzing bag and immersed in distilled water. Which substances will be found outside the bag in the distilled water?
a. KCl solution and glucose solutions
b. an albumin solution (colloid)
c. an albumin solution (colloid), KCl solution, and glucose solution
d. urea solution and NaCl solution

CONCEPT MAP

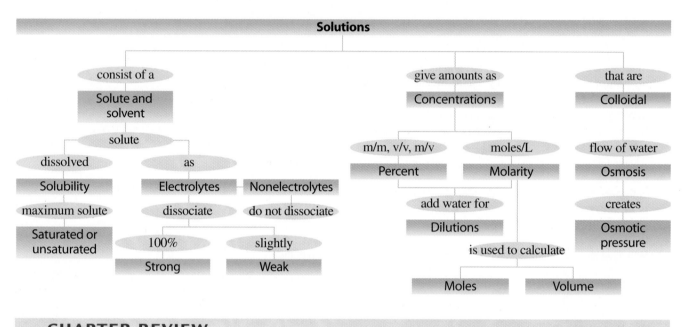

CHAPTER REVIEW

7.1 Solutions
A solution forms when a solute dissolves in a solvent. In a solution, the particles of solute are evenly distributed in the solvent. The solute and solvent may be solid, liquid, or gas. The polar O—H bond leads to hydrogen bonding in water molecules. An ionic solute dissolves in water, a polar solvent, because the polar water molecules attract and pull the ions into solution, where they become hydrated. The expression "like dissolves like" means that a polar ionic solute dissolves in a polar solvent while a nonpolar solute requires a nonpolar solvent.

7.2 Electrolytes and Nonelectrolytes
Substances that release ions in water are called electrolytes because the solution will conduct an electrical current. Strong electrolytes are completely ionized, whereas weak electrolytes are only partially ionized. Nonelectrolytes are substances that dissolve in water to produce molecules and cannot conduct electrical currents. An equivalent is the amount of a electrolyte that carries one mole of positive or negative charge. One mole of Na^+ is 1 equivalent. One mole of Ca^{2+} has 2 equivalents. In fluid replacement solutions, the concentrations of electrolytes are expressed as mEq/L of solution.

7.3 Solubility
A solution that contains the maximum amount of dissolved solute is a saturated solution. The solubility of a solute is the maximum amount of a solute that can dissolve in 100 g of solvent. A solution containing less than the maximum amount of dissolved solute is unsaturated. An increase in temperature increases the solubility of most solids in water, but decreases the solubility of gases in water.

7.4 Percent Concentration
The concentration of a solution is the amount of solute dissolved in a certain amount of solution. Mass percent expresses the ratio of the mass of solute to the mass of solution multiplied by 100. Percent concentration is also expressed as volume/volume and mass/volume ratios. In calculations of grams or milliliters of solute or solution, the percent concentration is used as a conversion factor.

7.5 Molarity and Dilution
Molarity is the moles of solute per liter of solution. Units of molarity, moles/liter, are used in conversion factors to solve for moles of solute or volume of solution. In dilution, the volume of solvent increases and the solute concentration decreases.

7.6 Solutions in Chemical Reactions
When solutions are involved in chemical reactions, the moles of a substance in solution can be determined from the volume and molarity of the solution. If the mass or solution volume and molarity of substances in a reaction are given, the balanced equation can be used to determine the quantities or concentrations of any of the other substances in the reaction.

7.7 Properties of Solutions
Colloids contain particles that do not settle out and pass through filters but not through semipermeable membranes. Suspensions have very large particles that settle out of solution.

   In osmosis, solvent (water) passes through a semipermeable membrane from a solution of a lower solute concentration to a solution of a higher concentration. Isotonic solutions have osmotic pressures equal to that of body fluids. A red blood cell maintains its volume in an isotonic solution, but swells and may burst (hemolyzes) in a hypotonic solution, and shrinks (crenates) in a hypertonic solution. In dialysis, water and small solute particles pass through a dialyzing membrane, while larger particles are retained.

KEY TERMS

colloid A mixture having particles that are moderately large. Colloids pass through filters but cannot pass through semipermeable membranes.

concentration A measure of the amount of solute that is dissolved in a specified amount of solution.

crenation The shriveling of a cell due to water leaving the cell when the cell is placed in a hypertonic solution.

dialysis A process in which water and small solute particles pass through a semipermeable membrane.

dilution A process by which water (solvent) is added to a solution to increase the volume and decrease (dilute) the concentration of the solute.

electrolyte A substance that produces ions when dissolved in water; its solution conducts electricity.

equivalent (Eq) The amount of a positive or negative ion that supplies 1 mole of electrical charge.

hemodialysis A mechanical cleansing of the blood by an artificial kidney using the principle of dialysis.

hemolysis A swelling and bursting of red blood cells in a hypotonic solution due to an increase in fluid volume.

Henry's law The solubility of a gas in a liquid is directly related to the pressure of that gas above the liquid.

hydration The process of surrounding dissolved ions by water molecules.

hydrogen bond The attraction between a partially positive hydrogen atom in one molecule and a highly electronegative atom such as oxygen in another molecule.

hypertonic solution A solution that has a higher osmotic pressure than the red blood cells of the body.

hypotonic solution A solution that has a lower osmotic pressure than the red blood cells of the body.

isotonic solution A solution that has the same osmotic pressure as that of the red blood cells of the body.

mass percent The grams of solute in exactly 100 grams of solution.

mass/volume percent The grams of solute in exactly 100 mL of solution.

molarity (M) The number of moles of solute in exactly 1 L of solution.

nonelectrolyte A substance that dissolves in water as molecules; its solution will not conduct an electrical current.

osmosis The flow of a solvent, usually water, through a semipermeable membrane into a solution of higher solute concentration.

osmotic pressure The pressure that prevents the flow of water into the more concentrated solution.

physiological solution A solution that exerts the same osmotic pressure as normal body fluids.

saturated solution A solution containing the maximum amount of solute that can dissolve at a given temperature. Any additional solute will remain undissolved in the container.

semipermeable membrane A membrane that permits the passage of certain substances while blocking or retaining others.

solubility The maximum amount of solute that can dissolve in exactly 100 g of solvent, usually water, at a given temperature.

solute The component in a solution that changes state upon dissolving; if no change in state occurs, it is the component present in the smaller quantity.

solution A homogeneous mixture in which the solute is made up of small particles (ions or molecules) that can pass through filters and semipermeable membranes.

solvent The substance in which the solute dissolves; usually the component present in greatest amount.

strong electrolyte A polar or ionic compound that ionizes completely when it dissolves in water. Its solution is a good conductor of electricity.

suspension A mixture in which the solute particles are large enough and heavy enough to settle out and be retained by both filters and semipermeable membranes.

unsaturated solution A solution that contains less solute than can be dissolved.

volume percent A percent concentration that relates the volume of the solute to the volume of the solution.

weak electrolyte A substance that produces only a few ions along with many molecules when it dissolves in water. Its solution is a weak conductor of electricity.

ADDITIONAL QUESTIONS AND PROBLEMS

7.73 Why does iodine dissolve in hexane, but not in water?

7.74 How does temperature and pressure affect the solubility of solids and gases in water?

7.75 Calculate the mass percent of a solution containing 15.5 g of Na_2SO_4 and 75.5 g of H_2O.

7.76 How many grams of K_2CO_3 are in 750 mL of a 3.5% (m/v) K_2CO_3 solution?

7.77 A patient receives all her nutrition from fluids given through the vena cava. Every 12 hours, 750 mL of a solution that is 4% (m/v) amino acids (protein) and 25% (m/v) glucose (carbohydrate) is given along with 500 mL of a 10% (m/v) lipid (fat).

a. In 1 day, how many grams of amino acids, glucose, and lipid are given to the patient?
b. How many kilocalories does she obtain in 1 day?

7.78 An 80-proof brandy is 40.0% (v/v) ethyl alcohol. The "proof" is twice the percent concentration of alcohol in the beverage. How many milliliters of alcohol are present in 750 mL of brandy?

7.79 How many milliliters of a 12% (v/v) propyl alcohol solution would you take to obtain 4.5 mL of propyl alcohol?

7.80 How many liters of a 5.0% (m/v) glucose solution would you take to obtain 75 g of glucose?

7.81 If you were in the laboratory, how would you prepare 0.250 L of a 2.00 M KCl solution?

7.82 What is the molarity of a solution containing 15.6 g of KCl in 274 mL of solution?

7.83 A solution is prepared with 70.0 g of HNO_3 and 130.0 g of H_2O. It has a density of 1.21 g/mL.
  a. What is the mass percent of the HNO_3 solution?
  b. What is the total volume of the solution?
  c. What is the mass/volume percent?
  d. What is its molarity (M)?

7.84 What is the molarity of a 15% (m/v) NaOH solution?

7.85 How many grams of solute are in each of the following solutions?
  a. 2.5 L of 3.0 M $Al(NO_3)_3$
  b. 75 mL of 0.50 M $C_6H_{12}O_6$

7.86 Calcium carbonate $CaCO_3$ reacts with stomach acid (HCl, hydrochloric acid) according to the following equation:

$$CaCO_3(s) + 2HCl(aq) \longrightarrow CaCl_2(aq) + H_2O(l) + CO_2(g)$$

Tums, an antacid, contains $CaCO_3$. If Tums is added to 20.0 mL of 0.400 M HCl, how many grams of CO_2 gas are produced?

7.87 The antacid Amphogel contains aluminum hydroxide $Al(OH)_3$. How many milliliters of 6.00 M HCl are required to react with 60.0 mL of 1.00 M $Al(OH)_3$?

$$Al(OH)_3(s) + 3HCl(aq) \longrightarrow AlCl_3(aq) + 3H_2O(aq)$$

7.88 Why would a dialysis unit (artificial kidney) use isotonic concentrations of NaCl, KCl, $NaHCO_3$, and glucose in the dialysate?

7.89 Why would solutions with high salt content be used to prepare dried flowers?

7.90 A patient on dialysis has a high level of urea, a high level of sodium, and a low level of potassium in the blood. Why is the dialyzing solution prepared with a high level of potassium but no sodium or urea?

7.91 Why can't you drink seawater even if you are stranded on a desert island?

UNDERSTANDING THE CONCEPTS

7.92 Match the diagrams with
  a. a polar solute and a polar solvent
  b. a nonpolar solute and a polar solvent
  c. a nonpolar solute and a nonpolar solvent

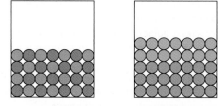

7.93 Select the diagram that represents the solution formed by a solute 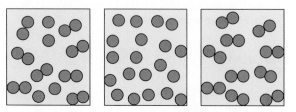 that is a
  a. nonelectrolyte
  b. weak electrolyte
  c. strong electrolyte

7.94 Do you think solution (1) has undergone heating or cooling to give the solid shown in (2) and (3)?

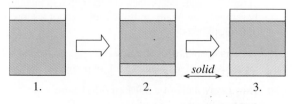

7.95 A pickle is made by soaking a cucumber in a *brine*, a salt–water solution. What makes the smooth cucumber become wrinkled like a prune?

7.96 Why do the lettuce leaves in a salad wilt after a vinaigrette dressing containing salt is added?

7.97 Select the container that represents the dilution of a 4% (m/v) KCl solution to each of the following:
  a. 2% (m/v) KCl
  b. 1% (m/v) KCl

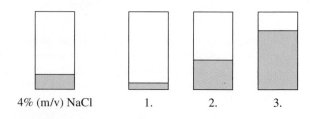

4% (m/v) NaCl   1.   2.   3.

7.98 Osmolarity (Osm) is the molar concentration of the particles—molecules and/or ions—in a solution. In physiological solutions, 5% (m/v) glucose ($C_6H_{12}O_6$) and 0.9% (m/v) NaCl, the osmolarity must be the same as for the blood, which is 0.3 Osm. Calculate the osmolarity of these physiological solutions.

7.99 A semipermeable membrane separates two compartments, A and B. If the levels of solutions in A and B are equal initially, select the diagram that illustrates the final levels for each of the following:

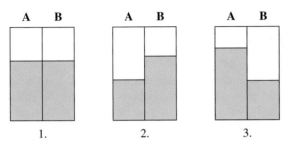

a. 2% (m/v) starch	8% (m/v) starch
b. 1% (m/v) starch	1% (m/v) starch
c. 5% (m/v) sucrose	1% (m/v) sucrose
d. 0.1% (m/v) sucrose	1% (m/v) sucrose

7.100 Select the diagram that represents the shape of a red blood cell when placed in each of the following solutions:

Normal red blood cell

a. 0.90% (m/v) NaCl
b. 10% (m/v) glucose
c. 0.01% (m/v) NaCl
d. 5.0% (m/v) glucose
e. 1% (m/v) glucose

QUESTIONS FOR TEAMWORK

7.101 In a laboratory experiment, a 10.0-mL sample of NaCl solution is poured into an evaporating dish with a mass of 24.10 g. The combined mass of the evaporating dish and NaCl is 36.15 g. After heating, the evaporating dish and dry NaCl have a combined mass of 25.50 g.
  a. What is the % (m/m) of the NaCl solution?
  b. What is the molarity (M) of the NaCl solution?
  c. If water is added to 10.0 mL of the initial NaCl solution to give a final volume of 60.0 mL, what is the molarity of the dilute NaCl solution?

7.102 A solution contains 4.56 g KCl in 175 mL of solution. If the density of the KCl solution is 1.12 g/mL, what are the percent (m/m) and molarity, M, for the potassium chloride solution?

7.103 A solution is prepared by dissolving 22.0 g of NaOH in 118.0 g of water. The NaOH solution has a density of 1.15 g/mL.
  a. What is the mass percent of the NaOH solution?
  b. What is the total volume of the solution?
  c. What is its molarity (M)?

7.104 A solution is prepared by dissolving 22.0 g NaOH in 118.0 g water. The NaOH solution has a density of 1.15 g/mL.

  a. What is the % (m/m) concentration of the NaOH solution?
  b. What is the total volume (mL) of the solution?
  c. What is the molarity (M) of the solution?

7.105 How many milliliters of a 1.75 M LiCl solution contain 15.2 g of LiCl?

7.106 How many grams of NaBr are contained in 75.0 mL of a 1.50 M NaBr solution?

7.107 Magnesium reacts with HCl to produce magnesium chloride and hydrogen gas

$$Mg(s) + 2HCl(aq) \longrightarrow MgCl_2(aq) + H_2(g)$$

What is the molarity of the HCl solution if 250. mL of the HCl solution reacts with magnesium to produce 4.20 L of H_2 gas measured at STP?

7.108 How many g NO gas can be produced from 80.0 mL of 4.00 M HNO_3 and excess Cu?

$$3Cu(s) + 8HNO_3(aq) \longrightarrow$$
$$3Cu(NO_3)_2(aq) + 4H_2O(l) + 2NO(g)$$

ANSWERS

Answers to Study Checks

7.1 Iodine is the solute, and ethyl alcohol is the solvent.

7.2 Yes. Both the solute and solvent are nonpolar substances; "like dissolves like."

7.3 A solution of a weak electrolyte will contain both molecules and ions.

7.4 0.194 mole Cl^-

7.5 57 g $NaNO_3$

7.6 The value of 65 g/100 g H_2O is more likely because the solubility of most solids increases when the temperature increases.

7.7 3.4% (m/m) NaCl solution

7.8 4.8% (m/v) Br_2 in CCl_4

7.9 18 g KCl and 207 g H_2O

7.10 2.12 M KNO_3

7.11 123 g $NaHCO_3$

7.12 4.5 moles HCl

7.13 4.8% (m/v) HCl

7.14 120. mL

7.15 1.47 g Zn

7.16 17.5 mL

7.17 colloids

7.18 5% glucose

7.19 The red blood cell will shrink (crenate).

Answers to Selected Questions and Problems

7.1 **a.** NaCl, solute; water, solvent
b. water, solute; ethanol, solvent
c. oxygen, solute; nitrogen, solvent

7.3 The polar water molecules pull the K^+ and I^- ions away from the solid and into solution, where they are hydrated.

7.5 **a.** water   **b.** CCl_4 **c.** water   **d.** CCl_4

7.7 In a solution of KF, only the ions of K^+ and F^- are present in the solvent. In an HF solution, there are a few ions of H^+ and F^- present but mostly dissolved HF molecules.

7.9 **a.** $KCl(s) \xrightarrow{H_2O} K^+(aq) + Cl^-(aq)$
b. $CaCl_2(s) \xrightarrow{H_2O} Ca^{2+}(aq) + 2Cl^-(aq)$
c. $K_3PO_4(s) \xrightarrow{H_2O} 3K^+(aq) + PO_4^{3-}(aq)$
d. $Fe(NO_3)_3(s) \xrightarrow{H_2O} Fe^{3+}(aq) + 3NO_3^-(aq)$

7.11 **a.** mostly molecules and a few ions   **b.** ions only
c. molecules only

7.13 **a.** strong electrolyte   **b.** weak electrolyte
c. nonelectrolyte

7.15 **a.** 1 Eq   **b.** 2 Eq   **c.** 2 Eq   **d.** 6 Eq

7.17 0.154 mole Na^+, 0.154 mole Cl^-

7.19 55 mEq/L

7.21 **a.** saturated   **b.** unsaturated

7.23 **a.** unsaturated   **b.** unsaturated   **c.** saturated

7.25 **a.** 68.0 g KCl   **b.** 12.0 g KCl

7.27 **a.** The solubility of solid solutes typically increases as temperature increases.
b. The solubility of a gas is less at a higher temperature.
c. Gas solubility is less at a higher temperature and the CO_2 pressure in the can is increased.

7.29 5% (m/m) is 5 g of glucose in 100 g of solution, whereas 5% (m/v) is 5 g of glucose in 100 mL of solution.

7.31 **a.** 17% (m/m)   **b.** 5.3% (m/m)

7.33 **a.** 30.% (m/v)   **b.** 11% (m/v)

7.35 **a.** 2.5 g KCl   **b.** 50. g NH_4Cl

7.37 79.9 mL of alcohol

7.39 **a.** 20 g mannitol   **b.** 300 g mannitol

7.41 2 L

7.43 **a.** 0.50 M glucose   **b.** 0.036 M KOH

7.45 **a.** 3.0 moles NaCl   **b.** 0.40 mole KBr

7.47 **a.** 120 g NaOH   **b.** 60. g KCl

7.49 **a.** 1.5 L   **b.** 10. L

7.51 Adding water (solvent) to the soup increases the volume and dilutes the tomato concentration.

7.53 **a.** 2.0 M HCl   **b.** 2.0 M NaOH
c. 2.5% (m/v) KOH   **d.** 3.0 (m/v) H_2SO_4

7.55 **a.** 0.60 L   **b.** 0.25 L   **c.** 6.0 L

7.57 **a.** 10.4 g $PbCl_2$ **b.** 18.8 mL $Pb(NO_3)_2$ solution

7.59 **a.** 206 mL HCl solution   **b.** 0.500 mole H_2 gas

7.61 **a.** solution   **b.** suspension

7.63 **a.** Water flows from wet soil through plant membranes toward the higher solute concentration in the plants.
b. Because a brine solution (salts in water) has a higher solute concentration, water flows from the cucumber to the brine and the cucumber becomes a pickle.

7.65 **a.** starch   **b.** from pure water into the starch
c. starch

7.67 **a.** B 10% (m/v) glucose   **b.** B 8% (m/v) albumin
c. B 10% (m/v) NaCl

7.69 **a.** hypotonic   **b.** hypotonic
c. isotonic   **d.** isotonic

7.71 **a.** NaCl   **b.** alanine   **c.** NaCl   **d.** urea

7.73 Because iodine is a nonpolar molecule, it will dissolve in hexane, a nonpolar solvent. Iodine does not dissolve in water because water is a polar solvent.

7.75 17.0% (m/m)

7.77 **a.** 60 g of amino acids, 380 g of glucose, and 100 g of lipids
b. 2700 kcal

7.79 38 mL of solution

7.81 To make a 2.00 M KCl solution, weigh out 37.3 g KCl (0.500 mole) and place into a volumetric flask. Add water to dissolve the KCl and give a final volume of 0.250 liter.

7.83 **a.** 35.0% (m/m) HNO_3 **b.** 165 mL
c. 42.4% (m/v)   **d.** 6.73 M

7.85 **a.** 1600 g $Al(NO_3)_3$ **b.** 6.8 g of $C_6H_{12}O_6$

7.87 30.0 mL HCl solution

7.89 The solution will dehydrate the flowers because water will flow out of the cells of the flowers into the more concentrated salt solution.

7.91 Drinking seawater will cause water to flow out of the body cells and further dehydrate a person.

7.93 **a.** 3   **b.** 1   **c.** 2

7.95 The skin of the cucumber acts like a semipermeable membrane and the more dilute solution inside flows into the brine solution.

7.97 **a.** 2   **b.** 3

7.99 **a.** 2   **b.** 1   **c.** 3   **d.** 2

8 Acids and Bases

"In a stat lab, we are sent blood samples of patients in emergency situations," says Audrey Trautwein, clinical laboratory technician, Stat Lab, Santa Clara Valley Medical Center. "We may need to assess the status of a trauma patient in ER or a patient who is in surgery. For example, an acidic blood pH diminishes cardiac function, and affects the actions of certain drugs. In a stat situation, it is critical that we obtain our results fast. This is done using a blood gas analyzer. As I put a blood sample into the analyzer, a small probe draws out a measured volume, which is tested simultaneously for pH, P_{O_2}, and P_{CO_2} as well as electrolytes, glucose, and hemoglobin. In about one minute we have our test results, which are sent to the doctor's computer."

the **Chemistry** place

Visit **www.aw-bc.com/chemplace** or use the CD-ROM in the book for extra quizzes, interactive tutorials, career resources, PowerPoint slides, math help, and case studies.

Lemons, grapefruit, and vinegar taste sour because they contain acids. We have acid in our stomach that helps us digest food; we produce lactic acid in our muscles when we exercise. Acid from bacteria turns milk sour to make cottage cheese or yogurt. Bases are solutions that neutralize acids. Sometimes we take antacids such as milk of magnesia to offset the effects of too much stomach acid.

The pH of a solution describes its acidity. The pH of body fluids, including blood and urine, is regulated primarily by the lungs and the kidneys. Major changes in the pH of the body fluids can severely affect biological activities within the cells. Buffers are present to prevent large fluctuations in pH.

LEARNING GOAL

Describe and name acids and bases.

WEB TUTORIAL
Nature of Acids and Bases

■ 8.1 Acids and Bases

The term *acid* comes from the Latin word *acidus*, which means "sour." We are familiar with the sour tastes of vinegar and lemons and other common acids in foods.

In the nineteenth century, Arrhenius was the first to describe **acids** as substances that produce hydrogen ions (H^+) when they dissolve in water. For example, hydrogen chloride ionizes in water to give hydrogen ions, H^+, and chloride ions, Cl^-. The hydrogen ions, H^+, give acids a sour taste, change blue litmus to red, and corrode some metals.

$$HCl(g) \xrightarrow{\text{H}_2\text{O}} H^+(aq) + Cl^-(aq)$$

Polar covalent compound   Ionization in water

■ Naming Acids

When an acid dissolves in water to produce hydrogen ion and a simple nonmetal anion, the prefix *hydro* is used before the name of the nonmetal and its *ide* ending is changed to *ic acid*. For example, hydrogen chloride (HCl) dissolves in water to form HCl(*aq*), which is named hydrochloric acid.

When an acid contains a polyatomic ion, the name of the acid comes from the name of the polyatomic ion. The *ate* in the name is replaced with *ic acid*. If the acid contains a polyatomic ion with an *ite* ending, its name ends with *ous acid*. The names of some common acids and their anions are listed in Table 8.1.

■ Bases

You may be familiar with some bases such as antacids, drain openers, and oven cleaners. According to the Arrhenius theory, **bases** are ionic compounds that dissociate into a metal ion and hydroxide ions (OH^-) when they dissolve in water. For example, sodium hydroxide is an Arrhenius base that dissociates in water to give sodium ions, Na^+, and hydroxide ions, OH^-.

NaOH(*s*)

● OH⁻
⊕ Na⁺

Water

$$NaOH(s) \xrightarrow{\text{H}_2\text{O}} Na^+(aq) + OH^-(aq)$$

Ionic compound   Dissociation   Hydroxide ion

TABLE 8.1

Naming Common Acids

Acid	Name of Acid	Anion	Name of Anion
HCl	**hydro**chlor**ic acid**	Cl^-	chlor**ide**
HBr	**hydro**brom**ic acid**	Br^-	brom**ide**
HNO_3	nitr**ic acid**	NO_3^-	nitr**ate**
HNO_2	nitr**ous acid**	NO_2^-	nitr**ite**
H_2SO_4	sulfur**ic acid**	SO_4^{2-}	sulf**ate**
H_2SO_3	sulfur**ous acid**	SO_3^{2-}	sulf**ite**
H_2CO_3	carbon**ic acid**	CO_3^{2-}	carbon**ate**
H_3PO_4	phosphor**ic acid**	PO_4^{3-}	phosph**ate**
$HClO_3$	chlor**ic acid**	ClO_3^-	chlor**ate**
$HClO_2$	chlor**ous acid**	ClO_2^-	chlor**ite**
CH_3COOH	acet**ic acid**	CH_3COO^-	acet**ate**

Most Arrhenius bases are formed from Groups 1A (1) and 2A (2) metals, such as NaOH, KOH, LiOH, and $Ca(OH)_2$. There are other bases such as $Al(OH)_3$ and $Fe(OH)_3$, but they are fairly insoluble. The hydroxide ions (OH^-) give Arrhenius bases common characteristics such as a bitter taste and soapy feel. Table 8.2 compares some characteristics of acids and bases.

Naming Bases

Typical Arrhenius bases are named as hydroxides.

Bases	Name
NaOH	sodium **hydroxide**
KOH	potassium **hydroxide**
$Ca(OH)_2$	calcium **hydroxide**
$Al(OH)_3$	aluminum **hydroxide**

TABLE 8.2

Some Characteristics of Acids and Bases

Characteristic	Acids	Bases
Reaction, Arrhenius	Produce H^+	Produce OH^-
Reaction, Brønsted–Lowry	Donate H^+	Accept H^+
Electrolytes	Yes	Yes
Taste	Sour	Bitter, chalky
Feel	May sting	Soapy, slippery
Litmus	Red	Blue
Phenolphthalein	Colorless	Red
Neutralization	Neutralize bases	Neutralize acids

SAMPLE PROBLEM 8.1

■ Names of Acids and Bases

Name the following as acids or bases.

a. H_3PO_4 **b.** NaOH   **c.** HNO_2 **d.** HBr

SOLUTION

a. phosphoric acid   **b.** sodium hydroxide
c. nitrous acid   **d.** hydrobromic acid

STUDY CHECK

Give the name for H_2SO_4 and KOH.

SAMPLE PROBLEM 8.2

■ **Dissociation of an Arrhenius Base**

Write an equation for the dissociation of LiOH(s) in water.

SOLUTION

LiOH dissociates in water to give a basic solution of lithium ions (Li^+) and hydroxide ions (OH^-).

$$LiOH(s) \xrightarrow{H_2O} Li^+(aq) + OH^-(aq)$$

STUDY CHECK

Calcium hydroxide is used in some antacids to counteract excess stomach acid. Write an equation for dissociation of calcium hydroxide.

QUESTIONS AND PROBLEMS

■ **Acids and Bases**

8.1 Indicate whether each of the following statements is characteristic of an acid or a base:
a. has a sour taste   **b.** neutralizes bases
c. produces H^+ ions in water   **d.** is named potassium hydroxide

8.2 Indicate whether each of the following statements is characteristic of an acid or a base:
a. neutralizes acids   **b.** produces OH^- in water
c. has a soapy feel   **d.** turns litmus red

8.3 Name each of the following as an acid or base:
a. HCl   **b.** $Ca(OH)_2$ **c.** H_2CO_3
d. HNO_3 **e.** H_2SO_3

8.4 Name each of the following as an acid or base:
a. $Al(OH)_3$ **b.** HBr   **c.** H_2SO_4 **d.** KOH   **e.** HNO_2

8.5 Write formulas for the following acids and bases:
a. magnesium hydroxide   **b.** hydrofluoric acid
c. phosphoric acid   **d.** lithium hydroxide
e. copper(II) hydroxide

8.6 Write formulas for the following acids and bases:
a. barium hydroxide   **b.** hydriodic acid   **c.** nitric acid
d. iron(III) hydroxide   **e.** sodium hydroxide

LEARNING GOAL

Identify conjugate acid–base pairs for Brønsted-Lowry acids and bases.

■ 8.2 Brønsted–Lowry Acids and Bases

Early in the twentieth century, Brønsted and Lowry expanded the definition of acids and bases. A **Brønsted–Lowry acid** donates a proton (hydrogen ion, H^+) to another substance, and a **Brønsted–Lowry base** accepts a proton.

A Brønsted–Lowry acid is a proton (H^+) donor.

A Brønsted–Lowry base is a proton (H^+) acceptor.

A free proton (H^+) does not actually exist in water. Its attraction to polar water molecules is so strong that the proton bonds to the water molecule and forms a **hydronium ion, H_3O^+.**

We can now write the formation of a hydrochloric acid solution as a transfer of a proton from hydrogen chloride to water. By accepting a proton in the reaction, water is acting as a base according to the Brønsted–Lowry concept.

In another reaction ammonia (NH_3) reacts with water. Because the nitrogen atom of NH_3 has a stronger attraction for a proton, water acts as an acid by donating a proton.

SAMPLE PROBLEM 8.3

■ **Acids and Bases**

In each of the following equations, identify the reactant that is an acid (H^+ donor) and the reactant that is a base (H^+ acceptor).

a. $HBr(aq) + H_2O(l) \longrightarrow H_3O^+(aq) + Br^-(aq)$

b. $H_2O(l) + CN^-(aq) \rightleftharpoons HCN(aq) + OH^-(aq)$

SOLUTION

a. HBr, acid; H_2O, base   **b.** H_2O, acid; CN^-, base

STUDY CHECK

When HNO_3 reacts with water, water acts as a base (H^+ acceptor). Write the equation for the reaction.

Conjugate Acid–Base Pairs

According to the Brønsted–Lowry theory, the reaction between an acid and base involves proton transfer. When molecules or ions are related by the loss or gain of one H^+, they make up a **conjugate acid–base pair**. Because protons are transferred in both a forward and reverse reaction, each acid–base reaction contains two conjugate acid–base pairs. In this general reaction, the acid HA has a conjugate base A^- and the base B has a conjugate acid BH^+.

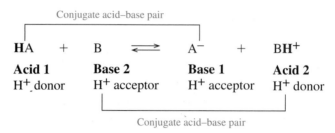

Now we can identify the conjugate acid–base pairs in a reaction such as hydrofluoric acid and water. Because the reaction is reversible, the conjugate acid H_3O^+ can transfer a proton to the conjugate base F^- and re-form the acid HF. Using the relationship of loss and gain of one H^+, we identify the conjugate acid–base pairs as HF and F^- along with H_3O^+ and H_2O.

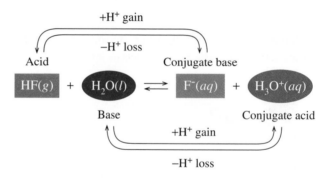

In another proton-transfer reaction, ammonia, NH_3, accepts H^+ from H_2O to form the conjugate acid NH_4^+ and conjugate base OH^-. Each of these conjugate acid–base pairs, NH_3 and NH_4^+ as well as H_2O and OH^-, are related by the loss and gain of one H^+.

SAMPLE PROBLEM 8.4

Conjugate Acid–Base Pairs

Write the formula of the conjugate base of each of the following Brønsted–Lowry acids.

a. $HClO_3$ **b.** H_2CO_3

SOLUTION

The conjugate base forms when the acid donates a proton.

a. ClO_3^- is the conjugate base of $HClO_3$
b. HCO_3^- is the conjugate base of H_2CO_3

STUDY CHECK

Write the conjugate acid of each of the following Brønsted–Lowry bases.

a. HS^- **b.** NO_2^-

SAMPLE PROBLEM 8.5

■ **Identifying Conjugate Acid–Base Pairs**

Identify the conjugate acid–base pairs in the following reaction:

$$HBr(aq) + NH_3(aq) \longrightarrow Br^-(aq) + NH_4^+(aq)$$

SOLUTION

In the reaction, HBr donates H^+ to NH_3. The resulting Br^- is the conjugate base and NH_4^+ is the conjugate acid. The conjugate acid–base pairs are HBr and Br^- along with NH_3 and NH_4^+.

STUDY CHECK

In the following reaction, identify the conjugate acid–base pairs.

$$HCN(aq) + SO_4^{2-}(aq) \rightleftharpoons CN^-(aq) + HSO_4^-(aq)$$

QUESTIONS AND PROBLEMS

■ **Brønsted–Lowry Acids and Bases**

8.7 In each of the following equations, identify the acid (proton donor) and base (proton acceptor) for the reactants:
  a. $HI(aq) + H_2O(l) \longrightarrow H_3O^+(aq) + I^-(aq)$
  b. $F^-(aq) + H_2O(l) \rightleftharpoons HF(aq) + OH^-(aq)$

8.2 In each of the following equations, identify the acid (proton donor) and base (proton acceptor) for the reactants:
  a. $CO_3^{2-}(aq) + H_2O(l) \rightleftharpoons HCO_3^-(aq) + OH^-(aq)$
  b. $H_2SO_4(aq) + H_2O(l) \longrightarrow H_3O^+(aq) + HSO_4^-(aq)$

8.9 Write the formula and name of the conjugate base for each of the following acids:
  a. HF   **b.** H_2O **c.** H_2CO_3 **d.** HSO_4^-

8.10 Write the formula and name of the conjugate base for each of the following acids:
  a. HCO_3^- **b.** H_3O^+ **c.** HPO_4^{2-} **d.** HNO_2

8.11 Write the formula and name of the conjugate acid for each of the following bases:
  a. CO_3^{2-} **b.** H_2O **c.** $H_2PO_4^-$ **d.** Br^-

8.12 Write the formula and name of the conjugate acid for each of the following bases:
  a. SO_4^{2-} **b.** CN^- **c.** OH^- **d.** ClO_2^-, chlorite ion

8.13 Identify the acid and base on the left side of the following equations and identify their conjugate species on the right side.
  a. $H_2CO_3(aq) + H_2O(l) \rightleftharpoons H_3O^+(aq) + HCO_3^-(aq)$
  b. $NH_4^+(aq) + H_2O(l) \rightleftharpoons H_3O^+(aq) + NH_3(aq)$
  c. $HCN(aq) + NO_2^-(aq) \rightleftharpoons CN^-(aq) + HNO_2(aq)$

8.14 Identify the acid and base on the left side of the following equations and identify their conjugate species on the right side.

a. $H_3PO_4(aq) + H_2O(l) \rightleftharpoons H_3O^+(aq) + H_2PO_4^-(aq)$

b. $CO_3^{2-}(aq) + H_2O(l) \rightleftharpoons OH^-(aq) + HCO_3^-(aq)$

c. $H_3PO_4(aq) + NH_3(aq) \rightleftharpoons NH_4^+(aq) + H_2PO_4^-(aq)$

LEARNING GOAL

■ Write equations for the dissociation of strong and weak acids.

8.3 Strengths of Acids and Bases

Acids are typically classified according to their ability to donate protons. Strong acids give up protons easily, whereas in a weak acid most of the molecules keep their protons and just a few give them up. Bases are classified in terms of their ability to accept protons. Strong bases have a strong attraction for protons, whereas weak bases have little attraction for protons.

Strong and Weak Acids

The six **strong acids** listed in Table 8.3 dissociate almost completely (≈100%) to give H_3O^+ ions and anions.

$$HCl(g) + H_2O(l) \longrightarrow H_3O^+(aq) + Cl^-(aq)$$

Other acids are **weak acids** that dissociate only slightly and produce small amounts of H_3O^+ ions. Even at high concentrations, weak acids produce low

Increasing acid strength

TABLE 8.3

Common Strong and Weak Acids

Strong acids

perchloric acid	$HClO_4$
sulfuric acid	H_2SO_4
hydroiodic acid	HI
hydrobromic acid	HBr
hydrochloric acid	HCl
nitric acid	HNO_3

Weak acids

hydronium ion	H_3O^+
hydrogen sulfate ion	HSO_4^-
nitrous acid	HNO_2
phosphoric acid	H_3PO_4
acetic acid	CH_3COOH
hydrofluoric acid	HF
carbonic acid	H_2CO_3
hydrosulfuric acid	H_2S
ammonium ion	NH_4^+
hydrocyanic acid	HCN
bicarbonate ion	HCO_3^-
hydrogen sulfide ion	HS^-
water	H_2O

FIGURE 8.1

A strong acid such as HCl is completely dissociated ($\approx 100\%$), whereas a weak acid such as CH_3COOH contains mostly molecules and a few ions.

Q **What is the difference between a strong acid and a weak acid?**

concentrations of hydronium ions. (See Figure 8.1.) The acidic products you drink or use at home are weak acids. In carbonated soft drinks, CO_2 is dissolved in water to form carbonic acid, H_2CO_3. Citric acid is a weak acid found in fruits and fruit juices such as lemons, oranges, and grapefruit. The vinegar we put on a salad is a 5% acetic acid solution. In the equation for acetic acid in water, a longer arrow indicates that the reaction favors the reactants.

$$CH_3COOH(l) \; + \; H_2O(l) \; \underset{\longleftarrow}{\overset{\longrightarrow}{}} \; H_3O^+(aq) \; + \; CH_3COO^-(aq)$$
acetic acid   acetate ion

Therefore an aqueous solution of a strong acid, HA, consists of mostly ions H_3O^+ and A^-. If an acid (HA) is a weak acid, the concentrations of H_3O^+ and A^- are small, which means that the aqueous solution consists of mostly reactants. (See Figure 8.2.)

$$HA \; + \; H_2O \; \underset{\longleftarrow}{\overset{\longrightarrow}{}} \; H_3O^+ \; + \; A^-$$

Strong and Weak Bases

The Arrhenius bases such as LiOH, KOH, NaOH, and $Ca(OH)_2$ are **strong bases** that dissociate completely (100%). Because these strong bases are ionic compounds, they dissociate in water to give an aqueous solution of a metal ion and hydroxide ion.

$$KOH(s) \; \xrightarrow{\; H_2O \;} \; K^+(aq) \; + \; OH^-(aq)$$

FIGURE 8.2

(a) A strong acid dissociates in water to give H_3O^+ and A^- ions. (b) A weak acid in water dissociates only slightly to form a solution containing only a few H_3O^+ and A^- ions and mostly undissociated HA molecules.

Q How does the height of the H_3O^+ and A^- in the bar diagram change for a strong acid compared to a weak acid?

Strong bases, such as NaOH (also known as lye), are used in household products to remove grease in ovens and to clean drains. Because high concentrations of hydroxide ions cause severe damage to the skin and eyes, directions must be followed carefully when such products are used in the home and use in the chemistry laboratory should be carefully supervised. If you spill an acid or a base on your skin or get some in your eyes, be sure to flood the area immediately with water.

The **weak bases** are poor acceptors of protons. A typical weak base, ammonia, NH_3, is found in window cleaners. In water, a few ammonia molecules accept protons to form ammonium hydroxide.

$$NH_3(g) + H_2O(l) \rightleftharpoons NH_4^+(aq) + OH^-(aq)$$

SAMPLE PROBLEM 8.6

■ **Strengths of Acids and Bases**

For the following questions select from one of the following:

$$H_2CO_3 \qquad H_2SO_4 \qquad HNO_2$$

a. Which is the strongest acid?
b. Which acid is the weakest?

SOLUTION

As derived from the information in Table 8.3:

a. The strongest acid in this group is H_2SO_4.
b. The weakest acid in this group is carbonic acid, H_2CO_3.

STUDY CHECK

Which is the stronger base: KOH or NH_3?

QUESTIONS AND PROBLEMS

■ Strengths of Acids and Bases

8.15 Identify the stronger acid in each pair:
    a. HBr or HNO_2
    b. H_3PO_4 or HSO_4^-
    c. HCN or H_2CO_3

8.16 Identify the stronger acid in each pair:
    a. NH_4^+ or H_3O^+
    b. H_2SO_4 or HCN
    c. H_2O or H_2CO_3

8.17 Identify the weaker acid in each pair:
    a. HCl or HSO_4^-
    b. HNO_2 or HF
    c. HCO_3^- or NH_4^+

8.18 Identify the weaker acid in each pair:
    a. HNO_3 or HCO_3^-
    b. HSO_4^- or H_2O
    c. H_2SO_4 or H_2CO_3

■ 8.4 Ionization of Water

We have seen that in some acid–base reactions water acts as an acid and in other reactions as a base. Does that mean water can be both an acid and a base? Yes, this is exactly what happens with water molecules in pure water. Let's see how this happens. In water, one water molecule donates a proton to another to produce H_3O^+ and OH^-, which means that water can behave as both an acid and a base. Let's take a look at the conjugate acid–base pairs in water.

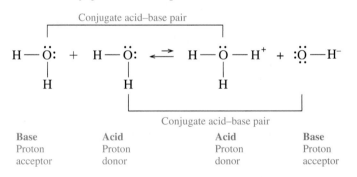

In the ionization of water, there is both a forward and reverse reaction.

$$H_2O(l) + H_2O(l) \rightleftharpoons H_3O^+(aq) + OH^-(aq)$$

In pure water, the transfer of a proton between two water molecules produces equal numbers of H_3O^+ and OH^-. Experiments have determined that, in pure water, the concentrations of H_3O^+ and OH^- at 25°C are each 1.0×10^{-7} M. Square brackets around the symbols indicate their concentrations in moles per liter (M).

Pure water $[H_3O^+] = [OH^-] = 1.0 \times 10^{-7}$ M

When we multiply these concentrations, it forms the **ion-product constant of water, K_w**, which is 1.0×10^{-14}. The concentration units are omitted in the K_w value.

$$K_w = [H_3O^+] \times [OH^-]$$
$$= (1.0 \times 10^{-7} \text{ M})(1.0 \times 10^{-7} \text{ M}) = 1.0 \times 10^{-14}$$

The K_w value of 1.0×10^{-14} is important because it applies to any aqueous solution: all aqueous solutions have H_3O^+ and OH^-.

When the $[H_3O^+]$ and $[OH^-]$ in a solution are equal, the solution is **neutral**. However, most solutions are not neutral and have different concentrations of $[H_3O^+]$ and $[OH^-]$. If acid is added to water, there is an increase in $[H_3O^+]$ and a decrease in $[OH^-]$, which makes an acidic solution. If base is added, $[OH^-]$ increases and $[H_3O^+]$ decreases, which makes a basic solution. (See Figure 8.3.) However, for any aqueous solution, whether it is neutral, acidic, or basic, the product $[H_3O^+ \times OH^-]$ is equal to K_w (1.0×10^{-14}). Therefore, if the $[H_3O^+]$ is given, K_w can be used to calculate the $[OH^-]$. Or if the $[OH^-]$ is given, K_w can be used to calculate the $[H_3O^+]$. (See Table 8.4.)

$$K_w = [H_3O^+] \times [OH^-]$$

$$[OH^-] = \frac{K_w}{[H_3O^+]} \qquad [H_3O^+] = \frac{K_w}{[OH^-]}$$

To illustrate these calculations, let us calculate the $[H_3O^+]$ for a solution that has an $[OH^-] = 1.0 \times 10^{-6}$ M.

TABLE 8.4

Examples of $[H_3O^+]$ and $[OH^-]$ in Neutral, Acidic, and Basic Solutions

Type of Solution	$[H_3O^+]$	$[OH^-]$	K_w
Neutral	1.0×10^{-7} M	1.0×10^{-7} M	1.0×10^{-14}
Acidic	1.0×10^{-2} M	1.0×10^{-12} M	1.0×10^{-14}
Acidic	2.5×10^{-5} M	4.0×10^{-10} M	1.0×10^{-14}
Basic	1.0×10^{-8} M	1.0×10^{-6} M	1.0×10^{-14}
Basic	5.0×10^{-11} M	2.0×10^{-4} M	1.0×10^{-14}

STEP 1 **Write the K_w for water.**

$$K_w = \boxed{[H_3O^+]}\,[OH^-] = 1.0 \times 10^{-14}$$

STEP 2 **Arrange the K_w to solve for the unknown.**   Dividing through by the $[OH^-]$ gives

$$\frac{K_w}{[OH^-]} = \frac{[H_3O^+] \times [\cancel{OH^-}]}{[\cancel{OH^-}]} = \frac{1.0 \times 10^{-14}}{[OH^-]}$$

$$\boxed{[H_3O^+]} = \frac{1.0 \times 10^{-14}}{[OH^-]}$$

STEP 3 **Substitute the $[OH^-]$ and calculate the $[H_3O^+]$.**

$$\boxed{[H_3O^+]} = \frac{1.0 \times 10^{-14}}{[1.0 \times 10^{-6}]} = 1.0 \times 10^{-8}\,M$$

Because the $[OH^-]$ of 1.0×10^{-6} M is larger than the $[H_3O^+]$ of 1.0×10^{-8} M, the solution is basic.

Guide to Calculating $[H_3O^+]$ and $[OH^-]$ in Aqueous Solutions

STEP 1
Write the K_w constant for water.

STEP 2
Solve the K_w for the unknown $[H_3O^+]$ or $[OH^-]$.

STEP 3
Substitute the known $[H_3O^+]$ or $[OH^-]$ and calculate.

SAMPLE PROBLEM 8.7

■ **Calculating $[H_3O^+]$ and $[OH^-]$ in Solution**

A vinegar solution has a $[H_3O^+] = 2.0 \times 10^{-3}$ M at 25°C. What is the $[OH^-]$ of the vinegar solution? Is the solution acidic, basic, or neutral?

SOLUTION

STEP 1 **Write the K_w for water.**

$$K_w = [H_3O^+] \times \boxed{[OH^-]} = 1.0 \times 10^{-14}$$

STEP 2 **Arrange the K_w to solve for the unknown.**   Rearranging the K_w for OH^- gives

$$\frac{K_w}{[H_3O^+]} = \frac{[\cancel{H_3O^+}] \times [OH^-]}{[\cancel{H_3O^+}]} = \frac{1.0 \times 10^{-14}}{[H_3O^+]}$$

$$[OH^-] = \frac{1.0 \times 10^{-14}}{[H_3O^+]}$$

STEP 3 **Substitute the known $[H_3O^+]$ or $[OH^-]$ and calculate.**

$$\boxed{[OH^-]} = \frac{1.0 \times 10^{-14}}{[2.0 \times 10^{-3}]} = 5.0 \times 10^{-12}\,M$$

Because the $[H_3O^+]$ of 2.0×10^{-3} M is much larger than the $[OH^-]$ of 5.0×10^{-12} M, the solution is acidic.

STUDY CHECK

What is the $[H_3O^+]$ of an ammonia cleaning solution with an $[OH^-] = 4.0 \times 10^{-4}$ M? Is the solution acidic, basic, or neutral?

■

QUESTIONS AND PROBLEMS

■ Ionization of Water

8.19 Why are the concentrations of H_3O^+ and OH^- equal in pure water?

8.20 What is the meaning and value of K_w?

8.21 In an acidic solution, how does the concentration of H_3O^+ compare to the concentration of OH^-?

8.22 If a base is added to pure water, why does the $[H_3O^+]$ decrease?

8.23 Indicate whether the following are acidic, basic, or neutral solutions:
- **a.** $[H_3O^+] = 2.0 \times 10^{-5}$ M *A*
- **b.** $[H_3O^+] = 1.4 \times 10^{-9}$ M *B*
- **c.** $[OH^-] = 8.0 \times 10^{-3}$ M *Bo*
- **d.** $[OH^-] = 3.5 \times 10^{-10}$ M *A*

8.24 Indicate whether the following are acidic, basic, or neutral solutions:
- **a.** $[H_3O^+] = 6.0 \times 10^{-12}$ M
- **b.** $[H_3O^+] = 1.4 \times 10^{-4}$ M
- **c.** $[OH^-] = 5.0 \times 10^{-12}$ M
- **d.** $[OH^-] = 4.5 \times 10^{-2}$ M

8.25 Calculate the $[OH^-]$ of each aqueous solution with the following $[H_3O^+]$:
- **a.** coffee, 1.0×10^{-5} M
- **b.** soap, 1.0×10^{-8} M
- **c.** cleanser, 5.0×10^{-10} M
- **d.** lemon juice, 2.5×10^{-2} M

8.26 Calculate the $[OH^-]$ of each aqueous solution with the following $[H_3O^+]$:
- **a.** NaOH, 1.0×10^{-12} M
- **b.** aspirin, 6.0×10^{-4} M
- **c.** milk of magnesia, 1.0×10^{-9} M
- **d.** stomach acid, 5.2×10^{-2} M

8.27 Calculate the $[OH^-]$ of each aqueous solution with the following $[H_3O^+]$:
- **a.** vinegar, 1.0×10^{-3} M
- **b.** urine, 5.0×10^{-6} M
- **c.** ammonia, 1.8×10^{-12} M
- **d.** NaOH, 4.0×10^{-13} M

8.28 Calculate the $[OH^-]$ of each aqueous solution with the following $[H_3O^+]$:
- **a.** baking soda, 1.0×10^{-8} M
- **b.** orange juice, 2.0×10^{-4} M
- **c.** milk, 5.0×10^{-7} M
- **d.** bleach, 4.8×10^{-12} M

LEARNING GOAL

Calculate pH from $[H_3O^+]$; given the pH, calculate $[H_3O^+]$ and $[OH^-]$ of a solution.

the
Chemistry place

CASE STUDY
Hyperventilation and Blood pH

■ 8.5 The pH Scale

Many kinds of careers such as respiratory therapy, wine and beer making, medicine, agriculture, spa cleaning, and soap manufacturing require personnel to measure the $[H_3O^+]$ and $[OH^-]$ of solutions. The proper levels of acidity are necessary for soil to support plant growth and prevent algae in swimming pool water. Measuring the acidity levels of blood and urine checks the function of the kidneys.

On the pH scale, a number between 0 and 14 represents the H_3O^+ concentration. A pH value less than 7 corresponds to an acidic solution; a pH value greater than 7 indicates a basic solution. (See Figure 8.4.)

Acidic solution	$pH < 7$	$[H_3O^+] > 1.0 \times 10^{-7}$ M
Neutral solution	$pH = 7$	$[H_3O^+] = 1.0 \times 10^{-7}$ M
Basic solution	$pH > 7$	$[H_3O^+] < 1.0 \times 10^{-7}$ M

In the laboratory, a pH meter is commonly used to determine the pH of a solution. There are also indicators and pH papers that turn specific colors when placed in solutions of different pH values. The pH is found by comparing the colors to a color chart. (See Figure 8.5.)

0 1.0 M HCl

Gastric juice 1.6

Lemon juice 2.2

Vinegar 2.8   Carbonated beverage 3.0

Orange 3.5

Acidic

Apple juice 3.8
Tomato 4.2

Coffee 5.0

Soil 5.5   Bread 5.5

Potato 5.8   Urine 6.0

Milk 6.4

Neutral
$[H^+] = [OH^-]$ Water (pure) 7.0

Drinking water 7.2

Blood 7.4

Bile 8.0   Detergents 8.0–9.0

Seawater 8.5

Basic

Milk of magnesia 10.5

Ammonia 11.0

Bleach 12.0

1.0 M NaOH (lye) 14.0

FIGURE 8.4

On the pH scale, values below 7 are acidic, a value of 7 is neutral, and values above 7 are basic.

Q **Is apple juice an acidic, basic, or a neutral solution?**

Calculating the pH of Solutions

the **Chemistry place**

WEB TUTORIAL
The pH Scale

The pH scale is a log scale that corresponds to the hydrogen-ion concentrations of aqueous solutions. Mathematically, **pH** is the negative logarithm (base 10) of the H_3O^+ concentration.

$$pH = -\log[H_3O^+]$$

Essentially, the negative powers of ten in the concentrations are converted to positive numbers. For example, a lemon juice solution with $[H_3O^+] = 1.0 \times 10^{-2}$ M

(a)

(b)

(c)

FIGURE 8.5

The pH of a solution can be determined using **(a)** a pH meter, **(b)** pH paper, and **(c)** indicators that turn different colors corresponding to different pH values.

Q **If a pH meter reads 4.00, is the solution acidic, basic, or neutral?**

has a pH of 2.00. This can be calculated using the pH equation. For whole numbers in the $[H_3O^+]$, remember to add the correct number of significant zeros to the resulting pH obtained on your calculator.

$$pH = -\log [1.0 \times 10^{-2}]$$
$$pH = -(-2.00)$$
$$= 2.00$$

Let's look at how we determine the number of significant figures in the pH. For a logarithm, the number of decimal places in the pH value is equal to the number of significant figures in the $[H_3O^+]$. The number to the left of the decimal point is the power of ten.

$$[H_3O^+] = 1.0 \times 10^{-2} \qquad pH = 2.00$$

2 significant figures   2 decimal places

Steps for a pH Calculation

The pH of a solution is determined using the *log* key and *changing sign*. For example, to calculate the pH of a vinegar solution with $[H_3O^+] = 2.4 \times 10^{-3}$ M you can use the following steps:

Operation	Display Shows
STEP 1 Enter the $[H_3O^+]$ value. 2.4 [EE or EXP] 3	2.4^{03} or $2.4\ 03$
Press the [+/−] to change the power to −3. (For calculators without a change sign key, consult the instructions for the calculator.)	2.4^{-03} or $2.4-03$
STEP 2 Press the [log] key.	−2.619789
Change the sign.	2.619789

Guide to Calculating pH of an Aqueous Solution

STEP 1
Enter the concentration of $[H_3O^+]$.

STEP 2
Press the *log* key and *change the sign.*

STEP 3
Adjust significant figures to the *right* of the decimal point to equal SFs in the coefficient.

The steps can be combined to give the calculator sequence as follows:

$$pH = -\log[2.4 \times 10^{-3}] = 2.4 \boxed{\text{EE or EXP}} \ 3 \ \boxed{+/-} \ \boxed{\log} \ \boxed{+/-}$$
$$= 2.619789$$

Be sure to check the instructions for your calculator. On some calculators, the log key is used first, followed by the concentration.

STEP 3 Adjust significant figures. In a pH value, the number to the *left* of the decimal point is an *exact* number derived from the power of ten. The number of digits to the *right* of the decimal point is equal to the number of significant figures in the coefficient.

Coefficient   **Power of ten**

$[H_3O^+] = 2.4 \quad \times \quad 10^{-3} \, M \qquad pH = -\log[2.4 \times 10^{-3}] = 2.62$

2 Significant figures (2SFs)   Exact   Exact   2 decimal places

Because pH is a log scale, a change of one pH unit corresponds to a ten-fold change in $[H_3O^+]$. It is important to note that the pH decreases as the $[H_3O^+]$ increases. For example, a solution with a pH of 2.00 has a $[H_3O^+]$ 10 times higher than a solution with a pH of 3.00 and 100 times higher than a solution with a pH of 4.00.

SAMPLE PROBLEM 8.8

■ Calculating pH

Determine the pH for the following solutions:

a. $[H_3O^+] = 1.0 \times 10^{-5} \, M$ **b.** $[H_3O^+] = 5 \times 10^{-8} \, M$

SOLUTION

a. STEP 1 Enter the concentration of $[H_3O^+]$ and the *change sign* key.

                                                 Display

$pH = -\log[1.0 \times 10^{-5}] = 1.0 \boxed{\text{EE or EXP}} \ 5 \ \boxed{+/-}$ $\boxed{1.0^{-05}}$ or $\boxed{1.0 - 05}$

STEP 2 Press the *log* key and *change sign* key.

     $\boxed{\log}$ $\boxed{+/-}$ $\boxed{5}$

STEP 3 Adjust significant figures to the *right* of the decimal point to equal the number of significant figures in the coefficient.

         $1.0 \times 10^{-5} \, M$ $pH = 5.00$

         2 SFs   ⟶ 2 SFs to the *right* of the decimal point

b. STEP 1 Enter the concentration and press the *change sign* key.

     $pH = -\log[5 \times 10^{-8}] = 5 \boxed{\text{EE or EXP}} \ \boxed{+/-}$ $\boxed{5^{-08}}$ or $\boxed{5 - 08}$

STEP 2 Press the *log* key, and then the *change sign* key.

     $\boxed{\log}$ $\boxed{+/-}$ $\boxed{7.301029}$

STEP 3 **Adjust significant figures to the *right* of the decimal point to equal the significant figures in the coefficient.**

$$5 \times 10^{-8} \text{ M} \qquad \text{pH} = 7.3$$

1 SF ⟶ 1 SF to the *right* of the decimal point

STUDY CHECK

What is the pH of bleach with $[H_3O^+] = 4.2 \times 10^{-12}$ M?

SAMPLE PROBLEM 8.9

■ **Calculating pH from [OH⁻]**

What is the pH of an ammonia solution with $[OH^-] = 3.7 \times 10^{-3}$ M?

SOLUTION

STEP 1 **Enter the $[H_3O^+]$ and press the *change sign* key.** Because $[OH^-]$ is given for the ammonia solution, we have to calculate $[H_3O^+]$ using the ion product of water, K_w. Dividing through by $[OH^-]$ gives $[H_3O^+]$.

$$\frac{K_w}{[OH^-]} = \frac{[H_3O^+]\ [\cancel{OH^-}]}{[\cancel{OH^-}]} = \frac{1.0 \times 10^{-14}}{[OH^-]}$$

$$[H_3O^+] = \frac{1.0 \times 10^{-14}}{[3.7 \times 10^{-3} \text{ M}]} = 2.7 \times 10^{-12} \text{ M}$$

$$\text{pH} = -\log[2.7 \times 10^{-12}] = 2.7 \boxed{\text{EE or EXP}}\ 12\ \boxed{+/-} = \boxed{2.7^{-12}} \text{ or } \boxed{2.7-12}$$

STEP 2 **Press the *log* key, and then the *change sign* key.**

$$\boxed{\log} \quad \boxed{+/-} \hspace{4cm} \boxed{11.56863}$$

STEP 3 **Adjust significant figures to the *right* of the decimal point to equal the SFs in the coefficient.**

$$2.7 \times 10^{-12} \text{ M} \qquad \text{pH} = 11.57$$

2 SFs ⟶ 2 SFs *after* the decimal point

STUDY CHECK

Calculate the pH of a sample of acid rain that has $[OH^-] = 2 \times 10^{-10}$ M.

Sometimes, we need to determine the $[H_3O^+]$ from the pH. For whole number pH values, the negative pH value is the power of 10 in the H_3O^+ concentration.

$$[H_3O^+] = 1 \times 10^{-\text{pH}}$$

SAMPLE PROBLEM 8.10

■ **Calculating $[H_3O^+]$ from pH**

Determine $[H_3O^+]$ for solutions having the following pH values:

a. pH = 3.0 **b.** pH = 12.0

SOLUTION

For pH values that are whole numbers, the $[H_3O^+]$ can be written 1.0×10^{-pH}.

a. $[H_3O^+] = 1 \times 10^{-3}$ M   b. $[H_3O^+] = 1 \times 10^{-12}$ M

STUDY CHECK

What is the $[H_3O^+]$ and $[OH^-]$ of beer if it has a pH of 5.0?

A comparison of $[H_3O^+]$, $[OH^-]$, and their corresponding pH values is given in Table 8.5.

TABLE 8.5

A Comparison of $[H_3O^+]$, $[OH^-]$, and Corresponding pH Values

$[H_3O^+]$	pH	$[OH^-]$	
10^0	0	10^{-14}	
10^{-1}	1	10^{-13}	
10^{-2}	2	10^{-12}	
10^{-3}	3	10^{-11}	Acidic
10^{-4}	4	10^{-10}	
10^{-5}	5	10^{-9}	
10^{-6}	6	10^{-8}	
10^{-7}	7	10^{-7}	Neutral
10^{-8}	8	10^{-6}	
10^{-9}	9	10^{-5}	
10^{-10}	10	10^{-4}	
10^{-11}	11	10^{-3}	Basic
10^{-12}	12	10^{-2}	
10^{-13}	13	10^{-1}	
10^{-14}	14	10^0	

QUESTIONS AND PROBLEMS

■ The pH Scale

8.29 Why does a neutral solution have a pH of 7.00?

8.30 If you know the $[OH^-]$, how can you determine the pH of a solution?

8.31 State whether each of the following solutions is acidic, basic, or neutral:
   a. blood, pH 7.38   b. vinegar, pH 2.8
   c. drain cleaner, pH 11.2   d. coffee, pH 5.5

8.32 State whether each of the following solutions is acidic, basic, or neutral:
   a. soda, pH 3.2   b. shampoo, pH 5.7
   c. laundry detergent, pH 9.4   d. rain, pH 5.8

8.33 Calculate the pH of each solution given the following $[H_3O^+]$ or $[OH^-]$ values.
   a. $[H_3O^+] = 1 \times 10^{-4}$ M   b. $[H_3O^+] = 3 \times 10^{-9}$ M
   c. $[OH^-] = 1 \times 10^{-5}$ M   d. $[OH^-] = 2.5 \times 10^{-11}$ M

8.34 Calculate the pH of each solution given the following $[H_3O^+]$ or $[OH^-]$ values.
   a. $[H_3O^+] = 1 \times 10^{-8}$ M   b. $[H_3O^+] = 5 \times 10^{-6}$ M
   c. $[OH^-] = 4 \times 10^{-2}$ M   d. $[OH^-] = 8 \times 10^{-3}$ M

◾ Using Vegetables and Flowers as pH Indicators

Many flowers and vegetables with strong color, especially reds and purples, contain compounds that change color with changes in pH. Some examples are red cabbage, cranberry juice, and cranberry drinks.

MATERIALS NEEDED

Red cabbage, water, and a saucepan; or cranberry juice or drinks

Several glasses or small glass containers and some tape and a pen or pencil to mark the containers

Several colorless household solutions such as vinegar, lemon juice, other fruit juices, baking soda, antacids, aspirin, window cleaners, soaps, shampoos, and detergents

PROCEDURE

1. Obtain a bottle of cranberry juice or cranberry drink, or use a red cabbage to prepare the red cabbage pH indicator, as follows: Tear up several red cabbage leaves and place them in a saucepan and cover with water. Boil for about 5 minutes. Cool and collect the purple solution.

2. Place small amounts of each household solution into separate clear glass containers and mark what each one is. If the sample is a solid or a thick liquid, add a small amount of water. Add some cranberry juice or some red cabbage indicator until you obtain a color.

3. Observe the colors of the various samples. The colors that indicate acidic solutions are the pink and orange colors (pH 1–4) and the pink to lavender colors (5–6). A neutral solution has about the same purple color as the indicator. Bases will give blue to green color (pH 8–11) or a yellow color (pH 12–13).

4. Arrange your samples by color and pH. Classify each of the solutions as acidic (1–6), neutral (7), or basic (8–13).

5. Try to make an indicator using other colorful fruits or flowers.

QUESTIONS

1. Which products that tested acidic listed an acid on their labels?
2. Which products that tested basic listed a base on their labels?
3. How many products were neutral?
4. Which flowers or vegetables behaved as indicators?

8.35 Complete the following table:

$[H_3O^+]$	$[OH^-]$	pH	Acidic, Basic, or Neutral?
	1×10^{-6} M		
		3.0	
2×10^{-5} M			
1×10^{-12} M			

8.36 Complete the following table:

$[H_3O^+]$	$[OH^-]$	pH	Acidic, Basic, or Neutral?
		10.0	
			Neutral
	1×10^{-5} M		
1×10^{-2} M			

◾ Write balanced equations for reactions of acids and bases.

◾ 8.6 Reactions of Acids and Bases

Typical reactions of acids and bases include the reactions of acids with metals, bases, and carbonate or bicarbonate ions. For example, when you drop an antacid tablet in water, the bicarbonate ion and citric acid in the tablet react to produce carbon dioxide bubbles, a salt, and water.

◾ Acids and Metals

Acids react with certain metals known as *active metals* to produce hydrogen gas (H_2) and the salt of that metal. Active metals include potassium, sodium, calcium, magnesium, aluminum, zinc, iron, and tin. In these single replacement reactions, the metal loses electrons and the metal ion replaces the hydrogen in the acid.

$$Mg(s) + 2HCl(aq) \longrightarrow MgCl_2(aq) + H_2(g)$$

metal   acid   salt   hydrogen

$$Zn(s) + 2HCl(aq) \longrightarrow ZnCl_2(aq) + H_2(g)$$

metal   acid   salt   hydrogen

Acids, Carbonates, and Bicarbonates

When strong acids are added to a carbonate or bicarbonate (also known as hydrogen carbonate), the reaction produces bubbles of carbon dioxide gas, a salt, and water. In the reaction, H^+ is transferred to the carbonate to give carbonic acid, H_2CO_3, which breaks down rapidly to CO_2 and H_2O. A simpler equation called a net ionic equation is written by omitting the metal ions and chloride ions that are not reacting.

$$HCl(aq) + NaHCO_3(aq) \longrightarrow CO_2(g) + H_2O + NaCl(aq)$$

$$H^+(aq) + HCO_3^-(aq) \longrightarrow CO_2(g) + H_2O$$

$$2HCl(aq) + Na_2CO_3(aq) \longrightarrow CO_2(g) + H_2O + 2NaCl(aq)$$

$$2H^+(aq) + CO_3^{2-}(aq) \longrightarrow CO_2(g) + H_2O$$

ENVIRONMENTAL NOTE

Acid Rain

Rain typically has a pH of 6.2. It is slightly acidic because carbon dioxide in the air combines with water to form carbonic acid. However, in many parts of the world, rain has become considerably more acidic, with pH values as low as 3 being reported. One cause of acid rain is the sulfur dioxide (SO_2) gas produced when coal that contains sulfur is burned.

In the air, the SO_2 gas reacts with oxygen to produce SO_3, which then combines with water to form sulfuric acid, H_2SO_4, a strong acid.

$$S(s) + O_2(g) \longrightarrow SO_2(g)$$

$$2SO_2(g) + O_2(g) \longrightarrow 2SO_3(g)$$

$$SO_3(g) + H_2O(l) \longrightarrow H_2SO_4(aq)$$

In parts of the United States, acid rain has made lakes so acidic they are no longer able to support fish and plant life. Limestone ($CaCO_3$) is sometimes added to these lakes to neutralize the acid. In Eastern Europe, acid rain has brought about an environmental disaster. Nearly 40% of the forests in Poland have been severely damaged, and some parts of the land are so acidic that crops will not grow. Throughout Europe and the United States, monuments made of marble (a form of $CaCO_3$) are deteriorating as acid rain dissolves the marble.

$$2H^+(aq) + CaCO_3(s) \longrightarrow Ca^{2+}(aq) + H_2O(l) + CO_2(g)$$

1935   1994

Marble statue in Washington Square Park

Efforts to slow or stop the damaging effects of acid rain include the reduction of sulfur emissions. This will require installation of expensive equipment in coal-burning plants to absorb more SO_2 before it is emitted. In some outdated plants, this may be impossible, and they will need to be closed. It is a difficult problem for engineers and scientists, but one that must be solved.

Acids and Hydroxides: Neutralization

Neutralization is a reaction between an acid and a base to produce a salt and water. The cation in the salt comes from the base and the anion comes from the acid. In the neutralization reactions of strong acids and strong bases, water is always one of the products. We can write the following equation for the neutralization of HCl and NaOH.

$$\underset{\text{acid}}{HCl(aq)} + \underset{\text{base}}{NaOH(aq)} \longrightarrow \underset{\text{salt}}{NaCl(aq)} + \underset{\text{water}}{H_2O(l)}$$

To see the actual reaction, we write the strong acid and strong base as individual ions.

$$\mathbf{H^+}(aq) + Cl^-(aq) + Na^+(aq) + \mathbf{OH^-}(aq) \longrightarrow$$
$$Na^+(aq) + Cl^-(aq) + \mathbf{H_2O}(l) \quad \text{Ionic equation}$$

In the neutralization reaction, H^+ from the acid reacts with OH^- from the base to form water, leaving the spectator ions from the salt (Na^+ and Cl^-) in solution. When we omit the spectator ions from the ionic equation, we obtain the *net ionic* equation, which allows us to see that the net reaction for neutralization is the reaction of H^+ and OH^-.

$$\mathbf{H^+}(aq) + \cancel{Cl^-(aq)} + \cancel{Na^+(aq)} + \mathbf{OH^-}(aq) \longrightarrow$$
$$\cancel{Na^+(aq)} + \cancel{Cl^-(aq)} + \mathbf{H_2O}(l) \quad \text{Ionic equation}$$

$$H^+(aq) + OH^-(aq) \longrightarrow H_2O(l) \quad \text{Net ionic equation}$$

Balancing Neutralization Equations

In a neutralization reaction, one H^+ always reacts with one OH^-. Therefore, the coefficients in the neutralization equation must be chosen so that the H^+ from the acid is balanced by the OH^- provided by the base. We balance the neutralization of HCl and $Ba(OH)_2$ as follows.

Guide to Balancing an Equation for Neutralization

STEP 1
Write the reactants and products.

STEP 2
Balance the H^+ in the acid with the OH^- in the base.

STEP 3
Balance the H_2O with the H^+ and the OH^-.

STEP 4
Write the salt from the remaining ions.

STEP 1 Write the reactants and products.

$$HCl(aq) + Ba(OH)_2(s) \longrightarrow H_2O(l) + \text{salt}$$

STEP 2 Balance the H^+ in the acid with the OH^- in the base. Placing a 2 in front of the HCl provides $2 H^+$ for the $2 OH^-$ in $Ba(OH)_2$.

$$2HCl(aq) + Ba(OH)_2(s) \longrightarrow H_2O(l) + \text{salt}$$

STEP 3 Balance the H_2O with the H^+ and OH^-.

$$2HCl(aq) + Ba(OH)_2(s) \longrightarrow 2H_2O(l) + \text{salt}$$

STEP 4 Write the salt from the remaining ions in the acid and base.

$$2HCl(aq) + \mathbf{Ba(OH)_2}(s) \longrightarrow 2H_2O(l) + \mathbf{BaCl_2}(aq)$$

SAMPLE PROBLEM 8.11

Reactions of Acids

Write a balanced equation for the reaction of HCl(*aq*) with each of the following:

a. Al(*s*)
b. K_2CO_3(*ag*)
c. $Mg(OH)_2$(*s*)

SOLUTION

a. Al

> **STEP 1 Write the reactants and products.**   When a metal reacts with an acid, the products are H_2 gas and a salt.
>
> $$Al(s) + HCl(aq) \longrightarrow H_2(g) + salt$$
>
> **STEP 2 Determine the formula of the salt.**   When $Al(s)$ dissolves, it forms Al^{3+}, which is balanced by 3 Cl^- from HCl.
>
> $$Al(s) + HCl(aq) \longrightarrow H_2(g) + AlCl_3(aq)$$
>
> **STEP 3 Balance the equation.**
>
> $$2Al(s) + 6HCl(aq) \longrightarrow 3H_2(g) + 2AlCl_3(aq)$$

b. K_2CO_3

> **STEP 1 Write the reactants and products.**   When a carbonate reacts with an acid, the products are $CO_2(g)$, $H_2O(l)$, and a salt.
>
> $$K_2CO_3(s) + HCl(aq) \longrightarrow CO_2(g) + H_2O(l) + salt$$
>
> **STEP 2 Determine the formula of the salt.**   When $K_2CO_3(s)$ dissolves, it forms K^+, which is balanced by 1 Cl^- from HCl.
>
> $$\textbf{K}_2\textbf{CO}_3(s) + \textbf{HCl}(aq) \longrightarrow \textbf{CO}_2(g) + \textbf{H}_2\textbf{O}(l) + \textbf{KCl}(aq)$$
>
> **STEP 3 Balance the equation.**
>
> $$K_2CO_3(aq) + \textbf{2}HCl(aq) \longrightarrow CO_2(g) + H_2O(l) + \textbf{2}KCl(aq)$$

c. $Mg(OH)_2$

> **STEP 1 Write the reactants and products.**   When a base reacts with an acid, the products are $H_2O(l)$ and a salt.
>
> $$Mg(OH)_2(s) + HCl(aq) \longrightarrow H_2O(l) + salt$$
>
> **STEP 2 Balance the H^+ in the acid with the OH^- in the base.**   Placing a 2 in front of the HCl provides 2 H^+ for 2 OH^- in $Mg(OH)_2$.
>
> $$Mg(OH)_2(s) + \textbf{2}HCl(aq) \longrightarrow H_2O(l) + salt$$
>
> **STEP 3 Balance the H_2O with the H^+ and OH^-.**
>
> $$Mg(OH)_2(s) + 2HCl(aq) \longrightarrow \textbf{2}H_2O(l) + salt$$
>
> **STEP 4 Write the salt from the remaining ions in the acid and base.**
>
> $$\textbf{Mg(OH)}_2(s) + \textbf{2HCl}(aq) \longrightarrow \textbf{2H}_2\textbf{O}(l) + \textbf{MgCl}_2(aq)$$

STUDY CHECK

Write the balanced equation for the reaction between H_2SO_4 and $NaHCO_3$.

Acid–Base Titration

Suppose we need to find the molarity of an HCl solution of unknown concentration. We can do this by a laboratory procedure called **titration** in which we neutralize an acid sample with a known amount of base. In our titration, we first place

HEALTH NOTE

Antacids

Antacids are substances used to neutralize excess stomach acid (HCl). Some antacids are mixtures of aluminum hydroxide and magnesium hydroxide. These hydroxides are not very soluble in water, so the levels of available OH^- are not damaging to the intestinal tract. However, aluminum hydroxide has the side effects of producing constipation and binding phosphate in the intestinal tract, which may cause weakness and loss of appetite. Magnesium hydroxide has a laxative effect. These side effects are less likely when a combination of the antacids is used.

$$Al(OH)_3(aq) + 3HCl(aq) \longrightarrow AlCl_3(aq) + 3H_2O(l)$$

$$Mg(OH)_2(s) + 2HCl(aq) \longrightarrow MgCl_2(aq) + 2H_2O(l)$$

Some antacids use calcium carbonate to neutralize excess stomach acid. About 10% of the calcium is absorbed into the bloodstream, where it elevates the levels of serum calcium. Calcium carbonate is not recommended for patients who have peptic ulcers or a tendency to form kidney stones.

$$CaCO_3(s) + 2HCl(aq) \longrightarrow$$
$$H_2O(l) + CO_2(g) + CaCl_2(aq)$$

Still other antacids contain sodium bicarbonate. This type of antacid has a tendency to increase blood pH and elevate sodium levels in the body fluids. It also is not recommended in the treatment of peptic ulcers.

$$NaHCO_3(s) + HCl(aq) \longrightarrow NaCl(aq) + CO_2(g) + H_2O(l)$$

The neutralizing substances in some antacid preparations are given in Table 8.6.

TABLE 8.6

Basic Compounds in Some Antacids

Antacid	Base(s)
Amphojel	$Al(OH)_3$
Milk of magnesia	$Mg(OH)_2$
Mylanta, Maalox, Di-Gel, Gelusil, Riopan	$Mg(OH)_2$, $Al(OH)_3$
Bisodol	$CaCO_3$, $Mg(OH)_2$
Titralac, Tums, Pepto-Bismol	$CaCO_3$
Alka-Seltzer	$NaHCO_3$, $KHCO_3$

a measured volume of the acid in a flask and add a few drops of an *indicator* such as phenolphthalein. In an acidic solution, phenolpthalein is colorless. Then we fill a buret with a NaOH solution of known molarity and carefully add NaOH to the acid in the flask, as shown in Figure 8.6.

In the titration, we neutralize the acid by adding a volume of base that contains a matching number of moles of OH^-. We know that neutralization has taken place when the phenolphthalein in the solution changes from colorless to pink. This is called the neutralization *endpoint*. From the volume added and molarity of the NaOH, we can calculate the number of moles of NaOH and then the concentration of the acid.

SAMPLE PROBLEM 8.12

Titration of an Acid

A 25.0 mL sample of an HCl solution is placed in a flask with a few drops of phenolphthalein (indicator). If 32.6 mL of a 0.185 M NaOH is needed to reach the endpoint, what is the concentration (M) of the HCl solution?

$$NaOH(aq) + HCl(aq) \longrightarrow NaCl(aq) + H_2O(l)$$

FIGURE 8.6

The titration of an acid. A known volume of an acid is placed in a flask with an indicator and titrated with a measured volume of a base, such as NaOH, to the neutralization point.

Q What data is needed to determine the molarity of the acid in the flask?

SOLUTION

STEP 1 **Given:**   32.6 mL of 0.185 M NaOH; 25.0 mL HCl = 0.0250 L HCl
        Need: Molarity of HCl

STEP 2 **Plan**

32.6 mL [Metric factor] L [Molarity factor] moles NaOH [Mole factor] moles HCl [Divide by liters.] M HCl

STEP 3 **Equalities/Conversion Factors**

Guide to Calculations
for an Acid-Base Titration

$$1 \text{ L NaOH} = 1000 \text{ mL NaOH} \qquad 1 \text{ L NaOH} = 0.185 \text{ mole NaOH}$$

$$\frac{1 \text{ L}}{1000 \text{ mL}} = \frac{1000 \text{ mL}}{1 \text{ L}} \qquad \frac{1 \text{ L}}{0.185 \text{ mole}} = \frac{0.185 \text{ mole}}{1 \text{ L}}$$

STEP 1
State the given and needed
quantities and concentrations.

$$1 \text{ mole HCl} = 1 \text{ mole NaOH}$$

$$\frac{1 \text{ mole HCl}}{1 \text{ mole NaOH}} = \frac{1 \text{ mole NaOH}}{1 \text{ mole HCl}}$$

STEP 2
Write a plan to calculate
molarity or volume.

STEP 4 **Set up problem.**

$$32.6 \; \cancel{\text{mL NaOH}} \times \frac{1 \; \cancel{\text{L NaOH}}}{1000 \; \cancel{\text{mL NaOH}}} \times \frac{0.185 \; \cancel{\text{mole NaOH}}}{1 \; \cancel{\text{L NaOH}}} \times$$

STEP 3
State equalities and conversion
factors including concentration.

$$\frac{1 \text{ mole HCl}}{1 \; \cancel{\text{mole NaOH}}} = 0.0603 \text{ mole HCl}$$

STEP 4
Set up problem to
calculate needed quantity.

$$\text{Molarity HCl} = \frac{0.0603 \text{ mole HCl}}{0.0250 \text{ L HCl}} = 0.241 \text{ M}$$

STUDY CHECK

What is the molarity of an HCl solution, if 28.6 mL of a 0.175 M NaOH solution is
needed to neutralize a 25.0 mL sample of the HCl solution?

QUESTIONS AND PROBLEMS

▪ Reactions of Acids and Bases

8.37 Complete and balance the equation for the following reactions:
   a. $ZnCO_3(s) + HBr(aq) \longrightarrow$
   b. $Zn(s) + HCl(aq) \longrightarrow$
   c. $HCl(aq) + NaHCO_3(s) \longrightarrow$
   d. $H_2SO_4(aq) + Mg(OH)_2(s) \longrightarrow$

8.38 Complete and balance the equations for the following reactions:
   a. $KHCO_3(s) + HCl(aq) \longrightarrow$
   b. $Ca(s) + H_2SO_4(aq) \longrightarrow$
   c. $H_2SO_4(aq) + Al(OH)_3(s) \longrightarrow$
   d. $Na_2CO_3(s) + H_2SO_4(aq) \longrightarrow$

8.39 Balance each of the following neutralization reactions:
   a. $HCl(aq) + Mg(OH)_2(s) \longrightarrow MgCl_2(aq) + H_2O(l)$
   b. $H_3PO_4(aq) + LiOH(aq) \longrightarrow Li_3PO_4(aq) + H_2O(l)$

8.40 Balance each of the following neutralization reactions:
   a. $HNO_3(aq) + Ba(OH)_2(s) \longrightarrow Ba(NO_3)_2(aq) + H_2O(l)$
   b. $H_2SO_4(aq) + Al(OH)_3(aq) \longrightarrow Al_2(SO_4)_3(aq) + H_2O(l)$

8.41 Write a balanced equation for the neutralization of each of the following:
  a. $H_2SO_4(aq)$ and $NaOH(aq)$
  b. $HCl(aq)$ and $Fe(OH)_3(aq)$
  c. $H_2CO_3(aq)$ and $Mg(OH)_2(s)$

8.42 Write a balanced equation for the neutralization of each of the following:
  a. $H_3PO_4(aq)$ and $NaOH(aq)$
  b. $HI(aq)$ and $LiOH(aq)$
  c. $HNO_3(aq)$ and $Ca(OH)_2(s)$

8.43 What is the molarity of an HCl solution if 5.00 mL HCl solution is titrated with 28.6 mL of 0.145 M NaOH solution?

$$HCl(aq) + NaOH(aq) \longrightarrow NaCl(aq) + H_2O(l)$$

8.44 If 29.7 mL of 0.205 M KOH is required to completely neutralize 25.0 mL of a CH_3COOH solution, what is the molarity of the acetic acid solution?

$$CH_3COOH(aq) + KOH(aq) \longrightarrow CH_3COOK(aq) + H_2O(l)$$

8.45 If 38.2 mL of 0.163 M KOH is required to neutralize completely 25.0 mL H_2SO_4 solution, what is the molarity of the acid solution?

$$H_2SO_4(aq) + 2KOH(aq) \longrightarrow K_2SO_4(aq) + 2H_2O(l)$$

8.46 A solution of 0.162 M NaOH is used to neutralize 25.0 mL H_2SO_4 solution. If 32.8 mL NaOH solution is required to reach the endpoint, what is the molarity of the H_2SO_4 solution?

$$H_2SO_4(aq) + 2NaOH(aq) \longrightarrow Na_2SO_4(aq) + 2H_2O(l)$$

LEARNING GOAL

Describe the role of buffers in maintaining the pH of a solution.

the
Chemistry
place

WEB TUTORIAL
pH and Buffers

8.7 Buffers

When a small amount of acid or base is added to pure water, the pH changes drastically. However, if a solution is buffered, there is little change in pH. A **buffer solution** resists a change in pH when small amounts of acid or base are added. For example, blood is buffered to maintain a pH of about 7.4. If the pH of the blood goes even slightly above or below this value, changes in our uptake of oxygen and our metabolic process can be drastic enough to cause death. Even though we are constantly obtaining acids and bases from foods and biological processes, the buffers in the body so effectively absorb those compounds that blood pH remains essentially unchanged. (See Figure 8.7.)

In a buffer, an acid must be present to react with any OH^- that is added, and a base must be available to react with any added H_3O^+. However, that acid and base must not be able to neutralize each other. Therefore, a combination of a weak acid–base conjugate pair is used in buffers. Most buffer solutions consist of nearly equal concentrations of a weak acid and a salt containing its conjugate base. Buffers may also contain a weak base and the salt of the weak base.

For example, a typical buffer contains acetic acid (CH_3COOH) and a salt such as sodium acetate (CH_3COONa). As a weak acid, acetic acid dissociates slightly in water to form H_3O^+ and a very small amount of CH_3COO^-. The presence of the salt provides a much larger concentration of acetate ion (CH_3COO^-), which is necessary for its buffering capability.

$$CH_3COOH(aq) + H_2O(l) \rightleftharpoons H_3O^+(aq) + CH_3COO^-(aq)$$
Large amount   Large amount

Let's see how this buffer solution maintains the H_3O^+ concentration. When a small amount of acid is added, it will combine with the acetate ion (anion) as the

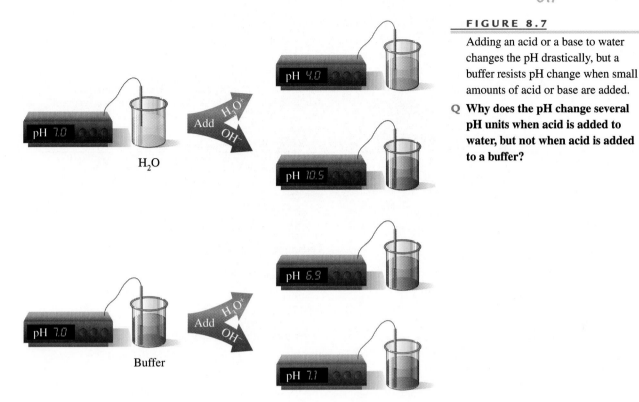

equilibrium shifts to the reactant acetic acid. There will be a small decrease in the $[CH_3COO^-]$ and a small increase in $[CH_3COOH]$, but the $[H_3O^+]$ will not change very much.

$$CH_3COOH(aq) + H_2O(l) \longleftarrow H_3O^+(aq) + CH_3COO^-(aq)$$

If a small amount of base is added to this buffer solution, it is neutralized by the acetic acid and water is produced. The concentration of $[CH_3COOH]$ decreases slightly and the $[CH_3COO^-]$ increases slightly, but again the $[H_3O^+]$ does not change very much. (See Figure 8.8.)

$$CH_3COOH(aq) + OH^-(aq) \longleftarrow H_2O(aq) + CH_3COO^-(aq)$$

FIGURE 8.8

The buffer described here consists of about equal concentrations of acetic acid (CH_3COOH) and its conjugate base, acetate ion (CH_3COO^-). Adding H_3O^+ to the buffer uses up some CH_3COO^-, whereas adding OH^- neutralizes some CH_3COOH. The pH of the solution is maintained as long as the added amounts of acid or base are small compared to the concentrations of the buffer components.

Q **How does this acetic acid/acetate ion buffer maintain pH?**

▪ Buffers in the Blood

The arterial blood has a normal pH of 7.35–7.45. If changes in H_3O^+ lower the pH below 6.8 or raise it above 8.0, cells cannot function properly and death may result. In our cells, CO_2 is continually produced as an end product of cellular metabolism. Some CO_2 is carried to the lungs for elimination, and the rest dissolves in body fluids such as plasma and saliva, forming carbonic acid. As a weak acid, carbonic acid dissociates to give bicarbonate and H_3O^+. More of the anion HCO_3^- is supplied by the kidneys to give an important buffer system in the body fluid: the H_2CO_3/HCO_3^- buffer.

$$CO_2 + H_2O \rightleftharpoons H_2CO_3 \overset{H_2O}{\rightleftharpoons} H_3O^+ + HCO_3^-$$

Excess H_3O^+ entering the body fluids reacts with the HCO_3^- and excess OH^- reacts with the carbonic acid.

$$H_2CO_3 + H_2O \longleftarrow H_3O^+ + HCO_3^-$$

$$H_2CO_3 + OH^- \longrightarrow H_2O + HCO_3^-$$

In the body, the concentration of carbonic acid is closely associated with the partial pressure of CO_2. Table 8.7 lists the normal values for arterial blood. If the CO_2 level rises, it produces more H_2CO_3 and more H_3O^+, which lowers the pH. This condition is called **acidosis**. Difficulty with ventilation or gas diffusion can lead to respiratory acidosis, which can happen in emphysema or when the medulla of the brain is affected by an accident or depressive drugs.

A lowering of the CO_2 level leads to a high blood pH, a condition called **alkalosis**. Excitement, trauma, or a high temperature may cause a person to hyperventilate, which expels large amounts of CO_2. As the partial pressure of CO_2 in the blood falls below normal, H_2CO_3 forms CO_2 and H_2O, which decreases the $[H_3O^+]$ and raises the pH. Table 8.8 lists some of the conditions that lead to changes in the blood pH and some possible treatments. The kidneys also regulate H_3O^+ and HCO_3^- components, but more slowly than the adjustment made by the lungs through ventilation.

TABLE 8.7

Normal Values for Blood Buffer in Arterial Blood

P_{CO_2}	40 mm Hg
H_2CO_3	2.4 mmoles/L of plasma
HCO_3^-	24 mmoles/L of plasma
pH	7.35–7.45

TABLE 8.8

Acidosis and Alkalosis: Symptoms, Causes, and Treatments

Respiratory Acidosis: CO_2 ↑ pH ↓

Symptoms:	Failure to ventilate, suppression of breathing, disorientation, weakness, coma
Causes:	Lung disease blocking gas diffusion (e.g., emphysema, pneumonia, bronchitis, and asthma); depression of respiratory center by drugs, cardiopulmonary arrest, stroke, poliomyelitis, or nervous system disorders
Treatment:	Correction of disorder, infusion of bicarbonate

Respiratory Alkalosis: CO_2 ↓ pH ↑

Symptoms:	Increased rate and depth of breathing, numbness, light-headedness, tetany
Causes:	Hyperventilation due to anxiety, hysteria, fever, exercise; reaction to drugs such as salicylate, quinine, and antihistamines; conditions causing hypoxia (e.g., pneumonia, pulmonary edema, and heart disease)
Treatment:	Elimination of anxiety-producing state, rebreathing into a paper bag

Metabolic Acidosis: H^+ ↑ pH ↓

Symptoms:	Increased ventilation, fatigue, confusion
Causes:	Renal disease, including hepatitis and cirrhosis; increased acid production in diabetes mellitus, hyperthyroidism, alcoholism, and starvation; loss of alkali in diarrhea; acid retention in renal failure
Treatment:	Sodium bicarbonate given orally, dialysis for renal failure, insulin treatment for diabetic ketosis

Metabolic Alkalosis: H^+ ↓ pH ↑

Symptoms:	Depressed breathing, apathy, confusion
Causes:	Vomiting, diseases of the adrenal glands, ingestion of excess alkali
Treatment:	Infusion of saline solution, treatment of underlying diseases

SAMPLE PROBLEM 8.13

▪ Identifying Buffer Solutions

Indicate whether each of the following would make a buffer solution:

a. HCl, a strong acid, and NaCl
b. H_3PO_4, a weak acid
c. HF, a weak acid, and NaF

SOLUTION

a. No. A solution of a strong acid and its salt is completely ionized.
b. No. A weak acid is not sufficient for a buffer; the salt of the weak acid is also needed.
c. Yes. This mixture contains a weak acid and its salt.

STUDY CHECK

Will a mixture of NaCl and Na_2CO_3 make a buffer solution? Explain.

QUESTIONS AND PROBLEMS

▪ Buffers

8.47 Which of the following represent a buffer system? Explain.
   a. NaOH and NaCl
   b. H_2CO_3 and $NaHCO_3$
   c. HF and KF
   d. KCl and NaCl

8.48 Which of the following represent a buffer system? Explain.
   a. H_3PO_4
   b. $NaNO_3$
   c. CH_3COOH and CH_3COONa
   d. HCl and NaOH

8.49 Consider the buffer system of hydrofluoric acid, HF, and its salt, NaF.

$$HF(aq) \rightleftharpoons H^+(aq) + F^-(aq)$$

   a. What is the purpose of the buffer system?
   b. Why is a salt of the acid needed?
   c. How does the buffer react when some H^+ is added?
   d. How does the buffer react when some OH^- is added?

8.50 Consider the buffer system of nitrous acid, HNO_2, and its salt, $NaNO_2$.

$$HNO_2(aq) \rightleftharpoons H^+(aq) + NO_2^-(aq)$$

   a. What is the purpose of a buffer system?
   b. What is the purpose of $NaNO_2$ in the buffer?
   c. How does the buffer react when some H^+ is added?
   d. How does the buffer react when some OH^- is added?

CONCEPT MAP

CHAPTER REVIEW

8.1 Acids and Bases
According to the Arrhenius theory, an acid produces H^+, and a base produces OH^- in aqueous solutions.

8.2 Brønsted–Lowry Acids and Bases
According to the Brønsted–Lowry theory, acids are proton (H^+) donors, and bases are proton acceptors. Two conjugate acid–base pairs are present in an acid–base reaction. Each acid–base pair is related by the loss or gain of one H^+. For example, when the acid HF donates a H^+, the F^- it forms is its conjugate base because F^- is capable of accepting a H^+. The other acid–base pair would be H_2O and H_3O^+.

8.3 Strengths of Acids and Bases
In strong acids, all the H^+ in the acid is donated to H_2O; in a weak acid, only a small percentage of acid molecules produce H_3O^+. Strong bases are hydroxides of Groups 1A (1) and 2A (2) that dissociate completely in water. An important weak base is ammonia, NH_3.

8.4 Ionization of Water
In pure water, a few molecules transfer protons to other water molecules, producing small, but equal, amounts of $[H_3O^+]$ and $[OH^-]$, such that each has a concentration of 1×10^{-7} mole/L. The ion product, K_w, $[H_3O^+][OH^-] = 1 \times 10^{-14}$, applies to all aqueous

solutions. In acidic solutions, the $[H_3O^+]$ is greater than the $[OH^-]$. In basic solutions, the $[OH^-]$ is greater than the $[H_3O^+]$.

8.5 The pH Scale
The pH scale is a range of numbers from 0 to 14 related to the $[H_3O^+]$ of the solution. A neutral solution has a pH of 7. In acidic solutions, the pH is below 7, and in basic solutions the pH is above 7. Mathematically, pH is the negative logarithm of the hydronium ion concentration ($-\log[H_3O^+]$).

8.6 Reactions of Acids and Bases
When an acid reacts with a metal, hydrogen gas and a salt are produced. The reaction of an acid with a carbonate or bicarbonate produces carbon dioxide, a salt, and water. In neutralization, an acid reacts with a base to produce a salt and water. In titration, an acid sample is neutralized with a known amount of a base. From the volume and molarity of the base, the concentration of the acid is calculated.

8.7 Buffers
A buffer solution resists changes in pH when small amounts of acid or base are added. A buffer contains either a weak acid and its salt or a weak base and its salt. The weak acid picks up added OH^-, and the anion of the salt picks up added H^+. Buffers are important in maintaining the pH of the blood.

KEY TERMS

acid A substance that dissolves in water and produces hydrogen ions (H^+), according to the Arrhenius theory. All acids are proton donors, according to the Brønsted–Lowry theory.

acidosis A physiological condition in which the blood pH is lower than 7.35.

alkalosis A physiological condition in which the blood pH is higher than 7.45.

base A substance that dissolves in water and produces hydroxide ions (OH^-) according to the Arrhenius theory. All bases are proton acceptors, according to the Brønsted–Lowry theory.

Brønsted–Lowry acids and bases An acid is a proton donor, and a base is a proton acceptor.

buffer solution A mixture of a weak acid or a weak base and its salt that resists changes in pH when small amounts of an acid or base are added.

conjugate acid–base pair An acid and base that differ by one H^+. When an acid donates a proton, the product is its conjugate base, which is capable of accepting a proton in the reverse reaction.

hydronium ion, H_3O^+ The ion formed by the attraction of a proton (H^+) to an H_2O molecule.

ion-product constant of water, K_w The product of $[H_3O^+]$ and $[OH^-]$ in solution; $K_w = [H_3O^+][OH^-]$.

neutral The term that describes a solution with equal concentrations of $[H_3O^+]$ and $[OH^-]$.

neutralization A reaction between an acid and a base to form a salt and water.

pH A measure of the $[H_3O^+]$ in a solution; $pH = -\log[H_3O^+]$.

strong acid An acid that completely ionizes in water.

strong base A base that completely ionizes in water.

titration The addition of base to an acid sample to determine the concentration of the acid.

weak acid An acid that ionizes only slightly in solution.

weak base A base that ionizes only slightly in solution.

ADDITIONAL QUESTIONS AND PROBLEMS

8.51 Name each of the following:
   a. H_2SO_4
   b. KOH
   c. $Ca(OH)_2$
   d. HCl
   e. HNO_2

8.52 Are the following examples of body fluids acidic, basic, or neutral?
   a. saliva, pH 6.8
   b. urine, pH 5.9
   c. pancreatic juice, pH 8.0
   d. bile, pH 8.4
   e. blood, pH 7.45

8.53 What are some similarities and differences between strong and weak acids?

8.54 What are some ingredients found in antacids? What do they do?

8.55 One ingredient in some antacids is $Mg(OH)_2$.
   a. If the base is not very soluble in water, why is it considered a strong base?
   b. What is the neutralization reaction of $Mg(OH)_2$ with stomach acid, HCl?

8.56 Acetic acid, which is the acid in vinegar, is a weak acid. Why?

8.57 Determine the pH for the following solutions:
   a. $[H_3O^+] = 2.0 \times 10^{-8}$ M
   b. $[H_3O^+] = 5.0 \times 10^{-2}$ M
   c. $[OH^-] = 3.5 \times 10^{-4}$ M
   d. $[OH^-] = 0.005$ M

8.58 Are the solutions in problem 8.57 acidic, basic, or neutral?

8.59 What are the $[H_3O^+]$ and $[OH^-]$ for a solution with the following pH values?
   a. 3.0   b. 6.0
   c. 8.0   d. 11.0

8.60 Solution A has a pH of 4.0 and solution B has a pH of 6.0.
   a. Which solution is more acidic?
   b. What is the $[H_3O^+]$ in each?
   c. What is the $[OH^-]$ in each?

8.61 A buffer is made by dissolving H_3PO_4 and NaH_2PO_4 in water.
   a. Write an equation that shows how this buffer neutralizes small amounts of acids.
   b. Write an equation that shows how this buffer neutralizes small amounts of base.

8.62 A buffer is made by dissolving CH_3COOH and CH_3COONa in water.
   a. Write an equation that shows how this buffer neutralizes small amounts of acid.
   b. Write an equation that shows how this buffer neutralizes small amounts of base.

8.63 Calculate the volume (mL) of a 0.150 M NaOH solution that will completely neutralize the following:
   a. 25.0 mL of a 0.288 M HCl solution
   b. 10.0 mL of a 0.560 M H_2SO_4 solution

8.64 How many milliliters of 0.215 M NaOH solution are needed to completely neutralize 2.50 mL of 0.825 M H_2SO_4 solution?

8.65 A solution of 0.205 M NaOH is used to neutralize 20.0 mL H_2SO_4. If 45.6 mL NaOH is required to reach the endpoint, what is the molarity of the H_2SO_4 solution?

$$H_2SO_4(aq) + 2NaOH(aq) \longrightarrow Na_2SO_4(aq) + 2H_2O(l)$$

8.66 A 10.0 mL sample of vinegar, which is an aqueous solution of acetic acid, CH_3COOH, requires 16.5 mL of 0.500 M NaOH to reach the endpoint in a titration. What is the molarity of the acetic acid solution?

$$CH_3COOH(aq) + NaOH(aq) \longrightarrow$$
$$CH_3COONa(aq) + H_2O(l)$$

UNDERSTANDING THE CONCEPTS

8.67 In each of the following diagrams of acid solutions, determine if each diagram represents a strong acid or a weak acid. The acid has the formula HX.

8.68 Adding a few drops of a strong acid to water will lower the pH appreciably. However, adding the same number of drops to a buffer does not appreciably alter the pH. Why?

Water   Buffer

8.69 Sometimes, during stress or trauma, a person can start to hyperventilate. Then the person might breathe into a paper bag to avoid fainting.
  a. What changes occur in the blood pH during hyperventilation?
  b. How does breathing into a paper bag help return blood pH to normal?

8.70 In the blood plasma, pH is maintained by the carbonic acid–bicarbonate buffer system.
  a. How is pH maintained when acid is added to the buffer system?
  b. How is pH maintained when base is added to the buffer system?

QUESTIONS FOR TEAMWORK

8.71 Consider the following:
  1. H_2S **2.** H_3PO_4 **3.** HCO_3^-
  a. For each, write the formula of the conjugate base.
  b. Write the formula of the weakest acid.
  c. Write the formula of the strongest acid.

8.72 Identify the conjugate acid–base pairs in each of the following equations.
  a. $NH_3(aq) + HNO_3(aq) \rightleftharpoons NH_4^+(aq) + NO_3^-(aq)$
  b. $H_2O(l) + HBr(aq) \rightleftharpoons H_3O^+(aq) + Br^-(aq)$
  c. $HNO_2(aq) + HS^-(aq) \rightleftharpoons H_2S(g) + NO_2^-(aq)$
  d. $Cl^-(aq) + H_2O(l) \rightleftharpoons OH^-(aq) + HCl(aq)$

8.73 Complete and balance each of the following:
  a. $ZnCO_3(s) + H_2SO_4(aq) \longrightarrow$
  b. $Al(s) + HCl(aq) \longrightarrow$
  c. $H_3PO_4(aq) + Ca(OH)_2(s) \longrightarrow$
  d. $KHCO_3(s) + HNO_3(aq) \longrightarrow$

8.74 Determine each of the following for a 0.050 M KOH solution.
  a. $[H_3O^+]$
  b. pH
  c. products when allowed to react with H_3PO_4
  d. milliliters required to neutralize 40.0 mL of 0.035 M H_2SO_4

8.75 What is the pH of a solution prepared by dissolving 2.5 g HCl in water to make 425 mL of solution?

8.76 Consider the reaction of KOH and HNO_2.
  a. Write the balanced chemical equation.
  b. Calculate the milliliters of 0.122 M KOH required to neutralize 36.0 mL of 0.250 M HNO_2.

ANSWERS

Answers to Study Checks

8.1 Sulfuric acid; potassium hydroxide

8.2 $Ca(OH)_2(s) \xrightarrow{H_2O} Ca^{2+}(aq) + 2OH^-(aq)$

8.3 $HNO_3(aq) + H_2O(l) \longrightarrow H_3O^+(aq) + NO_3^-(aq)$

8.4 A base accepts a proton to form its conjugate acid.
a. H_2S
b. HNO_2

8.5 The conjugate acid/base pairs are HCN/CN^- and SO_4^{2-}/HSO_4^-.

8.6 KOH

8.7 $[H_3O^+] = 2.5 \times 10^{-11}$ M; basic

8.8 11.38

8.9 4.3

8.10 $[H_3O^+] = 1 \times 10^{-5}$ M; $[OH^-] = 1 \times 10^{-9}$ M

8.11 $H_2SO_4(aq) + 2NaHCO_3(s) \longrightarrow$
$$Na_2SO_4(aq) + 2CO_2(g) + 2H_2O(l)$$

8.12 0.200 M HCl

8.13 No. Both substances are salts; the mixture has no weak acid present.

Answers to Selected Questions and Problems

8.1 **a.** acid **b.** acid
  c. acid **d.** base

8.3 **a.** hydrochloric acid **b.** calcium hydroxide
  c. carbonic acid **d.** nitric acid
  e. sulfurous acid

8.5 **a.** $Mg(OH)_2$ **b.** HF **c.** H_3PO_4
  d. LiOH **e.** $Cu(OH)_2$

8.7 **a.** HI is the acid (proton donor) and H_2O is the base (proton acceptor).
  b. H_2O is the acid (proton donor) and F^- is the base (proton acceptor).

8.9 **a.** F^-, fluoride ion
  b. OH^-, hydroxide ion
  c. HCO_3^-, bicarbonate ion *or* hydrogen carbonate ion
  d. SO_4^{2-}, sulfate ion

8.11 **a.** HCO_3^-, bicarbonate ion *or* hydrogen carbonate ion
  b. H_3O^+, hydronium ion
  c. H_3PO_4, phosphoric acid
  d. HBr, hydrobromic acid

8.13 **a.** acid H_2CO_3; conjugate base HCO_3^-
    base H_2O; conjugate acid H_3O^+
  b. acid NH_4^+; conjugate base NH_3
    base H_2O; conjugate acid H_3O^+
  c. acid HCN; conjugate base CN^-
    base NO_2^-; conjugate acid HNO_2

8.15 **a.** HBr **b.** HSO_4^- **c.** H_2CO_3

8.17 **a.** HSO_4^- **b.** HF **c.** HCO_3^-

8.19 In pure water, $[H_3O^+] = [OH^-]$ because one of each is produced every time a proton transfers from one water to another.

8.21 In an acidic solution, the $[H_3O^+]$ is greater than the $[OH^-]$.

8.23 **a.** acidic **b.** basic **c.** basic **d.** acidic

8.25 **a.** 1.0×10^{-9} M
  b. 1.0×10^{-6} M
  c. 2.0×10^{-5} M
  d. 4.0×10^{-13} M

8.27 **a.** 1.0×10^{-11} M
  b. 2.0×10^{-9} M
  c. 5.6×10^{-3} M
  d. 2.5×10^{-2} M

8.29 In a neutral solution, the $[H_3O^+]$ is 1.0×10^{-7} M and the pH is 7.00, which is the negative value of the power of 10.

8.31 **a.** basic **b.** acidic **c.** basic **d.** acidic

8.33 **a.** 4.0 **b.** 8.5 **c.** 9.0 **d.** 3.40

8.35

$[H_3O^+]$	$[OH^-]$	pH	Acidic, Basic, or Neutral?
1×10^{-8} M	1×10^{-6} M	8.0	Basic
1×10^{-3} M	1×10^{-11} M	3.0	Acidic
2×10^{-5} M	5×10^{-10} M	4.7	Acidic
1×10^{-12} M	1×10^{-2} M	12.0	Basic

8.37 **a.** $ZnCO_3(s) + 2HBr(aq) \longrightarrow$
      $ZnBr_2(aq) + CO_2(g) + H_2O(l)$
  b. $Zn(s) + 2HCl(aq) \longrightarrow ZnCl_2(aq) + H_2(g)$
  c. $HCl(g) + NaHCO_3(s) \longrightarrow$
      $NaCl(aq) + H_2O(l) + CO_2(g)$
  d. $H_2SO_4(aq) + Mg(OH)_2(s) \longrightarrow MgSO_4(aq) + 2H_2O(l)$

8.39 **a.** $2HCl(aq) + Mg(OH)_2(s) \longrightarrow MgCl_2(aq) + 2H_2O(l)$
  b. $H_3PO_4(aq) + 3LiOH(aq) \longrightarrow Li_3PO_4(aq) + 3H_2O(l)$

8.41 **a.** $H_2SO_4(aq) + 2NaOH(aq) \longrightarrow$
      $Na_2SO_4(aq) + 2H_2O(l)$
  b. $3HCl(aq) + Fe(OH)_3(aq) \longrightarrow FeCl_3(aq) + 3H_2O(l)$
  c. $H_2CO_3(aq) + Mg(OH)_2(s) \longrightarrow MgCO_3(s) + 2H_2O(l)$

8.43 0.829 M HCl

8.45 0.125 M H_2SO_4

8.47 (b) and (c) are buffer systems. (b) contains the weak acid H_2CO_3 and its salt $NaHCO_3$. (c) contains HF, a weak acid, and its salt KF.

8.49 **a.** A buffer system keeps the pH constant.
  b. To neutralize any H^+ added.
  c. The added H^+ reacts with F^- from NaF.
  d. The added OH^- is neutralized by the HF.

8.51 **a.** sulfuric acid
 b. potassium hydroxide
 c. calcium hydroxide
 d. hydrochloric acid
 e. nitrous acid

8.53 Both strong and weak acids produce H_3O^+ in water. Weak acids are only slightly ionized, whereas a strong acid exists as ions in solution.

8.55 **a.** The $Mg(OH)_2$ that dissolves is completely ionized, making it a strong base.
 b. $Mg(OH)_2(aq) + 2HCl(aq) \longrightarrow$
$$MgCl_2\,(aq) + 2H_2O(l)$$

8.57 **a.** pH 7.70   **b.** pH 1.30
 c. pH 10.54   **d.** pH 11.7

8.59 **a.** $[H_3O^+] = 1 \times 10^{-3}$ M; $[OH^-] = 1 \times 10^{-11}$ M
 b. $[H_3O^+] = 1 \times 10^{-6}$ M; $[OH^-] = 1 \times 10^{-8}$ M

c. $[H_3O^+] = 1 \times 10^{-8}$ M; $[OH^-] = 1 \times 10^{-6}$ M
 d. $[H_3O^+] = 1 \times 10^{-11}$ M; $[OH^-] = 1 \times 10^{-3}$ M

8.61 **a.** acid: $H_2PO_4^-\,(aq) + H_3O^+\,(aq) \longrightarrow$
$$H_3PO_4(aq) + H_2O(l)$$
 b. base: $H_3PO_4(aq) + OH^-(aq) \longrightarrow$
$$H_2PO_4^-(aq) + H_2O(l)$$

8.63 **a.** 48.0 mL NaOH   **b.** 74.7 mL NaOH

8.65 0.234 M H_2SO_4

8.67 **a.** weak acid   **b.** strong acid   **c.** weak acid

8.69 **a.** Hyperventilation will lower the CO_2 level in the blood, which lowers the H_2CO_3 concentration, which decreases the H_3O^+ and increases the blood pH.
 b. Breathing into a bag will increase the CO_2 level, increase the H_2CO_3, increase H_3O^+, and lower the blood pH.

9 Nuclear Radiation

"Everything we do in this department involves radioactive materials," says Julie Goudak, nuclear medicine technologist at Kaiser Hospital. "The radioisotopes are given in several ways. The patient may ingest an isotope, breathe it in, or receive it by an IV injection. We do many diagnostic tests, particularly of the heart function, to determine if a patient needs a cardiac CAT scan."

A nuclear medicine technologist administers isotopes that emit radiation to determine the level of function of an organ such as the thyroid or heart, to detect the presence and size of a tumor, or to treat disease. A radioisotope locates in a specific organ and its radiation is used by a computer to create an image of that organ. From this data, a physician can make a diagnosis and design a treatment program.

the Chemistry place

Visit **www.aw-bc.com/chemplace** or use the CD-ROM in the book for extra quizzes, interactive tutorials, career resources, PowerPoint slides, math help, and case studies.

A female patient, aged 50, complains of nervousness, irritability, increased perspiration, brittle hair, and muscle weakness. Her hands are shaky at times and her heart often beats rapidly. She has been experiencing weight loss. The doctor decides to test for hyperthyroidism. To get a detailed look at the thyroid, a thyroid scan is ordered. The patient is given a small amount of an iodine radioisotope, which will be taken up by the thyroid. The scan shows a higher than normal rate of uptake of the radioactive iodine indicating an overactive thyroid gland, a condition called hyperthyroidism. Treatment for hyperthyroidism includes the use of drugs to lower the level of thyroid hormone, the use of radioactive iodine to destroy thyroid cells, or surgical removal of part or the entire thyroid. In our case, the nuclear physician decides to use radioactive iodine. To begin treatment, the patient drinks a solution containing radioactive iodine. In the following few weeks, the cells that take up the radioactive iodine are destroyed by the radiation. Tests show that the patient's thyroid is smaller, and that the blood level of thyroid hormone is normal once again.

With the discovery of X rays in 1895 and the production of artificial radioactive substances in 1934, the field of nuclear medicine was established. In 1937, the first radioactive isotope was used to treat a patient with leukemia at the University of California at Berkeley. Major strides in the use of radioactivity in medicine occurred in 1946, when a radioactive iodine isotope was successfully used to diagnose thyroid function and to treat hyperthyroidism and thyroid cancer. In the 1970s and 1980s, a variety of radioactive substances were used to produce an image of an organ such as liver, spleen, thyroid gland, kidney, and brain, and to detect heart disease. Today, procedures in nuclear medicine provide information about the function and structure of every organ in the body, which allows the nuclear physician to diagnose and treat diseases early.

LEARNING GOAL

■ Describe alpha, beta, and gamma radiation.

■ 9.1 Natural Radioactivity

In Chapter 3, we learned that samples of most elements contain isotopes that differ only in the number of neutrons. While most isotopes contain nuclei that are stable, there are some that contain unstable nuclei. An unstable nucleus is **radioactive** because it spontaneously emits energy to become more stable. The energy emitted from a nucleus, called **radiation**, may be high-energy particles, such as alpha (α) or beta (β) particles, or gamma (γ) rays, which are pure energy. An isotope that emits radiation is called a **radioisotope**. While there are several naturally occurring radioisotopes, many more are produced artificially in nuclear laboratories. You may be familiar with X rays, which are also a type of radiation, although the energy of X rays is lower than that of nuclear radiation. X rays are not emitted from nuclei of unstable atoms but are produced in a different way.

In the symbol for an isotope, the mass number is written in the upper left corner of the symbol, and the atomic number is written in the lower left corner. For

TABLE 9.1

Stable and Radioactive Isotopes of Some Elements		
Magnesium	**Iodine**	**Uranium**
Stable Isotopes		
$^{24}_{12}\text{Mg}$	$^{127}_{53}\text{I}$	None
Magnesium-24	Iodine-127	
Radioactive Isotopes		
$^{23}_{12}\text{Mg}$	$^{125}_{53}\text{I}$	$^{235}_{92}\text{U}$
Magnesium-23	Iodine-125	Uranium-235
$^{27}_{12}\text{Mg}$	$^{131}_{53}\text{I}$	$^{238}_{92}\text{U}$
Magnesium-27	Iodine-131	Uranium-238

example, a radioactive isotope of iodine used in the diagnosis and treatment of thyroid conditions has a mass number of 131 and an atomic number of 53. We can write its symbol as shown below.

Mass number (protons and neutrons)
Element
Atomic number (protons)

$$^{131}_{53}\text{I}$$

This isotope is also called iodine-131 or I-131. Radioactive isotopes are named by writing the mass number after the element's name or symbol. When necessary, we can obtain the atomic number from the periodic table. Table 9.1 compares some stable, nonradioactive isotopes with some radioactive isotopes.

Types of Radiation

Different forms of radiation are emitted from an unstable nucleus when a change takes place among its protons and neutrons. Symbols for different forms of radiation can include the charge and mass number, written in the lower and upper left corner. The release of radiation of any type produces a more stable, lower energy nucleus. One type of radiation consists of alpha particles. An **alpha particle** contains 2 protons and 2 neutrons, which gives it a mass number of 4 and an atomic number of 2. Because it has 2 protons, an alpha particle has a charge of 2+. That makes it identical to a helium nucleus. In equations, it is written as the Greek letter alpha (α) or as the symbol for helium.

α or $^{4}_{2}\text{He}$
Alpha particle

Another type of radiation occurs when a radioisotope emits **beta particles**. A beta particle, which is a high-energy electron, has a charge of 1− and, because its mass is so much less than the mass of a proton, it is given a mass number of 0. It is represented by the Greek letter beta (β) or by the symbol for the electron, with the charge, −1, written on the lower left.

β or $^{0}_{-1}e$ ●
Beta particle

Beta particles are produced from an unstable nucleus when a neutron is transformed into a proton and an electron. The high-energy electron is emitted from the nucleus as beta radiation. This electron or beta particle did not exist in the nucleus until the following change occurred:

$$^{0}_{1}n \longrightarrow ^{1}_{1}H \quad + \quad ^{0}_{-1}e \text{ or } \beta$$

Neutron in the nucleus · New proton remains in the nucleus · New electron formed and emitted as a beta particle

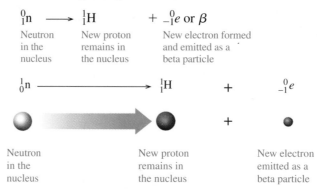

Neutron in the nucleus · New proton remains in the nucleus · New electron emitted as a beta particle

A **positron** is similar to an electron except that a positron has a positive (+1) charge. A positron is produced by an unstable nucleus when a proton is transformed into a neutron and a positron. The electron and positron each have a mass number of 0. Instead of the atomic number, the charge is shown when we write the symbols of an electron and a positron as follows.

	Electron	**Positron**
Mass number Charge	$^{0}_{-1}e$	$^{0}_{+1}e$

$$^{1}_{1}p \longrightarrow ^{1}_{0}n \quad + \quad ^{0}_{+1}e \text{ (or } \beta^{+})$$

Positron

A positron is an example of *antimatter*, a term physicists use to describe a particle that is the exact opposite of a particle, in this case, an electron. When an electron and a positron collide, their minute masses are completely converted to energy in the form of gamma rays.

$$^{0}_{-1}e \quad + \quad ^{0}_{+1}e \longrightarrow 2 \, ^{0}_{0}\gamma$$

When the symbol β is used with no charge, it is taken to mean a beta particle rather than a positron.

Gamma rays are high-energy radiation, released as an unstable nucleus undergoes a rearrangement of its particles to give a more stable, lower energy nucleus. A gamma ray is shown as the Greek letter gamma (γ). Because gamma rays are energy only, there is no mass or charge associated with their symbol.

γ
Gamma ray

Table 9.2 summarizes the types of radiation we will use in nuclear equations.

TABLE 9.2

Some Common Forms of Radiation

Type of Radiation	Symbol		Mass Number	Charge
Alpha particle	α	$^{4}_{2}He$	4	2+
Beta particle	β	$^{0}_{-1}e$	0	1−
Positron	β^{+}	$^{0}_{+1}e$	0	1+
Gamma ray	γ	$^{0}_{0}\gamma$	0	0
Proton	$^{1}_{1}H$	$^{1}_{1}p$	1	1+
Neutron	$^{1}_{0}n$	n	1	0

■ **Writing Symbols for Radiation Particles**

Write the symbol for an alpha particle.

SOLUTION

The alpha particle contains 2 protons and 2 neutrons. It has a mass number of 4 and an atomic number of 2.

α or ^4_2He

STUDY CHECK

What is the symbol used for beta radiation?

■ **Radiation Protection**

Nuclear radiation is harmful because the particles have tremendous energy. Therefore, it is important that the radiologist, doctor, and nurse working with radioactive isotopes use proper radiation protection. Proper **shielding** is necessary to prevent exposure. Alpha particles are the heaviest of the radiation particles; they travel only a few centimeters in the air before they collide with air molecules, acquire electrons, and become helium atoms. A piece of paper, clothing, and our skin are protection against alpha particles. Lab coats and gloves will also provide sufficient shielding. However, if ingested or inhaled, alpha particles can bring about serious internal damage because their mass and high charge causes much ionization in a short distance.

Beta particles have a very small mass and move much faster and farther than alpha particles, traveling as much as several meters through air. They can pass through paper and penetrate as far as 4–5 mm into body tissue. External exposure to beta particles can burn the surface of the skin, but they are stopped before they can reach the internal organs. Heavy clothing such as lab coats and gloves are needed to protect the skin from beta particles.

Gamma rays travel great distances through the air and pass through many materials, including body tissues. Only very dense shielding, such as lead or concrete, will stop them. (See Figure 9.1.) Because they can penetrate so deeply, exposure to gamma rays can be extremely hazardous. Even the syringe used to give an injection of a gamma-emitting radioactive isotope is placed inside a special lead-glass cover.

When preparing radioactive materials, the radiologist wears special gloves and works behind leaded windows. Long tongs are used within the work area to pick up vials of radioactive material, keeping them away from the hands and body. (See Figure 9.1.) Table 9.3 summarizes the shielding materials required for the various types of radiation.

Try to keep the time you must spend in a radioactive area to a minimum. A certain amount of radiation is emitted every minute. Remaining in a radioactive area twice as long exposes a person to twice as much radiation.

Keep your distance! The greater the distance from the radioactive source, the lower the intensity of radiation received. If you double your distance from the radiation source, the intensity of radiation drops to $\left(\frac{1}{2}\right)^2$ or one-fourth of its previous value.

WEB TUTORIAL
Radiation and Its Biological Effects

FIGURE 9.1

A person working with radioisotopes wears protective clothing and gloves and stands behind a lead shield.

Q What types of radiation does the lead shield block?

▪ Biological Effects of Radiation

When high-energy radiation strikes molecules in its path, electrons may be knocked away. The result of this *ionizing radiation* is the formation of unstable ions or radicals. A free radical is a particle that has an unpaired electron. For example, when radiation passes through the human body, it may interact with water molecules, removing electrons and producing H_2O^+ ions or free radicals.

When ionizing radiation strikes the cells of the body, the unstable ions or free radicals that form can cause undesirable chemical reactions. The cells in the body most sensitive to radiation are the ones undergoing rapid division—those of the bone marrow, skin, reproductive organs, and intestinal lining, as well as all cells of growing children. Damaged cells may lose their ability to produce necessary materials. For example, if radiation damages cells of the bone marrow, red blood cells may no longer be produced. If sperm cells or ova or the cells of a fetus are damaged, birth defects may result. In contrast, cells of the nerves, muscles, liver, and adult bones are much less sensitive to radiation because they undergo little or no cellular division.

Cancer cells are another example of rapidly dividing cells. Because cancer cells are highly sensitive to radiation, large doses of radiation are used to destroy them. The surrounding normal tissue, dividing at a slower rate, shows a greater resistance to radiation and suffers less damage. In addition, normal tissue is able to repair itself more readily than cancerous tissue. However, this repair is not always complete. Possible long-term effects of ionizing radiation include a shortened life span, malignant tumors, leukemia, anemia, and genetic mutations.

TABLE 9.3
Properties of Ionizing Radiation and Shielding Required

Property	Alpha	Beta	Gamma
Characteristics	Helium nucleus	Electron	High-energy rays
Symbols	α, ^4_2He	β, $^0_{-1}e$	γ, $^0_0\gamma$
Travel distance in air	2–4 cm	200–300 cm	500 m
Tissue depth	0.05 mm	4–5 mm	50 cm or more
Shielding	Paper, clothing	Heavy clothing, lab coats, gloves,	Lead, thick concrete
Typical source	Radium-226	Carbon-14	Technetium-99m

SAMPLE PROBLEM 9.2

▪ Radiation Protection

How does the type of shielding for alpha radiation differ from that used for gamma radiation?

SOLUTION

Alpha radiation is stopped by paper and clothing. However, lead or concrete is needed for protection from gamma radiation.

STUDY CHECK

Besides shielding, what other methods help reduce exposure to radiation?

QUESTIONS AND PROBLEMS

▪ Natural Radioactivity

9.1 **a.** How are an alpha particle and a helium nucleus similar? different?
   b. What symbols are used for alpha particles?

9.2 **a.** How are a beta particle and an electron similar? different?
   b. What symbols are used for beta particles?

9.3 Naturally occurring potassium consists of three isotopes, potassium-39, potassium-40, and radioactive potassium-41.
   a. Write the atomic symbols for each isotope.
   b. In what ways are the isotopes similar, and in what ways do they differ?

9.4 Naturally occurring iodine is iodine-127. Medically, radioactive isotopes of iodine-125 and iodine-130 are used.
   a. Write the atomic symbols for each isotope.
   b. In what ways are the isotopes similar, and in what ways do they differ?

9.5 Supply the missing information in the following table:

Medical Use	Atomic Symbol	Mass Number	Number of Protons	Number of Neutrons
Heart imaging	$^{201}_{81}Tl$	201	81	120
Radiation therapy	60/27 CO	60	27	33
Abdominal scan	67/31 GA	67	31	36
Hyperthyroidism	$^{131}_{53}I$ I	131	53	78
Leukemia treatment	32/15 P	32	15	17

9.6   Supply the missing information in the following table:

Medical Use	Atomic Symbol	Mass Number	Number of Protons	Number of Neutrons
Cancer treatment	$^{60}_{27}\text{Co}$			
Brain scan		99	43	
Blood flow		141	58	
Bone scan		85		47
Lung function	$^{133}_{54}\text{Xe}$			

9.7   Write a symbol for the following:
   a. alpha particle   **b.** neutron
   c. beta particle   **d.** nitrogen-15
   e. iodine-125

9.8   Write a symbol for the following:
   a. proton   **b.** gamma ray
   c. electron   **d.** positron
   e. cobalt-60

9.9   Identify the symbol for X in each of the following:
   a. $^{0}_{-1}\text{X}$ **b.** $^{4}_{2}\text{X}$ **c.** $^{1}_{0}\text{X}$
   d. $^{24}_{11}\text{X}$ **e.** $^{14}_{6}\text{X}$

9.10 Identify the symbol for X in each of the following:
   a. $^{1}_{1}\text{X}$ **b.** $^{32}_{15}\text{X}$ **c.** $^{0}_{0}\text{X}$
   d. $^{59}_{26}\text{X}$ **e.** $^{0}_{+1}\text{X}$

9.11 a. Why does beta radiation penetrate further in solid material than alpha radiation?
   b. How does ionizing radiation cause damage to cells of the body?
   c. Why does the x-ray technician leave the room when you receive an X ray?
   d. What is the purpose of wearing gloves when handling radioisotopes?

9.12 a. As a nurse in an oncology unit, you sometimes give an injection of a radioisotope. What are three ways you can minimize your exposure to radiation?
   b. Why are cancer cells more sensitive to radiation than nerve cells?
   c. What is the purpose of placing a lead apron on a patient who is receiving routine dental X rays?
   d. Why are the walls in a radiology office built of thick concrete blocks?

■ 9.2 Nuclear Equations

When a nucleus spontaneously breaks down by emitting radiation, the process is called *radioactive decay*. It can be shown as a *nuclear equation* using the symbols for the original radioactive nucleus, the new nucleus, and the radiation emitted.

Radioactive nucleus $\longrightarrow$ new nucleus + radiation ($\alpha, \beta, \gamma, \beta^{+}$)

A nuclear equation is balanced when the sum of the mass numbers and the sum of the atomic numbers of the particles and atoms on one side of the equation are equal to their counterparts on the other side.

The changes in mass number and atomic number of an atom that emits a radioactive particle are shown in Table 9.4.

LEARNING GOAL

Write an equation showing mass numbers and atomic numbers for radioactive decay.

the
**Chemistry
place**

WEB TUTORIAL
Radiation and Its Biological Effects

TABLE 9.4

Mass Number and Atomic Number Changes due to Radiation

Decay Process	Radiation Symbol	Change in Mass Number	Change in Atomic Number	Change in Neutron Number
Alpha emission	^4_2He or α	-4	-2	-2
Beta emission	$^0_{-1}e$ or β	0	$+1$	-1
Positron emission	$^0_{+1}e$ or β^+	0	-1	$+1$
Gamma emission	$^0_0\gamma$	0	0	0

Radioactive nucleus

$^{238}_{92}\text{U}$

Radiation

New nucleus

Alpha particle

^4_2He

○ Neutron

● Proton

$^{234}_{90}\text{Th}$

146 neutrons
92 protons

$^{238}_{92}\text{U}$

Radioactive nucleus

144 neutrons
90 protons

2 neutrons
2 protons

$^{238}_{92}\text{U} \longrightarrow \ ^{234}_{90}\text{Th} \ + \ ^4_2\text{He}$

New nucleus   Alpha particle

Alpha Emitters

Alpha emitters are radioisotopes that decay by emitting alpha particles. For example, uranium-238 decays to thorium-234 by emitting an alpha particle.

The alpha particle emitted contains 2 protons, which means that the new nucleus has 2 fewer protons, or 90 protons. That means that the new nucleus has an atomic number of 90 and is therefore thorium (Th). Because the alpha particle has a mass number of 4, the mass number of the thorium isotope is 234, 4 less than that of the original uranium nucleus.

Guide to Completing a Nuclear Equation

In another example of radioactive decay, radium-226 emits an alpha particle to form a new isotope whose mass number, atomic number, and identity we must determine.

STEP 1 **Write the incomplete nuclear equation.**

$$^{226}_{88}\text{Ra} \longrightarrow \ ? \ + \ ^4_2\text{He}$$

STEP 2 **Determine the missing mass number.** In the equation, the mass number, 226, of the radium is equal to the combined mass numbers of the alpha particle and the new nucleus. We can calculate the mass number of the new nucleus by subtracting:

$$226 \quad = ? + 4$$
$$226 - 4 = ?$$
$$222 \quad = ? \text{ (mass number of new nucleus)}$$

STEP 3 **Determine the missing atomic number.** The atomic number of radium, 88, must equal the sum of the atomic number of the alpha particle and the new nucleus. We can calculate the atomic number of the new nucleus by subtracting:

$$88 \quad = ? + 2$$
$$88 - 2 = ?$$
$$86 \quad = ? \text{ (atomic number of new nucleus)}$$

STEP 4 **Determine the symbol of the new nucleus.** On the periodic table, the element that has atomic number 86 is radon, Rn. The nucleus of this isotope of Rn is written as

$$^{222}_{86}\text{Rn}$$

Guide to Completing
a Nuclear Equation

STEP 1
Write the incomplete nuclear equation.

STEP 2
Determine the missing mass number.

STEP 3
Determine the missing atomic number.

STEP 4
Determine the symbol of the new nucleus.

STEP 5
Complete the nuclear equation.

^4_2He

86	87	88
Rn	Fr	Ra

STEP 5 **Complete the nuclear equation.**

$$^{226}_{88}\text{Ra} \longrightarrow {}^{222}_{86}\text{Rn} + {}^{4}_{2}\text{He}$$

In this nuclear reaction, a radium-226 nucleus decays by releasing an alpha particle and produces a radon-222 nucleus.

ENVIRONMENTAL NOTE

SAMPLE PROBLEM 9.3

■ **Writing an Equation for Alpha Decay**

Smoke detectors, required in homes and apartments, contain an alpha emitter such as americium-241. The alpha particles ionize air molecules, producing a constant stream of electrical current. However, when smoke particles enter the detector, they interfere with the formation of ions in the air, and the electric current is interrupted. This causes the alarm to sound and warns the occupants of the danger of fire. Complete the following nuclear equation for the decay of americium-241:

$$^{241}_{95}\text{Am} \longrightarrow ? + {}^{4}_{2}\text{He}$$

SOLUTION

STEP 1 **Write the incomplete nuclear equation.**

$$^{241}_{95}\text{Am} \longrightarrow ? + {}^{4}_{2}\text{He}$$

STEP 2 **Determine the missing mass number.** In the equation, the mass number, 241, of the americium is equal to the sum of the mass numbers of the alpha particle and the new nucleus. We calculate the mass number of the new nucleus as follows:

$$241 \quad = ? + 4$$
$$241 - 4 = ?$$
$$237 \quad = ? \text{ (mass number of new nucleus)}$$

STEP 3 **Determine the missing atomic number.** The atomic number of americium-95 must equal the sum of the atomic number of the alpha particle and the new nucleus. We can calculate the atomic number of the new nucleus as follows:

$$95 \quad = ? + 2$$
$$95 - 2 = ?$$
$$93 \quad = ? \text{ (atomic number of new nucleus)}$$

STEP 4 **Determine the symbol of the new nucleus.** On the periodic table, the element that has atomic number 93 is neptunium, Np. The symbol of this isotope of Np is written as $^{237}_{93}\text{Np}$.

STEP 5 **Complete the nuclear equation.**

$$^{241}_{95}\text{Am} \longrightarrow {}^{237}_{93}\text{Np} + {}^{4}_{2}\text{He}$$

In this nuclear reaction, an americium-241 nucleus decays by releasing an alpha particle and produces a neptunium-237 nucleus.

■ **Radon in our Homes**

The presence of radon has become a much publicized environmental and health issue because of radiation danger. Radioactive isotopes that produce radon, such as radium-226 and uranium-238, are naturally present in many types of rocks and soils. Radium 226 emits an alpha particle and is converted into radon gas, which diffuses out of the rocks and soil.

$$^{226}_{88}\text{Ra} \longrightarrow {}^{222}_{86}\text{Rn} + {}^{4}_{2}\text{He}$$

As uranium-238 decays, it also forms radium-226, which in turn produces radon. Uranium-238 has been found in particularly high levels in an area between Pennsylvania and New England.

Outdoors, radon gas poses little danger because it dissipates in the air. However, if the source of radon is under a house or building, the gas can enter the house through cracks in the foundation or other openings, where the radon can be inhaled by those living or working there. Inside the lungs, radon emits alpha particles to form polonium-218, which is known to cause cancer when present in the lungs.

$$^{222}_{86}\text{Rn} \longrightarrow {}^{218}_{84}\text{Po} + {}^{4}_{2}\text{He}$$

Some researchers have estimated that 10% of all lung cancer deaths in the United States are due to radon. The Environmental Protection Agency (EPA) recommends that the maximum level of radon not exceed 4 picocuries (pCi) per liter of air in a home. One (1) picocurie (pCi) is equal to 10^{-12} curies (Ci): curies are described in Section 9.3. In California, 1% of all the houses surveyed exceeded the EPA's recommended maximum radon level.

STUDY CHECK

Write a balanced nuclear equation for the alpha emitter polonium-214.

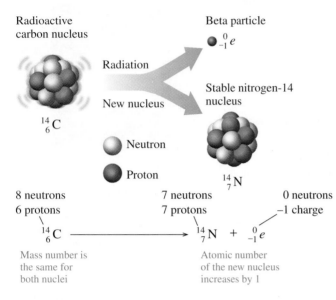

Radioactive carbon nucleus

Beta particle

$^{0}_{-1}e$

Radiation

New nucleus

Stable nitrogen-14 nucleus

$^{14}_{6}C$

Neutron

Proton

$^{14}_{7}N$

8 neutrons
6 protons

7 neutrons
7 protons

0 neutrons
−1 charge

$^{14}_{6}C \longrightarrow ^{14}_{7}N + ^{0}_{-1}e$

Mass number is the same for both nuclei

Atomic number of the new nucleus increases by 1

Beta Emitters

A beta emitter is a radioactive isotope that decays by emitting beta particles. To form a beta particle, the unstable nucleus converts a neutron into a proton. The newly formed proton adds to the number of protons already in the nucleus and increases the atomic number by 1. However, the mass number of the newly formed nucleus stays the same. For example, carbon-14 decays by emitting a beta particle and forming a nitrogen nucleus.

In the nuclear equation of a beta emitter, the mass number of the radioactive nucleus and the mass number of the new nucleus are the same, and the atomic number of the new nucleus increases by 1, indicating a change of one element into another. At left is the nuclear equation for the beta decay of carbon-14.

Beta Emitters in Medicine

The radioactive isotopes of several biologically important elements are beta emitters. When a radiologist wants to treat a malignancy within the body, a beta emitter may be used. The short range of penetration into the tissue by beta particles is advantageous for certain conditions. For example, some malignant tumors increase the fluid within the body tissues. A compound containing phosphorus-32, a beta emitter, is injected into the body cavity where the tumor is located. The beta particles travel only a few millimeters through the tissue, so only the malignancy and any tissue within that range are affected. The growth of the tumor is slowed or stopped, and the production of fluid decreases. Phosphorus-32 is also used to treat leukemia, polycythemia vera (excessive production of red blood cells), and lymphomas.

$^{32}_{15}P \longrightarrow ^{32}_{16}S + ^{0}_{-1}e$

Another beta emitter, iron-59, is used in blood tests to determine the level of iron in the blood and the rate of production of red blood cells by the bone marrow.

$^{59}_{26}Fe \longrightarrow ^{59}_{27}Co + ^{0}_{-1}e$

SAMPLE PROBLEM 9.4

Writing an Equation for Beta Decay

Cobalt-60 decays by emitting a beta particle. Write the nuclear equation for its decay.

SOLUTION

STEP 1 **Write the incomplete nuclear equation.**

$$^{60}_{27}Co \longrightarrow ? + ^{0}_{-1}e$$

STEP 2 **Determine the missing mass number.** In the equation, the mass number of cobalt-60 is equal to the sum of the mass numbers of the beta particle and the new nucleus. We can calculate the mass number of the new nucleus as follows:

$60 \quad = ? + 0$

$60 - 0 = ?$

$60 \quad = ?$ (mass number of new nucleus)

STEP 3 **Determine the missing atomic number.** The atomic number of cobalt-60 must equal the sum of the atomic number of the beta particle and the new nucleus. We calculate the atomic number of the new nucleus as follows:

$27 \quad = ? - 1$

$27 + 1 = ?$

$28 \quad = ?$ (atomic number of new nucleus)

STEP 4 Determine the symbol of the new nucleus. On the periodic table, the element that has atomic number 28 is nickel (Ni). The nucleus of this isotope of Ni is written as

$$^{60}_{28}\text{Ni}$$

STEP 5 Complete the nuclear equation.

$$^{60}_{27}\text{Co} \longrightarrow {}^{60}_{28}\text{Ni} + {}^{0}_{-1}e$$

In this nuclear reaction, a cobalt-60 nucleus decays by releasing a beta particle and produces a nickel-60 nucleus.

STUDY CHECK

Write the nuclear equation for iodine-131, a beta emitter.

Gamma Emitters

There are very few pure gamma emitters, although gamma radiation accompanies most alpha and beta radiation. In radiology, one of the most commonly used gamma emitters is technetium (Tc). The excited state called metastable technetium is written as technetium-99m, Tc-99m, or ^{99m}Tc. By emitting energy in the form of gamma rays, the excited nucleus becomes more stable.

$$^{99m}_{43}\text{Tc} \longrightarrow {}^{99}_{43}\text{Tc} + \gamma$$

Figure 9.2 summarizes the changes in the nucleus for alpha, beta, positron, and gamma radiation.

Producing Radioactive Isotopes

Today, more than 1500 radioisotopes are produced by converting stable, nonradioactive isotopes into radioactive ones. To do this, a stable atom is bombarded by fast-moving alpha particles, protons, or neutrons. When one of these particles is absorbed by a stable nucleus, the nucleus becomes unstable and the atom is now a radioactive isotope, or *radioisotope*. The process of changing one element into another is called *transmutation*.

When boron-10, a nonradioactive isotope, is bombarded by an alpha particle, it is converted to nitrogen-13, a radioisotope. In this bombardment reaction, a neutron is emitted.

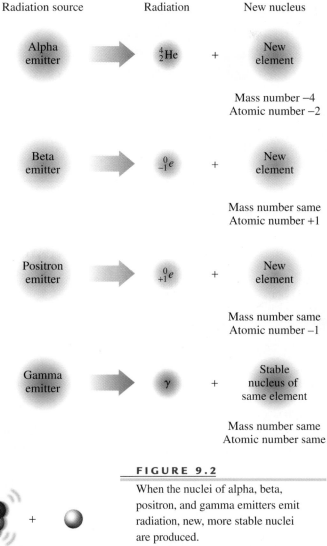

Radiation source	Radiation	New nucleus
Alpha emitter	$^{4}_{2}\text{He}$ +	New element
		Mass number −4 Atomic number −2
Beta emitter	$^{0}_{-1}e$ +	New element
		Mass number same Atomic number +1
Positron emitter	$^{0}_{+1}e$ +	New element
		Mass number same Atomic number −1
Gamma emitter	γ +	Stable nucleus of same element
		Mass number same Atomic number same

FIGURE 9.2

When the nuclei of alpha, beta, positron, and gamma emitters emit radiation, new, more stable nuclei are produced.

Q What changes occur in the number of protons and neutrons when an alpha emitter gives off radiation?

$$^{4}_{2}\text{He} \quad + \quad {}^{10}_{5}\text{B} \quad \longrightarrow \quad {}^{13}_{7}\text{N} \quad + \quad {}^{1}_{0}\text{n}$$

Bombarding particle   Stable nucleus   New radioactive nucleus   Neutron

All of the known elements that have atomic numbers greater than 92 have been produced by bombardment; none of these elements occurs naturally. Most have been produced in only small amounts and exist for such a short time that it is difficult to study their properties. An example is element 105, dubnium, which is produced when californium-249 is bombarded with nitrogen-15.

$$^{249}_{98}\text{Cf} + {}^{15}_{7}\text{N} \longrightarrow {}^{260}_{105}\text{Db} + 4\,{}^{1}_{0}\text{n}$$

Technetium-99m is a radioisotope used in nuclear medicine for several diagnostic procedures, including the detection of brain tumors and examinations of the liver and spleen. The source of technetium-99m is molybdenum-99, which is produced in a nuclear reactor by neutron bombardment of molybdenum-98.

$$^{98}_{42}\text{Mo} + {}^{1}_{0}\text{n} \longrightarrow {}^{99}_{42}\text{Mo}$$

Many radiology laboratories have a small generator containing the radioactive molybdenum-99, which decays to give the technetium-99m radioisotope.

$$^{99}_{42}\text{Mo} \longrightarrow {}^{99m}_{43}\text{Tc} + {}^{0}_{-1}e$$

The technetium-99m radioisotope decays by emitting gamma rays. Gamma emission is desirable for diagnostic work because the gamma rays pass through the body to the detection equipment.

$$^{99m}_{43}\text{Tc} \longrightarrow {}^{99}_{43}\text{Tc} + \gamma$$

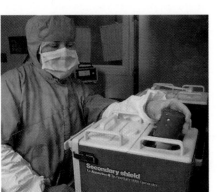

SAMPLE PROBLEM 9.5

■ Writing an Equation for Proton Bombardment

A new isotope is produced by the bombardment of zinc-66 by protons. Write the equation for this nuclear bombardment.

SOLUTION

STEP 1 **Write the incomplete nuclear equation.**

$$^{66}_{30}\text{Zn} + {}^{1}_{1}\text{H} \longrightarrow ?$$

STEP 2 **Determine the missing mass number.**   In the equation, the sum of the mass numbers of zinc (66) and hydrogen (1) must equal the mass number of the new nucleus. We calculate the mass number of the new nucleus as follows:

$$66 + 1 = ?$$
$$67 \quad = ? \text{ (mass number of new nucleus)}$$

STEP 3 **Determine the missing atomic number.**   The sum of the atomic numbers of zinc (30) and hydrogen (1) must equal the atomic number of the new nucleus. We calculate the atomic number of the new nucleus as follows:

$$30 + 1 = ?$$
$$31 \quad = ? \text{ (atomic number of new nucleus)}$$

STEP 4 **Determine the symbol of the new nucleus.**   On the periodic table, the element that has atomic number 31 is gallium, Ga. The symbol of this isotope of Ga is written as

$$^{67}_{31}\text{Ga}$$

STEP 5 Complete the nuclear equation.

$$^{66}_{30}Zn + {}^{1}_{1}H \longrightarrow {}^{67}_{31}Ga$$

Proton   New isotope

STUDY CHECK

Write the balanced nuclear equation for the bombardment of aluminum-27 by an alpha particle to produce a new isotope and one neutron.

■

QUESTIONS AND PROBLEMS

■ Nuclear Equations

9.13 Write a balanced nuclear equation for the alpha decay of each of the following radioactive isotopes:
   a. $^{208}_{84}Po$ **b.** $^{232}_{90}Th$ **c.** $^{251}_{102}No$

9.14 Write a balanced nuclear equation for the alpha decay of each of the following radioactive isotopes:
   a. $^{243}_{96}Cm$ **b.** $^{252}_{99}Es$ **c.** $^{251}_{98}Cf$

9.15 Write a balanced nuclear equation for the beta decay of each of the following radioactive isotopes:
   a. $^{25}_{11}Na$ **b.** $^{20}_{8}O$ **c.** $^{92}_{38}Sr$

9.16 Write a balanced nuclear equation for the beta decay of each of the following radioactive isotopes:
   a. potassium-42
   b. iron-59
   c. iron-60

9.17 Complete each of the following nuclear equations:
   a. $^{28}_{13}Al \longrightarrow ? + {}^{0}_{-1}e$
   b. $? \longrightarrow {}^{86}_{36}Kr + {}^{1}_{0}n$
   c. $^{66}_{29}Cu \longrightarrow {}^{66}_{30}Zn + ?$

9.18 Complete each of the following nuclear equations:
   a. $^{11}_{6}C \longrightarrow {}^{7}_{4}Be + ?$
   b. $^{35}_{16}S \longrightarrow ? + {}^{0}_{-1}e$
   c. $? \longrightarrow {}^{90}_{39}Y + {}^{0}_{-1}e$

9.19 Complete each of the following bombardment reactions:
   a. $^{9}_{4}Be + {}^{1}_{0}n \longrightarrow ?$
   b. $^{32}_{16}S + ? \longrightarrow {}^{32}_{15}P$
   c. $? + {}^{1}_{0}n \longrightarrow {}^{24}_{11}Na + {}^{4}_{2}He$

9.20 Complete each of the following bombardment reactions:
   a. $^{40}_{18}Ar + ? \longrightarrow {}^{43}_{19}K + {}^{1}_{1}H$
   b. $^{238}_{92}U + {}^{1}_{0}n \longrightarrow ?$
   c. $? + {}^{1}_{0}n \longrightarrow {}^{14}_{6}C + {}^{1}_{1}H$

■ 9.3 Radiation Measurement

LEARNING GOAL

■ Describe the measurement of radiation.

One of the most common instruments for detecting beta and gamma radiation is the Geiger counter. It consists of a metal tube filled with a gas such as argon. When radiation enters a window on the end of the tube, it produces charged particles in

WEB TUTORIAL
Nuclear Chemistry

the gas, which produce an electrical current. Each burst of current is amplified to give a click and a readout on a meter.

$$Ar + radiation \longrightarrow Ar^+ + e^-$$

Radiation is measured in several different ways. We can measure the activity of a radioactive sample or determine the impact of radiation on biological tissue.

Activity

When a radiology laboratory obtains a radioisotope, the activity of the sample is measured in terms of the number of nuclear disintegrations per second. The **curie (Ci)**, the original unit of activity, was defined as the number of disintegrations that occur in one second for 1 g of radium, which is equal to 3.7×10^{10} disintegrations per second. The curie was named for Marie Curie, a Polish scientist, who along with her husband Pierre discovered the radioactive elements radium and polonium. A newer unit of radiation activity is the **becquerel (Bq)**, which is one disintegration per second.

Biological Effect

After we measure the activity of a radioisotope, we often want to know how much radiation the tissues in the body absorb. The **rad (radiation absorbed dose)** is a unit that measures the amount of radiation absorbed by a gram of material such as body tissue. The newer unit for absorbed dose is the **gray (Gy)**, which is equal to 100 rads.

The **rem (radiation equivalent in humans)** measures the biological effects of different kinds of radiation. Alpha particles don't penetrate the skin. But if they should enter the body by some other route, they cause a lot of damage even though the particles travel a short distance in tissue. High-energy radiation such as beta particles and high-energy protons and neutrons that penetrate the skin and travel into tissue cause more damage. Gamma rays are damaging because they travel a long way through tissue and create a great deal of ionization.

To determine the **equivalent dose** or rem dose, the absorbed dose (rads) is multiplied by a factor that adjusts for biological damage caused by a particular form of radiation. For beta and gamma radiation the factor is 1, so the biological damage in rems is the same as the absorbed radiation (rads). For high-energy protons and neutrons, the factor is about 10, and for alpha particles it is 20.

Biological damage (rem) = absorbed dose (rad) × factor

Often the measurement for an equivalent dose will be in units of millirems (mrem). One rem is equal to 1000 mrem. The newer unit is the **sievert (Sv)**. One sievert is equal to 100 rems. Table 9.5 summarizes the units used to measure radiation.

■ Radiation and Food

Foodborne illnesses caused by pathogenic bacteria such as *Salmonella, Listeria*, and *Escherichia coli* have become a major health concern in the United States. The Centers for Disease Control and Prevention estimates that each year *E. coli* in contaminated foods infects 20,000 people in the United States, and that 500 people die. *E. coli* has been responsible for outbreaks of illness from contaminated ground beef, fruit juices, lettuce, and alfalfa sprouts.

The Food and Drug Administration (FDA) has approved the use of 0.3 kilogray (0.3 kGy) to 1 kGy of ionizing radiation produced by cobalt-60 or cesium-137 for the treatment of foods. The irradiation technology is much like that used to sterilize medical supplies. Cobalt pellets are placed in stainless steel tubes, which are arranged in racks. When food moves through the series of racks, the gamma rays pass through the food and kill the bacteria.

It is important for consumers to understand that when food is irradiated, it never comes in contact with the radioactive source. The gamma rays pass through the food to kill bacteria, but that does not make the food radioactive. The radiation kills bacteria because it stops their ability to divide and grow. We cook or heat food thoroughly for the same purpose. Radiation, as well as heat, has little effect on the food itself because its cells are no longer dividing or growing. Thus irradiated food is not harmed although a small amount of vitamin B_1 and C may be lost.

Currently, tomatoes, blueberries, strawberries, and mushrooms are being irradiated to allow them to be harvested when completely ripe and extend their shelf life (see Figure 9.3). The FDA has also approved the irradiation of pork, poultry, and beef in order to decrease potential infections and to extend shelf life. Currently, irradiated vegetable and meat products are available in retail markets in South

(a)

(b)

FIGURE 9.3

(a) The FDA requires this symbol to appear on irradiated retail foods. (b) After 2 weeks, the irradiated strawberries on the right show no spoilage. Mold is starting to grow on the nonirradiated ones on the left.

Q **Why are irradiated foods used on space ships and in nursing homes?**

Africa. Apollo 17 astronauts ate irradiated foods on the moon, and some U.S. hospitals and nursing homes now use irradiated poultry to reduce the possibility of infections among patients. The extended shelf life of irradiated food also makes it useful for campers and military personnel. Soon consumers concerned about food safety will have a choice of irradiated meats, fruits, and vegetables at the market.

TABLE 9.5

Some Units of Radiation Measurement

Measurement	Common Unit	SI Unit
Activity	curie (Ci) = 3.7×10^{10} disintegrations/s	becquerel (Bq) = 1 disintegrations/s
Absorbed dose	rad	gray (Gy)
Biological damage	rem = rad × factor	1 sievert (Sv)

the
Chemistry
place

CASE STUDY
Food Irradiation

SAMPLE PROBLEM 9.6

■ Radiation Measurement

One treatment of bone pain involves intravenous administration of the radioisotope phosphorous-32, which is primarily incorporated into bone. A typical dose of 7 mCi can produce up to 450 rads in the bone. What is the difference between the units mCi and rads?

SOLUTION

The millicuries (mCi) indicate the activity of the P-32 in terms of nuclei that break down in 1 second. The radiation absorbed dose (rads) is a measure of amount of radiation absorbed by the bone.

STUDY CHECK

If P-32 is a beta emitter, how do the number of rems compare to the rads?

Exposure to Radiation

We are all exposed to low levels of radiation every day. Naturally occurring radioactive isotopes are part of the atoms of wood, brick, and concrete in our homes and the buildings where we work and go to school. This radioactivity, called background radiation, is present in the soil, in the food we eat, in the water we drink, and in the air we breathe. For example, one of the naturally occurring isotopes of potassium, potassium-40, is radioactive. It is found in the body because it is always present in any potassium-containing food. Other naturally occurring radioisotopes in air and food are carbon-14, radon-222, strontium-90, and iodine-131. Table 9.6 lists some common sources of radiation.

We are constantly exposed to radiation (cosmic rays) produced in space by the sun. At higher elevations, the amount of radiation from outer space is greater because there are fewer air molecules to absorb the radiation. People living at high altitudes or flying in an airplane receive more radiation from cosmic rays than those who live at sea level. For example, a person living in Denver receives about twice the cosmic radiation as a person living in Los Angeles.

A person living close to a nuclear power plant normally does not receive much additional radiation, perhaps 0.1 millirem (mrem) in 1 year. (One rem equals 1000 mrem.) However, in the accident at the Chernobyl nuclear power plant in 1986, it is estimated that people in a nearby town received as much as 1 rem/hr.

In addition to naturally occurring radiation from construction materials in our homes, we receive radiation from television. In the medical clinic, dental and chest X rays also add to our radiation exposure. The average person in the United States receives about 0.17 rem or 170 mrem of radiation annually.

Radiation Sickness

The larger the dose of radiation received at one time, the greater the effect on the body. Exposure to radiation under 25 rem usually cannot be detected. Whole-body exposure of 100 rem produces a temporary decrease in the number of white blood cells. If the exposure to radiation is 100 rem or higher, the person suffers the symptoms of radiation sickness: nausea, vomiting, fatigue, and a reduction in white-cell count. A whole-body dosage greater than 300 rem can lower the white-cell count to zero. The victim suffers diarrhea, hair loss, and infection. Exposure to radiation of about 500 rem is expected to cause death in 50% of the people receiving that dose. This amount of radiation is called the lethal dose for one-half the population, or the LD_{50}. The LD_{50} varies for different life-forms, as Table 9.7 shows. Radiation dosages of about 600 rem would be fatal to all humans within a few weeks.

TABLE 9.6

Average Annual Radiation Received by a Person in the United States

Source	Dose (mrem)
Natural	
The ground	15
Air, water, food	30
Cosmic rays	40
Wood, concrete, brick	50
Medical	
Chest X ray	50
Dental X ray	20
Upper gastrointestinal tract X ray	200
Other	
Television	2
Air travel	1
Radon	200[a]
Cigarette smoking	35

[a]Varies widely.

TABLE 9.7

Lethal Doses of Radiation for Some Life-Forms

Life-Form	LD_{50} (rem)
Insect	100,000
Bacterium	50,000
Rat	800
Human	500
Dog	300

■ **Radiation Measurement**

9.21 a. How does a Geiger counter detect radiation?
   b. What are the SI unit and the older unit that describe the activity of a radioactive sample?
   c. What are the SI unit and the older unit that describe the radiation dose absorbed by tissue?
   d. What is meant by the term kilogray?

9.22 a. What is background radiation?
   b. What are the SI unit and the older unit that describe the biological effect of radiation?
   c. What is meant by the terms mCi and mrem?
   d. Why is a factor used to determine the dose equivalent?

9.23 A sample of iodine-131 has an activity of 3.0 Ci. The recommended dosage of iodine-131 is 4.20 μCi/kg of body weight. How many microcuries of iodine-131 are needed for a 70.0-kg patient with hyperthyroidism?

9.24 a. The dosage of technetium-99m for a lung scan is 20 μCi/kg of body weight. How many millicuries should be given to a 50.0-kg patient? (1 mCi = 1000 μCi)
   b. Suppose a person absorbed 50 mrads of alpha radiation. What would be the dose equivalent in mrems?

9.25 Why would an airline pilot be exposed to more background radiation than the person who works at the ticket counter?

9.26 In radiation therapy, a patient receives high doses of radiation. What symptoms of radiation sickness might the patient exhibit?

■ 9.4 Half-Life of a Radioisotope

The **half-life** of a radioisotope is the amount of time it takes for one-half of a sample to decay. For example, ^{131}I has a half-life of 8.0 days. As ^{131}I decays it produces a beta particle and the nonradioactive isotope ^{131}Xe.

$$^{131}_{53}\text{I} \longrightarrow {}^{131}_{54}\text{Xe} + {}^{0}_{-1}\beta$$

Suppose we have a sample that contains 20.0 grams of ^{131}I. We don't know which specific nucleus will emit radiation, but we do know that in 8 days, one-half of all the nuclei in the sample will decay to give ^{131}Xe. That means that after 8.0 days, there are one-half of the number of ^{131}I atoms in the sample, or 10.0 g of ^{131}I, remaining. The decay process has also produced 10.0 g of the product ^{131}Xe. After another half-life or 8.0 days passes, 5.00 g of the 10.0 g of ^{131}I will decay to ^{131}Xe. Now there are 5.00 g of ^{131}I left, while there is a total of 15.0 g of ^{131}Xe.

$$20 \text{ g }{}^{131}\text{I} \xrightarrow{\text{1 half-life}} 10 \text{ g }{}^{131}\text{I} \xrightarrow{\text{2 half-lives}} 5 \text{ g }{}^{131}\text{I}$$

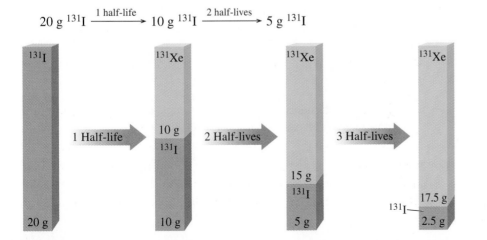

LEARNING GOAL

■ Given the half-life of a radioisotope, calculate the amount of radioisotope remaining after one or more half-lives.

the
Chemistry
place

WEB TUTORIAL
Nuclear Chemistry

FIGURE 9.4

The decay curve for iodine-131 shows that one-half of the radioactive sample decays and one-half remains radioactive after each half-life of 8 days.

Q **How many grams of the 20-g sample remain radioactive after 2 half-lives?**

A third half-life, or another 8.0 days, results in 2.50 g of the ^{131}I decaying to give ^{131}Xe, which leaves 2.50 g of ^{131}I still capable of producing radiation. This information can be summarized in Table 9.8 as follows:

TABLE 9.8

Activity of an ^{131}I Sample with Time

Time elapsed	0 days	8.0 days	16.0 days	24.0 days
Half-lives	0	1	2	3
^{131}I remaining	20.0 g	10.0 g	5.00 g	2.50 g
^{131}Xe produced	0 g	10.0 g	15.0 g	17.5 g

A **decay curve** is a diagram of the decay of a radioactive isotope. Figure 9.4 shows such a curve for the ^{131}I we have discussed.

SAMPLE PROBLEM 9.7

■ **Using Half-Lives of a Radioisotope**

Phosphorus-32, a radioisotope used in the treatment of leukemia, has a half-life of 14 days. If a sample contains 8.0 g of phosphorus-32, how many grams of phosphorus-32 remain after 42 days?

SOLUTION

STEP 1 **Given** 8.0 g ^{32}P; 42 days; 14 days/half-life
Need g ^{32}P remaining

STEP 2 **Plan**

42 days ▸ Half-life ▸ Number of half-lives

8.0 g ^{32}P ▸ Number of half-lives ▸ g ^{32}P remaining

STEP 3 **Equalities/Conversion Factors**

$$1 \text{ half-life} = 14 \text{ d}$$
$$\frac{14 \text{ d}}{1 \text{ half-life}} \quad \text{and} \quad \frac{1 \text{ half-life}}{14 \text{ d}}$$

STEP 4 **Set Up Problem** We can do this problem with two calculations. First, we determine the number of half-lives in the amount of time that has elapsed.

$$\text{Number of half-lives} = 42 \text{ d} \times \frac{1 \text{ half-life}}{14 \text{ d}} = 3 \text{ half-lives}$$

Now, we determine how much of the sample decays in 3 half-lives and how many grams of the phosphorus remain.

$$8.0 \text{ g } ^{32}\text{P} \xrightarrow{14 \text{ d}} 4.0 \text{ g } ^{32}\text{P} \xrightarrow{14 \text{ d}} 2.0 \text{ g } ^{32}\text{P} \xrightarrow{14 \text{ d}} 1.0 \text{ g } ^{32}\text{P}$$

STUDY CHECK

Iron-59 has a half-life of 46 days. If the laboratory received a sample of 8.0 g iron-59, how many grams are still active after 184 days?

■

Naturally occurring isotopes of the elements usually have long half-lives, as shown in Table 9.9. They disintegrate slowly and produce radiation over a long period of time, even hundreds or millions of years. In contrast, many of the radioisotopes used in nuclear medicine have much shorter half-lives. They disintegrate rapidly and produce almost all their radiation in a short period of time. For example, technetium-99m emits half of its radiation in the first 6 hr. This means that a small amount of the radioisotope given to a patient is essentially gone within 2 days. The decay products of technetium 99m are totally eliminated by the body.

TABLE 9.9

Half-Lives of Some Radioisotopes

Element	Radioisotope	Half-Life
Naturally Occurring Radioisotopes		
Carbon	^{14}C	5730 yr
Potassium	^{40}K	1.3×10^9 yr
Radium	^{226}Ra	1600 yr
Uranium	^{238}U	4.5×10^9 yr
Some Medical Radioisotopes		
Chromium	^{51}Cr	28 days
Iodine	^{131}I	8 days
Iron	^{59}Fe	46 days
Technetium	^{99m}Tc	6.0 hr

SAMPLE PROBLEM 9.8

■ **Dating Using Half-Lives**

In Los Angeles, the remains of ancient animals have been unearthed at La Brea tar pits. Suppose a bone sample from the tar pits is subjected to the carbon-14 dating method. If the sample shows that two half-lives have passed, when did the animal live?

SOLUTION

We can calculate the age of the bone sample by using the half-life of carbon-14 (5730 years).

$$2 \text{ half-lives} \times \frac{5730 \text{ years}}{1 \text{ half-life}} = 11{,}500 \text{ years}$$

We would estimate that the animal lived 11,500 years ago, or about 9500 B.C.E.

STUDY CHECK

Suppose that a piece of wood found in a tomb had $\frac{1}{8}$ (3 half-lives) of its original C-14 activity. About how many years ago was the wood part of a living tree?

■

EXPLORE YOUR WORLD

■ **Modeling Half-Lives**

Obtain a piece of paper and a licorice stick or celery stalk. Draw a vertical and a horizontal axis on the paper. Label the vertical axis as radioactive atoms and the horizontal axis as minutes. Place the licorice stick or celery against the vertical axis and mark its height for zero minutes. In the next minute, cut the licorice stick or celery in two. (You can eat the half if you are hungry.) Place the shortened licorice stick or celery at 1 minute on the horizontal axis and mark its height. Every minute cut the licorice stick or celery in half again and mark the shorter height at the corresponding time. Keep reducing the length by half until you cannot divide the licorice or celery in half any more. Connect the points you made for each minute. What does the curve look like? How does this curve represent the concept of a half-life for a radioisotope?

Dating Ancient Objects

A technique known as radiological dating is used by geologists, archaeologists, and historians as a way to determine the age of ancient objects. The age of an object derived from plants or animals (such as wood, fiber, natural pigments, bone, and cotton or woolen clothing) is determined by measuring the amount of carbon-14, a naturally occurring radioactive form of carbon. In 1960, Willard Libby received the Nobel Prize for the work he did developing carbon-14 dating techniques during the 1940s. Carbon-14 is produced in the upper atmosphere by the bombardment of $^{14}_{7}N$ by high-energy neutrons from cosmic rays.

$$^{1}_{0}n \quad + \quad ^{14}_{7}N \quad \longrightarrow \quad ^{14}_{6}C \quad + \quad ^{1}_{1}H$$

Neutron from   Nitrogen in   Radioactive   Proton
cosmic rays   atmosphere   carbon-14

The carbon-14 reacts with oxygen to form radioactive carbon dioxide, $^{14}CO_2$. Because carbon dioxide is continuously absorbed by living plants during the process of photosynthesis, some carbon-14 will be taken into the plant. After the plant dies, no more carbon-14 is taken up, and the amount of carbon-14 contained in the plant steadily decreases as it undergoes radioactive β decay.

$$^{14}_{6}C \longrightarrow ^{14}_{7}N + ^{0}_{-1}e$$

Scientists use the half-life of carbon-14 (5730 years) to calculate the amount of time that has passed since the plant died, a process called **carbon dating**. The smaller the amount of carbon-14 remaining in the sample, the greater the number of half-lives that have passed. Thus, the approximate age of the sample can be determined. For example, a wooden beam found in an ancient Indian dwelling might have one-half of the carbon-14 found in living plants today. Thus, the dwelling was probably constructed about 5730

years ago, one half-life of carbon-14. Carbon-14 dating was used to determine that the Dead Sea Scrolls are about 2000 years old.

A radiological dating method used for determining the age of rocks is based on the radioisotope uranium-238, which decays through a series of reactions to lead-206. The uranium-238 isotope has a very long half-life, about 4×10^9 (4 billion) years. Measurements of the amounts of uranium-238 and lead-206 enable geologists to determine the age of rock samples. The older rocks will have a higher percentage of lead-206 because more of the uranium-238 has decayed. The age of rocks brought back from the moon by the *Apollo* missions was determined using uranium-238. They were found to be about 4×10^9 years old, approximately the same age calculated for Earth.

QUESTIONS AND PROBLEMS

Half-Life of a Radioisotope

9.27 What is meant by the term half-life?

9.28 Why are radioisotopes with short half-lives used for diagnosis in nuclear medicine?

9.29 Technetium-99m is an ideal radioisotope for scanning organs because it has a half-life of 6.0 hr and is a pure gamma emitter. Suppose that 80.0 mg were prepared in the technetium generator this morning. How many milligrams would remain after the following intervals?
a. one half-life   **b.** two half-lives   **c.** 18 hr   **d.** 24 hr

9.30 A sample of sodium-24 with an activity of 12 mCi is used to study the rate of blood flow in the circulatory system. If sodium-24 has a half-life of 15 hr, what is the activity of the sodium after $2\frac{1}{2}$ days?

9.31 Strontium-85, used for bone scans, has a half-life of 64 days. How long will it take for the radiation level of strontium-85 to drop to one-fourth of its original level? To one-eighth?

9.32 Fluorine-18, which has a half-life of 110 min, is used in PET scans. If 100 mg of fluorine-18 is shipped at 8:00 A.M., how many milligrams of the radioisotope are still active if the sample arrives at the radiology laboratory at 1:30 P.M.?

9.5 Medical Applications Using Radioactivity

LEARNING GOAL

Describe the use of radioisotopes in medicine.

When a radiologist wants to determine the condition of an organ in the body, a patient is given a radioisotope that is known to concentrate in that organ. The cells in the body cannot differentiate between a nonradioactive atom and a radioactive one. However, radioactive atoms can be detected because they emit radiation and the nonradioactive atoms do not. Some radioisotopes used in nuclear medicine are listed in Table 9.10.

Scans with Radioisotopes

After a patient receives a radioisotope, the radiologist determines the level and location of radioactivity emitted by the radioisotope. An apparatus called a scanner is used to produce an image of the organ. The scanner moves slowly across the patient's body above the region where the organ containing the radioisotope is located. The gamma rays emitted from the radioisotope in the organ can be used to expose a photographic plate, producing a **scan** of the organ. On a scan, an area of decreased or increased radiation can indicate such conditions as a disease of the organ, a tumor, a blood clot, or edema.

A common method of determining thyroid function is the use of radioactive iodine uptake (RAIU). Taken orally, the radioisotope iodine-131 mixes with the iodine already present in the thyroid. Twenty-four hours later, the amount of iodine taken up by the thyroid is determined. A detection tube held up to the area of the thyroid gland detects the radiation coming from the iodine-131 that has located there. (See Figure 9.5.)

A patient with a hyperactive thyroid will have a higher than normal level of radioactive iodine, whereas a patient with a hypoactive thyroid will record low values. If the patient has hyperthyroidism, treatment is begun to lower the activity of the thyroid. One treatment involves giving the patient a therapeutic dosage of radioactive iodine, which has a higher radiation count than the diagnostic dose. The radioactive iodine goes to the thyroid where its radiation destroys some of the thyroid cells. The thyroid produces less thyroid hormone, bringing the hyperthyroid condition under control.

TABLE 9.10

Medical Applications of Radioisotopes

Isotope	Half-Life	Medical Application
Ce-141	32.5 days	Gastrointestinal tract diagnosis; measuring myocardial blood flow
Ga-67	78 hr	Abdominal imaging; tumor detection
Ga-68	68 min	Detect pancreatic cancer
P-32	4.3 days	Treatment of leukemia, polycythemia vera (excess red blood cells), pancreatic cancer
I-125	60 days	Treatment of brain cancer; osteoporosis detection
I-131	8 days	Imaging thyroid; treatment of Graves' disease, goiter, and hyperthyroidism; treatment of thyroid and prostate cancer
Sr-85	65 days	Detection of bone lesions; brain scans
Tc-99m	6 hr	Imaging of skeleton and heart muscle, brain, liver, heart, lungs, bone, spleen, kidney, and thyroid; *most widely used radioisotope in nuclear medicine*

(b)

(a)

FIGURE 9.5

(a) A scanner is used to detect radiation from a radioisotope that has accumulated in an organ. (b) A scan of the thyroid shows the accumulation of radioactive iodine-131 in the thyroid.

Q **What type of radiation would move through body tissues to create a scan?**

Positron Emission Tomography (PET)

Radioisotopes that emit a positron are used in an imaging method called positron emission tomography (PET). A positron is a particle emitted from the nucleus that has the same mass as an electron but has a positive charge.

$$\beta^+ \quad \text{or} \quad _{+1}^{0}e$$
Positron

Carbon-11 is an example of a radioisotope that emits a positron when it decays.

$$_{6}^{11}C \longrightarrow _{+1}^{0}e + _{5}^{11}B$$
Positron

A positron is produced when a proton changes to a neutron.

$$_{1}^{1}H \longrightarrow _{+1}^{0}e + _{0}^{1}n$$

The new nucleus retains the same mass number but has a lower atomic number. The positron exists for only a moment before it collides with an electron producing gamma rays, which can be detected.

$$_{+1}^{0}e + _{-1}^{0}e \longrightarrow 2\gamma$$
Positron   Electron   Gamma rays
         in an atom   produced

FIGURE 9.6

These PET scans of the brain show a normal brain on the left and a brain affected by Alzheimer's disease on the right.

Q **When positrons collide with electrons, what type of radiation is produced that gives an image of an organ?**

Medically, positron emitters such as carbon-11, oxygen-15, and nitrogen-13 are used to diagnose conditions involving blood flow, metabolism, and particularly brain function. Glucose labeled with carbon-11 is used to detect damage in the brain from epilepsy, stroke, and Parkinson's disease. The gamma rays from the positrons emitted by carbon-11 are detected by computerized equipment to create a three-dimensional image of the organ. (See Figure 9.6.)

■ Medical Application of Radioactivity

In hyperthyroidism, the thyroid produces an overabundance of thyroid hormone. Why would a radioisotope such as iodine-131 with an activity of 10 μCi be used to image the thyroid?

SOLUTION

The thyroid takes up the iodine-131 because the thyroid uses most of the iodine in the body. An activity level of 10 μCi is high enough to determine the level of thyroid function, but low enough to keep cell damage to a minimum.

STUDY CHECK

The evaluation of the thyroid image determined that the patient had an enlarged thyroid. A few weeks later, that same patient was given 100 mCi of iodine-131 intravenously. Why was the activity of the iodine-131 sample used for treatment so much greater than the activity of the iodine-131 used in the diagnostic procedure?

QUESTIONS AND PROBLEMS

■ Medical Applications Using Radioactivity

9.33 Bone and bony structures contain calcium and phosphorus.
  a. Why would the radioisotopes of calcium-47 and phosphorus-32 be used in the diagnosis and treatment of bone diseases?
  b. During nuclear tests, scientists were concerned that strontium-85, a radioactive product, would be harmful to the growth of bone in children. Explain.

9.34 a. Technetium-99m emits only gamma radiation. Why would this type of radiation be used in diagnostic imaging rather than an isotope that also emits beta or alpha radiation?
  b. A patient with polycythemia vera (excess production of red blood cells) receives radioactive phosphorus-32. Why would this treatment reduce the production of red blood cells in the bone marrow of the patient?

9.35 In a diagnostic test for leukemia, a patient receives 4.0 mL of a solution containing selenium-75. If the activity of the selenium-75 is 45 μCi/mL, what is the dose received by the patient?

9.36 A vial contains radioactive iodine-131 with an activity of 2.0 mCi per milliliter. If the thyroid test requires 3.0 mCi in an "atomic cocktail," how many milliliters are used to prepare the iodine-131 solution?

HEALTH NOTE

■ Radiation Doses in Diagnostic and Therapeutic Procedures

We can compare the levels of radiation exposure commonly used during diagnostic and therapeutic procedures in nuclear medicine. In diagnostic procedures, the radiologist minimizes radiation damage by using only enough radioisotope to evaluate the condition of an organ or tissue. The doses used in **radiation therapy** are much greater than those used for diagnostic procedures. For example, a therapeutic dose would be used to destroy the cells in a malignant tumor. Although there will be some damage to surrounding tissue, the healthy cells are more resistant to radiation and can repair themselves. (See Table 9.11.)

TABLE 9.11

Radiation Doses Used for Diagnostic and Therapeutic Procedures

Organ/Condition	Dose (rem)
Diagnostic	
Liver	0.3
Thyroid	50.0
Lung	2.0
Therapeutic	
Lymphoma	4500
Skin cancer	5000–6000
Lung cancer	6000
Brain tumor	6000–7000

9.6 Nuclear Fission and Fusion

LEARNING GOAL

Describe the processes of nuclear fission and fusion.

In the 1930s, scientists bombarding uranium-235 with neutrons discovered that the U-235 nucleus splits into two medium-weight nuclei and produces a great amount of energy. This was the discovery of nuclear **fission**. The energy generated by splitting the atom was called atomic energy. When uranium-235 absorbs a neutron, it breaks apart to form two smaller nuclei, several neutrons, and a great amount of energy. A typical equation for nuclear fission is

$$\,^{1}_{0}n \;+\; \,^{235}_{92}U \;\longrightarrow\; \,^{91}_{36}Kr \;+\; \,^{142}_{56}Ba \;+\; 3\,^{1}_{0}n \;+\; \text{Energy}$$

If we could weigh these products with great accuracy, we would find that their total mass is slightly less than the mass of the starting materials. The missing

HEALTH NOTE

Other Imaging Methods

COMPUTED TOMOGRAPHY (CT)

Another imaging method used to detect changes within the body is computed tomography (CT). A computer monitors the degree of absorption of 30,000 x-ray beams directed at the brain at successive layers. The differences in absorption based upon the densities of the tissues and fluids in the brain provide a series of images of the brain. This technique is successful in the identification of brain hemorrhages, tumors, and atrophy. (See Figure 9.7.)

MAGNETIC RESONANCE IMAGING (MRI)

Magnetic resonance imaging (MRI) is a powerful imaging technique that does not involve x-ray radiation. It is the least invasive imaging method available. MRI is based on the absorption of energy when the protons in hydrogen atoms are excited by a strong magnetic field. Hydrogen atoms make up 63% of all the atoms in the body. In the nucleus, the spin of the single proton acts like a tiny magnet. With no external field, the spins of the protons have random orientations. However, when placed within a large magnet, the spins of the protons align with the magnetic field. A magnet aligned with the field has a lower energy than one that is aligned against the field. As the MRI scan proceeds, radiofrequency pulses of energy are applied. When a nucleus absorbs certain energy, its proton "flips" and becomes aligned against the field. Because hydrogen atoms in the body are in different chemical environments, frequencies of different energies are absorbed. The energies absorbed are calculated and converted to color images of the body. MRI is particularly useful to image soft tissues because soft tissues contain large amounts of water. (See Figure 9.8.)

FIGURE 9.7

A CT scan shows a brain tumor (yellow area) in the center of the right side of the brain.

Q **What is the type of radiation used to give a CT scan?**

FIGURE 9.8

An MRI scan of the heart and lungs, with the left ventricle shown in red.

Q **What is the source of energy in MRI?**

mass has been converted into energy, consistent with the famous equation derived by Albert Einstein:

$$E = mc^2$$

E is the energy released, m is the mass lost, and c is the speed of light, 3×10^8 m/s. Even though the mass loss is very small, when it is multiplied by the speed of light squared the result is a large value for the energy released. The fission of 1 g of uranium-235 produces about as much energy as the burning of 3 tons of coal.

Chain Reaction

Fission begins when a neutron collides with the nucleus of a uranium atom. The resulting nucleus is unstable and splits into smaller nuclei. This fission process also releases several neutrons and large amounts of gamma radiation and energy. The neutrons emitted have high energies and bombard more uranium-235 nuclei.

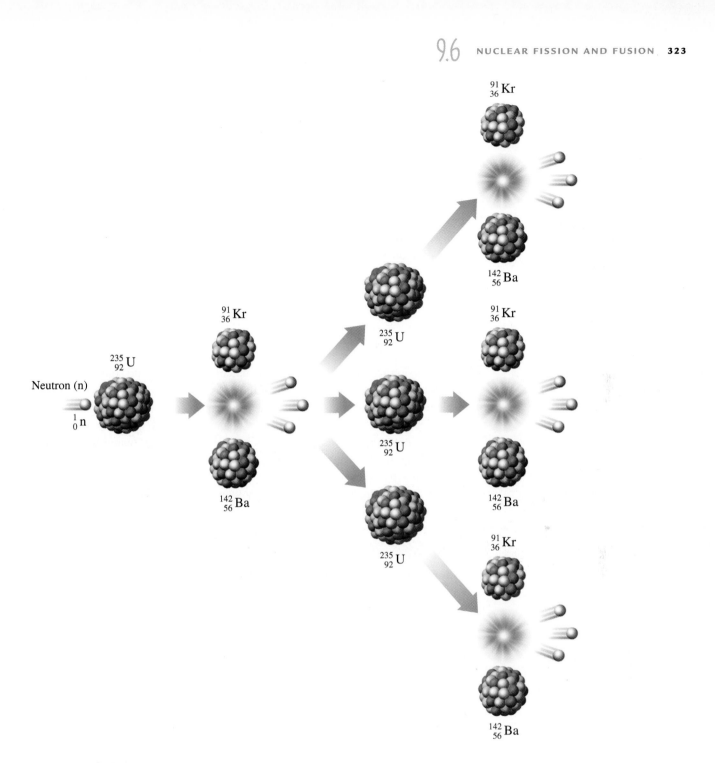

FIGURE 9.9

In a nuclear chain reaction, the fission of each ^{235}U atom produces three neutrons that cause the nuclear fission of more and more ^{235}U atoms.

Q Why is the fission of ^{235}U called a chain reaction?

As fission continues, there is a rapid increase in the number of high-energy neutrons capable of splitting more uranium atoms, a process called a **chain reaction**. To sustain a nuclear chain reaction, sufficient quantities of uranium-235 must be brought together to provide a critical mass in which almost all the neutrons immediately collide with more uranium-235 nuclei. So much heat and energy are released that an atomic explosion occurs. (See Figure 9.9).

Nuclear Fusion

In **fusion**, two small nuclei such as those in hydrogen combine to form a larger nucleus. Mass is lost, and a tremendous amount of energy is released, even more than the energy released from nuclear fission. However, a very high temperature (100 000 000°C) is required to overcome the repulsion of the hydrogen nuclei and cause them to undergo fusion. Fusion reactions occur continuously in the sun and other stars, providing us with heat and light. The huge amounts of energy produced by our sun come from the fusion of 6×10^{11} kilograms of hydrogen every second. The following fusion reaction involves the combination of two isotopes of hydrogen.

$$\begin{array}{ccccccc} {}^{3}_{1}H & + & {}^{2}_{1}H & \longrightarrow & {}^{4}_{2}He & + & {}^{1}_{0}n & + & \text{Energy} \end{array}$$

The fusion reaction has tremendous potential as a possible source for future energy needs. One of the advantages of fusion as an energy source is that hydrogen is plentiful in the oceans. Although scientists expect some radioactive waste from fusion reactors, the amount is expected to be much less than that from fission, and the waste products should have shorter half-lives. However, fusion is still in the experimental stage because the extremely high temperatures needed have been difficult to reach and even more difficult to maintain. Research groups around the world are attempting to develop the technology needed to make the fusion reaction a reality in our lifetime.

SAMPLE PROBLEM 9.10

▪ Identifying Fission and Fusion

Classify the following as pertaining to nuclear fission, nuclear fusion, or both:

a. Small nuclei combine to form larger nuclei.
b. Large amounts of energy are released.
c. Very high temperatures are needed for reaction.

SOLUTION

a. fusion   **b.** both fusion and fission   **c.** fusion

STUDY CHECK

Would the following reaction be an example of a fission or fusion reaction?

$${}^{2}_{1}H + {}^{1}_{1}H \longrightarrow {}^{3}_{2}He$$

QUESTIONS AND PROBLEMS

▪ Nuclear Fission and Fusion

9.37 What is nuclear fission?

9.38 How does a chain reaction occur in nuclear fission?

9.39 Complete the following fission reaction:

$${}^{235}_{92}U + {}^{1}_{0}n \longrightarrow {}^{131}_{50}Sn + ? + 2\,{}^{1}_{0}n + \text{energy}$$

ENVIRONMENTAL NOTE

▪ Nuclear Power Plants

In a nuclear power plant, the quantity of uranium-235 is held below a critical mass, so it cannot sustain a chain reaction. The fission reactions are slowed by placing control rods, which absorb some of the fast-moving neutrons, among the uranium samples. In this way, less fission occurs, and there is a slower, controlled production of energy. The heat from the controlled fission is used to produce steam. The steam drives a generator, which produces electricity. Approximately 10% of the electrical energy produced in the United States is generated in nuclear power plants.

Although nuclear power plants help meet some of our energy needs, there are some problems. One of the most serious problems is the production of radioactive by-products that have very long half-lives. It is essential that these waste products be stored safely for a very long time in a place where they do not contaminate the environment. Early in 1990, the Environmental Protection Agency gave its approval for the storage of radioactive hazardous wastes in chambers 2150 ft underground. In 1998 the Waste Isolation Pilot Plant (WIPP) repository site in New Mexico was ready to receive plutonium waste from former U.S. bomb factories. Although authorities claim the caverns are safe, some people are concerned with the safe transport of the radioactive waste by trucks on the highways.

9.40 In another fission reaction, U-235 bombarded with a neutron produces Sr-94, another small nucleus, and 3 neutrons. Write the complete equation for the fission reaction.

9.41 Indicate whether each of the following are characteristic of the fission or fusion process or both:
   a. Neutrons bombard a nucleus. *Fis*
   b. The nuclear process occurring in the sun. *Fus*
   c. A large nucleus splits into smaller nuclei. *Fis*
   d. Small nuclei combine to form larger nuclei. *Fus*

9.42 Indicate whether each of the following are characteristic of the fission or fusion process or both:
   a. Very high temperatures are required to initiate the reaction.
   b. Less radioactive waste is produced.
   c. Hydrogen nuclei are the reactants.
   d. Large amounts of energy are released when the nuclear reaction occurs.

CONCEPT MAP

CHAPTER REVIEW

9.1 Natural Radioactivity
Radioactive isotopes have unstable nuclei that break down (decay), spontaneously emitting alpha (α), beta (β), and gamma (γ) radiation. Because radiation can damage the cells in the body, proper protection must be used: shielding, limiting the time of exposure, and distance.

9.2 Nuclear Equations
A balanced equation is used to represent the changes that take place in the nuclei of the reactants and products. The new isotopes and the type of radiation emitted can be determined from the symbols that show the mass numbers and atomic numbers of the isotopes in the nuclear reaction. A radioisotope is produced artificially when a nonradioactive isotope is bombarded by a small particle. Many radioactive isotopes used in nuclear medicine are produced in this way.

9.3 Radiation Measurement
In a Geiger counter, radiation ionizes gas in a metal tube, which produces an electrical current. The curie (Ci) measures the number of nuclear transformations of a radioactive sample. Activity is also measured in becquerel (Bq) units. The amount of radiation absorbed by a substance is measured in rads or the gray (Gy). The rem and the sievert (Sv) are units used to determine the biological damage from the different types of radiation.

9.4 Half-Life of a Radioisotope
Every radioisotope has its own rate of emitting radiation. The time it takes for one-half of a radioactive sample to decay is called its half-life. For many medical radioisotopes, such as Tc-99m and I-131, half-lives are short. For other isotopes, usually naturally occurring ones such as C-14, Ra-226, and U-238, half-lives are extremely long.

9.5 Medical Applications Using Radioactivity
In nuclear medicine, radioisotopes are given that go to specific sites in the body. By detecting the radiation they emit, an evaluation can be made about the location and extent of an injury, disease, tumor, or the level of function of a particular organ. Higher levels of radiation are used to treat or destroy tumors.

9.6 Nuclear Fission and Fusion
In fission, a large nucleus breaks apart into smaller pieces, releasing one or more types of radiation and a great amount of energy. In fusion, small nuclei combine to form a larger nucleus while great amounts of energy are released.

KEY TERMS

alpha particle A nuclear particle identical to a helium ($^{4}_{2}He$, or α) nucleus (2 protons and 2 neutrons).

becquerel (Bq) A unit of activity of a radioactive sample equal to one disintegration per second.

beta particle A particle identical to an electron ($^{0}_{-1}e$, or β) that forms in the nucleus when a neutron changes to a proton and an electron.

carbon dating A technique used to date ancient specimens that contain carbon. The age is determined by the amount of active C-14 that remains in the sample.

chain reaction A fission reaction that will continue once it has been initiated by a high-energy neutron bombarding a heavy nucleus such as U-235.

curie (Ci) A unit of radiation equal to 3.7×10^{10} disintegrations/s.

decay curve A diagram of the decay of a radioactive element.

equivalent dose The measure of biological damage from an absorbed dose that has been adjusted for the type of radiation.

fission A process in which large nuclei are split into smaller pieces, releasing large amounts of energy.

fusion A reaction in which large amounts of energy are released when small nuclei combine to form larger nuclei.

gamma ray High-energy radiation (γ) emitted to make a nucleus more stable.

gray (Gy) A unit of absorbed dose equal to 100 rads.

half-life The length of time it takes for one-half of a radioactive sample to decay.

positron A particle with no mass and a positive charge produced when a proton is transformed into a neutron and a positron.

rad (radiation absorbed dose) A measure of an amount of radiation absorbed by the body.

radiation Energy or particles released by radioactive atoms.

radiation therapy The use of high doses of radiation to destroy harmful tissues in the body.

radioactive The process by which an unstable nucleus breaks down with the release of high-energy radiation.

radioisotope A radioactive atom of an element.

rem (radiation equivalent in humans) A measure of the biological damage caused by the various kinds of radiation (rad $\times$ radiation biological factor).

scan The image of a site in the body created by the detection of radiation from radioactive isotopes that have accumulated in that site.

shielding Materials used to provide protection from radioactive sources.

sievert (Sv) A unit of biological damage (equivalent dose) equal to 100 rems.

ADDITIONAL QUESTIONS AND PROBLEMS

9.43 Carbon-12 is a nonradioactive isotope of carbon, and carbon-14 is a radioactive isotope. What is similar and different about the two carbon isotopes?

9.44 Give the number of protons, neutrons, and electrons in atoms of the following isotopes:
a. boron-10   **b.** zinc-72
c. iron-59   **d.** gold-198

9.45 Describe alpha, beta, and gamma radiation in terms of the following:
a. type of radiation
b. symbols
c. depth of tissue penetration
d. type of shielding needed for protection

9.46 When you have dental X rays at the dentist's office, the technician places a heavy lead apron over you, and then leaves the room to take your X rays. What is the purpose of these activities?

9.47 Write the balanced nuclear equations for each of the following emitters:
a. thorium-225 (α)   **b.** bismuth-210 (α)
c. cesium-137 (β)   **d.** tin-126 (β)

9.48 Write the balanced nuclear equations for each of the following emitters:
a. potassium-40 (β)   **b.** sulfur-35 (β)
c. platinum-190 (α)   **d.** radium-210 (α)

9.49 Complete each of the following nuclear equations:
a. $^{14}_{7}\text{N} + ^{4}_{2}\text{He} \longrightarrow ? + ^{1}_{1}\text{H}$
b. $^{27}_{13}\text{Al} + ^{4}_{2}\text{He} \longrightarrow ^{30}_{14}\text{Si} + ?$
c. $^{235}_{92}\text{U} + ^{1}_{0}\text{n} \longrightarrow ^{90}_{38}\text{Sr} + 3^{1}_{0}\text{n} + ?$

9.50 Complete each of the following nuclear equations:
a. $^{59}_{27}\text{Co} + ? \longrightarrow ^{56}_{25}\text{Mn} + ^{4}_{2}\text{He}$
b. $? \longrightarrow ^{14}_{7}\text{N} + ^{0}_{-1}e$
c. $^{76}_{36}\text{Kr} + ^{0}_{-1}e \longrightarrow ?$

9.51 Write the atomic symbols and a complete nuclear equation for the following:
a. When two oxygen-16 atoms collide, one of the products is an alpha particle. What is the other product?
b. When californium-249 is bombarded by oxygen-18, a new element, seaborgium-263, and four neutrons are produced.
c. Radon-222 emits an alpha particle, and the product emits another alpha particle. Write the two nuclear reactions.

9.52 Write the symbols and complete the following nuclear equations:
a. When polonium-210 emits an alpha particle, the product is lead-206.
b. Bismuth-211 decays by emitting an alpha particle. The product is another radioisotope that emits a beta particle. Write equations for the two nuclear changes.

9.53 If the amount of radioactive phosphorus-32 in a sample decreases from 1.2 g to 0.30 g in 28 days, what is the half-life of phosphorus-32?

9.54 If the amount of radioactive iodine-123 in a sample decreases from 0.4 g to 0.1 g in 26.2 hours, what is the half-life of iodine-123?

9.55 Iodine-131, a beta emitter, has a half-life of 8.0 days.
a. Write the nuclear equation for the beta-decay of iodine-131.
b. How many grams of a 12.0-g sample of iodine-131 would remain after 40 days?
c. How many days have passed if 48 g of iodine-131 decayed to 3.0 grams of iodine-131?

9.56 Cesium-137, a beta emitter, has a half-life of 30 years.
a. Write the nuclear equation for the beta-decay of cesium-137.
b. How many grams of a 16 g sample of cesium-137 would remain after 90 years?
c. How many years will be needed for 28 grams of cesium-137 to decay to 3.5 g of cesium-137?

9.57 A nurse was accidentally exposed to potassium-42 while doing some brain scans for possible tumors. The error was not discovered until 36 hours later when the activity of the potassium-42 sample was 2.0 μCi. If potassium-42 has a half-life of 12 hr, what was the activity of the sample at the time the nurse was exposed?

9.58 A wooden object from the site of an ancient temple has a carbon-14 activity of 10 counts per minute compared with a reference piece of wood cut today that has an activity of 40 counts per minute. If the half-life for carbon-14 is 5730 years, what is the age of the ancient wood object?

9.59 A 120-mg sample of technetium-99 m is used for a diagnostic test. If technetium-99 m has a half-life of 6.0 hr, how much of the technetium-99 m sample remains 24 hr after the test?

9.60 The half-life of oxygen-15 is 124 seconds. If a sample of oxygen-15 has an activity of 4000 becquerels, how many minutes will elapse before it reaches an activity of 500 becquerels?

9.61 What is the purpose of irradiating meats, fruits, and vegetables?

9.62 The irradiation of foods was approved in the United States in the 1980s.
  a. Why have we not seen many irradiated products in our markets?
  b. Would you buy foods that have been irradiated? Why or why not?

9.63 What is the difference between fission and fusion?

9.64 **a.** What are the products in the fission of uranium-235 that make possible a nuclear chain reaction?
  b. What is the purpose of placing control rods among uranium samples in a nuclear reactor?

9.65 Where does fusion occur naturally?

9.66 Why are scientists continuing to try to build a fusion reactor even though very high temperatures have been difficult to reach and maintain?

UNDERSTANDING THE CONCEPTS

9.67 Consider the following nucleus of a radioactive isotope.

○ proton
○ neutron

  a. What is the nuclear symbol for this isotope?
  b. If this isotope decays by emitting a positron, what does the resulting nucleus look like?

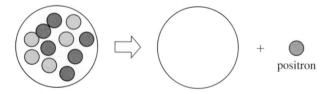

positron

9.68 Draw in the radioactive nucleus that emits a beta particle to form the following nucleus.

+ beta particle

9.69 Draw in the nucleus of the atom to complete the following nuclear reaction.

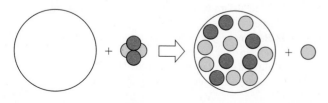

9.70 Complete the following equation by drawing the nucleus of the atom produced.

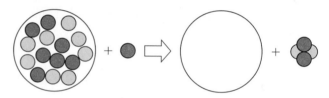

9.71 Carbon dating of small bits of charcoal used in cave paintings has determined that some of the paintings are from 10 000 to 30 000 years old. Carbon-14 has a half-life of 5730 years. In a 1 μg sample of carbon from a live tree, the activity of ^{14}C is 6.4 μCi. If researchers determine that 1 μg charcoal from a prehistoric cave painting in France has an activity of 0.80 μC, what is the age of the painting?

9.72

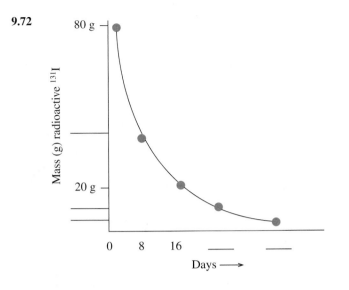

a. Complete the values for the mass of radioactive ^{131}I on the vertical axis.

b. Complete the number of days on the horizontal axis.

c. What is the half-life in days of ^{131}I?

QUESTIONS FOR TEAMWORK

9.73 Identify each of the following nuclear reactions as alpha decay, beta decay, positron emission, or gamma radiation.

a. $^{27m}_{13}\text{Al} \longrightarrow ^{27}_{13}\text{Al} + ^{0}_{0}\gamma$

b. $^{8}_{5}\text{B} \longrightarrow ^{8}_{4}\text{Be} + ^{0}_{+1}e$

c. $^{90}_{38}\text{Sr} \longrightarrow ^{90}_{39}\text{Y} + ^{0}_{-1}e$

d. $^{218}_{85}\text{At} \longrightarrow ^{214}_{83}\text{Bi} + ^{4}_{2}\text{He}$

9.74 Complete and balance each of the following nuclear equations:

a. $^{23m}_{12}\text{Mg} \longrightarrow \underline{\hspace{1cm}} + ^{0}_{0}\gamma$

b. $^{61}_{30}\text{Zn} \longrightarrow ^{61}_{29}\text{Cu} + \underline{\hspace{1cm}}$

c. $^{241}_{95}\text{Am} + ^{4}_{2}\text{He} \longrightarrow \underline{\hspace{1cm}} + 2\,^{1}_{0}\text{n}$

d. $^{126}_{50}\text{Sn} \longrightarrow \underline{\hspace{1cm}} + ^{0}_{-1}e$

9.75 The half-life for the radioactive decay of calcium-47 is 4.5 days. If a sample has an activity of 4.0 μCi after 18 days, what was the initial activity of the sample?

9.76 A 16 μg sample of sodium-24 decays to 2.0 μg in 45 hr. What is the half-life of ^{24}Na?

ANSWERS

Answers to Study Checks

9.1 β or $^{0}_{-1}e$

9.2 Limiting the time one spends near a radioactive source and staying as far away as possible will reduce exposure to radiation.

9.3 $^{214}_{84}\text{Po} \longrightarrow ^{210}_{82}\text{Pb} + ^{4}_{2}\text{He}$

9.4 $^{131}_{53}\text{I} \longrightarrow ^{131}_{54}\text{Xe} + ^{0}_{-1}e$

9.5 $^{27}_{13}\text{Al} + ^{4}_{2}\text{He} \longrightarrow ^{30}_{15}\text{P} + ^{1}_{0}\text{n}$

9.6 For β, the factor is 1; rads and rems are equal.

9.7 0.50 g

9.8 17,200 years

9.9 An iodine-131 sample with a higher activity is used in radiation therapy when radiation is needed to destroy some of the cells in the thyroid gland.

9.10 fusion

Answers to Selected Questions and Problems

9.1 **a.** Both an alpha particle and a helium nucleus have 2 protons and 2 neutrons. However, an α-particle is emitted from a nucleus during radioactive decay.

b. α, $^{4}_{2}\text{He}$

9.3 **a.** $^{39}_{19}\text{K}$, $^{40}_{19}\text{K}$, $^{41}_{19}\text{K}$

b. They all have 19 protons and 19 electrons, but they differ in the number of neutrons.

9.5

Medical Use	Isotope Symbol	Mass Number	Number of Protons	Number of Neutrons
Heart imaging	$^{201}_{81}\text{Tl}$	201	81	120
Radiation therapy	$^{60}_{27}\text{Co}$	60	27	33
Abdominal scan	$^{67}_{31}\text{Ga}$	67	31	36
Hyperthyroidism	$^{131}_{53}\text{I}$	131	53	78
Leukemia treatment	$^{32}_{15}\text{P}$	32	15	17

9.7 **a.** α, ^4_2He **b.** ^1_0n **c.** β, $^0_{-1}e$
d. $^{15}_7\text{N}$ **e.** $^{125}_{53}\text{I}$

9.9 **a.** β or e^- **b.** α or He **c.** n
d. Na **e.** C

9.11 **a.** Because β particles move faster than α particles, they can penetrate further into tissue.
b. Ionizing radiation forms reactive species that cause undesirable reactions in the cells.
c. X-ray technicians leave the room to increase the distance between them and the radiation. Also a wall that contains lead shields them.
d. Wearing gloves shields the skin from α and β radiation.

9.13 **a.** $^{208}_{84}\text{Po} \longrightarrow {}^{204}_{82}\text{Pb} + {}^4_2\text{He}$
b. $^{232}_{90}\text{Th} \longrightarrow {}^{228}_{88}\text{Ra} + {}^4_2\text{He}$
c. $^{251}_{102}\text{No} \longrightarrow {}^{247}_{100}\text{Fm} + {}^4_2\text{He}$

9.15 **a.** $^{25}_{11}\text{Na} \longrightarrow {}^{25}_{12}\text{Mg} + {}^0_{-1}e$
b. $^{20}_8\text{O} \longrightarrow {}^{20}_9\text{F} + {}^0_{-1}e$
c. $^{92}_{38}\text{Sr} \longrightarrow {}^{92}_{39}\text{Y} + {}^0_{-1}e$

9.17 **a.** $^{28}_{14}\text{Si}$ **b.** $^{87}_{36}\text{Kr}$ **c.** $^0_{-1}e$

9.19 **a.** $^{10}_4\text{Be}$ **b.** $^0_{-1}e$ **c.** $^{27}_{13}\text{Al}$

9.21 **a.** When radiation enters the Geiger counter, charged particles are produced which create a burst of current that is detected by the instrument.
b. becquerel (Bq), curie (Ci)
c. gray (Gy), rad
d. 1000 Gy

9.23 294 μCi

9.25 When pilots are flying at high altitudes, there is less atmosphere to protect them from cosmic radiation.

9.27 A half-life is the time it takes for one-half of a radioactive sample to decay.

9.29 **a.** 40.0 mg **b.** 20.0 mg
c. 10.0 mg **d.** 5.00 mg

9.31 128 days, 192 days

9.33 **a.** Since the elements Ca and P are part of bone, their radioactive isotopes will also become part of the bony structures of the body where their radiation can be used to diagnose or treat bone diseases.
b. Strontium (Sr) acts much like calcium (Ca) because both are Group 2A (2) elements. The body will accumulate radioactive strontium in bones in the same way that it incorporates calcium. Radioactive strontium is harmful to children because the radiation it produces causes more damage in cells that are dividing rapidly.

9.35 180 μCi

9.37 Nuclear fission is the splitting of a large atom into smaller fragments with the release of large amounts of energy.

9.39 $^{103}_{42}\text{Mo}$

9.41 **a.** fission **b.** fusion **c.** fission **d.** fusion

9.43 Each has 6 protons and 6 electrons, but carbon-12 has 6 neutrons and carbon-14 has 8 neutrons. Carbon-12 is a stable isotope, but carbon-14 is radioactive and will emit radiation.

9.45 **a.** α radiation consists of a helium nucleus emitted from the nucleus of a radioisotope. β radiation is an electron and γ radiation is high-energy radiation emitted from the nucleus of a radioisotope.
b. α particle: ^4_2He
β particle: $^0_{-1}e$
γ radiation: γ
c. α particles penetrate 0.05 mm into tissue, β particles 4–5 mm, and γ rays 50 cm or more.
d. Paper or clothing will shield you from α particles; heavy clothing, lab coats, and gloves will shield you from β particles; and lead and concrete are needed for shielding from γ radiation.

9.47 **a.** $^{225}_{90}\text{Th} \longrightarrow {}^{221}_{88}\text{Ra} + {}^4_2\text{He}$
b. $^{210}_{83}\text{Bi} \longrightarrow {}^{206}_{81}\text{Tl} + {}^4_2\text{He}$
c. $^{137}_{55}\text{Cs} \longrightarrow {}^{137}_{56}\text{Ba} + {}^0_{-1}e$
d. $^{126}_{50}\text{Sn} \longrightarrow {}^{126}_{51}\text{Sb} + {}^0_{-1}e$

9.49 **a.** $^{17}_8\text{O}$ **b.** ^1_1H **c.** $^{143}_{54}\text{Xe}$

9.51 **a.** $^{16}_8\text{O} + {}^{16}_8\text{O} \longrightarrow {}^4_2\text{He} + {}^{28}_{14}\text{Si}$
b. $^{249}_{98}\text{Cf} + {}^{18}_8\text{O} \longrightarrow {}^{263}_{106}\text{Sg} + 4\,{}^1_0\text{n}$
c. $^{222}_{86}\text{Rn} \longrightarrow {}^{218}_{84}\text{Po} + {}^4_2\text{He}$
$^{218}_{84}\text{Po} \longrightarrow {}^{214}_{82}\text{Pb} + {}^4_2\text{He}$

9.53 14 days

9.55 **a.** $^{131}_{53}\text{I} \longrightarrow {}^0_{-1}e + {}^{131}_{54}\text{Xe}$ **b.** 0.375 g
c. 32 days

9.57 16 μCi

9.59 7.5 mg

9.61 The irradiation of meats, fruits, and vegetables kills bacteria such as *E. coli* that can cause foodborne illnesses. In addition, spoilage is deterred, and shelf life is extended.

9.63 In the fission process, an atom splits into smaller nuclei. In fusion, small nuclei combine (fuse) to form a larger nucleus.

9.65 Fusion occurs naturally in the sun and other stars.

9.67 **a.** $^{11}_6\text{C}$
b.

positron

9.69

9.71 17,200 years old

10

Introduction to Organic Chemistry: Alkanes

"When we have a hazardous materials spill, the first thing we do is isolate it," says Don Dornell, assistant fire chief, Burlingame Fire Station. "Then our technicians and a county chemist identify the product from its flammability and solubility in water so we can use the proper materials to clean up the spill. We use different methods for alcohol, which mixes with water, than for gasoline, which floats. Because hydrocarbons are volatile, we use foam to cover them and trap the vapors. At oil refineries, we will use foams, but many times we squirt water on the tanks to cool the contents below their boiling points, too. By knowing the boiling point of the product and its density and vapor density, we know if it floats or sinks in water and where its vapors will go."

the **Chemistry** place

Visit **www.aw-bc.com/chemplace** or use the CD-ROM in the book for extra quizzes, interactive tutorials, career resources, PowerPoint slides, math help, and case studies.

O rganic chemistry is the chemistry of carbon compounds that contain, primarily, carbon and hydrogen. The element carbon has a special role in chemistry because it bonds with other carbon atoms to give a vast array of molecules. The variety of molecules is so great that we find organic compounds in many common products we use such as gasoline, medicine, shampoos, plastic bottles, and perfumes. The food we eat is composed of different organic compounds that supply us with fuel for energy and the carbon atoms needed to build and repair the cells of our bodies.

Although many organic compounds occur in nature, chemists have synthesized even more. The cotton, wool, or silk in your clothes contain naturally occurring organic compounds, whereas materials such as polyester, nylon, or plastic have been synthesized through organic reactions. Sometimes it is convenient to synthesize a molecule in the lab even though that molecule is also found in nature. For example, vitamin C synthesized in a laboratory has the same structure as the vitamin C in oranges or lemons. In these chapters, you will learn about the structures and reactions of organic molecules, which will provide a foundation for understanding the more complex molecules of biochemistry.

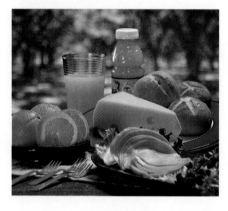

LEARNING GOAL

■ Identify properties characteristic of organic or inorganic compounds.

WEB TUTORIAL
Introduction to Organic Molecules

■ 10.1 Organic Compounds

At the beginning of the nineteenth century, scientists classified chemical compounds as inorganic and organic. An inorganic compound was a substance that was composed of minerals, and an organic compound was a substance that came from an organism, thus the use of the word "organic." It was thought that some type of "vital force," which could only be found in living cells, was required to synthesize an organic compound. This perception was shown to be incorrect in 1828 when the German chemist Friedrick Wöhler synthesized urea, a product of protein metabolism, by heating an inorganic compound, ammonium cyanate.

$$NH_4CNO \xrightarrow{\text{Heat}} H_2N - \overset{\overset{\displaystyle O}{\|}}{C} - NH_2$$

Ammonium cyanate (inorganic)   Urea, organic

We now define organic chemistry as the study of carbon compounds. **Organic compounds** form nonpolar molecules, and as a result, the attractions between molecules are weak, which accounts for the low melting and boiling points of most carbon compounds. Typically, organic compounds are not soluble in water. For example, vegetable oil, which is a mixture of organic compounds, does not dissolve in water, but floats on top.

Many organic compounds undergo combustion and burn vigorously in air. In contrast, many of the inorganic compounds are ionic, which leads to high melting and boiling points. Inorganic compounds that are ionic or polar covalent are usually soluble in water. Most inorganic substances do not burn in air. Table 10.1 contrasts some of the properties associated with organic and inorganic compounds such as propane, C_3H_8, and sodium chloride, NaCl. (See Figure 10.1.)

TABLE 10.1

Some Properties of Organic and Inorganic Compounds

Property	Organic	Example: C_3H_8	Inorganic	Example: NaCl
Bonding	Mostly covalent	Covalent	Many are ionic, some covalent	Ionic
Polarity of bonds	Nonpolar, unless a more electronegative atom is present	Nonpolar	Most are ionic or polar covalent, a few are nonpolar covalent	Ionic
Melting point	Usually low	$-188°C$	High	$801°C$
Boiling point	Usually low	$-42°C$	High	$1413°C$
Flammability	High	Burns in air	Low	Does not usually burn
Solubility in water	Not soluble, unless a polar group is present	No	Most are soluble, unless nonpolar	Yes

SAMPLE PROBLEM 10.1

■ Properties of Organic Compounds

Indicate whether the following properties are characteristic of organic or inorganic compounds:

a. not soluble in water   **b.** high melting point   **c.** burns in air

SOLUTION

a. Many organic compounds are not soluble in water.
b. Inorganic compounds are most likely to have high melting points.
c. Organic compounds are most likely to be flammable.

FIGURE 10.1

Propane, C_3H_8, is an organic compound, whereas sodium chloride, NaCl, is an inorganic compound.

Q Why is propane used as a fuel?

STUDY CHECK

Octane is not soluble in water. What type of compound is octane?

Bonding in Organic Compounds

The **hydrocarbons** are organic compounds that consist of only carbon and hydrogen. In the simplest hydrocarbon, methane (CH_4), the four valence electrons of carbon are shared with four hydrogen atoms to form an octet. In the electron-dot formula, each shared pair of electrons represents a single bond. In organic molecules, every carbon atom always has four bonds. An **expanded structural formula** is written when we show the bonds between all of the atoms.

methane

The Tetrahedral Structure of Carbon

The VSEPR theory (Chapter 4) predicts that when four bonds are arranged as far apart as possible, they have a tetrahedral shape. In CH_4 the bonds to hydrogen are directed to the corners of a tetrahedron with bond angles of 109.5°. The three-dimensional structure of methane can be illustrated as a ball-and-stick model or a space-filling model. (See Figure 10.2.)

An organic compound with two atoms of carbon is ethane, C_2H_6. Each carbon atom is bonded to another carbon and three hydrogen atoms. When there are two or more carbon atoms in a molecule, each carbon retains the tetrahedral shape if bonded to four other atoms. In the ball-and-stick model of ethane C_2H_6, two tetrahedra are attached to each other. (See Figure 10.3.) All the bond angles are close to 109.5°.

FIGURE 10.2

Three-dimensional representations of methane, CH_4: **(a)** tetrahedron, **(b)** ball-and-stick model, **(c)** space-filling model, **(d)** expanded structural formula.

Q **Why does methane have a tetrahedral shape and not a flat shape?**

Methane

FIGURE 10.3

Three-dimensional representations of ethane, C_2H_6: **(a)** tetrahedral shape of each carbon, **(b)** ball-and-stick model, **(c)** space-filling model, **(d)** expanded structural formula.

Q **How is the tetrahedral shape maintained in a molecule with two carbon atoms?**

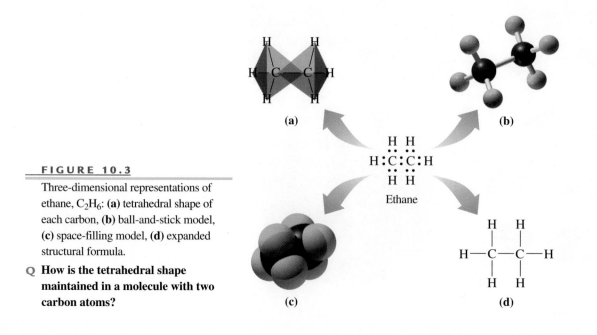

Ethane

QUESTIONS AND PROBLEMS

■ Organic Compounds

10.1 Identify the following as formulas of organic or inorganic compounds:
  a. KCl   **b.** C_4H_{10} **c.** CH_3CH_2OH
  d. H_2SO_4 **e.** $CaCl_2$ **f.** CH_3CH_2Cl

10.2 Identify the following as formulas of organic or inorganic compounds:
  a. $C_6H_{12}O_6$ **b.** Na_2SO_4 **c.** I_2
  d. C_2H_5Cl **e.** $C_{10}H_{22}$ **f.** CH_4

10.3 Identify the following properties as most typical of organic or inorganic compounds:
  a. soluble in water   **b.** low boiling point
  c. burns in air   **d.** solid at room temperature

10.4 Identify the following properties as most typical of organic or inorganic compounds:
  a. high melting point   **b.** gas at room temperature
  c. covalent bonds   **d.** produces ions in water

10.5 Match the following physical and chemical properties with the compounds ethane, C_2H_6, or sodium bromide, NaBr.
  a. boils at $-89°C$ **b.** burns vigorously
  c. solid at $250°C$ **d.** dissolves in water

10.6 Match the following physical and chemical properties with the compounds cyclohexane, C_6H_{12}, or calcium nitrate, $Ca(NO_3)_2$.
  a. melts at $500°C$ **b.** insoluble in water
  c. produces ions in water   **d.** is a liquid at room temperature

10.7 Why is the structure of the CH_4 molecule three-dimensional rather than two-dimensional?

10.8 In a propane molecule with three carbon atoms, what is the geometry around each carbon atom?

Propane

■ 10.2 Alkanes

The **alkanes** are a class or family of organic compounds that contain only carbon and hydrogen atoms and only form single bonds. One of the most common uses of alkanes is as fuels. Methane, used in gas heaters and gas cooktops, is an alkane with one carbon atom. The alkanes ethane, propane, and butane contain two, three, and four carbon atoms connected in a row or a *continuous* chain. These names are part of the **IUPAC** (International Union of Pure and Applied Chemistry) **system**, which chemists use to name organic compounds. Alkanes with five or more carbon atoms in a chain are named using Greek prefixes: *pent*(5), *hex*(6), *hept*(7), *oct*(8), *non*(9), and *dec*(10). (See Table 10.2.)

■ Condensed Structural Formulas

In a **condensed structural formula**, we write each carbon atom and its attached hydrogen atoms as a group. A subscript indicates the number of hydrogen atoms bonded to each carbon atom.

| Expanded | Condensed | Expanded | Condensed |

LEARNING GOAL

Write the IUPAC names and structural formulas for alkanes.

the Chemistry place

WEB TUTORIAL
Introduction to Organic Molecules

TABLE 10.2

IUPAC Names for the First Ten Continuous-Chain Alkanes

Number of Carbon Atoms	Prefix	Name	Molecular Formula	Condensed Structural Formula
1	meth	methane	CH_4	CH_4
2	eth	ethane	C_2H_6	$CH_3{-}CH_3$
3	prop	propane	C_3H_8	$CH_3{-}CH_2{-}CH_3$
4	but	butane	C_4H_{10}	$CH_3{-}CH_2{-}CH_2{-}CH_3$
5	pent	pentane	C_5H_{12}	$CH_3{-}CH_2{-}CH_2{-}CH_2{-}CH_3$
6	hex	hexane	C_6H_{14}	$CH_3{-}CH_2{-}CH_2{-}CH_2{-}CH_2{-}CH_3$
7	hept	heptane	C_7H_{16}	$CH_3{-}CH_2{-}CH_2{-}CH_2{-}CH_2{-}CH_2{-}CH_3$
8	oct	octane	C_8H_{18}	$CH_3{-}CH_2{-}CH_2{-}CH_2{-}CH_2{-}CH_2{-}CH_2{-}CH_3$
9	non	nonane	C_9H_{20}	$CH_3{-}CH_2{-}CH_2{-}CH_2{-}CH_2{-}CH_2{-}CH_2{-}CH_2{-}CH_3$
10	dec	decane	$C_{10}H_{22}$	$CH_3{-}CH_2{-}CH_2{-}CH_2{-}CH_2{-}CH_2{-}CH_2{-}CH_2{-}CH_2{-}CH_3$

FIGURE 10.4

A ball-and-stick model of hexane.

Q **Why do the carbon atoms in hexane appear to be arranged in a zigzag chain?**

The molecular formula gives the total number of each kind of atom, but does not indicate the arrangement of the atoms in the molecule. Table 10.3 shows the molecular, expanded, and condensed structural formulas for alkanes with one, two, and three carbon atoms.

In **continuous alkanes**, the carbon atoms are connected in a row. However, in chains of three carbon atoms or more, the carbon atoms do not lie in a straight line. The tetrahedral shape of carbon arranges the carbon bonds in a zigzag pattern, which is seen in the ball-and-stick model of hexane. (See Figure 10.4.)

In an alkane, the groups attached to each carbon are not in fixed positions, but rotate freely about the bond connecting the two carbon atoms. This motion is analogous to the independent rotation of the wheels of a toy car. Thus different arrangements occur during the rotation about a single bond.

Suppose we could look at butane, C_4H_{10}, as it rotates. Sometimes the CH_3 groups line up in front of each other, and at other times they are opposite each other. As the CH_3 groups turn around the single bond, many arrangements are possible. Butane can be depicted by a variety of two-dimensional structural formulas as shown in Table 10.4. It is important to recognize that all of these structural formulas represent the same continuous chain of four carbon atoms.

TABLE 10.3

Writing Structural Formulas for Some Alkanes

Alkane	Methane	Ethane	Propane
Molecular formula	CH_4	C_2H_6	C_3H_8
Structural formulas			

Condensed	CH_4	$CH_3{-}CH_3$	$CH_3{-}CH_2{-}CH_3$

or

$$CH_3{\diagup}^{CH_2}{\diagdown}CH_3$$

TABLE 10.4

Some Structural Formulas for Butane, C_4H_{10}

Expanded structural formula

$$
\begin{array}{ccccc}
& H & H & H & H \\
& | & | & | & | \\
H- & C- & C- & C- & C-H \\
& | & | & | & | \\
& H & H & H & H
\end{array}
$$

Condensed structural formulas

$CH_3-CH_2-CH_2-CH_3$

$$
\begin{array}{cc}
CH_2-CH_2 \\
| \quad\quad | \\
CH_3 \quad CH_3
\end{array}
$$

$$
\begin{array}{c}
CH_3 \\
| \\
CH_2-CH_2 \\
\quad\quad | \\
\quad\quad CH_3
\end{array}
$$

$$
\begin{array}{c}
CH_3 \\
| \\
CH_2 \\
| \\
CH_2 \\
| \\
CH_3
\end{array}
$$

$$
\begin{array}{c}
CH_3-CH_2 \\
\quad\quad | \\
\quad CH_2-CH_3
\end{array}
$$

$$
\begin{array}{c}
CH_3 \\
| \\
CH_2-CH_2-CH_3
\end{array}
$$

$$
\begin{array}{cc}
CH_3 & CH_2 \\
\diagdown & \diagup \quad \diagdown \\
CH_2 & CH_3
\end{array}
$$

SAMPLE PROBLEM 10.2

■ Drawing Expanded and Condensed Structural Formulas for Alkanes

A molecule of butane, C_4H_{10}, has four carbon atoms in a row. What are its expanded and condensed structural formulas?

SOLUTION

In the expanded structural formula, four carbon atoms are connected to each other and to hydrogen atoms using single bonds to give each carbon atom a total of four bonds. In the condensed structural formula, each carbon atom and its attached hydrogen atoms are written as CH_3-, or $-CH_2-$.

H—C—C—C—C—H   Expanded structural formula

$CH_3-CH_2-CH_2-CH_3$ Condensed structural formula

STUDY CHECK

Write the expanded and condensed structural formulas of pentane, C_5H_{12}.

■ Cycloalkanes

Hydrocarbons can also form cyclic structures called **cycloalkanes**, which have two fewer hydrogen atoms than the corresponding alkanes. Thus, the simplest cycloalkane, cyclopropane, C_3H_6, has a ring of three carbon atoms bonded to six hydrogen atoms. A simplified formula, which omits the hydrogen atoms and looks like a geometric figure, is a convenient way to show cyclic structures. Each corner of the triangle represents a carbon atom with four bonds to other carbon and hydrogen atoms.

The ball-and-stick models and their structural formulas for several cycloalkanes are shown in Table 10.5. A cycloalkane is named by adding the prefix *cyclo* to the name of the alkane with the same number of carbon atoms.

TABLE 10.5

Formulas of Some Common Cycloalkanes

Ball and Stick Models

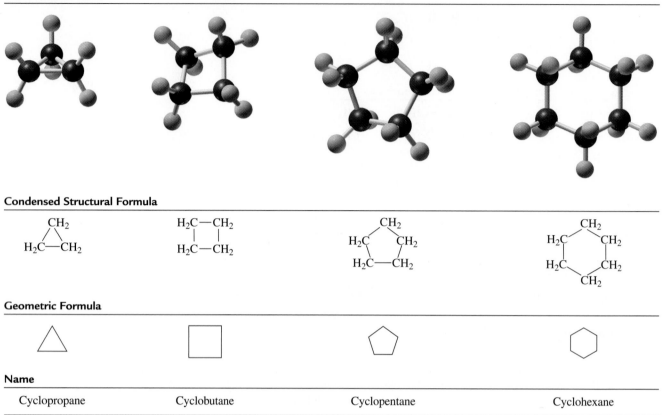

Condensed Structural Formula

| CH_2 H_2C-CH_2 | H_2C-CH_2 H_2C-CH_2 | CH_2 H_2C CH_2 H_2C-CH_2 | CH_2 H_2C CH_2 H_2C CH_2 CH_2 |

Geometric Formula

| △ | □ | ⬠ | ⬡ |

Name

| Cyclopropane | Cyclobutane | Cyclopentane | Cyclohexane |

SAMPLE PROBLEM 10.3

■ **Naming Alkanes**

Give the IUPAC name for each of the following:

a. $CH_3-CH_2-CH_2-CH_2-CH_3$ **b.** ⬡   **c.** $CH_3-CH_2-CH_3$

SOLUTION

a. A chain with five carbon atoms is pentane.
b. The ring of six carbon atoms is named cyclohexane.
c. This alkane is named propane because it has three carbon atoms.

STUDY CHECK

What is the IUPAC name of the following compound?

□

QUESTIONS AND PROBLEMS

■ Alkanes

10.9 Give the IUPAC name for each of the following alkanes.

a. CH$_3$
    |
  CH$_2$—CH$_2$—CH$_2$
          |
          CH$_3$

b. CH$_3$—CH$_3$

                                          CH$_2$—CH$_3$
                                         |
                                         CH$_2$
                                         |

c. CH$_3$—CH$_2$—CH$_2$

10.10 Give the IUPAC name for each of the following alkanes.

a. CH$_4$

b. CH$_3$—CH$_2$—CH$_2$—CH$_3$

                              CH$_3$
                              |
                              CH$_2$
                              |

c. CH$_3$

10.11 Write the condensed structural formula or geometric figure for each of the following:
    a. methane   **b.** ethane   **c.** pentane   **d.** cyclopropane

10.12 Write the condensed structural formula or geometric figure for each of the following:
    a. propane   **b.** hexane   **c.** heptane   **d.** cyclopentane

---

■ 10.3 Alkanes with Substituents

LEARNING GOAL

■
Write the IUPAC names for alkanes with substituents.

When an alkane has four or more carbon atoms, the atoms can be arranged so that a side group called a **branch** or **substituent** is attached to a carbon chain. For example, we can write two different structural formulas for the molecular formula C$_4$H$_{10}$. One formula contains the four carbon atoms in a continuous chain. In the other formula, a carbon atom in a *branch* is attached to a chain of three atoms. (See Figure 10.5.) An alkane with at least one branch is called a **branched alkane**. When two compounds have the same molecular formula but different arrangements of atoms, they are called **isomers**.

In another example, we can write the structural formulas of three different isomers with the molecular formula C$_5$H$_{12}$ as follows:

Isomers of C$_5$H$_{12}$

Continuous Chain	Branched Chains	
CH$_3$—CH$_2$—CH$_2$—CH$_2$—CH$_3$	CH$_3$ | CH$_3$—CH—CH$_2$—CH$_3$	CH$_3$ | CH$_3$—C—CH$_3$ | CH$_3$

SAMPLE PROBLEM 10.4

■ Isomers

Identify each pair of structural formulas as isomers or the same molecule.

a. CH$_3$ CH$_3$
     | |
   CH$_2$—CH$_2$ and CH$_2$—CH$_2$—CH$_3$
                       |
                       CH$_3$

FIGURE 10.5

The isomers of C$_4$H$_{10}$ have the same number and type of atoms, but bonded in a different order.

Q **What makes these molecules isomers?**

$$CH_3$$
$$|$$
b. $CH_3—CH—CH_2—CH_2—CH_3$ and

$$CH_3 \quad CH_3$$
$$| \quad\quad |$$
$$CH_3—CH—CH—CH_3$$

SOLUTION

a. The structural formulas represent the same molecule because each has four atoms in a continuous chain.

b. These are isomers because the molecular formula C_6H_{14} is identical but the atoms are bonded in a different order. One has one CH_3 side group attached to a five-carbon chain, and the other has two CH_3 side groups attached to a four-carbon chain.

STUDY CHECK

The following compound has the same molecular formula as the compounds in part b of the preceding Sample Problem. Is the following structural formula an isomer or identical to one of the molecules in part b?

$$CH_3$$
$$|$$
$$CH_3—CH_2—CH—CH_2—CH_3$$

Substituents in Alkanes

In the IUPAC names for alkanes, a carbon branch is named as an **alkyl group**, which is an alkane that is missing one hydrogen atom. The alkyl group is named by replacing the *ane* ending of the corresponding alkane name with *yl*. Alkyl groups cannot exist on their own: they must be attached to a carbon chain. When a halogen atom is attached to a carbon chain, it is named as a *halo* group: fluoro (F), chloro (Cl), bromo (Br), or iodo (I). Some of the common groups attached to carbon chains are illustrated in Table 10.6.

TABLE 10.6

Names and Formulas of Some Common Substituents

Substituent	Name
$CH_3—$	methyl
$CH_3—CH_2—$	ethyl
$CH_3—CH_2—CH_2—$	propyl
$CH_3—\overset{\displaystyle |}{C}H—CH_3$	isopropyl
$F—, Cl—, Br—, I—$	fluoro, chloro, bromo, iodo

Rules for Naming Alkanes with Substituents

In the IUPAC system of naming, the longest carbon chain is numbered to give the location of one or more substituents attached to it. Let's take a look at how we use the IUPAC system to name the following alkane:

$$CH_3$$
$$|$$
$$CH_3—CH—CH_2—CH_2—CH_3$$

STEP 1 **Write the alkane name of the longest continuous chain of carbon atoms.** In this alkane, the longest chain has five carbon atoms, which is *pentane*.

$$CH_3$$
$$|$$
$$CH_3—CH—CH_2—CH_2—CH_3 \quad\quad \text{pentane}$$

STEP 2 **Number the carbon atoms starting from the end nearest a substituent.** Once you start numbering, continue in that same direction.

$$CH_3$$
$$|$$
$$CH_3—CH—CH_2—CH_2—CH_3 \quad\quad \text{pentane}$$
$$\;\,1\quad\;\, 2\quad\;\;\; 3\quad\;\;\;\, 4\quad\;\;\, 5$$

Guide to Naming Alkanes

STEP 1
Write the alkane name of the longest continuous chain of carbon atoms.

STEP 2
Number the carbon atoms starting from the end nearest a substituent.

STEP 3
Give the location and name of each substituent (alphabetical order) as a prefix to the name of the main chain.

STEP 3 **Give the location and name of each substituent as a prefix to the alkane name.** Place a hyphen between the number and the substituent name.

$$CH_3—CH—CH_2—CH_2—CH_3$$
$$\overset{CH_3}{|}$$
1 2 3 4 5

2-methyl pentane

List the substituents in alphabetical order.

$$CH_3—CH—CH—CH_2—CH_3$$
$$\overset{CH_3}{|}\ \overset{Cl}{|}$$
1 2 3 4 5

3-chloro-2-methylpentane

Use a prefix (di-, tri-, tetra-) to indicate a group that appears more than once. Use commas to separate two or more numbers.

$$CH_3—CH—CH—CH_2—CH_3$$
$$\overset{CH_3}{|}\ \overset{CH_3}{|}$$
1 2 3 4 5

2,3-dimethylpentane

When there are two or more substituents, the main chain is numbered in the direction that gives the lowest set of numbers.

$$CH_3—CH—CH_2—C—CH_3$$
$$\overset{Br}{|}\qquad\overset{CH_3}{|}$$
$$\underset{Br}{|}$$
5 4 3 2 1

2,4-dibromo-2-methylpentane

When a substituent is attached to a cycloalkane, its name is listed in front of the cycloalkane name. No number in needed for a single substituent.

methylcyclopentane

SAMPLE PROBLEM 10.5

■ **Writing IUPAC Names**

Give the IUPAC name for the following alkane:

$$CH_3—CH—CH_2—C—CH_2—CH_3$$
$$\overset{CH_3}{|}\qquad\overset{Br}{|}$$
$$\underset{CH_3}{|}$$

SOLUTION

STEP 1 **Write the alkane name of the longest continuous chain of carbon atoms.** In this alkane, the longest chain has six carbon atoms, which is *hexane*.

$$CH_3—CH—CH_2—C—CH_2—CH_3$$
$$\overset{CH_3}{|}\qquad\overset{Br}{|}$$
$$\underset{CH_3}{|}$$

hexane

STEP 2 **Number the carbon atoms starting from the end nearest a substituent.**

$$CH_3-\overset{\displaystyle CH_3}{\underset{\displaystyle CH_3}{\overset{|}{C}}}H-CH_2-\overset{\displaystyle Br}{\underset{\displaystyle CH_3}{\overset{|}{C}}}-CH_2-CH_3 \qquad \text{hexane}$$

1   2   3   4   5   6

STEP 3 **Give the location and name of each substituent in front of the name of the longest chain. List the names of different substituents in alphabetical order.** Place a hyphen between the number and the substituent names and commas to separate two or more numbers. A prefix (di-, tri-, tetra-) indicates a group that appears more than once.

$$CH_3-\overset{\displaystyle CH_3}{\overset{|}{C}}H-CH_2-\overset{\displaystyle Br}{\underset{\displaystyle CH_3}{\overset{|}{C}}}-CH_2-CH_3 \qquad \text{4-bromo- 2,4-dimethyl hexane}$$

1   2   3   4   5   6

STUDY CHECK

Give the IUPAC name for the following compound:

$$CH_3-CH_2-\overset{\displaystyle CH_3}{\overset{|}{C}}H-CH_2-\overset{\displaystyle CH_3}{\overset{|}{C}}H-CH_2-Cl$$

Drawing Structural Formulas for Alkanes

The IUPAC name gives all the information needed to draw the condensed structural formula of an alkane. Suppose you are asked to draw the condensed structural formula of 2,3-dimethylbutane. The alkane name gives the number of carbon atoms in the longest chain. The names in the beginning indicate the substituents and where they are attached. We can break down the name in the following way.

2,3-dimethylbutane

2,3-	di	methyl	but	ane
Substituents on carbons 2 and 3	Two identical groups	CH_3- alkyl groups	4 Carbon atoms in the main chain	Single C—C bonds

Guide to Drawing Alkane Formulas

STEP 1
Draw the main chain of carbon atoms.

STEP 2
Number the chain and place the substituents on the carbons indicated by the numbers.

STEP 3
Add the correct number of hydrogen atoms to give four bonds to each C atom.

SAMPLE PROBLEM 10.6

■ **Drawing Structures from IUPAC Names**

Write the condensed structural formula for 2,3-dimethylbutane.

SOLUTION

We can use the following guide to draw the condensed structural formula.

STEP 1 **Draw the main chain of carbon atoms.** For butane, we draw a chain of four carbon atoms.

C—C—C—C

STEP 2 **Number the chain and place the substituents on the carbons indicated by the numbers.** The first part of the name indicates two methyl groups (CH$_3$—), one on carbon 2 and one on carbon 3.

$$
\begin{array}{cccc}
& \overset{\text{methyl}}{\text{CH}_3} & \overset{\text{methyl}}{\text{CH}_3} & \\
& | & | & \\
\text{C}-\!\!&\text{C}-\!\!&\text{C}-\!\!&\text{C} \\
1 & 2 & 3 & 4
\end{array}
$$

STEP 3 **Add the correct number of hydrogen atoms to give four bonds to each C atom.**

$$
\begin{array}{ccc}
\text{CH}_3 & \text{CH}_3 & \\
| & | & \\
\text{CH}_3-\text{CH}-\text{CH}-\text{CH}_3
\end{array}
$$

2,3-dimethylbutane

STUDY CHECK

What is the structural formula for 2,4-dimethylpentane?

Haloalkanes

In a **haloalkane**, halogen atoms replace hydrogen atoms in an alkane. The halo substituents are numbered and arranged alphabetically just as we did with the alkyl groups. Many times chemists use the common, traditional name for these compounds rather than the systematic IUPAC name. Simple haloalkanes are commonly named as alkyl halides; the carbon group is named as an alkyl group followed by the halide name.

	CH$_3$—Cl	CH$_3$—CH$_2$—Br	$\overset{\text{F}}{\underset{|}{\text{CH}_3-\text{CH}-\text{CH}_3}}$
IUPAC:	chloromethane	bromoethane	2-fluoropropane
Common:	methyl chloride	ethyl bromide	isopropyl fluoride

SAMPLE PROBLEM 10.7

Naming Haloalkanes

Freon 11 and Freon 12 are compounds, known as chlorofluorocarbons (CFCs), that were used as refrigerants and aerosol propellants. What are their IUPAC names?

Freon 11   Freon 12

SOLUTION

Freon 11, trichlorofluoromethane; Freon 12, dichlorodifluoromethane.

HEALTH NOTE

Common Uses of Haloalkanes

Some common uses of haloalkanes include solvents and anesthetics. For many years, carbon tetrachloride was widely used in dry cleaners and in home spot removers to take oils and grease out of clothes. However, this use was discontinued when carbon tetrachloride was found to be toxic to the liver, where it can cause cancer. Today, dry cleaners use other halogenated compounds such as dichloromethane, 1,1,1-trichloroethane, and 1,1,2-trichloro-1,2,2-trifluoroethane.

CH_2Cl_2 Cl_3C-CH_3 $FCl_2C-CClF_2$

Dichloro-   1,1,1-trichloroethane   1,1,2-trichloro-
methane   1,2,2-trifluoroethane

General anesthetics are compounds that are inhaled or injected to cause a loss of sensation so that surgery or other procedures can be done without causing pain to the patient. As nonpolar compounds, they are soluble in the nonpolar nerve membranes, where they decrease the ability of the nerve cells to conduct the sensation of pain. Chloroform $CHCl_3$ was once used as an anesthetic, but it is toxic and may be carcinogenic. One of the most widely used general anesthetics is halothane, also called Fluothane. It has a pleasant odor, is nonexplosive, has few side effects, undergoes few reactions within the body, and is eliminated quickly.

halothane (Fluothane)

For minor surgeries, a local anesthetic such as chloroethane (ethyl chloride) CH_3-CH_2-Cl is applied to an area of the skin. Chloroethane evaporates quickly, which cools the skin and causes a loss of sensation.

STUDY CHECK

Ethylene dibromide is the common name of a haloalkane used as a fumigant. What is its IUPAC name?

Ethylene dibromide

QUESTIONS AND PROBLEMS

Alkanes with Substituents

10.13 Indicate whether each of the following pairs of structural formulas represent isomers or the same molecule.

          CH$_3$ CH$_3$ CH$_3$ CH$_3$
          | | | |
c. CH$_2$—CH—CH$_2$—CH$_3$ and CH$_3$—CH—CH—CH$_3$

10.14 Indicate whether each of the following pairs of structural formulas represent isomers or the same molecule.

             CH$_3$ CH$_3$
             | |
a. CH$_3$—C—CH$_3$ and   CH—CH$_2$—CH$_3$
             | |
             CH$_3$ CH$_3$

           CH$_3$ CH$_3$ CH$_3$ CH$_3$ CH$_3$
           | | | | |
b. CH$_3$—CH—CH—CH$_2$ and CH$_3$—CH—CH$_2$—CH—CH$_3$

         CH$_3$ CH$_3$
         | |
c. CH$_3$—CH—CH$_2$—CH$_3$ and CH$_3$—CH$_2$—CH—CH$_3$

10.15 Give the IUPAC name for each of the following alkanes:

          CH$_3$ CH$_3$
         | |
a. CH$_3$—CH—CH$_2$—CH$_3$ **b.** CH$_3$—C—CH$_3$
                                    |
                                    CH$_3$

                CH$_3$ CH$_3$
                | |
c. CH$_3$—CH$_2$—CH—CH—CH$_3$

          CH$_3$ CH$_2$—CH$_3$
         | |
d. CH$_3$—C—CH$_2$—CH—CH$_2$—CH$_3$
         |
         CH$_3$

10.16 Give the IUPAC name for each of the following alkanes:

         CH$_3$
        |
a. CH$_3$—CH—CH$_2$—CH$_2$—CH$_3$

         CH$_3$ CH$_3$
        | |
b. CH$_3$—CH—CH—CH$_3$

             CH$_3$ CH$_3$
            | |
c. CH$_3$—CH$_2$—CH—CH$_2$—CH—CH$_3$

                 CH$_2$—CH$_3$
                 |
d. CH$_3$—CH$_2$—CH—CH—CH$_2$—CH$_3$
                |
               CH$_2$—CH$_3$

10.17 Give the IUPAC name for each of the following cycloalkanes:

  a.   **b.** CH$_3$ **c.** CH$_2$CH$_3$

10.18 Give the IUPAC name for each of the following cycloalkanes.

  a. CH$_3$ **b.**   **c.**
                                       CH$_3$

10.19 Draw a condensed structural formula for each of the following alkanes:
  a. 2-methylbutane
  b. 3,3-dimethylpentane
  c. 2,3,5-trimethylhexane
  d. 3-ethyl-2,5-dimethyloctane

10.20 Draw a condensed structural formula for each of the following alkanes:
  a. 3-ethylpentane
  b. 3-ethyl-2-methylpentane
  c. 2,2,3,5-tetramethylhexane
  d. 4-ethyl-2,2-dimethyloctane

10.21 Draw the structural formula for each of the following cycloalkanes.
  a. methylcyclopentane
  b. cyclobutane

10.22 Draw the structural formula for each of the following cycloalkanes.
  a. cyclopropane
  b. ethylcyclohexane

10.23 Give the IUPAC name for each of the following compounds:
  a. CH_3-CH_2-Br
  b. $CH_3-CH_2-CH_2-F$

  c. $CH_3-\overset{\overset{\displaystyle CH_3}{|}}{CH}-Cl$
  d. $CHCl_3$

10.24 Give the IUPAC name for each of the following compounds:

  a. $CH_3-CH_2-\overset{\overset{\displaystyle Cl}{|}}{CH}-CH_3$
  b. CCl_4

  c. $CH_3-\overset{\overset{\displaystyle CH_3}{|}}{\underset{\underset{\displaystyle CH_3}{|}}{C}}-I$
  d. CH_3F

10.25 Write the condensed structural formula for each of the following compounds:
  a. 2-chloropropane
  b. 2-bromo-3-chlorobutane
  c. methyl bromide
  d. tetrabromomethane

10.26 Write the condensed structural formula for each of the following compounds:
  a. 1,1,2,2-tetrabromopropane
  b. 2-bromopropane
  c. 2,3-dichloro-2-methylbutane
  d. dibromodichloromethane

LEARNING GOAL

Identify the properties of alkanes and write a balanced equation for combustion.

the
Chemistry
place

CASE STUDY
Hazardous Materials

10.4 Properties of Alkanes

Useful alkanes include the gasoline and diesel fuels that power our cars and the heating oils that heat our homes. You may have used a mixture of hydrocarbons such as mineral oil or petrolatum as a laxative or to soften your skin. The differences in uses of many of the alkanes result from their physical properties including solubility, density, and boiling point.

Solubility and Density

Alkanes are nonpolar, which makes them insoluble in water. However, they are soluble in nonpolar solvents such as other alkanes. Alkanes have densities from 0.65 g/mL to about 0.70 g/mL, which is less dense than water (1.0 g/mL). If there is an oil spill in the ocean, the alkanes in the crude oil remain on the surface and spread over a large area. In the *Exxon Valdez* oil spill in 1989, 40 million liters of oil covered over 25,000 square kilometers of water in Prince William Sound, Alaska. (See Figure 10.6.) If the crude oil reaches the beaches and inlets, there can be considerable damage to beaches, shellfish, fish, birds, and wildlife habitats. Even today there is oil on the surface, or just beneath the surface, in some areas of Prince William Sound. Cleanup includes both mechanical and chemical methods. In one method, a nonpolar compound that is "oil-attracting" is used to pick up oil, which is then scraped off into recovery tanks.

Some Uses of Alkanes

The first four alkanes—methane, ethane, propane, and butane—are gases at room temperature and are widely used as heating fuels. Tanks of liquid propane and butane are used to provide fuels for heating homes and cooking on barbecues. As chlorofluorocarbons (CFCs) are phased out, butane has replaced Freon as a propellant in aerosol containers.

Alkanes having 5–8 carbon atoms (pentane, hexane, heptane, and octane) are liquids at room temperature. They are highly volatile, which makes them useful in fuels such as gasoline. Liquid alkanes with 9–17 carbon atoms have higher boiling points and are found in kerosene, diesel, and jet fuels. Motor oil is a mixture of high-molecular-weight liquid hydrocarbons and is used to lubricate the internal components of engines. Mineral oil is a mixture of liquid hydrocarbons and is used as a laxative and a lubricant. Alkanes with 18 or more carbon atoms are waxy solids at room temperature. The high-molecular-weight alkanes, known as paraffins, are used in waxy coatings of fruits and vegetables to retain moisture, inhibit mold growth, and enhance appearance. (See Figure 10.7.) Petrolatum, or Vaseline, is a mixture of liquid hydrocarbons, which have low boiling points, that are encapsulated in solid hydrocarbons. It is used in ointments and cosmetics and as a lubricant and a solvent.

Combustion

Carbon–carbon single bonds are difficult to break, which makes alkanes the least reactive family of organic compounds. However, alkanes burn readily in oxygen.

The growth of civilization was strongly influenced by the discovery of fire. Heat from burning wood was used to make pottery and glass, extract metals, make weapons, and forge tools. An alkane undergoes **combustion** when it reacts with oxygen to produce carbon dioxide, water, and energy.

$$Alkane + O_2 \longrightarrow CO_2 + H_2O + energy$$

Methane is the gas we use to cook our foods and heat our homes. Propane is the gas used in portable heaters and gas barbecues. (See Figure 10.8.) Gasoline, a mixture of liquid hydrocarbons, is the fuel that powers our cars, lawn mowers, and snow blowers. As alkanes, they all undergo combustion. The equations for the combustion of methane (CH_4) and propane (C_3H_8) follow:

$$CH_4 + 2O_2 \longrightarrow CO_2 + 2H_2O + energy$$
$$C_3H_8 + 5O_2 \longrightarrow 3CO_2 + 4H_2O + energy$$

In the cells of our bodies, energy is produced by the combustion of glucose. Although a series of reactions is involved, we can write the overall combustion of glucose in our cells as follows:

$$C_6H_{12}O_6 + 6O_2 \xrightarrow{\text{enzymes}} 6CO_2 + 6H_2O + energy$$

FIGURE 10.6

In oil spills, large quantities of oil spread over the water.

Q **What physical properties cause oil to remain on the surface of water?**

FIGURE 10.7

The solid alkanes that make up waxy coatings on fruits and vegetables help retain moisture, inhibit mold, and enhance appearance.

Q **Why does the waxy coating help the fruits and vegetables retain moisture?**

■ Combustion

Write a balanced equation for the complete combustion of butane.

SOLUTION

The balanced equation for the complete combustion of butane can be written

$$2C_4H_{10} + 13O_2 \longrightarrow 8CO_2 + 10H_2O$$

STUDY CHECK

Write a balanced equation for the complete combustion of the following:

$$\begin{array}{c} CH_3 \\ | \\ CH_3-CH-CH_2-CH_3 \end{array}$$

FIGURE 10.8

The propane fuel in the tank undergoes combustion, which provides energy.

Q **What is the balanced equation for the combustion of propane?**

QUESTIONS AND PROBLEMS

■ Properties of Alkanes

10.27 Heptane, C_7H_{16}, has a density of 0.68 g/mL and boils at 98°C.
   a. What is the structural formula of heptane?
   b. Is it a solid, liquid, or gas at room temperature?
   c. Is it soluble in water?
   d. Will it float or sink in water?

10.28 Nonane, C_9H_{20}, has a density of 0.79 g/mL and boils at 151°C.
   a. What is the structural formula of nonane?
   b. Is it a solid, liquid, or gas at room temperature?
   c. Is it soluble in water?
   d. Will it float or sink in water?

10.29 Write a balanced equation for the complete combustion of each of the following compounds:
   a. ethane, C_2H_6 **b.** cyclopropane, C_3H_6
   c. octane   **d.** cyclohexane

10.30 Write a balanced equation for the complete combustion of each of the following compounds:
   a. hexane, C_6H_{14} **b.** cyclopentane, C_5H_{10}
   c. nonane   **d.** 2-methylbutane

HEALTH NOTE

■ Incomplete Combustion

You may already know that it is dangerous to burn natural gas, oil, or wood in a closed room where ventilation and fresh air are not adequate. A gas heater, fireplace, or wood stove must have proper ventilation. If the supply of oxygen is limited, incomplete combustion produces carbon monoxide. The incomplete combustion of methane in natural gas is written as:

$$2CH_4(g) + 3O_2(g) \longrightarrow 2CO(g) + 4H_2O(g) + heat$$

Limited oxygen supply   Carbon monoxide

Carbon monoxide (CO) is a colorless, odorless, poisonous gas. When inhaled, CO passes into the bloodstream, where it attaches to hemoglobin (as you saw in the Health Note on Oxygen, Hemoglobin, and Carbon Monoxide Poisoning in Chapter 5). When CO binds to the hemoglobin, it reduces the amount of oxygen (O_2) reaching the organs and cells. As a result, a healthy person can experience a reduction in exercise capability, visual perception, and manual dexterity.

When the amount of hemoglobin bound to CO (COHb) is 10% or less, a person may experience shortness of breath, mild headache, and drowsiness, which are symptoms that may be mistaken for the flu. Heavy smokers can have as high as 9% COHb in their blood. When as much as 30% of the hemoglobin is bound to CO, a person may experience more severe symptoms including dizziness, mental confusion, severe headache, and nausea. If 50% or more of the hemoglobin is bound to CO, a person could become unconscious and die if not treated immediately with oxygen.

Crude Oil

Crude oil or petroleum contains a wide variety of hydrocarbons. At an oil refinery, the components in crude oil are separated by fractional distillation, a process that removes groups or fractions of hydrocarbons by continually heating the mixture to higher temperatures. (See Table 10.7.) Fractions containing alkanes with longer carbon chains require higher temperatures before they reach their boiling temperature and form gases. The gases are removed and passed through a distillation column where they cool and condense back to liquids. The major use of crude oil is to obtain gasoline, but a barrel of crude oil is only about 35% gasoline. To increase the production of gasoline, heating oils are broken down to give the lower weight alkanes.

TABLE 10.7

Typical Alkane Mixtures Obtained by Distillation of Crude Oil

Distillation Temperatures (°C)	Number of Carbon Atoms	Product
Below 30	1–4	Natural gas
30–200	5–12	Gasoline
200–250	12–16	Kerosene, jet fuel
250–350	15–18	Diesel fuel, heating oil
350–450	18–25	Lubricating oil
Nonvolatile residue	Over 25	Asphalt, tar

Crude oil Distillation tower

T

Natural gas

< 30°C

Crude oil

30–200°C — Gasoline

200–250°C — Kerosene, jet fuel

250–350°C — Diesel fuel, heating oil

350–450°C — Lubricating oil

Heating burner

Asphalt, tar

Nonvolatile residue

■ 10.5 Functional Groups

In organic compounds, carbon atoms are most likely to bond with hydrogen, oxygen, nitrogen, sulfur, and halogens such as chlorine. Hydrogen with one valence electron forms a single convalent bond. An octet is achieved by nitrogen forming three covalent bonds and oxygen and sulfur each forming two covalent bonds. The halogens with seven valence electrons form one covalent bond. Table 10.8 lists the number of covalent bonds most often formed by elements found in organic compounds.

Organic compounds number in the millions and more are synthesized every day. It might seem that the task of learning organic chemistry would be overwhelming. However, within this vast number of compounds, there are characteristic structural features called **functional groups**, which are a certain group of atoms that react in a predictable way. Compounds with the same functional group undergo similar chemical reactions. The identification of functional groups allows us to classify organic compounds according to their structure and to name compounds within each family. For now we will focus on recognizing the patterns of atoms that make up the functional groups.

■ Alkenes, Alkynes, and Aromatic Compounds

Earlier in section 10.2, we learned that alkanes contain only carbon–carbon single bonds. The **alkenes** contain a functional group that is a double bond between two adjacent carbon atoms; **alkynes** contain a triple bond. **Aromatic compounds** contain benzene, a molecule that has a ring of six carbon atoms with one hydrogen atom attached to each carbon. The benzene structure is represented as a hexagon with a circle in the center.

	An alkene	An alkyne	An aromatic
Functional group	$-\overset{\displaystyle |}{C}=\overset{\displaystyle |}{C}-$	$-C\equiv C-$	
Condensed structural formula	$CH_2=CH_2$	$HC\equiv CH$	

TABLE 10.8

Covalent Bonds for Elements in Organic Compounds			
Element	Group	Covalent Bonds	Structure of Atoms
H	1A (1)	1	H—
C	4A (14)	4	$-\overset{\displaystyle |}{\underset{\displaystyle |}{C}}-$
N	5A (15)	3	$-\overset{\displaystyle |}{\underset{\displaystyle \cdot\cdot}{N}}-$
O, S	6A (16)	2	$-\overset{\cdot\cdot}{\underset{\cdot\cdot}{O}}-$ $-\overset{\cdot\cdot}{\underset{\cdot\cdot}{S}}-$
F, Cl, Br, I	7A (17)	1	$-\overset{\cdot\cdot}{\underset{\cdot\cdot}{X}}:$ (X = F, Cl, Br, I)

Alcohols and Ethers

The characteristic functional group in **alcohols** is the **hydroxyl (—OH) group** bonded to a carbon atom. In **ethers**, the characteristic structural feature is an oxygen atom bonded to two carbon atoms. The oxygen atom also has two unshared pairs of electrons, but they are not shown in the structural formulas.

$$CH_3-CH_2-\textbf{OH} \qquad CH_3-\textbf{O}-CH_3$$
An alcohol   An ether

Functional group $-O-H$ $-O-$

Aldehydes and Ketones

The aldehydes and ketones contain a **carbonyl group** (C=O), which is a carbon with a double bond to oxygen. In an **aldehyde**, the carbon atom of the carbonyl group is bonded to another carbon and one hydrogen atom. Only the simplest aldehyde, CH_2O, has a carbonyl group attached to two hydrogen atoms. In a **ketone**, the carbonyl group is bonded to two other carbon atoms.

An aldehyde   A ketone

Functional group

▪ Classifying Organic Compounds

Classify the following organic compounds according to their functional groups:

a. $CH_3-CH_2-CH_2-OH$ **b.** $CH_3-CH=CH-CH_3$

c. $CH_3-CH_2-\overset{\overset{\textstyle O}{\|}}{C}-CH_2-CH_3$

SOLUTION

a. alcohol   **b.** alkene   **c.** ketone

STUDY CHECK

Why is $CH_3-CH_2-O-CH_3$ an ether, but $CH_3-\overset{\overset{\textstyle OH}{|}}{CH}-CH_3$ is an alcohol?

Carboxylic Acids and Esters

In **carboxylic acids**, the functional group is the *carboxyl group*, which is a combination of the *carb*onyl and hydro*xyl* groups.

An alkene

An alkyne

An aromatic

An alcohol

An ether

An aldehyde

A ketone

A carboxylic acid

An ester

An amine

An amide

$$CH_3-\overset{\overset{\textstyle O}{\|}}{C}-O-H \quad \text{or} \quad CH_3COOH \quad \text{or} \quad CH_3CO_2H$$

A carboxylic acid

Functional group $\quad -\overset{\overset{\textstyle O}{\|}}{C}-O-H \quad$ or $\quad -COOH \quad$ or $\quad -CO_2H$

An **ester** is similar to a carboxylic acid, except the oxygen of the carboxyl group is attached to a carbon and not to hydrogen.

$$CH_3-\overset{\overset{\textstyle O}{\|}}{C}-O-CH_3 \quad \text{or} \quad CH_3COOCH_3 \quad \text{or} \quad CH_3CO_2CH_3$$

An ester

Functional group $\quad -\overset{\overset{\textstyle O}{\|}}{C}-O- \quad$ or $\quad -COO- \quad$ or $\quad -CO_2-$

Amines

In **amines**, the central atom is a nitrogen atom. Amines are derivatives of ammonia, NH_3, in which carbon atoms replace one, two, or three of the hydrogen atoms.

$$NH_3 \qquad CH_3-NH_2 \qquad CH_3-\underset{\underset{\textstyle CH_3}{|}}{NH} \qquad CH_3-\underset{\underset{\textstyle CH_3}{|}}{N}-CH_3$$

Ammonia   Examples of amines

In an **amide**, the hydroxyl group of a carboxylic acid is replaced by a nitrogen group.

$$CH_3-\overset{\overset{\textstyle O}{\|}}{C}-NH_2$$

An amide

A list of the common functional groups in organic compounds is shown in Table 10.9.

SAMPLE PROBLEM 10.10

■ **Identifying Functional Groups**

Classify the following organic compounds according to their functional groups:

a. $CH_3-CH_2-NH-CH_3$

b. $CH_3CO_2CH_2CH_3$

c. $CH_3-CH_2-\overset{\overset{\textstyle O}{\|}}{C}-OH$

SOLUTION

a. amine   **b.** ester   **c.** carboxylic acid

STUDY CHECK

How does a carboxylic acid differ from an ester?

TABLE 10.9

Classification of Organic Compounds

Class	Example	Functional Group	Characteristic
Alkene	$H_2C{=}CH_2$	$\diagdown C{=}C \diagup$	Carbon–carbon double bond
Alkyne	$HC{\equiv}CH$	$-C{\equiv}C-$	Carbon–carbon triple bond
Aromatic	(benzene ring with 6 H)	(benzene ring)	Six-atom carbon ring with six hydrogen atoms
Haloalkane	CH_3-Cl	$-F,-Cl,-Br,-I$	One or more halogen atoms
Alcohol	CH_3-CH_2-OH	$-OH$	Hydroxyl group ($-OH$)
Ether	CH_3-O-CH_3	$-O-$	Oxygen atom bonded to two carbons
Thiol	CH_3-SH	$-SH$	A $-SH$ group bonded to carbon
Aldehyde	$CH_3-\overset{\overset{\displaystyle O}{\|}}{C}-H$	$-\overset{\overset{\displaystyle O}{\|}}{C}-H$	Carbonyl group (carbon–oxygen double bond) with $-H$
Ketone	$CH_3-\overset{\overset{\displaystyle O}{\|}}{C}-CH_3$	$-\overset{\overset{\displaystyle O}{\|}}{C}-$	Carbonyl group (carbon–oxygen double bond) between carbon atoms
Carboxylic acid	$CH_3-\overset{\overset{\displaystyle O}{\|}}{C}-O-H$	$-\overset{\overset{\displaystyle O}{\|}}{C}-O-H$	Carboxyl group (carbon–oxygen double bond and $-OH$)
Ester	$CH_3-\overset{\overset{\displaystyle O}{\|}}{C}-O-CH_3$	$-\overset{\overset{\displaystyle O}{\|}}{C}-O-$	Carboxyl group with $-H$ replaced by a carbon
Amine	CH_3-NH_2	$-\overset{\displaystyle \|}{N}-$	Nitrogen atom with one or more carbon groups
Amide	$CH_3-\overset{\overset{\displaystyle O}{\|}}{C}-NH_2$	$-\overset{\overset{\displaystyle O}{\|}}{C}-\overset{\displaystyle \|}{N}-$	Carbonyl group bonded to nitrogen

QUESTIONS AND PROBLEMS

■ Functional Groups

10.31 Identify the class of compounds that contains each of the following functional groups.
   a. hydroxyl group attached to a carbon chain
   b. carbon–carbon double bond
   c. carbonyl group attached to a hydrogen atom
   d. carboxyl group attached to two carbon atoms

10.32 Identify the class of compounds that contains each of the following functional groups:
   a. a nitrogen atom attached to one or more carbon atoms
   b. carboxyl group
   c. oxygen atom bonded to two carbon atoms
   d. a carbonyl group between two carbon atoms

10.33 Classify the following molecules according to their functional groups. The possibilities are alcohol, ether, ketone, carboxylic acid, or amine:
   a. $CH_3-CH_2-O-CH_2-CH_3$
   b. $CH_3-\overset{\overset{\displaystyle OH}{\|}}{CH}-CH_3$ **c.** $CH_3-\overset{\overset{\displaystyle O}{\|}}{C}-CH_2-CH_3$

ENVIRONMENTAL NOTE

■ Functional Groups in Familiar Compounds

The flavors and odors of foods and many household products can be attributed to the functional groups of organic compounds. As we discuss these familiar products, look for the functional groups we have described.

Ethyl alcohol is the alcohol found in alcoholic beverages. Isopropyl alcohol is another alcohol commonly used to disinfect skin before giving injections and to treat cuts.

$$CH_3—CH_2—OH \qquad CH_3—\overset{\displaystyle OH}{\underset{|}{CH}}—CH_3$$
ethyl alcohol   isopropyl alcohol

Ketones and aldehydes are in many items we use or eat each day. Acetone or dimethyl ketone is produced in great amounts commercially. Acetone is used as an organic solvent because it dissolves a wide variety of organic substances. You may be familiar with acetone as fingernail polish remover. Ketones and aldehydes used in the food industry are found in flavorings such as vanilla, cinnamon, and spearmint. When we buy a small bottle of liquid flavorings, the aldehyde or ketone is dissolved in alcohol because the compounds are not very soluble in water. The aldehyde butyraldehyde adds a "buttery" taste to foods and margarine.

$$CH_3—CH_2—CH_2—\overset{\displaystyle O}{\overset{||}{C}}—H$$
butyraldehyde "butter" flavoring

The sour tastes of vinegar and fruit juices and the pain from ant stings are all due to carboxylic acids. Acetic acid is the carboxylic acid that makes up vinegar. Aspirin also contains a carboxylic acid group. Esters found in fruits produce the pleasant aromas and tastes of bananas, oranges, pears, and pineapples. Esters are also used as solvents in many household cleaners, polishes, and glues.

One of the characteristics of fish is their odor, which is due to amines. Amines produced when proteins decay have a particularly pungent and offensive odor.

$$H_2N—CH_2—CH_2—CH_2—CH_2—NH_2$$
putrescine

$$CH_3—\overset{\displaystyle O}{\overset{||}{C}}—OH$$
Acetic acid (in vinegar)

$$CH_3—\overset{\displaystyle O}{\overset{||}{C}}—O—CH_2—CH_2—CH_3$$
Propyl acetate (pears)

$$CH_3 — NH_2$$
Methyl amine

$$CH_3 — \overset{\displaystyle O}{\overset{||}{C}} — O—CH_2 — CH_2 — CH_2 — CH_2 — CH_3$$
Pentyl acetate (bananas)

$$H_2N—CH_2—CH_2—CH_2—CH_2—CH_2—NH_2$$
cadaverine

Alkaloids are biologically active amines synthesized by plants to ward off insects and animals. Some typical alkaloids include caffeine, nicotine, histamine, and the decongestant epinephrine. Many are painkillers and hallucinogens such as morphine, LSD, marijuana, and cocaine. Certain parts of our neurons have receptor sites that respond to the various alkaloids. By modifying the structures of certain alkaloids to eliminate side effects, chemists have synthesized painkillers and drugs such as Novocain, codeine, and Valium.

d. $CH_3—CH_2—CH_2—COOH$

e. $CH_3—CH_2—NH_2$

10.34 Classify the following molecules according to their functional groups. The possibilities are alkene, aldehyde, carboxylic acid, ester, or amide:

a. $CH_3CH_2COOCH_2CH_3$

b. $CH_3—\overset{\displaystyle O}{\overset{||}{C}}—NH_2$

c. $CH_3—CH_2—CH_2—\overset{\displaystyle O}{\overset{||}{C}}—H$

d. $CH_3—CH_2—CH_2—COOH$

e. $CH_3—CH{=}CH—CH_3$

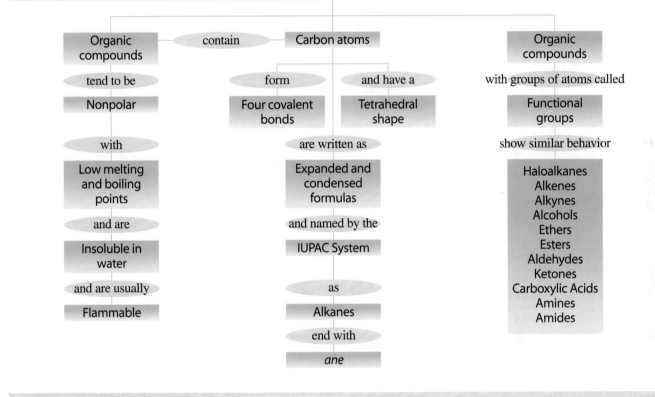

CHAPTER REVIEW

10.1 Organic Compounds

Most organic compounds have covalent bonds and form nonpolar molecules. Often they have low melting points and low boiling points, are not very soluble in water, produce molecules in solutions, and burn vigorously in air. In contrast, many inorganic compounds are ionic or contain polar covalent bonds and form polar molecules. Many have high melting and boiling points, are usually soluble in water, produce ions in water, and do not burn in air. Carbon atoms share four valence electrons to form four covalent bonds. In the simplest organic molecule, methane, CH_4, the four bonds that bond hydrogen to the carbon atom are directed out to the corners of a tetrahedron with bond angles of $109.5°$.

10.2 Alkanes

Alkanes are hydrocarbons that have only C—C single bonds. In the expanded structural formula, a separate line is drawn for every bonded atom. A condensed structural formula depicts groups composed of each carbon atom and its attached hydrogen atoms. In cycloalkanes, the carbon atoms form a ring or cyclic structure. The name is written by placing the prefix *cyclo* before the alkane name with the same number of carbon atoms.

The IUPAC system is used to name organic compounds in a systematic manner. The IUPAC name indicates the number of carbon atoms.

10.3 Alkanes with Substituents

For a continuous alkane, the carbon atoms are connected in a chain and bonded to hydrogen atoms. Substituents such as alkyl groups can replace hydrogen atoms on an alkane. A haloalkane contains one or more F, Cl, Br, or I atoms. In the IUPAC system, halogen atoms are named as fluoro, chloro, bromo, or iodo substituents attached to the main chain. In the common name, alkyl halides, the name of the alkyl group precedes the halide, for example methyl chloride.

10.4 Properties of Alkanes

As nonpolar molecules, alkanes are not soluble in water. They are less dense than water. With only weak attractions, they have low melting and boiling points. Although the C—C bonds in alkanes resist most reactions, alkanes undergo combustion. In combustion, or burning, alkanes react with oxygen to produce carbon dioxide and water.

10.5 Functional Groups

An organic molecule contains a characteristic group of atoms called a functional group that determines the molecule's family name and chemical reactivity. Functional groups are used to classify organic compounds, act as reactive sites in the molecule, and provide a system of naming for organic compounds. Some common functional groups include the hydroxyl group (—OH) in alcohols, the carbonyl group (C=O) in aldehydes and ketones, and a nitrogen atom $\left(-\overset{|}{\underset{|}{N}}-\right)$ in amines, and a nitrogen and carbonyl group in amides.

SUMMARY OF NAMING

Type	Example	Characteristic	Structure
Alkane	Propane	single C—C, C—H bonds	$CH_3-CH_2-CH_3$
	Methylpropane		$CH_3-\overset{\overset{\textstyle CH_3}{\textstyle |}}{CH}-CH_3$
Haloalkane	1-Chloropropane	halogen atom	$CH_3-CH_2-CH_2-Cl$
Cycloalkane	Cyclobutane	carbon ring	□

SUMMARY OF REACTIONS

COMBUSTION

$$Alkane + O_2 \longrightarrow CO_2 + H_2O + energy$$

KEY TERMS

alcohols A class of organic compounds that contains the hydroxyl (—OH) group bonded to a carbon atom.

aldehydes A class of organic compounds that contains a carbonyl group (C=O) bonded to at least one hydrogen atom.

alkanes Hydrocarbons containing only single bonds between carbon atoms.

alkenes Hydrocarbons that contain carbon–carbon double bonds (C=C).

alkyl group An alkane minus one hydrogen atom. Alkyl groups are named like the alkanes except a *yl* ending replaces *ane*.

alkynes Hydrocarbons that contain carbon–carbon triple bonds (C≡C).

amide A class of organic compounds in which the hydroxyl group of a carboxylic acid is replaced by a nitrogen group.

amines A class of organic compounds that contains a nitrogen atom bonded to one or more carbon atoms.

aromatic A compound that contains benzene. Benzene has a six-carbon ring with a hydrogen attached to each carbon.

branch A carbon group or halogen bonded to the main carbon chain.

branched alkane A hydrocarbon containing a substituent bonded to the main chain.

carbonyl group A functional group that contains a double bond between a carbon atom and an oxygen atom (C=O).

carboxylic acids A class of organic compounds that contains the functional group (—COOH).

combustion A chemical reaction in which an alkane reacts with oxygen to produce CO_2, H_2O, and energy.

condensed structural formula A structural formula that shows the arrangement of the carbon atoms in a molecule but groups each carbon atom with its bonded hydrogen atoms (CH_3, CH_2, or CH).

continuous alkane An alkane in which the carbon atoms are connected in a row, one after the other.

cycloalkane An alkane that is a ring or cyclic structure.

esters A class of organic compounds that contains a —COO— group with an oxygen atom bonded to carbon.

ethers A class of organic compounds that contains an oxygen atom bonded to two carbon atoms.

expanded structural formula A type of structural formula that shows the arrangement of the atoms by drawing each bond in the hydrocarbon as C—H or C—C.

functional group A group of atoms that determine the physical and chemical properties and naming of a class of organic compounds.

haloalkane A type of alkane that contains one or more halogen atoms.

hydrocarbons Organic compounds consisting of only carbon and hydrogen.

hydroxyl group The group of atoms (—OH) characteristic of alcohols.

isomers Organic compounds in which identical molecular formulas have different arrangements of atoms.

IUPAC system An organization known as International Union of Pure and Applied Chemistry that determines the system for naming organic compounds.

ketones A class of organic compounds in which a carbonyl group is bonded to two carbon atoms.

organic compounds Compounds made of carbon that typically have covalent bonds, nonpolar molecules, low melting and boiling points, are insoluble in water, and flammable.

substituent Groups of atoms such as an alkyl group or a halogen bonded to the main chain or ring of carbon atoms.

ADDITIONAL QUESTIONS AND PROBLEMS

10.35 Compare organic and inorganic compounds in terms of:
a. types of bonds   b. solubility in water
c. melting points   d. flammability

10.36 Identify each of the following compounds as organic or inorganic:
a. Na_2SO_4 b. $CH_2\!\!=\!\!CH_2$
c. Cr_2O_3 d. $C_{12}H_{22}O_{11}$

10.37 Match the following physical and chemical properties with the compounds butane, C_4H_{10}, or potassium chloride, KCl.
a. melts at $-138°C$
b. burns vigorously in air
c. melts at $770°C$
d. produces ions in water
e. is a gas at room temperature

10.38 Match the following physical and chemical properties with the compounds cyclohexane, C_6H_{12}, or calcium nitrate, $Ca(NO_3)_2$.
a. contains only covalent bonds
b. melts above $500°C$
c. insoluble in water
d. liquid at room temperature
e. produces ions in water

10.39 Complete the following structures by adding the correct number of hydrogen atoms:
a. C—C   b. C—C—O

c. C—C—C (with O double-bonded to middle C)   d. C—O—C—C

10.40 Identify the functional groups in each.
a. $CH_3—NH_2$
b. $CH_3—\overset{\overset{\displaystyle O}{\|}}{C}—CH_3$
c. $CH_3—\overset{\overset{\displaystyle OH}{|}}{CH}—CH_3$
d. $CH_3—C\!\!\equiv\!\!CH$
e. $CH_3—O—CH_2—CH_3$
f. ⬡—CH_3

10.41 Write the name of each of the following alkyl groups:
a. $CH_3—$
b. $CH_3—CH_2—CH_2—$
c. $CH_3—\overset{\overset{\displaystyle CH_3}{|}}{CH}—$
d. $CH_3—CH_2—$

10.42 Give the IUPAC names for each of the following molecules:
a. $CH_3—CH_2—\overset{\overset{\displaystyle CH_3}{|}}{\underset{\underset{\displaystyle CH_3}{|}}{C}}—CH_3$

b. $CH_3—CH_2—Cl$

c. $CH_3—CH_2—CH—CH_2—CH—CH_3$ (with $CH_3—CH_2$ branch and Br branch)
d. ⬡

10.43 Write the condensed structural formula for each of the following molecules:
a. 3-ethylhexane
b. 2,3-dimethylpentane
c. 1,3-dichloro-3-methylheptane

10.44 Write the condensed structural formula for each of the following molecules:
a. ethylcyclopropane
b. methylcyclohexane
c. isopropylcyclopentane

10.45 In an automobile engine, "knocking" occurs when the combustion of gasoline occurs too rapidly. The octane number of gasoline represents the ability of a gasoline mixture to reduce knocking. A sample of gasoline is compared with heptane, rated 0 because it reacts with severe knocking, and 2,2,4-trimethylpentane, which has a rating of 100 because of its low knocking. Write the condensed structural formula, molecular formula, and equation for the complete combustion of 2,2,4-trimethylpentane.

10.46 Draw the structures of the following halogenated compounds, which are used as refrigerants.
a. Freon 14, tetrafluoromethane
b. Freon 114, 1,2-dichloro-1,1,2,2-tetrafluoroethane
c. Freon C318, octafluorocyclobutane

10.47 Write a balanced equation for the complete combustion of each of the following:
a. propane   b. C_5H_{12}
c. cyclobutane   d. octane

10.48 Write a balanced equation for the complete combustion of each of the following:
a. hexane   b. methylcyclohexane
c. cyclopentane   d. 2-methylpropane

10.49 For each of the definitions, find a corresponding term from the following list: alkane, alkene, alkyne, alcohol, ether, aldehyde, ketone, carboxylic acid, ester, amine, functional group, tetrahedral, isomers
a. An organic compound that contains a hydroxyl group bonded to a carbon.
b. A hydrocarbon that contains one or more carbon–carbon double bonds.
c. An organic compound in which the carbon of a carbonyl group is bonded to a hydrogen.
d. A hydrocarbon that contains only carbon–carbon single bonds.
e. An organic compound in which the carbon of a carbonyl group is bonded to a hydroxyl group.
f. An organic compound that contains a nitrogen atom bonded to one or more carbon atoms.

g. The three-dimensional shape of a carbon bonded to four hydrogen atoms.

10.50 For each of the definitions, find a corresponding term from the following list: alkane, alkene, alkyne, alcohol, ether, aldehyde, ketone, carboxylic acid, ester, amine, functional group, tetrahedral, isomers
a. Organic compounds with identical molecular formulas that differ in the order the atoms are connected.

b. An organic compound in which the hydrogen atom of a carboxyl group is replaced by a carbon atom.
c. An organic compound that contains an oxygen atom bonded to two carbon atoms.
d. A hydrocarbon that contains a carbon–carbon triple bond.
e. A charactistic group of atoms that make compounds behave and react in a particular way.
f. An organic compound in which the carbonyl group is bonded to two carbon atoms.

UNDERSTANDING THE CONCEPTS

10.51 Suncreens contain compounds that absorb UV light such as oxybenzone and 2-ethylhexyl p-methoxycinnamate.

Identify the functional groups in each of the following UV-absorbing compounds used in suncreens:
a. oxybenzone

b. 2-ethylhexyl p-methoxycinnamate

10.52 Oxymetazoline is a vasoconstrictor used in nasal decongestant sprays such as Afrin.

What functional groups are in oxymetazoline?

10.53 Decimemide is used as an anticonvulsant.

What functional groups are in decimemide?

10.54 The odor and taste of pineapples is from ethyl butyrate.

What functional group is in ethyl butyrate?

$$CH_3-CH_2-CH_2-\overset{\overset{\displaystyle O}{\|}}{C}-O-CH_2-CH_3$$

QUESTIONS FOR TEAMWORK

10.55 The density of pentane, a component of gasoline, is 0.63 g/mL. The heat of combustion for pentane is 845 kcal per mole.
  a. Write an equation for the complete combustion of pentane.
  b. What is the molar mass?
  c. How much heat is produced when 1 gallon of pentane is burned (1 gallon = 3.78 liters)?
  d. How many liters of CO_2 at STP are produced from the complete combustion of 1 gallon of pentane?

10.56 Write structural formulas of two esters and a carboxylic acid that each have molecular formula $C_3H_6O_2$.

10.57 Draw all the possible structures of an organic compound with 6 carbon atoms that has a 4-carbon chain.

10.58 Draw all the possible structures of an organic compound with 4 carbon atoms that has a 3-carbon ring and a hydroxyl group.

10.59 Consider the compound propane.
  a. Draw the chemical structure.
  b. Write the equation for the complete combustion of propane.
  c. How many grams O_2 are needed to react with 12.0 L propane at STP?
  d. How many grams of CO_2 would be produced from the reaction in part **c.**?

10.60 Consider the compound ethylcyclopentane.
  a. Draw the chemical structure.
  b. Write the equation for the complete combustion of ethylcyclopentane.
  c. Calculate the grams O_2 required for the reaction of 25.0 g ethylcyclopentane.
  d. How many liters of CO_2 would be produced at STP from the reaction in part **c.**?

ANSWERS

Answers to Study Checks

10.1 Octane is not soluble in water; it is an organic compound.

10.2

$$CH_3-CH_2-CH_2-CH_2-CH_3$$

10.3 cyclobutane

10.4 This is another isomer. There is a five-carbon chain with a carbon group bonded to the middle (third) carbon.

10.5 1-chloro-2,4-dimethylhexane

10.6

10.7 1,2-dibromoethane

10.8

$$C_5H_{12} + 8O_2 \longrightarrow 5CO_2 + 6H_2O$$

10.9 $CH_3-CH_2-O-CH_3$ contains the functional group $C-O-C$: it is an ether.

contains the $-OH$ functional group; it is an alcohol.

10.10 A carboxylic acid has a carboxyl group COOH. In an ester, the oxygen atom of the carboxyl group is attached to a carbon atom, not hydrogen.

Answers to Selected Questions and Problems

10.1 **a.** inorganic   **b.** organic
       c. organic   **d.** inorganic
       e. inorganic   **f.** organic

10.3 **a.** inorganic   **b.** organic
       c. organic   **d.** inorganic

10.5 **a.** ethane   **b.** ethane
       c. NaBr   **d.** NaBr

10.7 VSEPR theory predicts that the four bonds in CH_4 will be as far apart as possible, which means that the hydrogen atoms are at the corners of a tetrahedron.

10.9 **a.** pentane   **b.** ethane   **c.** hexane

10.11 a. CH_4 **b.** CH_3-CH_3
       c. $CH_3-CH_2-CH_2-CH_2-CH_3$
       d.

10.13 a. same molecule
       b. isomers of C_5H_{12}
       c. isomers of C_6H_{14}

10.15 a. 2-methylbutane   **b.** 2,2-dimethylpropane
       c. 2,3-dimethylpentane   **d.** 4-ethyl-2,2-dimethylhexane

10.17 a. cyclobutane
       b. methylcyclopentane
       c. ethylcyclohexane

10.19 a.

b.

c.
$$\underset{\begin{array}{ccc}CH_3 & CH_3 & CH_3\end{array}}{CH_3-CH-CH-CH_2-CH-CH_3}$$

d.
$$\underset{\begin{array}{cccc}CH_3 & CH_2-CH_3 & CH_3\end{array}}{CH_3-CH-CH-CH_2-CH-CH_2-CH_2-CH_3}$$

10.21 a. CH₃ (cyclopentane structure)   **b.** (square / cyclobutane)

10.23 a. bromoethane   **b.** 1-fluoropropane
c. 2-chloropropane   **d.** trichloromethane

10.25 a.
$$\underset{}{\overset{Cl}{CH_3-CH-CH_3}}$$

b.
$$\overset{\;\;Br\;\;\;\;\;Cl}{CH_3-CH-CH-CH_3}$$

c. CH₃Br   **d.** CBr₄

10.27 a. CH₃—CH₂—CH₂—CH₂—CH₂—CH₂—CH₃
b. liquid
c. insoluble in water
d. float

10.29 a. $2C_2H_6 + 7O_2 \longrightarrow 4CO_2 + 6H_2O$
b. $2C_3H_6 + 9O_2 \longrightarrow 6CO_2 + 6H_2O$
c. $2C_8H_{18} + 25O_2 \longrightarrow 16CO_2 + 18H_2O$
d. $C_6H_{12} + 9O_2 \longrightarrow 6CO_2 + 6H_2O$

10.31 a. alcohol   **b.** alkene
c. aldehyde   **d.** ester

10.33 a. ether   **b.** alcohol   **c.** ketone
d. carboxylic acid   **e.** amine

10.35 a. Organic compounds have covalent bonds; inorganic compounds have ionic as well as polar covalent and a few have nonpolar covalent bonds.
b. Most organic compounds are insoluble in water; many inorganic compounds are soluble in water.
c. Most organic compounds have low melting points; inorganic compounds have high melting points.
d. Most organic compounds are flammable; inorganic compounds are not usually flammable.

10.37 a. butane   **b.** butane
c. potassium chloride   **d.** potassium chloride
e. butane

10.39 a.
$$\underset{\begin{array}{cc}H & H\end{array}}{\overset{\begin{array}{cc}H & H\end{array}}{H-C-C-H}}$$

b.

c.
$$\underset{\begin{array}{cc}H & H\end{array}}{\overset{\begin{array}{ccc}H & H & O\end{array}}{H-C-C-C-H}}$$

d.
$$\underset{\begin{array}{ccc}H & H & H\end{array}}{\overset{\begin{array}{ccc}H & H & H\end{array}}{H-C-O-C-C-H}}$$

10.41 a. methyl   **b.** propyl
c. isopropyl   **d.** ethyl

10.43 a.
$$\overset{CH_2-CH_3}{CH_3-CH_2-CH-CH_2-CH_2-CH_3}$$

b.
$$\overset{\begin{array}{cc}CH_3 & CH_3\end{array}}{CH_3-CH-CH-CH_2-CH_3}$$

c.
$$\underset{CH_3}{\overset{Cl}{Cl-CH_2-CH_2-C-CH_2-CH_2-CH_2-CH_3}}$$

10.45 Condensed structural formula
$$\underset{CH_3}{\overset{\begin{array}{cc}CH_3 & CH_3\end{array}}{CH_3-C-CH_2-CH-CH_3}}$$

molecular formula of C_8H_{18}

The combustion reaction: $2C_8H_{18} + 25O_2 \longrightarrow 16CO_2 + 18H_2O$

10.47 a. $C_3H_8 + 5O_2 \longrightarrow 3CO_2 + 4H_2O$
b. $C_5H_{12} + 8O_2 \longrightarrow 5CO_2 + 6H_2O$
c. $C_4H_8 + 6O_2 \longrightarrow 4CO_2 + 4H_2O$
d. $2C_8H_{18} + 25O_2 \longrightarrow 16CO_2 + 18H_2O$

10.49 a. alcohol   **b.** alkene
c. aldehyde   **d.** alkane
e. carboxylic acid   **f.** amine
g. tetrahedral

10.51 a. aromatic, ether, alcohol, ketone
b. aromatic, ether, alkene, ketone

10.53 aromatic, ether, amide

11 Unsaturated Hydrocarbons

"Dentures replace natural teeth that are extracted due to cavities, bad gums, or trauma," says Dr. Irene Hilton, dentist, La Clinica De La Raza. "I make an impression of teeth using alginate, which is a polysaccharide extracted from seaweed. I mix the compound with water and place the gel-like material in the patient's mouth, where it becomes a hard, cement-like substance. I fill this mold with gypsum ($CaSO_4$) and water, which form a solid to which I add teeth made of plastic or porcelain. When I get a good match to the patient's own teeth, I prepare a preliminary wax denture. This is placed in the patient's mouth to check the bite and adjust the position of the replacement teeth. Then a permanent denture is made using a hard plastic polymer (methyl methacrylate)."

the Chemistry place

Visit **www.aw-bc.com/chemplace** or use the CD-ROM in the book for extra quizzes, interactive tutorials, career resources, PowerPoint slides, math help, and case studies.

n Chapter 10, we looked at saturated hydrocarbons, which contain all single bonds between their carbon atoms. Now we will investigate unsaturated hydrocarbons, carbon compounds containing one or more carbon–carbon double bonds or triple bonds. When we cook with vegetable oils such as corn oil, safflower oil, or olive oil, we are using unsaturated lipids that have double bonds in their long carbon chains. Saturated fats from animal sources also have long chains of carbon atoms, but they are connected only by the single bonds of alkanes. If we compare the two types of fats, we find considerable differences in their physical and chemical properties. Vegetable oils are liquid at room temperature whereas animal fats are solid. Because double bonds are reactive, unsaturated fats can be oxidized by oxygen in the air, especially at warm temperatures, forming products that have rancid, unpleasant odors. The saturated fats are more resistant to reactions.

LEARNING GOAL

Identify structural formulas as alkenes, cycloalkenes, and alkynes and write their IUPAC or common names.

Ethene

Ethyne

11.1 Alkenes and Alkynes

Alkenes and alkynes are families of hydrocarbons that contain double and triple bonds respectively. They are called **unsaturated hydrocarbons** because they have fewer hydrogen atoms than alkanes.

Identifying Alkenes and Alkynes

An **alkene** contains at least one double bond between carbons. The double bond forms when two adjacent carbon atoms share two pairs of valence electrons. The simplest alkene is ethene, C_2H_4, which is more likely to be called by its common name ethylene. In ethene, each carbon atom is attached to two H atoms and the other carbon atom in the double bond. The resulting molecule has a flat geometry because the carbon and hydrogen atoms all lie in the same plane. (See Figure 11.1.)

In an **alkyne**, a triple bond occurs when two carbon atoms share three pairs of valence electrons. The simplest alkyne is called ethyne, but is commonly known as acetylene. In ethyne each carbon atom in the triple bond is attached to two other atoms. The most common is acetylene, which is used in welding where it burns at a very high temperature.

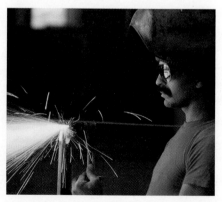

FIGURE 11.1

Ball-and-stick models of ethene and ethyne show the functional groups of double or triple bonds.

Q **Why are these compounds called unsaturated hydrocarbons?**

■ **Identifying Unsaturated Compounds**

Classify each of the following structural formulas as an alkane, alkene, or alkyne.

a. $CH_3-C{\equiv}C-CH_3$

b. $CH_3-CH_2-CH_3$

c.
$$CH_3-CH_2-\overset{\overset{\displaystyle CH_3}{|}}{C}{=}CH-CH_2-CH_3$$

SOLUTION

The structural formula with a double bond is an alkene, and the one with a triple bond is an alkyne.

a. alkyne   **b.** alkane   **c.** alkene

STUDY CHECK

Is the following structure a cycloalkane or cycloalkene?

Naming Alkenes and Alkynes

The IUPAC names for alkenes and alkynes are similar to those of alkanes. The simplest alkene, ethene (ethylene) is an important plant hormone involved in promoting the ripening of fruit. Commercially grown fruit, such as avocados, bananas, and tomatoes, are often picked before they are ripe. Before the fruit is brought to market, it is exposed to ethylene to accelerate the ripening process. Ethylene also accelerates the breakdown of cellulose in plants, which causes flowers to wilt and leaves to fall from trees.

The IUPAC name of the simplest alkyne is ethyne, although acetylene, its common name, is often used. See Table 11.1 for a comparison of the naming for alkanes, alkenes, and alkynes. For alkenes and alkynes, the longest carbon chain must contain the double or triple bond.

STEP 1 **Name the longest carbon chain that contains the double or triple bond.** Replace the corresponding alkane ending with *ene* for an alkene and *yne* for an alkyne.

STEP 2 **Number the longest chain from the end nearest the double or triple bond.** Indicate the position of the double or triple bond with the number of the first unsaturated carbon. Unsaturated compounds with two or three carbons do not need numbers.

$$\underset{4\quad\ \ 3\quad\ \ 2\quad\ \ 1}{CH_3-CH_2-CH{=}CH_2}\qquad\underset{1\quad\ \ 2\quad\ \ 3\quad\ \ 4}{CH_3-CH{=}CH-CH_3}\qquad\underset{1\quad\ \ 2\quad\ \ 3\quad\ \ 4}{CH_3-C{\equiv}C-CH_3}$$

1-butene   2-butene   2-butyne

TABLE 11.1

Comparison of Names for Alkanes, Alkenes, and Alkynes

Alkane	Alkene	Alkyne
H_3C-CH_3	$H_2C{=}CH_2$	$HC{\equiv}CH$
Ethane	Ethene (ethylene)	Ethyne (acetylene)
$CH_3-CH_2-CH_3$	$CH_3-CH{=}CH_2$	$CH_3-C{\equiv}CH$
Propane	Propene	Propyne

the **Chemistry** place

WEB TUTORIAL
Organic Molecules and Isomers
Introduction to Organic Molecules

EXPLORE YOUR WORLD

■ **Ripening Fruit**

Obtain two unripe green bananas. Place one in a plastic bag and seal it. Leave them on the counter. Check the bananas twice a day to observe any difference in the ripening process.

QUESTIONS

1. What compound helps ripen the bananas?
2. What are some possible reasons for any difference in the ripening rate?
3. If you wish to ripen an avocado, what procedure might you use?

STEP 3 Give the location and name of each substituent (alphabetical order) as a prefix to the alkene or alkyne name.

4-methyl-1-pentene   2,3-dimethyl-2-butene   3-chloro-1-butyne

---

SAMPLE PROBLEM 11.2

■ **Naming Alkenes and Alkynes**

Write the IUPAC name for each of the following:

$$\text{CH}_3$$
a. $\text{CH}_3\text{—CH—CH}\text{=}\text{CH—CH}_3$
b. $\text{CH}_3\text{—CH}_2\text{—C}\text{≡}\text{C—CH}_2\text{—CH}_3$

SOLUTION

Guide to Naming Alkenes and Alkynes

STEP 1
Name the longest carbon chain with a double or triple bond.

STEP 2
Number the carbon chain starting from the end nearest a double or triple bond.

STEP 3
Give the location and name of each substituent (alphabetical order) as a prefix to the name, if needed.

a. **STEP 1** **Name the longest carbon chain that contains the double or triple bond.** There are five carbon atoms in the longest carbon chain containing the double bond. Replacing the corresponding alkane ending with *ene* gives pentene.

STEP 2 **Number the longest chain from the end nearest the double or triple bond.** The number of the first carbon in the double bond is used to give the location of the double bond.

2-pentene

STEP 3 **Give the location and name of each substituent (alphabetical order) as a prefix to the alkene or alkyne name.** The methyl group is located on carbon 4.

$$\text{CH}_3$$
$\text{CH}_3\text{—CH—CH}\text{=}\text{CH—CH}_3$ 4-methyl-2-pentene
$\quad$ 5 $\quad$ 4 $\quad$ 3 $\quad$ 2 $\quad$ 1

b. **STEP 1** **Name the longest carbon chain that contains the double or triple bond.** There are six carbon atoms in the longest chain containing the triple bond. Replacing the corresponding alkane ending with *yne* gives hexyne.

STEP 2 **Number the main chain from the end nearest the double or triple bond.** The number of the first carbon in the triple bond is used to give the location of the double bond.

$\text{CH}_3\text{—CH}_2\text{—C}\text{≡}\text{C—CH}_2\text{—CH}_3$ 3-hexyne
$\quad$ 1 $\quad$ 2 $\quad$ 3 $\quad$ 4 $\quad$ 5 $\quad$ 6

STEP 3 **Give the location and name of each substituent (alphabetical order) as a prefix to the alkene or alkyne name.** There are no substituents in this formula.

STUDY CHECK

Draw the structural formulas for each of the following:

a. 2-pentyne   **b.** cyclopentene

QUESTIONS AND PROBLEMS

■ Alkenes and Alkynes

11.1 Identify the following as alkanes, alkenes, cycloalkenes, or alkynes:

b. $CH_3-CH_2-C{\equiv}C-H$

c.

11.2 Identify the following as alkanes, alkenes, cycloalkenes, or alkynes:

a.

b. $CH_3-\overset{\overset{\displaystyle CH_3}{|}}{C}{=}\underset{\underset{\displaystyle CH_3}{|}}{C}-CH_3$

c. $C{\equiv}CH$

11.3 Compare the structural formulas of propene and propyne.

11.4 Compare the structural formulas of 1-butyne and 2-butyne.

11.5 Give the IUPAC name for each of the following:

a. $CH_2{=}CH_2$ **b.** $CH_3-\overset{\overset{\displaystyle CH_3}{|}}{C}{=}CH_2$

c. $CH_3-CH_2-C{\equiv}C-CH_3$ **d.** ☐

11.6 Give the IUPAC name for each of the following:

a. $CH_2{=}CH-CH_2-CH_3$ **b.** $CH_3-C{\equiv}C-CH_2-CH_2-\overset{\overset{\displaystyle CH_3}{|}}{C}H-CH_3$

c. ⬡   **d.** $CH_3-CH_2-CH{=}CH-CH_3$

11.7 Draw the structural formula for each of the following compounds:
  a. propene   **b.** 1-pentene
  c. 2-methyl-1-butene   **d.** cyclohexene

11.8 Draw the structural formula for each of the following compounds:
  a. cyclopentene   **b.** 3-methyl-1-butyne
  c. 3,4-dimethyl-1-pentene   **d.** cyclobutene

■ 11.2 Cis–Trans Isomers

LEARNING GOAL

■ Write the structural formulas and names for cis-trans isomers of alkenes.

In alkenes, there is no rotation around the carbons in the rigid double bond. As a result, any groups connected to the double bond remain fixed on one side or the other. In a **cis isomer**, two similar groups are on the same side of the double bond.

■ Fragrant Alkenes

The odors you associate with lemons, oranges, roses, and lavender are due to volatile compounds that are synthesized by the plants. Often it is unsaturated compounds that are responsible for the pleasant flavors and fragrances of many fruits and flowers. They were some of the first kinds of compounds to be extracted from natural plant material. In ancient times, they were highly valued in their pure forms. Limonene and myrcene give the characteristic odors and flavors to lemons and oranges and bay leaves, respectively. Geraniol and citronellal give roses and lemon grass their distinct aromas. In the food and perfume industries, these compounds are extracted or synthesized and used as perfumes and flavorings.

$$CH_3-\underset{\underset{CH_3}{|}}{C}=CH-CH_2-CH_2-\underset{\underset{CH_3}{|}}{CH}-CH_2-CH_2OH$$

Geraniol, roses

$$CH_3-\underset{\underset{CH_3}{|}}{C}=CH-CH_2-CH_2-\underset{\underset{CH_2}{||}}{C}-CH=CH_2$$

Myrcene, bay leaves

$$CH_3-\underset{\underset{CH_3}{|}}{C}=CH-CH_2-CH_2-\underset{\underset{CH_3}{|}}{C}=CH-CHO$$

Citronellal, lemon grass

Limonene, lemons, and oranges

cis-2-butene

trans-2-butene

FIGURE 11.2

Ball-and-stick models of the cis and trans isomers of 2-butene.

Q What feature in 2-butene accounts for the cis and trans isomers?

In the **trans isomer**, the similar groups are on opposite sides of the double bond. For example, cis–trans isomers can be written for 2-butene. (See Figure 11.2.) In general, trans isomers are more stable than their cis counterparts because the large groups attached to the double bond are further apart. The cis–trans isomers of 2-butene are different compounds with different physical and chemical properties:

cis-2-butene
(mp – 139°C; bp 3.7°C)

trans-2-butene
(mp – 106°C; bp 0.3°C)

Not every alkene shows cis–trans isomerism. If one of the carbons in the double bond is attached to identical groups, the molecule does not have cis–trans isomers. This is the case of 1-butene and 2-methyl-1-propene, isomers of 2-butene. If the hydrogen atoms are interchanged on carbon 1, the same structure results. Alkynes do not have cis–trans isomers because the carbons in the triple bond are each attached to only one group.

1-butene

2-methyl-1-propene

As long as the groups attached to the double bond are different, an alkene will show cis–trans isomers. Another example of cis–trans isomers is the following:

cis-3-hexene   trans-3-hexene

the
Chemistry
place
WEB TUTORIAL
Geometric Isomers

SAMPLE PROBLEM 11.3

▪ Identifying Cis–Trans Isomers

Identify the following as cis or trans isomers.

SOLUTION

a. This isomer is a cis isomer because the Br and Cl atoms are on the same side of the double bond.

b. This isomer is a trans isomer because the CH_3— and —CH_2—CH_3 groups are on the opposite sides of the double bond.

STUDY CHECK

Is the following compound *cis*-3-hexene or *trans*-3-hexene?

QUESTIONS AND PROBLEMS

▪ Cis–Trans Isomers

11.9 Write the IUPAC name of each of the following using cis or trans prefixes:

11.10 Write the IUPAC name of each of the following using cis or trans prefixes:

▪ Modeling Cis–Trans Isomers

Because cis–trans isomerism is not easy to imagine, here are some things you can do to understand the difference in rotation around a single bond compared to a double bond and how it affects groups that are attached to the carbon atoms in the double bond.

Put the fingertips of your index fingers together. This is a model of a single bond. Consider the index fingers as a pair of carbon atoms and think of your thumbs and other fingers as other parts of a carbon chain. While your index fingers are touching, twist your hands and change the position of the thumbs relative to each other. Notice how the relationship of your other fingers changes.

Now place the tips of your index fingers and middle fingers together in a model of a double bond. As you did before, twist your hands to move the thumbs apart. What happens? Can you change the location of your thumbs relative to each other without breaking the double bond? The difficulty of moving your hands with two fingers touching represents the lack of rotation about a double bond. You have made a model of a cis isomer when both thumbs are on the same side. If you turn one hand over so one thumb points down and one points up, you have made a model of a trans isomer.

Cis-hands (*cis*-thumbs/fingers)

Trans-hands (*trans*-thumbs/fingers)

Pheromones in Insect Communication

Insects and many other organisms emit minute quantities of chemicals called pheromones. Insects use pheromones to send messages to individuals of the same species. Some pheromones warn of danger, others call for defense, mark a trail, or attract the opposite sex. In the last 40 years, the structures of many pheromones have been chemically determined. One of the most studied is bombykol, the sex pheromone produced by the female of the silkworm moth species. The bombykol molecule is a 16 carbon chain with one cis double bond, one trans double bond, and an alcohol group. A few molecules of synthetic bombykol will attract male silkworm moths from distances of over one kilometer. The effectiveness of many of these pheromones depends on the cis or trans configuration of the double bonds in the molecules. A certain species will respond to one isomer but not the other.

Scientists are interested in synthesizing pheromones for use as nontoxic alternatives to pesticides. When used in a trap, bombykol can be used to isolate male silkworm moths. When a synthetic pheromone is released in several areas of a field or crop, the males cannot locate the females, which disrupts the reproductive cycle. This technique has been successful with controlling the oriental fruit moth, the grapevine moth, and the pink bollworm.

Bombykol, sex attractant for the silkworm moth

Cis–Trans Isomers for Night Vision

The retinas of the eyes consist of two types of cells: rods and cones. The rods on the edge of the retina allow us to see in dim light, and the cones, in the center, produce our vision in bright light. In the rods, there is a substance called rhodopsin that absorbs light. Rhodopsin is composed of *cis*-11-retinal, an unsaturated compound, attached to a protein. When rhodopsin absorbs light, the *cis*-11-retinal isomer is converted to its trans isomer, which changes its shape. The trans form no longer fits the protein and it separates from the protein. The change from the cis to trans isomer and the separation from the protein generates an electrical signal that the brain converts into an image.

An enzyme (isomerase) converts the trans isomer back to the *cis*-11-retinal isomer and the rhodopsin re-forms. If there is a deficiency of rhodopsin in the rods of the retina, night blindness may occur. One common cause is a lack of vitamin A in the diet. In our diet, we obtain vitamin A from plant pigments containing β-carotene, which is found in foods such as carrots, squash, and spinach. In the small intestine, the β-carotene is converted to vitamin A, which can be converted to *cis*-11-retinal or stored in the liver for future use. Without a sufficient quantity of retinal, not enough rhodopsin is produced to enable us to see adequately in dim light.

CIS-TRANS ISOMERS OF RETINAL

11-*cis*-retinal → (Light) → 11-*trans*-retinal

b. CH₃, H
C=C
H CH₂—CH₂—CH₂—CH₃
c. CH₃—CH₂—CH₂ H
C=C
H CH₂—CH₃

11.11 Draw the structural formula for each of the following:
a. *trans*-2-butene   **b.** *cis*-2-pentene
c. *trans*-3-heptene

11.12 Draw the structural formula for each of the following:
a. *cis*-3-hexene   **b.** *trans*-2-pentene
c. *cis*-4-octene

11.3 Addition Reactions

For alkenes and alkynes, the most characteristic reaction is the **addition** of atoms or groups of atoms to the carbons of the double or triple bond. Addition occurs because double and triple bonds are easily broken, which provides electrons for new single bonds. The general equation for the addition of a reactant A—B to an alkene can be written as follows:

The addition reactions have different names that depend on the type of reactant we add to the alkene, as Table 11.2 shows.

Hydrogenation

In a reaction called **hydrogenation**, atoms of hydrogen add to the carbons in a double or triple bond to form alkanes. A catalyst such as platinum (Pt), nickel (Ni), or palladium (Pd) is added to speed up the reaction. The general equation for hydrogenation can be written as follows:

Double bond (unsaturated)   Single bond (saturated)

Some examples of the hydrogenation of alkenes and alkynes follow:

CH₃—CH=CH—CH₃ + H—H $\xrightarrow{Pt}$ CH₃—CH—CH—CH₃
2-butene   butane

cyclohexene   cyclohexane

TABLE 11.2

Reactants and Addition Reactions	
Reactant Added	**Name of Addition Reaction**
H₂	Hydrogenation
HOH (water)	Hydration

The hydrogenation of alkynes requires two molecules of hydrogen to form the alkane product.

$$CH_3—C{\equiv}C—CH_3 + 2\,H—H \xrightarrow{Pt} CH_3—\overset{\overset{\displaystyle H}{|}}{\underset{\underset{\displaystyle H}{|}}{C}}—\overset{\overset{\displaystyle H}{|}}{\underset{\underset{\displaystyle H}{|}}{C}}—CH_3$$

2-butyne   butane

SAMPLE PROBLEM 11.4

■ Writing Equations for Hydrogenation

Write the structural formula for the product of the following hydrogenation reactions:

a. $CH_3—CH{=}CH_2 + H_2 \xrightarrow{Pt}$

b. ⬠ $+ H_2 \xrightarrow{Pt}$ **c.** $HC{\equiv}CH + 2H_2 \xrightarrow{Ni}$

SOLUTION

In an addition reaction, hydrogen adds to the double or triple bond to give an alkane.

a. $CH_3—CH_2—CH_3$ **b.** ⬠   **c.** $H_3C—CH_3$

STUDY CHECK

Draw the structural formula of the product of the hydrogenation of 2-methyl-1-butene using a platinum catalyst.

HEALTH NOTE

■ Hydrogenation of Unsaturated Fats

Vegetable oils such as corn oil or safflower oil are unsaturated fats composed of fatty acids that contain double bonds. The process of hydrogenation is used commercially to convert the double bonds in the unsaturated fats in vegetable oils to saturated fats such as margarine, which are more solid. Adjusting the amount of added hydrogen produces partially hydrogenated fats such as soft margarine, solid margarine in sticks, and shortenings, which are used in cooking. For example, oleic acid is a typical unsaturated fatty acid in olive oil and has a cis-double bond at carbon 9. When oleic acid is hydrogenated, it is converted to stearic acid, a saturated fatty acid.

Oleic acid (found in olive oil and other unsaturated fats)   Stearic acid (found in saturated fats)

Hydration

Alkenes react with water (HOH) when the reaction is catalyzed by a strong acid such as H_2SO_4. In this reaction called **hydration**, H— attaches to one of the carbon atoms in the double bond, and —OH group to the other carbon. Hydration is used to prepare alcohols, which have the functional group —OH. In the general equation the acid is represented by H^+.

When water adds to a double bond in which the carbon atoms are attached to different groups, the H attaches to the carbon that already has the most H atoms.

SAMPLE PROBLEM 11.5

▪ Writing Products of Hydration

Write the structural formulas for the products that form in the following hydration reactions:

a. CH_3—CH_2—CH_2—CH=CH_2 + HOH $\xrightarrow{H^+}$

b. ☐ + HOH $\xrightarrow{H^+}$

SOLUTION

a. Water adds H— and —OH to the double bond. The H— adds to the CH_2 in the double bond, and the —OH bonds to the CH.

$$CH_3-CH_2-CH_2-\overset{\overset{OH}{\downarrow}}{CH}=\overset{\overset{H}{\downarrow}}{CH_2} \xrightarrow{H^+} CH_3-CH_2-CH_2-\overset{\overset{OH}{|}}{CH}-CH_3$$

b. In cyclobutene, the H— adds to one side of the double bond, and the —OH adds to the other side.

FIGURE 11.3

Synthetic polymers are used to replace diseased veins and arteries. A broken hip is repaired using a metal piece that fits into an artificial plastic cup socket.

Q Why are the substances in these plastic devices called polymers?

LEARNING GOAL

Draw structural formulas of monomers that form a polymer or a three-monomer section of a polymer.

WEB TUTORIAL
Polymers

STUDY CHECK

Draw the structural formula for the alcohol obtained by the hydration of 2-methyl-2-butene.

QUESTIONS AND PROBLEMS

▪ Addition Reactions

11.13 Give the condensed structural formulas of the products in each of the following reactions:

a. $CH_3—CH_2—CH_2—CH{=}CH_2 + H_2 \xrightarrow{Pt}$

b. $CH_3—CH{=}CH—CH_3 + HOH \xrightarrow{H^+}$

c. ☐ + HOH $\xrightarrow{H^+}$

d. cyclopentene + $H_2 \xrightarrow{Pt}$

e. 2-pentyne + $2H_2 \xrightarrow{Pt}$

11.14 Give the condensed structural formulas of the products in each of the following reactions:

a. $CH_3—CH_2—CH{=}CH_2 + HOH \xrightarrow{H^+}$

b. cyclohexene + $H_2 \xrightarrow{Pt}$

c. *cis*-2-butene + $H_2 \xrightarrow{Pt}$

d. $CH_3—\overset{\displaystyle CH_3}{\underset{\displaystyle CH_3}{C}}{=}CH—CH_2—CH_3 + H_2 \xrightarrow{Pt}$

e. $CH_3—\overset{\displaystyle CH_3}{C}H—C{\equiv}CH + 2H_2 \xrightarrow{Pt}$

▪ 11.4 Polymers of Alkenes

Polymers are large molecules that consist of small repeating units called **monomers**. In the past hundred years, the plastics industry has made synthetic polymers that are in many of the materials we use every day, such as carpeting, plastic wrap, nonstick pans, plastic cups, and rain gear. In medicine, synthetic polymers are used to replace diseased or damaged body parts such as hip joints, teeth, heart valves, and blood vessels. (See Figure 11.3.)

▪ Addition Polymers

Many of the synthetic polymers are made by addition reactions of monomers that are small alkenes. The conditions for many polymerization reactions require high temperatures and very high pressure (over 1000 atm). In an addition reaction, the polymer grows as monomers are added to the end of the chain. Polyethylene, a polymer made from ethylene $CH_2{=}CH_2$, is used in plastic bottles, film, and plastic dinnerware. (See Figure 11.4.) In the polymerization, a series of addition reactions

Polyethylene

Vinyl chloride

Polypropylene

Polytetrafluoroethylene (Teflon)

Polydichloroethylene (Saran)

Polystyrene

FIGURE 11.4

Synthetic polymers provide a wide variety of items that we use every day.

Q **What are some alkenes used to make the polymers in these plastic items?**

joins one monomer to the next until a long carbon chain forms that contains as many as 1000 monomers.

ethene (ethylene) monomers   polyethylene section

Table 11.3 lists several alkene monomers that are used to produce common synthetic polymers, and Figure 11.4 shows examples of each.

The alkane-like nature of these plastic synthetic polymers makes them unreactive. Thus, they do not decompose easily (they are nonbiodegradable) and have become contributors to pollution. Efforts are being made to make them more degradable. It is becoming increasingly important to recycle plastic material,

EXPLORE YOUR WORLD

■ **Polymers and Recycling Plastics**

1. Make a list of the items you use or have in your room or home that are made of polymers.
2. Recycling information on the bottom or side of a plastic bottle includes a triangle with a code number that identifies the type of polymer used to make the plastic. Make a collection of several different kinds of plastic bottles. Try to find plastic items with each type of polymer.

QUESTIONS

1. What are the most common types of plastics among the plastic containers in your collection?
2. What are the monomer units of some of the plastics you looked at?

TABLE 11.3

Some Alkenes and Their Polymers

Monomer	Polymer Section	Common Uses
$CH_2{=}CH_2$ Ethene (ethylene)	Polyethylene	Plastic bottles, film, insulation materials
$CH_2{=}CH$ with Cl Chloroethene (vinyl chloride)	Polyvinyl chloride (PVC)	Plastic pipes and tubing, garden hoses, garbage bags
$CH_2{=}CH$ with CH_3 Propene (propylene)	Polypropylene	Ski and hiking clothing, carpets, artificial joints
$F{-}C{=}C{-}F$ Tetrafluoroethene	Polytetrafluoroethylene (Teflon)	Nonstick coatings
$CH_2{=}C{-}Cl$ with Cl 1,1-Dichloroethene	Polydichloroethylene (Saran)	Plastic film and wrap
$H_2C{=}CH$ (phenyl) Phenylethene (styrene)	$-CH_2-CH-CH_2-CH-CH_2-CH-$ Polystyrene	Plastic coffee cups and cartons, insulation

rather than add to our growing landfills. You can identify the type of polymer used to manufacture a plastic item by looking for the recycle symbol (arrows in a triangle) found on the label or on the bottom of the plastic container. For example, either the number 5 or the letters PP inside the triangle is a code for a polypropylene plastic.

1	2	3	4	5	6
PETE	HDPE	PV	LDPE	PP	PS
Polyethylene terephthalate	High-density polyethylene	Polyvinyl chloride	Low-density polyethylene	Polypropylene	Polystyrene

SAMPLE PROBLEM 11.6

▪ Polymers

What are the starting monomers for the following polymers?

a. polypropylene   **b.** Saran

SOLUTION

a. propene (propylene) $CH_2\!=\!\overset{\displaystyle CH_3}{\underset{\displaystyle |}{CH}}$

b. 1,1-dichloroethene, $CH_2\!=\!\overset{\displaystyle Cl}{\underset{\displaystyle |}{C}}\!-\!Cl$

STUDY CHECK

What is the monomer for PVC?

QUESTIONS AND PROBLEMS

■ Polymers of Alkenes

11.15 What is a polymer?

11.16 What is a monomer?

11.17 Write an equation that represents the formation of a part of the Teflon polymer from three of the monomer units.

11.18 Write an equation that represents the formation of a part of the polystyrene polymer from three of the monomer units.

■ 11.5 Aromatic Compounds

LEARNING GOAL

■ Describe the bonding in benzene; name aromatic compounds, and write their structural formulas.

In 1825, Michael Faraday isolated a hydrocarbon called benzene, which had the molecular formula C_6H_6. Because many compounds containing benzene had fragrant odors, the family of benzene compounds became known as **aromatic compounds**. A molecule of **benzene** consists of a ring of six carbon atoms with one hydrogen atom attached to each carbon. Each carbon atom uses 3 valence electrons to bond to the hydrogen atom and two adjacent carbons. That leaves 1 valence electron to share in a double bond with an adjacent carbon. In 1865, August Kekulé proposed that the carbon atoms were arranged in a flat ring with alternating single and double bonds between the carbon atoms. This idea led to two ways of writing the benzene structure, as follows:

Structures for benzene

However, there is only one structure of benzene. Today we know that all the bonds in benzene are identical. In the benzene ring, scientists determined that the electrons are shared equally, a unique feature that makes aromatic compounds especially stable. To show this stability, the benzene structure is also represented as a hexagon with a circle in the center.

Naming Aromatic Compounds

Aromatic compounds that contain a benzene ring with a single substituent are usually named as benzene derivatives. However, many of these compounds have been important in chemistry for many years and still use their common names. Some widely used names such as toluene, aniline, and phenol are allowed by IUPAC rules.

toluene   ethylbenzene   aniline   phenol
(methylbenzene)   (benzenamine)   (hydroxybenzene)

When a benzene ring is a substituent, C_6H_5 —, it is named as a phenyl group.

phenyl group   3-phenyl-1-butene

When there are two or more substituents on a benzene ring, the ring is numbered to give the lowest numbers to the substituents.

1,2-dichlorobenzene   1,3-dichlorobenzene   1,4-dichlorobenzene

The substituents are named alphabetically.

1,3,5-trichlorobenzene   4-bromo-2-chlorotoluene   2,6-dibromo-4-chlorotoluene

---

SAMPLE PROBLEM 11.7

Naming Aromatic Compounds

Give the IUPAC name for each of the following aromatic compounds:

SOLUTION

a. chlorobenzene

b. 4-bromo-3-chlorotoluene

c. 1,2-dimethylbenzene

STUDY CHECK

Name the following compound.

note—Diethyl benzene

$$CH_2{-}CH_3$$

$$CH_2{-}CH_3$$

QUESTIONS AND PROBLEMS

■ Aromatic Compounds

11.19 Cyclohexane and benzene each have six carbon atoms. How are they different?

11.20 In the Health Note "Some Common Aromatic Compounds," what part of each molecule is the aromatic portion?

11.21 Give the IUPAC name for each of the following:

a. CH_3 / Cl

b. $CH_2{-}CH_3$

c. Cl / Cl Cl

d. CH_3 / Br Cl

11.22 Give the IUPAC name for each of the following:

a.

b. CH_3

c. Cl / Cl

d. Br / Cl Cl

11.23 Draw the structural formulas for each of the following compounds:

a. toluene

b. 1,3-dichlorobenzene

c. 4-ethyltoluene

d. 4-chlorotoluene

11.24 Draw the structural formulas for each of the following compounds:

a. benzene

b. 2-chlorotoluene

c. propylbenzene

d. 1,2,4-trichlorobenzene

▪ Polycyclic Aromatic Hydrocarbons (PAHs)

Large aromatic compounds known as polycyclic aromatic hydrocarbons are formed by fusing together two or more benzene rings edge-to-edge. In a fused ring compound, neighboring benzene rings share two carbon atoms. Naphthalene with two benzene rings is well known for its use in mothballs. Anthracene with three rings is used in the manufacture of dyes.

Napthalene   Anthracene   Phenanthrene

When a polycyclic compound contains phenanthrene, it may act as a carcinogen, a substance known to cause cancer. For example, some aromatic compounds in cigarette smoke cause cancer,

as seen in the lung tissue of a heavy smoker. Benz[a]pyrene, a product of combustion, has been identified in coal tar, tobacco smoke, barbecued meats, and automobile exhaust.

Benz[a]pyrene

Compounds containing five or more fused benzene rings such as benz[a]pyrene are potent carcinogens. The molecules interact with the DNA in the cells, causing abnormal cell growth and cancer. Increased exposure to carcinogens increases the chance of DNA alterations in the cells.

CONCEPT MAP

Unsaturated Hydrocarbons

include

Alkenes — with a — **Double bond** $-C=C-$

Alkynes — with a — **Triple bond** $-C\equiv C-$

Aromatic compounds — contain a — **Benzene ring**

Double bond:
- can exist as → **Cis–trans isomers**
- undergo → **Addition reactions**

such as:
- **Hydrogenation** — which adds H_2 to form → **Alkanes**
- **Hydration** — which adds H_2O to form → **Alcohols**

CHAPTER REVIEW

11.1 Alkenes and Alkynes

Alkenes are unsaturated hydrocarbons that contain carbon–carbon double bonds ($C=C$). Alkynes contain a triple bond ($-C\equiv C-$). The IUPAC names of alkenes end with *ene*, while alkyne names end with *yne*. The main chain is numbered from the end nearest the double or triple bond.

11.2 Cis–Trans Isomers

Isomers of alkenes occur when the carbon atoms in the double bond are connected to different atoms or groups. In the cis isomer the similar groups are on the same side of the double bond, whereas in the trans isomer they are connected on the opposite sides of the double bond.

11.3 Addition Reactions

The addition of small molecules to the double bond is a characteristic reaction of alkenes. Hydrogenation adds hydrogen atoms to the double bond of an alkene to yield an alkane. Water can also add to a double bond. When there are a different number of groups attached to the carbons in the double bond, the H from the HOH adds to the carbon with the greater number of hydrogen atoms.

11.4 Polymers of Alkenes

Polymers are long-chain molecules that consist of many repeating units of smaller carbon molecules called monomers. Many materials that we use every day are synthetic polymers, including carpeting, plastic wrap, nonstick pans, and nylon. These synthetic materials are often made by addition reactions in which a catalyst links the carbon atoms from various kinds of alkene molecules.

11.5 Aromatic Compounds

Most aromatic compounds contain benzene, C_6H_6, a cyclic structure represented as a hexagon with a circle in the center. Many aromatic compounds use the parent name benzene, although common names such as toluene are retained. The benzene ring is numbered and the branches are listed in alphabetical order.

SUMMARY OF NAMING

Type	Example	Characteristic	Structure
Alkene	Propene (propylene)	double bond	$CH_3-CH=CH_2$
Cycloalkene	Cyclopropene	double bond in a carbon ring	△
Alkyne	Propyne	triple bond	$CH_3-C\equiv CH$
Aromatic	Benzene	Aromatic ring of six carbons	⬡
	Methylbenzene or toluene		CH_3 ⬡

SUMMARY OF REACTIONS

HYDROGENATION

Alkene + H_2 $\xrightarrow{Pt}$ alkane

$$CH_2=CH-CH_3 + H_2 \xrightarrow{Pt} CH_3-CH_2-CH_3$$

Alkyne + $2H_2$ $\xrightarrow{Pt}$ alkane

$$CH_3-C\equiv CH + 2H_2 \xrightarrow{Pt} CH_3-CH_2-CH_3$$

HYDRATION OF ALKENES

Alkene + H—OH $\xrightarrow{H^+}$ alcohol

$$CH_2=CH-CH_3 + H-OH \xrightarrow{H^+} CH_3-\overset{\overset{\displaystyle OH}{|}}{CH}-CH_3$$

KEY TERMS

addition A reaction in which atoms or groups of atoms bond to a double bond or triple bond. Addition reactions include the addition of hydrogen (hydrogenation) and water (hydration).

alkene An unsaturated hydrocarbon containing a carbon–carbon double bond.

alkyne An unsaturated hydrocarbon containing a carbon–carbon triple bond.

aromatic compounds Compounds that contain the ring structure of benzene.

benzene A ring of six carbon atoms each of which is attached to a hydrogen atom, C_6H_6.

cis isomer A geometric isomer in which similar groups are connected on the same side of the double bond.

hydration An addition reaction in which the components of water, H— and —OH, bond to the carbon–carbon double bond to form an alcohol.

hydrogenation The addition of hydrogen (H_2) to the double bond of alkenes or alkynes to yield alkanes.

monomer The small organic molecule that is repeated many times in a polymer.

polymer A very large molecule that is composed of many small, repeating structural units that are identical.

trans isomer A geometric isomer in which similar groups are connected to opposite sides of the double bond in an alkene.

unsaturated hydrocarbons A compound of carbon and hydrogen in which the carbon chain contains at least one double (alkene) or triple carbon–carbon bond (alkyne). An unsaturated compound is capable of an addition reaction with hydrogen, which converts the double or triple bonds to single carbon–carbon bonds.

ADDITIONAL QUESTIONS AND PROBLEMS

11.25 Compare the formulas and bonding in propane, cyclopropane, propene, and propyne.

11.26 Compare the formulas and bonding in butane, cyclobutane, cyclobutene, and 2-butyne.

11.27 Give the IUPAC name for each of the following compounds:

b. CH_3—CH_2—$C{\equiv}C$—CH_3

11.28 Write the condensed structures of each of the following compounds:
   a. 2-pentyne
   b. *cis*-2-heptene
   c. *trans*-3-hexene
   d. 2,3-dichloro-1-butene

11.29 Indicate if the following pairs of structures represent isomers, cis–trans isomers, or the same molecule.

11.30 What role do cis–trans isomers play in night vision?

11.31 Write the cis and trans isomers for each of the following:
   a. 2-pentene
   b. 3-hexene
   c. 2-butene
   d. 2-hexene

11.32 Write the structures of the products for the following.

a. CH_3—$CH{=}CH$—CH_3 + H_2 $\xrightarrow{\text{Ni}}$

b. ⬠ + H_2 $\xrightarrow{\text{Ni}}$

c. CH_3—$CH{=}CH$—CH_3 + HOH $\xrightarrow{H^+}$

d. ⬠ + HOH $\xrightarrow{H^+}$

e. ⬡ + HOH $\xrightarrow{H^+}$

11.33 Name each of the following aromatic compounds:

11.34 Write the structural formulas for each of the following:
   a. ethylbenzene
   b. 1,3-dichlorobenzene
   c. 1,2,4-trimethylbenzene
   d. 1,4-dimethylbenzene

UNDERSTANDING THE CONCEPTS

11.35 Draw a part of the polymer (use four monomers) of Teflon made from 1,1,2,2-tetrafluoroethene.

11.36 A garden hose is made of polyvinylchloride (PVC) from chloroethene (vinyl chloride). Draw a part of the polymer (use four monomers) for PVC.

11.37 Explosives used in mining contain TNT or 2,4,6-trinitrotoluene.

a. If the functional group *nitro* is —NO_2, what is the structural formula of TNT?

b. TNT is actually a mixture of isomers of trinitrotoluene. Draw two possible isomers.

11.38 Margarine is produced from the hydrogenation of vegetable oils, which contain unsaturated fatty acids. How many grams of hydrogen are required to completely saturate 75.0 g oleic acid, $C_{18}H_{34}O_2$, which has one double bond?

QUESTIONS FOR TEAMWORK

11.39 Using each of the following carbon chains for C_5H_{10}, write and name all the possible alkenes, including those with cis and trans isomers.

C—C—C—C—C

C—C—C—C
      |
      C

11.40 How many grams of hydrogen are needed to hydrogenate 30.0 g 2-butene?

11.41 If a female silkworm moth secretes 50 ng Bombykol, a sex attractant, how many molecules did she secrete? (See Environmental Note "Pheromones in Insect Communication")

11.42 Acetylene gas reacts with oxygen and burns at high temperature in an acetylene torch.
   a. Write the balanced equation for the complete combustion of acetylene.
   b. How many grams of oxygen are needed to react with 8.5 L acetylene at STP?

c. How many liters of CO_2 (at STP) are produced when 30.0 g acetylene undergoes combustion?

ANSWERS

Answers to Study Checks

11.1 This is cylcobutene, a cycloalkene.

11.2 a. $CH_3-C\equiv C-CH_2-CH_3$ **b.**

11.3 *trans*-3-hexene

11.4
$$CH_3-\underset{\underset{CH_3}{|}}{CH}-CH_2-CH_3$$

11.5
$$CH_3-\underset{\underset{OH}{|}}{\overset{\overset{CH_3}{|}}{C}}-CH_2-CH_3$$

11.6 The monomer of PVC, polyvinyl chloride, is chloroethene:

$$\underset{\underset{H}{|}}{\overset{\overset{H}{|}}{C}}=\underset{\underset{H}{|}}{\overset{\overset{Cl}{|}}{C}}$$

11.7 1, 3-diethylbenzene

Answers to Selected Questions and Problems

11.1 a. An alkene has a double bond.
b. An alkyne has a triple bond.
c. A cycloalkene has a double bond in a ring.

11.3 Propene contains a double bond, and propyne has a triple bond.

11.5 a. ethene
b. 2-methyl-1-propene
c. 2-pentyne
d. cyclobutene

11.7 a. $CH_3-CH=CH_2$

b. $CH_2=CH-CH_2-CH_2-CH_3$

c. $CH_2=\underset{\underset{CH_3}{|}}{C}-CH_2-CH_3$ **d.**

11.9 a. *cis*-2-butene
b. *trans*-3-octene
c. *cis*-3-heptene

11.11 a.
$$\underset{\underset{H}{}}{\overset{\overset{CH_3}{}}{C}}=\underset{\underset{CH_3}{}}{\overset{\overset{H}{}}{C}}$$

b.
$$\underset{\underset{H}{}}{\overset{\overset{CH_3}{}}{C}}=\underset{\underset{H}{}}{\overset{\overset{CH_2-CH_3}{}}{C}}$$

c.
$$\underset{\underset{H}{}}{\overset{\overset{CH_3-CH_2}{}}{C}}=\underset{\underset{CH_2-CH_2-CH_3}{}}{\overset{\overset{H}{}}{C}}$$

11.13 a. $CH_3-CH_2-CH_2-CH_2-CH_3$

b.
$$CH_3-\underset{\underset{OH}{|}}{CH}-CH_2-CH_3$$

c.

d.

e. $CH_3-CH_2-CH_2-CH_2-CH_3$

11.15 A polymer is a very large molecule composed of small units that are repeated many times.

11.17

11.19 Cyclohexane, C_6H_{12}, is a cycloalkane in which six carbon atoms in a ring are linked by single bonds. In benzene, C_6H_6, electrons shared equally link the six carbon atoms.

11.21 a. 2-chlorotoluene
b. ethylbenzene
c. 1,3,5-trichlorobenzene
d. 3-bromo-5-chlorotoluene

11.23

11.25 All the compounds have three carbon atoms. Propane is a saturated alkane, and cyclopropane is a saturated cyclic hydrocarbon. Both propene and propyne are unsaturated hydrocarbons, but propene has a double bond and propyne has a triple bond.

11.27 a. 2-methyl-1-pentene
b. 2-pentyne
c. cyclopentene
d. *trans*-2-pentene
e. cyclohexene

11.29 a. isomers   **b.** cis–trans isomers
c. identical   **d.** isomers

SOLUTION

a. The parent chain is pentane; the alcohol is named *pentanol*. The carbon chain is numbered to give the position of the —OH group on carbon 2 and the methyl group on carbon 4. The compound is named *4-methyl-2-pentanol*.

b. The compound is a *phenol* because the —OH is attached to a benzene ring. The ring is numbered with carbon 1 attached to the —OH in the direction that gives the bromine the lower number. The compound is named *2-bromophenol*.

STUDY CHECK

Give the IUPAC name for the following:

$$\begin{array}{c} \text{Cl} \\ | \\ \text{CH}_3\text{—CH—CH}_2\text{—CH}_2\text{—OH} \end{array}$$

Classification of Alcohols

Alcohols are classified by the number of carbon groups attached to the carbon atom bonded to the hydroxyl (—OH) group. A **primary (1°) alcohol** has one alkyl group attached to the carbon atom bonded to the —OH, a **secondary (2°) alcohol** has two alkyl groups, and a **tertiary (3°) alcohol** has three alkyl groups.

Primary (1°) alcohol   **Secondary (2°) alcohol**   **Tertiary (3°) alcohol**

Carbon attached to OH group

SAMPLE PROBLEM 12.2

Classifying Alcohols

Classify each of the following alcohols as primary, secondary, or tertiary.

a. $\text{CH}_3\text{—CH}_2\text{—CH}_2\text{—OH}$

b. $\begin{array}{c} \text{OH} \\ | \\ \text{CH}_3\text{—CH}_2\text{—C—CH}_3 \\ | \\ \text{CH}_3 \end{array}$

SOLUTION

a. One alkyl group attached to the carbon atom bonded to the —OH makes this a primary alcohol.

b. Three alkyl groups attached to the carbon atom bonded to the —OH makes this a tertiary alcohol.

HEALTH NOTE

▪ Some Important Alcohols and Phenols

Methanol (*methyl alcohol*), the simplest alcohol, is found in many solvents and paint removers. If ingested, methanol is oxidized to formaldehyde, which can cause headaches, blindness, and death. Methanol is used to make plastics, medicines, and fuels. In car racing, it is used as a fuel because it is less flammable and has a higher octane rating than gasoline.

Ethanol (*ethyl alcohol*) has been known since prehistoric times as an intoxicating product formed by the fermentation of grains and starches.

$$C_6H_{12}O_6 \xrightarrow{\text{fermentation}} 2CH_3{-}CH_2{-}OH + 2CO_2$$

Today, ethanol for commercial uses is produced by allowing ethene and water to react at high temperatures and pressures. It is used as a solvent for perfumes, varnishes, and some medicines, such as tincture of iodine. "Gasohol" is a mixture of ethanol and gasoline used as a fuel.

$$H_2C{=}CH_2 + H_2O \xrightarrow{\text{300°C, 200 atm, catalyst}} CH_3{-}CH_2{-}OH$$

Several of the essential oils of plants, which produce the odor or flavor of the plant, are derivatives of phenol. Eugenol is found in cloves, vanillin in vanilla bean, isoeugenol in nutmeg, and thymol in thyme and mint. Thymol has a pleasant, minty taste and is used in mouthwashes and by dentists to disinfect a cavity before adding a filling compound.

1,2,3-Propanetriol (*glycerol or glycerin*), a trihydroxy alcohol, is a viscous liquid obtained from oils and fats during the production of soaps. The presence of several polar —OH groups makes it strongly attracted to water, a feature that makes glycerin useful as a skin softener in products such as skin lotions, cosmetics, shaving creams, and liquid soaps.

$$HO{-}CH_2{-}\overset{\displaystyle OH}{CH}{-}CH_2{-}OH$$
1,2,3-propanetriol (glycerol)

1,2-Ethanediol (*ethylene glycol*) is used as antifreeze in heating and cooling systems. It is also a solvent for paints, inks, and plastics, and is used in the production of synthetic fibers such as Dacron. If ingested, it is extremely toxic. In the body, it is oxidized to oxalic acid, which forms insoluble salts in the kidneys that cause renal damage, convulsions, and death. Because its sweet taste is attractive to pets and children, ethylene glycol solutions must be carefully stored.

$$HO{-}CH_2{-}CH_2{-}OH \xrightarrow{[O]} HO{-}\overset{\displaystyle O}{\underset{}{C}}{-}\overset{\displaystyle O}{\underset{}{C}}{-}OH$$
1,2-ethanediol (ethylene glycol)　　oxalic acid

Vanillin

Eugenol

Thymol

Isoeugenol

STUDY CHECK

Classify the following as primary, secondary, or tertiary:

$$CH_3-\underset{\underset{\displaystyle OH}{|}}{CH}-CH_3$$

Thiols

Thiols are a family of sulfur-containing organic compounds that have a **sulfhydryl** (—SH) **group**. They have structures similar to alcohols except that a —SH group takes the place of an —OH group. (See Figure 12.1.)

An important property of thiols is their strong, sometimes disagreeable, odor, which is characteristic of oysters, cheddar cheese, onions, and garlic. To help us detect natural gas (methane) leaks, a small amount of ethanethiol is added to the gas supply.

Ethers

An *ether* contains an oxygen atom that is attached by single bonds to two carbon groups that are alkyls or aromatic rings. Ethers have a bent structure like water and alcohols except both hydrogen atoms are replaced by alkyl groups.

CH_3-CH_2-SH
Ethanethiol

CH_3-SH
Methanethiol
Oysters and cheese

$CH_3-CH_2-CH_2-SH$
1-Propanethiol
Onions

$CH_2=CH-CH_2-SH$
2-Propene-1-thiol
Garlic

FIGURE 12.1

Thiols are sulfur-containing compounds with an —SH group.

Q Why do thiols have structures similar to alcohols?

Most ethers are named by their common names. Write the name of each alkyl or aromatic group attached to the oxygen atom in alphabetical order followed by the word *ether*. The IUPAC names are used only when the ether is more complex.

methyl group

CH_3—O—CH_2—CH_2—CH_3 propyl group

Common name: methyl propyl ether

Some examples of the common names for ethers follow:

CH_3—O—CH_3 CH_3—CH_2—O—CH_2—CH_3 CH_3—CH_2—O—⬡

dimethyl ether   diethyl ether

ethyl phenyl ether

SAMPLE PROBLEM 12.3

■ **Ethers**

Give the common name for the following ether:

CH_3—CH_2—O—CH_2—CH_2—CH_3

SOLUTION

The groups attached to the oxygen are an ethyl group and a propyl group. The common name is *ethyl propyl ether*.

STUDY CHECK

Draw the structure of methyl phenyl ether.

QUESTIONS AND PROBLEMS

■ **Alcohols, Ethers, and Thiols**

12.1 Give the IUPAC name for each of the following:

a. CH_3—CH_2—OH

    OH
    |
b. CH_3—CH_2—CH—CH_3

        OH
        |
c. CH_3—CH—CH_2—CH_2—CH_3

d.

12.2 Give the IUPAC name for each of the following alcohols:

    CH_2—CH_3
    |
a. ☐—OH

        CH_3
        |
b. CH_3—CH_2—CH—CH_2—OH

        CH_3 CH_3
        | |
c. CH_3—CH_2—CH—CH—CH_2—OH

                OH
                |
d. CH_3—CH_2—CH_2—CH—CH_3

12.3   Write the condensed structural formula of each of the following alcohols:
    a. 1-propanol   **b.** methyl alcohol
    c. 3-pentanol   **d.** 2-methyl-2-butanol

12.4   Write the condensed structural formula of each of the following alcohols:
    a. ethyl alcohol   **b.** 3-methyl-1-butanol
    c. 2,4-dichlorocyclohexanol   **d.** propyl alcohol

12.5   Name each of the following phenols:

12.6   Name each of the following phenols:

12.7   Classify each of the following as a primary, secondary, or tertiary alcohol:

    CH_3
    |
a. $CH_3-CH-CH_2-CH_2-OH$ **b.** $CH_3-CH_2-CH_2-CH_2-OH$

    OH   OH
    |
c. $CH_3-C-CH_2-CH_3$ **d.** $-CH_3$
    |
    CH_3

12.8   Classify each of the following as a primary, secondary, or tertiary alcohol:

    CH_3 CH_3
                                     |
a.   **b.** $CH_3-CH-CH_2-OH$
       OH

                                  CH_3
                                   |
c. CH_2-OH **d.** $CH_3-CH_2-CH_2-C-OH$
                                   |
                                   CH_3

12.9   Give a common name for each of the following ethers:
    a. $CH_3-O-CH_2-CH_3$
    b. $CH_3-CH_2-CH_2-O-CH_2-CH_2-CH_3$

    c.   **d.** $CH_3-O-CH_2-CH_2-CH_3$

12.10   Give the common name for each of the following ethers:
    a. $CH_3-CH_2-O-CH_2-CH_2-CH_3$

    b.   **c.**

    d. CH_3-O-CH_3

12.11   Write the condensed structural formula for each of the following ethers:
    a. ethyl propyl ether   **b.** cyclopropyl ethyl ether
    c. ethyl methyl ether

12.12   Write the condensed structural formula for each of the following ethers:
    a. diethyl ether   **b.** diphenyl ether
    c. cyclohexyl methyl ether

■ Ethers as Anesthetics

Anesthesia is the loss of sensation and consciousness. A general anesthetic is a substance that blocks signals to the awareness centers in the brain so the person has a loss of memory, a loss of feeling pain, and an artificial sleep. The term *ether* has been associated with anesthesia because diethyl ether was the most widely used anesthetic for more than a hundred years. Although it is easy to administer, ether is very volatile and highly flammable. A small spark in the operating room could cause an explosion. Since the 1950s, anesthetics such as Forane (isoflurane), Ethrane (enflurane),

and Penthrane (methoxyflurane) have been developed that are not as flammable and do not cause nausea. Most of these anesthetics retain the ether group, but the addition of halogen atoms reduces the volatility and flammability of the ethers. More recently, they have been replaced by halothane (1-bromo-1-chloro-2,2,2-trifluoroethane), discussed in Chapter 10, because of the side effects of the ether-type inhalation anesthetics.

Forane
(isoflurane)

Ethrane
(enflurane)

Penthrane
(methoxyflurane)

■ 12.2 Properties of Alcohols, Phenols, and Ethers

In Chapter 10, we learned that hydrocarbons, which are composed of only carbon and hydrogen, are nonpolar. In this chapter, we looked at compounds containing the element oxygen, which is strongly electronegative. As a result the oxygen and hydrogen atoms in the hydroxyl group, —OH, form hydrogen bonds. Although ethers also contain an oxygen atom, there is no hydrogen atom attached. Thus, ethers do not hydrogen bond with each other, but they do hydrogen bond with water.

Hydrogen bonds

No hydrogen bonds

Methyl alcohol

Dimethyl ether

■ Boiling Points

Higher temperatures are required to provide the energy needed to break the hydrogen bonds between alcohol molecules. Thus, alcohols have higher boiling points than alkanes and ethers of similar mass. The boiling points of ethers are close to those of alkanes of similar mass because hydrogen bonding does not occur.

Solubility in Water

The oxygen atom in alcohols and ethers influences their solubility in water. In alcohols, the polar —OH group can hydrogen bond with water, which makes alcohols with one to four carbon atoms very soluble in water.

Water

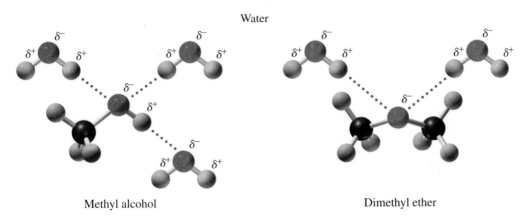

Methyl alcohol   Dimethyl ether

When there are more carbon atoms in the alkyl portion of the alcohol, the effect of the —OH group is diminished. The alkane portion does not participate in hydrogen bonding with water. Thus alcohols with five or more carbon atoms are not very soluble in water.

Ethers are more soluble in water than alkanes of similar mass because the oxygen can hydrogen bond with water, but ethers are not as soluble as alcohols. Ethers such as diethyl ether are very useful as solvents for hydrocarbons. However, ether vapors are highly flammable and react with oxygen to form explosive compounds. The utmost care must be taken when working with ethers.

Table 12.1 compares the boiling points and solubility of some alkanes, alcohols, and ethers by mass.

TABLE 12.1

Solubility and Boiling Points of Some Typical Alkanes, Alcohols, and Ethers of Similar Molar Mass

Compound	Structural Formula	Molar Mass (g/mole)	Boiling Point (°C)	Soluble in Water
Propane	CH_3—CH_2—CH_3	44	−42	No
Dimethyl ether	CH_3—O—CH_3	46	−23	Yes
Ethanol	CH_3—CH_2—OH	46	78	Yes
Butane	CH_3—CH_2—CH_2—CH_3	58	0	No
Ethyl methyl ether	CH_3—O—CH_2—CH_3	60	8	Yes
1-Propanol	CH_3—CH_2—CH_2—OH	60	97	Yes

▪ Solubility of Phenols

Phenol is soluble in water because the hydroxyl group ionizes slightly as a weak acid. In fact, an early name for phenol was *carbolic acid*. A concentrated solution of phenol is very corrosive and highly irritating to the skin; it can cause severe burns and ingestion can be fatal. Dilute solutions of phenol were previously used in hospitals as antiseptics, but they have generally been replaced.

phenol   phenoxide ion

SAMPLE PROBLEM 12.4

▪ Properties of Alcohols, Ethers, and Phenols

Predict which compounds in each pair will be more soluble in water.

a. propane or ethanol
b. 1-propanol or 1-hexanol

SOLUTION

a. The alcohol ethanol is more soluble because it can form hydrogen bonds with water.
b. The 1-propanol is more soluble because it has a shorter carbon chain.

STUDY CHECK

Dimethyl ether and ethanol both have molar masses of 46. However, ethanol has a much higher boiling point than dimethyl ether. How would you explain this difference in boiling points?

QUESTIONS AND PROBLEMS

▪ Properties of Alcohols, Phenols, and Ethers

12.13 Predict the compound with the higher boiling point in the following pairs:
  a. ethane or methanol
  b. diethyl ether or 1-butanol
  c. 1-butanol or pentane

12.14 Glycerol (1,2,3-propanetriol) has a boiling point of 290°C. 1-Pentanol, which has about the same molar mass as glycerol, boils at 138°C. Why is the boiling point of glycerol so much higher?

12.15 Are each of the following soluble in water? Explain.
  a. CH_3-CH_2-OH
  b. CH_3-O-CH_3
  c. $CH_3-CH_2-CH_2-CH_2-CH_2-CH_2-OH$
  d. $CH_3-CH_2-CH_3$

12.16 Give an explanation for the following observations:
  a. Ethanol is soluble in water, but propane is not.
  b. Dimethyl ether is soluble in water, but pentane is not.
  c. 1-Propanol is soluble in water, but 1-hexanol is not.

12.3 Reactions of Alcohols and Thiols

LEARNING GOAL

Write equations for the combustion, dehydration, and oxidation of alcohols.

In Chapter 10, we learned that hydrocarbons undergo combustion in the presence of oxygen. Alcohols burn with oxygen too. For example, in a restaurant, a dessert may be prepared by pouring a liquor on fruit or ice cream and lighting it. (See Figure 12.2.) The combustion of the ethanol in the liquor proceeds as follows:

$$CH_3—CH_2—OH + 3O_2 \longrightarrow 2CO_2 + 3H_2O + energy$$

Dehydration of Alcohols to Form Alkenes

Earlier we saw that alkenes can add water to yield alcohols. In a reverse reaction alcohols lose a water molecule when they are heated with an acid catalyst such as H_2SO_4. During the **dehydration** of an alcohol, a H— and —OH are removed from *adjacent carbon atoms of the same alcohol* to produce a water molecule. A double bond forms between the same two carbon atoms to produce an alkene product.

FIGURE 12.2

A flaming dessert is prepared using a liquor that undergoes combustion.

Q What is the equation for the combustion of the ethanol in the liquor?

Examples

SAMPLE PROBLEM 12.5

Dehydration of Alcohols

Draw the condensed structural formulas for the alkenes produced by the dehydration of the following alcohols:

SOLUTION

a. $CH_3—CH_2—CH=CH—CH_3$

b. The —OH of this alcohol is removed along with an H from an adjacent carbon. Remember that the hydrogens are not drawn in this type of geometric figure.

STUDY CHECK

What is the name of the alkene produced by the dehydration of cyclopentanol?

■ Oxidation of Alcohols

In organic chemistry, it is convenient to think of **oxidation** as a loss of hydrogen atoms or the addition of oxygen. An oxidation reaction occurs when there is an increase in the number of carbon–oxygen bonds. In a reduction, the product has fewer bonds between carbon and oxygen.

■ Oxidation of Primary and Secondary Alcohols

The oxidation of a primary alcohol produces an aldehyde, which contains a double bond between carbon and oxygen. The oxidation occurs by removing two hydrogen atoms, one from the —OH group and another from the carbon that is bonded to the —OH. To indicate the presence of an oxidizing agent, reactions are often written with the symbol [O].

Aldehydes oxidize further, this time by the addition of another oxygen to form a carboxylic acid. This step occurs so readily that it is often difficult to isolate the aldehyde product during oxidation. We will learn more about carboxylic acids in Chapter 13.

In the oxidation of secondary alcohols, the products are ketones. One hydrogen is removed from the —OH and another from the carbon bonded to the —OH group. The result is a ketone that has the carbon–oxygen double bond attached to alkyl groups on both sides.

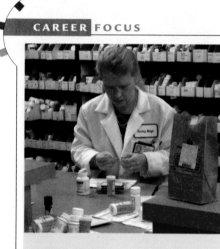

Tertiary alcohols do not oxidize readily because there are no hydrogen atoms on the carbon bonded to the —OH group. Because C—C bonds are usually too strong to oxidize, tertiary alcohols resist oxidation.

3° Alcohol

SAMPLE PROBLEM 12.6

■ **Oxidation of Alcohols**

Draw the structural formula of the aldehyde or ketone formed by the oxidation of each of the following:

a. CH_3—CH_2—$\overset{\overset{\textstyle OH}{|}}{CH}$—$CH_3$
b. CH_3—CH_2—CH_2—OH

SOLUTION

a. The oxidation of a secondary alcohol produces a ketone.

$$CH_3-CH_2-\overset{\overset{\textstyle O}{\|}}{C}-CH_3$$

b. The oxidation of a primary alcohol produces an aldehyde.

$$CH_3-CH_2-\overset{\overset{\textstyle O}{\|}}{C}-H$$

STUDY CHECK

Draw the structural formula of the aldehyde or ketone formed by the oxidation of 2-propanol.

■

Oxidation of Thiols

Thiols also undergo oxidation by a loss of hydrogen atoms from the —SH groups. The oxidized product is called a **disulfide**.

Example

$$CH_3-S-H + H-S-CH_3 \xrightarrow{[O]} CH_3-S-S-CH_3 + H_2O$$

methanethiol   dimethyl disulfide

the
**Chemistry
place**

CASE STUDY
Alcohol Toxicity

■ Oxidation of Alcohol in the Body

Ethanol is the most commonly abused drug in the United States. When ingested in small amounts, ethanol may produce a feeling of euphoria in the body although it is a depressant. In the liver, enzymes such as alcohol dehydrogenases oxidize ethanol to acetaldehyde, a substance that impairs mental and physical coordination. If the blood alcohol concentration exceeds 0.4%, coma or death may occur. Table 12.2 gives some of the typical behaviors exhibited at various levels of blood alcohol.

$$CH_3CH_2OH \xrightarrow{[O]} CH_3\overset{\displaystyle O}{\overset{\|}{C}}H \xrightarrow{[O]} 2CO_2 + H_2O$$

Ethanol (ethyl alcohol) Ethanal (acetaldehyde)

The acetaldehyde produced from ethanol in the liver is further oxidized to acetic acid, which is converted to carbon

dioxide and water in the citric acid (Krebs) cycle. Thus, the enzymes in the liver can eventually break down ethanol, but the aldehyde and carboxylic acid intermediates can cause considerable damage while they are present within the cells of the liver.

A person weighing 150 lb requires about one hour to completely metabolize 10 ounces of beer. However, the rate of metabolism of ethanol varies between nondrinkers and drinkers. Typically, nondrinkers and social drinkers can metabolize 12–15 mg of ethanol/dL of blood in one hour, but an alcoholic can metabolize as much as 30 mg of ethanol/dL in one hour. Some effects of alcohol metabolism include an increase in liver lipids (fatty liver), an increase in serum triglycerides, gastritis, pancreatitis, ketoacidosis, alcoholic hepatitis, and psychological disturbances.

When the Breathalyzer test is used for suspected drunk drivers, the driver exhales a volume of breath into a solution containing the orange Cr^{6+} ion. If there is ethyl alcohol present in the exhaled air, the alcohol is oxidized, and the Cr^{6+} is reduced to give a green solution of Cr^{3+}.

$$CH_3CH_2OH + Cr^{6+} \xrightarrow{[O]} CH_3\overset{\displaystyle O}{\overset{\|}{C}}OH + Cr^{3+}$$

ethanol orange acetic acid green

Sometimes alcoholics are treated with a drug called Antabuse (disulfiram), which prevents the oxidation of acetaldehyde to acetic acid. As a result, acetaldehyde accumulates in the blood, which causes nausea, profuse sweating, headache, dizziness, vomiting, and respiratory difficulties. Because of these unpleasant side effects, the patient is less likely to use alcohol.

TABLE 12.2

Typical Behaviors Exhibited by a 150-lb Person Consuming Alcohol

Number of Beers (12 oz) or Glasses of Wine (5 oz)	Blood Alcohol Level (w/v %)	Typical Behavior
1	0.025	Slightly dizzy, talkative
2	0.05	Euphoria, loud talking and laughing
4	0.10	Loss of inhibition, loss of coordination, drowsiness, legally drunk in most states
8	0.20	Intoxicated, quick to anger, exaggerated emotions
12	0.30	Unconscious
16–20	0.40–0.50	Coma and death

Much of the protein in the hair is cross-linked by disulfide bonds, which occur mostly between the thiol groups of the amino acid cysteine.

$$NH_2\!-\!\underset{\underset{cysteine}{}}{\overset{\overset{\displaystyle COOH}{|}}{CH}}\!-\!CH_2\!-\!SH + HS\!-\!CH_2\!-\!\overset{\overset{\displaystyle COOH}{|}}{CH}\!-\!NH_2$$

$$\xrightarrow{[O]}\ NH_2\!-\!\overset{\overset{\displaystyle COOH}{|}}{CH}\!-\!CH_2\!-\!S\!-\!S\!-\!CH_2\!-\!\overset{\overset{\displaystyle COOH}{|}}{CH}\!-\!NH_2\!-\!H_2O$$

When a person is given a "perm," a reducing substance is used to break the disulfide bonds. While the hair is still wrapped around the curlers, an oxidizing substance is applied that causes new disulfide bonds to form between different parts of the protein hair strands, which gives the hair a new shape.

QUESTIONS AND PROBLEMS

■ Reactions of Alcohols and Thiols

12.17 Draw the condensed structural formula of the alkene produced by each of the following dehydration reactions:

a. $CH_3-CH_2-CH_2=CH_2-OH \xrightarrow[\text{Heat}]{H^+}$

b.
OH
$\xrightarrow[\text{Heat}]{H^+}$

c.
OH
$CH_3-CH_2-CH-CH_2-CH_3 \xrightarrow[\text{Heat}]{H^+}$

12.18 Draw the condensed structural formula of the alkene produced by each of the following dehydration reactions:

a. CH_3-CH_2-OH

b.
CH$_3$
$CH_3-CH-CH_2-OH \xrightarrow[\text{Heat}]{H^+}$

c.
OH
$\xrightarrow[\text{Heat}]{H^+}$

12.19 Draw the condensed structural formula of the aldehyde or ketone when each of the following alcohols is oxidized [O] (if no reaction, write *none*):

a. $CH_3-CH_2-CH_2-CH_2-CH_2-OH$

b.
OH
$CH_3-CH_2-CH-CH_3$

c.
OH

d.
OH   CH$_3$
$CH_3-CH-CH_2-CH-CH_3$

12.20 Draw the condensed structural formula of the organic product when each of the following alcohols is oxidized [O] (if no reaction, write *none*):

a.
CH$_3$
$CH_3-CH-CH_2-CH_2-OH$

b.
OH
$CH_3-CH_2-C-CH_3$
CH$_3$

c.
OH
$CH_3-CH_2-CH-CH_2-CH_3$

d.
OH

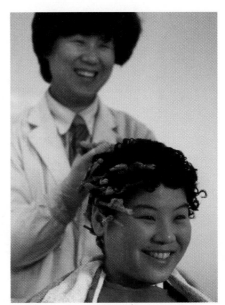

■ 12.4 Aldehydes and Ketones

As we learned in Chapter 10, the carbonyl group consists of a carbon–oxygen double bond. The double bond in the carbonyl group is similar to that of alkenes, except the carbonyl group has a dipole. The oxygen atom with two lone pairs of electrons is much more electronegative than the carbon atom. Therefore, the carbonyl group has a strong dipole with a partial negative charge (δ^-) on the oxygen, and a partial positive charge (δ^+) on the carbon. The polarity of the carbonyl group strongly influences the physical and chemical properties of aldehydes and ketones.

LEARNING GOAL

Identify compounds with the carbonyl group as aldehydes and ketones; write their IUPAC names.

the
Chemistry
place

WEB TUTORIAL
Aldehydes and Ketones

Carbonyl group

Aldehyde   Ketone

FIGURE 12.3

The carbonyl group in aldehydes and ketones.

Q If aldehydes and ketones both contain a carbonyl group, how can you differentiate between compounds from each family?

In an **aldehyde**, the carbon of the carbonyl group is bonded to at least one hydrogen atom. That carbon may also be bonded to another hydrogen, a carbon of an alkyl group, or an aromatic ring. (See Figure 12.3.) In a **ketone**, the carbonyl group is bonded to two alkyl groups or aromatic rings.

In the condensed structural formula, the aldehyde group may be drawn as separate atoms or it may be written as —CHO, with the double bond understood. The keto group (C=O) is sometimes written as CO.

Aldehyde

$$CH_3-CH_2-\overset{\overset{\text{O}}{\|}}{C}-H \ = \ CH_3-CH_2-CHO$$

Ketone

$$CH_3-\overset{\overset{\text{O}}{\|}}{C}-CH_3 \ = \ CH_3-CO-CH_3$$

SAMPLE PROBLEM 12.7

▪ Identifying Aldehydes and Ketones

Identify each of the following compounds as an aldehyde or ketone:

a. $CH_3-\overset{\overset{\text{CH}_3}{|}}{\underset{\overset{|}{\text{CH}_3}}{C}}-CH_2-\overset{\overset{\text{O}}{\|}}{C}-H$ **b.**   **c.**

SOLUTION

a. aldehyde   **b.** ketone   **c.** aldehyde

STUDY CHECK

Draw the condensed structural formula of a ketone that has a carbonyl group bonded to two ethyl groups.

▪ Naming Aldehydes

In the IUPAC names of aldehydes, the *e* of the alkane name is replaced with *al*.

STEP 1 **Name the longest carbon chain containing the carbonyl group by replacing the *e* in the corresponding alkane name by *al*.** No number is needed for the aldehyde group because it always appears at the end of the chain.

The IUPAC system names the aldehyde of benzene as benzaldehyde.

benzaldehyde

The carbonyl carbon is at the end of the chain

| IUPAC: | Methanal | Ethanal | Propanal | Butanal |
| Common: | **Form**aldehyde | **Acet**aldehyde | **Propion**aldehyde | **Butyr**aldehyde |

FIGURE 12.4

In the structures of aldehydes, the carbonyl group is always the end carbon.

Q Why is the carbon in the carbonyl group in aldehydes always at the end of the chain?

The first four unbranched aldehydes are often referred to by their common names, which end in *aldehyde*. (See Figure 12.4.) the roots of these common names are derived from Latin or Greek words that indicate the source of the corresponding carboxylic acid. We will look at carboxylic acids in the next section.

STEP 2 **Name and number any substituents on the carbon chain by counting the carbonyl carbon as carbon 1.**

2-methylpropanal   4-methylpentanal

SAMPLE PROBLEM 12.8

■ **Naming Aldehydes**

Give the IUPAC names for the following aldehydes:

SOLUTION

a. pentanal

b. 4-chlorobenzaldehyde

c. The longest unbranched chain has four atoms with a methyl group on the third carbon. The IUPAC name is 3-methylbutanal.

STUDY CHECK

What are the IUPAC and common names of the aldehyde with three carbon atoms?

ENVIRONMENTAL NOTE

▪Vanilla

Vanilla has been used as a flavoring for over a thousand years. After drinking a beverage made from powdered vanilla and cocoa beans with Emperor Montezuma in Mexico, Cortez took vanilla back to Europe where it became popular for flavoring and for scenting perfumes and tobacco. Thomas Jefferson introduced vanilla to the United States in the late 1700s. Today much of the vanilla we use in the world is grown in Mexico, Madagascar, Réunion, Seychelles, Tahiti, Ceylon, Java, the Philippines, and Africa.

The vanilla plant is a member of the orchid family. There are many species of *Vanilla*, but *Vanilla planifolia* (or *V. fragrans*) is considered to produce the best flavor. The vanilla plant grows like a vine and can grow to 100 feet in length. Its flowers are hand-pollinated to produce a green fruit that is picked in 8 or 9 months. It is sun-dried to form a long, dark brown pod, which is called "vanilla bean" because it looks like a string bean. The flavor and fragrance of the vanilla bean comes from the black seeds found inside the dried bean.

The seeds and pod are used to flavor desserts such as custards and ice cream. The extract of vanilla is made by chopping up vanilla beans and mixing them with a 35% alcohol–water mixture. The liquid, which contains the aldehyde vanillin, is drained from the bean residue and used for flavoring.

vanillin

▪ Naming Ketones

Aldehydes and ketones are some of the most important classes of organic compounds. Because they have played a major role in organic chemistry for more than a century, the common names for unbranched ketones are still in use. In the common names the alkyl groups bonded to the carbonyl group are named as substituents and listed alphabetically followed by *ketone*. Acetone, which is another name for propanone, has been retained by the IUPAC system.

In the IUPAC system, the name of a ketone is obtained by replacing the *e* in the corresponding alkane name with *one*.

STEP 1 **Name the longest carbon chain containing the carbonyl group by replacing the *e* in the corresponding alkane name by *one*.**

STEP 2 **Number the main chain starting from the end nearest the carbonyl group.** Place the number of the carbonyl carbon in front of the ketone name. (Propanone and butanone do not require numbers.)

propanone
(dimethyl ketone: acetone)

butanone
(ethyl methyl ketone)

3-pentanone
(diethyl ketone)

STEP 3 Name and number any substituents on the carbon chain.

3-methylbutanone   2-bromo-3-pentanone

STEP 4 For cyclic ketones, the prefix *cyclo* is used in front of the ketone name. Any substituents are located by numbering the ring starting with the carbonyl carbon as carbon 1. The ring is numbered so that the substituents have the lowest possible number.

cyclopentanone   3-methylcyclohexanone   2,3-dichlorocyclopentanone

---

SAMPLE PROBLEM 12.9

■ Names of Ketones

Give the IUPAC name for the following ketone:

SOLUTION

The longest chain is five carbon atoms. Counting from the right, the carbonyl group is on carbon 2 and a methyl group is on carbon 4. The IUPAC name is 4-methyl-2-pentanone.

STUDY CHECK

What is the common name of 3-hexanone?

QUESTIONS AND PROBLEMS

■ Aldehydes and Ketones

12.21 Identify the following compounds as aldehydes or ketones:

a. CH₃—CH₂—C—CH₃   **b.**   **c.**

12.22 Identify the following compounds as aldehydes or ketones:

a.   **b.** CH₃—CH—C—H

c.

Some Important Aldehydes and Ketones

Formaldehyde, the simplest aldehyde, is a colorless gas with a pungent odor. Industrially, it is a reactant in the synthesis of polymers used to make fabrics, insulation materials, carpeting, pressed wood products such as plywood, and plastics for kitchen counters. An aqueous solution called formalin, which contains 40% formalde-hyde, is used as a germicide and to preserve biological specimens. Exposure to formaldehyde fumes can irritate eyes, nose, upper respiratory tract, and cause skin rashes, headaches, dizziness, and general fatigue.

Acetone, or propanone (dimethyl ketone), which is the simplest ketone, is a colorless liquid with a mild odor that has wide use as a solvent in cleaning fluids, paint and nail-polish removers, and rubber cement. (See Figure 12.5.) It is extremely flammable and care must be taken when using acetone. In the body, acetone may be produced in uncontrolled diabetes, fasting, and high-protein diets when large amounts of fats are metabolized for energy.

Several naturally occurring aromatic aldehydes are used to flavor food and as fragrances in perfumes. Benzaldehyde is found in almonds, vanillin in vanilla beans, and cinnamaldehyde in cinnamon.

benzaldehyde (almond) vanillin (vanilla) cinnamaldehyde (cinnamon)

The flavor of butter or margarine is from butanedione, muscone is used to make musk perfumes, and oil of spearmint contains carvone.

butanedione (butter flavor) muscone (musk) carvone (spearmint oil)

O O
‖ ‖
CH₃—C—C—CH₃
Butanedione

FIGURE 12.5

Acetone is used as a solvent in paint and nail-polish removers.

Q **What is the IUPAC name of acetone?**

12.23 Give a common name for each of the following compounds:

a. CH₃—C(=O)—H

b. CH₃—C(=O)—CH₂—CH₂—CH₃

c. H—C(=O)—H

12.24 Give the common name for each of the following compounds:

a. $CH_3-\overset{\overset{O}{\|}}{C}-CH_2-CH_3$ b. $CH_3-CH_2-\overset{\overset{O}{\|}}{C}-CH_2-CH_3$

c. $CH_3-CH_2-\overset{\overset{O}{\|}}{C}-H$

12.25 Give the IUPAC name for each of the following compounds:

a. $CH_3-CH_2-\overset{\overset{O}{\|}}{C}-H$ b. $CH_3-CH_2-\overset{\overset{O}{\|}}{C}-\overset{\overset{CH_3}{|}}{CH}-CH_3$

c.

d. $\overset{\overset{O}{\|}}{C}-H$

12.26 Give the IUPAC name for each of the following compounds:

a. $CH_3-CH_2-CH_2-\overset{\overset{O}{\|}}{C}-H$ b. $CH_3-CH_2-CH_2-\overset{\overset{O}{\|}}{C}-CH_3$

c.

d. $\overset{\overset{O}{\|}}{C}-H$

12.27 Write the condensed structural formula for each of the following compounds:
   a. acetaldehyde   **b.** 2-pentanone
   c. butyl methyl ketone   **d.** 3-methylpentanal

12.28 Write the condensed structural formula for each of the following compounds:
   a. propionaldehyde   **b.** butanal
   c. 4-bromobutanone   **d.** acetone

12.5 Properties of Aldehydes and Ketones

LEARNING GOAL

Compare the boiling points and solubility of aldehydes and ketones to those of alkanes and alcohols.

At room temperature, formaldehyde (bp $-21°C$) and acetaldehyde (bp $21°C$) are gases. Aldehydes containing from 3 to 10 carbon atoms are liquids. The polar carbonyl group with a partially negative oxygen atom and a partially positive carbon atom has an influence on the boiling points and the solubility of aldehydes and ketones in water.

Boiling Points

The polar carbonyl group gives aldehydes and ketones higher boiling points than alkanes and ethers of similar mass. The increase in boiling points is due to dipole–dipole interactions.

Dipole–dipole interaction

$$>C^{\delta+}=O^{\delta-} \;||||||||\; >C^{\delta+}=O^{\delta-} \;||||||||\; >C^{\delta+}=O^{\delta-}$$

However, because there is no hydrogen on the oxygen atom, aldehydes and ketones cannot form hydrogen bonds with each other. Thus they have boiling points that are lower than alcohols.

	CH_3—CH_2—CH_2—CH_3	CH_3—CH_2—O—CH_3	CH_3—CH_2—$\overset{\overset{\displaystyle O}{\|}}{C}$—H	CH_3—$\overset{\overset{\displaystyle O}{\|}}{C}$—$CH_3$	CH_3—CH_2—CH_2—OH
Name	butane	ethyl methyl ether	propanal	propanone	1-propanol
Molar Mass	58	60	58	58	60
Family	alkane	ether	aldehyde	ketone	alcohol
bp	0°C	8°C	49°C	56°C	97°C

Increasing boiling point ⟶

Solubility of Aldehydes and Ketones in Water

Hydrogen bond

Acetaldehyde

Although aldehydes and ketones do not hydrogen bond with each other, the electronegative oxygen atom does hydrogen bond with water molecules. Carbonyl compounds with one to four carbons are very soluble in water. However, those with five carbon atoms or more are not very soluble because the alkyl portions diminish the effect of the polar carbonyl group.

SAMPLE PROBLEM 12.10

■ **Boiling Point and Solubility**

Would you expect ethanol CH_3CH_2OH to have a higher or lower boiling point than ethanal CH_3CHO? Explain.

SOLUTION

Ethanol would have a higher boiling point because its molecules can hydrogen bond with each other, while molecules of ethanal cannot.

STUDY CHECK

If acetone molecules cannot hydrogen bond with each other, why is acetone soluble in water?

Hydrogen bond

Acetone

Oxidation

Aldehydes oxidize readily to carboxylic acids. In contrast, ketones do not undergo further oxidation.

CH_3—$\overset{\overset{\displaystyle O}{\|}}{C}$—H $\xrightarrow{\text{Oxidation}}$ CH_3—$\overset{\overset{\displaystyle O}{\|}}{C}$—OH

acetaldehyde   acetic acid

CH_3—$\overset{\overset{\displaystyle O}{\|}}{C}$—$CH_3$ $\xrightarrow{\text{Oxidation}}$ no reaction

propanone

$$Ag^+ + 1e^- \longrightarrow Ag(s)$$

FIGURE 12.6

In Tollens' test, a silver mirror forms when the oxidation of an aldehyde reduces silver ion to metallic silver. The silvery surface of a mirror is formed in a similar way.

Q What is the product of the oxidation of an aldehyde?

Tollens' Test

Tollens' test, which uses a solution of Ag^+ ($AgNO_3$) and ammonia, oxidizes aldehydes, but not ketones. The silver ion is reduced and forms a "silver mirror" on the inside of the container. Commercially, a similar process is used to make mirrors by applying a mixture of $AgNO_3$ and ammonia on glass with a spray gun. (See Figure 12.6.)

$$\underset{\text{Acetaldehyde}}{CH_3-\overset{\overset{\displaystyle O}{\|}}{C}-H} + \underset{\substack{\text{Tollens}\\\text{reagent}}}{2Ag^+} \xrightarrow{[O]} \underset{\substack{\text{Silver}\\\text{mirror}}}{2Ag(s)} + \underset{\text{Acetic acid}}{CH_3-\overset{\overset{\displaystyle O}{\|}}{C}-OH}$$

Another test, called **Benedict's test**, gives a positive test with compounds that have an aldehyde functional group and an adjacent hydroxyl group. When Benedict's reagent containing Cu^{2+} ($CuSO_4$) ions is added to this type of aldehyde, a brick-red solid of Cu_2O forms. (See Figure 12.7.) The test is negative with simple aldehydes and ketones.

$$\underset{\text{2-hydroxypropanal}}{CH_3-\overset{\overset{\displaystyle OH}{|}}{CH}-\overset{\overset{\displaystyle O}{\|}}{C}-H} + \underset{\substack{\text{Benedict's}\\\text{reagent}}}{Cu^{2+}} \xrightarrow{[O]} \underset{\text{brick-red solid}}{Cu_2O(s)} + \underset{\text{2-hydroxypropanoic acid}}{CH_3-\overset{\overset{\displaystyle OH}{|}}{CH}-\overset{\overset{\displaystyle O}{\|}}{C}-OH}$$

Cu^{2+} $Cu_2O(s)$

FIGURE 12.7

The blue Cu^{2+} in Benedict's solution forms a brick-red solid of Cu_2O in a positive test for many sugars and aldehydes with adjacent hydroxyl groups.

Q Which test tube contains an aldehyde with an adjacent hydroxyl group?

Because many sugars such as glucose contain this type of aldehyde grouping, Benedict's reagent can be used to determine the presence of glucose in blood or urine.

D-glucose   D-gluconic acid

QUESTIONS AND PROBLEMS

■ Properties of Aldehydes and Ketones

12.29 Which compound in each of the following pairs would have the higher boiling point? Explain.

a. $CH_3-CH_2-CH_3$ or $CH_3-CH_2-\overset{\overset{\displaystyle O}{\|}}{C}-H$

b. propanal or pentanal   **c.** butanal or 1-butanol

12.30 Which compound in each of the following pairs would have the higher boiling point? Explain.

a. $CH_3-CH_2-\overset{\overset{\displaystyle OH}{|}}{CH}-CH_3$ or $CH_3-CH_2-\overset{\overset{\displaystyle O}{\|}}{C}-CH_3$

b. butane or butanone   **c.** propanone or pentanone

12.31 Which compound in each of the following pairs would be more soluble in water? Explain.

a. $CH_3-\overset{\overset{\displaystyle O}{\|}}{C}-CH_2-CH_3$ or $CH_3-\overset{\overset{\displaystyle O}{\|}}{C}-\overset{\overset{\displaystyle O}{\|}}{C}-CH_3$

b. acetone or 2-pentanone

12.32 Which compound in each of the following pairs would be more soluble in water? Explain.
a. $CH_3-CH_2-CH_3$ or CH_3-CH_2-CHO
b. propanone or 3-hexanone

12.33 Draw the condensed structural formula of the organic product when each of the following alcohols is oxidized [O] (if no reaction, write *none*):
a. $CH_3-CH_2-CH_2-CH_2-CH_2-OH$

b. $CH_3-CH_2-\overset{\overset{\displaystyle OH}{|}}{CH}-CH_3$ **c.**

d. $CH_3-\overset{\overset{\displaystyle OH}{|}}{CH}-CH_2-\overset{\overset{\displaystyle CH_3}{|}}{CH}-CH_3$ **e.** $CH_3-\overset{\overset{\displaystyle CH_3}{|}}{CH}-CH_2-CH_2-OH$

12.34 Draw the condensed structural formula of the organic product when each of the following alcohols is oxidized [O] (if no reaction, write *none*):

c. $CH_3-CH_2-\overset{\overset{\displaystyle OH}{|}}{\underset{\underset{\displaystyle CH_3}{|}}{C}}-CH_3$ d. $CH_3-CH-\overset{\overset{\displaystyle OH}{|}}{CH}-CH_2-CH_3$
                                                                $\underset{\underset{\displaystyle OH}{|}}{}$

e.

12.6 Chiral Molecules

LEARNING GOAL

Identify chiral and achiral carbon atoms in an organic molecule.

In the preceding chapters, we have looked at some types of isomers. Let's review those now. Molecules are structural isomers when they have the same molecular formula, but different bonding arrangements.

Isomers

C_2H_6O CH_3-CH_2-OH CH_3-O-CH_3
                ethanol   dimethyl ether

C_3H_6O $CH_3-CH_2-\overset{\overset{\displaystyle O}{\|}}{C}-H$ $CH_3-\overset{\overset{\displaystyle O}{\|}}{C}-CH_3$
                propanal   propanone

Another group of isomers called **stereoisomers** have identical molecular formulas, too, but they are not structural isomers. In stereoisomers the atoms are bonded in the same sequence, but differ in the way they are arranged in space.

Chirality

When stereoisomers have mirror images that are different, they are said to have "handedness." If you look at the palms of your hands, your thumbs are on opposite sides. If you turn your palms toward each other, you have mirror images. (See Figure 12.8.) The left hand is the mirror image of the right hand. Your hands are not

Left hand   Right hand

Mirror image of right hand

FIGURE 12.8

A pair of hands are chiral because they have mirror images that cannot be superimposed on each other.

Q **Why are your shoes chiral objects?**

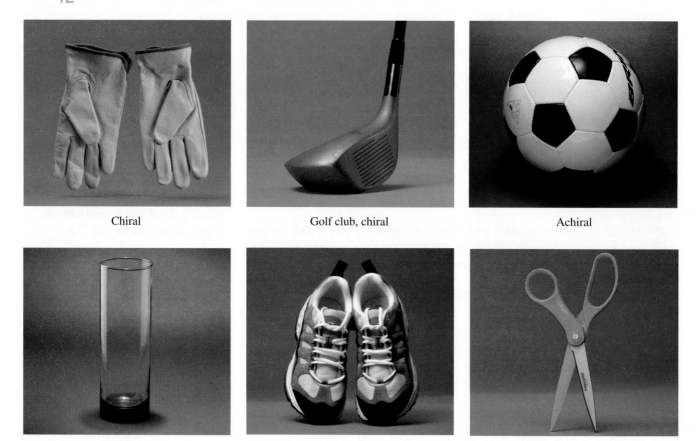

| Chiral | Golf club, chiral | Achiral |

| Achiral | Chiral | Right-handed scissors, chiral |

FIGURE 12.9

Everyday objects can be chiral or achiral.

Q Why are some of the above objects chiral and others are achiral?

superimposable. Objects such as hands that have nonsuperimposable mirror images are **chiral**. Left and right shoes are chiral; left- and right-handed golf clubs are chiral. When one mirror image can be superimposed on the other, the object is *achiral*. The object has no "handedness."

Molecules in nature also have mirror images, and often the "left-handed" stereoisomer has a different biological effect than does the "right-handed" one. For some compounds, one isomer has a certain odor, and the mirror image has a completely different odor. For example, one enantiomer of limonene smells like oranges, while its mirror image has the odor of lemons. If we try to put a left-hand glove on our right hand, or a right shoe on our left foot, or use left-handed scissors if we are right handed, we begin to realize that certain properties of mirror images can be very different. (See Figure 12.9.)

Chiral Carbon Atoms

A carbon compound is chiral if it has at least one carbon atom bonded to four different atoms or groups. This type of carbon atom is called a **chiral carbon** because there are two different ways that it can bond to four atoms or groups of atoms. The resulting structures are mirror images of each other. Let's look at the mirror images of a carbon bonded to four different atoms. (See Figure 12.10.) If

FIGURE 12.10

(a) The enantiomers of a chiral molecule are mirror images. (b) The enantiomers of a chiral molecule cannot be superimposed on each other.

Q Why is the carbon atom in this compound chiral?

we line up the hydrogen and iodine atoms in the mirror images, the bromine and chorine atoms appear on opposite sides. No matter how we turn the models, we cannot align all four atoms at the same time. When stereoisomers cannot be superimposed, they are called **enantiomers**. If two or more atoms are the same, the atoms can be aligned (superimposed) and the mirror images represent the same structure. (See Figure 12.11 on the following page.)

SAMPLE PROBLEM 12.11

■ **Chiral Carbons**

Indicate whether the carbon in red is chiral or not chiral.

SOLUTION

a. Not chiral. Two of the substituents on the carbon are the same (Cl). A chiral carbon must be bonded to four different groups or atoms.

b. Chiral. Carbon 2 is bonded to four different groups: one OH, one CH_3, one CH_2—CH_3, and one H.

c. Not chiral. Carbon 2 is bonded to only three groups, not four.

STUDY CHECK

Circle the two chiral carbons in the structural formula of the carbohydrate erythrose.

$$HO-CH_2-\overset{\overset{\displaystyle OH}{|}}{CH}-\overset{\overset{\displaystyle OH}{|}}{CH}-\overset{\overset{\displaystyle O}{||}}{C}-H$$

erythrose

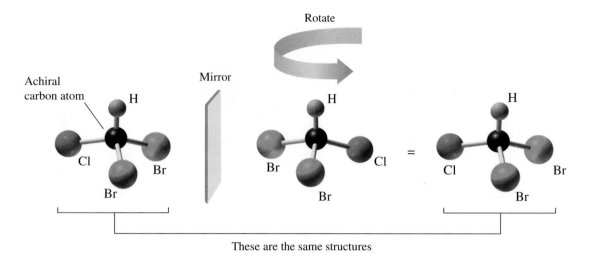

These are the same structures

FIGURE 12.11

The mirror images of an achiral compound can be superimposed on each other.

Q **Why do the mirror images have the same structure?**

Drawing Fischer Projections

Emil Fischer devised a simplified system for drawing stereoisomers that shows the arrangements of the atoms. Fischer received the Nobel Prize in 1902 for his contributions to carbohydrate and protein chemistry. In a **Fischer projection**, the bonds to a chiral atom are drawn as intersecting lines, with the chiral carbon being at the center where the lines cross. The horizontal lines represent the bonds that come forward in the three-dimensional structure, and the vertical lines represent the bonds that point away.

By convention the carbon chain in the Fischer projection is written vertically with the most highly oxidized carbon at the top. For glyceraldehyde, the simplest sugar, the carbonyl group, which is the most highly oxidized group in the molecule, is written at the top. The letter L is assigned to the stereoisomer that has the —OH group on the left of the chiral carbon. The letter D is assigned to the structure where the —OH is on the right of the chiral carbon. Let's look at how glyceraldehyde is converted from a three-dimensional view to a Fischer projection. (See Figure 12.12.)

FIGURE 12.12

In a Fischer projection, the chiral carbon atom is at the center with horizontal lines for bonds that extend toward the viewer and vertical lines for bonds that point away.

Q **Why does glyceraldehyde have only one chiral carbon atom?**

Dash-wedge structures of glyceraldehyde

Fischer projections of glyceraldehyde

L-glyceraldehyde

D-glyceraldehyde

Fischer projections can also be written for larger compounds that have two or more chiral carbons. For example, the mirror images below are different. Each chiral atom is bonded to four different groups. To draw the mirror image of a Fischer projection, the positions of the substituents on the horizontal lines are reversed, while the groups on the vertical line are left unchanged.

We can also draw the mirror image of the carbohydrate erythrose, which has two chiral carbons.

L-erythrose   D-erythrose

SAMPLE PROBLEM 12.12

▪ Fischer Projections

Determine if each Fischer projection is a chiral compound. If so, identify it as the D or L isomer and draw the mirror image.

a.
$$CH_2OH$$
$$HO{-}|{-}H$$
$$CH_3$$

b.
$$CH_3$$
$$HO{-}|{-}H$$
$$CH_3$$

c.
$$CHO$$
$$H{-}|{-}OH$$
$$CH_3$$

SOLUTION

a. This chiral compound with four different substituents is the L isomer. The mirror image is written by reversing the H and OH on the horizontal lines.

$$CH_2OH$$
$$H{-}|{-}OH$$
$$CH_3$$

b. The compound is achiral because it has two identical groups (CH_3).

c. This chiral compound with four different substituents is the D isomer. The mirror image is written by reversing the H and OH on the horizontal lines.

$$CHO$$
$$HO{-}|{-}H$$
$$CH_3$$

STUDY CHECK

Draw the Fischer projections for the D and L stereoisomers of 2-hydroxypropanal.

HEALTH NOTE

■ Enantiomers in Biological Systems

Most stereoisomers that are active in biological systems consist of only one enantiomer. Rarely are both enantiomers of biological molecules active. This happens because the enzymes and cell surface receptors on which metabolic reactions take place also have "handedness." Thus, only one of the enantiomers of a reactant or a drug interacts with its enzymes or receptors; the other is inactive. At the target site, the chiral receptor fits the arrangement of the substituents in only one enantiomer. Its mirror image, which is inactive, does not fit properly. This is similar to the idea that your right hand will only fit into a right-handed glove. (See Figure 12.13.)

A substance called carvone exists as two enantiomers. One enantiomer gives the odor of spearmint oil, while the other enantiomer produces the flavor of caraway seeds. Such differences in odor are due to the chiral receptor sites in the nose that fit the shape of one enantiomer, but not the other. Thus, our senses of smell and also taste are sensitive to the chirality of molecules.

Enantiomers of Carvone

From spearmint oil   From caraway seeds

In the brain, one stereoisomer of LSD causes hallucinations because it affects the production of serotonin, a chemical that is important in sensory perception. However, its enantiomer LSD produces little effect in the brain. The behavior of nicotine and epinephrine (adrenaline) also depends upon one of their enantiomers. One enantiomer of nicotine is more toxic than the other.

In epinephrine (adrenaline), one enantiomer is responsible for the constriction of blood vessels.

Nicotine   Adrenalin (epinephrine)

A substance used to treat Parkinson's disease is L-dopa, which is converted to dopamine in the brain, where it raises the serotonin level. However, the D-dopa enantiomer has no biological effect.

L-Dopa, anti-Parkinsonian drug   D-Dopa has no biological effect

For many drugs, only one of the enantiomers is biologically active. However, for many years, drugs have been produced that were mixtures of their enantiomers. Today, drug researchers are using *chiral technology* to produce the active enantiomers of chiral drugs. Chiral catalysts are being designed that direct the formation of just one enantiomer rather than both. The benefits of producing only the active enantiomer include using a lower dose, enhancing activity, reducing interactions with other drugs, and eliminating possible harmful side effects from the nonactive enantiomer. Several active enantiomers are now being produced such as L-dopa and the active enantiomer of the popular analgesic ibuprofen used in Advil, Motrin, and Nuprin.

Ibuprofen

FIGURE 12.13

(a) The substituents on the biologically active enantiomer bind to all the sites on a chiral receptor; **(b)** its enantiomer does not bind properly and is not active biologically.

Q Why don't all the substituents of the mirror image of the active enantiomer fit into a chiral receptor site?

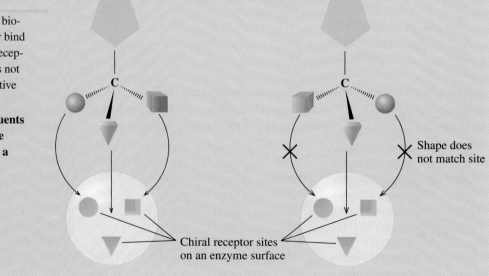

Shape does not match site

Chiral receptor sites on an enzyme surface

QUESTIONS AND PROBLEMS

■ Chiral Molecules

12.35 Identify each of the following structures as chiral or achiral. If chiral, indicate the chiral carbon.

$$\textbf{a.}\ CH_3-\underset{\underset{OH}{|}}{CH}-CH_3 \qquad \textbf{b.}\ CH_3-\underset{\underset{Br}{|}}{CH}-CH_2-CH_3$$

$$\textbf{c.}\ CH_3-\underset{\underset{Br}{|}}{CH}-\underset{\overset{O}{\|}}{C}-H \qquad \textbf{d.}\ CH_3-CH_2-\underset{\overset{O}{\|}}{C}-CH_3$$

12.36 Identify each of the following structures as chiral or achiral. If chiral, indicate the chiral carbon.

$$\textbf{a.}\ CH_3-\underset{\underset{CH_3}{|}}{\overset{\overset{Cl}{|}}{C}}-CH_2-\underset{\underset{}{\overset{\overset{Cl}{|}}{CH}}}-CH_3 \qquad \textbf{b.}\ CH_3-\underset{\overset{Br}{|}}{C}=CH-CH_3$$

$$CH_3-\underset{\underset{CH_3}{|}}{\overset{\overset{OH}{|}}{C}}-\overset{\overset{OH}{|}}{CH}-CH_3 \qquad \textbf{d.}\ Br-CH_2-\underset{\overset{Cl}{|}}{CH}-CH_3$$

12.37 Identify the chiral carbon in each of the following naturally occurring compounds.
   a. citronellol; one enantiomer has the geranium odor

$$CH_3-\underset{\overset{CH_3}{|}}{C}=CH-CH_2-CH_2-\underset{\overset{CH_3}{|}}{CH}-CH_2-CH_2-OH$$

   b. alanine, amino acid

$$H_2N-\underset{\overset{CH_3}{|}}{CH}-\underset{\overset{O}{\|}}{C}-OH$$

12.38 Identify the chiral carbon in each of the following naturally occurring compounds.
   a. amphetamine (Benzedrine), stimulant, treatment of hyperactivity

   b. Norepinephrine, increases blood pressure and nerve transmission

12.39 Draw Fischer projections for each of the following dash-wedge structures.

12.40 Draw Fischer projections for each of the following dash-wedge structures.

a.
```
        Br
        ┊
        C
     ╱  ┊  ╲
  HO    ┊   Br
      CH₂OH
```
b.
```
        CH₃
        ┊
        C
     ╱  ┊  ╲
   H    ┊   OH
      CH₂OH
```
c.
```
        CHO
        ┊
        C
     ╱  ┊  ╲
  HO    ┊   H
      CH₂OH
```

12.41 Indicate whether each pair of Fischer projections represent enantiomers or identical structures.

a.
CH₃
Br——Cl and Cl——Br
CH₃
CH₃

b.
CHO
HO——H and H——OH
CH₃
CH₃

c.
H
Br——Cl and Br——Cl
CH₃
CH₃ / H

d.
COOH
H——OH and HO——H
CH₃
CH₃

12.42 Indicate whether each pair of Fischer projections represent enantiomers or identical structures.

a.
CH₂OH
Br——Cl and Cl——Br
CH₃
CH₃

b.
CHO
H——H and H——H
CH₃
CH₃

c.
CH₃
H——OH and HO——H
CH₂CH₃
CH₂CH₃

d.
COOH
H——NH₂ and H₂N——H
CH₃
CH₃

CONCEPT MAP

CHAPTER REVIEW

12.1 Alcohols, Ethers, and Thiols
The functional group of an alcohol is the hydroxyl group —OH bonded to a carbon chain. In a phenol, the hydroxyl group is bonded to an aromatic ring. In the IUPAC system, the names of alcohols have *ol* endings, and the location of the —OH group is given by numbering the carbon chain. Simple alcohols are generally named by their common names with the alkyl name preceding the term *alcohol*. A cyclic alcohol is named as a cycloalkanol. An aromatic alcohol is named as a phenol. Alcohols are classified according to the number of alkyl or aromatic groups bonded to the carbon that holds the —OH group. In a primary (1°) alcohol one alkyl group is attached to the hydroxyl carbon. In a secondary (2°) alcohol two alkyl groups are attached, and in a tertiary (3°) alcohol there are three alkyl groups bonded to the hydroxyl carbon. In thiols, the functional group is —SH, which is analogous to the —OH group of alcohols. In an ether, an oxygen atom is connected by single bonds to two alkyl or aromatic groups. In the common names of ethers, the alkyl groups are listed alphabetically followed by the name *ether*.

12.2 Properties of Alcohols, Phenols, and Ethers
The —OH group allows alcohols to hydrogen bond, which causes alcohols to have higher boiling points than alkanes and ethers of similar mass. Short-chain alcohols and ethers can hydrogen bond with water, which makes them soluble in water.

12.3 Reactions of Alcohols and Thiols
At high temperatures, alcohols dehydrate in the presence of an acid to yield alkenes. Primary alcohols are oxidized to aldehydes, which can oxidize further to carboxylic acids. Secondary alcohols are oxidized to ketones. Tertiary alcohols do not oxidize.

12.4 Aldehydes and Ketones
Aldehydes and ketones contain a carbonyl group (C=O), which consists of a double bond between a carbon and an oxygen atom. In aldehydes, the carbonyl group appears at the end of carbon chains. In ketones, the carbonyl group occurs between two alkyl groups. In the IUPAC system, the *e* in the corresponding alkane is replaced with *al* for aldehydes and *one* for ketones. For ketones with more than four carbon atoms in the main chain, the carbonyl group is numbered to show its location. Many of the simple aldehydes and ketones use common names.

12.5 Properties of Aldehydes and Ketones
Because they contain a polar carbonyl group, aldehydes and ketones have higher boiling points than alkanes and ethers. However, their boiling points are lower than alcohols because aldehydes and ketones cannot hydrogen bond with each other. Aldehydes and ketones can hydrogen bond with water molecules, which makes carbonyl compounds with one to four carbon atoms soluble in water. Aldehydes are easily oxidized to carboxylic acids, but ketones do not oxidize further. Aldehydes, but not ketones, react with Tollens' reagent to give silver mirrors. In Benedict's test, aldehydes with adjacent hydroxyl groups reduce blue Cu^{2+} to give a brick-red Cu_2O solid.

12.6 Chiral Molecules
Chiral molecules are molecules with mirror images that cannot be superimposed on each other. These types of stereoisomers are called enantiomers. A chiral molecule must have at least one chiral carbon, which is a carbon bonded to four different atoms or groups of atoms. The Fischer projection is a simplified way to draw the arrangements of atoms by placing the chiral carbons at the center of crossed lines. The names of the mirror images are labeled D or L to differentiate between the enantiomers.

SUMMARY OF NAMING

Structure	Family	IUPAC Name	Common Name
CH_3—OH	Alcohol	Methanol	Methyl alcohol
⬡—OH	Phenol	Phenol	Phenol
CH_3—SH	Thiol	Methanethiol	
CH_3—O—CH_3	Ether		Dimethyl ether
CH_3—S—S—CH_3	Disulfide	Dimethyldisulfide	
H—$\overset{\overset{\displaystyle O}{\|}}{C}$—H	Aldehyde	Methanal	Formaldehyde
CH_3—$\overset{\overset{\displaystyle O}{\|}}{C}$—$CH_3$	Ketone	Propanone	Acetone; Dimethyl ketone

SUMMARY OF REACTIONS

COMBUSTION OF ALCOHOLS

$$CH_3—CH_2—OH + 3O_2 \longrightarrow 2CO_2 + 3H_2O$$

ethanol   oxygen   carbon dioxide water

DEHYDRATION OF ALCOHOLS TO FORM ALKENES

$$CH_3—CH_2—CH_2—OH \xrightarrow[\text{Heat}]{H^+}$$

1-propanol

$$CH_3—CH=CH_2 + H_2O$$

propene

OXIDATION OF PRIMARY ALCOHOLS TO FORM ALDEHYDES

$$CH_3—\underset{\text{ethanol}}{\overset{\overset{\displaystyle OH}{|}}{CH_2}} \xrightarrow{[O]} CH_3—\underset{\text{acetaldehyde}}{\overset{\overset{\displaystyle O}{\|}}{C}}—H + H_2O$$

OXIDATION OF SECONDARY ALCOHOLS TO FORM KETONES

$$CH_3—\underset{\text{2-propanol}}{\overset{\overset{\displaystyle OH}{|}}{CH}}—CH_3 \xrightarrow{[O]} CH_3—\underset{\text{propanone}}{\overset{\overset{\displaystyle O}{\|}}{C}}—CH_3 + H_2O$$

OXIDATION OF THIOLS TO FORM DISULFIDES

methanethiol   dimethyl disulfide

OXIDATION OF ALDEHYDES TO CARBOXYLIC ACIDS

$$CH_3—\underset{\text{acetaldehyde}}{\overset{\overset{\displaystyle O}{\|}}{C}}—H \xrightarrow{[O]} CH_3—\underset{\text{acetic acid}}{\overset{\overset{\displaystyle O}{\|}}{C}}—OH$$

KEY TERMS

alcohols Organic compounds that contain the hydroxyl (—OH) functional group, attached to a carbon chain.

aldehyde An organic compound with a carbonyl functional group and at least one hydrogen.

$$\overset{\overset{\displaystyle O}{\|}}{—C}—H \;=\; —CHO$$

Benedict's test A test for aldehydes with adjacent hydroxyl groups in which Cu^{2+} ($CuSO_4$) ions in Benedict's reagent are reduced to a brick-red solid of Cu_2O.

carbonyl group A functional group that contains a carbon–oxygen double bond ($C=O$).

chiral Objects or molecules that have mirror images that cannot be superimposed on each other.

chiral carbon A carbon atom that is bonded to four different atoms or groups of atoms.

dehydration A reaction that removes water from an alcohol in the presence of an acid to form alkenes.

disulfides Compounds formed from thiols; disulfides contain the —S—S— functional group.

enantiomers Stereoisomers that are mirror images that cannot be superimposed on each other.

ether An organic compound, —O—, in which an oxygen atom is bonded to two alkyl or two aromatic groups, or a mix of the two.

Fischer projection A system for drawing stereoisomers that shows horizontal lines for bonds coming forward, and

vertical lines for bonds going back with the chiral atom at the center.

ketone An organic compound in which the carbonyl functional group is bonded to two alkyl groups.

$$\overset{\overset{\displaystyle O}{\|}}{—C}— \;=\; —CO—$$

oxidation The loss of two hydrogen atoms from a reactant to give a more oxidized compound, e.g., primary alcohols oxidize to aldehydes, secondary alcohols oxidize to ketones. An oxidation can also be the addition of an oxygen atom as in the oxidation of aldehydes to carboxylic acids.

phenol An organic compound that has an —OH group attached to a benzene ring.

primary (1°) alcohol An alcohol that has one alkyl group bonded to the alcohol carbon atom.

secondary (2°) alcohol An alcohol that has two alkyl groups bonded to the carbon atom with the —OH group.

stereoisomers Isomers that have atoms bonded in the same order, but with different arrangements in space.

sulfhydryl group The —SH functional group.

tertiary (3°) alcohol An alcohol that has three alkyl groups bonded to the carbon atom with the —OH.

thiols Organic compounds that contain a thiol group (—SH).

Tollens' test A test for aldehydes in which Ag^+ in Tollens' reagent is reduced to metallic silver, which forms a "silver mirror" on the walls of the container.

ADDITIONAL QUESTIONS AND PROBLEMS

12.43 Classify each of the following as primary, secondary, or tertiary alcohols:

a.

b.

c. CH₃—CH—CH₂—OH
 with CH₃ above

d. CH₃—C—CH₂—CH—CH₃
 with CH₃ (top), CH₃ (bottom) on second C, and OH on fourth C

e. HO—CH₂—CH₂—CH₃

f.

12.44 Classify each of the following as primary, secondary, or tertiary alcohols:

a.

b.

c. CH₃—CH—CH₂—CH₃
 with CH₂—OH above

d. CH₃—C—CH₂—CH—CH₃
 with OH (top left), CH₃ (bottom left), CH₃ (top right)

e. CH₃—CH₂—CH₂—CH₂—OH

f.

12.45 Identify each of the following as an alcohol, a phenol, an ether, or a thiol:

a.

b.

c. CH₃—CH—CH₃
 with SH above

d. CH₃—C—CH₂—CH—CH₃
 with OH (top left), CH₃ (bottom left), CH₃ (top right)

e. CH₃—CH₂—CH₂—O—CH₃

f. CH₃—CH—CH₂—CH—CH₃
 with Br (top left), OH (top right)

g.

12.46 Identify each of the following as an alcohol, a phenol, an ether, or a thiol:

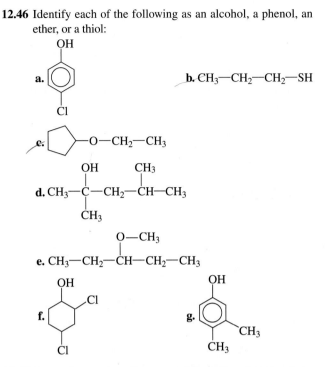

a.

b. CH₃—CH₂—CH₂—SH

c. —O—CH₂—CH₃ (cyclopentane)

d. CH₃—C—CH₂—CH—CH₃
 with OH (top left), CH₃ (bottom left), CH₃ (top right)

e. CH₃—CH₂—CH—CH₂—CH₃
 with O—CH₃ above

f.

g.

12.47 Draw the condensed structural formula of each of the following compounds:
a. 3-methylcyclopentanol
b. 4-chlorophenol
c. 2-methyl-3-pentanol
d. ethyl phenyl ether
e. 3-pentanone
f. 2,4-dibromophenol

12.48 Draw the condensed structural formula of each of the following compounds:
a. 3-pentanol
b. 2-pentanol
c. methyl propyl ether
d. 3-methyl-2-butanol
e. cyclohexanol

12.49 Which compound in each pair would you expect to have the higher boiling point? Why?
a. butane or 1-propanol
b. 1-propanol or ethyl methyl ether
c. ethanol or 1-butanol

12.50 Which compound in each pair would you expect to have the higher boiling point? Why?
a. propane or ethyl alcohol
b. 2-propanol or 2-pentanol
c. diethyl ether or 1-butanol

12.51 Explain why each of the following compounds would be soluble or insoluble in water.
a. 2-propanol
b. dimethyl ether
c. 1-hexanol

12.52 Explain why each of the following compounds would be soluble or insoluble in water.
a. glycerol
b. butane
c. 1,3-hexanediol

12.53 Draw the condensed structural formula for the product of each of the following reactions:
a. CH₃—CH₂—CH₂—OH $\xrightarrow{H^+,\ heat}$
b. CH₃—CH₂—CH₂—OH $\xrightarrow{[O]}$

c. CH₃—CH₂—CH(OH)—CH₃ $\xrightarrow{[O]}$

d. [cyclohexanol] $\xrightarrow{H^+, \text{ heat}}$ **e.** [cyclohexanol] $\xrightarrow{[O]}$

12.54 Draw the condensed structural formula for the product of each of the following reactions:

a. CH₃—CH(CH₃)—CH₂—OH $\xrightarrow{H^+, \text{ heat}}$

b. CH₃—CH(CH₃)—CH(OH)—CH₃ $\xrightarrow{[O]}$

c. [cyclopentanol] $\xrightarrow{H^+, \text{ heat}}$ **d.** [cyclopentanol] $\xrightarrow{[O]}$

e. CH₃—CH₂—CH₂—CH(OH)—CH₃ $\xrightarrow{[O]}$

12.55 Identify the functional groups in the following molecule:

testosterone

12.56 Identify the functional groups in the following molecule:

tetrahydrocannabinol (THC)

12.57 Write the condensed structural formulas for each of the following naturally occurring compounds:
   a. 2,5-Dichlorophenol, a defense pheromone of a grasshopper.
   b. Skunk scent, which contains 3-methyl-1-butanethiol.
   c. Pentachlorophenol, which is a wood preservative.

12.58 Dimethyl ether and ethyl alcohol both have the molecular formula C₂H₆O. One has a boiling point of −23°C, and the other, 79°C. Draw the condensed structural formulas of each compound. Decide which boiling point goes with which compound and explain. Check the boiling points in a chemistry handbook.

12.59 Give the IUPAC and common names (if any) for each of the following compounds:

a. **b.**

c. Cl—CH₂—CH₂—C(=O)—H

d. CH₃—CH(Cl)—C(=O)—CH₂—CH₃ **e.** [3-methylcyclohexanone]

12.60 Give the IUPAC and common names (if any) for each of the following compounds:

a. CH₃—CH₂—C(=O)—CH₃

b. [3,5-dichlorobenzaldehyde]

c. [2-chlorocyclopentanone]

d. CH₃—CH(CH₃)—CH(OH)—CH₂—C(=O)—H

e. [1-phenylethanone]

12.61 Draw the condensed structural formulas of each of the following:
   a. 3-methylcyclopentanone
   b. 4-chlorobenzaldehyde
   c. 3-chloropropionaldehyde
   d. ethyl methyl ketone
   e. 3-methylhexanal
   f. 2-heptanone, an alarm pheromone of bees

12.62 Draw the condensed structural formulas of each of the following:
   a. propionaldehyde
   b. 2-chlorobutanal
   c. 2-methylcyclohexanone
   d. 3,5-dimethylhexanal
   e. 3-bromocyclopentanone
   f. *trans*-2-hexenal, an alarm pheromone of an ant

12.63 Which of the following compounds are soluble in water?
   a. CH₃—CH₂—CH₂—CH₃

   b. CH₃—CH₂—C(=O)—H **c.** CH₃—C(=O)—CH₃

   d. CH₃—CH₂—CH₂—OH

   e. CH₃—CH₂—C(=O)—CH₂—CH₂—CH₃

12.64 Which of the following compounds are soluble in water?

   a. CH₃—CH₂—C(=O)—CH₃ **b.** H—C(=O)—H

c. CH$_3$—$\overset{\displaystyle O}{\overset{\|}{C}}$—H   **d.** CH$_3$—CH$_2$—CH$_3$

e. CH$_3$—CH$_2$—$\overset{\displaystyle CH_3}{\underset{|}{CH}}$—CH$_2$—CH$_2$—$\overset{\displaystyle O}{\overset{\|}{C}}$—H

12.65 In each of the following pairs of compounds, select the compound with the higher boiling point:

a. CH$_3$—CH$_2$—OH or CH$_3$—$\overset{\displaystyle O}{\overset{\|}{C}}$—H

b. CH$_3$—CH$_2$—CH$_2$—CH$_3$ or CH$_3$—CH$_2$—$\overset{\displaystyle O}{\overset{\|}{C}}$—H

12.66 In each of the following pairs of compounds, select the compound with the higher boiling point:

a. CH$_3$—CH$_2$—CH$_2$—OH or CH$_3$—$\overset{\displaystyle O}{\overset{\|}{C}}$—CH$_3$

b. CH$_3$—CH$_2$—CH$_2$—CH$_3$ or CH$_3$—$\overset{\displaystyle O}{\overset{\|}{C}}$—CH$_3$

12.67 Identify the chiral carbons, if any, in each of the following compounds.

a. H—$\overset{\displaystyle Cl}{\underset{\displaystyle Cl}{\overset{|}{\underset{|}{C}}}}$—$\overset{\displaystyle Cl}{\underset{\displaystyle H}{\overset{|}{\underset{|}{C}}}}$—O—H

b. CH$_3$—$\overset{\displaystyle H}{\overset{|}{C}}$=$\overset{\displaystyle CH_3}{\overset{|}{C}}$—CH$_3$

c. HO—CH$_2$—$\overset{\displaystyle OH}{\overset{|}{CH}}$—CH$_2$—OH

d. CH$_3$—$\overset{\displaystyle NH_2}{\overset{|}{CH}}$—$\overset{\displaystyle O}{\overset{\|}{C}}$—H

e. CH$_3$—CH$_2$—$\overset{\displaystyle Br}{\overset{|}{CH}}$—CH$_2$—CH$_2$—CH$_3$

12.68 Identify the chiral carbons, if any, in each of the following compounds.

a. CH$_3$—$\overset{\displaystyle OCH_3}{\overset{|}{CH}}$—CH$_3$ **b.** CH$_3$—$\overset{\displaystyle OH}{\overset{|}{CH}}$—$\overset{\displaystyle O}{\overset{\|}{C}}$—CH$_3$

c. CH$_3$—$\overset{\displaystyle OH}{\underset{\displaystyle OH}{\overset{|}{\underset{|}{C}}}}$—CH$_3$ **d.** CH$_3$—$\overset{\displaystyle CH_3}{\overset{|}{CH}}$—$\overset{\displaystyle O}{\overset{\|}{C}}$—CH$_3$

e. CH$_3$—$\overset{\displaystyle Br}{\underset{\displaystyle OH}{\overset{|}{\underset{|}{C}}}}$—CH$_2$—CH$_3$

12.69 Identify each of the following pairs of Fischer projections as enantiomers or identical compounds:

a.
CH$_2$OH   CH$_2$OH
H——OH and HO——H
CH$_2$OH   CH$_2$OH

b.
CHO   CHO
H——OH and HO——H
CH$_2$OH   CH$_2$OH

c.
CH$_3$ CH$_2$OH
H——Cl and H——Cl
CH$_2$OH   CH$_3$

d.
OH   OH
H——OH and HO——H
CH$_3$ CH$_3$

12.70 Identify each of the following pairs of Fischer projections as enantiomers or identical compounds:

a.
CH$_2$CH$_3$ CH$_2$CH$_3$
H——Cl and Cl——H
CH$_2$OH   CH$_2$OH

b.
CH$_2$OH   CH$_2$OH
H——OH and HO——H
CH$_2$OH   CH$_2$OH

c.
CH$_3$ CH$_3$
H——Cl and H——Cl
CH$_2$OH   CH$_2$OH

d.
CH$_3$ CH$_3$
H——OH and HO——H
CH$_2$OH   CH$_2$OH

12.71 Draw the structural formula of the organic product when each of the following is oxidized:
a. CH$_3$—CH$_2$—CH$_2$—OH

b. CH$_3$—$\overset{\displaystyle OH}{\overset{|}{CH}}$—CH$_2$—CH$_2$—CH$_3$

c. CH$_3$—CH$_2$—CH$_2$—$\overset{\displaystyle O}{\overset{\|}{C}}$—H   **d.**

12.72 Draw the structural formula of the organic product when each of the following is oxidized:

a. CH$_3$—CH$_2$—$\overset{\displaystyle OH}{\overset{|}{CH}}$—CH$_2$OH

b. CH$_3$—CH$_2$—$\overset{\displaystyle OH}{\overset{|}{CH}}$—CH$_3$

c. CH$_3$—$\overset{\displaystyle CH_3}{\overset{|}{CH}}$—CH$_2$—$\overset{\displaystyle O}{\overset{\|}{C}}$—H

d.

UNDERSTANDING THE CONCEPTS

12.73 A compound called butylated hydroxytoluene or BHT is a common preservative added to cereal. It is named as 1-hydroxy-2,6-dimethylethyl-4-methylbenzene. Draw the structure of BHT.

12.74 The compound 1,3-dihydroxy-2-propanone or dihydroxy-acetone (DHA) used in "sunless" tanning lotions darkens the skin without sun. What is its structural formula?

12.75 Identify the functional groups in each of the following:
a. Urushiol, a substance in poison ivy and poison oak that causes itching and blistering of the skin.

b. Menthol gives a peppermint taste and odor used in candy and throat lozenges.

12.76 Identify the functional groups in each of the following:
a. BHA, an anti-oxidant used as a preservative in foods such as baked goods, butter, meats, and snack foods.

b. Vanillin, a flavoring, obtained from the seeds of the vanilla bean.

QUESTIONS FOR TEAMWORK

12.77 Draw the structures and give the IUPAC names of all the alcohols that have the formula $C_5H_{12}O$.

12.78 Draw the structures and give the IUPAC names of all the aldehydes and ketones that have the formula $C_5H_{10}O$.

12.79 A compound with the formula C_4H_8O is synthesized from 2-methyl-1-propanol and oxidizes easily to give a carboxylic acid. What is the structure of the compound?

12.80 Methyl *tert*-butyl ether (MTBE) or methyl 2-methyl-2-propyl ether is a fuel additive for gasoline to boost the

octane rating. It increases the oxygen content, which reduces CO emissions to an acceptable level determined by the Clean Air Act.

a. If fuel mixtures are required to contain 2.7% oxygen by mass, how many grams MTBE must be added to each 100 g gasoline?

b. How many liters of MTBE would be in a liter of fuel if the density of both gasoline and MTBE is 0.740 g/mL?

c. Write the equation for the complete combustion of MTBE.

d. How many liters of air containing 21% (v/v) O_2 are required at STP to completely react (combust) 1.00 L liquid MTBE?

12.81 Compound A has the formula C_3H_8O. When A is heated with acid, compound B (C_3H_6) forms. When A is oxidized, compound C (C_3H_6O) forms that oxidizes further to a carboxylic acid. Another compound D has the formula C_3H_8O. When D is heated with acid, compound B (C_3H_6) forms. When D is oxidized, compound E (C_3H_6O) forms that cannot be oxidized further. What are the structures and names of compounds A, B, C, D, and E?

ANSWERS

Answers to Study Checks

12.1 3-chloro-l-butanol

12.2 secondary

12.3

12.4 Ethanol molecules can hydrogen bond with each other, but dimethyl ether molecules cannot. Thus, a higher temperature is required to break the hydrogen bonds between ethanol molecules.

12.5 cyclopentene

12.6 2-propanol loses two hydrogen atoms to form a ketone.

12.7 $CH_3-CH_2-\overset{O}{\overset{\|}{C}}-CH_2-CH_3$

12.8 propanal (IUPAC), propionaldehyde (common)

12.9 ethyl propyl ketone

12.10 The oxygen atom in the carbonyl group of acetone hydrogen bonds with water molecules.

12.11

12.12

D-2-hydroxypropanal   L-2-hydroxypropanal

Answers to Selected Questions and Problems

12.1 a. ethanol
b. 2-butanol
c. 2-pentanol
d. 4-methylcyclohexanol

12.3 a. $CH_3-CH_2-CH_2-OH$
b. CH_3-OH
c. $CH_3-CH_2-\overset{OH}{\underset{|}{CH}}-CH_2-CH_3$
d. $CH_3-\overset{OH}{\underset{\underset{CH_3}{|}}{\overset{|}{C}}}-CH_2-CH_3$

12.5 a. phenol
b. 2-bromophenol
c. 3-bromophenol

12.7 a. 1°   **b.** 1°   **c.** 3°   **d.** 3°

12.9 a. ethyl methyl ether
b. dipropyl ether
c. cyclohexyl methyl ether
d. methyl propyl ether

12.11 a. $CH_3-CH_2-O-CH_2-CH_2-CH_3$

b. $CH_3-CH_2-O-\triangle$

c. $CH_3-CH_2-O-CH_3$

12.13 a. methanol   **b.** 1-butanol
c. 1-butanol

12.15 a. yes; hydrogen bonding
b. yes; hydrogen bonding
c. no; long carbon chain diminishes effect of —OH group
d. no; alkanes are nonpolar

12.17 a. $CH_3-CH_2-CH=CH_2$

b. (cyclopentene structure)

c. $CH_3-CH=CH-CH_2-CH_3$

12.19 a. $CH_3-CH_2-CH_2-CH_2-\overset{\overset{\displaystyle O}{\|}}{C}-H$

b. $CH_3-CH_2-\overset{\overset{\displaystyle O}{\|}}{C}-CH_3$

c. (cyclohexanone structure with O)

d. $CH_3-\overset{\overset{\displaystyle O}{\|}}{C}-CH_2-\overset{\overset{\displaystyle CH_3}{|}}{CH}-CH_3$

12.21 a. ketone   **b.** ketone   **c.** aldehyde

12.23 a. acetaldehyde
b. methyl propyl ketone
c. formaldehyde

12.25 a. propanal   **b.** 2-methyl-3-pentanone
c. 3-methylcyclohexanone   **d.** benzaldehyde

12.27 a. $CH_3-\overset{\overset{\displaystyle O}{\|}}{C}-H$

b. $CH_3-\overset{\overset{\displaystyle O}{\|}}{C}-CH_2-CH_2-CH_3$

c. $CH_3-\overset{\overset{\displaystyle O}{\|}}{C}-CH_2-CH_2-CH_2-CH_3$

d. $CH_3-CH_2-\overset{\overset{\displaystyle CH_3}{|}}{CH}-CH_2-\overset{\overset{\displaystyle O}{\|}}{C}-H$

12.29 a. $CH_3-CH_2-\overset{\overset{\displaystyle O}{\|}}{C}-H$ has a higher boiling point because it has a polar carbonyl group.

b. Pentanal has a higher molar mass and thus a higher boiling point.
c. 1-Butanol has a higher boiling point because it can hydrogen bond with other 1-butanol molecules.

12.31 a. $CH_3-\overset{\overset{\displaystyle O}{\|}}{C}-\overset{\overset{\displaystyle O}{\|}}{C}-CH_3$; more hydrogen bonding
b. acetone; lower number of carbon atoms

12.33 a. $CH_3-CH_2-CH_2-CH_2-\overset{\overset{\displaystyle O}{\|}}{CH}$

b. $CH_3-CH_2-\overset{\overset{\displaystyle O}{\|}}{C}-CH_3$

c. (cyclohexanone structure with O)

d. $CH_3-\overset{\overset{\displaystyle O}{\|}}{C}-CH_2-\overset{\overset{\displaystyle CH_3}{|}}{CH}-CH_3$

e. $CH_3-\overset{\overset{\displaystyle CH_3}{|}}{CH}-CH_2-\overset{\overset{\displaystyle O}{\|}}{CH}$

12.35 a. achiral
b. chiral $CH_3-\overset{\overset{\displaystyle Br}{|}}{CH}-CH_2CH_3$ chiral carbon

chiral carbon $CH_3-\overset{\overset{\displaystyle Br}{|}}{CH}-\overset{\overset{\displaystyle O}{\|}}{C}-H$

c. chiral
d. achiral

12.37

a. $CH_3-\overset{\overset{\displaystyle CH_3}{|}}{C}=CH-CH_2-CH_2-\overset{\overset{\displaystyle CH_3}{|}}{CH}-CH_2-CH_2-OH$ chiral carbon

b. $H_2N-\overset{\overset{\displaystyle CH_3}{|}}{CH}-\overset{\overset{\displaystyle O}{\|}}{C}-OH$
chiral carbon

12.39 a. $\underset{\displaystyle CH_3}{\overset{\displaystyle H}{HO-\!\!|\!\!-Br}}$ **b.** $\underset{\displaystyle OH}{\overset{\displaystyle CH_3}{Cl-\!\!|\!\!-Br}}$

c. $\underset{\displaystyle CH_2CH_3}{\overset{\displaystyle CHO}{HO-\!\!|\!\!-H}}$

12.41 a. identical   **b.** enantiomers
c. enantiomers   **d.** enantiomers

12.43 a. 2°   **b.** 1°   **c.** 1°
d. 2°   **e.** 1°   **f.** 3°

12.45 a. alcohol   **b.** ether   **c.** thiol
d. alcohol   **e.** ether   **f.** alcohol
g. phenol

12.47 a.

b. OH

Cl

c. CH₃—CH—CH—CH₂—CH₃
with CH₃ and OH above

d. phenyl—O—CH₂—CH₃

e. CH₃—CH₂—C(=O)—CH₂—CH₃

f. OH with Br (ortho) and Br (para) on benzene ring

12.49 a. 1-propanol; hydrogen bonding
   b. 1-propanol; hydrogen bonding
   c. 1-butanol; higher molar mass

12.51 a. soluble; hydrogen bonding
   b. soluble; hydrogen bonding
   c. insoluble; long carbon chain diminishes effect
       of polar —OH on hydrogen bonding

12.53 a. CH₃—CH=CH₂

b. CH₃—CH₂—C(=O)—H

c. CH₃—CH₂—C(=O)—CH₃

d. cyclohexene

e. cyclohexanone

12.55 Testosterone contains cycloalkene, alcohol, and ketone
functional groups.

12.57 a. OH with Cl and two Cl substituents on benzene ring

b. CH₃—CH—CH₂—CH₂—SH with CH₃ above

c. OH with Cl, Cl, Cl, Cl, Cl on benzene ring

12.59 a. 2-bromo-4-chlorocyclopentanone
   b. 4-chloro-3-hydroxybenzaldehyde
   c. 3-chloropropanal

d. 2-chloro-3-pentanone
e. 3-methylcyclohexanone

12.61 a. O (cyclopentanone with CH₃)

b. CHO with Cl on benzene ring

c. Cl—CH₂—CH₂—C(=O)—H

d. CH₃—CH₂—C(=O)—CH₃

e. CH₃—CH₂—CH₂—CH—CH₂—C(=O)—H with CH₃ above

f. CH₃—CH₂—CH₂—CH₂—CH₂—C(=O)—CH₃

12.63 b, c, and d

12.65 a. CH₃—CH₂—OH

b. CH₃—CH₂—C(=O)—H

12.67 a. H—C(Cl)(Cl)—C(Cl)(H)—O—H

b. none   **c.** none

d. CH₃—CH(NH₂)—C(=O)—H

e. CH₃—CH₂—CH(Br)—CH₂—CH₂—CH₃

12.69 a. identical   **b.** enantiomers
   c. enantiomers (turn 180°)   **d.** identical

12.71 a. CH₃—CH₂—C(=O)—OH

b. CH₃—C(=O)—CH₂—CH₂—CH₃

c. CH₃—CH₂—CH₂—C(=O)—OH

d. cyclohexanone

12.73 OH with (CH₃)₃C—, —C(CH₃)₃, and CH₃ on benzene ring

12.75 a. aromatic, alcohol
   b. cycloalkane, alcohol

13 Carboxylic Acids, Esters, Amines, and Amides

"There are many carboxylic acids, including the alpha hydroxy acids, that are found today in skin products," says Dr. Ken Peterson, pharmacist and cosmetic chemist, Oakland. "When you take a carboxylic acid called a fatty acid and react it with a strong base, you get a salt called soap. Soap has a high pH because the weak fatty acid and the strong base won't have a neutral pH of 7. If you take soap and drop its pH down to 7, you will convert the soap to the fatty acid. When I create fragrances, I use my nose and my chemistry background to identify and break down the reactions that produce good scents. Many fragrances are esters, which form when an alcohol reacts with a carboxylic acid. For example, the ester that smells like pineapple is made from ethanol and butyric acid."

the Chemistry place

Visit **www.aw-bc.com/chemplace** or use the CD-ROM in the book for extra quizzes, interactive tutorials, career resources, PowerPoint slides, math help, and case studies.

arboxylic acids are similar to the weak acids we studied in Chapter 8. They have a sour or tart taste, produce hydronium ions in water, and neutralize bases. You encounter carboxylic acids when you use a vinegar salad dressing, which is a solution of acetic acid and water, or experience the sour taste of citric acid in a grapefruit or lemon. When a carboxylic acid combines with an alcohol, an ester is produced. Aspirin is an ester as well as a carboxylic acid. Fats known as triglycerides are esters of glycerol and fatty acids, which are long-chain carboxylic acids. Many fruits and flavorings including bananas, oranges, and strawberries contain esters, which produce their pleasant aromas and flavors.

Amines and amides are organic compounds that contain nitrogen. Many nitrogen-containing compounds are important to life as components of amino acids, proteins, and nucleic acids: DNA and RNA. Many amines that exhibit strong physiological activity are used in medicine as decongestants, anesthetics, and sedatives. Examples include dopamine, histamine, epinephrine, and amphetamine.

Alkaloids such as caffeine, nicotine, cocaine, and digitalis, which demonstrate powerful physiological activity, are naturally occurring amines obtained from plants. Amides, which are derived from carboxylic acids, are important in biology as the types of bonds that link amino acids in proteins. Some amides used medically include acetaminophen (Tylenol) used to reduce fever; phenobarbital, a sedative and anticonvulsant medication; and penicillin, an antibiotic.

13.1 Carboxylic Acids

LEARNING GOAL
Give the common names, IUPAC names, and condensed structural formulas of carboxylic acids.

In Chapter 12, we described the carbonyl group (C=O) as the functional group in aldehydes and ketones. In a **carboxylic acid**, a hydroxyl group is attached to the carbonyl group, forming a **carboxyl group**. The carboxyl functional group may be attached to an alkyl group or an aromatic group.

carbonyl group
hydroxyl group
carboxyl group

acetic acid

stearic acid (a fatty acid in fats and oils)

benzoic acid (aromatic acid)

The carboxyl group can be written in several different ways. For example, the condensed structural formula for propanoic acid can be written as follows:

Some condensed structural formulas for propanoic acid

WEB TUTORIAL
Carboxylic Acids

Naming Carboxylic Acids

The IUPAC names of carboxylic acids use the alkane names of the corresponding carbon chains.

> **STEP 1 Identify the longest carbon chain containing the carboxyl group and replace the *e* of the alkane name by *oic acid*.**

> **STEP 2 Number the carbon chain beginning with the carboxyl carbon as carbon 1.**

> **STEP 3 Give the location and names of substituents on the main chain.** The carboxyl function group takes priority over all the functional groups we have discussed.

methanoic acid   2-methylpropanoic acid   3-hydroxybutanoic acid

> **STEP 4 The aromatic carboxylic acid is called benzoic acid.** With the carboxyl carbon bonded to carbon 1, the ring is numbered in the direction that gives substituents the smallest possible numbers.

benzoic acid   4-aminobenzoic acid   3,4-dichlorobenzoic acid

Many carboxylic acids are still named by their common names, which are derived from their natural sources. In Chapter 12, we named aldehydes using the prefixes that represent the typical sources of carboxylic acids.

Formic acid is injected under the skin from bee or red ant stings and other insect bites. Acetic acid is the oxidation product of the ethanol in wines and apple cider. The resulting solution of acetic acid and water is known as vinegar. Butyric acid gives the foul odor to rancid butter. (See Table 13.1.)

TABLE 13.1

Names and Natural Sources of Carboxylic Acids

Condensed Structural Formulas	IUPAC Name	Common Name	
H—C—OH (=O)	methanoic acid	formic acid	
CH_3—C—OH (=O)	ethanoic acid	acetic acid	
CH_3—CH_2—C—OH (=O)	propanoic acid	propionic acid	
CH_3—CH_2—CH_2—C—OH (=O)	butanoic acid	butyric acid	

SAMPLE PROBLEM 13.1

■**Naming Carboxylic Acids**

Give the IUPAC and common name, if any, for each of the following carboxylic acids:

a. CH_3—CH_2—C—OH (=O)

b. CH_3—CH_2—CH(CH_3)—C—OH (=O)

c. C—OH (=O) attached to benzene ring

SOLUTION

a. This carboxylic acid has 3 carbon atoms. In the IUPAC system, the *e* in propane is replaced by *oic acid*, to give the name, *propanoic acid*. Its common name is *propionic acid*.

b. This carboxylic acid has a methyl group on the second carbon. It has the IUPAC name *2-methylbutanoic acid*.

c. An aromatic carboxylic acid is named as benzoic acid.

STUDY CHECK

Write the condensed structural formula of pentanoic acid.

HEALTH NOTE

Alpha Hydroxy Acids

Alpha hydroxy acids (AHAs) are naturally occurring carboxylic acids found in fruits, milk, and sugarcane. Cleopatra reportedly bathed in sour milk to smooth her skin. Dermatologists have been using products with a high concentration of AHAs to remove acne scars and reduce irregular pigmentation and age spots. Now lower concentrations (8–10%) of AHAs have been added to skin care products for the purpose of smoothing fine lines, improving skin texture, and cleansing pores. Several alpha hydroxy acids may be found in skin care products singly or in combination. Glycolic acid and lactic acid are most frequently used.

Recent studies indicate that products with AHAs increase sensitivity of the skin to sun and UV radiation. It is recommended that a sunscreen with a sun protection factor (SPF) of at least 15 be used when treating the skin with products that include AHAs. Products containing AHAs at concentrations under 10% and pH values greater than 3.5 are generally considered safe. However, the Food and Drug Administration has reports of AHAs causing skin irritation including blisters, rashes, and discoloration of the skin. The FDA does not require product safety reports from cosmetic manufacturers, although they are responsible for marketing safe products. The FDA advises that you test any product containing AHAs on a small area of skin before you use it on a large area.

Alpha Hydroxy Acid (Source)	Structure
Glycolic acid (Sugarcane, sugar beet)	$HO-CH_2-\overset{\overset{O}{\|\|}}{C}-OH$
Lactic acid (Sour milk)	$CH_3-\overset{\overset{OH}{\|}}{CH}-\overset{\overset{O}{\|\|}}{C}-OH$
Tartaric acid (Grapes)	$HO-\overset{\overset{O}{\|\|}}{C}-\overset{\overset{OH}{\|}}{CH}-\overset{\overset{OH}{\|}}{CH}-\overset{\overset{O}{\|\|}}{C}-OH$
Malic acid (Apples, grapes)	$HO-\overset{\overset{O}{\|\|}}{C}-CH_2-\overset{\overset{OH}{\|}}{CH}-\overset{\overset{O}{\|\|}}{C}-OH$
Citric acid (Citrus fruits: lemons, oranges, grapefruit)	$HO-\overset{\overset{CH_2-COOH}{\|}}{\underset{\underset{CH_2-COOH}{\|}}{C}}-COOH$

QUESTIONS AND PROBLEMS

Carboxylic Acids

13.1 What carboxylic acid is responsible for the pain of an ant sting?

13.2 What carboxylic acid solution is found in vinegar?

13.3 Explain the differences in the condensed structural formulas of propanal and propanoic acid.

13.4 Explain the differences in the condensed structural formulas of benzaldehyde and benzoic acid.

13.5 Give the IUPAC and common names (if any) for the following carboxylic acids:

a. $CH_3-\overset{\overset{O}{\|\|}}{C}-OH$ b. $CH_3-CH_2-\overset{\overset{O}{\|\|}}{C}-OH$

c.

d. $CH_3-\overset{\overset{Br}{\|}}{CH}-CH_2-CH_2-\overset{\overset{O}{\|\|}}{C}-OH$

13.6 Give the IUPAC and common names (if any) for the following carboxylic acids:

a. $H-\overset{\overset{O}{\|\|}}{C}-OH$ b.

c.

d. CH_3—$\overset{\overset{\displaystyle CH_3}{|}}{CH}$—$CH_2$—$\overset{\overset{\displaystyle O}{||}}{C}$—OH

13.7 Draw the condensed structural formulas of each of the following carboxylic acids:
   a. propionic acid   **b.** benzoic acid
   c. chloroethanoic acid   **d.** 3-hydroxypropanoic acid
   e. butyric acid   **f.** heptanoic acid

13.8 Draw the condensed structural formulas of each of the following carboxylic acids:
   a. pentanoic acid   **b.** 3-ethylbenzoic acid
   c. 2-hydroxyacetic acid   **d.** 2,4-dibromobutanoic acid
   e. 3-methylbenzoic acid   **f.** hexanoic acid

■ 13.2 Properties of Carboxylic Acids

Carboxylic acids are among the most polar organic compounds because the functional group consists of two polar groups: a hydroxyl (—OH) group and a carbonyl (C=O) group. This C=O double bond is similar to that of the aldehydes and ketones.

LEARNING GOAL

Describe the boiling points, solubility, and ionization of carboxylic acids in water.

WEB TUTORIAL
Carboxylic Acids

Therefore, carboxylic acids form hydrogen bonds with other carboxylic acid molecules and water. This ability to form hydrogen bonds has a major influence on both their boiling points and solubility in water.

■ Boiling Points

Carboxylic acids have higher boiling points than alcohols, ketones, and aldehydes of similar mass.

	CH_3—CH_2—$\overset{\overset{\displaystyle O}{||}}{C}$—H	CH_3—CH_2—CH_2—OH	CH_3—$\overset{\overset{\displaystyle O}{||}}{C}$—OH
Compound	propanal	1-propanol	acetic acid
Molar mass	58	60	60
bp	49°C	97°C	118°C

One reason for the high boiling points of carboxylic acids is the formation of hydrogen bonds between two carboxylic acids to give a dimer. Because the dimers are stable as gases, the mass of the carboxylic acid is effectively doubled, which means that higher temperatures are required to reach the boiling point and form a gas.

Two hydrogen bonds

A dimer of two acetic acid molecules

FIGURE 13.1

Acetic acid forms hydrogen bonds with water molecules.

Q **Why do the atoms in the carboxyl group hydrogen bond with water molecules?**

TABLE 13.2

Properties of Selected Carboxylic Acids

IUPAC name	bp (°C)	Soluble in water?
methanoic acid	101	Yes
ethanoic acid	118	Yes
propanoic acid	141	Yes
butanoic acid	164	Yes
pentanoic acid	187	Slightly
hexanoic acid	205	Slightly
benzoic acid	250	Slightly

Solubility in Water

Carboxylic acids with one to four carbons are very soluble in water because the carboxyl group forms hydrogen bonds with several water molecules. (See Figure 13.1.) However, as the length of the carbon chain increases, the nonpolar portion reduces solubility. Carboxylic acids having five or more carbons are not very soluble in water. Table 13.2 lists boiling points and solubilities for some selected carboxylic acids.

SAMPLE PROBLEM 13.2

■ **Properties of Carboxylic Acids**

Place the following organic compounds in order of increasing boiling points: butanoic acid, pentane, and 2-butanol.

SOLUTION

The boiling point increases when the molecules of a compound can form hydrogen bonds. The alkane has the lowest boiling point because alkanes cannot hydrogen bond. Alcohols and carboxylic acids have higher boiling points because they hydrogen bond via the —OH. The higher boiling points of carboxylic acids are due to the formation of stable dimers, which increase the effective mass and therefore the boiling point.

Pentane < 2-butanol < butanoic acid

STUDY CHECK

At its boiling point, acetic acid forms gas molecules that have a mass of 120 rather than the molar mass of 60 g/mole for CH_3COOH. Explain.

Acidity of Carboxylic Acids

One of the most important properties of carboxylic acids is their ionization in water as weak acids. In the ionization, a carboxylic acid donates a proton to a water molecule to produce an anion called a **carboxylate ion** and a hydronium ion. Carboxylic acids are more acidic than other organic compounds including phenols, but only a small percentage (~1%) of the carboxylic acid molecules are ionized, which means that most of the acid is not ionized.

$$CH_3-\overset{\overset{\displaystyle O}{\|}}{C}-OH + H_2O \rightleftarrows CH_3-\overset{\overset{\displaystyle O}{\|}}{C}-O^- + H_3O^+$$

Ethanoic acid (acetic acid)   Ethanoate ion (acetate ion)   Hydronium ion

SAMPLE PROBLEM 13.3

■ **Ionization of Carboxylic Acids in Water**

Write the equation for the ionization of propionic acid in water.

SOLUTION

The ionization of propionic acid produces a carboxylate ion and a hydronium ion.

$$CH_3-CH_2-\overset{\overset{\displaystyle O}{\|}}{C}-OH + H_2O \rightleftarrows CH_3-CH_2-\overset{\overset{\displaystyle O}{\|}}{C}-O^- + H_3O^+$$

STUDY CHECK

Write an equation for the ionization of formic acid in water.

Neutralization of Carboxylic Acids

Although carboxylic acids are weak acids, they are completely neutralized by strong bases such as NaOH and KOH. The products are water and a **carboxylic acid salt**, which is a carboxylate ion and the metal ion from the base. The carboxylate ion is named by replacing the *oic acid* ending of the acid name with *ate*.

Sodium propionate, a preservative, is added to bread, cheeses, and bakery items to inhibit the spoilage of the food by microorganisms. Sodium benzoate, an inhibitor of mold and bacteria, is added to juices, margarine, relishes, salads, and jams. Monosodium glutamate (MSG) is added to meats, fish, vegetables, and bakery items to enhance flavor, although it causes headaches in some people.

The carboxylic acid salts are solids at room temperature and have high melting points. Because they are ionic compounds, carboxylic acid salts of the alkali metals (Li^+, Na^+, and K^+) and NH_4^+ are usually soluble in water.

SAMPLE PROBLEM 13.4

■ Neutralization of a Carboxylic Acid

Write the equation for the neutralization of propionic acid with sodium hydroxide.

SOLUTION

The neutralization of an acid with a base produces the salt of the acid and water.

STUDY CHECK

What carboxylic acid will produce potassium butyrate when it is neutralized by KOH?

_____ ■

QUESTIONS AND PROBLEMS

■ **Properties of Carboxylic Acids**

13.9 Identify the compound in each pair that has the higher boiling point.
 a. acetic acid or butanoic acid
 b. 1-propanol or propanoic acid
 c. butanone or butanoic acid

13.10 Identify the compound in each pair that has the higher boiling point.
 a. acetone or propanoic acid
 b. propanoic acid or hexanoic acid
 c. ethanol or acetic acid

13.11 Select the compound in each group that is the most soluble in water:
 a. propanoic acid, hexanoic acid, benzoic acid
 b. pentane, 1-butanol, propionic acid

13.12 Select the compound in each group that is the most soluble in water:
 a. butanone, butanoic acid, butane
 b. acetic acid, pentanoic acid, octanoic acid

13.13 Write equations for the ionization of each of the following carboxylic acids in water:

$$\text{a. } H-\overset{\overset{\displaystyle O}{\|}}{C}-OH \qquad \text{b. } CH_3-CH_2-\overset{\overset{\displaystyle O}{\|}}{C}-OH \qquad \text{c. acetic acid}$$

13.14 Write equations for the ionization of each of the following carboxylic acids in water:

$$\text{a. } CH_3-\overset{\overset{\displaystyle CH_3}{|}}{CH}-\overset{\overset{\displaystyle O}{\|}}{C}-OH \qquad \text{b. hydroxyethanoic acid} \qquad \text{c. butanoic acid}$$

13.15 Write equations for the reaction of each of the following carboxylic acids with NaOH:
 a. formic acid   **b.** propanoic acid   **c.** benzoic acid

13.16 Write equations for the reaction of each of the following carboxylic acids with KOH:
 a. acetic acid   **b.** 2-methylbutanoic acid   **c.** 4-chlorobenzoic acid

---

LEARNING GOAL

Name an ester; write equations for the formation and hydrolysis of an ester.

■ 13.3 Esters

A carboxylic acid reacts with an alcohol to form an **ester**. In an ester, the —H of the carboxylic acid is replaced by an alkyl group. Aspirin is an ester as well as a carboxylic acid. Fats and oils in our diets contain esters of glycerol and fatty acids, which are long-chain carboxylic acids. The aromas and flavors of many fruits including bananas, oranges, and strawberries are due to esters.

Carboxylic acid	**Ester**
$CH_3-\overset{\overset{\displaystyle O}{\|}}{C}-O-H$	$CH_3-\overset{\overset{\displaystyle O}{\|}}{C}-O-CH_3$
ethanoic acid (acetic acid)	methyl ethanoate (methyl acetate)

Esterification

In a reaction called **esterification**, a carboxylic acid reacts with an alcohol when heated in the presence of an acid catalyst (usually H_2SO_4). In the reaction, water is produced from the —OH removed from the carboxylic acid and an —H lost by the alcohol.

$$CH_3-\overset{O}{\overset{\|}{C}}-O-H \ + \ H-O-CH_3 \ \underset{}{\overset{H^+,\ heat}{\rightleftarrows}} \ CH_3-\overset{O}{\overset{\|}{C}}-O-CH_3 \ + \ H-O-H$$

acetic acid   methyl alcohol   methyl acetate

If we use acetic acid and 1-propanol, we can write an equation for the formation of the ester that is responsible for the flavor and odor of pears.

$$CH_3-\overset{O}{\overset{\|}{C}}-OH \ + \ H-O-CH_2-CH_2-CH_3 \ \underset{}{\overset{H^+,\ heat}{\rightleftarrows}} \ CH_3-\overset{O}{\overset{\|}{C}}-O-CH_2-CH_2-CH_3 \ + \ H_2O$$

acetic acid   1-propanol   propyl acetate
                                                              (pears)

HEALTH NOTE

▪ Salicylic Acid and Aspirin

Chewing on a piece of willow bark was used as a way of relieving pain for many centuries. By the 1800s, chemists discovered that salicylic acid was the agent in the bark responsible for the relief of pain. However, salicylic acid, which has both a carboxylic group and a hydroxyl group, irritates the stomach lining. A less irritating ester of salicylic acid and acetic acid, called acetylsalicylic acid or "aspirin," was prepared in 1899 by the Bayer chemical company in Germany. In some aspirin preparations, a buffer is added to neutralize the carboxylic acid group and lessen its irritation of the stomach. Aspirin is used as an analgesic (pain reliever), antipyretic (fever reducer), and anti-inflammatory agent.

salicylic acid   methyl alcohol

methyl salicylate
(oil of wintergreen)

Oil of wintergreen, or methyl salicylate, has a spearmint odor and flavor. Because it can pass through the skin, methyl salicylate is used in skin ointments where it acts as a counterirritant, producing heat to soothe sore muscles.

SAMPLE PROBLEM 13.5

■ Writing Esterification Equations

The ester that gives the flavor and odor of apples can be synthesized from butyric acid and methyl alcohol. What is the equation for the formation of the ester in apples?

SOLUTION

$$CH_3—CH_2—CH_2—\overset{\overset{\displaystyle O}{\|}}{C}—OH \ + \ H—O—CH_3 \ \underset{}{\overset{H^+,\ heat}{\rightleftarrows}}$$

butyric acid   methyl alcohol

$$CH_3—CH_2—CH_2—\overset{\overset{\displaystyle O}{\|}}{C}—O—CH_3 \ + \ H_2O$$

methyl butyrate

STUDY CHECK

What carboxylic acid and alcohol are needed to form the following ester, which gives the flavor and odor to apricots?

$$CH_3—CH_2—\overset{\overset{\displaystyle O}{\|}}{C}—O—CH_2—CH_2—CH_2—CH_2—CH_3$$

Naming Esters

Methyl ethanoate
(methyl acetate)

The name of an ester consists of two words taken from the names of the alcohol and the acid. The first word indicates the *alkyl* part of the alcohol. The second word is the *carboxylate* name of the carboxylic acid. The IUPAC names of esters use the IUPAC names for the acid, while the common names of esters use the common names for the acid. Let's take a look at the following ester and break it into two parts, one from the alcohol and one from the acid. By writing and naming the alcohol and carboxylic acid that produced the ester, we can determine the name of the ester.

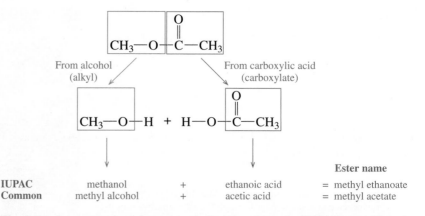

					Ester name
IUPAC	methanol	+	ethanoic acid	=	methyl ethanoate
Common	methyl alcohol	+	acetic acid	=	methyl acetate

The following examples of some typical esters show the IUPAC as well as the common names of esters.

ethyl ethanoate
(ethyl acetate)

methyl propanoate
(methyl propionate)

ethyl benzoate

TABLE 13.3

Some Esters in Fruits and Flavorings

Condensed Structural Formula and Name	Flavor/Odor

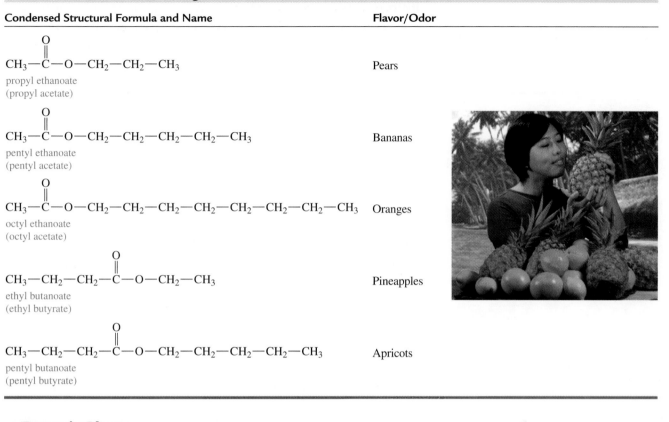

$$CH_3-\overset{\overset{\displaystyle O}{\|}}{C}-O-CH_2-CH_2-CH_3$$

propyl ethanoate
(propyl acetate)

Pears

$$CH_3-\overset{\overset{\displaystyle O}{\|}}{C}-O-CH_2-CH_2-CH_2-CH_2-CH_3$$

pentyl ethanoate
(pentyl acetate)

Bananas

$$CH_3-\overset{\overset{\displaystyle O}{\|}}{C}-O-CH_2-CH_2-CH_2-CH_2-CH_2-CH_2-CH_2-CH_3$$

octyl ethanoate
(octyl acetate)

Oranges

$$CH_3-CH_2-CH_2-\overset{\overset{\displaystyle O}{\|}}{C}-O-CH_2-CH_3$$

ethyl butanoate
(ethyl butyrate)

Pineapples

$$CH_3-CH_2-CH_2-\overset{\overset{\displaystyle O}{\|}}{C}-O-CH_2-CH_2-CH_2-CH_2-CH_3$$

pentyl butanoate
(pentyl butyrate)

Apricots

Esters in Plants

Many of the fragrances of perfumes and flowers and the flavors of fruits are due to esters. Small esters are volatile so we can smell them and soluble in water so we can taste them. Several of these are listed in Table 13.3.

SAMPLE PROBLEM 13.6

Naming Esters

Write the IUPAC and common names of the following ester:

$$CH_3-CH_2-\overset{\overset{\displaystyle O}{\|}}{C}-O-CH_2-CH_2-CH_3$$

SOLUTION

The alcohol part of the ester is propyl, and the carboxylic acid part is propanoic (propionic) acid.

From propanoic acid → propanoate   From propyl alcohol → propyl
   (or propionic acid → propionate)

$$CH_3-CH_2-\overset{\overset{\displaystyle O}{\|}}{C}\!-\!O-CH_2-CH_2-CH_3$$

IUPAC name: propyl propanoate
Common name: propyl propionate

STUDY CHECK

Draw the condensed structural formula of pentyl acetate.

Acid Hydrolysis of Esters

In **hydrolysis**, esters are split apart with water when heated in the presence of a strong acid, usually H_2SO_4 or HCl. The products of acid hydrolysis are the carboxylic acid and alcohol. Therefore, hydrolysis is the reverse of the esterification reaction. When hydrolysis of biological compounds occurs in the cells, an enzyme replaces the acid as the catalyst. In the hydrolysis reaction, the —OH from a water molecule bonds to the carbonyl group of the ester to form the carboxylic acid.

$$\underset{\text{methyl acetate}}{CH_3-\overset{\overset{\textstyle O}{\|}}{C}-O-CH_3} \ + \ \underset{\text{water}}{H-OH} \ \xrightleftharpoons{H^+} \ \underset{\text{acetic acid}}{CH_3-\overset{\overset{\textstyle O}{\|}}{C}-O-H} \ + \ \underset{\text{methyl alcohol}}{CH_3-OH}$$

SAMPLE PROBLEM 13.7

■ Acid Hydrolysis of Esters

Aspirin that has been stored for a long time may undergo hydrolysis in the presence of water and heat. What are the hydrolysis products of aspirin? Why does a bottle of old aspirin smell like vinegar?

aspirin, acetylsalicylic acid

SOLUTION

To write the hydrolysis products, separate the compound at the ester bond. Complete the formula of the carboxylic acid by adding —OH (from water) to the carbonyl group and an —H to complete the alcohol. The acetic acid in the products gives the vinegar odor to a sample of aspirin that has hydrolyzed.

STUDY CHECK

What are the names of the products from the acid hydrolysis of ethyl propionate?

Base Hydrolysis of Esters

When an ester undergoes hydrolysis with a strong base such as NaOH or KOH, the products are the carboxylic acid salt and the corresponding alcohol. The base hydrolysis reaction is also called **saponification**, which refers to the reaction of a long-chain fatty acid with NaOH to make soap. The carboxylic acid, which is produced in acid hydrolysis, is converted in an irreversible reaction to its carboxylate ion by the strong base.

$$CH_3-\overset{\overset{\displaystyle O}{\|}}{C}-O-CH_3 \ + \ NaOH \ \xrightarrow{\text{Heat}} \ CH_3-\overset{\overset{\displaystyle O}{\|}}{C}-O^-Na^+ \ + \ CH_3-OH$$

methyl ethanoate   sodium hydroxide   sodium acetate   methanol
(methyl acetate)   (sodium ethanoate)   (methyl alcohol)

SAMPLE PROBLEM 13.8

■ **Base Hydrolysis of Esters**

Ethyl acetate is a solvent widely used for fingernail polish, plastics, and lacquers. Write the equation of the hydrolysis of ethyl acetate by NaOH.

SOLUTION

The hydrolysis of ethyl acetate by NaOH gives the salt of acetic acid and ethyl alcohol.

$$CH_3-\overset{\overset{\displaystyle O}{\|}}{C}-O-CH_2-CH_3 \ + \ NaOH \ \xrightarrow{\text{Heat}} \ CH_3-\overset{\overset{\displaystyle O}{\|}}{C}-O^-Na^+ \ + \ HO-CH_2-CH_3$$

ethyl acetate   sodium acetate   ethyl alcohol

STUDY CHECK

Write the condensed structural formulas of the products from the hydrolysis of methyl benzoate by KOH.

QUESTIONS AND PROBLEMS

■ **Esters**

13.17 Identify each of the following as an aldehyde, a ketone, a carboxylic acid, or an ester:

a. $CH_3-\overset{\overset{\displaystyle O}{\|}}{C}-H$ b. $CH_3-\overset{\overset{\displaystyle O}{\|}}{C}-O-CH_3$

c. $CH_3-CH_2-\overset{\overset{\displaystyle O}{\|}}{C}-CH_3$ d. $CH_3-CH_2-\overset{\overset{\displaystyle O}{\|}}{C}-O-H$

13.18 Identify each of the following as an aldehyde, a ketone, a carboxylic acid, or an ester:

a. $CH_3-\overset{\overset{\displaystyle O}{\|}}{C}-OH$ b. $CH_3-\overset{\overset{\displaystyle O}{\|}}{C}-O-CH_2-CH_3$

c. $CH_3-CH_2-\overset{\overset{\displaystyle O}{\|}}{C}-H$ d. $CH_3-\overset{\displaystyle CH}{\underset{\displaystyle |}{CH_3}}-\overset{\overset{\displaystyle O}{\|}}{C}-O-CH_2-CH_3$

13.19 Write the condensed structural formula of the ester formed when each of the following react with methyl alcohol:
   a. acetic acid   b. butyric acid   c. benzoic acid

13.20 Write the condensed structural formula of the ester formed when each of the following react with methyl alcohol:
   a. formic acid   b. propionic acid   c. 2-methylpentanoic acid

13.21 Draw the condensed structural formulas of the ester formed when each of the following carboxylic acids and alcohols react:

a. $CH_3-CH_2-\overset{\overset{\displaystyle O}{\|}}{C}-OH \ + \ HO-CH_2-CH_2-CH_3 \ \overset{H^+}{\underset{}{\rightleftharpoons}}$

b. $CH_3-CH_2-CH_2-CH_2-\overset{\overset{\displaystyle O}{\|}}{C}-OH \ + \ HO-\overset{\displaystyle CH}{\underset{\displaystyle |}{}}\overset{\displaystyle CH_3}{}-CH_3 \ \overset{H^+}{\underset{}{\rightleftharpoons}}$

▪ Cleaning Action of Soaps

For many centuries, soaps were made by heating a mixture of animal fats (tallow) with lye, a basic solution obtained from wood ashes. In the soap-making process, fatty acids, which are long-chain carboxylic acids, undergo saponification with the strong base in lye.

Fatty acid

$$CH_3CH_2CH_2CH_2CH_2CH_2CH_2CH_2CH_2CH_2CH_2CH_2CH_2CH_2CH_2CH_2CH_2 \overset{O}{\overset{\|}{-C}} -OH \ + \ NaOH \longrightarrow$$

Carboxylic acid salt, "soap"

$$CH_3CH_2CH_2CH_2CH_2CH_2CH_2CH_2CH_2CH_2CH_2CH_2CH_2CH_2CH_2CH_2CH_2 \overset{O}{\overset{\|}{-C}} -O^- Na^+$$

Nonpolartail   Polar head
(hydrophobic)   (hydrophilic)

Today soaps are also prepared from fats such as coconut oil. Perfumes are added to give a pleasant-smelling soap. Because a soap is the salt of a long-chain fatty acid, the two ends of a soap molecule have different polarities. The long carbon chain end is nonpolar and *hydrophobic* (water-fearing). It is soluble in non-polar substances such as oil or grease; but it is not soluble in water. The carboxylate salt end is ionic and *hydrophilic* (water-loving). It is very soluble in water but not in oils or grease.

When a soap is used to clean grease or oil, the nonpolar ends of the soap molecules dissolve in the nonpolar fats and oils that accompany dirt. The water-loving salt ends of the soap molecules extend outside where they can dissolve in water. The soap molecules coat the oil or grease, forming clusters called *micelles*. The ionic ends of the soap molecules provide polarity to the micelles, which makes them soluble in water. As a result, small globules of oil and fat coated with soap molecules are pulled into the water and rinsed away.

One of the problems of using soaps is that the carboxylate end reacts with ions in water such as Ca^{2+} and Mg^{2+} and forms insoluble substances.

$$2CH_3(CH_2)_{16}COO^- + Mg^{2+} \longrightarrow [CH_3(CH_2)_{16}COO^-]_2Mg^{2+}$$

Stearate ion Magnesium Magnesium stearate
      ion   (insoluble)

Soap molecule

Fatty acid
chain

H_2O

Hydrophilic

Oil
and
grease

Hydrophobic

13.22 Draw the condensed structural formula of the ester formed when each of the following carboxylic acids and alcohols react:

a. $CH_3-CH_2-\overset{O}{\overset{\|}{C}}-OH \ + \ HO-CH_3 \overset{H^+}{\rightleftharpoons}$

b. $\overset{O}{\overset{\|}{C}}-OH \ + \ HO-CH_2-CH_2-CH_2-CH_3 \overset{H^+}{\rightleftharpoons}$

13.23 Name each of the following esters:

a. CH₃—O—C(=O)—H

b. CH₃—O—C(=O)—CH₃

c. CH₃—O—C(=O)—CH₂—CH₂—CH₃

d. CH₃—CH₂—CH₂—C(=O)—O—CH₂—CH₃

13.24 Name each of the following esters:

a. CH₃—CH₂—O—C(=O)—CH₂—CH₂—CH₃

b. CH₃—O—C(=O)—CH₂—CH₂—CH₂—CH₂—CH₃

c. CH₃—O—C(=O)—CH₂—CH₂—CH₃

d. CH₃—CH₂—C(=O)—O—CH₂—CH₂—CH₂—CH₃

13.25 Draw the condensed structural formulas of each of the following esters:
a. methyl acetate   **b.** butyl formate
c. ethyl pentanoate   **d.** propyl propanoate

13.26 Draw the condensed structural formulas of each of the following esters:
a. hexyl acetate   **b.** propyl propionate
c. ethyl butanoate   **d.** methyl benzoate

13.27 What is the ester responsible for the flavor and odor of the following fruit?
a. banana   **b.** orange   **c.** apricot

13.28 What flavor would you notice if you smelled or tasted the following?
a. ethyl butanoate   **b.** propyl acetate   **c.** pentyl acetate

13.29 What are the products of the acid hydrolysis of an ester?

13.30 What are the products of the base hydrolysis of an ester?

13.31 Draw the condensed structural formulas of the products from the acid- or base-catalyzed hydrolysis of each of the following compounds:

a. CH₃—CH₂—C(=O)—O—CH₃ + NaOH ⟶

b. CH₃—C(=O)—O—CH₂—CH₂—CH₃ + H₂O $\xrightarrow{H^+}$

c. CH₃—CH₂—CH₂—C(=O)—O—CH₂—CH₃ + H₂O $\xrightarrow{H^+}$

d. C₆H₅—C(=O)—O—CH₂—CH₃ + NaOH ⟶

13.32 Draw the condensed structural formulas of the products from the acid- or base-catalyzed hydrolysis of each of the following compounds:

a. CH₃—CH₂—C(=O)—O—[CH₂—CH₂—CH₂—CH₃ + H₂O $\xrightarrow{H^+}$

b. H—C(=O)—O—[CH₂—CH₃ + NaOH ⟶

c. CH₃—CH₂—C(=O)—O—[CH₃ + H₂O $\xrightarrow{H^+}$

$$\text{d. } CH_3—CH_2—\overset{\overset{\textstyle O}{\|}}{C}—O—\langle\bigcirc\rangle + H_2O \xrightarrow{H^+}$$

LEARNING GOAL

Classify amines as primary, secondary, or tertiary; name amines using common names. Describe some properties of amines.

13.4 Amines

Amines are considered as derivatives of ammonia (NH_3), in which one or more hydrogen atoms is replaced with alkyl or aromatic groups. For example, in methylamine, a methyl group replaces one hydrogen atom in ammonia. The bonding of two methyl groups gives dimethylamine, and the three methyl groups in trimethylamine replace all the hydrogen atoms in ammonia.

Classification of Amines

WEB TUTORIAL
Amine and Amide Functional Groups

When we classified alcohols in Chapter 12, we looked at the number of carbon atoms bonded to the alcohol carbon. Amines are classified in a similar way by counting the number of carbon atoms directly bonded to a nitrogen atom. In a *primary (1°) amine*, one carbon is bonded to a nitrogen atom. In a *secondary (2°) amine*, two carbons are bonded to the nitrogen atom, and a *tertiary (3°) amine* has three carbon atoms bonded to the nitrogen.

Ammonia	Primary (1°) amine	Secondary (2°) amine	Tertiary (3°) amine

| Ammonia | Methylamine | Dimethylamine | Trimethylamine |

SAMPLE PROBLEM 13.9

■ Classifying Amines

Classify the following amines as primary, secondary, or tertiary:

a. NH_2 (cyclohexyl)

b. $CH_3—\overset{\overset{\textstyle H}{|}}{N}—CH_2—CH_3$

c. $\langle\bigcirc\rangle—\overset{}{\underset{\overset{|}{H}}{N}}—CH_3$

SOLUTION

a. This is a primary amine because there is one alkyl group (cyclohexyl) attached to a nitrogen atom.

b. This is a secondary amine. There are two alkyl groups (methyl and ethyl) attached to the nitrogen atom.

c. This is a secondary amine with two carbon groups, methyl and phenyl, bonded to the nitrogen atom.

STUDY CHECK

Classify the following amine as primary, secondary, or tertiary:

$$CH_3-CH_2-\underset{\underset{CH_3}{|}}{N}-CH_2-CH_3$$

Naming Amines

For simple amines, the common names are often used. In the common name, the alkyl groups bonded to the nitrogen atom are listed in alphabetical order. The prefixes *di* and *tri* are used to indicate two and three identical substituents.

| methylamine | dimethylamine | ethylmethylpropylamine |

Aromatic Amines

The aromatic amines use the name *aniline*. Alkyl groups attached to the nitrogen of aniline are named with the prefix *N-* followed by the alkyl name.

aniline   4-bromoaniline   *N*-methylaniline

SAMPLE PROBLEM 13.10

■ **Naming Amines**

Give the common name for each of the following amines:

a. $CH_3-CH_2-NH_2$ **b.** $CH_3-\underset{\underset{CH_3}{|}}{N}-CH_3$ **c.** (aniline structure with NH_2)

SOLUTION

a. This amine has one ethyl group attached to the nitrogen atom; its name is *ethylamine*.

b. This amine has three methyl groups attached to the nitrogen atom; its name is *trimethylamine*.

c. This aromatic amine is *aniline*.

STUDY CHECK

Draw the structure of ethylpropylamine.

Amines in Health and Medicine

In response to allergic reactions or injury to cells the body increases the production of histamine, which causes blood vessels to dilate and increases the permeability of the cells. Redness and swelling occur in the area. Administering an antihistamine such as diphenhydramine helps block the effects of histamine.

Histamine

Diphenhydramine

In the body, hormones called biogenic amines carry messages between the central nervous system and nerve cells. Epinephrine (adrenaline) and norepinephrine (noradrenaline) are released by the adrenal medulla in "fight or flight" situations to raise the blood glucose level and move the blood to the muscles. Used in remedies for colds, hay fever, and asthma, the norepinephrine contracts the capillaries in the mucous membranes of the respiratory passages. The prefix *nor* in a drug name means there is one less CH₃— group on the nitrogen atom. Parkinson's disease is a result of a deficiency in another biogenic amine called dopamine.

Epinephrine (adrenaline)

Norepinephrine (noradrenaline)

Dopamine

Produced synthetically, amphetamines (known as "uppers") are stimulants of the central nervous system much like epinephrine, but they also increase cardiovascular activity and depress the appetite. They are sometimes used to bring about weight loss, but they can cause chemical dependency. Benzedrine and Neo-Synephrine (phenylephrine) are used in medications to reduce respiratory congestion from colds, hay fever, and asthma. Sometimes, Benzedrine is taken internally to combat the desire to sleep, but it has side effects. Methedrine is used to treat depression and in the illegal form is known as "speed" or "crank." The prefix *meth* means that there is one more methyl group on the nitrogen atom.

Benzedrine (amphetamine)

Neo-Synephrine (phenylephrine)

Methamphetamine (methedrine)

Properties of Amines

Hydrogen bond

Amines have higher boiling points than hydrocarbons of similar mass, but lower than the alcohols. Because amines contain a polar N—H bond, they form hydrogen bonds. However, the nitrogen atom in amines is not as electronegative as the oxygen in alcohols, which makes the hydrogen bonds weaker in amines. In primary amines, —NH₂ can form more hydrogen bonds, which gives them higher boiling points than the secondary amines of the same mass. It is not possible for tertiary amines to hydrogen bond with each other (no N—H bonds), which makes their boiling points much lower and similar to those of alkanes and ethers.

$$CH_3-CH_2-CH_2-OH$$
propanol
bp 97°C

$$CH_3-CH_2-CH_2-NH_2$$
propylamine (1°)
bp 48°C

$$CH_3-CH_2-NH-CH_3$$
ethylmethylamine (2°)
bp 36°C

trimethylamine (3°)
bp 3°C

Solubility in Water

Like alcohols, the smaller amines, including tertiary ones, are soluble in water because they form hydrogen bonds with water. However, in amines with more than six carbon atoms, the effect of hydrogen bonding is diminished, and their solubility in water decreases. Large amines are not soluble in water.

Hydrogen bonding

$$CH_3 - \overset{\overset{\displaystyle H}{|}}{\underset{\underset{\displaystyle H}{|}}{\overset{..}{N}}} : \overset{\delta^-}{\cdots\cdots} H \overset{\delta^+}{-} O^\delta$$

■ Properties of Amines

If the compounds trimethylamine and ethylmethylamine have the same molar mass, why is the boiling point of trimethylamine (3°C) lower than that of ethyl-methylamine (36°C)?

SOLUTION

With a polar N—H bond, hydrogen bonds form between ethylmethylamine molecules. Thus, a higher temperature is required to break the hydrogen bonds and form a gas. However, trimethylamine, which is a tertiary amine, has no N—H bond and cannot form hydrogen bonds between the amine molecules.

STUDY CHECK

Why is $CH_3-CH_2-NH_2$ more soluble in water than
$CH_3-CH_2-CH_2-CH_2-CH_2-CH_2-CH_2-NH_2$?

Amines React as Bases

In Chapter 8, we saw that ammonia (NH_3) acts as a Brønsted–Lowry base because it accepts a proton (H^+) from water to produce an ammonium ion (NH_4^+) and a hydroxide ion (OH^-). Let's review that equation:

$$\overset{..}{N}H_3 + H_2O \rightleftharpoons NH_4^+ + OH^-$$

ammonia   ammonium ion   hydroxide ion

In water, amines also act as Brønsted–Lowry bases because the lone electron pair on the nitrogen atom accepts a proton from water and produces hydroxide ions.

$$CH_3-\overset{..}{N}H_2 + H_2O \rightleftharpoons CH_3-\overset{+}{N}H_3 + OH^-$$

methylamine   methylammonium ion   hydroxide ion

$$CH_3-\underset{\underset{\displaystyle CH_3}{|}}{\overset{..}{N}H} + H_2O \rightleftharpoons CH_3-\underset{\underset{\displaystyle CH_3}{|}}{\overset{+}{N}H_2} + OH^-$$

dimethylamine   dimethylammonium ion   hydroxide ion

Amine Salts

When you squeeze lemon juice on fish, the "fishy odor" of the amines is removed by converting them to amine salts. In a neutralization reaction, an amine acts as a base and reacts with an acid to form an **amine salt**. The lone pair of electrons on the nitrogen atom accepts a proton H^+ from an acid to give an amine salt; no water is formed. An amine salt is named by replacing the *amine* part of the amine name with *ammonium* followed by the name of the negative ion.

WEB TUTORIAL
Amines as Bases

the Chemistry place

CASE STUDY
Death By Chocolate?

Neutralization of an Amine

Properties of Amine Salts

As ionic compounds, amine salts are solids at room temperature, odorless, and soluble in water and body fluids. For this reason, amines used as drugs are converted to their amine salts. The amine salt of ephedrine is used as a bronchodilator and in decongestant products such as Sudafed. The amine salt of diphenhydramine is used in products such as Benadryl for relief of itching and pain from skin irritations and rashes. (See Figure 13.2.) In pharmaceuticals, the naming of the amine salt follows an older method of giving the amine name followed by the name of the acid.

ephedrine hydrochloride
ephedrine HCl
Sudafed

diphenhydramine hydrochloride
diphenhydramine HCl
Benadryl

When an amine salt reacts with a strong base such as NaOH, it is converted back to the amine, which is also called the free amine or free base.

$$CH_3-NH_3^+Cl^- + NaOH \longrightarrow CH_3-NH_2 + NaCl + H_2O$$

The narcotic cocaine is typically extracted from coca leaves using an acidic HCl solution to give a white, solid amine salt, which is cocaine hydrochloride. This is the form in which cocaine is smuggled and sold illegally on the street to be snorted or injected. "Crack cocaine" is the free amine or free base of the amine obtained by treating the cocaine hydrochloride with NaOH and ether, a process known as "free-basing." The solid product is known as "crack cocaine" because it makes a crackling noise when heated. The free amine is rapidly absorbed when smoked and gives stronger highs than the cocaine hydrochloride. Unfortunately, these effects of crack cocaine have caused a rise in addiction to cocaine.

cocaine hydrochloride

cocaine ("free base")

HEALTH NOTE

■ Alkaloids: Amines in Plants

Alkaloids are physiologically active nitrogen-containing compounds produced by plants. The term *alkaloid* refers to the "alkali-like" or basic characteristics we have seen for amines. Certain alkaloids are used in anesthetics, in antidepressants, and as stimulants, although many are habit forming.

As a stimulant, nicotine increases the level of adrenaline in the blood, which increases the heart rate and blood pressure. It is well known that smoking cigarettes can damage the lungs and that exposure to tars and other carcinogens in cigarette smoke can lead to lung cancer. However, nicotine is responsible for the addiction of smoking. Coniine, which is obtained from hemlock, is an extremely toxic alkaloid.

nicotine   coniine

Caffeine is a stimulant of the central nervous system. Present in coffee, tea, soft drinks, chocolate, and cocoa, caffeine increases alertness, but may cause nervousness and insomnia. Caffeine is also used in certain pain relievers to counteract the drowsiness caused by an antihistamine.

Several alkaloids are used in medicine. Quinine obtained from the bark of the cinchona tree has been used in the treatment of malaria since the 1600s. Atropine from belladonna is used in low concentrations to accelerate slow heart rates and as an anesthetic for eye examinations.

quinine   atropine

For many centuries morphine and codeine, alkaloids found in the oriental poppy plant, have been used as effective painkillers. Codeine, which is structurally similar to morphine, is used in some prescription painkillers and cough syrups. Heroin, obtained by a chemical modification of morphine, is strongly addicting and is not used medically.

caffeine   Morphine   Codeine

SAMPLE PROBLEM 13.12

■ **Reactions of Amines**

Write an equation that shows ethylamine

a. ionizing as a weak base in water
b. neutralized by HCl

SOLUTION

a. In water, ethylamine acts as a weak base by accepting a proton from water to produce ethylammonium hydroxide.

$$CH_3—CH_2—NH_2 + H—OH \rightleftharpoons CH_3—CH_2—NH_3^+ + OH^-$$

b. $CH_3—CH_2—NH_2 + HCl \longrightarrow CH_3—CH_2—NH_3^+Cl^-$

STUDY CHECK

What is the condensed structural formula of the salt formed by the reaction of trimethylamine and HCl?

■

QUESTIONS AND PROBLEMS

■ **Amines**

13.33 Classify each of the following amines as primary, secondary, or tertiary:

a. $CH_3—CH_2—CH_2—NH_2$

b. $CH_3—\overset{\overset{\displaystyle H}{|}}{N}—CH_2—CH_3$

c.

d. $CH_3—\overset{\overset{\displaystyle CH_3}{|}}{CH}—\overset{\overset{\displaystyle CH_3}{|}}{N}—CH_2—CH_3$

13.34 Classify each of the following amines as primary, secondary, or tertiary:

a. $CH_3—CH_2—\overset{\overset{\displaystyle NH_2}{|}}{CH}—CH_3$

b. $CH_3—CH_2—\overset{\overset{\displaystyle CH_3}{|}}{N}—CH_2—CH_3$

c.

d. $CH_3—\overset{\overset{\displaystyle H}{|}}{N}—\overset{\overset{\displaystyle CH_3}{|}}{\underset{\underset{\displaystyle CH_3}{|}}{C}}—CH_3$

13.35 Write the common names for each of the following:
a. $CH_3—CH_2—NH_2$ **b.** $CH_3—NH—CH_2—CH_2—CH_3$
c. $CH_3—CH_2—\overset{\overset{\displaystyle CH_3}{|}}{N}—CH_2—CH_3$ **d.** $CH_3—\overset{\overset{\displaystyle NH_2}{|}}{CH}—CH_3$

13.36 Write the common names for each of the following:
a. $CH_3—CH_2—CH_2—NH_2$ **b.** $CH_3—NH—CH_2—CH_3$
c. $CH_3—CH_2—CH_2—CH_2—NH_2$ **d.** $CH_3—CH_2—\overset{\overset{\displaystyle CH_2—CH_3}{|}}{N}—CH_2—CH_3$

13.37 Draw the condensed structural formulas for each of the following amines:
a. ethylamine   **b.** N-methylaniline   **c.** butylpropylamine

13.38 Draw the condensed structural formulas for each of the following amines:
a. dimethylamine   **b.** 4-chloroaniline   **c.** N,N-diethylaniline

13.39 Indicate the compound in each pair that would have the higher boiling point:
    a. CH_3—CH_2—NH_2 or CH_3—CH_2—OH
    b. CH_3—NH_2 or CH_3—CH_2—CH_2—NH_2

               CH_3
               |
    c. CH_3—N—CH_3 or CH_3—CH_2—CH_2—NH_2

13.40 Indicate the compound in each pair that would have the higher boiling point:
    a. CH_3—CH_2—CH_2—CH_3 or CH_3—CH_2—CH_2—NH_2
    b. CH_3—NH_2 or CH_3—CH_2—NH_2

                                 NH_2
                                 |
    c. CH_3—CH_2—CH_2—OH or CH_3—CH—CH_3

13.41 Indicate if each of the following is soluble in water:
    a. CH_3—CH_2—NH_2
    b. CH_3—NH—CH_3

                        CH_2—CH_2—CH_3
                        |
    c. CH_3—CH_2—CH_2—N—CH_2—CH_2—CH_3

               NH_2
               |
    d. CH_3—CH—CH_2—CH_3

13.42 Indicate if each of the following is soluble in water:
    a. CH_3—CH_2—CH_2—NH_2
    b. CH_3—CH_2—CH_2—NH—CH_2—CH_3

               CH_3
               |
    c. CH_3—N—CH_3

              H
              |
              N
    d.

13.43 Write an equation for the ionization of each of the following amines in water:
    a. methylamine   **b.** dimethylamine   **c.** aniline

13.44 Write an equation for the ionization of each of the following amines in water:
    a. ethylamine   **b.** propylamine   **c.** *N*-methylaniline

---

■ **13.5 Amides**

The **amides** are derivatives of carboxylic acids in which a nitrogen group replaces the hydroxyl group. An amide is produced when a carboxylic acid reacts with ammonia or an amine. A molecule of water is eliminated, and the fragments of the carboxylic acid and amine molecules join to form the amide, much like the formation of ester.

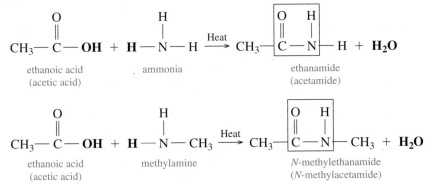

LEARNING GOAL

■ Write common and IUPAC names of amides and the products of formation and hydrolysis.

Ethanamide
(Acetamide)

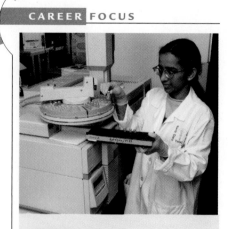
SAMPLE PROBLEM 13.13

■ Formation of Amides

Give the structural formula of the amide product in each of the following reactions:

a.

b.

$$CH_3 - \overset{\overset{\displaystyle O}{\|}}{C} - OH \ + \ NH_2 - CH_2 - CH_3 \xrightarrow{\text{Heat}}$$

SOLUTION

a. The structural formula of the amide product can be written by attaching the carbonyl group from the acid to the nitrogen atom of the amine. —OH is removed from the acid and —H from the amine to form water.

b.

$$CH_3 - \overset{\overset{\displaystyle O}{\|}}{C} - \overset{\overset{\displaystyle H}{|}}{N} - CH_2 - CH_3$$

STUDY CHECK

What are the condensed structural formulas of the carboxylic acid and amine needed to prepare the following amide?

$$H - \overset{\overset{\displaystyle O}{\|}}{C} - \overset{\overset{\displaystyle CH_3}{|}}{N} - CH_3$$

Naming Simple Amides

In both the common and IUPAC names, simple amides are named by dropping the *ic acid* or *oic acid* from the carboxylic acid names and adding the suffix *amide*. An alkyl group attached to the nitrogen of an amide is named with the prefix *N-* followed by the alkyl name.

methanamide (formamide)   ethanamide (acetamide)   butanamide (butyramide)   benzamide

■ Naming Amides

Give the common and IUPAC names for

$$CH_3 - CH_2 - \overset{\overset{\displaystyle O}{\|}}{C} - NH_2$$

SOLUTION

The IUPAC name of the carboxylic acid is propanoic acid; the common name is propionic acid. Replacing the *oic acid* or *ic acid* ending with *amide* gives the IUPAC name of *propanamide* and common name of *propionamide*.

STUDY CHECK

Draw the condensed structural formula of benzamide.

■ Physical Properties of Amides

The amides do not have the properties of bases that we saw for the amines. Only formamide is a liquid at room temperature, while the other amides are solids. For primary amides, the —NH_2 group can form several hydrogen bonds, which gives primary amides high melting points. The melting points of the secondary amides are lower because the number of hydrogen bonds decreases. Tertiary amides have even lower melting points because they cannot form hydrogen bonds with other tertiary amides.

WEB TUTORIAL
Amine and Amide Functional Groups

Hydrogen bonding between amide molecules

The amides with one to five carbon atoms are soluble in water because they can hydrogen bond with water molecules.

Hydrogen bonding of amides with water

Amides in Health and Medicine

The simplest natural amide is urea, an end product of protein metabolism in the body. The kidneys remove urea from the blood and provide for its excretion in urine. If the kidneys malfunction, urea is not removed and builds to a toxic level, a condition called uremia. Urea is also used as a component of fertilizer, to increase nitrogen in the soil.

Urea

Many barbiturates are cyclic amides of barbituric acid that act as sedatives in small dosages or sleep inducers in larger dosages. They are often habit forming. Barbiturate drugs include phenobarbital (Luminal) and pentobarbital (Nembutal).

Luminal (phenobarbital)

Nembutal (pentobarbital)

Aspirin substitutes contain phenacetin or acetaminophen, which is used in Tylenol. Like aspirin, acetaminophen reduces fever and pain, but it has little anti-inflammatory effect.

phenacetin

acetaminophen

Hydrolysis of Amides

Amides undergo hydrolysis when water is added back to the amide bond to split the molecule. When an acid is used, the hydrolysis products of an amide are the carboxylic acid and the ammonium salt. In base hydrolysis, the amide produces the salt of the carboxylic acid and ammonia or an amine.

Acid Hydrolysis of Amides

Base Hydrolysis of Amides

CH_3—CH_2—$\overset{\overset{\displaystyle O}{\|}}{C}$—$NH$—$CH_3$ + NaOH ⟶ CH_3—CH_2—$\overset{\overset{\displaystyle O}{\|}}{C}$—$O^-Na^+$ + NH_2—CH_3

N-methylpropanamide   sodium propanoate, a salt   methanamine
(*N*-methylpropionamide)   (sodium propionate)   (methylamine)

SAMPLE PROBLEM 13.15

Hydrolysis of Amides

Write the structural formulas for the products for the hydrolysis of *N*-methylpentanamide with NaOH.

SOLUTION

Hydrolysis of the amide with a base produces a carboxylate salt (sodium pentanoate) and the corresponding amine (methylamine).

$$CH_3-CH_2-CH_2-CH_2-\overset{\overset{\displaystyle O}{\|}}{C}O^- Na^+ \ + \ NH_2-CH_3$$

STUDY CHECK

What are the structures of the products from the hydrolysis of *N*-methylbutyramide with HBr?

QUESTIONS AND PROBLEMS

■ Amides

13.45 Give the IUPAC and common names (if any) for each of the following amides:

a. $CH_3-\overset{\overset{\displaystyle O}{\|}}{C}-NH_2$

b. $CH_3-CH_2-CH_2-\overset{\overset{\displaystyle O}{\|}}{C}-NH_2$

c. $H-\overset{\overset{\displaystyle O}{\|}}{C}-NH_2$

13.46 Give the IUPAC and common names (if any) for each of the following amides:

a. $CH_3-CH_2-\overset{\overset{\displaystyle O}{\|}}{C}-NH_2$

b. $CH_3-CH_2-CH_2-CH_2-CH_2-\overset{\overset{\displaystyle O}{\|}}{C}-NH_2$

c. ⬡$-\overset{\overset{\displaystyle O}{\|}}{C}-NH_2$

13.47 Draw the condensed structural formulas for each of the following amides:
   a. propionamide   **b.** 2-methylpentanamide   **c.** methanamide

13.48 Draw the condensed structural formulas for each of the following amides:
   a. formamide   **b.** benzamide   **c.** 3-methylbutyramide

13.49 Write the condensed structural formulas for the products of the acid hydrolysis of each of the following amides with HCl:

a. $CH_3-\overset{\overset{\displaystyle O}{\|}}{C}-NH_2$

b. $CH_3-CH_2-\overset{\overset{\displaystyle O}{\|}}{C}-NH_2$

c. $CH_3-CH_2-CH_2-\overset{\overset{\displaystyle O}{\|}}{C}-NH-CH_3$

d. ⬡$-\overset{\overset{\displaystyle O}{\|}}{C}-NH_2$

   e. *N*-ethylpentanamide

13.50 Write the condensed structural formulas for the products of the base hydrolysis of each of the following amides with NaOH:

a. CH₃—CH₂—CH—C—NH₂
        (CH₃) (O)

b. CH₃—CH₂—CH₂—C—N—CH₂—CH₃
        (O) (CH₂—CH₃)

c. C—N—CH₂—CH₂—CH₂—CH₃
    (O) (CH₃)

d. CH₃—CH—C—N—CH₂—CH₃
    (Cl) (O) (CH₃)

e. C—N—CH₂—CH₂—CH₃
    (O) (H)

CONCEPT MAP

Carboxylic Acids, Esters, Amines, and Amides

Carboxylic acids — form — Amides — hydrolyze — Amines

Carboxylic acids have a Carboxyl group

are Polar

ionize as Weak acids

High boiling points; soluble in water up to 5 C atoms

react with Alcohols — to form — Esters — hydrolyze — Alcohols

Amines have a Nitrogen atom

are Soluble in water up to 6 C atoms

ionize as Weak bases

CHAPTER REVIEW

13.1 Carboxylic Acids
A carboxylic acid contains the carboxyl functional group, which is a hydroxyl group connected to the carbonyl group.

13.2 Properties of Carboxylic Acids
The carboxyl group contains polar bonds of O—H and C=O, which makes a carboxylic acid with one to four carbon atoms very soluble in water. As weak acids, carboxylic acids ionize slightly by donating a proton to water to form carboxylate and hydronium ions. Carboxylic acids are neutralized by base, producing the carboxylate salt and water.

13.3 Esters
In an ester, an alkyl or aromatic group has replaced the H of the hydroxyl group of a carboxylic acid. In the presence of a strong acid, a carboxylic acid reacts with an alcohol to produce an ester. A molecule of water is removed: —OH from the carboxylic acid and —H from the alcohol molecule. The names of esters consist of two words, one from the alcohol and the other from the carboxylic acid with the *ic* ending replaced by *ate*. Esters undergo acid hydrolysis by adding water to yield the carboxylic acid and alcohol (or phenol). Base hydrolysis, or saponification, of an ester produces the carboxylate salt and an alcohol.

13.4 Amines
A nitrogen atom attached to one, two, or three alkyl or aromatic groups forms a primary, secondary, or tertiary amine. Many amines, synthetic or naturally occurring, have physiological activity. In the common names of simple amines, the alkyl groups are listed alphabetically followed by the suffix *amine*. Primary and secondary amines form hydrogen bonds, which make their boiling points higher than alkanes of similar mass. Small amines are soluble in water. In water, amines act as weak bases to produce ammonium and hydroxide ions. When amines react with acids, they form amine salts. As ionic compounds, amine salts are solids, soluble in water, and odorless compared to the amines.

13.5 Amides
Amides are derivatives of carboxylic acids in which the hydroxyl group is replaced by —NH_2. Amides are named by replacing the *ic acid* or *oic acid* with *amide*. Hydrolysis of an amide by an acid produces an amine salt. Hydrolysis by a base produces the salt of the carboxylic acid.

SUMMARY OF NAMING

Family	Condensed Structural Formula	IUPAC Name	Common Name
Carboxylic acid	CH_3—C(=O)—OH	Ethanoic acid	Acetic acid
Ester	CH_3—C(=O)—O—CH_3	Methyl ethanoate	Methyl acetate
Amine	CH_3—CH_2—NH_2		Ethylamine
Amide	CH_3—C(=O)—NH_2	Ethanamide	Acetamide

SUMMARY OF REACTIONS

IONIZATION OF A CARBOXYLIC ACID IN WATER

$$CH_3-\overset{O}{\overset{\|}{C}}-OH + H_2O \rightleftharpoons CH_3-\overset{O}{\overset{\|}{C}}-O^- + H_3O^+$$

ethanoic acid   ethanoate ion   hydronium
(acetic acid)   (acetate ion)   ion

NEUTRALIZATION OF A CARBOXYLIC ACID

$$CH_3-CH_2-\overset{O}{\overset{\|}{C}}-OH + NaOH \longrightarrow CH_3-CH_2-\overset{O}{\overset{\|}{C}}-O^-Na^+ + H_2O$$

propanoic acid   sodium   sodium propanoate
(propionic acid)   hydroxide   (sodium propionate)

ESTERIFICATION: CARBOXYLIC ACID AND AN ALCOHOL

ethanoic acid   methanol   methyl ethanoate
(acetic acid)   (methyl alcohol)   (methyl acetate)

ACID HYDROLYSIS OF AN ESTER

methyl ethanoate   ethanoic acid   methanol (methyl
(methyl acetate)   (acetic acid)   alcohol)

BASE HYDROLYSIS OF AN ESTER

methyl propanoate   sodium   sodium propanoate   methanol
(methyl propionate)   hydroxide   (sodium propionate)   (methyl
                                                                   alcohol)

IONIZATION OF AMINES IN WATER

methylamine   methylammonium hydroxide

FORMATION OF AMINE SALTS

methylamine   methylammonium chloride

FORMATION OF AMIDES

propanoic acid   ammonia   propanamide
(propionic acid)   (propionamide)

ACID HYDROLYSIS OF AMIDES

ethanamide   ethanoic acid   ammonium
(acetamide)   (acetic acid)   chloride

img_4

BASE HYDROLYSIS OF AMIDES

N-methylpropanamide
(N-methylpropionamide)

sodium propanoate, a salt
(sodium propionate)

methylamine

KEY TERMS

alkaloids Amines having physiological activity that are produced in plants.

amides Organic compounds containing the carbonyl group attached to an amino group or a substituted nitrogen atom:

$$\underset{\|}{\overset{O}{C}}-NH_2 \qquad \underset{\|}{\overset{O}{C}}-\underset{|}{\overset{H}{N}}-$$

amines Organic compounds containing a nitrogen atom attached to one, two, or three hydrocarbon groups.

amine salt An ionic compound produced from an amine and an acid.

carboxyl group A functional group found in carboxylic acids composed of carbonyl and hydroxyl groups.

$$\underset{\|}{\overset{O}{C}}-OH \qquad \text{carboxyl group}$$

carboxylate ion The anion produced when a carboxylic acid donates a proton to water.

carboxylic acids A family of organic compounds containing the carboxyl group.

$$\underset{\|}{\overset{O}{C}}-OH \qquad \text{carboxylic acid}$$

carboxylic acid salt The product of neutralization of a carboxylic acid; a carboxylate ion and the metal ion from the base.

esterification The formation of an ester from a carboxylic acid and an alcohol with the elimination of a molecule of water in the presence of an acid catalyst.

esters A family of organic compounds in which an alkyl group replaces the hydrogen atom in a carboxylic acid.

$$\underset{\|}{\overset{O}{C}}-O-C \qquad \text{Ester}$$

hydrolysis The splitting of a molecule by the addition of water. Esters hydrolyze to produce a carboxylic acid and an alcohol. Amides yield the corresponding carboxylic acid and amine or their salts.

saponification The hydrolysis of an ester with a strong base to produce a salt of the carboxylic acid and an alcohol.

ADDITIONAL QUESTIONS AND PROBLEMS

13.51 Give the IUPAC and common names (if any) for each of the following compounds:

13.52 Give the IUPAC and common names (if any) for each of the following compounds:

a. $CH_3-\underset{\underset{CH_3}{|}}{CH}-CH_2-CH_2-\underset{\|}{\overset{O}{C}}-OH$

13.53 Draw the condensed structural formulas of each of the following:
a. methyl acetate
b. ethyl butanoate
c. 3-methylpentanoic acid
d. ethyl benzoate

13.54 Draw the condensed structural formulas of each of the following:
a. ethyl butyrate
b. 2-methylpentanoic acid
c. 3,5-dimethylhexanoic acid
d. propyl acetate

13.55 In each of the following pairs of compounds, select the compound that would have the higher boiling point:

a. $CH_3-CH_2-CH_2-OH$ or $CH_3-\overset{\overset{O}{\|}}{C}-OH$

b. $CH_3-CH_2-CH_2-CH_3$ or $CH_3-CH_2-\overset{\overset{O}{\|}}{C}-OH$

c. $CH_3-\overset{\overset{O}{\|}}{C}-OH$ or $CH_3-CH_2-\overset{\overset{O}{\|}}{C}-OH$

13.56 In each of the following pairs of compounds, select the compound that would have the higher boiling point:

a. $CH_3-CH_2-CH_2-OH$ or $CH_3-\overset{\overset{O}{\|}}{C}-O-CH_3$

b. $CH_3-O-\overset{\overset{O}{\|}}{C}-CH_3$ or $CH_3-CH_2-\overset{\overset{O}{\|}}{C}-OH$

c. $CH_3-\overset{\overset{O}{\|}}{C}-O-CH_3$ or $CH_3-CH_2-CH_2-CH_3$

13.57 Write the products of the following reactions:

a. $CH_3-CH_2-\overset{\overset{O}{\|}}{C}-OH + KOH \longrightarrow$

b. $CH_3-CH_2-\overset{\overset{O}{\|}}{C}-OH + CH_3OH \overset{H^+}{\rightleftharpoons}$

c.

$\overset{\overset{O}{\|}}{C}-OH$ (benzoic acid)

$+ CH_3-CH_2-OH \overset{H^+}{\rightleftharpoons}$

13.58 Write the products of the following reactions:

a. $CH_3-\overset{\overset{O}{\|}}{C}-OH + NaOH \longrightarrow$

b. $CH_3-\overset{\overset{CH_3}{|}}{CH}-\overset{\overset{O}{\|}}{C}-OH + KOH \longrightarrow$

c. $CH_3-\overset{\overset{CH_3}{|}}{CH}-\overset{\overset{O}{\|}}{C}-OH + CH_3OH \overset{H^+}{\rightleftharpoons}$

13.59 Write the products of the following reactions:

a. $CH_3-CH_2-\overset{\overset{O}{\|}}{C}-O-\overset{\overset{CH_3}{|}}{CH}-CH_3 + H_2O \overset{H^+}{\rightleftharpoons}$

b. $CH_3-\overset{\overset{CH_3}{|}}{CH}-\overset{\overset{O}{\|}}{C}-O-CH_2-CH_2-CH_3 + H_2O \overset{H^+}{\longrightarrow}$

13.60 Write the products of the following reactions:

a. $CH_3-CH_2-\overset{\overset{O}{\|}}{C}-O-\overset{\overset{CH_3}{|}}{CH}-CH_3 + NaOH \longrightarrow$

b. $CH_3-\overset{\overset{CH_3}{|}}{CH}-\overset{\overset{O}{\|}}{C}-O-CH_2-CH_2-CH_3 + NaOH \longrightarrow$

13.61 Draw the structure of each of the following compounds:
a. dimethylamine
b. cyclohexylamine
c. dimethylammonium chloride
d. triethylamine
e. N-methylaniline

13.62 In each pair, indicate the compound that has the higher boiling point:
a. 1-butanol or butylamine
b. trimethylamine or propylamine
c. butylamine or diethylamine
d. butane or propylamine

13.63 In each pair, indicate the compound that is more soluble in water:
a. ethylamine or butylamine
b. trimethylamine or tripropylamine
c. butylamine or pentane
d. $NH_2-CH_2-CH_2-CH_2-NH_2$ or $CH_3-CH_2-CH_2-NH_2$

13.64 Give the IUPAC name for each of the following amides:

a. $H-\overset{\overset{O}{\|}}{C}-NH_2$

b. $CH_3-CH_2-\overset{\overset{O}{\|}}{C}-NH_2$

c. $CH_3-\overset{\overset{O}{\|}}{C}-NH_2$

d. $CH_3-CH_2-CH_2-\overset{\overset{O}{\|}}{C}-NH_2$

13.65 Write the structure of each product of the following reactions:
a. $CH_3-CH_2-NH_2 + H_2O \longrightarrow$
b. $CH_3-CH_2-NH_2 + HCl \longrightarrow$
c. $CH_3-CH_2-NH-CH_3 + H_2O \longrightarrow$
d. $CH_3-CH_2-NH-CH_3 + HCl \longrightarrow$
e. $CH_3-CH_2-CH_2-NH_3^+Cl^- + NaOH \longrightarrow$

f. $CH_3-CH_2-\overset{\overset{CH_3}{|}}{NH_2^+}Cl^- + NaOH \longrightarrow$

13.66 Toradol is used in dentistry to relieve pain. Name the functional groups in this molecule.

13.67 Voltaren is indicated for acute and chronic treatment of the symptoms of rheumatoid arthritis. Name the functional groups in this molecule.

13.68 Using a reference book such as the *Merck Index* or *Physicians' Desk Reference*, look up the structural formula of the following medicinal drugs and list the functional groups in the compounds. You may need to refer to the cross-index of names in the back of the reference book.

a. Keflex, an antibiotic

b. Inderal, a β-channel blocker used to treat heart irregularities
c. ibuprofen, an anti-inflammatory agent
d. Aldomet (methyldopa)
e. Percodan, a narcotic pain reliever
f. triamterene, a diuretic

UNDERSTANDING THE CONCEPTS

13.69 The sweetener aspartame is made from two amino acids, aspartic acid and phenylalanine. Identify the functional groups in aspartame.

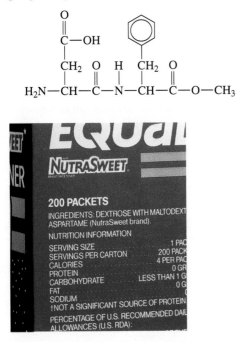

13.70 Some aspirin substitutes contain phenacetin to reduce fever. Identify the functional groups in phenacetin.

$$CH_3-CH_2-O-\text{⟨benzene⟩}-NH-\overset{\displaystyle O}{\overset{\displaystyle \|}{C}}-CH_3$$

13.71 Propyl acetate is the ester that gives the odor and smell of pears.

a. What is the structure of propyl acetate?
b. Write an equation for the formation of propyl acetate.
c. Write an equation for the acid hydrolysis of propyl acetate.
d. Write an equation for the base hydrolysis of propyl acetate with NaOH.
e. How many mL of 0.208 M NaOH are needed to complete hydrolyze (saponify) 1.58 g propyl acetate?

13.72 Ethyl octanoate is a flavor component of mangos.

a. What is the structure of ethyl octanoate?
b. Write an equation for the formation of ethyl octanoate.
c. Write an equation for the acid hydrolysis of ethyl octanoate.
d. Write an equation for the base hydrolysis of ethyl octanoate with NaOH.
e. How many mL of 0.315 M NaOH are needed to complete hydrolyze (saponify) 2.840 g ethyl octanoate?

QUESTIONS FOR TEAMWORK

13.73 For the following:
   a. Give the common name of each.
   b. Identify isomers.
   c. Explain the following order of increasing boiling points:

$$CH_3-O-CH_3 \qquad CH_3-CH_2-NH_2 \qquad CH_3-CH_2-OH \qquad \underset{118°C}{CH_3-\overset{\overset{\displaystyle O}{\|}}{C}-OH}$$

$$-25°C \qquad\qquad 17°C \qquad\qquad 79°C$$

13.74 Why is the boiling point of trimethylamine (3°C) lower than propylamine (48°C) when the two compounds are isomers and have the same molar mass?

13.75 Novocain, a local anesthetic, is the hydrochloride salt of procaine.

$$H_2N-\bigcirc-\overset{\overset{\displaystyle O}{\|}}{C}-O-CH_2-CH_2-\underset{\underset{\displaystyle CH_2-CH_3}{|}}{\overset{\overset{\displaystyle CH_2-CH_3}{|}}{N}}$$

procaine

   a. What is the formula of the amine salt formed when procaine reacts with HCl?
   b. Why is Novocain, the amine salt, used rather than the amine procaine?

13.76 Lidocaine (xylocaine) is used as a local anesthetic and cardiac depressant.

Lidocaine (xylocaine)

   a. What is the formula of the amine salt formed when lidocaine reacts with HCl?
   b. Why is the amine salt of lidocaine used rather than the amine?

ANSWERS

Answers to Study Checks

13.1 $CH_3-CH_2-CH_2-CH_2-\overset{\overset{\displaystyle O}{\|}}{C}-OH$

13.2 Two carboxylic acid molecules hydrogen bond forming a dimer, which has a mass twice that of the single acid molecule.

13.3 $H-\overset{\overset{\displaystyle O}{\|}}{C}-OH + H_2O \rightleftharpoons H-\overset{\overset{\displaystyle O}{\|}}{C}-O^- + H_3O^+$

13.4 butanoic acid, butyric acid

13.5 propanoic (propionic) acid and 1-pentanol

13.6 $CH_3-\overset{\overset{\displaystyle O}{\|}}{C}-O-CH_2-CH_2-CH_2-CH_2-CH_3$

13.7 propanoic (propionic) acid and ethanol

13.8 $+ CH_3-OH$

13.9 tertiary (3°)

13.10 $CH_3-CH_2-NH-CH_2-CH_2-CH_3$

13.11 Hydrogen bonding makes amines with six or fewer carbon atoms soluble in water.

13.12 $CH_3-\underset{\underset{\displaystyle CH_3}{|}}{\overset{\overset{\displaystyle CH_3}{|}}{\overset{+}{N}}}-HCl^-$

13.13 $H-\overset{\overset{\displaystyle O}{\|}}{C}-OH$ and $H-\underset{\underset{\displaystyle CH_3}{|}}{\overset{\overset{\displaystyle CH_3}{|}}{N}}$

13.14

$$\bigcirc-\overset{\overset{\displaystyle O}{\|}}{\underset{\underset{\displaystyle NH_2}{}}{C}}$$

13.15 $CH_3-CH_2-CH_2-\overset{\overset{\displaystyle O}{\|}}{C}-OH$ and $CH_3-NH_3{}^+Br^-$

Answers to Selected Problems

13.1 methanoic acid (formic acid)

13.3 Each compound contains three carbon atoms. They differ because propanal, an aldehyde, contains a carbonyl group bonded to a hydrogen. In propanoic acid, the carbonyl group connects to a hydroxyl group, forming a carboxyl group.

13.5 **a.** ethanoic acid (acetic acid)
b. propanoic acid (propionic acid)
c. 4-hydroxybenzoic acid
d. 4-bromopentanoic acid

13.7 **a.** CH₃—CH₂—C(=O)—OH

b.

(benzene ring with C(=O)—OH)

c. Cl—CH₂—C(=O)—OH

d. HO—CH₂—CH₂—C(=O)—OH

e. CH₃—CH₂—CH₂—C(=O)—OH

f. CH₃—CH₂—CH₂—CH₂—CH₂—CH₂—C(=O)—OH

13.9 **a.** Butanoic acid has a higher molar mass and would have a higher boiling point.
b. Propanoic acid can form more hydrogen bonds and would have a higher boiling point.
c. Butanoic acid can form hydrogen bonds and would have a higher boiling point.

13.11 **a.** Propanoic acid
b. Propionic acid

13.13 **a.** H—C(=O)—OH + H₂O ⇌ H—C(=O)—O⁻ + H₃O⁺

b. CH₃—CH₂—C(=O)—OH + H₂O ⇌

CH₃—CH₂—C(=O)—O⁻ + H₃O⁺

c. CH₃—C(=O)—OH + H₂O ⇌

CH₃—C(=O)—O⁻ + H₃O⁺

13.15 **a.** H—C(=O)—OH + NaOH ⟶ H—C(=O)—O⁻Na⁺ + H₂O

b. CH₃—CH₂—C(=O)—OH + NaOH ⟶

CH₃—CH₂—C(=O)—O⁻Na⁺ + H₂O

c. (benzene ring with C(=O)—OH) + NaOH ⟶ (benzene ring with C(=O)—O⁻Na⁺) + H₂O

13.17 **a.** aldehyde
b. ester
c. ketone
d. carboxylic acid

13.19 **a.** CH₃—C(=O)—O—CH₃

b. CH₃—CH₂—CH₂—C(=O)—O—CH₃

c. (benzene ring with) C(=O)—O—CH₃

13.21 **a.** CH₃—CH₂—C(=O)—O—CH₂—CH₂—CH₃

b. CH₃—CH₂—CH₂—CH₂—C(=O)—O—CH(CH₃)—CH₃

13.23 **a.** methyl formate (methyl methanoate)
b. methyl acetate (methyl ethanoate)
c. methyl butyrate (methyl butanoate)
d. ethyl butyrate (ethyl butanoate)

13.25 **a.** CH₃—C(=O)—O—CH₃

b. H—C(=O)—O—CH₂—CH₂—CH₂—CH₃

c. CH₃—CH₂—CH₂—CH₂—C(=O)—O—CH₂—CH₃

d. CH₃—CH₂—C(=O)—O—CH₂—CH₂—CH₃

13.27 **a.** pentyl ethanoate (pentyl acetate)
b. octyl ethanoate (octyl acetate)
c. pentyl butanoate (pentyl butyrate)

13.29 The products of the acid hydrolysis of an ester are an alcohol and a carboxylic acid.

13.31 **a.** CH₃—CH₂—C(=O)—O⁻Na⁺   and   CH₃—OH

b. CH₃—C(=O)—OH   and   CH₃—CH₂—CH₂—OH

c. CH₃—CH₂—CH₂—C(=O)—OH   and   CH₃—CH₂—OH

d. and CH_3-CH_2-OH

13.33 a. primary   **b.** secondary
c. tertiary   **d.** tertiary

13.35 a. ethylamine   **b.** methylpropylamine
c. diethylmethylamine   **d.** isopropylamine

13.37 a. $CH_3-CH_2-NH_2$

b. [benzene ring with NH—CH₃ group]

c. $CH_3-CH_2-CH_2-CH_2-\overset{\overset{H}{|}}{N}-CH_2-CH_2-CH_3$

13.39 a. CH_3-CH_2-OH
b. $CH_3-CH_2-CH_2-NH_2$
c. $CH_3-CH_2-CH_2-NH_2$

13.41 a. yes   **b.** yes   **c.** no   **d.** yes

13.43 a. $CH_3-NH_2 + H_2O \rightleftharpoons CH_3-NH_3^+ + OH^-$
b. $CH_3-NH-CH_3 + H_2O \rightleftharpoons$
$$CH_3-\overset{+}{N}H_2-CH_3 + OH^-$$

c. [benzene ring with NH₂] $+ H_2O \rightleftharpoons$ [benzene ring with NH₃⁺] $+ OH^-$

13.45 a. ethanamide (acetamide)
b. butanamide (butyramide)
c. methanamide (formamide)

13.47 a. $CH_3-CH_2-\overset{\overset{O}{\|}}{C}-NH_2$

b. $CH_3-CH_2-CH_2-\overset{\overset{CH_3}{|}}{CH}-\overset{\overset{O}{\|}}{C}-NH_2$

c. $H-\overset{\overset{O}{\|}}{C}-NH_2$

13.49 a. $CH_3-\overset{\overset{O}{\|}}{C}-OH + NH_4^+Cl^-$

b. $CH_3-CH_2-\overset{\overset{O}{\|}}{C}-OH + NH_4^+Cl^-$

c. $CH_3-CH_2-CH_2-\overset{\overset{O}{\|}}{C}-OH + CH_3-NH_3^+Cl^-$

d. [benzene ring]$-\overset{\overset{O}{\|}}{C}-OH + NH_4^+Cl^-$

e. $CH_3-CH_2-CH_2-CH_2-\overset{\overset{O}{\|}}{C}-OH + CH_3-CH_2-NH_3^+Cl^-$

13.51 a. 3-methylbutanoic acid
b. ethyl benzoate
c. ethyl propanoate; ethyl propionate
d. 2-chlorobenzoic acid
e. pentanoic acid

13.53 a. $CH_3-O-\overset{\overset{O}{\|}}{C}-CH_3$

b. $CH_3-CH_2-O-\overset{\overset{O}{\|}}{C}-CH_2-CH_2-CH_3$

c. $CH_3-CH_2-\overset{\overset{CH_3}{|}}{CH}-CH_2-\overset{\overset{O}{\|}}{C}-OH$

d. [benzene ring]$-\overset{\overset{O}{\|}}{C}-O-CH_2-CH_3$

13.55 a. $CH_3-\overset{\overset{O}{\|}}{C}-OH$

b. $CH_3-CH_2-\overset{\overset{O}{\|}}{C}-OH$

c. $CH_3-CH_2-\overset{\overset{O}{\|}}{C}-OH$

13.57 a. $CH_3-CH_2-\overset{\overset{O}{\|}}{C}-O^-K^+ + H_2O$

b. $CH_3-CH_2-\overset{\overset{O}{\|}}{C}-O-CH_3 + H_2O$

c. [benzene ring]$-\overset{\overset{O}{\|}}{C}-O-CH_2-CH_3 + H_2O$

13.59 a. $CH_3-CH_2-\overset{\overset{O}{\|}}{C}-OH$ and $HO-\overset{\overset{CH_3}{|}}{CH}-CH_3$

b. $CH_3-\overset{\overset{CH_3}{|}}{CH}-\overset{\overset{O}{\|}}{C}-OH$ and $HO-CH_2-CH_2-CH_3$

13.61 a. $CH_3-\overset{\overset{CH_3}{|}}{NH}$

b. [cyclohexane ring with NH₂]

c. $CH_3-\overset{\overset{CH_3}{|}}{\overset{+}{N}H_2} \ Cl^-$

d. $CH_3-CH_2-\overset{\overset{\displaystyle CH_2-CH_3}{|}}{N}-CH_2-CH_3$

e. $\underset{\bigcirc}{\overset{\displaystyle HN-CH_3}{|}}$

13.63 a. ethylamine
b. trimethylamine
c. butylamine
d. $NH_2-CH_2-CH_2-CH_2-NH_2$

13.71 a. $CH_3-CH_2-CH_2-O-\overset{\overset{\displaystyle O}{\|}}{C}-CH_3$

b. $CH_3-CH_2-CH_2-OH + HO-\overset{\overset{\displaystyle O}{\|}}{C}-CH_3 \underset{\longleftarrow}{\overset{H^+, \text{ heat}}{\longrightarrow}} CH_3-CH_2-CH_2-O-\overset{\overset{\displaystyle O}{\|}}{C}-CH_3 + H_2O$

c. $CH_3-CH_2-CH_2-O-\overset{\overset{\displaystyle O}{\|}}{C}-CH_3 + H_2O \underset{\longleftarrow}{\overset{H^+}{\longrightarrow}} CH_3-CH_2-CH_2-OH + HO-\overset{\overset{\displaystyle O}{\|}}{C}-CH_3$

d. $CH_3-CH_2-CH_2-O-\overset{\overset{\displaystyle O}{\|}}{C}-CH_3 + NaOH \overset{\text{Heat}}{\longrightarrow} CH_3-CH_2-CH_2-OH + Na^+ {}^-O-\overset{\overset{\displaystyle O}{\|}}{C}-CH_3$
e. 74.4 mL of 0.208 M NaOH

13.65 a. $CH_3-CH_2-NH_3{}^+OH^-$
b. $CH_3-CH_2-NH_3{}^+Cl^-$
c. $CH_3-CH_2-\overset{+}{N}H_2-CH_3\,OH^-$
d. $CH_3-CH_2-\overset{+}{N}H_2-CH_3\,Cl^-$
e. $CH_3-CH_2-CH_2-NH_2 + NaCl + H_2O$
f. $CH_3-CH_2-\overset{\overset{\displaystyle CH_3}{|}}{N}H + NaCl + H_2O$

13.67 carboxylic acid salt, aromatic, amine

13.69 amine, carboxylic acid, amide, aromatic, ester

14 Carbohydrates

"We use a refractometer to measure sugar content in a small sample of juices from the grapes in different areas of the vineyard," says Leslie Bucher, laboratory director at Bouchaine Winery. "We also measure the alcohol content during fermentation and run tests for sulfur, pH, and total acid."

As grapes ripen, there is an increase in the sugars, which are the monosaccharides fructose and glucose. The sugar content is affected by soil conditions and the amount of sun and water. When the grapes are ripe and sugar content is at a desirable level, they are harvested. During fermentation, enzymes from yeast convert about half the sugar to ethanol, and half to carbon dioxide. Grapes harvested with 22.5% sugar will ferment to give a wine with 12.5–13.5% alcohol content.

the Chemistry place

Visit **www.aw-bc.com/chemplace** or use the CD-ROM in the book for extra quizzes, interactive tutorials, career resources, PowerPoint slides, math help, and case studies.

Of all the organic compounds in nature, carbohydrates are the most abundant. In plants, energy from the sun converts carbon dioxide and water into the carbohydrate glucose. Many of the glucose molecules are made into long-chain polymers of starch that store energy or into cellulose to build the structural framework of the plant. About 65% of the foods in our diet consist of carbohydrates. Each day you enjoy carbohydrates in foods such as bread, pasta, potatoes, and rice. During digestion and cellular metabolism, the starches are broken down into glucose, which is oxidized further in our cells to provide our bodies with energy and to provide the cells with carbon atoms for building molecules of protein, lipids, and nucleic acids. Cellulose has other important uses, too. The wood in our furniture, the pages in this book, and the cotton in our clothing are made of cellulose.

14.1 Carbohydrates

Carbohydrates such as table sugar, lactose in milk, and cellulose are all made of carbon, hydrogen, and oxygen. Simple sugars, which have formulas of $C_n(H_2O)_n$, were once thought to be hydrates of carbon, thus the name *carbohydrate*. In a series of reactions called photosynthesis, energy from the sun is used to combine the carbon atoms from carbon dioxide (CO_2) and the hydrogen and oxygen atoms of water into the carbohydrate glucose.

$$6CO_2 \ + \ 6H_2O \ + \ energy \xrightleftharpoons[\text{Respiration}]{\text{Photosynthesis}} \underset{\text{glucose}}{C_6H_{12}O_6} \ + \ 6O_2$$

In the body, glucose is oxidized in a series of metabolic reactions known as respiration, which releases chemical energy to do work in the cells. Carbon dioxide and water are produced and returned to the atmosphere. The combination of photosynthesis and respiration is called the carbon cycle, in which energy from the sun is stored in plants by photosynthesis and made available to us when the carbohydrates in our diets are metabolized. (See Figure 14.1.)

Types of Carbohydrates

The simplest carbohydrates are the **monosaccharides**. A monosaccharide cannot be split or hydrolyzed into smaller carbohydrates. One of the most common carbohydrates, glucose, $C_6H_{12}O_6$, is a monosaccharide. **Disaccharides** consist of two monosaccharide units joined together. Likewise, a disaccharide can be split into two monosaccharide units. For example, ordinary table sugar, sucrose, $C_{12}H_{22}O_{11}$, is a disaccharide that can be hydrolyzed in the presence of an acid or an enzyme to give one molecule of glucose and one molecule of another monosaccharide, fructose.

$$\underset{\text{sucrose}}{C_{12}H_{22}O_{11}} \ + \ H_2O \xrightarrow{\text{H}^+ \text{ or enzyme}} \underset{\text{glucose}}{C_6H_{12}O_6} \ + \ \underset{\text{fructose}}{C_6H_{12}O_6}$$

FIGURE 14.1

During photosynthesis, energy from the sun combines CO_2 and H_2O to form glucose $C_6H_{12}O_6$ and O_2. During respiration in the body, carbohydrates are oxidized to CO_2 and H_2O, while energy is produced.

Q **What are the reactants and products of respiration?**

Polysaccharides are carbohydrates that are naturally occurring polymers containing many monosaccharide units. In the presence of an acid or an enzyme, a polysaccharide can be completely hydrolyzed to yield many molecules of monosaccharide.

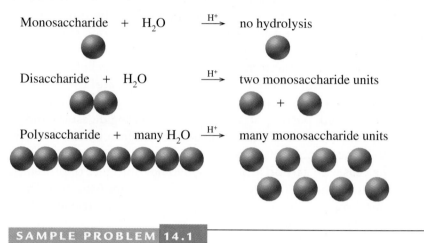

SAMPLE PROBLEM 14.1

▪ Carbohydrates

Classify the following carbohydrates as mono-, di-, or polysaccharides:

a. When lactose, milk sugar, is hydrolyzed, two monosaccharide units are produced.
b. Cellulose, a carbohydrate in cotton, yields thousands of monosaccharide units when completely hydrolyzed.

SOLUTION

a. A disaccharide contains two monosaccharide units.
b. A polysaccharide contains many monosaccharide units.

STUDY CHECK

Fructose found in fruits does not undergo hydrolysis. What type of carbohydrate is it?

Monosaccharides

Monosaccharides are simple sugars that have an unbranched chain of three to eight carbon atoms, one in a carbonyl group and the rest attached to hydroxyl groups. There are two types of monosaccharide structures. In an **aldose**, the carbonyl group is on the first carbon (—CHO); a **ketose** contains the carbonyl group on the second carbon atom as a ketone (C=O).

erythrose, a
polyhydroxy
aldehyde

erythrulose, a
polyhydroxy
ketone

A monosaccharide with three carbon atoms is a *triose*, one with four carbon atoms is a *tetrose*, a *pentose* has five carbons, and a *hexose* contains six carbons. We can use both classification systems to indicate the type of carbonyl group and the number of carbon atoms. An aldopentose is a five-carbon monosaccharide that is an aldehyde; a ketohexose would be a six-carbon monosaccharide that is a ketone. Some examples:

glyceraldehyde
(aldotriose)

threose
(aldotetrose)

ribose
(aldopentose)

fructose
(ketohexose)

SAMPLE PROBLEM 14.2

■ Monosaccharides

Classify each of the following monosaccharides to indicate their carbonyl group and number of carbon atoms:

a. ribulose

b. glucose

SOLUTION

a. The structural formula has a ketone group; ribulose is a ketose. Because there are five carbon atoms, it is a pentose. Combining these classifications makes it a ketopentose.

b. The structural formula has an aldehyde group; glucose is an aldose. Because there are six carbon atoms, it is an aldohexose.

STUDY CHECK

The simplest ketose is a triose named dihydroxyacetone. Draw its structural formula.

QUESTIONS AND PROBLEMS

■ **Carbohydrates**

14.1 What reactants are needed for photosynthesis and respiration?

14.2 What is the relationship between photosynthesis and respiration?

14.3 What is a monosaccharide? A disaccharide?

14.4 What is a polysaccharide?

14.5 What functional groups are found in all monosaccharides?

14.6 What is the difference between an aldose and a ketose?

14.7 What are the functional groups and number of carbons in a ketopentose?

14.8 What are the functional groups and number of carbons in an aldohexose?

14.9 Classify each of the following monosaccharides as an aldose or ketose.

14.10 Classify each of the monosaccharides in problem 14.9 according to the number of carbon atoms in the chain.

■ 14.2 Structures of Monosaccharides

In Chapter 12 we learned that chiral compounds exist as mirror images that cannot be superimposed. The monosaccharides, which contain chiral carbons, exist as mirror images.

■ Fischer Projections

Let's take a look again at the Fischer projection for the simplest aldose, glyceraldehyde. By convention the carbon chain is written vertically with the aldehyde group (most oxidized carbon) at the top. The letter L is assigned to the stereoisomer if the —OH group is on the left of the chiral carbon. In D-glyceraldehyde, the —OH is on the right. The very bottom carbon atom in the Fischer projection of a carbohydrate, —CH_2OH, is not chiral because it does not have four different groups bonded to it.

L-glyceraldehyde   D-glyceraldehyde

Most of the carbohydrates we will study have carbon chains with five or six carbon atoms, which means that they have several chiral carbons. Then we use the chiral carbon furthest from the carbonyl group to determine the D or L isomer. The following are the isomers of ribose, which is a five-carbon monosaccharide, and glucose, a six-carbon monosaccharide.

SAMPLE PROBLEM 14.3

▪ Identifying D and L Isomers of Sugars

Is the following structure the D or L enantiomer of ribose?

SOLUTION

In ribose, carbon 4 is the chiral atom furthest from the carbonyl group. Because the hydroxyl group on carbon 4 is on the left, this enantiomer is L-ribose.

Chiral carbon furthest from carbonyl group

STUDY CHECK

Draw the Fischer projection for D-ribose.

Some Important Monosaccharides

The hexoses glucose, galactose, and fructose are important monosaccharides. Although structural formulas for both the D and L isomers can be written for these hexoses, the D isomers are the more common form of carbohydrates found in nature. We can write their open-chain structures as follows:

The most common hexose, D-**glucose**, $C_6H_{12}O_6$, also known as dextrose and blood sugar, is found in fruits, vegetables, corn syrup, and honey. (See Figure 14.2.) It is a building block of the disaccharides sucrose, lactose, and maltose and poly-saccharides such as starch, cellulose, and glycogen.

Galactose is an aldohexose that does not occur in the free form in nature. It is obtained as a hydrolysis product of the disaccharide lactose, a sugar found in milk and milk products. Galactose is important in the cellular membranes of the brain and nervous system. The only difference in the structures of D-glucose and D-galactose is the arrangement of the —OH group on carbon 4.

In a condition called *galactosemia*, an enzyme needed to convert galactose to glucose is missing. The accumulation of galactose in the blood and tissues can lead to cataracts, mental retardation, and cirrhosis. The treatment for galactosemia is the removal of all galactose-containing foods, mainly milk and milk products, from the diet. If this is done for an infant immediately after birth, the damaging effects of galactose accumulation can be avoided.

FIGURE 14.2

The sweet taste of honey is due to the monosaccharides of D-glucose and D-fructose.

Q **What are some differences in the structures of D-glucose and D-fructose?**

HEALTH NOTE

Hyperglycemia and Hypoglycemia

In the body, glucose normally occurs at a concentration of 70–90 mg/dL (1 dL = 100 mL) of blood. However, the amount of glucose depends on the time that has passed since eating. In the first hour after a meal, the level of glucose rises to about 130 mg/dL of blood, and then decreases over the next 2–3 hours as it is used in the tissues.

A doctor may order a glucose tolerance test to evaluate the body's ability to return to normal glucose concentration in response to the ingestion of a specified amount of glucose. The patient fasts for 12 hours and then drinks a solution containing glucose. A blood sample is taken immediately, followed by more blood samples each half-hour for 2 hours, and then every hour for a total of 5 hours. If the blood glucose exceeds 140 mg/dL in plasma and remains high, hyperglycemia may be indicated. The term *glyc* or *gluco* refers to "sugar." The prefix *hyper* means above or over, and *hypo* is below or under. Thus the blood sugar level in *hyperglycemia* is above normal and below normal in *hypoglycemia*.

An example of a disease that can cause hyperglycemia is diabetes mellitus, which occurs when the pancreas is unable to produce suffi-cient quantities of insulin. As a result, glucose levels in the body fluids can rise as high as 350 mg/dL plasma. Symptoms of diabetes in people under the age of 40 include thirst, excessive urination, increased appetite, and weight loss. In older persons, diabetes is sometimes a consequence of excessive weight gain.

When a person is hypoglycemic, the blood glucose level rises and then decreases rapidly to levels as low as 40 mg/dL plasma. In some cases, hypoglycemia is caused by overproduction of insulin by the pancreas. Low blood glucose can cause dizziness, general weakness, and muscle tremors. A diet may be prescribed that consists of several small meals high in protein and low in carbohydrate. Some hypoglycemic patients are finding success with diets that include more complex carbohydrates rather than simple sugars.

In contrast to glucose and galactose, **fructose** is a ketohexose. The structure of fructose differs from glucose at carbons 1 and 2 by the location of the carbonyl group.

D-glucose   D-fructose

Fructose is the sweetest of the carbohydrates, twice as sweet as sucrose (table sugar). This makes fructose popular with dieters because less fructose, and therefore fewer calories, are needed to provide a pleasant taste. After fructose enters the bloodstream, it is converted to its isomer, glucose. (See Figure 14.2.) Fructose is found in fruit juices and honey; it is also called levulose and fruit sugar. Fructose is also obtained as one of the hydrolysis products of sucrose, the disaccharide known as table sugar.

■ **Monosaccharides**

Ribulose has the following structure.

```
        CH2OH
         |
         C=O
         |
  H—C—OH
         |
  H—C—OH
         |
        CH2OH
```

a. Identify the above compound D- or L-ribulose.

b. Write the structure of its mirror image.

SOLUTION

a. D-ribulose

b. The mirror image, L-ribulose, has the following structural formula:

```
         CH2OH
          |
          C=O
          |
  HO—C—H
          |
  HO—C—H
          |
         CH2OH
```

STUDY CHECK

What type of carbohydrate is ribulose?

QUESTIONS AND PROBLEMS

■ **Structures of Monosaccharides**

14.11 What is a Fischer projection?

14.12 Write the Fischer projection formula for D-glyceraldehyde and L-glyceraldehyde.

14.13 State whether each of the following sugars is the D or L isomer:

a.
```
       CHO
  HO——┼——H
   H——┼——OH
      CH2OH
     threose
```

b.
```
      CH2OH
       |
       C=O
  HO——┼——H
   H——┼——OH
      CH2OH
     xylulose
```

c.
```
       CHO
   H——┼——OH
   H——┼——OH
  HO——┼——H
  HO——┼——H
      CH2OH
     mannose
```

d.
```
       CHO
   H——┼——OH
   H——┼——OH
   H——┼——OH
   H——┼——OH
      CH2OH
      allose
```

14.14 State whether each of the following sugars is the D or L isomer:

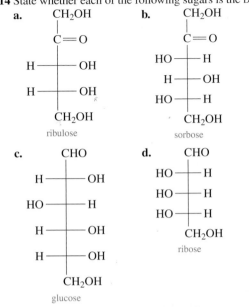

a.
CH₂OH
|
C=O
|
H —— OH
|
H —— OH
|
CH₂OH

ribulose

b.
CH₂OH
|
C=O
|
HO —— H
|
H —— OH
|
HO —— H
|
CH₂OH

sorbose

c.
CHO
|
H —— OH
|
HO —— H
|
H —— OH
|
H —— OH
|
CH₂OH

glucose

d.
CHO
|
HO —— H
|
HO —— H
|
HO —— H
|
CH₂OH

ribose

14.15 Write the mirror images for a–d in problem 14.13.

14.16 Write the mirror images for a–d in problem 14.14.

14.17 Draw the open-chain structure of D-glucose and L-glucose.

14.18 Draw the open-chain structure of D-fructose and L-fructose.

14.19 How does the open-chain structure of D-galactose differ from D-glucose?

14.20 How does the open-chain structure of D-fructose differ from D-glucose?

14.21 Identify a monosaccharide that fits each of the following descriptions:
    a. also called blood sugar
    b. not metabolized in galactosemia
    c. also called fruit sugar

14.22 Identify a monosaccharide that fits each of the following descriptions:
    a. high blood levels in diabetes
    b. obtained as a hydrolysis product of lactose
    c. the sweetest of the monosaccharides

■ 14.3 Cyclic Structures of Monosaccharides

LEARNING GOAL

■ Draw and identify the cyclic structures of monosaccharides.

Most of the time, monosaccharides exist as ring structures formed when a carbonyl group and a hydroxyl group in the *same* molecule react. The product is a ring structure that is the most stable form of aldopentoses and aldohexoses. While the carbonyl group in an aldohexose could react with several of the —OH groups, the aldohexoses typically form six-atom rings. Let's look at how we draw the cyclic structure for D-glucose starting with the Fischer projection.

■ Drawing Cyclic Structures for Glucose

STEP 1 Think of turning the open chain of glucose clockwise to the right. Then the —OH groups written on the right, other than the one on carbon 6, are drawn down, and the —OH group on the left is up.

D-glucose (open chain)

STEP 2 Fold the chain into a hexagon placing the —CH₂OH up and the —OH group close to the carbonyl carbon. Form the cyclic structure by bonding the oxygen in the —OH group to the carbonyl carbon. This structure is known as a **Haworth structure**.

Carbon-5 oxygen bonds to carbonyl   Cyclic structure

STEP 3 In the cyclic structure, carbon 1 is bonded to a new —OH group. There are two ways to place the —OH, either up or down, which gives two isomers. The —OH group on carbon 1 is down in the α (alpha) form and up in the β (beta) form.

Such differences in structural forms may seem trivial. However, we can digest starch products such as pasta to obtain glucose because the polysaccharide contains the α isomers of glucose. We cannot digest paper or wood because cellulose consists of only β-D-glucose units. Humans have an α-amylase, an enzyme needed for the digestion of starches, but not a β-amylase for the digestion of cellulose.

Sometimes, the cyclic structure is simplified to show only the position of the hydroxyl groups on the six-atom ring structure.

α-D-glucose (simplified structure)   β-D-glucose

In solution, the ring of α-D-glucose opens and closes to form β-D-glucose. Although the open chain is essential, only a small amount of the open chain is present at any given time.

α-D-glucose
(36% in mixture)

D-glucose
open-chain (trace)

β-D-glucose
(64% in mixture)

Cyclic Structures of Galactose

Galactose is an aldohexose that differs from glucose only in the arrangement of the —OH group on carbon 4. Thus, its cyclic structure is also similar to glucose, except that in galactose the —OH on carbon 4 is up. With the formation of a new hydroxyl group on carbon 1, galactose also exists in α and β forms.

D-galactose

α-D-galactose

β-D-galactose

Cyclic Structures of Fructose

In contrast to glucose and galactose, fructose is a ketohexose. It forms a cyclic structure when a hydroxyl group on carbon 5 reacts with the ketone group. The cyclic structure for fructose is a five-atom ring with carbon 2 at the right corner. A new hydroxyl group is on carbon 2 in addition to the carbon 1 from the CH$_2$OH group. There are also α and β forms of fructose.

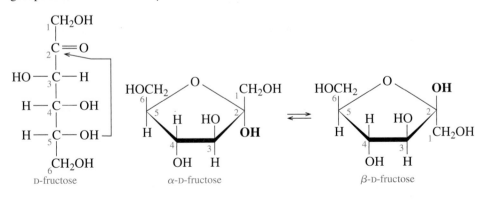

D-fructose

α-D-fructose

β-D-fructose

SAMPLE PROBLEM 14.5

▪ **Cyclic Structures for Sugars**

Identify the cyclic structure as α- or β-D-mannose, a carbohydrate found in immunoglobulins.

SOLUTION

The cyclic structure is β-D-mannose because the hydroxyl group on carbon 1 is written above the carbon.

STUDY CHECK

Draw the cyclic structure for α-D-mannose.

QUESTIONS AND PROBLEMS

▪ **Cyclic Structures of Monosaccharides**

14.23 What are the kind and number of atoms in the ring portion of the cyclic structure of glucose?

14.24 What are the kind and number of atoms in the ring portion of the cyclic structure of fructose?

14.25 Draw the cyclic structures for the α and β forms of D-glucose.

14.26 Draw the cyclic structures for the α and β forms of D-fructose.

14.27 Identify each of the following cyclic structures as the α or β form.

14.28 Identify each of the following cyclic structures as the α or β form.

b.

14.4 Chemical Properties of Monosaccharides

LEARNING GOAL

Identify the products of oxidation or reduction of monosaccharides; determine whether a carbohydrate is a reducing sugar.

Monosaccharides contain functional groups that can undergo chemical reactions. In an aldose, the aldehyde group can be oxidized to a carboxylic acid. The carbonyl group in both an aldose and ketose can be reduced to give a hydroxyl group. The hydroxyl groups can react with other compounds to form a variety of derivatives that are important in biological structures.

Oxidation of Monosaccharides

Although monosaccharides exist mostly in their cyclic forms, the aldehyde group of the open-chain structure oxidizes easily. When the monosaccharide oxidizes, a carboxylic group is produced. At the same time the Cu^{2+} in Benedict's reagent is reduced to Cu^+, forming a brick-red precipitate of Cu_2O. Monosaccharides that reduce another substance such as Benedict's reagent are called **reducing sugars**.

Open chain of D-glucose, a reducing sugar

D-gluconic acid

Fructose is also a reducing sugar. In the open-chain form, a rearrangement between the hydroxyl group on carbon 1 and the ketone group provides an aldehyde group that can be oxidized.

D-fructose (ketose)

D-glucose (aldose)

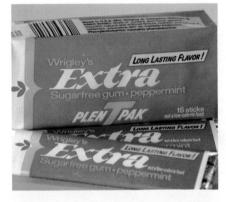

D-Sorbitol

Reduction of Monosaccharides

The reduction of the carbonyl group in monosaccharides produces sugar alcohols, which are also called *alditols*. D-glucose is reduced to D-glucitol, better known as sorbitol. D-Mannose is reduced to give D-mannitol.

D-glucose   →   D-glucitol or D-sorbitol

Sugar alcohols such as sorbitol, xylitol from xylose, and mannitol from mannose are used as sweeteners in many sugar-free products such as diet drinks and sugarless

Testing for Glucose in Urine

Normally, blood glucose flows through the kidneys and is reabsorbed into the bloodstream. However, if the blood level exceeds about 160 mg of glucose/dL of blood, the kidneys cannot reabsorb it all, and glucose spills over into the urine, a condition known as glucosuria. A symptom of diabetes mellitus is a high level of glucose in the urine.

Benedict's test can be used to determine the presence of glucose in urine. The amount of cuprous oxide (Cu_2O) formed is proportional to the amount of reducing sugar present in the urine. Low to moderate levels of reducing sugar turn the solution green; solutions with high glucose levels turn Benedict's yellow or brick-red. Table 14.1 lists some colors associated with the concentration of glucose in the urine.

In another clinical test that is more specific for glucose, the enzyme glucose oxidase is used. The oxidase enzyme converts glucose to gluconic acid and hydrogen peroxide, H_2O_2. The peroxide produced reacts with a dye in the test strip to give different colors. The level of glucose present in the urine is found by matching the color produced to a color chart on the container.

TABLE 14.1

Glucose Test Results

Color	Glucose Present in Urine	
	%	mg/dL
Blue	0	0
Blue-green	0.25	250
Green	0.50	500
Yellow	1.00	1000
Brick-red	2.00	2000

gum as well as products for people with diabetes. However, there are some side effects of these sugar substitutes. Some people experience some discomfort such as gas and diarrhea from the ingestion of sugar alcohols. The development of cataracts in diabetics is attributed to the accumulation of sorbitol in the lens of the eye.

CASE STUDY
Diabetes and Blood Glucose

SAMPLE PROBLEM 14.6

▪ Reducing Sugars

Why is D-glucose called a *reducing sugar*?

SOLUTION

D-Glucose is easily oxidized by Benedict's reagent. A carbohydrate that reduces Cu^{2+} to Cu^{+} is called a reducing sugar.

STUDY CHECK

A test using Benedict's reagent turns brick-red with a urine sample. According to Table 14.1, what might this result indicate?

QUESTIONS AND PROBLEMS

▪ Chemical Properties of Monosaccharides

14.29 Draw the product xylitol produced from the reduction of D-xylose.

D-xylose

14.30 Draw the product mannitol produced from the reduction of D-mannose.

D-mannose

14.31 Write the oxidation and reduction products of D-arabinose. What is the name of the sugar alcohol produced?

D-arabinose

14.32 Write the oxidation and reduction products of D-ribose. What is the name of the sugar alcohol produced?

D-ribose

LEARNING GOAL

Describe the monosaccharide units and linkages in disaccharides.

■ 14.5 Disaccharides

A disaccharide is composed of two monosaccharides linked together. The most common disaccharides are maltose, lactose, and sucrose. Their hydrolysis, by an acid or an enzyme, gives the following monosaccharides.

$$\text{Maltose} + \text{H}_2\text{O} \xrightarrow{\text{H}^+} \text{glucose} + \text{glucose}$$

$$\text{Lactose} + \text{H}_2\text{O} \xrightarrow{\text{H}^+} \text{glucose} + \text{galactose}$$

$$\text{Sucrose} + \text{H}_2\text{O} \xrightarrow{\text{H}^+} \text{glucose} + \text{fructose}$$

Maltose, or malt sugar, is a disaccharide obtained from starch. When maltose in barley and other grains is hydrolyzed by yeast enzymes, glucose is obtained, which can undergo fermentation to give ethanol. Maltose is used in cereals, candies, and the brewing of beverages.

When a hydroxyl group in one monosaccharide reacts with a hydroxyl group in another monosaccharide, a **glycosidic bond** forms and the product is a disaccharide. In maltose, the glycosidic bond that joins the two glucose molecules is designated an α-1,4 linkage to show that —OH on carbon 1 of α-D-glucose is joined to carbon 4 of the second glucose. Because the second glucose molecule has a free —OH on carbon 1, there are α and β forms of maltose. Because carbon 1 can open to give a free aldehyde group to oxidize, maltose is a reducing sugar.

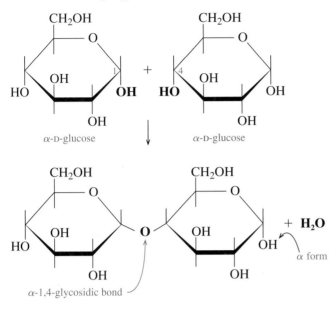

Lactose, milk sugar, is a disaccharide found in milk and milk products. (See Figure 14.3.) It makes up 6–8% of human milk and about 4–5% of cow's milk and is used in products that attempt to duplicate mother's milk. Some people do not produce sufficient quantities of the enzyme needed to hydrolyze lactose, and the sugar remains undigested, causing abdominal cramps and diarrhea. In some commercial milk products, an enzyme called lactase is added to break down lactose. The bond in lactose is a β-1,4-glycosidic bond because the β form of galactose forms a bond with a hydroxyl group on carbon 4 of glucose. The hydroxyl group

FIGURE 14.3

Lactose, a disaccharide found in milk and milk products, contains galactose and glucose.

Q What type of glycosidic bond links galactose and glucose in lactose?

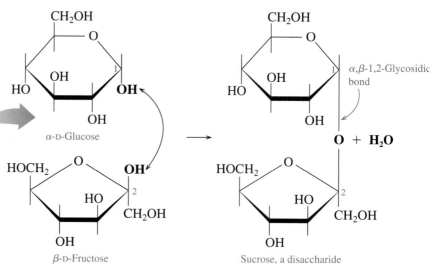

FIGURE 14.4

Sucrose, a disaccharide obtained from sugar beets and sugar cane, contains glucose and fructose.

Q **Why is sucrose a nonreducing sugar?**

Sugar and Sweeteners

Add a tablespoon of sugar to a glass of water and stir. Taste. Add more tablespoons of sugar, stir, and taste. If you have other carbohydrates such as fructose, honey, cornstarch, arrowroot, or flour, add some of each to separate glasses of water and stir. If you have some artificial sweeteners, add a few drops of the sweetener or a package, if solid, to a glass of water. Taste each.

QUESTIONS

1. Which substance is the most soluble in water?
2. Place the substances in order from the one that tastes least sweet to the sweetest.
3. How does your list compare to the values in Table 14.2?
4. How does the sweetness of sucrose compare with the artificial sweeteners?
5. Check the labels of food products in your kitchen. Look for sugars such as sucrose, fructose, or artificial sweeteners such as aspartame or sucralose on the label. How many grams of sugar are in a serving of the food?

on carbon 1 of glucose gives both α- and β-lactose. Because the open chain has an aldehyde group that can be oxidized, lactose is a reducing sugar.

Sucrose consists of an α-D-glucose and β-D-fructose molecule joined by an α, β-1,2-glycosidic bond. (See Figure 14.4.) The structure of sucrose differs from the other disaccharides because the glycosidic bond between carbon 1 of glucose and carbon 2 of fructose cannot open to give an aldehyde. Sucrose cannot react with Benedict's reagent; sucrose is not a reducing sugar.

You already know that ordinary table sugar is sucrose, a disaccharide that is the most abundant carbohydrate in the world. Most of the sucrose for table sugar comes from sugar cane (20% by mass) or sugar beets (15% by mass). Both the raw and refined forms of sugar are sucrose. Some estimates indicate that each person in the United States consumes an average of 45 kg (100 lb) of sucrose every year either by itself or in a variety of food products.

SAMPLE PROBLEM 14.7

Glycosidic Bonds in Disaccharides

Melebiose is a disaccharide that has a sweetness of about 30 compared with sucrose (= 100).

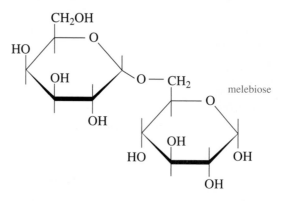

a. What are the monosaccharide units in melebiose?
b. What type of glycosidic bond links the monosaccharides?
c. Is the compound drawn as α- or β-melebiose?

HEALTH NOTE

How Sweet Is My Sweetener?

Although many of the monosaccharides and disaccharides taste sweet, they differ considerably in their degree of sweetness. Dietetic foods contain sweeteners that are noncarbohydrate or carbohydrates that are sweeter. Some examples of sweeteners compared with sucrose are shown in Table 14.2.

Sucralose is made from sucrose by replacing some of the hydroxyl groups with chlorine atoms.

sucralose

Aspartame, which is marketed as Nutra-Sweet, is used in a large number of sugar-free products. It is a noncarbohydrate sweetener made of aspartic acid and a methyl ester of phenylalanine. It does have some caloric value, but it is so sweet that a very small quantity is needed. However, one of the breakdown products phenylalanine, poses a danger to anyone who cannot metabolize it properly, a condition called phenylketonuria (PKU).

From aspartic acid   From phenylalanine   methyl ester
Aspartame (Nutra-Sweet)

TABLE 14.2

Relative Sweetness of Sugars and Artificial Sweeteners

	Sweetness Relative to Sucrose (= 100)
Monosaccharides	
galactose	30
sorbitol	36
glucose	75
fructose	175
Disaccharides	
lactose	16
maltose	33
sucrose	100 ← reference standard
Artificial Sweeteners (Noncarbohydrate)	
sucralose	60 000
aspartame	18 000
saccharin	45 000

Saccharin has been used as a noncarbohydrate artificial sweetener for the past 25 years. The use of saccharin has been banned in Canada because studies indicate that it may cause bladder tumors. However, it is still approved for use by the FDA in the United States.

saccharin

SOLUTION

a. The monosaccharide on the left side is α-D-galactose; on the right is α-D-glucose.

b. The monosaccharide units are linked by an α-1,6-glycosidic bond.

c. The downward position of the hydroxyl group on carbon 1 makes it α-melebiose.

STUDY CHECK

Cellobiose is a disaccharide composed of two β-D-glucose molecules linked by a β-1,4-glycosidic linkage. Draw a structural formula for β-cellobiose.

QUESTIONS AND PROBLEMS

■ **Disaccharides**

14.33 For each of the following disaccharides, give the monosaccharide units produced by hydrolysis, the type of glycosidic bond, and the identity of the disaccharide including the α or β form:

14.34 For each of the following disaccharides, give the monosaccharide units produced by hydrolysis, the type of glycosidic bond, and the identity of the disaccharide including the α or β form:

14.35 Indicate whether the sugars in problem 14.33 will undergo oxidation.

14.36 Indicate whether the sugars in problem 14.34 will undergo oxidation.

14.37 Identify disaccharides that fit each of the following descriptions:
   a. ordinary table sugar
   b. found in milk and milk products
   c. also called malt sugar
   d. hydrolysis gives galactose and glucose

14.38 Identify disaccharides that fit each of the following descriptions:

a. not a reducing sugar
b. composed of two glucose units
c. also called milk sugar
d. hydrolysis gives glucose and fructose

Blood Types and Carbohydrates

Every individual's blood can be typed as one of four blood groups, A, B, AB, and O. Although there is some variation among ethnic groups in the United States, the incidence of blood types in the general population is about 43% O, 40% A, 12% B, and 5% AB.

The blood types are determined by three or four monosaccharides that are attached to the red blood cells. All the blood types include *N*-acetylglucosamine, galactose, and fucose. In type A blood, a fourth monosaccharide, *N*-acetylgalactosamine, is bonded to the galactose. The structures of these monosaccharides are as follows:

N-acetylglucosamine (*N*-AcGlu)   D-galactose (Gal)

L-fucose (Fuc)   *N*-acetylgalactosamine (*N*-AcGal)

In type B blood, there is a second molecule of galactose. Type AB blood contains both A and B blood types. A person with type A blood produces antibodies against type B, whereas a person with type B blood produces antibodies against A. Type AB blood produces no antibodies, whereas type O produces both. Thus, if a person with type A blood receives a transfusion of type B blood, factors in the recipient's blood will agglutinate the donor's red blood cells. If you become a blood donor, your blood is screened to make sure that an exact match is made with the blood type of the recipient. Table 14.3 summarizes the compatibility of blood groups for transfusion.

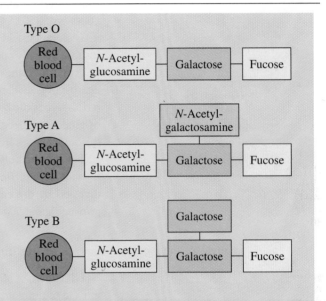

Type O: Red blood cell — *N*-Acetyl-glucosamine — Galactose — Fucose

Type A: Red blood cell — *N*-Acetyl-glucosamine — Galactose — Fucose; Galactose bonded to *N*-Acetyl-galactosamine

Type B: Red blood cell — *N*-Acetyl-glucosamine — Galactose — Fucose; Galactose bonded to Galactose

TABLE 14.3

Compatibility of Blood Groups

Blood Group	Can Receive Blood Types	Cannot Receive Blood Types
A	A, O	B, AB
B	B, O	A, AB
AB[a]	A, B, AB, O	Can receive all blood types
O[b]	O	A, B, AB

[a]AB universal recipient
[b]O universal donor

the
Chemistry place

WEB TUTORIAL
Polymers

■ 14.6 Polysaccharides

A polysaccharide is a polymer of many monosaccharides joined together. Four biologically important polysaccharides—amylose, amylopectin, cellulose, and glycogen—are all polymers of D-glucose that differ only in the type of glycosidic bonds and the amount of branching in the molecule.

Starch, a storage form of glucose in plants, is found as insoluble granules in rice, wheat, potatoes, beans, and cereals. Starch is composed of two kinds of polysaccharides, amylose and amylopectin. **Amylose**, which makes up about 20% of starch, consists of 250 to 4000 α-D-glucose molecules connected by α-1,4-glycosidic bonds in a continuous chain. Sometimes called a straight-chain polymer, polymers of amylose are actually coiled in helical fashion.

Amylopectin, which makes up as much as 80% of plant starch, is a branched-chain polysaccharide. Like amylose, the glucose molecules are connected by α-1,4-glycosidic bonds. However, at about every 25 glucose units, there is a branch of glucose molecules attached by an α-1,6-glycosidic bond between carbon 1 of the branch and carbon 6 in the main chain. (See Figure 14.5.)

Starches hydrolyze easily in water and acid to give dextrins, which then hydrolyze to maltose and finally glucose. In our bodies, these complex carbohydrates are digested by the enzymes amylase (in saliva) and maltase. The glucose obtained provides about 50% of our nutritional calories.

$$\text{Amylose, amylopectin} \xrightarrow{\text{H}^+ \text{ or amylase}} \text{dextrins} \xrightarrow{\text{H}^+ \text{ or amylase}} \text{maltose} \xrightarrow{\text{H}^+ \text{ or maltase}} \text{many D-glucose units}$$

Glycogen, or animal starch, is a polymer of glucose that is stored in the liver and muscle of animals. It is hydrolyzed in our cells at a rate that maintains the blood level of glucose and provides energy between meals. The structure of glycogen is very similar to that of amylopectin found in plants except that glycogen is more highly branched. In glycogen, the glucose units are joined by α-1,4-glycosidic bonds, and branches occurring about every 10–15 glucose units are attached by α-1,6-glycosidic bonds.

Cellulose is the major structural material of wood and plants. Cotton is almost pure cellulose. In **cellulose**, glucose molecules form a long unbranched chain similar to that of amylose. However, the glucose units in cellulose are linked by β-1,4-glycosidic bonds. The β isomers do not form coils like the α isomers but are aligned in parallel rows that are held in place by hydrogen bonds between hydroxyl groups in adjacent chains, which makes cellulose insoluble in water. This gives a rigid structure to the cell walls in wood and fiber that is more resistant to hydrolysis than the starches. (See Figure 14.6.)

Enzymes in our saliva and pancreatic juices hydrolyze the α-1,4-glycosidic bonds of the starches. However, there are no enzymes in humans that are able to hydrolyze the β-1,4-glycosidic bonds of cellulose; we cannot digest cellulose. Some animals such as goats and cows and insects like termites are able to obtain glucose from cellulose. Their digestive systems contain bacteria and protozoa with enzymes such as cellulase that can hydrolyze β-1,4-glycosidic bonds.

SAMPLE PROBLEM 14.8

■ **Structures of Polysaccharides**

Identify the polysaccharide described by each of the following:

a. A polysaccharide that is stored in the liver and muscle tissues

(a) Unbranched chain of amylose

Amylose (20%)

Glucose monomers

Amylopectin (80%)

(b) Branched-chain of amylopectin

FIGURE 14.5

The structure of (a) amylose is a straight-chain polysaccharide of glucose units, and (b) amylopectin is a branched chain of glucose.

Q **What are the two types of glycosidic bonds that link glucose molecules in amylopectin?**

b. An unbranched polysaccharide containing β-1,4-glycosidic bonds
c. A starch containing α-1,4- and α-1,6-glycosidic bonds

SOLUTION

a. glycogen   **b.** cellulose   **c.** amylopectin, glycogen

STUDY CHECK

Cellulose and amylose are both unbranched glucose polymers. How do they differ?

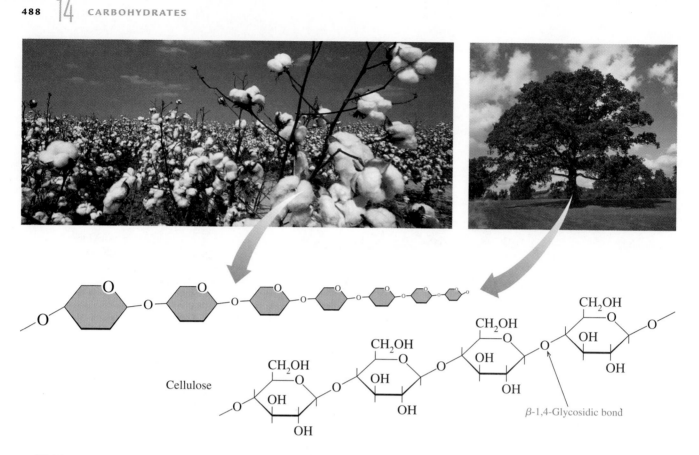

Cellulose

β-1,4-Glycosidic bond

FIGURE 14.6

The polysaccharide cellulose is composed of β-1,4-glycosidic bonds.

Q **Why are humans unable to digest cellulose?**

QUESTIONS AND PROBLEMS

■ Polysaccharides

14.39 Describe the similarities and differences in the following polysaccharides:
a. amylose and amylopectin   **b.** amylopectin and glycogen

14.40 Describe the similarities and differences in the following polysaccharides:
a. amylose and cellulose   **b.** cellulose and glycogen

14.41 Give the name of one or more polysaccharides that matches each of the following descriptions:
a. not digestible by humans
b. the storage form of carbohydrates in plants
c. contains only α-1,4-glycosidic bonds
d. the most highly branched polysaccharide

14.42 Give the name of one or more polysaccharides that matches each of the following descriptions:
a. the storage form of carbohydrates in animals
b. contains only β-1,4-glycosidic bonds
c. contains both α-1,4- and α-1,6-glycosidic bonds
d. produces maltose during digestion

CONCEPT MAP

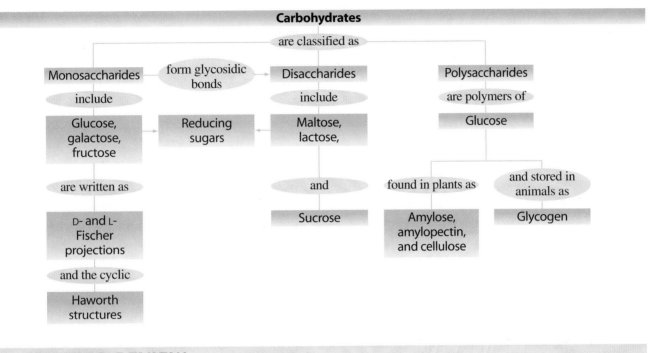

CHAPTER REVIEW

14.1 Carbohydrates

Carbohydrates are classified as monosaccharides (simple sugars), disaccharides (two monosaccharide units), and polysaccharides (many monosaccharide units). Monosaccharides are polyhydroxy aldehydes (aldoses) or ketones (ketoses). Monosaccharides are also classified by their number of carbon atoms: *triose*, *tetrose*, *pentose*, or *hexose*.

14.2 Structures of Monosaccharides

Chiral molecules can exist in two different forms, which are mirror images of each other. In a Fischer projection (straight chain), the prefixes D- and L- are used to distinguish between the mirror images. In D-monosaccharides, the —OH is on the right of the chiral carbon farthest from the carbonyl carbon; it is on the left in L-monosaccharides. Important monosaccharides are the aldohexoses glucose and galactose and the ketohexose fructose.

14.3 Cyclic Structures of Monosaccharides

The predominant form of monosaccharides is the cyclic form of five or six atoms. The cyclic structure forms by a reaction between an OH (usually the one on carbon 5 in hexoses) with the carbonyl group of the same molecule. The formation of a new hydroxyl group on carbon 1 (or 2 in fructose) gives α and β forms of the cyclic monosaccharide.

14.4 Chemical Properties of Monosaccharides

The aldehyde group in an aldose can be oxidized to a carboxylic acid, while the carbonyl group in an aldose or a ketose can be reduced to give a hydroxyl group. Monosaccharides are reducing sugars because the open-chain aldehyde group (also available in ketoses) can be oxidized by a metal ion such as Cu^{2+}.

14.5 Disaccharides

Disaccharides are two monosaccharide units joined together by a glycosidic bond. In the most common disaccharides, maltose, lactose, and sucrose, there is at least one glucose unit.

14.6 Polysaccharides

Polysaccharides are polymers of monosaccharide units. Starch consists of amylose, an unbranched chain of glucose, and amylopectin, a branched polymer of glucose. Glycogen, the storage form of glucose in animals, is similar to amylopectin with more branching. Cellulose is also a polymer of glucose, but in cellulose the glycosidic bonds are β bonds rather than α bonds as in the starches. Humans can digest starches, but not cellulose, to obtain energy. However, cellulose is important as a source of fiber in our diets.

SUMMARY OF CARBOHYDRATES

Carbohydrate	Food Sources	
Monosaccharides		
glucose	Fruit juices, honey, corn syrup	
galactose	Lactose hydrolysis	
fructose	Fruit juices, honey, sucrose hydrolysis	
Disaccharides		**Monosaccharides**
maltose	Germinating grains, starch hydrolysis	glucose + glucose
lactose	Milk, yogurt, ice cream	glucose + galactose
sucrose	Sugar cane, sugar beets	glucose + fructose
Polysaccharides		
amylose	Rice, wheat, grains, cereals	Unbranched polymer of glucose joined by α-1,4-glycosidic bonds
amylopectin	Rice, wheat, grains, cereals	Branched polymer of glucose joined by α-1,4- and α-1,6-glycosidic bonds
glycogen	Liver, muscles	Highly branched polymer of glucose joined by α-1,4- and α-1,6-glycosidic bonds
cellulose	Plant fiber, bran, beans, celery	Unbranched polymer of glucose joined by β-1,4-glycosidic bonds

SUMMARY OF REACTIONS

FORMATION OF DISACCHARIDES

monosaccharide + monosaccharide ⟶ disaccharide + H_2O

glycosidic bond

OXIDATION AND REDUCTION OF MONOSACCHARIDES

D-glucitol ⟵ reduction ─ D-glucose ─ oxidation ⟶ D-gluconic acid

HYDROLYSIS OF DISACCHARIDES

Sucrose + H_2O ⟶ glucose + fructose

Lactose + H_2O ⟶ glucose + galactose

Maltose + H_2O ⟶ glucose + glucose

KEY TERMS

aldose A monosaccharide that contains an aldehyde group.

amylopectin A branched-chain polymer of starch composed of glucose units joined by α-1,4- and α-1,6-glycosidic bonds.

amylose An unbranched polymer of starch composed of glucose units joined by α-1,4-glycosidic bonds.

carbohydrate A simple or complex sugar composed of carbon, hydrogen, and oxygen.

cellulose An unbranched polysaccharide composed of glucose units linked by β-1,4-glycosidic bonds that cannot be hydrolyzed by the human digestive system.

disaccharides Carbohydrates composed of two monosaccharides joined by a glycosidic bond.

fructose A monosaccharide found in honey and fruit juices; it is combined with glucose in sucrose. Also called levulose and fruit sugar.

galactose A monosaccharide that occurs combined with glucose in lactose.

glucose The most prevalent monosaccharide in the diet. An aldohexose that is found in fruits, vegetables, corn syrup, and honey. Also known as blood sugar and dextrose. Combines in glycosidic bonds to form most of the polysaccharides.

glycogen A polysaccharide formed in the liver and muscles for the storage of glucose as an energy reserve. It is composed of glucose in a highly branched polymer joined by α-1,4- and α-1,6-glycosidic bonds.

glycosidic bond The bond that forms when the hydroxyl group of one monosaccharide reacts with the hydroxyl group of another monosaccharide. It is the type of bond that links monosaccharide units in di- or polysaccharides.

Haworth structure The cyclic structure that represents the closed chain of a monosaccharide.

ketose A monosaccharide that contains a ketone group.

lactose A disaccharide consisting of glucose and galactose found in milk and milk products.

maltose A disaccharide consisting of two glucose units; it is obtained from the hydrolysis of starch and in germinating grains.

monosaccharide A polyhydroxy compound that contains an aldehyde or ketone group.

polysaccharides Polymers of many monosaccharide units, usually glucose. Polysaccharides differ in the types of glycosidic bonds and the amount of branching in the polymer.

reducing sugar A carbohydrate with a free aldehyde group capable of reducing the Cu^{2+} in Benedict's reagent.

sucrose A disaccharide composed of glucose and fructose; a nonreducing sugar, commonly called table sugar or "sugar."

ADDITIONAL QUESTIONS AND PROBLEMS

14.43 What are the structural differences in D-glucose and D-galactose?

14.44 What are the structural differences in D-glucose and D-fructose?

14.45 How do D-galactose and L-galactose differ?

14.46 How do α-D-glucose and β-D-glucose differ?

14.47 Consider the sugar D-gulose.

D-Gulose

a. What is the Fisher projection for L-gulose?
b. Draw the Haworth structure for α- and β-D-gulose.

14.48 Consider the structures for D-gulose in question 14.47.
a. What is the structure and name of the product formed by the reduction of D-gulose?
b. Write the structure and name of the product formed by the oxidation of D-gulose.

14.49 D-Sorbitol, a sweetener found in seaweed and berries, contains only hydroxyl functional groups. When D-sorbitol is oxidized, it forms D-glucose. What is the structural formula of D-sorbitol?

14.50 Raffinose is a trisaccharide found in Australian manna and in cottonseed meal. It is composed of three different monosaccharides. Identify the monosaccharides in raffinose.

14.51 If α-galactose is dissolved in water, β-galactose is eventually present. Explain how this occurs.

14.52 Why are lactose and maltose reducing sugars, but sucrose is not?

14.53 β-Cellobiose is a disaccharide obtained from the hydrolysis of cellulose. It is quite similar to maltose except it has a β-1,4-glycosidic bond. What is the structure of β-cellobiose?

14.54 The disaccharide trehalose found in mushrooms is composed of two α-D-glucose molecules joined by an α-1,1-glycosidic bond. Draw the structure of trehalose.

UNDERSTANDING THE CONCEPTS

14.55 Isomaltose, obtained from the breakdown of starch, has the following structure.

    a. Is isomaltose a mono-, di-, or polysaccharide?
    b. What are the monosaccharides in isomaltose?
    c. What is the glycosidic link in isomaltose?
    d. Is this the α or β form of isomaltose?
    e. Would isomaltose be a reducing sugar?

14.56 Sophorose, a saccharide found in certain types of beans, has the following structure:

    a. Is sophorose a mono-, di-, or polysaccharide?
    b. What are the monosaccharides in sophorose?
    c. What is the glycosidic link in sophorose?
    d. Is this the α or β form of sophorose?
    e. Would sophorose be a reducing sugar?

14.57 Melezitose is a saccharide with the following structure:

    a. Is melezitose a mono-, di-, tri-, or polysaccharide?
    b. What are the monosaccharides in melezitose?

14.58 What are the disaccharides and polysaccharides present in each of the following?

    (a)   (b)

    (c)   (d)

QUESTIONS FOR TEAMWORK

14.59 Gentiobiose is found in saffron.
    a. Gentiobiose contains two glucose molecules linked by a β-1,6-glycoside bond. Draw the structure of α-gentiobiose.
    b. Would gentiobiose be a reducing sugar? Why or why not?

14.60 From the compounds shown that follow, select those that meet the following statements:
    a. is the L-enantiomer of mannose
    b. a ketopentose
    c. an aldopentose
    d. a ketohexose

A   B   C   D

ANSWERS

Answers to Study Checks

14.1 a monosaccharide

14.2

14.3

H
   \\
     C=O
H——OH
H——OH
H——OH
   CH₂OH

14.4 Ribulose is a ketopentose.

14.5

14.6 This indicates a high level of reducing sugar (probably glucose) in the urine. One common cause of this condition is diabetes mellitus.

14.7

14.8 Cellulose contains glucose units connected by β-1,4-glycosidic bonds, whereas the glucose units in amylose are connected by α-1,4-glycosidic bonds.

Answers to Selected Questions and Problems

14.1 Photosynthesis requires CO_2, H_2O, and the energy from the sun. Respiration requires O_2 from the air and glucose from our foods.

14.3 A monosaccharide cannot be split into smaller carbohydrates. A disaccharide is composed of two monosaccharide units.

14.5 Hydroxyl groups are found in all monosaccharides along with a carbonyl on the first or second carbon.

14.7 A ketopentose contains hydroxyl and ketone functional groups and has five carbon atoms.

14.9 a. ketose   **b.** aldose   **c.** ketose
d. aldose   **e.** aldose

14.11 A Fischer projection is a two-dimensional representation of the three-dimensional structure of a molecule.

14.13 a. D   **b.** D   **c.** L   **d.** D

14.15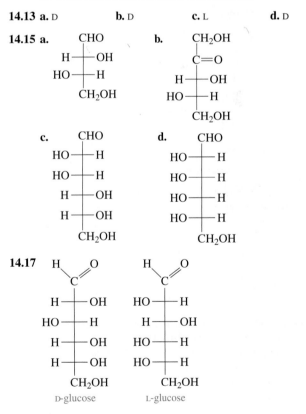

14.17

H   O   H   O
 \\   //   \\   //
   C   C
H——OH   HO——H
HO——H   H——OH
H——OH   HO——H
H——OH   HO——H
   CH₂OH   CH₂OH
  D-glucose   L-glucose

14.19 In D-galactose the hydroxyl on carbon four extends to the left. In D-glucose this hydroxyl goes to the right.

14.21 a. glucose   **b.** galactose   **c.** fructose

14.23 In the cyclic structure of glucose, there are five carbon atoms and an oxygen.

14.25

α-D-glucose

β-D-glucose

14.27 a. α form   **b.** α form

14.29

xylitol

14.31 Oxidation product:

Reduction product (sugar alcohol):

CH₂OH

HO—C—H

H—C—OH

H—C—OH

CH₂OH

D-arabitol

14.33 a. galactose and glucose; β-1,4 bond; β-lactose
b. glucose and glucose; α-1,4 bond; α-maltose

14.35 a. Can be oxidized
b. Can be oxidized

14.37 a. sucrose   **b.** lactose
c. maltose   **d.** lactose

14.39 a. Amylose is an unbranched polymer of glucose units joined by α-1,4 bonds; amylopectin is a branched polymer of glucose joined by α-1,4 and α-1,6 bonds.
b. Amylopectin, which is produced in plants, is a branched polymer of glucose, joined by α-1,4 and α-1,6 bonds.

Glycogen, which is produced in animals, is a highly branched polymer of glucose, joined by α-1,4 and α-1,6 bonds.

14.41 a. cellulose   **b.** amylose, amylopectin
c. amylose   **d.** glycogen

14.43 They differ only at carbon 4 where the —OH in D-glucose is on the right side and in D-galactose it is on the left side.

14.45 D-galactose is the mirror image of L-galactose. In D-galactose, the —OH group on carbon 5 is on the right side whereas in L-galactose, the —OH group on carbon 5 is on the left side.

14.47 a.

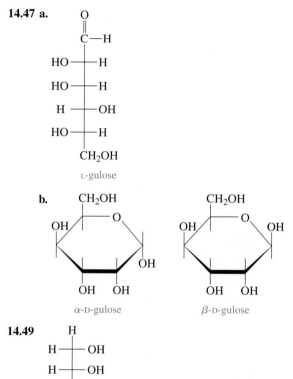

L-gulose

b.

α-D-gulose   β-D-gulose

14.49

14.51 The α-galactose forms an open-chain structure, and when the chain closes, it can form both α- and β-galactose.

14.53

14.55 a. disaccharide   **b.** α-glucose   **c.** α-1,6
d. α   **e.** Yes.

14.57 a. trisaccharide   **b.** glucose and fructose

15 Lipids

"In our toxicology lab, we measure the drugs in samples of urine or blood," says Penny Peng, assistant supervisor of chemistry, Toxicology Lab, Santa Clara Valley Medical Center. "But first we extract the drugs from the fluid and concentrate them so they can be detected in the machine we use. We extract the drugs by using different organic solvents such as methanol, ethyl acetate, or methylene chloride, and by changing the pH. We evaporate most of the organic solvent to concentrate any drugs it may contain. A small sample of the concentrate is placed into a machine called a gas chromatograph. As the gas moves over a column, the drugs in it are separated. From the results, we can identify as many as 10 to 15 different drugs from one urine sample."

the Chemistry place

Visit **www.aw-bc.com/chemplace** or use the CD-ROM in the book for extra quizzes, interactive tutorials, career resources, PowerPoint slides, math help, and case studies.

When we talk of fats and oils, waxes, steroids, cholesterol, and fat-soluble vitamins, we are discussing lipids. All the lipids are naturally occurring compounds that vary considerably in structure but share a common feature of being soluble in nonpolar solvents, but not in water. Fats, which are one family of lipids, have many functions in the body; they store energy and protect and insulate internal organs. Other types of lipids are found in nerve fibers and in hormones, which act as chemical messengers. Because they are not soluble in water, a major function of lipids is to build the cell membranes that separate the internal contents of cells from the surrounding aqueous environment.

Many people are concerned about the amounts of saturated fats and cholesterol in our diets. Researchers suggest that saturated fats and cholesterol are associated with diseases such as diabetes, cancers of the breast, pancreas, and colon, and atherosclerosis, a condition in which deposits of lipid materials accumulate in the coronary blood vessels. These plaques restrict the flow of blood to the tissue, causing necrosis (death) of the tissue. In the heart, this could result in a *myocardial infarction* (heart attack).

The American Institute for Cancer Research has recommended that we increase the fiber and starch content of our diets by adding more vegetables, fruits, and whole grains with moderate amounts of foods with low levels of fat and cholesterol such as fish, poultry, lean meats, and low-fat dairy products. They also suggest that we limit our intake of foods high in fat and cholesterol such as eggs, nuts, fatty or organ meats, cheeses, butter, and coconut and palm oil.

LEARNING GOAL

Describe the classes of lipids.

WEB TUTORIAL
Fats

■ 15.1 Lipids

Lipids are a family of biomolecules that have the common property of being soluble in organic solvents but not in water. The word "lipid" comes from the Greek word *lipos*, meaning "fat" or "lard." Typically, the lipid content of a cell can be extracted using an organic solvent such as ether or chloroform. Lipids are an important feature in cell membranes, fat-soluble vitamins, and steroid hormones.

■ Types of Lipids

Within the lipid family, there are distinct structures that distinguish the different types of lipids. Lipids such as waxes, fats, oils, and glycerophospholipids are esters that can be hydrolyzed to give fatty acids along with other products including an alcohol. Steroids are characterized by the steroid nucleus of four fused carbon rings. They do not contain fatty acids and cannot be hydrolyzed. Figure 15.1 illustrates the general structure of lipids we will discuss in this chapter.

FIGURE 15.1

Structures for some classes of lipids that are naturally occurring compounds in cells and tissues.

Q **What chemical property do waxes, triacylglycerols, and steroids have in common?**

SAMPLE PROBLEM 15.1

■ Classes of Lipids

What type of lipid does not contain fatty acids?

SOLUTION

The steroids are a group of lipids with no fatty acids.

STUDY CHECK

What are the components of a glycerophospholipid?

QUESTIONS AND PROBLEMS

■ Lipids

15.1   What are some functions of lipids in the body?

15.2   What are some of the different kinds of lipids?

15.3   Lipids are not soluble in water. Are lipids polar or nonpolar molecules?

15.4   Which of the following solvents might be used to dissolve an oil stain?
  a. water   **b.** CCl_4 **c.** diethyl ether
  d. benzene   **e.** NaCl solution

EXPLORE YOUR WORLD

■ Solubility of Fats and Oils

Place some water in a small bowl. Add a drop of a vegetable oil. Then add a few more drops of the oil. Now add a few drops of liquid soap and mix. Record your observations.

Place a small amount of fat such as margarine, butter, shortening, or vegetable oil on a dish or plate. Run water over it. Record your observations. Mix some soap with the fat substance and run water over it again. Record your observations.

QUESTIONS

1. Do the drops of oil in the water separate or do they come together? Explain.
2. How does the soap affect the oil layer?
3. Why don't the fats on the dish or plate wash off with water?
4. In general what is the solubility of lipids in water?
5. Why does soap help to wash the fats off the plate?

15.2 Fatty Acids

The fatty acids are the simplest type of lipids and are found as components in more complex lipids. A **fatty acid** contains a long carbon chain attached to a carboxylic acid group at one end. Although the carboxylic acid part is hydrophilic, the long hydrophobic carbon chain makes long-chain fatty acids insoluble in water. Naturally occurring fatty acids have an even number of carbon atoms, usually between 10 and 20. An example of a fatty acid is lauric acid, a 12-carbon acid found in coconut oil. In a simplified structure of a fatty acid called a line-bond formula, the carbon chain is written as a zig-zag line that indicates the bonds between carbon atoms. The structural formula of lauric acid can be written in several forms as follows.

Writing Formulas for Lauric Acid

$$CH_3-(CH_2)_{10}-\overset{\overset{\displaystyle O}{\|}}{C}-OH \qquad CH_3-(CH_2)_{10}-COOH$$

$$CH_3-CH_2-CH_2-CH_2-CH_2-CH_2-CH_2-CH_2-CH_2-CH_2-CH_2-C\overset{\displaystyle O}{\underset{\displaystyle OH}{<}}$$
Condensed structural formula

Line-bond structural formula

Saturated fatty acids such as lauric acid contain only single bonds between carbons. **Monounsaturated fatty acids** have one double bond in the carbon chain, and **polyunsaturated fatty acids** have two or more double bonds. Table 15.1 lists some of the typical fatty acids in lipids.

In Chapter 11 we saw that compounds with double bonds can have cis and trans isomers. This is also true of unsaturated fatty acids. For example, oleic acid, a monounsaturated fatty acid found in olives and corn, has one double bond at carbon 9. We can show its cis and trans structural formulas using the line-bond notation. The cis configuration is most prevalent in naturally occurring unsaturated fatty acids. In the cis isomer, the geometry of the carbon chain is not linear, but has a "kink" at the double bond site. As we will see, the cis bond has a major impact on the physical properties and uses of unsaturated fatty acids.

cis-Oleic acid
Cis double bond

trans-Oleic acid
Trans double bond

The human body is capable of synthesizing most fatty acids from carbohydrates or other fatty acids. However, humans do not synthesize sufficient amounts of fatty acids that have more than one double bond, such as linoleic acid, linolenic acid, and arachidonic acid. These fatty acids are called *essential* fatty acids because they must be provided by the diet. A deficiency of essential fatty acids can cause skin dermatitis in infants. However, the role of fatty acids in adult nutrition is not well understood. Adults do not usually have a deficiency of essential fatty acids.

TABLE 15.1

Structures and Melting Points of Common Fatty Acids

Name	Carbon Atoms	Source	Melting Point (°C)	Structures
Saturated Fatty Acids				
lauric acid	12	coconut	43	CH_3—$(CH_2)_{10}$—COOH
myristic acid	14	nutmeg	54	CH_3—$(CH_2)_{12}$—COOH
palmitic acid	16	palm	62	CH_3—$(CH_2)_{14}$—COOH
stearic acid	18	animal fat	69	CH_3—$(CH_2)_{16}$—COOH
Monounsaturated Fatty Acids				
palmitoleic acid	16	butter	0	CH_3—$(CH_2)_5$—CH=CH—$(CH_2)_7$—COOH
oleic acid	18	olives, corn	13	CH_3—$(CH_2)_7$—CH=CH—$(CH_2)_7$—COOH
Polyunsaturated Fatty Acids				
linoleic acid	18	soybean, safflower, sunflower	−9	CH_3—$(CH_2)_4$—CH=CH—CH_2—CH=CH—$(CH_2)_7$—COOH
linolenic acid	18	corn	−17	CH_3—CH_2—CH=CH—CH_2—CH=CH—CH_2—CH=CH—$(CH_2)_7$—COOH

Physical Properties of Fatty Acids

The saturated fatty acids fit close together in a regular pattern, which allows strong attractions to occur between the carbon chains. As a result, a significant amount of energy and high temperatures are required to separate the fatty acids and melt the fat. As the length of the carbon chain increases, more interactions occur between the carbon chains, requiring higher melting points. Saturated fatty acids are usually solids at room temperature.

In unsaturated fatty acids, the cis double bonds cause the carbon chain to bend, which gives the molecules an irregular shape. As a result, fewer interactions occur between carbon chains. Consequently, the melting points of unsaturated fats are lower than those of saturated fats. (See Figure 15.2.) Most unsaturated fats are liquid oils at room temperature.

SAMPLE PROBLEM 15.2

■ Structures and Properties of Fatty Acids

Consider the structural formula of oleic acid.

$$CH_3(CH_2)_7CH{=}CH(CH_2)_7\overset{\displaystyle O}{\overset{\displaystyle \|}{C}}OH$$

Stearic acid, mp 69°C

(a)

Oleic acid, mp 13°C

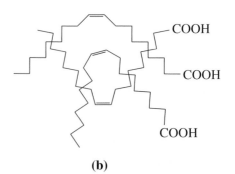

(b)

FIGURE 15.2

(a) In saturated fatty acids, the molecules fit closely together to give high melting points. **(b)** In unsaturated fatty acids, molecules cannot pack closely together, resulting in lower melting points.

Q Why does the cis double bond affect the melting points of unsaturated fatty acids?

a. Why is the substance called an acid?

b. How many carbon atoms are in oleic acid?

c. Is it a saturated or unsaturated fatty acid?

d. Is it most likely to be solid or liquid at room temperature?

e. Would it be soluble in water?

SOLUTION

a. Oleic acid contains a carboxylic acid group.

b. It contains 18 carbon atoms.

c. It is an unsaturated fatty acid.

d. It is liquid at room temperature.

e. No, its long hydrocarbon chain makes it insoluble in water.

STUDY CHECK

Palmitoleic acid is a fatty acid with the following formula:

$$CH_3(CH_2)_5CH{=}CH(CH_2)_7\overset{\overset{\displaystyle O}{\|}}{C}OH$$

a. How many carbon atoms are in palmitoleic acid?

b. Is it a saturated or unsaturated fatty acid?

c. Is it most likely to be solid or liquid at room temperature?

Prostaglandins

Prostaglandins are hormone-like substances produced in low amounts in most cells of the body. The variety of prostaglandins is formed from the unsaturated fatty acid arachidonic acid with 20 carbons. Prostaglandins are sometimes referred to as eicosanoids (*eicos* is the Greek word for 20). Most prostaglandins have a hydroxyl group on carbon 11 and carbon 15, and a trans double bond at carbon 13. Those with a ketone group on carbon 9 are designated as PGE, and as PGF when there is an hydroxyl group on carbon 9.

Although prostaglandins are broken down quickly, they have potent physiological effects. Some prostaglandins increase blood pressure, and others lower blood

HEALTH NOTE

▪ Omega-3 Fatty Acids in Fish Oils

Because unsaturated fats are now recognized as being more beneficial to health than saturated fats, Americans have been changing their diets to include more unsaturated fats. Unsaturated fats contain two types of unsaturated fatty acids, omega-3 and omega-6. Counting from the CH_3— end, the first double bond occurs at carbon 6 in an omega-6 fatty acid, whereas in the omega-3 type, the first double bond occurs at the third carbon.

The omega-6 fatty acids are mostly found in grains, oils from plants, and eggs, and the omega-3 fatty acids are mostly found in cold-water fish such as tuna and salmon. The benefits of omega-3 fatty acids were first recognized when a study was conducted of the diets of the Inuit peoples of Alaska who have a high-fat diet and high levels of blood cholesterol, but a very low occurrence of coronary heart disease. The fats in the Inuit diet are omega-3 fatty acids obtained primarily from fish.

In coronary heart disease, cholesterol forms plaque that adhere to the walls of the blood vessels. Blood pressure rises as blood has to squeeze through a smaller opening in the blood vessel. As more plaque forms, there is also a possibility of blood clots blocking the blood vessels and causing a heart attack. Omega-3 fatty acids lower the tendency of blood platelets to stick together, thereby reducing the possibility of blood clots. However, high levels of omega-3 fatty acids can increase bleeding if the ability of the platelets to form blood clots is reduced too much. It does seem that a diet that includes fish such as salmon, tuna, and herring can provide higher amounts of the omega-3 fatty acids, which help lessen the possibility of developing heart disease.

Omega-6 Fatty Acids

Linoleic acid $\quad CH_3-(CH_2)_4-CH=CH-CH_2-CH=CH-(CH_2)_7-COOH$
$\qquad\qquad\quad\ 1 \qquad\qquad\qquad\quad 6$

Arachidonic acid $\quad CH_3-(CH_2)_4-(CH=CH-CH_2)_4-(CH_2)_2-COOH$
$\qquad\qquad\qquad 1 \qquad\qquad\quad 6$

Omega-3 Fatty Acids

Linolenic acid $\qquad\qquad\qquad\quad CH_3-CH_2-(CH=CH-CH_2)_3-(CH_2)_6-COOH$
$\qquad\qquad\qquad\qquad\qquad\qquad\ 1 \quad\ 2 \qquad 3$

Eicosapentaenoic acid (EPA) $\quad CH_3-CH_2-(CH=CH-CH_2)_5-(CH_2)_2-COOH$
$\qquad\qquad\qquad\qquad\qquad\qquad\ 1 \quad\ 2 \qquad 3$

Docosahexaenoic acid (DHA) $\quad CH_3-CH_2-(CH=CH-CH_2)_6-CH_2-COOH$
$\qquad\qquad\qquad\qquad\qquad\qquad\ 1 \quad\ 2 \qquad 3$

pressure. Other prostaglandins stimulate contraction and relaxation in the smooth muscle of the uterus. When tissues are injured, arachidonic acid is converted to prostaglandins such as PGE and PGF that produce inflammation and pain in the area.

Arachidonic acid → (Analgesics ✗) → PGF₂ → Pain, fever, inflammation

The treatment of pain, fever, and inflammation is based on inhibiting the enzymes that convert arachidonic acid to prostaglandins. Several nonsteroidal anti-inflammatory drugs (NSAIDs), such as aspirin, block the production of prostaglandins, and in doing so decrease pain and inflammation and reduce fever (antipyretics). Ibuprofen has similar anti-inflammatory and analgesic effects. Other NSAIDs include naproxen (Aleve and Naprosyn), ketoprofen (Actron), and nabumetone (Relafen). Long-term use of such products can result in liver, kidney, and gastrointestinal damage. Some forms of PGE are being tested as inhibitors of gastric secretion for use in the treatment of stomach ulcers.

aspirin (acetylsalicyclic acid)   ibuprofen (Advil, Motrin)   naproxen (Aleve, Naprosyn)

QUESTIONS AND PROBLEMS

■ Fatty Acids

15.5 Describe some similarities and differences in the structures of a saturated fatty acid and an unsaturated fatty acid.

15.6 Stearic acid and linoleic acid both have 18 carbon atoms. Why does stearic acid melt at 69°C, but linoleic acid melts at −9°C?

15.7 Write the line-bond structure of the following fatty acids:
 a. palmitic acid **b.** oleic acid

15.8 Write the line-bond structure of the following fatty acids:
 a. stearic acid **b.** linoleic acid

15.9 Which of the following fatty acids are saturated, and which are unsaturated?
 a. lauric acid **b.** linolenic acid
 c. palmitoleic acid **d.** stearic acid

15.10 Which of the following fatty acids are saturated, and which are unsaturated?
 a. linoleic acid **b.** palmitic acid
 c. myristic acid **d.** oleic acid

15.11 How does the structure of a fatty acid with a cis double bond differ from the structure of a fatty acid with a trans double bond?

15.12 In each pair, identify the fatty acid with the lower melting point. Explain.
 a. myristic acid and stearic acid
 b. stearic acid and linoleic acid
 c. oleic acid and linolenic acid

15.13 Describe the position of the first double bond in an omega-3 and an omega-6 fatty acid. (See Health Note "Omega-3 Fatty Acids in Fish Oils.")

15.14 a. What are some sources of omega-3 and omega-6 fatty acids? (See Health Note "Omega-3 Fatty Acids in Fish Oils.")
 b. How may omega-3 fatty acids help in lowering the risk of heart disease?

15.15 What are some structural and functional differences in arachidonic acid and prostaglandins such as PGE_1?

15.16 What is the structural difference in PGE and PGF?

15.17 What are some functions of prostaglandins in the body?

15.18 How does an anti-inflammatory drug reduce inflammation?

Write the structural formula of a wax, fat, or oil produced by the reaction of a fatty acid and an alcohol or glycerol.

Ester bond

fatty acid — long-chain alcohol

Wax

the
Chemistry
place

WEB TUTORIAL
Triacylglycerols

15.3 Waxes, Fats, and Oils

Waxes are found in many plants and animals. Coatings of carnauba wax on fruits and the leaves and stems of plants help to prevent loss of water and damage from pests. Waxes on the skin, fur, and feathers of animals and birds provide a waterproof coating. A **wax** is an ester of a saturated fatty acid and a long-chain alcohol, each containing from 14 to 30 carbon atoms.

The formulas of some common waxes are given in Table 15.2. Beeswax obtained from honeycombs and carnauba wax obtained from palm trees are used to give a protective coating to furniture, cars, and floors. Jojoba wax is used in making candles and cosmetics such as lipstick. Lanolin, a mixture of waxes obtained from wool, is used in hand and facial lotions to aid retention of water, which softens the skin.

Fats and Oils: Triacylglycerols

In the body, fatty acids are stored as fats and oils known as **triacylglycerols**. These substances, also called *triglycerides*, are triesters of glycerol (a trihydroxy alcohol) and fatty acids. The general formula of a triacylglycerol follows.

triacylglycerol

A triacylglycerol is produced by **esterification**, a reaction in which the hydroxyl groups of glycerol form ester bonds with the carboxyl groups of fatty acids.

TABLE 15.2

Some Typical Waxes

Type	Structural Formula	Source	Uses
Beeswax	$CH_3(CH_2)_{14}-\overset{\overset{O}{\|\|}}{C}-O-(CH_2)_{29}CH_3$	Honeycomb	Candles, shoe polish, wax paper
Carnauba wax	$CH_3(CH_2)_{24}-\overset{\overset{O}{\|\|}}{C}-O-(CH_2)_{29}CH_3$	Brazilian palm tree	Waxes for furniture, cars, floors, shoes
Jojoba wax	$CH_3(CH_2)_{18}-\overset{\overset{O}{\|\|}}{C}-O-(CH_2)_{19}CH_3$	Jojoba	Candles, soaps, cosmetics

For example, glycerol and three molecules of stearic acid form glyceryl tristearate which is commonly named tristearin.

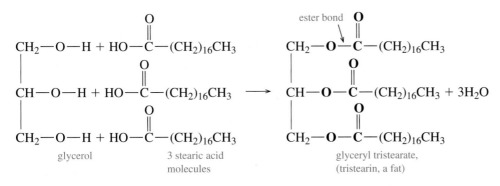

$$CH_2-O-H + HO-\overset{\overset{O}{\|}}{C}-(CH_2)_{16}CH_3$$

$$CH-O-H + HO-\overset{\overset{O}{\|}}{C}-(CH_2)_{16}CH_3 \longrightarrow$$

$$CH_2-O-H + HO-\overset{\overset{O}{\|}}{C}-(CH_2)_{16}CH_3$$

glycerol   3 stearic acid
                  molecules

ester bond

$$CH_2-O-\overset{\overset{O}{\|}}{C}-(CH_2)_{16}CH_3$$

$$CH-O-\overset{\overset{O}{\|}}{C}-(CH_2)_{16}CH_3 + 3H_2O$$

$$CH_2-O-\overset{\overset{O}{\|}}{C}-(CH_2)_{16}CH_3$$

glyceryl tristearate,
(tristearin, a fat)

Most fats and oils are mixed triacylglycerols that contain two or three different fatty acids. For example, a mixed triacylglycerol might be made from lauric acid, myristic acid, and palmitic acid. One possible structure for the mixed triacylglycerol follows:

$$CH_2-O-\overset{\overset{O}{\|}}{C}-(CH_2)_{10}CH_3 \quad \text{lauric acid}$$

$$CH-O-\overset{\overset{O}{\|}}{C}-(CH_2)_{12}CH_3 \quad \text{myristic acid}$$

$$CH_2-O-\overset{\overset{O}{\|}}{C}-(CH_2)_{14}CH_3 \quad \text{palmitic acid}$$

A mixed triacylglycerol

Triacylglycerols are the major form of energy storage for animals. Animals that hibernate eat large quantities of plants, seeds, and nuts that contain high levels of fats and oils. They gain as much as 14 kilograms a week. As the external temperature drops, the animal goes into hibernation. The body temperature drops to nearly freezing, and there is a dramatic reduction in cellular activity, respiration, and heart rate. Animals who live in extremely cold climates will hibernate for 4–7 months. During this time, stored fat is the only source of energy.

SAMPLE PROBLEM 15.3

■ Writing Structures for a Triacylglycerol

Draw the structural formula of triolein, a simple triacylglycerol that uses oleic acid.

SOLUTION

Triolein is the triacylglycerol of glycerol and three oleic acid molecules. Each fatty acid is attached by an ester bond to one of the hydroxyl groups in glycerol.

$$CH_2-O-\overset{\overset{O}{\|}}{C}-(CH_2)_7CH=CH(CH_2)_7CH_3$$

$$CH-O-\overset{\overset{O}{\|}}{C}-(CH_2)_7CH=CH(CH_2)_7CH_3$$

$$CH_2-O-\overset{\overset{O}{\|}}{C}-(CH_2)_7CH=CH(CH_2)_7CH_3$$

glyceryl trioleate (triolein)

STUDY CHECK

Write the structure of the triacylglycerol containing 3 molecules of myristic acid.

Melting Points of Fats and Oils

A **fat** is a triacylglycerol that is solid at room temperature, such as fats in meat, whole milk, butter, and cheese. Most fats come from animal sources.

An **oil** is a triacylglycerol that is usually liquid at room temperature. The most commonly used oils come from plant sources. Olive oil and peanut oil are monounsaturated because they contain large amounts of oleic acid. Oils from corn, cottonseed, safflower, and sunflower are polyunsaturated because they contain large amounts of fatty acids with two or more double bonds. (See Figure 15.3.) A few oils such as palm oil and coconut oil are solid at room temperature because they consist mostly of saturated fatty acids.

The amounts of saturated, monounsaturated, and polyunsaturated fatty acids in some typical fats and oils are shown in Figure 15.4. Saturated fatty acids

Triolein

FIGURE 15.3

Vegetable oils such as olive oil, corn oil, and safflower oil contain unsaturated fats.

Q Why is olive oil a liquid at room temperature?

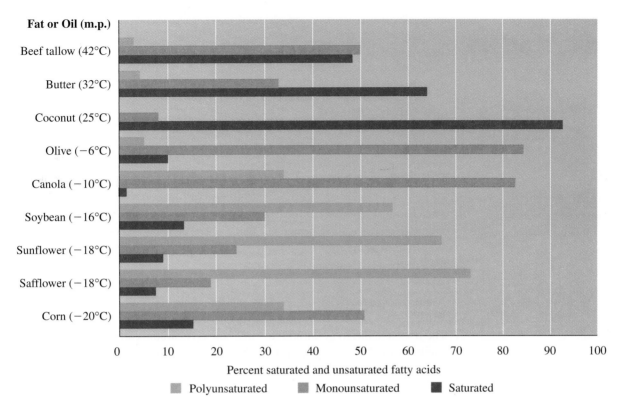

FIGURE 15.4

Fats and oils have low melting points because they have a higher percentage of unsaturated fatty acids than do animal fats.

Q   Why is the melting point of butter higher than olive or canola oil?

have higher melting points than unsaturated fatty acids because they pack together more tightly. Animal fats usually contain more saturated fatty acids than do vegetable oils. Therefore the melting points of animal fats are higher than those of vegetable oils.

QUESTIONS AND PROBLEMS

■ Waxes, Fats, and Oils

15.19 Draw the structure of an ester in beeswax formed from myricyl alcohol, $CH_3(CH_2)_{29}OH$, and palmitic acid.

15.20 Draw the structure of an ester in jojoba wax formed from arachidic acid, a 20-carbon saturated fatty acid, and 1-docosanol, $CH_3(CH_2)_{21}OH$.

15.21 Draw the structure of a triacylglycerol that contains stearic acid and glycerol.

15.22 A mixed triacylglycerol contains two palmitic acid molecules to every one oleic acid molecule. Write two possible structures (isomers) for the compound.

15.23 Draw the structure of tripalmitin.

15.24 Draw the structure of triolein.

15.25 Safflower oil is called a polyunsaturated oil, whereas olive oil is a monounsaturated oil. Explain.

15.26 Why does olive oil have a lower melting point than butter fat?

15.27 Why does coconut oil, a vegetable oil, have a melting point similar to fats from animal sources?

15.28 A label on a bottle of 100% sunflower seed oil states that it is lower in saturated fats than all the leading oils.
  a. How does the percentage of saturated fats in sunflower seed oil compare to that of safflower, corn, and canola oil? (See Figure 15.4.)
  b. Is the claim valid?

LEARNING GOAL

Draw the structure of the product from the reaction of a triacylglycerol with hydrogen, or an acid or base.

■ 15.4 Chemical Properties of Triacylglycerols

The chemical reactions of the triacylglycerols (fats and oils) are the same as we discussed for alkenes (Chapter 11), and carboxylic acids and esters (Chapter 13). We will look at the addition of hydrogen to the double bonds and the hydrolysis of the ester bonds of fats and oils.

■ Hydrogenation

The **hydrogenation** of unsaturated fats converts carbon–carbon double bonds to single bonds. The hydrogen gas is bubbled through the heated oil in the presence of a nickel catalyst.

$$-CH=CH- + H_2 \xrightarrow{Ni} -\overset{\overset{\displaystyle H}{|}}{\underset{\underset{\displaystyle H}{|}}{C}}-\overset{\overset{\displaystyle H}{|}}{\underset{\underset{\displaystyle H}{|}}{C}}-$$

For example, when hydrogen adds to all of the double bonds of triolein using a nickel catalyst, the product is the saturated fat tristearin.

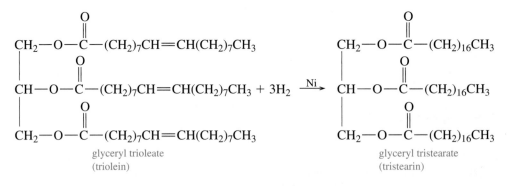

glyceryl trioleate
(triolein)

glyceryl tristearate
(tristearin)

In commercial hydrogenation, the addition of hydrogen is stopped before all the double bonds in an oil become completely saturated. Complete hydrogenation gives a very brittle product, whereas the partial hydrogenation of a liquid vegetable oil changes it to a soft, semisolid fat. As the oil becomes more saturated, the melting point increases and the fat becomes more solid at room temperature. Control of the degree of hydrogenation gives the various types of partially hydrogenated vegetable oil products on the market today—soft margarines, solid stick margarines, and solid shortenings. (See Figure 15.5.) Although these products now contain more saturated fatty acids than the original oils, they contain no cholesterol, unlike similar products from animal sources, such as butter and lard.

FIGURE 15.5

Many soft margarines, stick margarines, and solid shortenings are produced by the partial hydrogenation of vegetable oils.

Q How does hydrogenation change the structure of the fatty acids in the vegetable oils?

HEALTH NOTE

■ Olestra: A Fat Substitute

In 1968, food scientists designed an artificial fat called *olestra* as a source of nutrition for premature babies. However, olestra could not be digested and was never used for that purpose. Then scientists realized that olestra had the flavor and texture of a fat without the calories.

Olestra is manufactured by obtaining the fatty acids from the fats in cottonseed or soybean oils and bonding the fatty acids with the hydroxyl groups on sucrose. Chemically, olestra is composed of six to eight long-chain fatty acids attached by ester links to a sugar (sucrose) rather than to a glycerol molecule found in fats. This makes olestra a very large molecule, which cannot be absorbed through the intestinal walls. The enzymes and bacteria in the intestinal tract are unable to break down the olestra molecule and it travels through the intestinal tract undigested.

In 1996 the Food and Drug Administration approved olestra for use in potato chips, tortilla chips, crackers and fried snacks. In 1996, olestra snack products were test marketed in parts of Iowa, Wisconsin, Indiana, and Ohio. By 1997, there were reports of some adverse reactions, including diarrhea, abdominal cramps, and anal leakage, indicating that olestra may act as a laxative in some people. However, the manufacturers contend there is no direct proof that olestra is the cause of those effects.

The large molecule of olestra also combines with fat-soluble vitamins (A, D, E, and K) as well as the carotenoids from the foods we eat before they can be absorbed through the intestinal wall. Carotenoids are plant pigments in fruits and vegetables that protect against cancer, heart disease, and macular degeneration, a form of blindness in the elderly. The FDA now requires manufacturers to add the four vitamins, but not the carotenoids, to olestra products. Snack foods made with olestra are now in supermarkets nationwide. Since there are already low-fat snacks on the market, it remains to be seen whether olestra will have any significant effect on reducing the problem of obesity.

Fatty acids

$CH_3(CH_2)_6COOH$

$CH_3(CH_2)_8COOH$

Olestra

■ Trans Fatty Acids and Hydrogenation

In the early 1900s, margarine became a popular replacement for the highly saturated fats such as butter and lard. Margarine is produced by partially hydrogenating the unsaturated fats in vegetable oils such as safflower oil, corn oil, canola oil, cottonseed oil, and sunflower oil. Fats that are more saturated are more resistant to oxidation.

In vegetable oils, the unsaturated fats usually contain cis double bonds. As hydrogenation occurs, double bonds are converted to single bonds. However, some of the cis double bonds are converted to trans double bonds, which causes a change in the overall structure of the fatty acids. If the label on a product states that the oils have been "partially" or "fully hydrogenated," that product will also contain trans fatty acids. In the United States, it is estimated that 2–4% of our total calories come from trans fatty acids.

The concern about trans fatty acids is that their altered structure may make them behave like saturated fatty acids in the body. In the 1980s, research indicated that trans fatty acids have an effect on blood cholesterol similar to that of saturated fats, although study results vary. Several studies reported that trans fatty acids raise the levels of LDL-cholesterol, low-density lipoproteins containing cholesterol that can accumulate in the arteries. (LDLs and HDLs are described in the section on lipoproteins later in the chapter.) Some studies also report that trans fatty acids lower HDL-cholesterol, high-density lipoproteins that carry cholesterol to the liver to be excreted. But other studies did not report any decrease in HDL-cholesterol. In some American and European studies, an increased risk of breast cancer was associated with increased intake of trans fatty acids. However, these studies are not conclusive and not all studies have supported such findings. Current evidence does not yet indicate that the intake of trans fatty acids is a significant risk factor for heart disease. The trans fatty acids controversy will continue to be debated as more research is done.

Foods containing trans fatty acids include milk, bread, fried foods, ground beef, baked goods, stick margarine, butter, soft margarine, cookies, crackers, and vegetable shortening. The American Heart Association recommends that margarine should have no more than 2 grams of saturated fat per tablespoon and a liquid vegetable oil should be the first ingredient. They also recommend the use of soft margarine, which is lower in trans fatty acids because soft margarine is only slightly hydrogenated, and diet margarine because it has less fat and therefore fewer trans fatty acids.

Many health organizations agree that fat should account for less than 30% of daily calories (the current average for Americans is 34%) and saturated fat should be less than 10% of total calories. Lowering the overall fat intake would also decrease the amount of trans fatty acids. The Food and Drug Administration and the U.S. Department of Agriculture are encouraging the use of new food labels to inform consumers of the fat content of food. The best advice may be to reduce total fat in the diet by using fats and oils sparingly, cooking with little or no fat, substituting olive oil or canola oil for other oils, and limiting the use of coconut oil and palm oil, which are high in saturated fatty acids.

There are several products including peanut butter and butter-like spreads on the market that have 0% trans fatty acids. On the labels, they state that their products are nonhydrogenated, which avoids the production of the undesirable trans fatty acid. However, in the list of natural vegetable oils, such as soy and canola oil, there is also palm oil. Because palm oil has a melting point of 30°C, palm oil increases the overall melting point of the spread and gives a product that is solid at room temperature. However, palm oil contains high amounts of saturated fatty acids and has a similar effect in the body as does stearic acid (18 carbons) and fats derived from animal sources. Health experts recommend that we limit the amount of saturated fats, including palm oil, in our diets.

Hydrolysis

Triacylglycerols are hydrolyzed (split by water) in the presence of strong acids or digestive enzymes called *lipases*. The products of hydrolysis of the ester bonds are glycerol and three fatty acids. The polar glycerol is soluble in water, but the fatty acids with their long hydrocarbon chains are not.

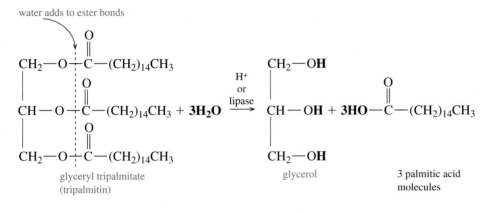

Saponification

When a fat is heated with a strong base such as sodium hydroxide, saponification of the fat gives glycerol and the sodium salts of the fatty acids, which are soaps. When NaOH is used, a solid soap is produced that can be molded into a desired shape; KOH produces a softer, liquid soap. Oils that are polyunsaturated produce softer soaps. Names like "coconut" or "avocado shampoo" tell you the sources of the oil used in the reaction.

Fat or oil + strong base ⟶ glycerol + salts of fatty acids (soaps)

SAMPLE PROBLEM 15.4

■ Reactions of Lipids

Write the equation for the reaction catalyzed by the enzyme lipase that hydrolyzes trilaurin during the digestion process.

Types of Fats

Read the labels on food products that contain fats such as butter, margarine, vegetable oils, peanut butter, and potato chips. Look for terms such as saturated, monounsaturated, polyunsaturated, and partially or fully hydrogenated.

QUESTIONS

1. How many grams of saturated, monounsaturated, and polyunsaturated fat are in one serving of the product?
2. What type(s) of fats or oils are in the product?
3. What percent of the total fat is saturated fat? Unsaturated fat?
4. The label on a container of peanut butter states that the cottonseed and canola oils used to make the peanut butter have been fully hydrogenated. What are the typical products that would form when hydrogen is added?
5. For each packaged food, determine the following:
   a. How many grams of fat are in one serving of the food?
   b. Using the caloric value for fat (9 kcal/gram of fat), how many calories (kilocalories) come from the fat in one serving?
   c. What is the percentage of fat in one serving?

SOLUTION

glyceryl trilaurate
(trilaurin)

glycerol   3 lauric acid
                  molecules

STUDY CHECK

What is the name of the product formed when a triacylglycerol containing oleic acid and linoleic acid is completely hydrogenated?

QUESTIONS AND PROBLEMS

Chemical Properties of Triacylglycerols

15.29 Write an equation for the hydrogenation of glyceryl trioleate, a fat containing glycerol and three oleic acid units.

15.30 Write an equation for the hydrogenation of glyceryl trilinolenate, a fat containing glycerol and three linolenic acid units.

15.31 a. Write an equation for the acid hydrolysis of glyceryl trimyristate (trimyristin).
   b. Write an equation for the NaOH saponification of glyceryl trimyristate (trimyristin).

15.32 a. Write an equation for the acid hydrolysis of glyceryl trioleate (triolein).
   b. Write an equation for the NaOH saponification of glyceryl trioleate (triolein).

15.33 Compare the structure of a triacylglycerol to the structure of olestra. (See Health Note "Olestra: A Fat Substitute.")

15.34 An oil is partially hydrogenated.
   a. Are all or just some of the double bonds converted to single bonds?
   b. What happens to many of the cis double bonds during hydrogenation?
   c. How can you reduce the amount of trans fatty acids in your diet?

15.35 Write the product of the hydrogenation of the following triacylglycerol.

15.36 Write all the products that would be obtained when the triacylglycerol in problem 15.35 undergoes complete hydrolysis.

15.5 Glycerophospholipids

The **glycerophospholipids** are a family of lipids similar to triacylglycerols except that one hydroxyl group of glycerol is replaced by the ester of phosphoric acid and an amino alcohol, bonded through a phosphodiester bond. We can compare the general structures of a triacylglycerol and a glycerophospholipid as follows:

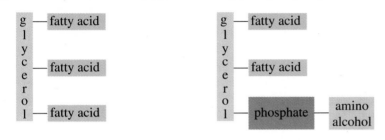

triacylglycerol (triglyceride)   glycerophospholipid

Three amino alcohols found in glycerophospholipids are choline, serine, and ethanolamine. Because these compounds exist in the body at physiological pH of 7.4, they are ionized.

$$HO-CH_2-CH_2-\overset{+}{\underset{CH_3}{\overset{CH_3}{N}}}-CH_3 \qquad HO-CH_2-\overset{\overset{+}{N}H_3}{\underset{}{CH}}-COO^- \qquad HO-CH_2-CH_2-\overset{+}{N}H_3$$

choline   serine   ethanolamine

Lecithins and **cephalins** are two types of glycerophospholipids that are particularly abundant in brain and nerve tissues as well as in egg yolks, wheat germ, and yeast. Lecithins contain choline, and cephalins contain ethanolamine and sometimes serine. In the following structural formulas, palmitic acid is used as an example of a fatty acid.

Glycerophospholipids contain both polar and nonpolar regions, which allow them to interact with both polar and nonpolar substances. The ionized alcohol and phosphate portion called "the head" is polar and can hydrogen bond with water. (See Figure 15.6.) The two fatty acids connected to the glycerol molecule represent the nonpolar "tails" of the phospholipid. The hydrocarbon chains that make up the "tails" are only soluble in other nonpolar substances, mostly lipids.

Polar head

(a) Chemical structure of a glycerophospholipid

Nonpolar tails

Polar head   Nonpolar tails

(b) Simplified way to draw a glycerophospholipid

FIGURE 15.6

(a) In a glycerophospholipid, a polar "head" contains the ionized amino alcohol and phosphoric acid groups, while the two fatty acids make up the nonpolar "tails."
(b) A simplified drawing indicates the polar region and the nonpolar region.
Q **Why are glycerophospholipids polar?**

Glycerophospholipids are the most abundant lipids in cell membranes, where they play an important role in cellular permeability. They make up much of the myelin sheath that protects nerve cells. In the body fluids, they combine with the less polar triglycerides and cholesterol to make them more soluble as they are transported in the body.

SAMPLE PROBLEM 15.5

■ Drawing Glycerophospholipid Structures

Draw the structure of a cephalin, using stearic acid for the fatty acids and serine for the amino alcohol. Describe each of the components in the glycerophospholipid.

SOLUTION

In general, glycerophospholipids are composed of a glycerol molecule in which two carbon atoms are attached to fatty acids such as stearic acid. The third carbon atom is attached in an ester bond to phosphate linked to an amino alcohol. In this example, the amino alcohol is serine.

STUDY CHECK

How do glycerophospholipids differ structurally from triacylgylcerols?

QUESTIONS AND PROBLEMS

▪ Glycerophospholipids

15.37 Describe the differences between triacylglycerols and glycerophospholipids.

15.38 Describe the differences between lecithins and cephalins.

15.39 Draw the structure of a glycerophospholipid containing two molecules of palmitic acid and ethanolamine. What is another name for this type of phospholipid?

15.40 Draw the structure of a glycerophospholipid that contains choline and palmitic acids.

15.41 Identify the type of the following glycerophospholipid and list its components:

15.42 Identify the type of the following glycerophospholipid and list its components:

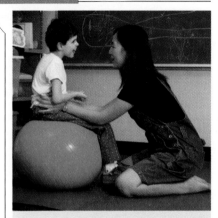
▪ 15.6 Steroids: Cholesterol and Steroid Hormones

Steroids are compounds containing the steroid nucleus, which consists of three cyclohexane rings and one cyclopentane ring fused together. The four rings in the steroid nucleus are designated A, B, C, and D. The carbon atoms are numbered beginning with the carbons in ring A and, in types of steroids like cholesterol, ending with two methyl groups.

Steroid

Cholesterol

Attaching other atoms and groups of atoms to the steroid structure forms a wide variety of steroid compounds. **Cholesterol**, which is one of the most important and abundant steroids in the body, is a *sterol* because it contains an oxygen atom as a hydroxyl (—OH) group on carbon 3. Like many steroids, cholesterol has methyl groups at carbon 10 and carbon 13 and a carbon chain at carbon 17 with a double bond between carbon 5 and carbon 6. In other steroids, the oxygen atom typically at carbon 3 forms a carbonyl (C═O) group.

Cholesterol

Cholesterol in the Body

Cholesterol is a component of cellular membranes, myelin sheath, and brain and nerve tissue. It is also found in the liver, bile salts, and skin, where it forms vitamin D. In the adrenal gland, it is used to synthesize steroid hormones. Cholesterol in the body is obtained from eating meats, milk, and eggs, and it is also synthesized by the liver from fats, carbohydrates, and proteins. There is no cholesterol in vegetable and plant products.

If a diet is high in cholesterol, the liver produces less. A typical daily American diet includes 400–500 mg of cholesterol, one of the highest in the world. The American Heart Association has recommended that we consume no more than 300 mg of cholesterol a day. The cholesterol contents of some typical foods are listed in Table 15.3.

When cholesterol exceeds its saturation level in the bile, gallstones may form. Gallstones are composed of almost 100% cholesterol with some calcium salts, fatty acids, and phospholipids. (See Figure 15.7.) High levels of cholesterol are also associated with the accumulation of lipid deposits (plaque) that line and narrow the coronary arteries. (See Figure 15.8.) Clinically, cholesterol levels are considered elevated if the total plasma cholesterol level exceeds 200 mg/dL.

Some research indicates that saturated fats in the diet may stimulate the production of cholesterol by the liver. A diet that is low in foods containing

FIGURE 15.7

Gallstones form in the gallbladder when cholesterol levels are high.

Q **What type of steroid is stored in the gallbladder?**

TABLE 15.3

Cholesterol Content of Some Foods

Food	Serving Size	Cholesterol (mg)
Liver (beef)	3 oz	370
Egg	1	250
Lobster	3 oz	175
Fried chicken	$3\frac{1}{2}$ oz	130
Hamburger	3 oz	85
Chicken (no skin)	3 oz	75
Fish (salmon)	3 oz	40
Butter	1 tablespoon	30
Whole milk	1 cup	35
Skim milk	1 cup	5
Margarine	1 tablespoon	0

(a)

(b)

FIGURE 15.8

Excess cholesterol forms plaque that can block an artery, resulting in a heart attack. **(a)** A normal, open artery shows no buildup of plaque. **(b)** An artery that is almost completely clogged by atherosclerotic plaque.

Q **What property of cholesterol would cause it to form deposits along the coronary arteries?**

cholesterol and saturated fats appears to be helpful in reducing the serum cholesterol level. Other factors that may also increase the risk of heart disease are family history, lack of exercise, smoking, obesity, diabetes, gender, and age.

SAMPLE PROBLEM 15.6

■ Cholesterol

Observe the structure of cholesterol for the following questions:

a. What part of cholesterol is the steroid nucleus?
b. What features have been added to the steroid nucleus in cholesterol?
c. What classifies cholesterol as a sterol?

SOLUTION

a. The four fused rings form the steroid nucleus.
b. The cholesterol molecule contains an alcohol group (—OH) on the first ring, one double bond in the second ring, and a branched carbon chain.
c. The alcohol group determines the sterol classification.

STUDY CHECK

Why is cholesterol in the lipid family?

■ Lipoproteins: Transporting Lipids

In the body, lipids must be transported through the bloodstream to tissues where they are stored, used for energy, or to make hormones. However, most lipids are nonpolar and insoluble in the aqueous environment of blood. They are made more soluble by combining them with phospholipids and proteins to form water-soluble complexes called **lipoproteins**. In general, lipoproteins are spherical particles with an outer surface of polar proteins and phospholipids that surround hundreds of nonpolar molecules of triacylglycerols and cholesteryl esters. (See Figure 15.9.)

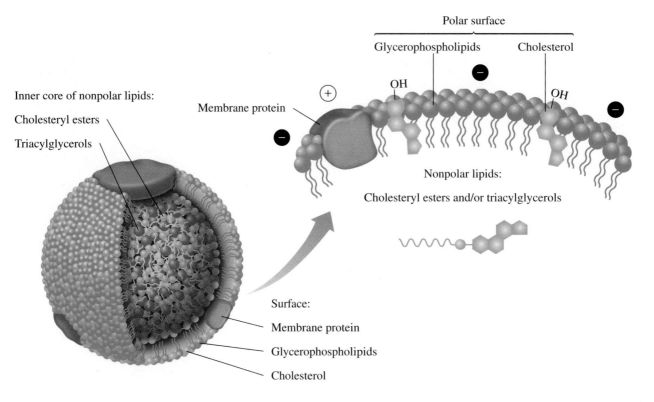

Inner core of nonpolar lipids:
Cholesteryl esters
Triacylglycerols

Surface:
Membrane protein
Glycerophospholipids
Cholesterol

FIGURE 15.9

A spherical lipoprotein particle surrounds nonpolar lipids with polar lipids and protein for transport to body cells.

Q **Why are the polar components on the surface of a lipoprotein particle and the nonpolar components at the center?**

Cholesteryl esters are the prevalent form of cholesterol in the blood. They are formed by the esterification of the hydroxyl group in cholesterol with a fatty acid.

cholesteryl ester

There are several types of lipoproteins that differ in density, lipid composition, and function. They include chylomicrons, very-low-density lipoprotein (VLDL), low-density lipoprotein (LDL), and high-density lipoprotein (HDL). (See Table 15.4.)

TABLE 15.4

Composition and Properties of Plasma Lipoproteins

	Chylomicron	VLDL	LDL	HDL
Density (g/mL)	0.94	0.950–1.006	1.006–1.063	1.063–1.210
Composition (% by mass)				
Triacylglycerol	86	55	6	4
Phospholipids	7	18	22	24
Cholesterol	2	7	8	2
Cholesteryl esters	3	12	42	15
Protein	2	8	22	55

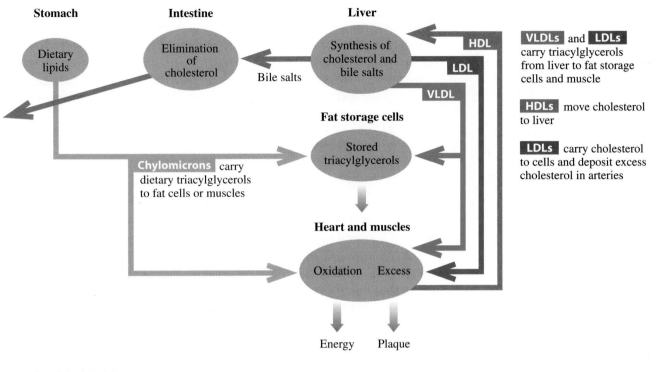

FIGURE 15.10

Lipoproteins such as HDLs and LDLs transport nonpolar lipids and cholesterol to cells and the liver.

Q What type of lipoprotein transports cholesterol to the liver?

The chylomicrons formed in the mucosal cells of the small intestine and the VLDLs formed in the liver transport triacylglycerols, phospholipids, and cholesterol to the tissues for storage or to the muscles for energy. (See Figure 15.10.) The LDLs transport cholesterol to tissues to be used for the synthesis of cell membranes, steroid hormones, and bile salts. When the level of LDL exceeds the amount of cholesterol needed by the tissues, the LDLs deposit cholesterol in the arteries, which can restrict blood flow and increase the risk of developing heart disease and/or myocardial infarctions (heart attacks). This is why LDL cholesterol is called "bad" cholesterol.

The HDLs remove excess cholesterol from the tissues and carry it to the liver where it is converted to bile salts and eliminated. When HDL levels are high, cholesterol that is not needed by the tissues is carried to the liver for elimination rather than deposited in the arteries, which gives the HDLs the name of "good" cholesterol. Most of the cholesterol in the body is synthesized in the liver, although some comes from the diet. However, a person on a high-fat diet reabsorbs cholesterol from the bile salts causing less cholesterol to be eliminated. In addition, higher levels of saturated fats stimulate the synthesis of cholesterol by the liver.

Because high cholesterol levels are associated with the onset of arteriosclerosis and heart disease, the serum levels of LDL and HDL are generally determined in a medical examination. For adults, recommended levels for total cholesterol are less than 200 mg/dL with LDL less than 130 mg/dL and HDL over 40 mg/dL. A lower level of serum cholesterol decreases the risk of heart disease. Higher HDL levels are found in people who exercise regularly and eat less saturated fat.

Steroid Hormones

The word *hormone* comes from the Greek "to arouse" or "to excite." Hormones are chemical messengers that serve as a kind of communication system from one part of the body to another. The *steroid* hormones, which include the sex hormones and the adrenocortical hormones, are closely related in structure to cholesterol and depend on cholesterol for their synthesis.

Two important male sex hormones, *testosterone* and *androsterone*, promote the growth of muscle and of facial hair and the maturation of the male sex organs and of sperm.

The *estrogens*, a group of female sex hormones, direct the development of female sexual characteristics: the uterus increases in size, fat is deposited in the breasts, and the pelvis broadens. *Progesterone* prepares the uterus for the implantation of a fertilized egg. If an egg is not fertilized, the levels of progesterone and estrogen drop sharply, and menstruation follows. Synthetic forms of the female sex hormones are used in birth-control pills. As with other kinds of steroids, side effects can include weight gain and a greater risk of forming blood clots. The structures of some steroid hormones follow:

Hormone	Biological Effects
testosterone (androgen) (produced in testes)	Development of male organs; male sexual characteristics including muscles and facial hair; sperm formation
estradiol (estrogen) (produced in ovaries)	Development of female sexual characteristics; ovulation
progesterone (produced in ovaries)	Prepares uterus for fertilized egg
norethindrone (synthetic progestin)	Contraceptive (birth control) pill

HEALTH NOTE

Anabolic Steroids

Some of the physiological effects of testosterone are to increase muscle mass and decrease body fat. Derivatives of testosterone called *anabolic steroids* that enhance these effects have been synthesized. Although they have some medical uses, anabolic steroids have been used in rather high dosages by some athletes in an effort to increase muscle mass. Such use is illegal.

Use of anabolic steroids in attempting to improve athletic strength can cause side effects including hypertension, fluid retention, increased hair growth, sleep disturbances, and acne. Over a long period of time, their use can be devastating and may cause irreversible liver damage and decreased sperm production.

SOME ANABOLIC STEROIDS

methandienone oxandrolone nandrolone stanozolol

Adrenal Corticosteroids

The adrenal glands, located on the top of each kidney, produce the corticosteroids. *Aldosterone*, a mineralocorticoid, is responsible for electrolyte and water balance by the kidneys. *Cortisone*, a glucocorticoid, increases the blood glucose level and stimulates the synthesis of glycogen in the liver from amino acids. Synthetic corticoids such as *prednisone* are derived from cortisone and used medically for reducing inflammation and treating asthma and rheumatoid arthritis, although precautions are given for long-term use.

Corticosteroids

cortisone
(produced in adrenal gland)

aldosterone (mineralocorticoid)
(produced in adrenal gland)

prednisone
(synthetic corticoid)

Biological Effects

Increases the blood glucose and glycogen levels from fatty acids and amino acids	Increases the reabsorption of Na^+ in kidneys; retention of water	Reduces inflammation; treatment of asthma and rheumatoid arthritis

SAMPLE PROBLEM 15.7

■ Steroid Hormones

What are the functional groups on the steroid nucleus in the sex hormones estradiol and testosterone?

SOLUTION

Estradiol contains a benzene ring and two alcohol groups. Testosterone contains a ketone group, a double bond, and an alcohol group.

STUDY CHECK

What are the similarities and differences in the structures of testosterone and the anabolic steroid nandrolone? (See Health Note "Anabolic Steroids.")

QUESTIONS AND PROBLEMS

■ Steroids: Cholesterol and Steroid Hormones

15.43 Draw the structure for the steroid nucleus.

15.44 Which of the following compounds are derived from cholesterol?
   a. glyceryl tristearate   **b.** cortisone
   c. testosterone   **d.** estradiol

15.45 What is the general structure of lipoproteins?

15.46 Why are lipoproteins needed to transport lipids in the bloodstream?

15.47 How do chylomicrons differ from very-low-density lipoproteins?

15.48 How do LDLs differ from HDLs?

15.49 Why are LDLs called "bad" cholesterol?

15.50 Why are HDLs called "good" cholesterol?

15.51 What are the similarities and differences between the sex hormones estradiol and testosterone?

15.52 What are the similarities and differences between the adrenal hormone cortisone and the synthetic corticoid prednisone?

15.53 Which of the following are male sex hormones?
   a. cholesterol   **b.** aldosterone
   c. estrogen   **d.** testosterone
   e. choline

15.54 Which of the following are adrenal steroids?
   a. cholesterol   **b.** aldosterone
   c. estrogen   **d.** testosterone
   e. choline

LEARNING GOAL

Describe the composition and function of the lipid bilayer in cell membranes.

WEB TUTORIAL
Membrane Structure

■ 15.7 Cell Membranes

The membrane of a cell separates the contents of a cell from the external fluids. It is semipermeable so that nutrients can enter the cell and waste products can leave. The main components of a cell membrane are the phospholipids. Earlier in this chapter we saw that the structures of such phospholipids consisted of a nonpolar region or tail with long-chain fatty acids and a polar region or head from phosphoric acid and amino alcohols that ionize at physiological pH.

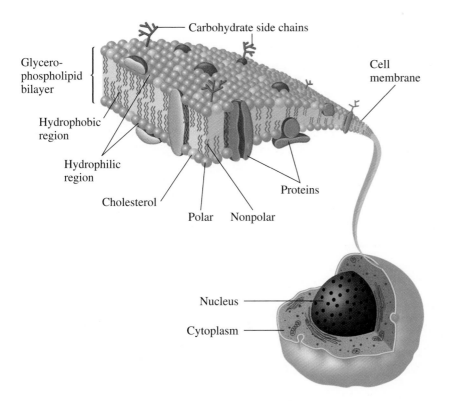

Carbohydrate side chains

Glycero-
phospholipid
bilayer

Hydrophobic
region

Hydrophilic
region

Cholesterol

Polar   Nonpolar

Proteins

Cell
membrane

Nucleus

Cytoplasm

FIGURE 15.11

The fluid mosaic model of a cell
membrane. Proteins and cholesterol
are embedded in a lipid bilayer of
phospholipids. The bilayer forms a
membrane-type barrier with polar
heads at the membrane surfaces and
the nonpolar tails in the center away
from the water.

Q **What types of fatty acids are
found in the phospholipids of the
lipid bilayer?**

In a cell membrane, two rows of phospholipids are arranged like a sandwich. Their nonpolar tails, which are hydrophobic ("water-fearing"), move to the center, while their polar heads, which are hydrophilic ("water-loving"), align on the outer edges of the membrane. This double row arrangement of phospholipids is called a **lipid bilayer**. (See Figure 15.11.) One row of the phospholipids forms the outside surface of the membrane, which is in contact with the external fluids, and the other row forms the inside surface or edge of the membrane, which is in contact with the internal contents of the cell.

Most of the phospholipids in the lipid bilayer contain unsaturated fatty acids. Due to the kinks in the carbon chains at the cis double bonds, the phospholipids do not fit closely together. As a result, the lipid bilayer is not a rigid, fixed structure, but one that is dynamic and fluid-like. In this liquid-like bilayer, there are also proteins, carbohydrates, and cholesterol molecules. For this reason, the model of biological membranes is referred to as the **fluid mosaic model** of membranes.

In the fluid mosaic model, proteins known as peripheral proteins emerge on just one of the surfaces, outer or inner. The integral proteins extend through the entire lipid bilayer and appear on both surfaces of membrane. Some proteins and some lipids on the outer surface of the cell membrane are attached to carbohydrates. These carbohydrate chains project into the surrounding fluid environment where they are responsible for cell recognition and communication with chemical messengers such as hormones and neurotransmitters. In animals, cholesterol makes up 20–25% of the lipid bilayer. Embedded among the nonpolar tails of the fatty acids, cholesterol molecules add strength and rigidity to the cell membrane.

Transport Through Cell Membranes

Ions and molecules flow in and out of the cell in several ways. In the simplest transport mechanism called diffusion or passive transport, ions and small molecules migrate from a higher concentration to a lower concentration. For example, some ions as well as small molecules such as O_2, urea, and water diffuse through cell membranes. If their concentration is greater outside the cell than inside, they diffuse into the cell. If water has a high concentration in the cell, it diffuses out of the cell.

Another type of transport called facilitated transport increases the rate of diffusion for substances that diffuse too slowly by passive diffusion to meet cell needs. This process utilizes the integral proteins that extend from one edge of the cell membrane to the other. Groups of integral proteins provide channels to transport chloride ion (Cl^-), bicarbonate ion (HCO_3^-), and glucose molecules in and out of the cell more rapidly.

Certain ions such as K^+, Na^+, and Ca^{2+}, move across a cell membrane against a concentration gradient. For example, the K^+ concentration is greater inside a cell, and the Na^+ concentration is greater outside. However, in the conduction of nerve impulses and contrac-

tion of muscles, K^+ moves into the cell, and Na^+ moves out. To move an ion from a lower to a higher concentration requires energy, which is accomplished by a process known as active transport. In active transport, a protein complex called a Na^+/K^+ pump breaks down ATP to ADP (Chapter 18), which releases energy to move Na^+ and K^+ against their concentration gradients.

SAMPLE PROBLEM 15.8

Lipid Bilayer in the Cell Membranes

Describe the role of phospholipids in the lipid bilayer.

SOLUTION

Phospholipids consist of polar and nonpolar parts. In a cell membrane, an alignment of the nonpolar sections toward the center with the polar sections on the outside produces a barrier that prevents the contents of a cell from mixing with the fluids on the outside of the cell.

STUDY CHECK

What is the function of cholesterol in the cell membrane?

QUESTIONS AND PROBLEMS

Cell Membranes

15.55 What types of lipids are found in cell membranes?

15.56 Describe the structure of a lipid bilayer.

15.57 What is the function of the lipid bilayer in a cell membrane?

15.58 How do the unsaturated fatty acids in the phospholipids affect the structure of cell membranes?

15.59 Where are proteins located in cell membranes?

15.60 How are carbohydrates attached to the lipid bilayer?

15.61 What is the function of the carbohydrates on a cell membrane?

15.62 Why is a cell membrane semipermeable?

CONCEPT MAP

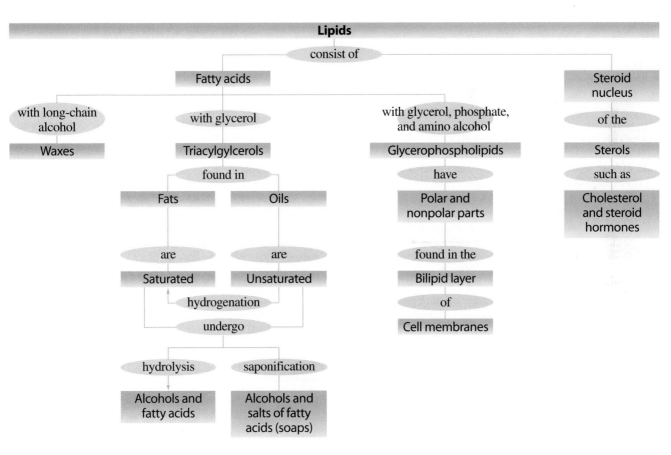

CHAPTER REVIEW

15.1 Lipids
Lipids are nonpolar compounds that are not soluble in water. Classes of lipids include waxes, fats and oils, glycerophospholipids, and steroids.

15.2 Fatty Acids
Fatty acids are unbranched carboxylic acids that typically contain an even number (10–20) of carbon atoms. Fatty acids may be saturated, monounsaturated with one double bond, or polyunsaturated with two or more double bonds. The double bonds in unsaturated fatty acids are almost always cis.

15.3 Waxes, Fats, and Oils
A wax is an ester of a long-chain fatty acid and a long-chain alcohol. The triacylglycerols of fats and oils are esters of glycerol with three long-chain fatty acids. Fats contain more saturated fatty acids and have higher melting points than most oils.

15.4 Chemical Properties of Triacylglycerols
The hydrogenation of unsaturated fatty acids converts double bonds to single bonds. The hydrolysis of the ester bonds in fats or oils produces glycerol and fatty acids. In saponification, a fat heated with a strong base produces glycerol and the salts of the fatty acids or soaps.

15.5 Glycerophospholipids

Glycerophospholipids are esters of glycerol with two fatty acids and a phosphate group attached to an amino alcohol.

15.6 Steroids: Cholesterol and Steroid Hormones

Steroids are lipids containing the steroid nucleus, which is a fused structure of four rings. Lipids, which are nonpolar, are transported through the aqueous environment of the blood by forming lipoproteins. Lipoproteins such as chylomicrons and LDL transport triacylglycerols from the intestines and the liver to fat cells for storage and muscles for energy. HDLs transport cholesterol from the tissues to the liver for elimination. The steroid hormones are closely related in structure to cholesterol and depend on cholesterol for their synthesis. The sex hormones, such as estrogen and testosterone, are responsible for sexual characteristics and reproduction. The adrenal corticosteroids such as aldosterone and cortisone regulate water balance and glucose levels in the cells.

15.7 Cell Membranes

All animal cells are surrounded by a semipermeable membrane that separates the cellular contents from the external fluids. The membrane is composed of two rows of glycerophospholipids in a lipid bilayer.

KEY TERMS

cephalins Phospholipids found in brain and nerve tissues that incorporate the amino alcohol serine or ethanolamine.

cholesterol The most prevalent of the steroid compounds found in cellular membranes; needed for the synthesis of vitamin D, hormones, and bile acids.

esterification The reaction of an alcohol such as glycerol with acids to form ester bonds.

fat Another term for solid triacylglycerols.

fatty acids Long-chain carboxylic acids found in fats.

fluid mosaic model The concept that cell membranes are lipid bilayer structures that contain an assortment of polar lipids and proteins in a dynamic, fluid arrangement.

glycerophospholipids Polar lipids of glycerol attached to two fatty acids and a phosphate group connected to an amino group such as choline, serine, or ethanolamine.

hydrogenation The addition of hydrogen to unsaturated fats.

lecithins Glycerophospholipids containing choline as the amino alcohol.

lipid bilayer A model of a cell membrane in which phospholipids are arranged in two rows interspersed with proteins arranged at different depths.

lipids A family of compounds that is nonpolar in nature and not soluble in water; includes fats, waxes, phospholipids, and steroids.

lipoprotein A combination of nonpolar lipids with glycerophospholipids and proteins to form a polar complex that can be transported through body fluids.

monounsaturated fatty acids Fatty acids with one double bond.

oil Another term for liquid triacylglycerols.

polyunsaturated fatty acids Fatty acids that contain two or more double bonds.

prostaglandins A number of compounds derived from arachidonic acid that regulate several physiological processes.

saturated fatty acids Fatty acids that have no double bonds; they have higher melting points than unsaturated lipids and are usually solid at room temperatures.

steroids Types of lipid composed of a multicyclic ring system.

triacylglycerols A family of lipids composed of three fatty acids bonded through ester bonds to glycerol, a trihydroxy alcohol.

wax The ester of a long-chain alcohol and a long-chain saturated fatty acid.

ADDITIONAL QUESTIONS AND PROBLEMS

15.63 Among the ingredients of a lipstick are beeswax, carnauba wax, hydrogenated vegetable oils, and glyceryl tricaprate (tricaprin). What types of lipids have been used? Draw the structure of glyceryl tricaprate (tricaprin). Draw the structure of this triacylglycerol using capric acid, a saturated 10-carbon fatty acid.

15.64 Because peanut oil floats on the top of peanut butter, many brands of peanut butter are hydrogenated. A solid product then forms that is mixed into the peanut butter and does not separate. If a triacylglycerol in peanut oil that contains one palmitic acid, one oleic acid, and one linoleic acid is completely hydrogenated, what is the product?

15.65 Trans fats are produced during the hydrogenation of polyunsaturated oils.
  a. What is the typical configuration of a monounsaturated fatty acid?
  b. How does a trans fat compare with a cis fat?
  c. Draw the structure of *trans*-oleic acid.

15.66 One mole of triolein is completely hydrogenated. What is the product? How many moles of hydrogen are required? How many grams of hydrogen? How many liters of hydrogen are needed if the reaction is run at STP?

15.67 On the list of ingredients of a cosmetic product are glyceryl stearate and lecithin. What are the structures of these compounds?

15.68 Some typical meals at fast-food restaurants are listed here. Calculate the number of kilocalories from fat and the percentage of total kilocalories due to fat (1 gram of fat = 9 kcal). Would you expect the fats to be mostly saturated or unsaturated? Why?
  a. a chicken dinner, 830 kcal, 46 g of fat
  b. a quarter-pound cheeseburger, 518 kcal, 29 g of fat
  c. pepperoni pizza (three slices), 560 kcal, 18 g of fat
  d. beef burrito, 470 kcal, 21 g of fat
  e. deep-fried fish (three pieces), 480 kcal, 28 g of fat

15.69 Identify the following as fatty acids, soaps, triacylglycerols, wax, glycerophospholipid, or steroid:
  a. beeswax   **b.** cholesterol
  c. lecithin   **d.** glyceryl tripalmitate (tripalmitin)
  e. sodium stearate **f.** safflower oil
  g. whale blubber   **h.** adipose tissue
  i. progesterone   **j.** cortisone
  k. stearic acid

15.70 Why would an animal that lives in a cold climate have more unsaturated triacylglycerols in its body fat than an animal that lives in a warm climate?

15.71 Identify the components contained in the following lipids a.-e.
  1. glycerol   **2.** fatty acid   **3.** phosphate
  4. amino alcohol   **5.** steroid nucleus
  a. estrogen   **b.** cephalin
  c. wax   **d.** triacylglycerol
  e. glycerophospholipid

15.72 Which of the following are found in cell membranes?
  a. cholesterol   **b.** triacylglycerols
  c. carbohydrates   **d.** proteins
  e. waxes   **f.** glycerophospholipids

15.73 Identify the lipoprotein in each description as
  1. chylomicrons   **2.** VLDL
  3. LDL   **4.** HDL
  a. "good" cholesterol
  b. transports most of the cholesterol to the cells
  c. carries triacylglycerols from the intestine to the fat cells
  d. transports cholesterol to the liver
  e. has the greatest abundance of protein
  f. "bad" cholesterol
  g. carries triacylglycerols synthesized in the liver to the muscles
  h. has the lowest density

UNDERSTANDING THE CONCEPTS

15.74 Jojoba wax in candles consists of stearic acid and a 22-carbon saturated alcohol. Write the structure of jojoba wax.

15.75 Palmitic acid is obtained from palm oil as glyceryl tripalmitate. Draw the structure of glyceryl tripalmitate.

15.76 Identify each of the following as saturated, monounsaturated, polyunsaturated, omega-3, or omega-6 fatty acids:

a.
$$CH_3-(CH_2)_4-CH=CH-CH_2-CH=CH-(CH_2)_7-COOH$$
b. linolenic acid
c. $CH_3-(CH_2)_{14}-COOH$
d. $CH_3-(CH_2)_7-CH=CH-(CH_2)_7-COOH$

15.77 Sunflower oil can be used to make margarine. A triacylglycerol in sunflower oil consists of 2 linoleic acids and 1 oleic acid.
  a. Write two isomers for the triacylglycerol in sunflower oil.
  b. Using one of the isomers, write the reaction that would be used when sunflower oil is used to make solid margarine.

QUESTIONS FOR TEAMWORK

15.78 Which of the following has the lower melting point? Explain.

CH$_3$—(CH$_2$)$_{16}$—COOH

        1

CH$_3$—(CH$_2$)$_7$—CH=CH—(CH$_2$)$_7$—COOH

              2

15.79 Draw the structure of glycerophospholipid that is made from stearic acid, palmitic acid, and a phosphate bonded to ethanolamine.

15.80 Olive oil consists of a high percentage of triolean.
a. Draw the structure for triolean.
b. How many liters of H$_2$ gas at STP are needed to completely saturate 100. g of triolean?
c. How many mL of 0.250 M NaOH are needed to completely saponify 100. g of triolean?

15.81 A sink drain can become clogged with solid fat such as glyceryl tristearate.
a. How would adding lye (NaOH) to the sink drain remove the blockage?
b. Write an equation for the reaction that occurs.

ANSWERS

Answers to Study Checks

15.1 Glycerol, two fatty acids, phosphate, and an amino alcohol.

15.2 **a.** 16   **b.** unsaturated   **c.** liquid

15.3

15.4 tristearin

15.5 Triacylglycerols contain three fatty acids connected by ester bonds to glycerol. Glycerophospholipids contain glycerol esterified to two fatty acids and a phosphate connected to an amino alcohol.

15.6 Cholesterol is not soluble in water; it is classified with the lipid family.

15.7 Testosterone and nandrolone both contain a steroid nucleus with one double bond and a ketone group in the first ring, and a methyl and alcohol group on the five-carbon ring. Nandrolone does not have a second methyl group at the first

and second ring fusion point, which is seen in the structure of testosterone.

15.8 Cholesterol adds strength and rigidity to the cell membrane.

Answers to Selected Questions and Problems

15.1 Lipids provide energy and protection and insulation for the organs in the body.

15.3 Because lipids are not soluble in water, a polar solvent, they are nonpolar molecules.

15.5 All fatty acids contain a long chain of carbon atoms with a carboxylic acid group. Saturated fats contain only carbon-to-carbon single bonds; unsaturated fats contain one or more double bonds. More saturated fats are found in animal fats, whereas vegetable oils contain more unsaturated fats.

15.7 **a.** palmitic acid

                         COOH

b. oleic acid

                         COOH

15.9 **a.** saturated   **b.** unsaturated
     c. unsaturated   **d.** saturated

15.11 In a cis fatty acid, the hydrogen atoms are on the same side of the double bond, which produces a bend in the carbon chain. In a trans fatty acid, the hydrogen atoms are on opposite sides of the double bond, which gives a carbon chain without any bend.

15.13 In an omega-3 fatty acid, there is a double bond on carbon 3 counting from the methyl group, whereas in an omega-6 fatty acid, there is a double bond beginning at carbon 6 counting from the methyl group.

15.15 Arachidonic acid contains four double bonds and no side groups. In PGE_1, a part of the chain forms cyclopentane and there are hydroxyl and ketone functional groups.

15.17 Prostaglandins affect blood pressure and stimulate contraction and relaxation of smooth muscle.

15.19

$$CH_3(CH_2)_{14}\overset{\overset{\displaystyle O}{\|}}{C}O(CH_2)_{29}CH_3$$

15.21

$$CH_2O\overset{\overset{\displaystyle O}{\|}}{C}(CH_2)_{16}CH_3$$
$$|$$
$$CHO\overset{\overset{\displaystyle O}{\|}}{C}(CH_2)_{16}CH_3$$
$$|$$
$$CH_2O\overset{\overset{\displaystyle O}{\|}}{C}(CH_2)_{16}CH_3$$

15.23

$$CH_2O\overset{\overset{\displaystyle O}{\|}}{C}(CH_2)_{14}CH_3$$
$$|$$
$$CHO\overset{\overset{\displaystyle O}{\|}}{C}(CH_2)_{14}CH_3$$
$$|$$
$$CH_2O\overset{\overset{\displaystyle O}{\|}}{C}(CH_2)_{14}CH_3$$

15.25 Safflower oil contains fatty acids with two or three double bonds; olive oil contains a large amount of oleic acid, which has only one (monounsaturated) double bond.

15.27 Although coconut oil comes from a vegetable, it has large amounts of saturated fatty acids and small amounts of unsaturated fatty acids.

15.29

$$CH_2O\overset{\overset{\displaystyle O}{\|}}{C}(CH_2)_7CH{=}CH(CH_2)_7CH_3$$
$$|$$
$$CHO\overset{\overset{\displaystyle O}{\|}}{C}(CH_2)_7CH{=}CH(CH_2)_7CH_3 + 3\ H_2 \xrightarrow{Ni}$$
$$|$$
$$CH_2O\overset{\overset{\displaystyle O}{\|}}{C}(CH_2)_7CH{=}CH(CH_2)_7CH_3$$

$$CH_2O\overset{\overset{\displaystyle O}{\|}}{C}(CH_2)_{16}CH_3$$
$$|$$
$$CHO\overset{\overset{\displaystyle O}{\|}}{C}(CH_2)_{16}CH_3$$
$$|$$
$$CH_2O\overset{\overset{\displaystyle O}{\|}}{C}(CH_2)_{16}CH_3$$

15.31 a.

$$CH_2O\overset{\overset{\displaystyle O}{\|}}{C}(CH_2)_{12}CH_3$$
$$|$$
$$CHO\overset{\overset{\displaystyle O}{\|}}{C}(CH_2)_{12}CH_3 + 3H_2O \xrightarrow{H^+}$$
$$|$$
$$CH_2O\overset{\overset{\displaystyle O}{\|}}{C}(CH_2)_{12}CH_3$$

$$CH_2OH$$
$$|$$
$$CHOH + 3CH_3(CH_2)_{12}\overset{\overset{\displaystyle O}{\|}}{C}OH$$
$$|$$
$$CH_2OH$$

b.

$$CH_2O\overset{\overset{\displaystyle O}{\|}}{C}(CH_2)_{12}CH_3$$
$$|$$
$$CHO\overset{\overset{\displaystyle O}{\|}}{C}(CH_2)_{12}CH_3 + 3\ NaOH \longrightarrow$$
$$|$$
$$CH_2O\overset{\overset{\displaystyle O}{\|}}{C}(CH_2)_{12}CH_3$$

$$CH_2OH$$
$$|$$
$$CHOH + 3CH_3(CH_2)_{12}\overset{\overset{\displaystyle O}{\|}}{C}O^-Na^+$$
$$|$$
$$CH_2OH$$

15.33 A triacylglycerol is composed of glycerol with three hydroxyl groups that form ester links with three long-chain fatty acids. In olestra, six to eight long-chain fatty acids form ester links with the hydroxyl groups on sucrose, a sugar. The olestra cannot be digested because our enzymes cannot break down the large olestra molecule.

15.35

$$CH_2O{-}\overset{\overset{\displaystyle O}{\|}}{C}{-}(CH_2)_{16}CH_3$$
$$|$$
$$CHO{-}\overset{\overset{\displaystyle O}{\|}}{C}{-}(CH_2)_{16}CH_3$$
$$|$$
$$CH_2O{-}\overset{\overset{\displaystyle O}{\|}}{C}{-}(CH_2)_{16}CH_3$$

15.37 A triacylglycerol consists of glycerol and three fatty acids. A glycerophospholipid consists of glycerol, two fatty acids, a phosphate group, and an amino alcohol.

15.39

$$CH_2O\overset{\overset{\displaystyle O}{\|}}{C}(CH_2)_{14}CH_3$$
$$|$$
$$CHO\overset{\overset{\displaystyle O}{\|}}{C}(CH_2)_{14}CH_3$$
$$|$$
$$CH_2OPOCH_2CH_2\overset{+}{N}H_3$$
$$|$$
$$O^-$$

This is a cephalin

15.41 This phospholipid is a cephalin. It contains glycerol, oleic acid, stearic acid, a phosphate, and ethanolamine.

15.43

15.45 Lipoproteins are large, spherically shaped molecules that transport lipids in the bloodstream. They consist of an outside layer of phospholipids and proteins surrounding an inner core of hundreds of nonpolar lipids and cholesterol esters.

15.47 Chylomicrons have a lower density than VLDLs. They pick up triacylglycerols from the intestine, whereas VLDLs transport triacylglycerols synthesized in the liver.

15.49 "Bad" cholesterol is the cholesterol carried by LDLs that can form deposits in the arteries called plaque, which narrows the arteries.

15.51 Both estradiol and testosterone contain the steroid nucleus and a hydroxyl group. Testosterone has a ketone group, a double bond, and two methyl groups. Estradiol has a benzene ring, a hydroxyl group in place of the ketone, and a methyl group.

15.53 d. Testosterone is a male sex hormone.

15.55 Phospholipids with smaller amounts of glycolipids and cholesterol.

15.57 The lipid bilayer in a cell membrane surrounds the cell and separates the contents of the cell from the external fluids.

15.59 The peripheral proteins in the membrane emerge on the inner or outer surface only, whereas the integral proteins extend through the membrane to both surfaces.

15.61 The carbohydrates attached to proteins and lipids on the surface of cells act as receptors for cell recognition and chemical messengers such as neurotransmitters.

15.63 Beeswax and carnauba are waxes. Vegetable oil and glyceryl tricaprate are triacylglycerols.

glyceryl tricaprate

15.77

a.

15.65 a. A typical fatty acid has a cis double bond.
b. A trans fatty acid has a trans double bond.
c.

15.67

glyceryl tristearate

lecithin

(Note: This lecithin structure is shown with choline and palmitic acid but ethanolamine or serine and other fatty acids are also possible in lecithin.)

15.69 a. wax **b.** steroid
c. glycerophospholipid **d.** triacylglycerol
e. soap **f.** triacylglycerol
g. triacylglycerol **h.** triacylglycerol
i. steroid **j.** steroid
k. fatty acid

15.71 a. 5 **b.** 1, 2, 3, 4 **c.** 2
d. 1, 2 **e.** 1, 2, 3, 4

15.73 a. 4 **b.** 3 **c.** 1
d. 4 **e.** 4 **f.** 3
g. 2 **h.** 1

15.75

$H_2C-O-\overset{\overset{\displaystyle O}{\|}}{C}-(CH_2)_{14}-CH_3$

$H-C-O-\overset{\overset{\displaystyle O}{\|}}{C}-(CH_2)_{14}-CH_3$

$H_2C-O-\overset{\overset{\displaystyle O}{\|}}{C}-(CH_2)_{14}-CH_3$

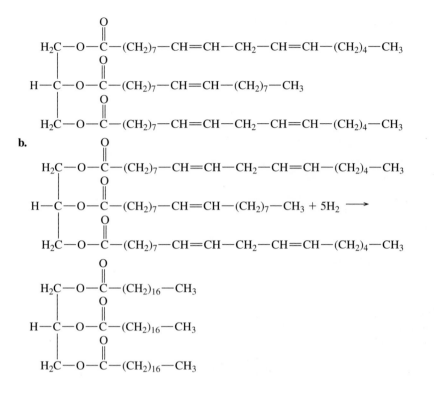

16 Amino Acids, Proteins, and Enzymes

"This lamb is fed with Lamb Lac, which is a chemically formulated replacement for ewe's milk," says part-time farmer Dennis Samuelson. "Its mother had triplets and didn't have enough milk to feed them all, so they weren't thriving the way the other lambs were. The Lamb Lac includes dried skim milk, dried whey, milk proteins, egg albumin, the amino acids methionine and lysine, vitamins, and minerals."

A veterinary technician diagnoses and treats diseases of animals, takes blood and tissue samples, and administers drugs and vaccines. Agricultural technologists assist in the study of farm crops to increase productivity and ensure a safe food supply. They look for ways to improve crop yields, develop safer methods of weed and pest control, and design methods to conserve soil and water.

the Chemistry place

Visit **www.aw-bc.com/chemplace** or use the CD-ROM in the book for extra quizzes, interactive tutorials, career resources, PowerPoint slides, math help, and case studies.

The word "protein" is derived from the Greek word *proteios*, meaning "first." All proteins in humans are polymers made up from 20 different amino acids. Each kind of protein is composed of amino acids arranged in a specific order that determines the characteristics of the protein and its biological action. Proteins provide structure in membranes, build cartilage and connective tissue, transport oxygen in blood and muscle, direct biological reactions as enzymes, defend the body against infection, and control metabolic processes as hormones. They can even be a source of energy.

The different functions of proteins depend on the structures and chemical behavior of amino acids, the building blocks of proteins. We will see how peptide bonds link amino acids and how the order of the amino acids in these protein polymers directs the formation of unique three-dimensional structures.

Every second, thousands of chemical reactions occur in the cells of the human body at rates that meet our physiological and metabolic needs. To make this happen, enzymes catalyze the chemical reactions in our cells, with a different enzyme for every reaction. Digestive enzymes in the mouth, stomach, and small intestine catalyze the hydrolysis of carbohydrates, fats, and proteins. Enzymes in the mitochondria extract energy from biomolecules to give us energy.

WEB TUTORIAL
Functions of Proteins

16.1 Functions of Proteins

The many kinds of proteins perform different functions in the body. There are proteins that form structural components such as cartilage, muscles, hair, and nails. Wool, silk, feathers, and horns are some other proteins made by animals. Proteins called enzymes regulate biological reactions such as digestion and cellular metabolism. Still other proteins, hemoglobin and myoglobin, carry oxygen in the blood and muscle. (See Figure 16.1). Table 16.1 gives examples of proteins that are classified by their functions in biological systems.

LEARNING GOAL

Classify proteins by their functions in the cells.

SAMPLE PROBLEM 16.1

Classifying Proteins by Function

Give the class of protein that would perform each of the following functions.

a. catalyzes metabolic reactions of lipids
b. carries oxygen in the bloodstream
c. stores amino acids in milk

SOLUTION

a. Enzymes catalyze metabolic reactions.
b. Transport proteins carry substances such as oxygen through the bloodstream.
c. A storage protein stores nutrients such as amino acids in milk.

STUDY CHECK

What proteins help regulate metabolism?

FIGURE 16.1

The horns of animals are made of proteins.

Q **What class of protein would be in horns?**

TABLE 16.1

Classification of Some Proteins and Their Functions

Class of Protein	Function in the Body	Examples
Structural	Provide structural components	*Collagen* is in tendons and cartilage. *Keratin* is in hair, skin, wool, and nails.
Contractile	Movement of muscles	*Myosin* and *actin* contract muscle fibers.
Transport	Carry essential substances throughout the body	*Hemoglobin* transports oxygen. *Lipoproteins* transport lipids.
Storage	Store nutrients	*Casein* stores protein in milk. *Ferritin* stores iron in the spleen and liver.
Hormone	Regulate body metabolism and nervous system	*Insulin* regulates blood glucose level. *Growth hormone* regulates body growth.
Enzyme	Catalyze biochemical reactions in the cells	*Sucrase* catalyzes the hydrolysis of sucrose. *Trypsin* catalyzes the hydrolysis of proteins.
Protection	Recognize and destroy foreign substances	*Immunoglobulins* stimulate immune responses.

QUESTIONS AND PROBLEMS

■ **Functions of Proteins**

16.1 Classify each of the following proteins according to its function:
   a. hemoglobin, oxygen carrier in the blood
   b. collagen, a major component of tendons and cartilage
   c. keratin, a protein found in hair
   d. amylase, an enzyme that hydrolyzes starch

16.2 Classify each of the following proteins according to its function:
   a. insulin, a protein needed for glucose utilization
   b. antibodies, proteins that disable foreign proteins
   c. casein, milk protein
   d. lipases that hydrolyze lipids

LEARNING GOAL

Draw the structure for an amino acid.

■ 16.2 Amino Acids

Proteins are composed of molecular building blocks called amino acids. An **amino acid** contains two functional groups, an amino group ($-NH_2$) and a carboxylic acid group ($-COOH$). In all of the 20 amino acids found in proteins, the amino group, the carboxylic group, and a hydrogen atom are bonded to a central carbon atom. Amino acids with this structure are called α (alpha) amino acids. Although there are many amino acids, only 20 different amino acids are present in the proteins in humans. The unique characteristics of the 20 amino acids are due to a side chain (R), which can be an alkyl, hydroxyl, thiol, amino, sulfide, aromatic, or cyclic group.

General Structure of an α-Amino Acid

Although it is convenient to write amino acids with carboxyl (—COOH) group and amino (—NH$_2$) group, they are usually ionized. At the pH of most body fluids, the carboxyl group loses a H$^+$, giving —COO$^-$, and the amino group accepts a H$^+$ to give an ammonium ion, —NH$_3$$^+$. The *dipolar* form of an amino acid, called a **zwitterion**, has a net charge of zero.

Classification of Amino Acids

Nonpolar amino acids, which have alkyl or aromatic side chains, are *hydrophobic* ("water-fearing"). **Polar amino acids**, which have polar side chains such as hydroxyl (—OH), thiol (—SH), and amide (—CONH$_2$) that form hydrogen bonds with water, are *hydrophilic* ("water attracting"). The **acidic amino acids** have side chains that contain a carboxylic acid group (—COOH) and can ionize as a weak acid. The side chains of the **basic amino acids** contain an amino group that can ionize as a weak base. The structures of the side chains (R), common names, three-letter abbreviations, and isoelectric points, pI (see section 16.3) of the 20 amino acids in proteins are listed in Table 16.2.

SAMPLE PROBLEM 16.2

Structural Formulas of Amino Acids

Write the structural formulas and abbreviations for the following amino acids:

a. alanine (R = —CH$_3$)   **b.** serine (R = —CH$_2$OH)

SOLUTION

a. The structure of the amino acids is written by attaching the side group (R) to the central carbon atom of the general structure of an amino acid.

alanine (Ala)

$$\overset{+}{\text{NH}}_3 - \text{CH} - \text{COO}^-$$ with CH$_3$ ← R group attached above CH

b. serine (Ser)

$$\overset{+}{\text{NH}}_3 - \text{CH} - \text{COO}^-$$ with OH—CH$_2$ attached above CH

STUDY CHECK

Classify the amino acids in the sample problem as polar or nonpolar.

Amino Acid Stereoisomers

All of the α-amino acids except for glycine are chiral because the α carbon is attached to four different atoms. Thus amino acids can exist as D and L enantiomers. We can write Fischer projections for amino acids as we did in Chapter 12 for aldehydes with the carboxylic acid group (the most highly oxidized carbon) at the top and the R group at the bottom. For the L isomer,

TABLE 16.2

The 20 Amino Acids in Proteins

Nonpolar Amino Acids

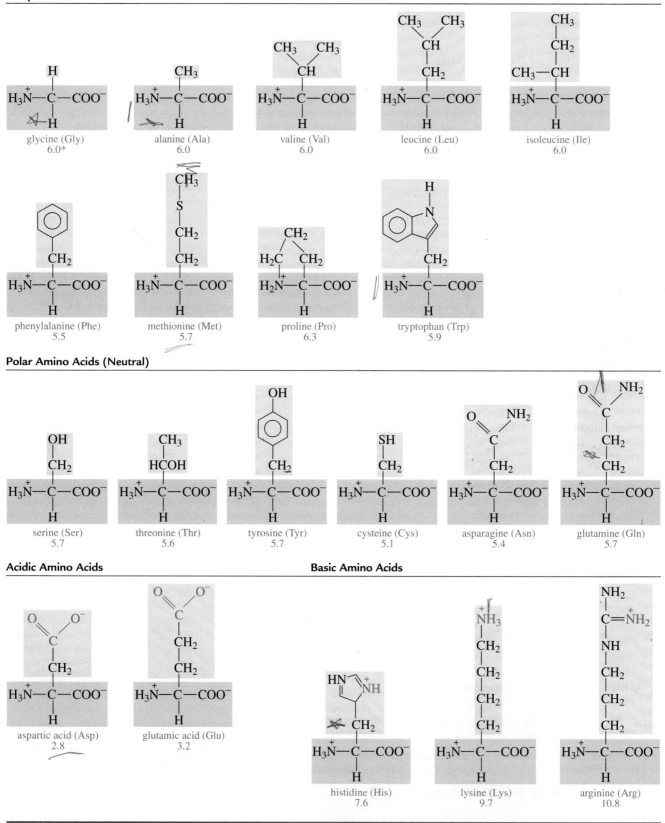

glycine (Gly)
6.0*

alanine (Ala)
6.0

valine (Val)
6.0

leucine (Leu)
6.0

isoleucine (Ile)
6.0

phenylalanine (Phe)
5.5

methionine (Met)
5.7

proline (Pro)
6.3

tryptophan (Trp)
5.9

Polar Amino Acids (Neutral)

serine (Ser)
5.7

threonine (Thr)
5.6

tyrosine (Tyr)
5.7

cysteine (Cys)
5.1

asparagine (Asn)
5.4

glutamine (Gln)
5.7

Acidic Amino Acids　　　　　　　　　　　**Basic Amino Acids**

aspartic acid (Asp)
2.8

glutamic acid (Glu)
3.2

histidine (His)
7.6

lysine (Lys)
9.7

arginine (Arg)
10.8

*Isoelectric points (pI)

the amino group, NH$_2$, is on the left, and in the D isomer, it is on the right. In biological systems, only L amino acids are incorporated into proteins. There are D amino acids found in nature, but not in proteins. Let's take a look at the enantiomers for glyceraldehyde and how the stereoisomers of alanine and cysteine are similar.

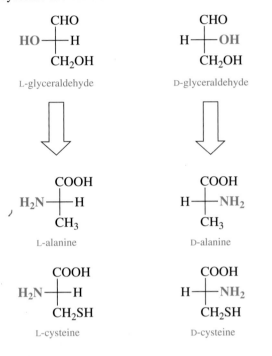

QUESTIONS AND PROBLEMS

■ Amino Acids

16.3 Describe the functional groups found in all α-amino acids.

16.4 How does the polarity of the side chain in leucine compare to the side chain in serine?

16.5 Draw the structural formula for each of the following amino acids:
    a. alanine
    b. threonine
    c. glutamic acid
    d. phenylalanine

16.6 Draw the structural formula for each of the following amino acids:
    a. lysine
    b. aspartic acid
    c. leucine
    d. tyrosine

16.7 Classify the amino acids in problem 16.5 as hydrophobic (nonpolar), hydrophilic (polar, neutral), acidic, or basic.

16.8 Classify the amino acids in problem 16.6 as hydrophobic (nonpolar), hydrophilic (polar, neutral), acidic, or basic.

16.9 Give the name of the amino acid represented by each of the following three-letter abbreviations:
    a. Ala   **b.** Val   **c.** Lys   **d.** Cys

16.10 Give the name of the amino acid represented by each of the following three-letter abbreviations:
    a. Trp   **b.** Met   **c.** Pro   **d.** Gly

16.11 Draw the Fischer projections for the following amino acids.
    a. L-valine
    b. D-cysteine

16.12 Draw the Fischer projections for the following amino acids.
    a. L-threonine
    b. D-valine

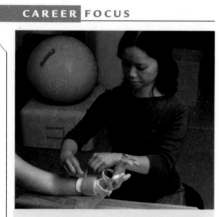
16.3 Amino Acids as Acids and Bases

At a certain pH known as the **isoelectric point (pI)**, the positive and negative charges are equal, which gives an overall charge of zero. However, in a solution that is more acidic than the pI, the $—COO^-$ group acts as a base and accepts an H^+, which gives an overall positive charge to the amino acid. In a solution more basic than the pI, the $—NH_3^+$ group acts as an acid and loses an H^+, which gives the amino acid an overall negative charge. Let's take a look at the changes in all the ionic forms of alanine from its zwitterion (pI = 6.0), to the positive ion in a more acidic solution, and the negative ion in basic solution.

alanine ion at
acidic pH < 6
(charge = 1+)

zwitterion of alanine
pH = 6.0
(charge = 0)

alanine ion at
a pH < 6
(charge = 1−)

The zwitterions for polar and nonpolar amino acids typically exist at pH values of 5.0 to 6.0. However, the zwitterions of the acidic amino acids form at pH values of about 3 because the carboxyl group in their side chain must pick up H^+. The basic amino acids with amino groups in their side chains form zwitterions at high pI values from 7.6 to 10.8. The pI values are included in the list of the amino acids in Table 16.2.

SAMPLE PROBLEM 16.3

Amino Acids in Acid or Base

Serine exists in its zwitterion form at a pH of 5.7. Draw the structural formula for the zwitterion of serine.

SOLUTION

As a zwitterion, both the carboxylic acid group and the amino group are ionized.

$$\overset{+}{NH_3}—\underset{\underset{\underset{OH}{|}}{\overset{|}{CH_2}}}{CH}—\overset{O}{\overset{||}{C}}—O^-$$ zwitterion of serine

STUDY CHECK

Draw the structure of serine at a pH of 2.0.

QUESTIONS AND PROBLEMS

■ Amino Acids as Acids and Bases

16.13 Write the zwitterion of each of the following amino acids:
  a. glycine
  b. cysteine
  c. serine
  d. alanine

16.14 Write the zwitterion of each of the following amino acids:
  a. phenylalanine
  b. methionine
  c. leucine
  d. valine

16.15 Write the positive ion (acidic ion) of each of the amino acids in problem 16.13 at a pH below 1.0.

16.16 Write the negative ion (basic ion) of each of the amino acids in problem 16.13 at a pH above 12.0.

16.17 Would the following ions of valine exist at a pH above, below, or at pI?

  a. H$_2$N—CH—COO$^-$ **b.** H$_3$N$^+$—CH—COOH   **c.** H$_3$N$^+$—CH—COO$^-$
                | | |
                CH   CH   CH
              /   \   /   \   /   \
          CH$_3$ CH$_3$ CH$_3$ CH$_3$ CH$_3$ CH$_3$

16.18 Would the following ions of serine exist at a pH above, below, or at pI?

  a. H$_3$N$^+$—CH—COO$^-$ **b.** H$_3$N$^+$—CH—COOH   **c.** H$_2$N—CH—COO$^-$
                  | | |
              CH$_2$OH   CH$_2$OH   CH$_2$OH

■ 16.4 Formation of Peptides

LEARNING GOAL

Draw the structure of a dipeptide.

The linking of two or more amino acids forms a **peptide**. A **peptide bond** is an amide bond that forms when the —COO$^-$ group of one amino acid reacts with the —NH$_3$$^+$ group of the next amino acid. Two amino acids linked together by a peptide bond form a *dipeptide*. We can write the formation of the dipeptide between glycine and alanine as follows. (See Figure 16.2.) In a peptide, the amino acid written on the left with the unreacted or free amino group (—NH$_3$$^+$) is called the **N terminal** amino acid. The **C terminal** amino acid is the last amino acid in the chain with the unreacted or free carboxyl group (—COO$^-$).

glycine   alanine   glycylalanine (Gly-Ala)

■ Naming Peptides

In naming a peptide, each amino acid beginning from the N terminal is named with a *yl* ending followed by the full name of the amino acid at the C terminal. For example, a tripeptide consisting of alanine, glycine, and serine is named as

Glycine   Alanine   Glycylalanine (Gly-Ala)   Water

FIGURE 16.2

A peptide bond between zwitterions of glycine and alanine form the dipeptide glycylalanine.

Q **What functional groups in glycine and alanine form the peptide bond?**

ala**nyl**gly**cyl**serine. For convenience, the order of amino acids in the peptide is often written as the sequence of three-letter abbreviations.

From alanine   From glycine   From serine
alanyl   glycyl   serine

alanylglycylserine
(Ala-Gly-Ser)

SAMPLE PROBLEM 16.4

■ Identifying a Tripeptide

Consider the following tripeptide.

a. What amino acid is the N terminal? What amino acid is the C terminal?
b. What is the three-letter abbreviation name for the order of amino acids in the tripeptide?

SOLUTION

a. Threonine is the N terminal; phenylalanine is the C terminal.
b. Thr-Leu-Phe

STUDY CHECK

What is the full name of the tripeptide in Sample Problem 16.4?

QUESTIONS AND PROBLEMS

■ Formation of Peptides

16.19 Draw the structural formula of each of the following peptides and give the abbreviation for their names:
a. alanylcysteine
b. serylphenylalanine
c. glycylalanylvaline
d. valylisoleucyltryptophan

16.20 Draw the structural formula of each of the following peptides and give the abbreviation for their names:
a. methionylaspartic acid
b. alanyltryptophan
c. methionylglutaminyllysine
d. histidylglycylglutamylalanine

■ 16.5 Levels of Protein Structure

When there are more than 50 amino acids in a chain, the polypeptide is usually called a **protein**. Each protein in our cells has a unique sequence of amino acids that determines its biological function.

■ Primary Structure

The **primary structure** is the order of the amino acids held together by peptide bonds.

The first protein to have its primary structure determined was insulin, which is a hormone that regulates the glucose level in the blood. In the primary structure of human insulin, there are two polypeptide chains. In chain A, there are 21 amino acids, and chain B has 30 amino acids. The polypeptide chains are held together by disulfide bonds formed by the side chains of the cysteine amino acids in each of the

WEB TUTORIAL
Structure of Proteins
Primary and Secondary Structure

HEALTH NOTE

■ Natural Opiates in the Body

Painkillers known as enkephalins and endorphins are produced naturally in the body. They are polypeptides that bind to receptors in the brain to give relief from pain. This effect appears to be responsible for the runner's high, for the temporary loss of pain when severe injury occurs, and for the analgesic effects of acupuncture.

The *enkephalins*, found in the thalamus and the spinal cord, are pentapeptides, the smallest molecules with opiate activity.

The amino acid sequence of an enkephalin is found in the longer amino acid sequence of the endorphins.

Four groups of *endorphins* have been identified: α-endorphin contains 16 amino acids, β-endorphin contains 31 amino acids, γ-endorphin has 17 amino acids, and δ-endorphin has 27 amino acids. Endorphins may produce their sedating effects by preventing the release of substance P, a polypeptide with 11 amino acids, which has been found to transmit pain impulses to the brain.

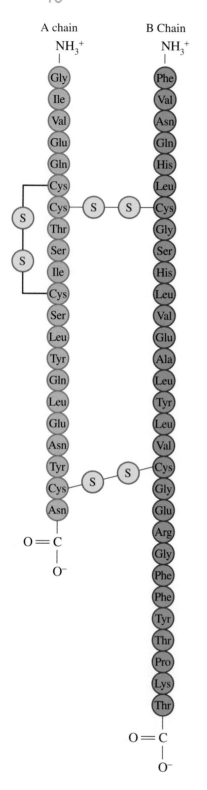

A chain   B Chain

chains. (See Figure 16.3.) Today human insulin produced through genetic engineering is used in the treatment of diabetes.

Secondary Structure

The **secondary structure** of a protein describes the way the amino acids next to or near to each other along the polypeptide are arranged in space. The three most common types of secondary structure are the *alpha helix*, the *beta-pleated sheet*, and the *triple helix* found in collagen.

The corkscrew shape of an **alpha helix (α helix)** is held in place by hydrogen bonds between each N—H group and the oxygen of a C=O group in the next turn of the helix, four amino acids down the chain. (See Figure 16.4.) Because

Carbon

Oxygen

Nitrogen

R group

Hydrogen

Peptide backbone of primary structure

Hydrogen bonds of secondary structure

FIGURE 16.3

The sequence of amino acids in human insulin is its primary structure.

Q What kinds of bonds occur in the primary structure of a protein?

FIGURE 16.4

The α (alpha) helix acquires a coiled shape from hydrogen bonds between the N—H of the peptide bond in one loop and the C=O of the peptide bond in the next loop.

Q What are the partial charges of the H in N—H and the O in C=O that permit hydrogen bonds to form?

Hydrogen bonds
between peptide
backbones

Carbon

Oxygen

Nitrogen

R group

Hydrogen

FIGURE 16.5

In a β (beta)-pleated sheet secondary structure, hydrogen bonds form between the peptide chains.

Q How do the hydrogen bonds differ in a β-pleated sheet from the alpha helix?

many hydrogen bonds form along the peptide backbone, this portion of the protein takes the shape of a strong, tight coil that looks like a telephone cord or a Slinky toy. All the side chains (R groups) of the amino acids are located on the outside of the helix.

Another type of secondary structure is known as the **beta-pleated sheet** (**β-pleated sheet**). In a β-pleated sheet, polypeptide chains are held together side by side by hydrogen bonds between the peptide chains. In a β-pleated sheet of silk fibroin, the small R groups of the prevalent amino acids, glycine, alanine, and serine, extend above and below the sheet. This results in a series of β-pleated sheets that are stacked close together. The hydrogen bonds holding the β-pleated sheets tightly in place account for the strength and durability of proteins such as silk. (See Figure 16.5.)

Collagen, the most abundant protein, makes up as much as one-third of all the protein in vertebrates. It is found in connective tissue, blood vessels, skin, tendons, ligaments, the cornea of the eye, and cartilage. The strong structure of collagen is a result of three polypeptides woven together like a braid to form a **triple helix**, as seen in Figure 16.6. Collagen has a high content of glycine (33%), proline (22%), alanine (12%), and smaller amounts of hydroxyproline, and hydroxylysine. The hydroxy forms of proline and lysine contain —OH groups that form hydrogen bonds across the peptide chains and give strength to the collagen triple helix.

hydroxyproline   hydroxylysine

When a diet is deficient in vitamin C, collagen fibrils are weakened because the enzymes needed to form hydroxyproline and hydroxylysine require vitamin C. Without the —OH groups of hydroxyproline and hydroxylysine, there is less hydrogen bonding between collagen fibrils. As a person ages, additional cross-links form between the fibrils, which make collagen less elastic. Bones, cartilage, and tendons become more brittle, and wrinkles are seen as the skin loses elasticity.

Tertiary Structure

The **tertiary structure** of a protein involves attractions and repulsions between the side chain groups of the amino acids in the polypeptide chain. As interactions occur between different parts of the peptide chain, segments of the chain twist and bend until the protein acquires a specific three-dimensional shape. The tertiary structure of a protein is stabilized by interactions between the R groups of the amino acids in one region of the polypeptide chain with R groups of amino acids in other regions of the protein. (See Figure 16.7.) Table 16.3 lists the stabilizing interactions of tertiary structures.

1. **Hydrophobic interactions** are interactions between two nonpolar R groups. Within the protein, the amino acids with nonpolar side chains push as far away from the aqueous environment as possible, which forms a hydrophobic center at the interior of the protein molecule.
2. **Hydrophilic interactions** are attractions between the external aqueous environment and amino acids that have polar or ionized side chains. The polar side chains pull toward the outer surface of globular proteins to hydrogen bond with water.
3. **Salt bridges** are ionic bonds between side chains of basic and acidic amino acids, which have positive and negative charges. The attraction of the oppositely charged side chains forms a strong bond called a salt bridge. If the pH changes, the basic and acidic side chains lose their ionic charges and cannot form salt bridges, which causes a change in the shape of the protein.

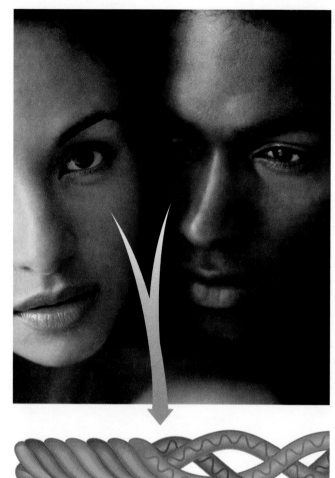

Triple helix   3 α-Helix peptide chains

FIGURE 16.6

Hydrogen bonds between polar R groups in three polypeptide chains form the triple helixes that combine to make fibers of collagen.

Q **What are some of the amino acids in collagen that form hydrogen bonds between the polypeptide chains?**

WEB TUTORIAL
Tertiary and Quaternary Structure

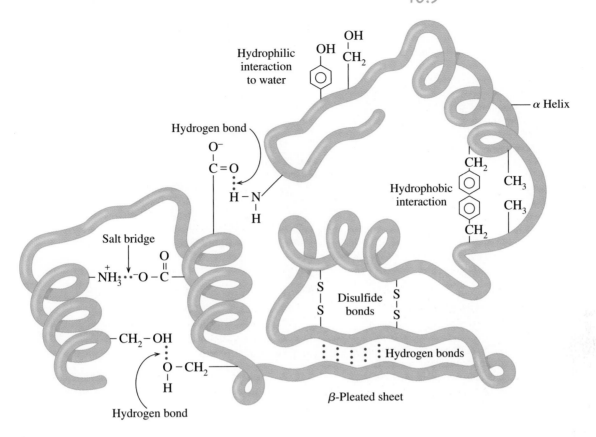

FIGURE 16.7

Interactions between amino acid R groups fold a protein into a specific three-dimensional shape called its tertiary structure.

Q **Why would one section of the protein chain move to the center while another section remains on the surface of the tertiary structure?**

4. **Hydrogen bonds** form between polar amino acids. For example, a hydrogen bond can occur between the —OH of serine and the —NH_2 of glutamine.
5. **Disulfide bonds** (—S—S—) are covalent bonds that form between the —SH groups of cysteines in the polypeptide chain.

TABLE 16.3

Some Cross-Links in Tertiary Structures

	Nature of Bonding
Hydrophobic interactions	Attractions between nonpolar groups
Hydrophilic interactions	Attractions between polar or ionized groups and water on the surface of the tertiary structure
Salt bridges	Ionic interactions between ionized acidic and basic amino acids
Hydrogen bonds	Occur between H and O or N
Disulfide bonds	Strong covalent links between sulfur atoms of two cysteine amino acids

---

SAMPLE PROBLEM 16.5

■ **Cross-Links in Tertiary Structures**

What type of interaction would you expect between the R groups of the following amino acids?

a. cysteine and cysteine   **b.** glutamic acid and lysine

SOLUTION

a. Because cysteine has an R group containing —SH, a disulfide bond could form.
b. An ionic bond (salt bridge) can form by the interaction of the —COO^- of glutamic acid and the —NH_3^+ of lysine.

STUDY CHECK

Would you expect to find valine and leucine on the outside or the inside of the tertiary structure? Why?

Essential Amino Acids

Of the 20 amino acids used to build the proteins in the body, only 10 can be synthesized in the body. The other 10 amino acids, listed in Table 16.4, are *essential amino acids* that cannot be synthesized and must be obtained from the proteins in the diet.

TABLE 16.4

Essential Amino Acids

arginine (Arg)*	methionine (Met)
histidine (His)*	phenylalanine (Phe)
isoleucine (Ile)	threonine (Thr)
leucine (Leu)	tryptophan (Trp)
lysine (Lys)	valine (Val)

*Required in diets of children, not adults.

Complete proteins, which contain all of the essential amino acids, are found in most animal products such as eggs, milk, meat, fish, and poultry. However, gelatin and plant proteins such as grains, beans, and nuts are *incomplete proteins* because they are deficient in one or more of the essential amino acids. Diets that rely on plant foods for protein must contain a variety of protein sources to obtain all the essential amino acids. For example, a diet of rice and beans contains all the essential amino acids because they are complementary protein sources. Rice contains the methionine and tryptophan deficient in beans, while beans contain the lysine that is lacking in rice. (See Table 16.5.)

TABLE 16.5

Amino Acid Deficiency in Selected Vegetables and Grains

Food Source	Amino Acids Missing
Eggs, milk, meat, fish, poultry	none
Wheat, rice, oats	lysine
Corn	lysine, tryptophan
Beans	methionine, tryptophan
Peas	methionine
Almonds, walnuts	lysine, tryptophan
Soy	low in methionine

Globular and Fibrous Proteins

A group of proteins known as **globular proteins** have compact, spherical shapes because sections of the polypeptide chain fold over on top of each other. It is the globular proteins that carry out the work of the cells: functions such as synthesis, transport, and metabolism.

Myoglobin is a globular protein that stores oxygen in skeletal muscle. High concentrations of myoglobin are found in the muscles of sea mammals, such as seals and whales, which allow them to stay under the water for long periods. Myoglobin contains 153 amino acids in a single polypeptide chain with about three-fourths of the chain in the α-helix secondary structure. The polypeptide chain, including its helical regions, forms a compact tertiary structure by folding upon itself. (See Figure 16.8.) Within the tertiary structure, a pocket of amino acids and a heme group binds and stores oxygen (O_2).

Pocket for oxygen (O_2)

Amino acids in protein chain

FIGURE 16.8

Myoglobin is a globular protein with a heme pocket in its tertiary structure that binds oxygen to be carried to the tissues.

Q **Would hydrophilic amino acids be found on the outside or inside of the myoglobin structure?**

The **fibrous proteins** are proteins that consist of long, thin, fiber-like shapes. They are typically involved in the structure of cells and tissues. Two types of fibrous protein are the α- and β-keratins. The α-keratins are the proteins that make up hair, wool, skin, and nails. In hair, three α-helixes coil together like a braid to form a fibril. Within the fibril, the α-helices are held together by disulfide (—S—S—) linkages between the R groups of the many cysteine amino acids in hair. Several fibrils bind together to form a strand of hair. (Figure 16.9.) The β-keratins are the type of proteins found in the feathers of birds and scales of reptiles. In β-keratins, the proteins consist of large amounts of β-pleated sheet structure.

Quaternary Structure: Hemoglobin

When a biologically active protein consists of two or more polypeptide subunits, the structural level is referred to as a **quaternary structure**. Hemoglobin, a globular protein that transports oxygen in blood, consists of four polypeptide chains or subunits, two α chains, and two β chains. (See Figure 16.10.) The subunits are held together in the quaternary structure by the same interactions that stabilize the tertiary structure, such as hydrogen bonds and salt bridges between side groups, disulfide links, and hydrophobic attractions. Each subunit of the hemoglobin contains a heme group that binds oxygen. In the adult hemoglobin molecule, all four subunits must be combined for the hemoglobin to properly function as an oxygen carrier. Therefore, the complete quaternary structure of hemoglobin can bind and transport four molecules of oxygen. Table 16.6 and Figure 16.11 summarize the structural levels of proteins.

α helix

Alpha keratin

FIGURE 16.9

The fibrous proteins of α-keratin wrap together to form fibrils that make up hair and wool.

Q **Why does hair have a large amount of cysteine amino acids?**

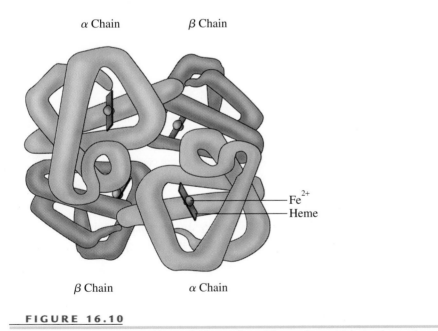

α Chain β Chain

Fe^{2+}
Heme

β Chain α Chain

FIGURE 16.10

The quaternary structure of hemoglobin consists of four polypeptide subunits, each containing a heme group that binds an oxygen molecule.

Q **What is the difference between a tertiary structure and a quaternary structure?**

TABLE 16.6

Summary of Structural Levels in Proteins	
Structural Level	**Characteristics**
Primary	The sequence of amino acids
Secondary	The coiled α-helix, β-pleated sheet, or a triple helix formed by hydrogen bonding between peptide bonds along the chain
Tertiary	A folding of the protein into a compact, three-dimensional shape stabilized by interactions between side R groups of amino acids
Quaternary	A combination of two or more protein subunits to form a larger, biologically active protein

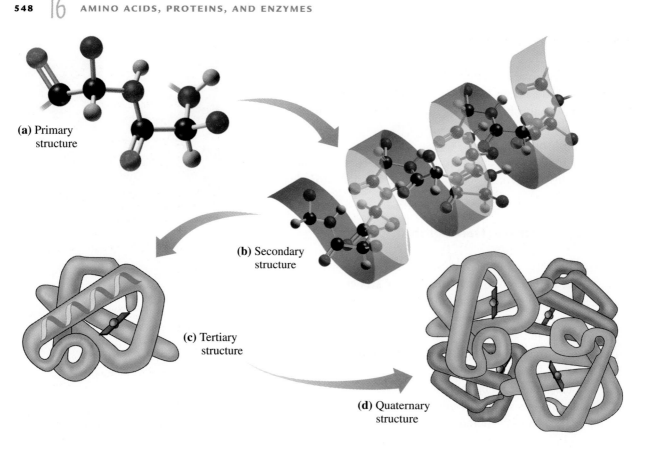

(a) Primary
structure

(b) Secondary
structure

(c) Tertiary
structure

(d) Quaternary
structure

FIGURE 16.11

Proteins consist of primary, secondary, tertiary, and sometimes quaternary structural levels.

Q What is the difference between a primary structure and a tertiary structure?

SAMPLE PROBLEM 16.6

■ **Identifying Protein Structure**

Indicate whether the following conditions are responsible for primary, secondary, tertiary, or quaternary protein structures:

a. Disulfide bonds form between portions of a protein chain.
b. Peptide bonds form a chain of amino acids.

SOLUTION

a. Disulfide bonds help to stabilize the tertiary structure of a protein.
b. The sequence of amino acids in a polypeptide is a primary structure.

STUDY CHECK

What structural level is represented by the grouping of two subunits in insulin?

Denaturation of Proteins

Denaturation of a protein occurs when there is a disruption of any of the bonds that stabilize the secondary, tertiary, or quaternary structure. However, the covalent amide bonds of the primary structure are not affected.

Heat, acid, base, heavy metal salts, agitation

Active protein   Denatured protein

When the interactions between the R groups are undone or altered, a globular protein unfolds like a loose piece of spaghetti. With the loss of its overall shape, the protein is no longer biologically active.

Denaturing agents include heat, acids and bases, organic compounds, heavy metal ions, and mechanical agitation. (See Table 16.7.)

▪ Denaturation of Milk Protein

Place some milk in five glasses. Add the following to the milk samples in glasses 1–4. The fifth glass of milk is a reference sample.
1. Vinegar, drop by drop. Stir.
2. One-half teaspoon of meat tenderizer. Stir.
2. One teaspoon of fresh pineapple juice. (Canned juice has been heated.)
4. One teaspoon of fresh pineapple juice that you have heated to boiling.

QUESTIONS

1. How did the appearance of the milk change in each of the samples?
2. What enzyme is listed on the package label of the tenderizer?
3. How does the effect of the heated pineapple juice compare with that of the fresh juice? Explain.
4. Why is cooked pineapple used when making gelatin (a protein) desserts?

TABLE 16.7

Examples of Protein Denaturation

Denaturing Agent	Bonds Disrupted	Examples
Heat above 50°C	Hydrogen bonds; hydrophobic attractions between nonpolar groups	Cooking food and autoclaving surgical items
Acids and bases	Hydrogen bonds between polar R groups; salt bridges	Lactic acid from bacteria, which denatures milk protein in the preparation of yogurt and cheese
Organic compounds	Hydrophobic interactions	Ethanol and isopropyl alcohol, which disinfect wounds and prepare the skin for injections
Heavy metal ions Ag^+, Pb^{2+}, and Hg^{2+}	Disulfide bonds in proteins, by forming ionic bonds	Mercury and lead poisoning
Agitation	Hydrogen bonds and hydrophobic interactions, by stretching polypeptide chains and disrupting stabilizing interactions	Whipped cream and meringue made from egg whites

Sickle-Cell Anemia

Sickle-cell anemia is a disease caused by an abnormality in the shape of one of the subunits of the hemoglobin protein. In the β chain, the sixth amino acid, glutamic acid, which is polar, is replaced by valine, a nonpolar amino acid.

Because valine has a hydrophobic side chain, it draws the hydrophobic pocket that binds to oxygen to the surface of the hemoglobin. The affected red blood cells change from a rounded shape to a crescent shape, like a sickle, which interferes with their ability to transport adequate quantities of oxygen. Hydrophobic attractions cause several sickle-cell hemoglobin molecules to stick together, which forms long fibers of sickle-cell hemoglobin. The clumps of insoluble fibers clog capillaries, where they cause inflammation, pain, and organ damage. Critically low oxygen levels may occur in the affected tissues.

In sickle-cell anemia, both genes for the altered hemoglobin must be inherited. However, a few sickled cells are found in persons who carry one gene for sickle-cell hemoglobin, a condition that is also known to provide protection from malaria.

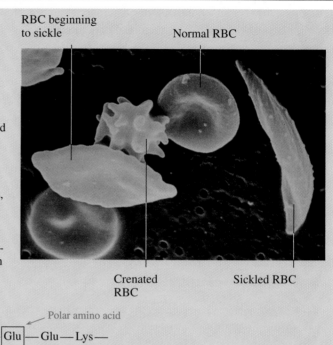

RBC beginning to sickle

Normal RBC

Crenated RBC

Sickled RBC

Polar amino acid

Normal β chain: Val — His — Leu — Thr — Pro — | Glu | — Glu — Lys —

Sickled β chain: Val — His — Leu — Thr — Pro — | Val | — Glu — Lys —

Nonpolar amino acid

QUESTIONS AND PROBLEMS

Levels of Protein Structure

16.21 What type of bonding occurs in the primary structure of a protein?

16.22 How can two proteins with exactly the same number and type of amino acids have different primary structures?

16.23 Two peptides each contain one molecule of valine and two molecules of serine. What are their possible primary structures?

16.24 What are three different types of secondary protein structure?

16.25 What happens to the primary structure of a protein when a protein forms a secondary structure?

16.26 In an α helix, how does bonding occur between the amino acids in the polypeptide chain?

16.27 What is the difference in bonding between an α helix and a β-pleated sheet?

16.28 How is the secondary structure of a β-pleated sheet different from that of a triple helix?

16.29 What type of interaction would you expect from the following R groups in a tertiary structure?
a. two cysteine residues
b. glutamic acid and lysine
c. serine and aspartic acid
d. two leucine residues

16.30 In myoglobin, about one-half of the 153 amino acids have nonpolar side chains.
a. Where would you expect those amino acids to be located in the tertiary structure?
b. Where would you expect the polar side chains to be?
c. Why is myoglobin more soluble in water than silk or wool?

16.31 A portion of a polypeptide chain contains the following sequence of amino acid residues:

-Leu-Val-Cys-Asp-

a. Which R groups can form a disulfide cross-link?
b. Which amino acid residues are likely to be found on the inside of the protein structure? Why?
c. Which amino acid residues would be found on the outside of the protein? Why?
d. How does the primary structure of a protein affect its tertiary structure?

16.32 State whether the following statements apply to primary, secondary, tertiary, or quaternary protein structure:
a. Side groups interact to form disulfide bonds or ionic bonds.
b. Peptide bonds join amino acids in a polypeptide chain.
c. Several polypeptides are held together by hydrogen bonds between adjacent chains.
d. Hydrogen bonding between carbonyl oxygen atoms and nitrogen atoms of amide groups causes a polypeptide to coil.
e. Hydrophobic side chains seeking a nonpolar environment move toward the inside of the folded protein.
f. Protein chains of collagen form a triple helix.
g. An active protein contains four tertiary subunits.

16.33 Indicate the changes in protein structure for each of the following:
a. An egg placed in water at 100°C is soft boiled in about 3 min.
b. Prior to giving an injection, the skin is wiped with an alcohol swab.
c. Surgical instruments are placed in a 120°C autoclave.
d. During surgery, a wound is closed by cauterization (heat).

16.34 Indicate the changes in protein structure for each of the following:
a. Tannic acid is placed on a burn.
b. Milk is heated to 60°C to make yogurt.
c. To avoid spoilage, seeds are treated with a solution of $HgCl_2$.
d. Hamburger is cooked at high temperatures to destroy *E. coli* bacteria that may cause intestinal illness.

16.6 Enzymes

As a **catalyst**, an enzyme increases the rate of a reaction by changing the way a reaction takes place, but is itself not changed at the end of the reaction. An uncatalyzed reaction in a cell may take place eventually, but not at a rate fast enough for survival. For example, the hydrolysis of proteins in our diet would eventually occur without a catalyst, but not fast enough to meet the body's requirements for amino acids. The chemical reactions in our cells must occur at incredibly fast rates under the mild conditions of pH 7.4 and a body temperature of 37°C. To do this, biological catalysts known as **enzymes** catalyze nearly all the chemical reactions that take place in the body.

As catalysts, enzymes lower the activation energy for the reaction. (See Figure 16.12.) As a result, less energy is required to convert reactant molecules to products, which allows more reacting molecules to form product.

Names of Enzymes

The names of enzymes describe the compound or the reaction that is catalyzed. The actual names of enzymes are derived by replacing the end of the name of the reaction or reacting compound with the suffix *ase*. For example, an *oxidase* catalyzes an oxidation reaction, and a *dehydrogenase* removes hydrogen atoms. The compound sucrose is hydrolyzed by the enzyme *sucrase*, and a lipid is hydrolyzed by

FIGURE 16.12

The activation energy needed for the reaction of CO_2 and H_2O is lowered by the enzyme carbonic anhydrase.

Q Why are enzymes used in biological reactions?

TABLE 16.8

Classes of Enzymes

Class	Reaction Catalyzed	Examples
Oxidoreductases	Oxidation–reduction reactions	*Oxidases* oxidize a substance. *Reductases* reduce a substance. *Dehydrogenases* remove 2 H atoms to form a double bond.
Transferases	Transfer a group between two compounds	*Transaminases* transfer amino groups. *Kinases* transfer phosphate groups.
Hydrolases	Hydrolysis reactions	*Proteases* hydrolyze peptide bonds. *Lipases* hydrolyze ester bonds in lipids. *Carbohydrases* hydrolyze glycosidic bonds in carbohydrates. *Phosphatases* hydrolyze phosphate–ester bonds. *Nucleases* hydrolyze nucleic acids.
Lyases	Add or remove groups involving a double bond without hydrolysis	*Carboxylases* add CO_2. *Deaminases* NH_3.
Isomerases	Rearrange atoms in a molecule to form an isomer	*Isomerases* convert cis to trans, or reverse. *Epimerases* convert D to L isomers, or reverse.
Ligases	Form bonds between molecules using ATP energy	*Synthetases* combine two molecules.

a *lipase*. Some early known enzymes use names that end in the suffix *in*, such as *papain* found in papaya, *rennin* found in milk, and *pepsin* and *trypsin*, enzymes that catalyze the hydrolysis of proteins.

More recently, a systematic method of classifying and naming enzymes has been established. The name and class of each indicates the type of reaction it catalyzes. There are six main classes of enzymes, as described in Table 16.8.

SAMPLE PROBLEM 16.7

■ **Naming Enzymes**

What chemical reaction do the following enzymes catalyze?

a. amino transferase
b. lactate dehydrogenase

SOLUTION

a. catalyzes the transfer of an amino group
b. catalyzes the removal of hydrogen from lactate

STUDY CHECK

What is the name of the enzyme that catalyzes the hydrolysis of lipids?

QUESTIONS AND PROBLEMS

■ **Enzymes**

16.35 Why do cellular reactions require enzymes?

16.36 How do enzymes make cellular reactions proceed at faster rates?

16.37 What is the reactant for each of the following enzymes?
   a. galactase   **b.** lipase   **c.** aspartase

16.38 What is the reactant for each of the following enzymes?
  a. peptidase
  b. cellulase
  c. lactase

16.39 What class of enzyme catalyzes each of the following reactions?
  a. hydrolysis of sucrose
  b. addition of oxygen
  c. converting glucose ($C_6H_{12}O_6$) to fructose ($C_6H_{12}O_6$)
  d. moving an amino group from one molecule to another

16.40 What class of enzyme catalyzes each of the following reactions?
  a. addition of water to a double bond
  b. removing hydrogen atoms
  c. splitting peptide bonds in proteins
  d. removing CO_2 from pyruvate

16.7 Enzyme Action

Nearly all enzymes are globular proteins. Each has a unique three-dimensional shape that recognizes and binds a small group of reacting molecules, which are called **substrates**. The tertiary structure of an enzyme plays an important role in how that enzyme catalyzes reactions.

Active Site

In a catalyzed reaction, an enzyme must first bind to a substrate in a way that favors catalysis. A typical enzyme is much larger than its substrate. However, within its large tertiary structure, there is a region called the **active site** where the enzyme binds a substrate or substrates and catalyzes the reaction. This active site is often a small pocket that closely fits the structure of the substrate. (See Figure 16.13.) Within the active site, the side chains of amino acids bind the substrate with hydrogen bonds, salt bridges, or hydrophobic attractions. The active site of a particular enzyme fits the shape of only a few types of substrates, which makes enzymes very specific about the type of substrate they bind.

Enzyme Catalyzed Reaction

The proper alignment of a substrate within the active site forms an **enzyme–substrate (ES) complex**. This combination of enzyme and substrate provides an alternative pathway for the reaction that has a lower activation energy. Within the active site, the amino acid side chains take part in catalyzing the chemical reaction. As soon as the catalyzed reaction is complete, the products are quickly released from the enzyme so it can bind to a new substrate molecule. We can write the catalyzed reaction of an enzyme (E) with a substrate (S) to form product (P) as follows:

Step 1	E + S	$\rightleftharpoons$	ES		
Step 2			ES	$\longrightarrow$	E + P
	E + S	$\rightleftharpoons$	ES $\longrightarrow$	E + P	
	Enzyme + substrate		ES complex	Enzyme + product	

Let's consider the hydrolysis of sucrose by sucrase. When sucrose binds to the active site of sucrase, the glycosidic bond of sucrose is placed into a geometry

LEARNING GOAL

Describe the role of an enzyme in an enzyme-catalyzed reaction.

FIGURE 16.13

On the surface of an enzyme, a small region called an active site binds a substrate and catalyzes a reaction of that substrate.

Q Why does an enzyme catalyze a reaction of only certain substrates?

WEB TUTORIAL
How Enzymes Work

favorable for reaction. The amino acid side chains catalyze the cleavage of the sucrose to give the products glucose and fructose.

Sucrase + sucrose $\rightleftharpoons$ sucrase–sucrose complex $\longrightarrow$ sucrase + glucose + fructose

$$E \quad + \quad S \qquad\qquad \text{ES complex} \qquad\qquad E \quad + \quad P_1 \quad + \quad P_2$$

Because the structures of the products are no longer attracted to the active site, they are released and the sucrase binds another sucrose substrate. (See Figure 16.14).

Lock-and-Key and Induced Fit Models

In an early theory of enzyme action called the **lock-and-key model**, the active site is described as having a rigid, nonflexible shape. Thus only those substrates with shapes that fit exactly into the active site are able to bind with that enzyme. The shape of the active site is analogous to a lock, and the proper substrate is the key that fits into the lock. (See Figure 16.15a.)

While the lock-and-key model explains the binding of substrates for many enzymes, certain enzymes have a broader range of activity than the lock and key model allows. In the **induced-fit model**, there is an interaction between both the enzyme and substrate. (See Figure 16.15b.) The active site adjusts to fit the shape of the substrate more closely. At the same time the substrate adjusts its shape to better adapt to the geometry of the active site. As a result, the reacting section of the substrate becomes aligned exactly with the groups in the active site that catalyze

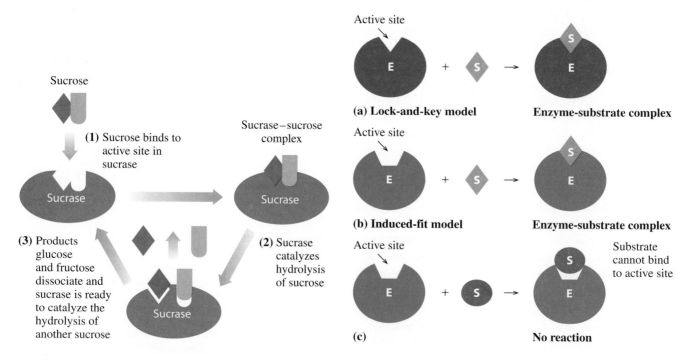

FIGURE 16.14

After sucrose binds to sucrase at the active site, it is properly aligned for the hydrolysis reaction by the active site. The monosaccharide products dissociate from the active site and the enzyme is ready to bind to another sucrose molecule.

Q **Why does the enzyme-catalyzed hydrolysis of sucrose go faster than the hydrolysis of sucrose in the chemistry laboratory?**

FIGURE 16.15

(a) In the lock-and-key model, a substrate fits the shape of the active site and forms an enzyme–substrate complex. (b) In the induced-fit model, a flexible active site and substrate adapt to provide a close fit to a substrate and proper orientation for reaction. (c) A substrate that does not fit or induce a fit in the active site cannot undergo catalysis by the enzyme.

Q **How does the induced-fit model differ from the lock-and-key model?**

the reaction. A different substrate could not induce these structural changes and no catalysis would occur. (See Figure 16.15c.)

SAMPLE PROBLEM 16.8

■ Enzyme Action

What is the function of the active site in an enzyme?

SOLUTION

The active site in an enzyme binds the substrate and contains the amino acid side chains that catalyze the reaction.

STUDY CHECK

How do the lock-and-key and the induced-fit models differ in their description of the active site in an enzyme?

HEALTH NOTE

■ Isoenzymes as Diagnostic Tools

Isoenzymes are different forms of an enzyme that catalyze the same reaction in different cells or tissues of the body. They consist of quaternary structures with slight variations in the amino acids in the polypeptide subunits. For example, there are five isoenzymes of *lactate dehydrogenase* (*LDH*) that catalyze the conversion between lactate and pyruvate.

$$CH_3-\underset{\underset{OH}{|}}{CH}-COO^- \underset{\text{dehydrogenase}}{\overset{\text{lactate}}{\rightleftharpoons}} CH_3-\underset{\underset{O}{||}}{C}-COO^- + 2H$$

lactate   pyruvate

Isoenzymes of lactate dehydrogenase	Highest levels found in the following:
H₄ (LDH₁)	Heart, kidneys
H₃M (LDH₂)	Red blood cells, heart, kidney, brain
H₂M₂ (LDH₃)	Brain, lung, white blood cells
HM₃ (LDH₄)	Lung, skeletal muscle
M₄ (LDH₅)	Skeletal muscle, liver

Isoenzymes as Diagnostic Tools   (Continued)

Each LDH isoenzyme contains a mix of polypeptide subunits, M and H. In the liver and muscle, lactate is converted to pyruvate by a LDH_5 isoenzyme with M subunits designated M_4. In the heart, the same reaction is catalyzed by a LDH_1 isoenzyme (H_4) containing four H subunits. Different combinations of the M and H subunits are found in the LDH isoenzymes of the brain, red blood cells, kidney, and white blood cells.

The different forms of an enzyme allow a medical diagnosis of damage or disease to a particular organ or tissue. In healthy tissues, isoenzymes function within the cells. However, when a disease damages a particular organ, cells die, which releases cell contents including the isoenzymes into the blood. Measurements of the elevated levels of specific isoenzymes in the blood serum

help to identify the disease and its location in the body. For example, an elevation in the serum LDH_5, which is the M_4 isoenzyme of lactate dehydrogenase, indicates liver damage or disease. When a myocardial infarction (MI), or heart attack, damages the cells in heart muscle, an increase in the level of LDH_1 (H_4) isoenzyme is detected in the blood serum. (See Table 16.9.)

Another isoenzyme used diagnostically is creatine kinase (CK), which consists of two types of polypeptide subunits. One subunit (B) is prevalent in the brain and the other predominates in skeletal muscle (M). Normally only the CK_3 is present in low amounts in the blood serum. However, in a patient who has suffered an MI, the levels of CK_2 will be elevated soon after the heart attack. Table 16.10 lists some enzymes used to diagnose tissue damage and diseases of certain organs.

TABLE 16.9

Isoenzymes of Lactate Dehydrogenase and Creatinine Kinase

Isoenzyme	Abundant in	Subunits
Lactate Dehydrogenase (LDH)		
LDH_1	Heart, kidneys	H_4
LDH_2	Red blood cells, heart, kidney, brain	MH_3
LDH_3	Brain, lung, white blood cells	M_2H_2
LDH_4	Lung, skeletal muscle	M_3H
LDH_5	Skeletal muscle, liver	M_4
Creatinine Kinanse (CK)		
CK_1	Brain, lung	BB
CK_2	Heart	MB
CK_3	Skeletal muscle, red blood cells	MM

TABLE 16.10

Serum Enzymes Used in Diagnosis of Tissue Damage

Condition	Diagnostic Enzymes Elevated
Heart attack, or liver disease (cirrhosis, hepatitis)	Lactate dehydrogenase (LDH) Aspartate transaminase (AST)
Heart attack	Creatine kinase (CK)
Hepatitis	Alanine transaminase (ALT)
Liver (carcinoma) or bone disease (rickets)	Alkaline phosphatase (ALP)
Pancreatic disease	Amylase, cholinesterase, lipase (LPS)
Prostate carcinoma	Acid phosphatase (ACP)

QUESTIONS AND PROBLEMS

Enzyme Action

16.41 Match the following three terms, (1) enzyme–substrate complex, (2) enzyme, and (3) substrate, with these phrases:
a. has a tertiary structure that recognizes the substrate
b. the combination of an enzyme with the substrate
c. has a structure that fits the active site of an enzyme

16.42 Match the following three terms, (1) active site, (2) lock-and-key model, and (3) induced-fit model, with these phrases:
a. the portion of an enzyme where catalytic activity occurs
b. an active site that adapts to the shape of a substrate
c. an active site that has a rigid shape

16.43 **a.** Write an equation that represents an enzyme-catalyzed reaction.
b. How is the active site different from the whole enzyme structure?

16.44 **a.** How does an enzyme speed up the reaction of a substrate?
b. After the products have formed, what happens to the enzyme?

16.45 What are isoenzymes?

16.46 How is the LDH isoenzyme in the heart different from LDH isoenzyme in the liver?

16.47 A patient arrives in emergency complaining of chest pains. What enzymes would you test for in the blood serum?

16.48 A patient who is an alcoholic has elevated levels of LDH and AST. What condition might be indicated?

■ 16.8 Factors Affecting Enzyme Activity

The **activity** of an enzyme describes how fast an enzyme catalyzes the reaction that converts a substrate to product. This activity is strongly affected by reaction conditions, which include the temperature, pH, concentration of the substrate and the presence of inhibitors.

LEARNING GOAL

■ Describe the effect of temperature, pH, concentration of substrate, and inhibitors on enzyme activity.

■ Temperature

Enzymes are very sensitive to temperature. At low temperatures, most enzymes show little activity because there is not a sufficient amount of energy for the catalyzed reaction to take place. At higher temperatures, enzyme activity increases as reacting molecules move faster to cause more collisions with enzymes. Enzymes are most active at **optimum temperature**, which is 37°C or body temperature for most enzymes. (See Figure 16.16.) At temperatures above 50°C, the tertiary structure and thus the shape of most proteins is destroyed, which causes a loss in enzyme activity. For this reason, equipment in hospitals and laboratories is sterilized in autoclaves where the high temperatures denature the enzymes in harmful bacteria.

■ pH

Enzymes are most active at their **optimum pH**, the pH that maintains the proper tertiary structure of the protein. (See Figure 16.17.) A pH value above or below

FIGURE 16.16

An enzyme attains maximum activity at its optimum temperature, usually 37°C. Lower temperatures slow the rate of reaction and temperatures above 50°C cause denaturation of the enzyme protein and loss of catalytic activity.

Q **Why is 37°C the optimum temperature for many enzymes?**

FIGURE 16.17

Enzymes are most active at their optimum pH. At a higher or lower pH, denaturation of the enzyme causes a loss of catalytic activity.

Q **Why does the digestive enzyme pepsin have an optimum pH of 2?**

TABLE 16.11

Optimum pH for Selected Enzymes

Enzyme	Location	Substrate	Optimum pH
pepsin	stomach	peptide bonds	2
urease	liver	urea	5
sucrase	small intestine	sucrose	6.2
pancreatic amylase	pancreas	amylose	7
trypsin	small intestine	peptide bonds	8
arginase	liver	arginine	9.7

the optimum pH causes a change in the three-dimensional structure of the enzyme that disrupts the active site. As a result the enzyme cannot bind substrate properly and no reaction occurs.

Enzymes in most cells have optimum pH values at physiological pH values around 7.4. However, enzymes in the stomach have a low optimum pH because they hydrolyze proteins at the acidic pH in the stomach. For example, pepsin, a digestive enzyme in the stomach, has an optimum pH of 2. Between meals, the pH in the stomach is 4 or 5 and pepsin shows little or no digestive activity. When food enters the stomach, the secretion of HCl lowers the pH to about 2, which activates pepsin.

If small changes in pH are corrected, an enzyme can regain its structure and activity. However, large variations from optimum pH permanently destroy the structure of the enzyme. Table 16.11 lists the optimum pH values for selected enzymes.

Substrate Concentration

In any catalyzed reaction, the substrate must first bind with the enzyme to form the substrate–enzyme complex. When enzyme concentration is kept constant, increasing the substrate concentration increases the rate of the catalyzed reaction as long as there are more enzyme molecules present than substrate molecules. At some point an increase in substrate concentration saturates the enzyme. With all the available enzyme molecules bonded to substrate, the rate of the catalyzed reaction reaches its maximum. Adding more substrate molecules cannot increase the rate further. (See Figure 16.18.)

FIGURE 16.18

Increasing the substrate concentration increases the rate of reaction until the enzyme molecules are saturated with substrate.

Q **What happens to the rate of reaction when the substrate saturates the enzyme?**

SAMPLE PROBLEM 16.9

■ **Factors Affecting Enzymatic Activity**

Describe what effect the changes in parts a and b would have on the rate of the reaction catalyzed by urease.

$$H_2N-\overset{\overset{\displaystyle O}{\|}}{C}-NH_2 \ + \ H_2O \ \xrightarrow{\text{urease}} \ 2NH_3 \ + \ CO_2$$

urea

a. increasing the urea concentration
b. lowering the temperature to 10°C

SOLUTION

a. An increase in urea concentration will increase the rate of reaction until all of the enzyme molecules are bound to the urea substrate. No further increase in rate occurs.

b. Because 10°C is lower than the optimum temperature of 37°C, the lower temperature will decrease the rate of the reaction.

STUDY CHECK

If urease has an optimum pH of 5, what is the effect of lowering the pH to 3?

Enzyme Inhibition

WEB TUTORIAL
Enzyme Inhibition

Many kinds of molecules called **inhibitors** cause enzymes to lose catalytic activity. Although inhibitors act differently, they all prevent the active site from binding with a substrate. Inhibition can be competitive or noncompetitive. In competitive inhibition, an inhibitor competes for the active site, whereas in noncompetitive inhibition, the inhibitor acts on a site that is not the active site.

A **competitive inhibitor** has a structure that is so similar to the substrate it competes for the active site on the enzyme. As long as the inhibitor occupies the active site, the substrate cannot bind to the enzyme and no reaction takes place. (See Figure 16.19.)

$$S \rightleftharpoons ES \longrightarrow E + P \quad \text{Favored by increasing [S]}$$

Enzyme–substrate complex

E

$$I \rightleftharpoons EI \quad \text{Favored when [S] is low}$$

Enzyme–inhibitor complex

As long as the concentration of the inhibitor is substantial, there is a loss of enzyme activity. However, increasing the substrate concentration displaces more of the inhibitor molecules. As more enzyme molecules bind to substrate (ES), enzyme activity is regained.

The structure of a **noncompetitive inhibitor** does not resemble the substrate and does not compete for the active site. Instead, a noncompetitive inhibitor binds to a site on the enzyme that is not the active site. When the noncompetitive inhibitor is bonded to the enzyme, the shape of the enzyme is distorted. Inhibition occurs because the substrate cannot fit in the active site, or it does not fit properly. Without the proper alignment of substrate with the amino acid side groups, no catalysis can take place. (See Figure 16.20.)

Because a noncompetitive inhibitor is not competing for the active site, the addition of more substrate does not reverse this type of inhibition. Examples of noncompetitive inhibitors are the heavy metal ions Pb^{2+}, Ag^+, and Hg^{2+} that bond with amino acid side groups such as $-COO^-$, or $-OH$. Catalytic activity is restored when chemical reagents remove the inhibitors.

Antibiotics produced by bacteria, mold, or yeast are inhibitors used to stop bacterial growth. For example, penicillin inhibits an enzyme needed for the formation of cell walls in bacteria, but not human cell membranes. With an incomplete cell wall, bacteria cannot survive, and the infection is stopped. However, some bacteria are resistant to penicillin because they produce penicillinase, an enzyme that breaks

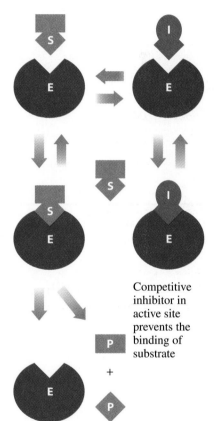

Competitive inhibitor in active site prevents the binding of substrate

FIGURE 16.19

With a structure similar to the substrate for an enzyme, a competitive inhibitor also fits the active site and competes with the substrate when both are present.

Q **Why does increasing the substrate concentration reverse the inhibition by a competitive inhibitor?**

FIGURE 16.20

A noncompetitive inhibitor binds to an enzyme at a site other than the active site, which distorts the enzyme and prevents the proper binding and catalysis of the substrate at the active site.

Q Why won't an increase in substrate concentration reverse the inhibition by a noncompetitive inhibitor?

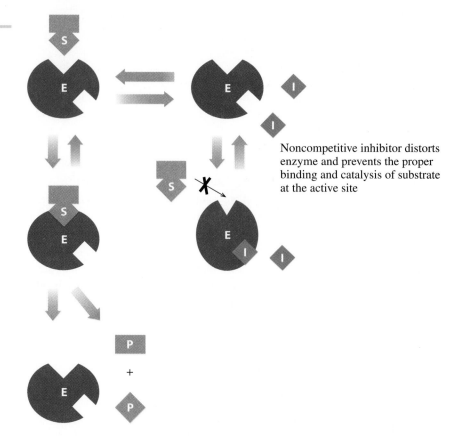

Noncompetitive inhibitor distorts enzyme and prevents the proper binding and catalysis of substrate at the active site

down penicillin. Over the years, derivatives of penicillin to which bacteria have not yet become resistant have been produced.

penicillin

R groups for penicillin derivatives

SAMPLE PROBLEM 16.10

■ Enzyme Inhibition

State the type of inhibition in the following:

a. The inhibitor has a structure that is similar to the substrate.
b. This inhibitor binds to the surface of the enzyme, changing its shape in such a way that it cannot bind to substrate.

SOLUTION

a. competitive inhibition   **b.** noncompetitive inhibition

STUDY CHECK

What type of enzyme inhibition can be reversed by adding more substrate?

QUESTIONS AND PROBLEMS

■ Factors Affecting Enzyme Activity

16.49 Trypsin, a peptidase that hydrolyzes polypeptides, functions in the small intestine at an optimum pH of 8. How is the rate of a trypsin-catalyzed reaction affected by each of the following conditions?
a. lowering the concentration of polypeptides
b. changing the pH to 3
c. running the reaction at 75°C

16.50 Pepsin, a peptidase that hydrolyzes proteins, functions in the stomach at an optimum pH of 2. How is the rate of a pepsin-catalyzed reaction affected by each of the following conditions?
a. increasing the concentration of proteins
b. changing the pH to 5
c. running the reaction at 0°C

16.51 The following graph shows the curves for pepsin, urease, and trypsin. Estimate the optimum pH for each.

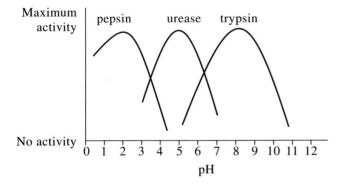

16.52 Refer to the graph in problem 16.51 to determine if the reaction rate in each condition will be at the optimum rate or not.
a. trypsin, pH 5   **b.** urease, pH 5   **c.** pepsin, pH 4
d. trypsin, pH 8   **e.** pepsin, pH 2

16.53 Indicate whether the following describe a competitive or a noncompetitive enzyme inhibitor:
a. The inhibitor has a structure similar to the substrate.
b. The effect of the inhibitor cannot be reversed by adding more substrate.
c. The inhibitor competes with the substrate for the active site.
d. The structure of the inhibitor is not similar to the substrate.
e. The addition of more substrate reverses the inhibition.

16.54 Oxaloacetate is an inhibitor of succinate dehydrogenase:

$$\underset{\text{succinate}}{\begin{array}{c} COO^- \\ | \\ CH_2 \\ | \\ CH_2 \\ | \\ COO^- \end{array}} \qquad \underset{\text{oxaloacetate}}{\begin{array}{c} COO^- \\ | \\ CH_2 \\ | \\ C=O \\ | \\ COO^- \end{array}}$$

a. Would you expect oxaloacetate to be a competitive or a noncompetitive inhibitor? Why?
b. Would oxaloacetate bind to the active site or elsewhere on the enzyme?
c. How would you reverse the effect of the inhibitor?

16.55 Methanol and ethanol are oxidized by alcohol dehydrogenase. In methanol poisoning, a high concentration of ethanol is given intravenously as an antidote.
a. Compare the structure of methanol and ethanol.
b. Would ethanol compete for the active site or bind to a different site?
c. Would ethanol be a competitive or noncompetitive inhibitor of methanol oxidation?

16.56 In humans, the antibiotic amoxycillin (a type of penicillin) is used to treat certain bacterial infections.
    a. Does the antibiotic inhibit enzymes in humans?
    b. Why does the antibiotic kill bacteria, but not humans?

LEARNING GOAL

Describe the types of cofactors found in enzymes.

16.9 Enzyme Cofactors

Enzymes are known as **simple enzymes** when their functional forms consist only of proteins with tertiary structures. However, many enzymes require small molecules or metal ions called **cofactors** to catalyze reactions properly. When the cofactor is a small organic molecule, it is known as a **coenzyme**. If an enzyme requires a cofactor, neither the protein structure nor the cofactor alone has catalytic activity.

Forms of Active Enzymes

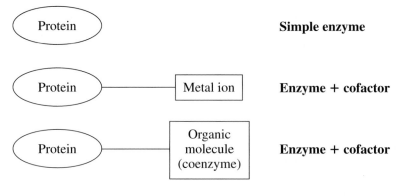

Metal Ions

Many enzymes must contain a metal ion to carry out their catalytic activity. The metal ions are bonded to one or more of the amino acid side chains. The metal ions from the minerals that we obtain from foods in our diet have various functions in catalysis. Ions such as Fe^{2+} and Cu^{2+} are used by oxidases because they lose or gain electrons in oxidation and reduction reactions. Other metals ions such as Zn^{2+} stabilize the amino acid side chains during hydrolysis reactions. Some metal cofactors required by enzymes are listed in Table 16.12.

TABLE 16.12

Enzymes and the Metal Ions Required as Cofactors

Metal Ion Cofactor	Function	Enzyme
Cu^{2+}	Oxidation–reduction	cytochrome oxidase
Fe^{2+}/Fe^{3+}	Oxidation–reduction	catalase
	Oxidation–reduction	cytochrome oxidase
Zn^{2+}	Used with NAD^+	alcohol dehydrogenase
		carbonic anhydrase
		carboxypeptidase A
Mg^{2+}	Hydrolyzes phosphate esters	glucose-6-phosphatase
Mn^{2+}	Removes electrons	arginase
Ni^{2+}	Hydrolyzes amides	urease

Vitamins and Coenzymes

Vitamins are organic molecules that are essential for normal health and growth. They are required in trace amounts and must be obtained from the diet because they are not synthesized in the body. Before vitamins were discovered, it was known that lime juice prevented the disease scurvy in sailors and that cod liver oil could prevent rickets. In 1912, scientists found that, in addition to carbohydrates, fats, and proteins, certain other factors called vitamins must be obtained from the diet.

Vitamins are classified into two groups by solubility: water-soluble and fat-soluble. **Water-soluble vitamins** have polar groups such as —OH and —COOH, which make them soluble in the aqueous environment of the cells. The **fat-soluble vitamins** are nonpolar compounds, which are soluble in the fat (lipid) components of the body such as fat deposits and cell membranes.

Most water-soluble enzymes are not stored in the body and excess amounts are eliminated in the urine each day. Therefore, the water-soluble vitamins must be in the foods of our daily diets. (See Figure 16.21.) Because many water-soluble vitamins are easily destroyed by heat, oxygen, and ultraviolet light, care must be taken in food preparation, processing, and storage. In the 1940s, a Committee on Food and Nutrition of the National Research Council began to recommend dietary enrichment of cereal grains. It was known that refining grains such as wheat caused a loss of vitamins. Thiamine (B_1), riboflavin (B_2), niacin, and iron were in the first group of added nutrients recommended. We now see the Recommended Daily Allowance (RDA) for many vitamins and minerals on food product labels such as cereals and bread.

The water-soluble vitamins are required by many enzymes as cofactors to carry out certain aspects of catalytic action. (See Table 16.13.) The coenzymes do not remain bonded to a particular enzyme, but are used over and over again by different enzymes to facilitate an enzyme-catalyzed reaction. (See Figure 16.22.) Thus, only small amounts of coenzymes are required in the cells.

The fat-soluble vitamins A, D, E, and K are not involved as coenzymes, but they are important in processes such as vision, formation of bone, protection from oxidation, and proper blood clotting. Because the fat-soluble vitamins are stored in

FIGURE 16.21

Oranges, lemons, peppers, and tomatoes contain vitamin C or ascorbic acid.

Q **What happens to the excess vitamin C that may be consumed in a day?**

TABLE 16.13

Vitamins and Function

Water-Soluble Vitamins	Coenzyme	Function
thiamine (vitamin B_1)	thiamine pyrophosphate	decarboxylation
riboflavin (vitamin B_2)	flavin adenine dinucleotide (FAD); flavin mononucleotide (FMN)	electron transfer
niacin (vitamin B_3)	nicotinamide adenine dinucleotide (NAD^+); nicotinamide adenine dinucleotide phosphate ($NADP^+$)	oxidation–reduction
pantothenic acid (vitamin B_5)	coenzyme A	acetyl group transfer
pyridoxine (vitamin B_6)	pyridoxal phosphate	transamination
cobalamin (vitamin B_{12})	methylcobalamin	methyl group transfer
ascorbic acid (vitamin C)	vitamin C	collagen synthesis, healing of wounds
biotin	biocytin	carboxylation
folic acid	tetrahydrofolate	methyl group transfer

Fat-Soluble Vitamins		
vitamin A	formation of visual pigments; development of epithelial cells	
vitamin D	absorption of calcium and phosphate; deposition of calcium and phosphate in bone	
vitamin E	antioxidant; prevents oxidation of vitamin A and unsaturated fatty acids	
vitamin K	synthesis of prothrombin for blood clotting	

FIGURE 16.22

The active forms of many enzymes require the combination of the protein with a coenzyme.

Q **What is the function of water-soluble vitamins in enzymes?**

FIGURE 16.23

Yellow and green fruits and vegetables contain vitamin A.

Q **Why is vitamin A called a fat-soluble vitamin?**

the body and not eliminated, it is possible to take too much, which could be toxic. (See Figure 16.23.)

SAMPLE PROBLEM 16.11

■ **Cofactors**

Indicate whether each of the following enzymes are active as a simple enzyme or require a cofactor.

a. a polypeptide that needs Mg^{2+} for catalytic activity
b. an active enzyme composed only of a polypeptide chain
c. an enzyme that consists of a quaternary structure attached to vitamin B_6

SOLUTION

a. The enzyme requires a cofactor.
b. An active enzyme that consists of only a polypeptide chain is a simple enzyme.
c. The enzyme requires a cofactor.

STUDY CHECK

Which of the nonprotein portions of the enzymes in Sample Problem 16.11 is a coenzyme?

QUESTIONS AND PROBLEMS

■ **Enzyme Cofactors**

16.57 Is the enzyme described in each of the following statements a simple enzyme or one that requires a cofactor?
　　a. requires vitamin B_1 (thiamine)
　　b. needs Zn^{2+} for catalytic activity
　　c. its active form consists of two polypeptide chains

16.58 Is the enzyme described in each of the following statements a simple enzyme or one that requires a cofactor?
　　a. requires vitamin B_2 (riboflavin)
　　b. its active form is composed of 155 amino acids
　　c. uses Cu^{2+} during catalysis

16.59 Identify a vitamin that is a component of each of the following coenzymes:
　　a. coenzyme A　　　　　　**b.** tetrahydrofolate (THF)　**c.** NAD^+

16.60 Identify a vitamin that is a component of each of the following coenzymes:
　　a. thiamine pyrophosphate　　**b.** FAD　　　　　　**c.** pyridoxal phosphate

CONCEPT MAP

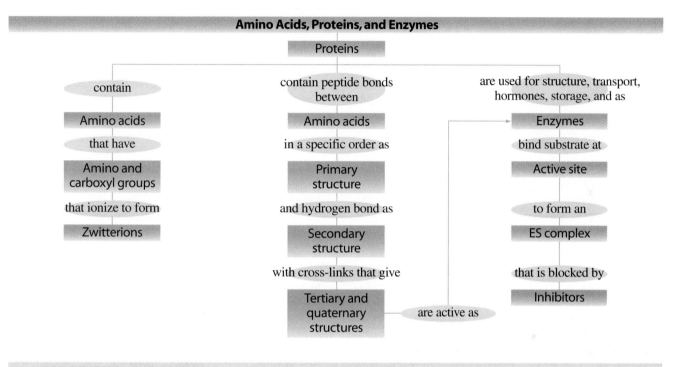

CHAPTER REVIEW

16.1 Functions of Proteins
Some proteins are enzymes or hormones, whereas others are important in structure, transport, protection, storage, and muscle contraction.

16.2 Amino Acids
A group of 20 amino acids provides the molecular building blocks of proteins. Attached to the central (alpha) carbon of each amino acid is an amino group, a carboxyl group, and a unique side group (R). The R group gives an amino acid the property of being nonpolar, polar, acidic, or basic.

16.3 Amino Acids as Acids and Bases
Amino acids exist as dipolar ions called zwitterions, as positive ions at low pH, and as negative ions at high pH levels. At the iso-electric point, zwitterions are neutral.

16.4 Formation of Peptides
Peptide bonds join amino acids to form peptides by forming an amide bond between the carboxyl group of one amino acid and the amino group of the second. Long chains of amino acids are called proteins.

16.5 Levels of Protein Structure
The primary structure of a protein is its sequence of amino acids. In the secondary structure, hydrogen bonds between peptide groups produce a characteristic shape such as an α helix, β-pleated sheet, or a triple helix.

In globular proteins, the polypeptide chain including its α-helical and β-pleated sheet regions folds upon itself to form a tertiary structure. A tertiary structure is stabilized by interactions that move hydrophobic R groups to the inside and hydrophilic R groups to the outside surface, and by attractions between R groups that form hydrogen bonds, disulfide bonds, and salt bridges. In a quaternary structure, two or more tertiary subunits must combine for biological activity. They are held together by the same interactions found in tertiary structures.

Denaturation of a protein occurs when heat or other denaturing agents destroy the structure of the protein (but not the primary structure) until biological activity is lost.

16.6 Enzymes
Enzymes act as biological catalysts by lowering activation energy and accelerating the rate of cellular reactions. The names of most enzymes ending in *ase* describe the compound or reaction catalyzed by the enzyme. Enzymes are classified by the main type of reaction they catalyze, such as oxidoreductase, transferase, or isomerase.

16.7 Enzyme Action
Within the tertiary structure of an enzyme, a small pocket called the active site binds the substrates. In the lock-and-key model, a substrate precisely fits the shape of the active site. In the induced-fit

model, substrates induce the active site to change structure to give an optimal fit by the substrate. In the enzyme–substrate complex, catalysis takes place when amino acid side chains react with a substrate. The products are released and the enzyme is available to bind another substrate molecule.

16.8 Factors Affecting Enzyme Activity

Enzymes are most effective at optimum temperature and pH, usually 37°C and 7.4. The rate of an enzyme-catalyzed reaction decreases considerably at temperature and pH values above or below the optimum. An increase in substrate concentration increases the reaction rate of an enzyme-catalyzed reaction. If an enzyme is saturated, adding more substrate will not increase the rate further. An inhibitor reduces the activity of an enzyme or makes it inactive. A competitive inhibitor has a structure similar to the sub-

strate and competes for the active site. When the active site is occupied, the enzyme cannot catalyze the reaction of the substrate. A noncompetitive inhibitor attaches elsewhere on the enzyme, changing the shape of both the enzyme and its active site. As long as the noncompetitive inhibitor is attached, the altered active site cannot bind with substrate.

16.9 Enzyme Cofactors

Simple enzymes are biologically active as a protein only, whereas other enzymes require small organic molecules or metals ions called cofactors. A cofactor may be a metal ion such as Cu^{2+} or Fe^{2+} or an organic molecule called a coenzyme. A vitamin is a small organic molecule needed for health and normal growth. Vitamins are obtained in small amounts through the foods in the diet. The water-soluble vitamins B and C function as coenzymes.

KEY TERMS

acidic amino acid An amino acid that has a carboxylic acid side chain (—COOH), which ionizes as a weak acid.

active site A pocket in a part of the tertiary enzyme structure that binds substrate and catalyzes a reaction.

activity The rate at which an enzyme catalyzes the reaction that converts substrate to product.

α (alpha) helix A secondary level of protein structure, in which hydrogen bonds connect the NH of one peptide bond with the C=O of a peptide bond later in the chain to form a coiled or corkscrew structure.

amino acid The building block of proteins, consisting of an amino group, a carboxylic acid group, and a unique side group attached to the alpha carbon.

basic amino acid An amino acid that contains an amino (—NH$_2$) group that can ionize as a weak base.

β (beta)-pleated sheet A secondary level of protein structure that consists of hydrogen bonds between peptide links in parallel polypeptide chains.

C terminal The amino acid that is the last amino acid in a peptide chain with an unreacted or free carboxyl group (—COO⁻).

catalyst A substance that takes part in a reaction to lower the activation energy, but is not changed.

coenzyme An organic molecule, usually a vitamin, required as a cofactor in enzyme action.

cofactor A nonprotein metal ion or an organic molecule that is necessary for a biologically functional enzyme.

collagen The most abundant form of protein in the body composed of fibrils of triple helixes with hydrogen bonding between —OH groups of hydroxyproline and hydroxylysine.

competitive inhibitor A molecule with a structure similar to the substrate that inhibits enzyme action by competing for the active site.

denaturation The loss of secondary and tertiary protein structure caused by heat, acids, bases, organic compounds, heavy metals, and/or agitation.

disulfide bonds Covalent —S—S— bonds that form between the —SH group of cysteines in a protein to stabilize the tertiary structure.

enzymes Globular proteins that catalyze biological reactions.

enzyme–substrate (ES) complex An intermediate consisting of an enzyme that binds to a substrate in an enzyme-catalyzed reaction.

fat-soluble vitamins Vitamins that are not soluble in water and can be stored in the liver and body fat.

fibrous protein A protein that is insoluble in water; consists of polypeptide chains with α helixes or β-pleated sheets, that make up the fibers of hair, wool, skin, nails, and silk.

globular proteins Proteins that acquire a compact shape from attractions between the R group of the amino acid residues in the protein.

hydrogen bonds Attractions between the polar R groups such as —OH, —NH$_2$, and —COOH of amino acids in a polypeptide chain.

hydrophilic interactions The interaction between polar R groups on the protein surface and water.

hydrophobic interactions The attraction between nonpolar R groups in a tertiary structure of a globular protein.

induced-fit model A model of enzyme action in which a substrate induces an enzyme to modify its shape to give an optimal fit with the substrate structure.

inhibitors Substances that make an enzyme inactive by interfering with its ability to react with a substrate.

isoelectric point (pI) The pH at which an amino acid is in the neutral zwitterion form.

isoenzyme Isoenzymes catalyze the same reaction, but have different combinations of polypeptide subunits.

lock-and-key model A model of an enzyme in which the substrate, like a key, exactly fits the shape of the lock, which is the specific shape of the active site.

N terminal The amino acid in a peptide written on the left with the unreacted or free amino group (—NH$_3$⁺).

noncompetitive inhibitor A substance that changes the shape of the enzyme, which prevents the active site from binding substrate properly.

nonpolar amino acids Amino acids that are not soluble in water because they contain a nonpolar R group.

optimum pH The pH at which an enzyme is most active.

optimum temperature The temperature at which an enzyme is most active.

peptide The combination of two or more amino acids joined by peptide bonds; dipeptide, tripeptide, and so on.

peptide bond The amide bond that joins amino acids in polypeptides and proteins.

polar amino acids Amino acids that are soluble in water because their R group is polar: hydroxyl (OH), thiol (SH), carbonyl (C=O), amino (NH_2), or carboxyl (COOH).

primary structure The sequence of the amino acids in a protein.

protein A term used for biologically active polypeptides that have many amino acids linked together by peptide bonds.

quaternary structure A protein structure in which two or more protein subunits form an active protein.

salt bridge The ionic bond formed between side groups of basic and acidic amino acids in a protein.

secondary structure The formation of an α helix, β-pleated sheet, or triple helix.

simple enzyme An enzyme that is active as a polypeptide only.

substrate The molecule that reacts in the active site in an enzyme-catalyzed reaction.

tertiary structure The folding of the secondary structure of a protein into a compact structure that is stabilized by the interactions of R groups.

triple helix The protein structure found in collagen consisting of three polypetide chains woven together like a braid.

vitamins Organic molecules, which are essential for normal health and growth, obtained in small amounts from the diet.

water-soluble vitamins Vitamins that are soluble in water, cannot be stored in the body, are easily destroyed by heat, ultraviolet light, and oxygen, and function as coenzymes.

zwitterion The dipolar form of an amino acid consisting of two oppositely charged ionic regions, $-NH_3^+$ and $-COO^-$.

ADDITIONAL QUESTIONS AND PROBLEMS

16.61 Seeds and vegetables are often deficient in one or more essential amino acids. Using the following table, state whether the following combinations would provide all the essential amino acids:

Source	Lysine	Tryptophan	Methionine
Oatmeal	No	Yes	Yes
Rice	No	Yes	Yes
Garbanzo beans	Yes	No	Yes
Lima beans	Yes	No	No
Cornmeal	No	Yes	Yes

    a. rice and garbanzo beans
    b. lima beans and cornmeal
    c. a salad of garbanzo beans and lima beans
    d. rice and lima beans
    e. rice and oatmeal
    f. oatmeal and lima beans

16.62 a. Draw the structure of Val-Ala-Leu.
    b. Would you expect to find this segment at the center or at the surface of a globular protein? Why?

16.63 What are some differences between the following pairs?
    a. secondary and tertiary protein structures
    b. essential and nonessential amino acids
    c. polar and nonpolar amino acids
    d. di- and tripeptides
    e. an ionic bond (salt bridge) and a disulfide bond
    f. fibrous and globular proteins
    g. α helix and β-pleated sheet
    h. tertiary and quaternary structures of proteins

16.64 a. Where is collagen found?
    b. What type of secondary structure is used to form collagen?

16.65 a. Draw the structure of Ser-Lys-Asp.
    b. Would you expect to find this segment at the center or at the surface of a globular protein? Why?

16.66 What type of interaction would you expect from the following R groups in a tertiary structure?
    a. threonine and asparagine
    b. valine and alanine
    c. arginine and aspartic acid

16.67 If serine were replaced by valine in a protein, how would the tertiary structure be affected?

16.68 If you eat rice, what other vegetable protein source(s) could you eat to ingest all essential amino acids?

16.69 How are enzymes different from the catalysts used in chemistry laboratories?

16.70 Why do enzymes function only under mild conditions?

16.71 Lactase is an enzyme that hydrolyzes lactose to glucose and galactose.
    a. What are the reactants and products of the reaction?
    b. Draw an energy diagram for the reaction with and without lactase.
    c. How does lactase make the reaction go faster?

16.72 Maltase is an enzyme that hydrolyzes maltose to two glucose molecules.
    a. What are the reactants and products of the reaction?
    b. Draw an energy diagram for the reaction with and without maltase.
    c. How does maltase make the reaction go faster?

16.73 Indicate whether each of the following would be a substrate (S) or an enzyme (E):
    a. lactose   **b.** lactase
    c. urease   **d.** trypsin
    e. pyruvate   **f.** transaminase

16.74 Indicate whether each of the following would be a substrate (S) or an enzyme (E):
    a. glucose   **b.** hydrolase
    c. maleate isomerase   **d.** alanine
    e. amylose   **f.** amylase

16.75 Give the substrate of each of the following enzymes:
    a. urease
    b. lactase
    c. aspartate transaminase
    d. tyrosinase

16.76 Give the substrate of each of the following enzymes:
    a. maltase
    b. fructose oxidase
    c. phenolase
    d. sucrase

16.77 How would the lock-and-key theory explain that sucrase hydrolyses sucrose, but not lactose?

16.78 How does the induced-fit model of enzyme action allow an enzyme to catalyze a reaction of a group of substrates?

16.79 If a blood test indicates a high level of LDH and CK, what could be the cause?

16.80 If a blood test indicates a high level of ALT, what could be the cause?

16.81 Indicate whether an enzyme is saturated or unsaturated in each of the following conditions:
a. adding more substrate does not increase the rate of reaction
b. doubling the substrate concentration doubles the rate of reaction

16.82 Indicate whether each of the following enzymes would be functional:
a. pepsin, a digestive enzyme, at pH 2
b. an enzyme at 37°C, if the enzyme is from a type of thermophilic bacteria that have an optimum temperature of 100°C

16.83 Are each of the following statements describing a simple enzyme or one that requires a cofactor?
a. contains Mg^{2+} in the active site
b. has catalytic activity as a tertiary protein structure
c. requires folic acid for catalytic activity

16.84 Are each of the following statements describing a simple enzyme or one that contains a cofactor?
a. contains riboflavin or vitamin B_2
b. has four subunits of polypeptide chains
c. requires Fe^{3+} in the active site for catalytic activity

UNDERSTANDING THE CONCEPTS

16.85 Ethylene glycol ($HO-CH_2-CH_2-OH$) is a major component of antifreeze. In the body, it is first converted to $HOOC-CHO$ (oxoethanoic acid) and then to $HOOC-COOH$ (oxalic acid), which is toxic.
a. What class of enzyme catalyzes both of the reactions of ethylene glycol?

b. The treatment for the ingestion of ethylene glycol is an intravenous solution of ethanol. How might this help prevent toxic levels of oxalic acid in the body?

16.86 Adults who are lactose intolerant cannot break down the disaccharide in milk products. To help digest dairy food,

a product known as Lactaid can be added to milk and the milk then refrigerated for 24 hours.
a. What enzyme is present in Lactaid and what is the major class?
b. What might happen to the enzyme if the digestion product were stored in a warm area?

16.87 Aspartame, which is used in artificial sweeteners, contains the following dipeptides.
a. What are the amino acids in aspartame?
b. How would you name the dipeptide in aspartame?

16.88 If cysteine, an amino acid prevalent in hair, has a pI of 5.0, which structure would it have in solutions with the following pH values?

(3)   (4)

a. 2   **b.** 5   **c.** 9

16.89 Identify the amino acids and type of crosslink that occur between the following side groups in tertiary protein structures:

a. $-CH_2-\overset{\overset{\text{O}}{\parallel}}{C}-NH_2$ and $HO-CH_2-$

b. $-CH_2-\overset{\overset{\text{O}}{\parallel}}{C}-O^-$ and $H_3\overset{+}{N}-(CH_2)_4-$

c. $-CH_2-SH$ and $HS-CH_2-$

d. $-CH_2-\overset{\overset{\text{CH}_3}{\mid}}{CH}-CH_3$ and CH_3-

16.90 Fresh pineapple contains the enzyme Bromelain that degrades proteins.

a. The directions on a Jello® package say not to add fresh pineapple. However, canned pineapple where pineapple is heated to high temperatures can be added. Why?

b. Fresh pineapple is used in a marinade to tenderize tough meat. Why?

QUESTIONS FOR TEAMWORK

16.91 Indicate the charge of each of the following amino acids at the following pH values as:

a. 0   **b.** +1   **c.** +2   **d.** -2   **e.** -1

1. serine at pH 6.0
2. glutamic acid at pH 10
3. arginine at pH 2
4. histidine at pH 7.6
5. isoleucine at pH 3
6. leucine at pH 9

16.92 Consider a mixture of the amino acids lysine, valine, and aspartic acid at pH 6.0 that is subjected to an electric current.

a. Indicate which amino acid would migrate toward the positive electrode (+), the negative electrode (−), or remain stationary.

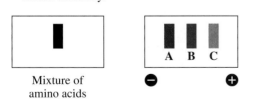

Mixture of amino acids

b. If the mixture is the result of hydrolyzing a tripeptide, what are the possible amino acid sequences containing one molecule of each?

c. In an enzyme, which of these amino acids would be found in

1. hydrophobic regions
2. hydrophilic regions
3. hydrogen bonds
4. disulfide bonds
5. salt bridges

16.93 What are some differences between the following pair?

a. secondary and tertiary protein structures
b. essential and nonessential amino acids
c. polar and nonpolar amino acids
d. di- and tripeptides
e. α-helix and β-pleated sheet
f. tertiary and quaternary protein structures

ANSWERS

Answers to Study Checks

16.1 hormones

16.2 a. nonpolar   **b.** polar

16.3

16.4 threonylleucylphenylalanine

16.5 Both are nonpolar and would be found on the inside of the tertiary structure.

16.6 quaternary

16.7 lipase

16.8 In the lock-and-key model, the shape of a substrate fits the shape of the active site exactly. In the induced-fit model, the substrate induces the active site to adjust its shape to fit the substrate.

16.9 At a pH lower than the optimum pH, denaturation will decrease the activity of urease.

16.10 competitive inhibition

16.11 vitamin B_6

Answers to Selected Questions and Problems

16.1 a. transport   **b.** structural
c. structural   **d.** enzyme

16.3 All amino acids contain a carboxylic acid group and an amino group on the α carbon.

16.5 a.   **b.**

c.   **d.**

16.7 a. hydrophobic nonpolar   **b.** hydrophilic polar, neutral
c. acidic   **d.** hydrophobic nonpolar

16.9 a. alanine   **b.** valine
c. lysine   **d.** cysteine

16.11 a.   **b.**

16.13 a.   **b.**

c.   **d.**

16.15 a.   **b.**

c.   **d.**

16.17 a. above pI   **b.** below pI   **c.** at pI

16.19

a. Ala-Cys

b. Ser-Phe

c. Gly-Ala-Val

d. Val-Ile-Trp

16.21 Amide bonds form to connect the amino acids that make up the protein.

16.23 Val-Ser-Ser, Ser-Val-Ser, or Ser-Ser-Val

16.25 The primary structure remains unchanged and intact as hydrogen bonds form between carbonyl oxygen atoms and amino hydrogen atoms.

16.27 In the α helix, hydrogen bonds form between the carbonyl oxygen atom and the amino hydrogen atom of the fourth amino acid in the sequence. In the β-pleated sheet, hydrogen bonds occur between parallel peptides or across sections of a long polypeptide chain.

16.29 a. a disulfide bond
b. salt bridge
c. hydrogen bond
d. hydrophobic interaction

16.31 a. cysteine
b. Leucine and valine will be found on the inside of the protein because they are hydrophobic.
c. The cysteine and aspartic acid would be on the outside of the protein because they are polar.
d. The order of the amino acids (the primary structure) provides the R groups, whose interactions determine the tertiary structure of the protein.

16.33 a. Placing an egg in boiling water disrupts hydrogen bonds and hydrophobic interactions, which changes secondary and tertiary structures and causes loss of overall shape.
b. Using an alcohol swab disrupts hydrophobic interactions and changes the tertiary structure.
c. The heat from an autoclave will disrupt hydrogen bonds and hydrophobic interactions, which changes secondary and tertiary structures of the protein in any bacteria present.
d. Heat will cause changes in the secondary and tertiary structure of surrounding protein, which results in the formation of solid protein that helps to close the wound.

16.35 The chemical reactions can occur without enzymes, but the rates are too slow. Catalyzed reactions, which are many times faster, provide the amounts of products needed by the cell at a particular time.

16.37 a. galactose   **b.** lipid   **c.** aspartic acid

16.39 a. hydrolase   **b.** oxidase
c. isomerase   **d.** transferase

16.41 a. enzyme   **b.** enzyme–substrate complex
c. substrate

16.43 a. E + S $\rightleftarrows$ ES $\longrightarrow$ E + P
b. The active site is a region or pocket within the tertiary structure of an enzyme that accepts the substrate, aligns the substrate for reaction, and catalyzes the reaction.

16.45 Isoenzymes are slightly different forms of an enzyme that catalyze the same reaction in different organs and tissues of the body.

16.47 A doctor might run tests for the enzymes CK, LDH, and AST to determine if the patient had a heart attack.

16.49 a. The rate would decrease.
b. The rate would decrease.
c. The rate would decrease.

16.51 pepsin, pH 2; urease, pH 5; trypsin, pH 8

16.53 a. competitive   **b.** noncompetitive
c. competitive   **d.** noncompetitive
e. competitive

16.55 a. methanol, CH_3OH; ethanol, CH_3CH_2OH
b. Ethanol has a similar structure to methanol and could compete for the active site.
c. Ethanol is a competitive inhibitor of methanol oxidation.

16.57 a. an enzyme that requires a cofactor
b. an enzyme that requires a cofactor
c. a simple enzyme

16.59 a. pantothenic acid (vitamin B_5)
b. folic acid
c. niacin (vitamin B_3)

16.61 a. yes   **b.** yes   **c.** no
d. yes   **e.** no   **f.** yes

16.63 a. The secondary structure of a protein depends on hydrogen bonds to form a helix or a pleated sheet; the tertiary structure is determined by the interaction of R groups and creates the three-dimensional structure of the protein.
b. Nonessential amino acids can be synthesized by the body; essential amino acids must be supplied by the diet.
c. Polar amino acids have hydrophilic side groups, whereas nonpolar amino acids have hydrophobic side groups.
d. Dipeptides contain two amino acids, whereas tripeptides contain three.
e. An ionic bond is an interaction between a basic and acidic side group; a disulfide bond links the sulfides of two cysteines.
f. Fibrous proteins consist of three to seven α helixes coiled like a rope. Globular proteins form a compact spherical shape.
g. The α helix is the secondary shape like a staircase or corkscrew. The β-pleated sheet is a secondary structure that is formed by many protein chains side by side like a pleated sheet.
h. The tertiary structure of a protein is its three-dimensional structure. The quaternary structure involves the grouping of two or more peptide units for the protein to be active.

16.65 a.

$$H_3\overset{+}{N}-\underset{\underset{CH_2OH}{|}}{CH}-\overset{\overset{O}{\|}}{C}-\underset{H}{N}-\underset{\underset{(CH_2)_4}{\underset{|}{|}}{\underset{NH_2}{}}}{CH}-\overset{\overset{O}{\|}}{C}-\underset{H}{N}-\underset{\underset{CH_2}{\underset{|}{|}}{\underset{COOH}{}}}{CH}-\overset{\overset{O}{\|}}{C}-O^-$$

b. This segment contains polar R groups, which would be found on the surface of a globular protein where they can hydrogen bond with water.

16.67 Serine is a polar amino acid, whereas valine is nonpolar. Valine would be in the center of the tertiary structure. However, serine would pull that part of the chain to the outside surface of the protein where it would form hydrophilic bonds with water.

16.69 When exposed to conditions of strong acids or bases or high temperatures, the proteins of enzymes are denatured rapidly causing a loss of tertiary structure and catalytic activity.

16.71 a. The reactant is lactose and the products are glucose and galactose.

b.

Reaction progress ⟶

c. By lowering the energy of activation, the enzyme furnishes a lower energy pathway by which the reaction can take place.

16.73 a. S   **b.** E
     c. E   **d.** E
     e. S   **f.** E

16.75 a. urea   **b.** lactose
     c. aspartate   **d.** tyrosine

16.77 Sucrose fits the shape of the active site in sucrase, but lactose does not.

16.79 A heart attack may be the cause.

16.81 a. saturated
     b. unsaturated

16.83 a. requires a cofactor
     b. simple enzyme
     c. requires a cofactor (coenzyme)

16.85 a. oxidoreductase
     b. At high concentration, ethanol, which acts as a competitive inhibitor of ethylene glycol, would saturate the alcohol dehydrogenase enzyme to allow ethylene glycol to be removed from the body without producing oxalic acid.

16.87 a. aspartic acid and phenylalanine
     b. aspartylphenylalanine

16.89 a. asparagine and serine; hydrogen bond
     b. aspartic acid and lysine; salt bridge
     c. cysteine and cysteine; disulfide bond
     d. leucine and alanine; hydrophobic interaction

17 Nucleic Acids and Protein Synthesis

"I run the Hepatitis C Clinic, where patients are often anxious when diagnosed," says Barbara Behrens, nurse practitioner, Hepatitis C Clinic, Kaiser Hospital. "The treatment for hepatitis C can produce significant reactions such as a radical drop in blood count. When this happens, I get help to them within 24 hours. I monitor our patients very closely and many call me whenever they need to."

Hepatitis C is an RNA virus or retrovirus that causes liver inflammation, often resulting in chronic liver disease. Unlike many viruses to which we eventually develop immunity, the hepatitis C virus undergoes mutations so rapidly that scientists have not been able to produce vaccines. People who carry the virus are contagious throughout their lives and are able to pass the virus to other people.

the Chemistry place

Visit **www.aw-bc.com/chemplace** or use the CD-ROM in the book for extra quizzes, interactive tutorials, career resources, PowerPoint slides, math help, and case studies.

Nucleic acids are the molecules in our cells that store and direct information for cellular growth and reproduction. Deoxyribonucleic acid (DNA), the genetic material in the nucleus of a cell, contains all the information needed for the development of a complete living system. The way you grow, your hair, your eyes, your physical appearance, the activities of the cells in your body are all determined by a set of directions contained in the DNA of your cells.

The nucleic acids are large molecules found in the nuclei of cells that contain all the information needed to direct the activities of a cell and its reproduction. All of the genetic information in the cell is called the *genome*. Every time a cell divides, the information in the genome is copied and passed on to the new cells. This replication process must duplicate the genetic instructions exactly. Some sections of DNA called *genes* contain the information to make a particular protein.

As a cell requires protein, another type of nucleic acid, RNA, translates the genetic information in DNA and carries that information to the ribosomes where the synthesis of protein takes place. However, mistakes sometimes occur that lead to mutations that affect the synthesis of a certain protein.

LEARNING GOAL

Describe the nitrogen bases and ribose sugars that make up the nucleic acids DNA and RNA.

FIGURE 17.1

A diagram of the general structure of a nucleotide found in nucleic acids.

Q **In a nucleotide, what types of groups are bonded to a five-carbon sugar?**

17.1 Components of Nucleic Acids

There are two closely related types of nucleic acids: *deoxyribonucleic acid* (**DNA**), and *ribonucleic acid* (**RNA**). Both are unbranched polymers of repeating monomer units known as *nucleotides*. A DNA molecule may contain several million nucleotides; smaller RNA molecules may contain up to several thousand. Each nucleotide has three components: a nitrogenous base, a five-carbon sugar, and a phosphate group. (See Figure 17.1.)

Nitrogen Bases

The **nitrogen-containing bases** in nucleic acids are derivatives of *pyrimidine* or *purine*.

pyrimidine   purine

In DNA, there are two purines, adenine (A) and guanine (G), and two pyrimidines, cytosine (C) and thymine (T). RNA contains the same bases, except thymine (5-methyluracil) is replaced by uracil (U). (See Figure 17.2.)

Pyrimidines

Cytosine (C)
(DNA and RNA)

Thymine (T)
(DNA only)

Uracil (U)
(RNA only)

Purines

Adenine (A)
(DNA and RNA)

Guanine (G)
(DNA and RNA)

FIGURE 17.2

DNA contains the nitrogen bases A, G, C, and T; RNA contains A, G, C, and U.

Q Which nitrogen bases are found in DNA?

Ribose and Deoxyribose Sugars

The nucleotides of RNA and DNA contain five-carbon pentose sugars. The atoms in the pentose sugars are numbered with primes ($1'$, $2'$, $3'$, $4'$, and $5'$) to differentiate them from the atoms in the nitrogen bases. (See Figure 17.3.) In RNA, the five-carbon sugar is *ribose*, which gives the letter R in the abbreviation RNA. In DNA, the five-carbon sugar is *deoxyribose*, which is similar to ribose except that there is no hydroxyl group (—OH) on C2′ of ribose. The *deoxy* prefix means "without oxygen" and provides the D in DNA.

Pentose sugars in RNA and DNA

Ribose in RNA

No oxygen
is bonded
to this carbon

Deoxyribose in DNA

FIGURE 17.3

The five-carbon pentose sugar found in RNA is ribose and deoxyribose in DNA.

Q What is the difference between ribose and deoxyribose?

SAMPLE PROBLEM 17.1

■ Components of Nucleic Acids

Identify each of the following bases as a purine or pyrimidine.

a.

b.

SOLUTION

a. Guanine is a purine.   **b.** Uracil is a pyrimidine.

STUDY CHECK

Indicate if the bases in Sample Problem 17.1 are found in RNA, DNA, or both.

■ Nucleosides and Nucleotides

A **nucleoside** is produced when a pyrimidine or a purine forms a glycosidic bond to C1′ of a sugar, either ribose or deoxyribose. For example, adenine, a purine, and ribose form a nucleoside called aden**osine**.

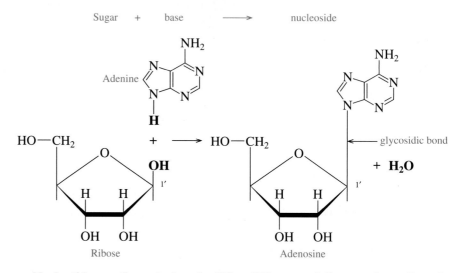

Sugar   +   base   ⟶   nucleoside

Ribose   Adenosine

Nucleotides are formed when the C5′ —OH group of ribose or deoxyribose in a nucleoside forms a phosphate ester. Other hydroxyl groups on ribose can form phosphate esters too, but only the 5′-monophosphate nucleotides are found in RNA and DNA. All the nucleotides in RNA and DNA are shown in Figure 17.4.

Adenosine 5'-monophosphate (AMP)
Deoxyadenosine 5'-monophosphate (dAMP)

Guanosine 5'-monophosphate (GMP)
Deoxyguanosine 5'-monophosphate (dGMP)

Cytidine 5'-monophosphate (CMP)
Deoxycytidine 5'-monophosphate (dCMP)

Uridine 5'-monophosphate (UMP)

Deoxythymidine 5'-monophosphate (dTMP)

FIGURE 17.4

The nucleotides of RNA are identical to those of DNA except in DNA the sugar is deoxyribose and deoxythymidine replaces uridine.

Q What are two differences in the nucleotides of RNA and DNA?

Naming Nucleotides

The name of a nucleotide is obtained from the name of the nucleoside followed by 5'-monophosphate. Nucleotides of DNA have the prefix *deoxy* added to the beginning of the nucleoside name. Although the letters A, G, C, U, and T represent the bases, they are often used in the abbreviations of the respective nucleotides as well. The names and abbreviations of the bases, nucleosides, and nucleotides in DNA and RNA are listed in Table 17.1.

TABLE 17.1

Nucleosides and Nucleotides in DNA and RNA

Base	Nucleosides	Nucleotides
RNA		
adenine (A)	adenosine (A)	adenosine 5'-monophosphate (AMP)
guanine (G)	guanosine (G)	guanosine 5'-monophosphate (GMP)
cytosine (C)	cytidine (C)	cytidine 5'-monophosphate (CMP)
uracil (U)	uridine (U)	uridine 5'-monophosphate (UMP)
DNA		
adenine (A)	deoxyadenosine (A)	deoxyadenosine 5'-monophosphate (dAMP)
guanine (G)	deoxyguanosine (G)	deoxyguanosine 5'-monophosphate (dGMP)
cytosine (C)	deoxycytidine (C)	deoxycytidine 5'-monophosphate (dCMP)
thymine (T)	deoxythymidine (T)	deoxythymidine 5'-monophosphate (dTMP)

SAMPLE PROBLEM 17.2

Nucleotides

Identify which nucleic acid (DNA or RNA) contains each of the following nucleotides; state the components of each nucleotide:

a. deoxyguanosine 5'-monophosphate (dGMP)
b. adenosine 5'-monophosphate (AMP)

SOLUTION

a. This DNA nucleotide consists of deoxyribose, guanine, and phosphate.
b. This RNA nucleotide contains ribose, adenine, and phosphate.

STUDY CHECK

What is the name and abbreviation of the DNA nucleotide of cytosine?

QUESTIONS AND PROBLEMS

Components of Nucleic Acids

17.1 Identify each of the following bases as a purine or pyrimidine:
a. thymine
b.

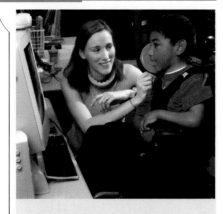

CAREER FOCUS

Occupational Therapist

"Occupational therapists teach children and adults the skills they need for the job of living," says occupational therapist Leslie Wakasa. "When working with the pediatric population, we are crucial in educating children with disabilities, their families, caregivers, and school staff in ways to help them be as independent as they can be in all aspects of their daily lives. It's rewarding when you can show children how to feed themselves, which is a huge self-esteem issue for them. The opportunity to help people become more independent is very rewarding."

A combination of technology and occupational therapy helps children who are nonverbal to communicate and interact with their environment. By leaning on a red switch, Alex is learning to use a computer.

17.2 Identify each of the following bases as a purine or pyrimidine:
  a. guanine
  b.

17.3 Identify the bases in problem 17.1 as present in RNA, DNA, or both.

17.4 Identify the bases in problem 17.2 as present in RNA, DNA, or both.

17.5 What are the names and abbreviations of the four nucleotides in DNA?

17.6 What are the names and abbreviations of the four nucleotides in RNA?

17.7 Identify each of the following as a nucleoside or nucleotide:
  a. adenosine
  b. deoxycytidine
  c. uridine
  d. cytidine 5'-monophosphate

17.8 Identify each of the following as a nucleoside or nucleotide:
  a. deoxythymidine
  b. guanosine
  c. adenosine
  d. uridine 5'-monophosphate

17.9 Draw the structure of deoxyadenosine 5'-monophosphate (dAMP).

17.10 Draw the structure of uridine 5'- monophosphate (UMP).

Describe the primary structures
of RNA and DNA.

the
Chemistry
place

WEB TUTORIAL
DNA and RNA Structure

17.2 Primary Structure of Nucleic Acids

The **nucleic acids** consist of polymers of many nucleotides in which the 3'-OH group of the sugar in one nucleotide bonds to the phosphate group on the 5'-carbon atom in the sugar of the next nucleotide. This phosphate link between the sugars in adjacent nucleotides is referred to as a **phosphodiester bond**. As more nucleotides are added through phosphodiester bonds, a backbone forms that consists of alternating sugar and phosphate groups.

RNA (ribonucleic acid)

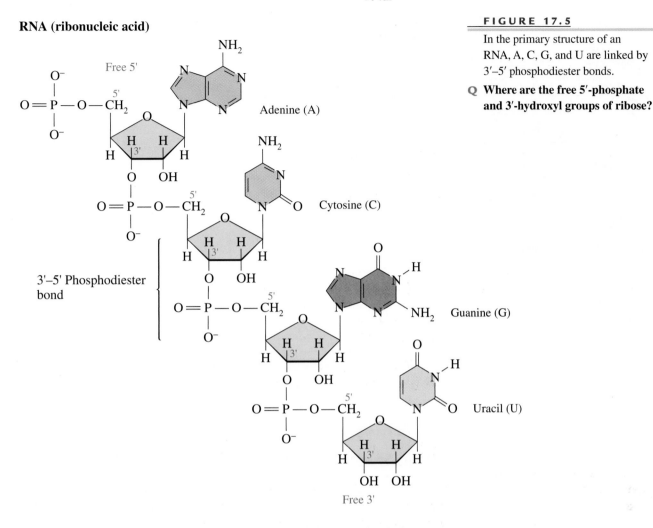

FIGURE 17.5

In the primary structure of an RNA, A, C, G, and U are linked by 3′–5′ phosphodiester bonds.

Q **Where are the free 5′-phosphate and 3′-hydroxyl groups of ribose?**

Each nucleic acid has its own unique sequence of bases, which is known as its **primary structure**. It is this sequence of bases that carries the genetic information from one cell to the next. Along a polynucleotide of a DNA or RNA chain, the bases attached to each of the sugars extend out from the nucleic acid backbone. In any nucleic acid, the sugar at the one end has an unreacted or free 5′-phosphate terminal end, and the sugar at the other end has a free 3′-hydroxyl group.

A nucleic acid sequence is read from the sugar with free 5′-phosphate to the sugar with the free 3′-hydroxyl group. The order of bases is often written using only the letters of the bases. Thus, the nucleotides in a section of RNA shown in Figure 17.5 is read as ACGU.

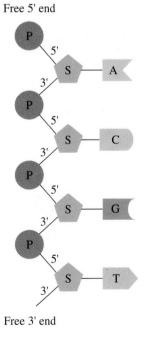

Free 5′ end

Free 3′ end

SAMPLE PROBLEM 17.3

▪ Bonding of Nucleotides

Draw the structure of an RNA dinucleotide formed by two cytidine monophosphates.

SOLUTION

STUDY CHECK

In the dinucleotide of cytidine shown in the solution to Sample Problem 17.3, identify the free 5′-phosphate group and the free 3′-hydroxyl (—OH) group.

FIGURE 17.6

A computer-generated model of a DNA molecule.

Q What is meant by the term *double helix*?

LEARNING GOAL

Describe the double helix of DNA.

QUESTIONS AND PROBLEMS

▪ Primary Structure of Nucleic Acids

17.11 How are the nucleotides held together in a nucleic acid chain?

17.12 How do the ends of a nucleic acid polymer differ?

17.13 Write the structure of the dinucleotide GC that would be in RNA.

17.14 Write the structure of the dinucleotide AT that would be in DNA.

▪ 17.3 DNA Double Helix

In the 1940s, biologists determined that DNA in a variety of organisms had a specific relationship between bases: adenine is paired only with thymine, and guanine is paired only with cytosine. In 1953, James Watson and Francis Crick proposed that DNA is a **double helix** that consists of two polynucleotide strands winding about each other like a spiral staircase. (See Figure 17.6.) The sugar–phosphate backbones are the outside railings with the nitrogen bases arranged like steps along the inside. The two strands run in opposite directions.

▪ Complementary Base Pairs

Each of the bases along one polynucleotide strand forms hydrogen bonds to a specific base on the opposite DNA strand. Adenine only bonds to thymine, and guanine only bonds to cytosine. (See Figure 17.7.) The pairs A—T and G—C are called **complementary base pairs**. The specificity of the base pairing is due to the

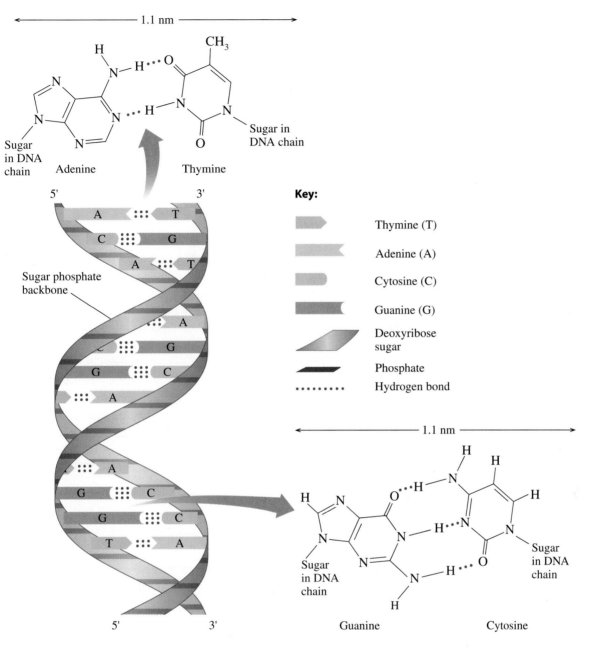

FIGURE 17.7

Hydrogen bonds between complementary base pairs hold the polynucleotide strands together in the double helix of DNA.

Q Why are G—C base pairs more stable than A—T base pairs?

fact that adenine and thymine form two hydrogen bonds, while cytosine and guanine form three hydrogen bonds. This explains why DNA has equal amounts of A and T bases and equal amounts of G and C.

SAMPLE PROBLEM 17.4

■ **Complementary Base Pairs**

Write the base sequence of the complementary segment for the following segment of a strand of DNA.

—A—C—G—A—T—C—T—

SOLUTION

In the complementary strand of DNA, the base A pairs with T, and G pairs with C.

Given segment of DNA:　　—A—C—G—A—T—C—T—
　　　　　　　　　　　　　: : : : : : :
Complementary segment:　—T—G—C—T—A—G—A—

STUDY CHECK

What is the sequence of bases that is complementary to a portion of DNA with a base sequence of —G—G—T—T—A—A—C—C—?

DNA Replication

In DNA **replication**, the strands in the parent DNA separate, which allows each of the original strands to make copies by synthesizing complementary strands. The replication process begins when an enzyme called *helicase* catalyzes the unwinding of a portion of the double helix by breaking the hydrogen bonds between the complementary bases. These single strands now act as templates for the synthesis of new complementary strands. (See Figure 17.8.)

After the base pairs are formed, *DNA polymerase* catalyzes the formation of phosphodiester bonds between the nucleotides. Eventually the entire double helix of the parent DNA is copied. In each new DNA molecule, one strand of the double helix is from the original DNA and one is a newly synthesized strand. This process produces two new DNAs called *daughter DNA* that are identical to each other and exact copies of the original parent DNA. In the process of DNA replication, complementary base pairing ensures the correct placements of bases in the new DNA strands.

SAMPLE PROBLEM 17.5

DNA Replication

In an original DNA strand, there is a portion with the bases —A—G—T—. What nucleotides are placed in a growing DNA daughter strand for this sequence?

SOLUTION

Complementary base pairing allows only one possible nucleotide to pair with each base on the original strand. Thymine will pair only with adenine, cytosine with guanine, and adenine with thymine to give the sequence —T—C—A—.

STUDY CHECK

What is the complementary section of DNA for the base sequence —G—C—A—A—T—C—?

QUESTIONS AND PROBLEMS

DNA Double Helix

17.15 How are the two strands of nucleic acid in DNA held together?

17.16 What is meant by complementary base pairing?

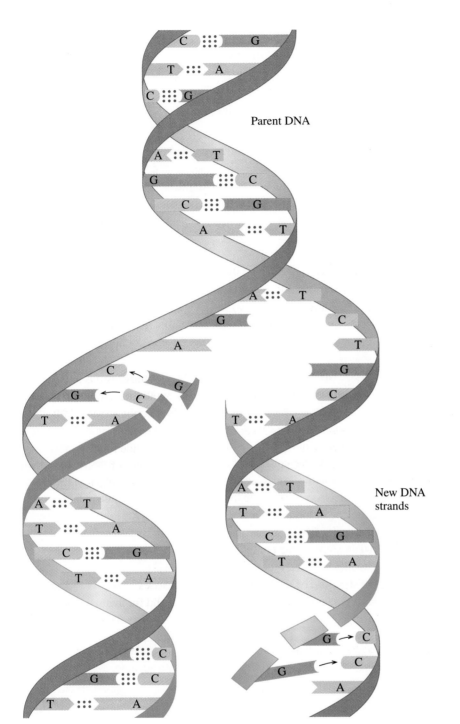

Parent DNA

New DNA strands

FIGURE 17.8

In DNA replication, the separate strands of the parent DNA are the templates for the synthesis of complementary strands, which produces two exact copies of DNA.

Q **How many strands of the parent DNA are in each of the new double-stranded copies of DNA?**

17.17 Complete the base sequence in a second DNA strand if a portion of one strand has the following base sequence:

a. —A—A—A—A—A—A—

b. —G—G—G—G—G—G—

c. —A—G—T—C—C—A—G—G—T—

d. —C—T—G—T—A—T—A—C—G—T—T—A—

17.18 Complete the base sequence in a second DNA strand if a portion of one strand has the following base sequence:

a. —T—T—T—T—T—T—

b. —C—C—C—C—C—C—C—C—C—C—

c. —A—T—G—G—C—A—

d. —A—T—A—T—G—C—G—C—T—A—A—A—

LEARNING GOAL

Identify the different RNAs;
describe the synthesis of mRNA.

17.4 RNA and the Genetic Code

Ribonucleic acid, RNA, which makes up most of the nucleic acid found in the cell, is involved with transmitting the genetic information needed to operate the cell. Similar to DNA, RNA molecules are unbranched polymers of nucleotides. However, there are several important differences.

1. The sugar in RNA is ribose rather than the deoxyribose found in DNA.
2. The nitrogen base uracil replaces thymine.
3. RNA molecules are single, not double stranded.
4. RNA molecules are much smaller than DNA molecules.

Types of RNA

There are three major types of RNA in the cells: *messenger RNA*, *ribosomal RNA*, and *transfer RNA*. Ribosomal RNA (**rRNA**), the most abundant type of RNA, is contained in the ribosomes. Ribosomes, which are the sites for protein synthesis, consist of two subunits, a large subunit and a small subunit. (See Figure 17.9.) Cells that synthesize large numbers of proteins have thousands of ribosomes in their cells.

Messenger RNA (**mRNA**) carries genetic information from the DNA in the nucleus to the ribosomes in the cytoplasm for protein synthesis. Each gene, a segment of DNA, produces a separate mRNA molecule when a certain protein is needed in the cell, but then the mRNA is broken down quickly. The size of an mRNA depends on the number of nucleotides in that particular gene.

Transfer RNA (**tRNA**), the smallest of the RNA molecules, interprets the genetic information in mRNA and brings specific amino acids to the ribosome for protein synthesis. Only the tRNAs can translate the genetic information into amino acids for proteins. There are one or more different tRNAs for each of the 20 amino acids. The structures of the transfer RNAs are similar, consisting of 70–90 nucleotides. Hydrogen bonds between some of the complementary bases in the chain produce loops that give some double-stranded regions.

Although the structure of tRNA is complex, we draw tRNA as a cloverleaf to illustrate its features. All tRNA molecules have a 3′ end with the nucleotide sequence ACC, which is known as the *acceptor stem*. An enzyme attaches an amino acid by forming an ester bond with the free —OH at the end of the acceptor stem.

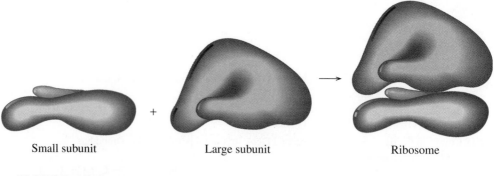

Small subunit   +   Large subunit   Ribosome

FIGURE 17.9

A typical ribosome consists of a small subunit and a large subunit.

Q **Why would there be many thousands of ribosomes in a cell?**

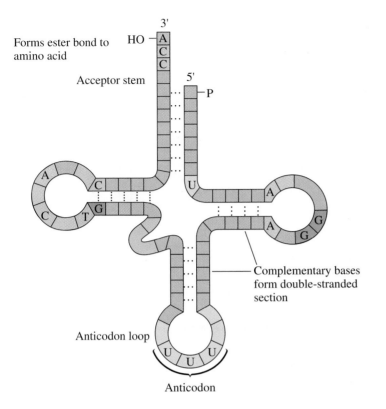

Forms ester bond to amino acid

Acceptor stem

Complementary bases form double-stranded section

Anticodon loop

Anticodon

FIGURE 17.10

A typical tRNA molecule has an acceptor stem at the 3′ end of the nucleic acid where an amino acid attaches, and an anticodon loop that complements three bases on mRNA.

Q Why will different tRNAs have different bases in the anticodon loop?

Each tRNA contains an **anticodon**, which is a series of three bases that complements three bases on an mRNA. (See Figure 17.10.) Table 17.2 summarizes the three types of RNA.

TABLE 17.2

Types of RNA Molecules

Type	Abbreviation	Percentage of Total RNA	Function in the Cell
Ribosomal RNA	rRNA	75	Major component of the ribosomes
Messenger RNA	mRNA	5–10	Carries information for protein synthesis from the DNA in the nucleus to the ribosomes
Transfer RNA	tRNA	10–15	Brings amino acids to the ribosomes for protein synthesis

SAMPLE PROBLEM 17.6

■ **Types of RNA**

What is the function of mRNA in a cell?

SOLUTION

mRNA carries the instructions for the synthesis of a protein from the DNA to the ribosomes.

STUDY CHECK

What is the function of tRNA in a cell?

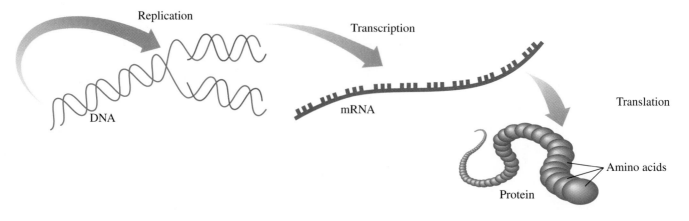

Replication

Translation

DNA

mRNA

Amino acids

Protein

FIGURE 17.11

The genetic information in DNA is replicated in cell division and used to produce messenger RNAs that are converted into amino acids for protein synthesis.

Q **What is the difference between transcription and translation?**

WEB TUTORIAL
Transcription
Overview of Protein Synthesis

Transcription: Synthesis of mRNA

We now look at the processes involved in transferring genetic information encoded in the DNA to the production of proteins. First, the genetic information is copied from a gene in DNA to make a messenger RNA (mRNA), a process called **transcription**. As a carrier molecule, the mRNA moves out of the nucleus and goes to the ribosomes. Then in a process called **translation**, tRNA molecules convert the information in the mRNA into amino acids, which are placed in the proper sequence to synthesize a protein. (See Figure 17.11.) The language that relates the series of nucleotides in mRNA to the amino acids specified is the *genetic code*.

$$\boxed{\text{DNA}} \xrightarrow{\text{Transcription}} \boxed{\text{mRNA}} \xrightarrow{\text{Translation}} \boxed{\text{protein}}$$

Transcription begins when the section of a DNA that contains the gene to be copied unwinds. Within this unwound DNA, a polymerase enzyme identifies a starting point that is the signal to begin mRNA synthesis. Just as in DNA synthesis, C is paired with G, T pairs with A, but A pairs with U (not T). The polymerase enzyme moves along the unwound DNA, forming bonds between the bases. Eventually the RNA polymerase reaches the termination point that signals the end of transcription, the new mRNA is released, and the DNA returns to its double helix structure. (See Figure 17.12.)

RNA polymerase

Section of bases on DNA template: —G—A—A—C—T—

Complementary base sequence in mRNA: —C—U—U—G—A—

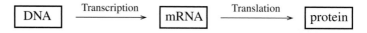

SAMPLE PROBLEM 17.7

■ RNA Synthesis

The sequence of bases in a part of the DNA template for mRNA is —CGATCA—. What is the corresponding mRNA produced?

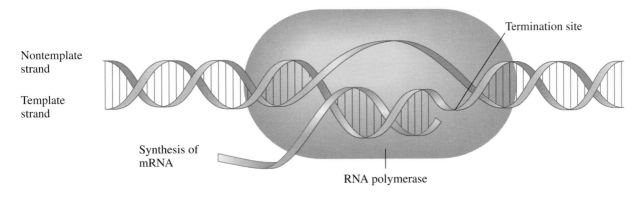

Nontemplate strand

Template strand

Synthesis of mRNA

RNA polymerase

Termination site

FIGURE 17.12

DNA undergoes transcription when RNA polymerase makes a complementary copy of a gene using only one of the DNA strands as the template.

Q **Why is the mRNA synthesized on only one DNA strand?**

SOLUTION

The nucleotides in DNA pair up with the ribonucleotides as follows: G ⟶ C, C ⟶ G, T ⟶ A, and A ⟶ U.

Portion of DNA template: —C—G—A—T—C—A—

Complementary bases in mRNA: —G—C—U—A—G—U—

STUDY CHECK

What is the DNA template that codes for the mRNA having the ribonucleotide sequence —G—G—G—U—U—U—A—A—A—?

The Genetic Code

The overall function of the RNAs in the cell is to facilitate the task of synthesizing protein. After the genetic information encoded in DNA is transcribed into mRNA molecules, the mRNAs move out of the nucleus to the ribosomes in the cytoplasm. At the ribosomes, the genetic information in the mRNAs is converted into a sequence of amino acids.

mRNA $\xrightarrow{\text{Translation at the ribosomes in the cytoplasm}}$ protein

Codons

Genetic information from DNA is encoded in the mRNA as a sequence of nucleotides. In the **genetic code**, a sequence of three bases (triplet), called a **codon**, specifies each amino acid in the protein. Early work on protein synthesis showed that repeating triplets of uracil (UUU) produced a polypeptide that contained only phenylalanine. Therefore, a sequence of —UUU—UUU—UUU— codes for three phenylalanines.

Codons in mRNA —UUU—UUU—UUU—

Translation

Amino acid sequence — Phe — Phe — Phe —

TABLE 17.3

mRNA Codons: The Genetic Code for Amino Acids

First Letter	Second Letter				Third Letter
	U	**C**	**A**	**G**	
U	UUU⎫ Phe UUC⎭ UUA⎫ Leu UUG⎭	UCU⎫ UCC⎪ Ser UCA⎪ UCG⎭	UAU⎫ Tyr UAC⎭ UAA STOP UAG STOP	UGU⎫ Cys UGC⎭ UGA STOP UGG Trp	U C A G
C	CUU⎫ CUC⎪ Leu CUA⎪ CUG⎭	CCU⎫ CCC⎪ Pro CCA⎪ CCG⎭	CAU⎫ His CAC⎭ CAA⎫ Gln CAG⎭	CGU⎫ CGC⎪ Arg CGA⎪ CGG⎭	U C A G
A	AUU⎫ AUC⎬ Ile AUA⎭ ᵃAUG Met/start	ACU⎫ ACC⎪ Thr ACA⎪ ACG⎭	AAU⎫ Asn AAC⎭ AAA⎫ Lys AAG⎭	AGU⎫ Ser AGC⎭ AGA⎫ Arg AGG⎭	U C A G
G	GUU⎫ GUC⎪ Val GUA⎪ GUG⎭	GCU⎫ GCC⎪ Ala GCA⎪ GCG⎭	GAU⎫ Asp GAC⎭ GAA⎫ Glu GAG⎭	GGU⎫ GGC⎪ Gly GGA⎪ GGG⎭	U C A G

ᵃCodon that signals the start of a peptide chain. STOP codons signal the end of a peptide chain.

The codons have now been determined for all 20 amino acids. A total of 64 codons are possible from the triplet combinations of A, G, C, and U. Three of these, UGA, UAA, and UAG, are stop signals that code for the termination of protein synthesis. All the other three-base codons shown in Table 17.3 specify amino acids, which means that one amino acid can have several codons. For example, glycine is the amino acid for the codons GGU, GGC, GGA, and GGG. The triplet AUG has two roles in protein synthesis. At the beginning of an mRNA, the codon AUG signals the start of protein synthesis. In the middle of a series of codons, the AUG codon specifies the amino acid methionine.

SAMPLE PROBLEM 17.8

■ **Codons**

What is the sequence of amino acids coded by the following codons in mRNA?

—GUC—AGC—CCA—

SOLUTION

According to Table 17.3, GUC codes for valine, AGC for serine, and CCA for proline. The sequence of amino acids is Val-Ser-Pro.

STUDY CHECK

The codon UGA does not code for an amino acid. What is its function in mRNA?

QUESTIONS AND PROBLEMS

■ **RNA and the Genetic Code**

17.19 What are the three different types of RNA?

17.20 What are the functions of each type of RNA?

17.21 What is meant by the term "transcription"?

17.22 What bases in mRNA are used to complement the bases A, T, G, and C in DNA?

17.23 Write the corresponding section of mRNA produced from the following section of DNA template:

—CCG—AAG—GTT—CAC—

17.24 Write the corresponding section of mRNA produced from the following section of DNA template:

—TAC—GGC—AAG—CTA—

17.25 What is a codon?

17.26 What is the genetic code?

17.27 What amino acid is coded for by each codon?
   a. CUU
   b. UCA
   c. GGU
   d. AGG

17.28 What amino acid is coded for by each codon?
   a. AAA
   b. GUC
   c. CGG
   d. GCA

17.29 When does the codon AUG signal the start of a protein, and when does it code for the amino acid methionine?

17.30 The codons UGA, UAA, and UAG do not code for amino acids. What is their role as codons in mRNA?

■ 17.5 Protein Synthesis: Translation

Once the mRNA is synthesized, it migrates out of the nucleus into the cytoplasm to the ribosomes. At the ribosomes, the *translation* process involves tRNA molecules, amino acids, and enzymes, all which convert the codons on mRNA into amino acids to make a protein.

We have seen that the genetic code carried by the mRNA consists of triplets of nucleotides that correspond to one of 20 different amino acids. Now the tRNAs are utilized to translate the codons into specific amino acids. This adaptor role of tRNA is possible because it reads both the triplet language of mRNA and picks up the amino acid that corresponds to that codon. The anticodon in the loop at the bottom of a tRNA contains a triplet of bases that complement a codon on mRNA. (See Figure 17.13.) The matching of a tRNA to the correct amino acid is essential. If the wrong amino acid is attached, it would be placed into the protein and make an incorrect protein.

■ Initiation of Protein Synthesis

Protein synthesis begins when an mRNA combines with the smaller subunit of a ribosome. Because the first codon in an mRNA is a *start* codon, AUG, a tRNA with anticodon UAC and carrying the amino acid methionine forms hydrogen bonds with the AUG codon. The larger ribosome completes the ribosomal unit and translation is ready to begin. (See Figure 17.14.) Protein synthesis takes place on

LEARNING GOAL

■ Describe the process of protein synthesis from mRNA.

WEB TUTORIAL
Translation

FIGURE 17.13

An activated tRNA with anticodon AGU bonds to serine at the acceptor stem.

Q **What is the codon for serine for this tRNA?**

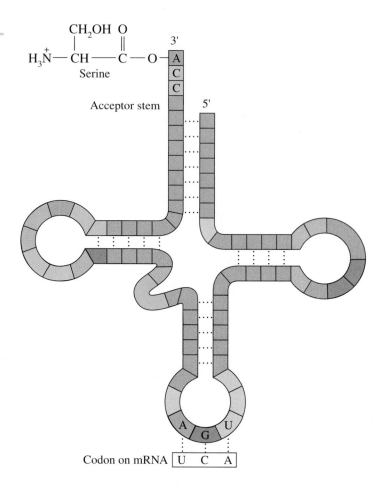

Serine

Acceptor stem

Codon on mRNA U C A

the ribosome at two adjacent sites on the larger subunit. A tRNA that has the anti-codon for the second codon with the second amino acid bonds to mRNA. A peptide bond forms between the amino acids and the first tRNA dissociates from the ribosome and returns to the cytoplasm. After the first tRNA detaches from the ribosome, the ribosome shifts to the next codon on the mRNA, a process called **translocation**. A new tRNA can now attach to the open binding site so that its amino acid forms a peptide bond with the previous amino acid. Again the earlier tRNA detaches and the ribosome shifts down the mRNA to read the next codon. Each time a peptide bond joins the new amino acid to the growing polypeptide chain. Sometimes several ribosomes, called a polysome, translate the same strand of mRNA at the same time to produce several copies of the peptide chain.

Termination

After all the amino acids for a particular protein have been linked together by peptide bonds, the ribosome encounters a stop codon. Because there are no tRNAs to complement the termination codon, protein synthesis ends. An enzyme releases the completed polypeptide chain from the ribosome. The initiating amino acid methionine is often removed from the beginning of the peptide chain. Now the amino acids in the chain form the three-dimensional structure that makes the polypeptide into a biologically active protein.

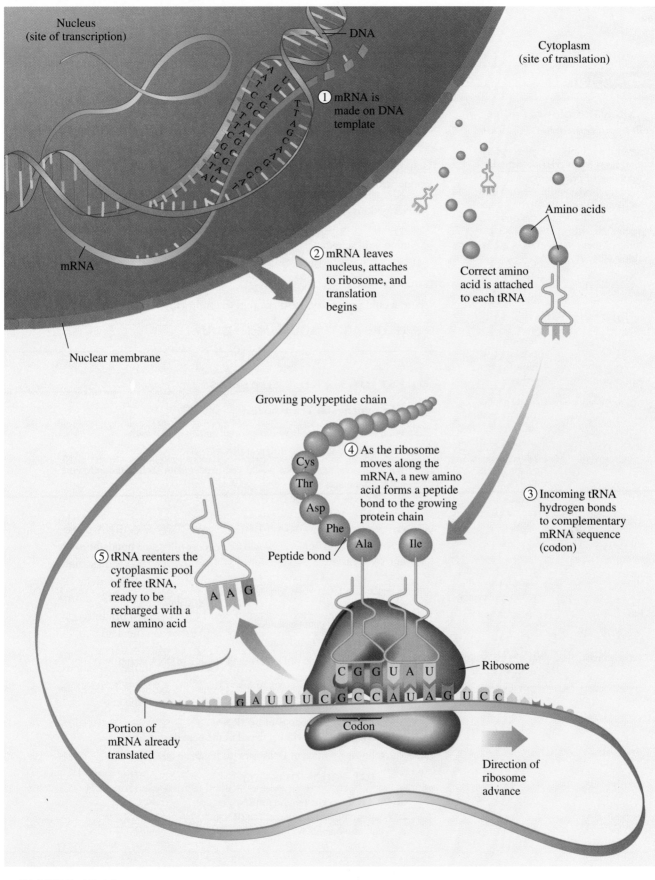

FIGURE 17.14

In the translation process, the mRNA synthesized by transcription attaches to a ribosome and tRNAs pick up their amino acids and place them in a growing peptide chain.

Q **How is the correct amino acid placed in the peptide chain?**

■ Many Antibiotics Inhibit Protein Synthesis

Several antibiotics stop bacterial infections by interfering with the synthesis of proteins needed by the bacteria. Some antibiotics act only on bacterial cells by binding to the ribosomes in bacteria but do not act on human cells. A description of some of these antibiotics is given in Table 17.4.

TABLE 17.4

Antibiotics That Inhibit Protein Synthesis in Bacterial Cells

Antibiotic	Effect on Ribosomes to Inhibit Protein Synthesis
Chloramphenicol	Inhibits peptide bond formation and prevents the binding of tRNA
Erythromycin	Inhibits peptide chain growth by preventing the translocation of the ribosome along the mRNA
Puromycin	Causes release of an incomplete protein by ending the growth of the polypeptide early
Streptomycin	Prevents the proper attachment of tRNAs
Tetracycline	Prevents the binding of tRNAs

SAMPLE PROBLEM 17.9

■ Protein Synthesis: Translation

What order of amino acids would you expect in a peptide for the mRNA sequence of —UCA—AAA—GCC—CUU—?

SOLUTION

Each of the codons specifies a particular amino acid. Using Table 17.3, we write a peptide with the following amino acid sequence:

RNA codons: —UCA—AAA—GCC—CUU—

Amino acid sequence: Ser — Lys — Ala — Leu

STUDY CHECK

Where would protein synthesis stop in the following series of bases in an mRNA?

—GGG—AGC—AGU—UAG—GUU—

QUESTIONS AND PROBLEMS

■ Protein Synthesis: Translation

17.31 What is the difference between a *codon* and an *anticodon*?

17.32 Why are there at least 20 different tRNAs?

17.33 What amino acid sequence would you expect from each of the following mRNA segments?
a. —AAA—AAA—AAA—
b. —UUU—CCC—UUU—CCC—
c. —UAC—GGG—AGA—UGU—

17.34 What amino acid sequence would you expect from each of the following mRNA segments?
a. —AAA—CCC—UUG—GCC—
b. —CCU—CGA—AGC—CCA—UGA—
c. —AUG—CAC—AAA—GAA—GUA—CUU—

17.35 How is a peptide chain extended?

17.36 What is meant by "translocation"?

17.37 The following portion of DNA is in the template DNA strand:

—GCT—TTT—CAA—AAA—

a. What is the corresponding mRNA section?
b. What are the anticodons of the tRNAs?
c. What amino acids will be placed in the peptide chain?

17.38 The following portion of DNA is in the template DNA strand:

—TGT—GGG—GTT—ATT—

a. What is the corresponding mRNA section?
b. What are the anticodons of the tRNAs?
c. What amino acids will be placed in the peptide chain?

17.6 Genetic Mutations

A **mutation** is a change in the DNA nucleotide sequence that alters the sequence of amino acids, which may alter the structure and function of a protein in a cell. Some mutations are known to result from X rays, overexposure to sun (ultraviolet or UV light), chemicals called mutagens, and possibly some viruses. If a change in DNA occurs in a somatic cell (a cell other than a reproductive cell) the altered DNA will be limited to that cell and its daughter cells. If there is uncontrolled growth, the mutation could lead to cancer. If the mutation occurs in germ cell DNA (egg or sperm), then all the DNA produced in a new individual will contain the same genetic change. If the genetic change greatly affects the catalysis of metabolic reactions or the formation of important structural proteins, the new cells may not survive or the person may exhibit a genetic disease.

Types of Mutations

Consider a triplet of bases CCG in the coding strand of DNA, which produces the codon GGC in mRNA. At the ribosome, tRNA would place the amino acid glycine in the peptide chain. (See Figure 17.15a.) Now, suppose that T replaces the first C in the DNA triplet, which gives TCG as the triplet. Then the codon produced in the mRNA is AGC, which brings the amino acid serine to the peptide chain. The replacement of one base in the coding strand of DNA with another is called a **substitution** mutation. The change in the codon can lead to the insertion of a different amino acid at that point in the polypeptide. Substitution is the most common way in which mutations occur. (See Figure 17.15b.)

In a **frameshift mutation**, a base is added to or deleted from the normal order of bases in the coding strand of DNA. Suppose now an A is deleted from the triplet AAA, which gives a new triplet of AAC. The next triplet becomes CGA rather than CCG and so on. All the triplets shift over by one base, which changes all the codons that follow and leads to a different sequence of amino acids from that point. Figure 17.15c illustrates a frameshift mutation by deletion.

Effect of Mutations

When a mutation causes a change in the amino acid sequence, the structure of the resulting protein can be altered severely and it may lose biological activity. If the protein is an enzyme, it may no longer bind to its substrate or react with the substrate at the active site. When an altered enzyme cannot catalyze a reaction, certain substances may accumulate until they act as poisons in the cell, or substances vital to survival may not be synthesized. If a defective enzyme occurs in a major metabolic pathway or in the building of a cell membrane, the mutation can be lethal. When a protein deficiency is genetic, the condition is called a **genetic disease**.

```
        X rays,
      UV sunlight,
        mutagens,
         viruses
DNA  ——————→  alteration of  ——→  defective  ——→  genetic disease (germ cells)
                    DNA               protein          or cancer (somatic cells)
```

FIGURE 17.15

An alteration in the DNA coding strand (template) produces a change in the sequence of amino acids in the protein, which may lead to a mutation. **(a)** A normal DNA leads to the correct amino order in a protein. **(b)** The substitution of a base in DNA leads to a change in the mRNA codon and a change in the amino acid. **(c)** The deletion of a base causes a frameshift mutation, which changes the amino acid order.

Q When would a substitution mutation cause protein synthesis to stop?

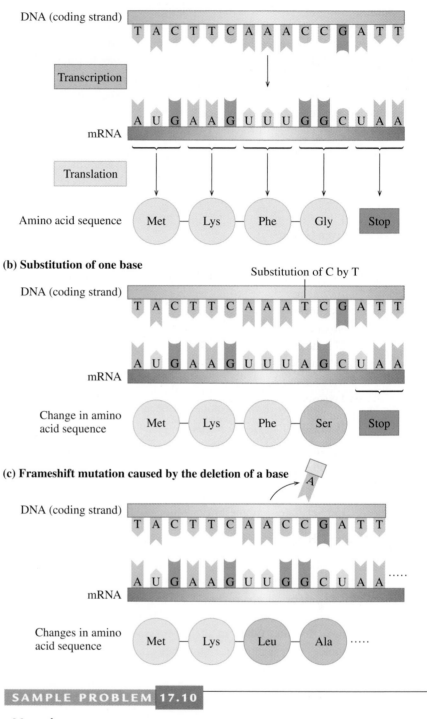

(a) Normal DNA and protein synthesis

DNA (coding strand)

T A C T T C A A A C C G A T T

Transcription

A U G A A G U U U G G C U A A

mRNA

Translation

Amino acid sequence   Met — Lys — Phe — Gly — Stop

(b) Substitution of one base

Substitution of C by T

DNA (coding strand)

T A C T T C A A A T C G A T T

A U G A A G U U U A G C U A A

mRNA

Change in amino acid sequence   Met — Lys — Phe — Ser   Stop

(c) Frameshift mutation caused by the deletion of a base

A

DNA (coding strand)

T A C T T C A A C C G A T T

A U G A A G U U G G C U A A .....

mRNA

Changes in amino acid sequence   Met — Lys — Leu — Ala .....

SAMPLE PROBLEM 17.10

■ **Mutations**

An mRNA has the sequence of codons —CCC—AGA—GGG—. If a base substitution in the DNA changes the mRNA codon of AGA to ACA, how is the amino acid sequence affected in the resulting protein?

SOLUTION

The initial mRNA sequence of —CCC—AGA—GGG— codes for the amino acids proline, arginine, and glycine. When the mutation occurs, the new sequence of mRNA codons CCC—ACA—GGG are for proline, threonine, and glycine. The amino acid arginine is replaced by threonine.

	Normal			*After Mutation*	
mRNA codons	—CCC—	AGA	—GGG—	—CCC—	ACA	—GGG—
Amino acids	— Pro —	Arg	— Gly —	— Pro —	Thr	— Gly —

STUDY CHECK

How might the protein made from this mRNA be affected by this mutation?

Genetic Diseases

A genetic disease is the result of a defective enzyme caused by a mutation in its genetic code. For example, phenylketonuria, PKU, results when DNA cannot direct the synthesis of the enzyme phenylalanine hydroxylase, required for the conversion of phenylalanine to tyrosine. In an attempt to break down the phenylalanine, other enzymes in the cells convert it to phenylpyruvate. The accumulation of phenylalanine and phenylpyruvate in the blood can lead to severe brain damage and mental retardation. If PKU is detected in a newborn baby, a diet is prescribed that eliminates all the foods that contain phenylalanine. Preventing the buildup of the phenylpyruvate ensures normal growth and development.

The amino acid tyrosine is needed in the formation of melanin, the pigment that gives the color to our skin and hair. If the enzyme that converts tyrosine to melanin is defective, no melanin is produced, a genetic disease known as albinism. Persons and animals with no melanin have no skin or hair pigment. (See Figure 17.16.) Table 17.5 lists some other common genetic diseases and the type of metabolism or area affected.

EXPLORE YOUR WORLD

■ A Model for DNA Replication and Mutation

1. Cut out 16 rectangular pieces of paper. For strand 1 of a segment of DNA, write on two pieces this nucleotide symbol with hydrogen bonds: A⚌, T⚌, G☰, and C☰. For strand 2, write on two pieces each: ⚌A, ⚌T, ☰G, and ☰C. Mix up the pieces for strand 1 and randomly place in vertical order. From the second group, strand 2 nucleotides, select the complementary bases to complete this section of DNA.

2. Using the nucleotides from experiment 1, make a strand 1 of

A—T—T—G—C—C

Arrange the corresponding bases to complete the DNA section. In the original column, change the G to an A. What does this do to the complementary strand? How could this change result in a mutation?

FIGURE 17.16

This peacock with albinism does not produce the melanin needed to make bright colors of its feathers.

Q Why are traits such as albinism related to the gene?

QUESTIONS AND PROBLEMS

■ Genetic Mutations

17.39 What is a substitution mutation?

17.40 How does a substitution mutation in the genetic code for an enzyme affect the order of amino acids in that protein?

17.41 What is the effect of a mutation on the amino acids in the polypeptide?

TABLE 17.5
Some Genetic Diseases

Genetic Disease	Result
Galactosemia	The transferase enzyme required for the metabolism of galactose-1-phosphate is absent. Accumulation of Gal-1-P leads to cataracts and mental retardation.
Cystic fibrosis	The most common inherited disease. Thick mucus secretions make breathing difficult and block pancreatic function.
Down syndrome	The leading cause of mental retardation, occurring in about 1 of every 800 live births. Mental and physical problems including heart and eye defects are the result of the formation of three chromosomes, usually chromosome 21, instead of a pair.
Familial hypercholesterolemia	A mutation of a gene on chromosome 19 results in high cholesterol levels that lead to early coronary heart disease in people 30–40 years old.
Muscular dystrophy (Duchenne)	One of 10 forms of MD. A mutation in the X chromosome results in the low or abnormal production of *dystrophin* by the X gene. This muscle-destroying disease appears at about age 5 with death by age 20 and occurs in about 1 of 10,000 males.
Huntington's disease (HD)	Appearing in middle age, HD affects the nervous system, leading to total physical impairment. It is the result of a mutation in a gene on chromosome 4, which can now be mapped to test people in families with HD.
Sickle-cell anemia	A defective hemoglobin from a mutation in a gene on chromosome 11 decreases the oxygen-carrying ability of red blood cells, which take on a sickled shape, causing anemia and plugged capillaries from red blood cell aggregation.
Hemophilia	One or more defective blood-clotting factors lead to poor coagulation, excessive bleeding, and internal hemorrhages.
Tay-Sachs disease	Hexosaminidase A is defective, causing an accumulation of gangliosides resulting in mental retardation, loss of motor control, and early death.

17.42 How can a mutation decrease the activity of a protein?

17.43 How is protein synthesis affected if the normal base sequence TTT in the DNA template is changed to TTC?

17.44 How is protein synthesis affected if the normal base sequence CCC is changed to ACC?

17.45 Consider the following portion of mRNA produced by the normal order of DNA nucleotides:

—ACA—UCA—CGG—GUA—

a. What is the amino acid order produced for normal DNA?
b. What is the amino acid order if a mutation changes UCA to ACA?
c. What is the amino acid order if a mutation changes CGG to GGG?
d. What happens to protein synthesis if a mutation changes UCA to UAA?
e. What happens if a G is added to the beginning of a chain?
f. What happens if the A is removed from the beginning of a chain?

17.46 Consider the following portion of mRNA produced by the normal order of DNA nucleotides:

—CUU—AAA—CGA—GUU—

a. What is the amino acid order produced for normal DNA?
b. What is the amino acid order if a mutation changes CUU to CCU?
c. What is the amino acid order if a mutation changes CGA to AGA?
d. What happens to protein synthesis if a mutation changes AAA to UAA?

17.47 a. A base substitution changes a codon for an enzyme from GCC to GCA. Why is there no change in the amino acid order in the protein?
b. In sickle-cell anemia, a base substitution in hemoglobin replaces glutamine (a polar amino acid) with valine. Why does the replacement of one amino acid cause such a drastic change in biological function?

17.48 a. A base substitution for an enzyme replaces leucine (a nonpolar amino acid) with alanine. Why does this change in amino acids have little effect on the biological activity of the enzyme?

b. A base substitution replaces cytosine in the codon UCA with adenine. How would this substitution affect the amino acids in the protein?

■ 17.7 Viruses

Viruses are small particles of 3 to 200 genes that cannot replicate without a host cell. A typical virus contains a nucleic acid, DNA or RNA, but not both, inside a protein coat. It does not have the necessary material such as nucleotides and enzymes to make proteins and grow. The only way a virus can replicate is to invade a host cell and take over the materials necessary for protein synthesis and growth. Some infections caused by viruses invading human cells are listed in Table 17.6. There are also viruses that attack bacteria, plants, and animals.

A viral infection begins when an enzyme in the protein coat makes a hole in the host cell, allowing the nucleic acids to enter and mix with the materials in the host cell. (See Figure 17.17.) If the virus contains DNA, the host cell begins to replicate

TABLE 17.6

Some Diseases Caused by Viral Infection	
Disease	**Virus**
Common cold	Coronavirus (over 100 types)
Influenza	Orthomyxovirus
Warts	Papovavirus
Herpes	Herpesvirus
Leukemia, cancers, AIDS	Retroviruses
Hepatitis	Hepatitis A virus (HAV), hepatitis B virus (HBV), hepatitis C (HCV)
Mumps	Paramyxovirus
Epstein-Barr	Epstein-Barr virus (EBV)

Head — DNA
Tail — Tail fiber

Uninfected cell

Infected cell

New viruses released

Viral DNA replicates

Cell bursts

Cellular enzymes → Viral protein + Viral DNA ← Cellular enzymes make viral proteins and viral DNA, which assemble into viruses.

FIGURE 17.17

After a virus attaches to the host cell, it injects its viral DNA and uses the host cell's amino acids to synthesize viral protein and uses the host cell's nucleic acids, enzymes, and ribosomes to make viral RNA. When the cell bursts, the new viruses are released to infect other cells.

Q Why does a virus need a host cell for replication?

FIGURE 17.18

After a retrovirus injects its viral RNA into a cell, it forms a DNA strand by reverse transcription. The DNA forms a double-stranded DNA called a provirus, which joins the host cell DNA. When the cell replicates, the provirus produces the viral RNA needed to produce more virus particles.

Q **What is reverse transcription?**

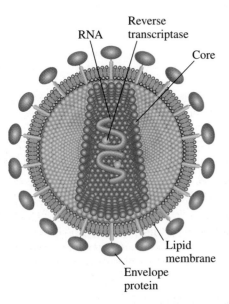

FIGURE 17.19

The HIV virus that causes AIDS syndrome destroys the immune system in the body.

Q **Is the HIV virus a DNA virus or an RNA retrovirus?**

WEB TUTORIAL
HIV Reproductive Cycle

the viral DNA in the same way it would replicate normal DNA. Viral DNA produces viral RNA, which proceeds to make the proteins for more coats. So many virus particles are synthesized that the cell bursts and releases new viruses to infect more cells.

Vaccines are inactive forms of viruses that boost the immune response by causing the body to produce antibodies to the virus. Several childhood diseases such as polio, mumps, chicken pox, and measles can be prevented through the use of vaccines. A protease produces a protein coat to form a viral particle that leaves the cell to infect other cells. (See Figure 17.18.)

Reverse Transcription

A virus that contains RNA as its genetic material is a **retrovirus**. Once inside the host cell, it must first make viral DNA using a process known as reverse transcription. A retrovirus contains a polymerase enzyme called *reverse transcriptase* that uses the viral RNA template to synthesize complementary strands of DNA. Once produced, the DNA strands form double-stranded DNA using the nucleotides present in the host cell. This newly formed viral DNA, called a *provirus*, joins the DNA of the host cell.

AIDS

In the early 1980s, a disease called AIDS (acquired immune deficiency syndrome) began to claim an alarming number of lives. An HIV-1 virus (human immunodeficiency virus type 1) is now known to be the AIDS-causing agent. (See Figure 17.19.) HIV is a retrovirus that infects and destroys T4 lymphocyte cells, which are involved in the immune response. After the HIV-1 virus binds to receptors on the surface of a T4 cell, the virus injects viral RNA into the host cell. As a retrovirus, the genes of the viral RNA direct the formation of viral DNA. The gradual depletion of T4 cells reduces the ability of the immune system to destroy harmful organisms. The AIDS syndrome is characterized by opportunistic infections such as *Pneumocystis carinii*, which causes pneumonia, and *Kaposi's sarcoma*, a skin cancer.

Treatment for AIDS is based on attacking the HIV-1 at different points in its life cycle including reverse transcription and protein synthesis. Nucleoside analogs mimic the structures of the nucleosides used for DNA synthesis. For example, the drug AZT (azidothymine) is similar to thymine, and ddI (dideoxyinosine) is similar to guanosine. Two other drugs include dideoxycytidine (ddC) and didehydro-3′-deoxythymidine (d4T). Such compounds are found in the "cocktails" that are providing extended remission of HIV infections. When a nucleoside analog is incorporated into viral DNA, the lack of a hydroxyl group on the 3′ carbon in the sugar prevents the formation of the sugar–phosphate bonds and stops the replication of the virus.

AZT
azidothymine

dideoxyinosine (ddI)

dideoxycytidine (ddC)

didehydro-3'-deoxythymidine (d4T)

The newest and most powerful anti-HIV drugs are the protease inhibitors such as saquinavir (Invirase), indinavir, and ritonavir. The inhibition of protease prevents the synthesis of proteins needed to make more copies of the virus. Researchers are not yet certain how long protease inhibitors will be beneficial for a person with AIDS, but they are encouraged by the current studies.

▪ Cancer

In an adult body, many cells do not continue to reproduce. When cells in the body begin to grow and multiply without control, they invade neighboring cells and appear as a tumor or growth (neoplasm). When tumors interfere with normal functions of the body, they are cancerous. If they are limited, they are benign. Cancer can be caused by chemical and environmental substances, by radiation, or by oncogenic viruses.

Some reports estimate that 70–80% of all human cancers are initiated by chemical and environmental substances. A carcinogen is any substance that increases the chance of inducing a tumor. Known carcinogens include dyes, cigarette smoke, and asbestos. More than 90% of all persons with lung cancer are smokers. A carcinogen causes cancer by reacting with molecules in a cell, probably DNA, and altering the growth of that cell. Some known carcinogens are listed in Table 17.7.

TABLE 17.7

Some Chemical and Environmental Carcinogens

Carcinogen	Tumor Site
Asbestos	Lung, respiratory tract
Arsenic	Skin, lung
Cadmium	Prostate, kidneys
Chromium	Lung
Nickel	Lung, sinuses
Aflatoxin	Liver
Nitrites	Stomach
Aniline dyes	Bladder
Vinyl chloride	Liver

Radiant energy from sunlight or medical radiation is another type of environmental factor. Skin cancer has become one of the most prevalent forms of cancer. It appears that DNA damage in the exposed areas of the skin causes mutations. The cells lose their ability to control protein synthesis, and uncontrolled cell division leads to cancer. The incidence of malignant melanoma, one of the most serious skin cancers, has been rapidly increasing. Some possible factors for this increase may be the popularity of suntanning as well as the reduction of the ozone layer, which absorbs much of the harmful radiation from sunlight.

Oncogenic viruses cause cancer when cells are infected. Several viruses associated with human cancers are listed in Table 17.8. Some cancers such as retinoblastoma and breast cancer appear to occur more frequently in families. There is some indication that a missing or defective gene may be responsible.

TABLE 17.8

Human Cancers Caused by Oncogenic Viruses

Virus	Disease
RNA viruses	
Human T-cell lymphotropic virus-type I (HTLV-I)	Leukemia
DNA viruses	
Epstein–Barr virus (EBV)	Epstein–Barr
	Burkitt's lymphoma (cancer of white blood B cells)
	Nasopharyngeal carcinoma
	Hodgkin's disease
Hepatitis B virus (HBV)	Liver cancer
Herpes simplex virus (type 2)	Cervical and uterine cancer
Papilloma virus	Cervical and colon cancer, genital warts

SAMPLE PROBLEM 17.11

▪ Viruses

Why are viruses unable to replicate on their own?

SOLUTION

Viruses only contain packets of DNA or RNA, but not the necessary replication machinery that includes enzymes and nucleosides.

STUDY CHECK

How do protease inhibitors affect the life cycle of the HIV-1 virus?

QUESTIONS AND PROBLEMS

■ Viruses

17.49 What type of genetic information is found in a virus?

17.50 Why do viruses need to invade a host cell?

17.51 A virus contains viral RNA.
   a. Why would reverse transcription be used in the life cycle of this type of virus?
   b. What is the name of this type of virus?

17.52 What is the purpose of a vaccine?

17.53 How do nucleoside analogs disrupt the life cycle of the HIV-1 virus?

17.54 How do protease inhibitors disrupt the life cycle of the HIV-1 virus?

CONCEPT MAP

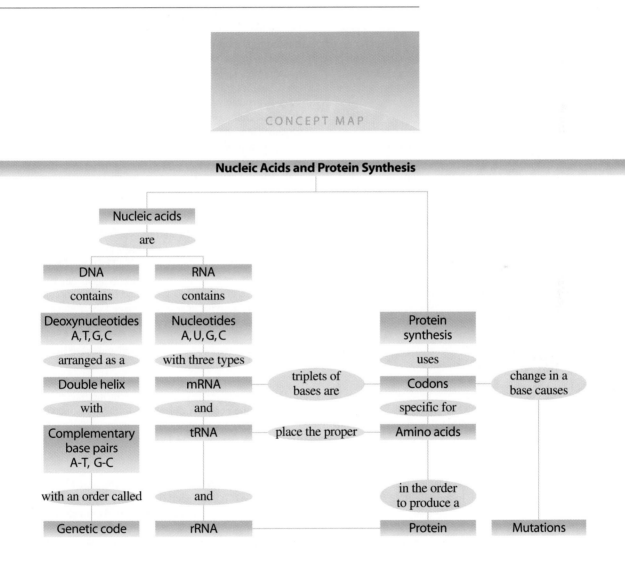

CHAPTER REVIEW

17.1 Components of Nucleic Acids

Nucleic acids, deoxyribonucleic acid (DNA) and ribonucleic acid (RNA), are polymers of nucleotides. A nucleoside is a combination of a pentose sugar and a nitrogen base. A nucleotide is composed of three parts: a nitrogen base, a sugar, and a phosphate group. In DNA, the sugar is deoxyribose and the nitrogen-containing base can be adenine, thymine, guanine, or cytosine. In RNA, the sugar is ribose and uracil replaces thymine.

17.2 Primary Structure of Nucleic Acids

Each nucleic acid has its own unique sequence of bases known as its primary structure. In a nucleic acid polymer, the 3′ OH of each ribose sugar in RNA or deoxyribose sugar in DNA forms a phosphodiester bond to the phosphate group of the 5′ carbon atom group of the sugar in the next nucleotide to give a backbone of alternating sugar and phosphate groups. There is a free 5′ phosphate at one end of the polymer, and a free 3′ OH group at the other end.

17.3 DNA Double Helix

A DNA molecule consists of two strands of nucleotides that are wound around each other like a spiral staircase. The two strands are held together by hydrogen bonds between complementary base pairs, A with T and G with C. During DNA replication, new DNA strands are made along each of the original DNA strands that serve as templates. Complementary base pairing ensures the correct pairing of bases to give identical copies of the original DNA.

17.4 RNA and the Genetic Code

The three types of RNA differ by function in the cell: ribosomal RNA makes up most of the structure of the ribosomes, messenger RNA carries genetic information from the DNA to the ribosomes, and transfer RNA places the correct amino acids in the protein.

Transcription is the process by which RNA polymerase produces mRNA from one strand of DNA. The bases in the mRNA are complementary to the DNA, except U is paired with A in RNA. The production of mRNA occurs when certain proteins are needed in the cell. The genetic code consists of a sequence of three bases (triplet) that specifies the order for the amino acids in a protein. The codon AUG signals the start of transcription and codons UAG, UGA, and UAA signal it to stop.

17.5 Protein Synthesis: Translation

Proteins are synthesized at the ribosomes in a translation process that includes three steps: initiation, translocation, and termination. During translation, tRNAs bring the appropriate amino acids to the ribosome and peptide bonds form. When the polypeptide is released, it takes on its secondary and tertiary structures and becomes a functional protein in the cell.

17.6 Genetic Mutations

A genetic mutation is a change of one or more bases in the DNA sequence that alters the structure and ability of the resulting protein to function properly. In a substitution, one base may be altered. A frameshift mutation inserts or deletes a base and changes all the codons after the base change.

17.7 Viruses

Viruses containing DNA or RNA invade host cells where they use the machinery within the cell to synthesize more viruses. For a retrovirus containing RNA, a viral DNA is synthesized by reverse transcription using the nucleotides and enzymes in the host cell. In the treatment of AIDS, nucleoside analogs inhibit the reverse transcriptase of the HIV-1 virus, and protease inhibitors disrupt the catalytic activity of protease needed to produce proteins for the synthesis of more viruses.

KEY TERMS

anticodon The triplet of bases in the center loop of tRNA that is complementary to a codon on mRNA.

codon A sequence of three bases in mRNA that specifies a certain amino acid to be placed in a protein. A few codons signal the start or stop of transcription.

complementary base pairs In DNA, adenine is always paired with thymine (A—T or T—A), and guanine is always paired with cytosine (G—C or C—G). In forming RNA, adenine is paired with uracil (A—U).

DNA Deoxyribonucleic acid; the genetic material of all cells containing nucleotides with deoxyribose sugar, phosphate, and the four nitrogen bases adenine, thymine, guanine, and cytosine.

double helix The helical shape of the double chain of DNA that is like a spiral staircase with a sugar–phosphate backbone on the outside and base pairs like stair steps on the inside.

frameshift mutation A mutation that inserts or deletes a base in a DNA sequence.

genetic code The sequence of codons in mRNA that specifies the amino acid order for the synthesis of protein.

genetic disease A physical malformation or metabolic dysfunction caused by a mutation in the base sequence of DNA.

mRNA Messenger RNA; produced in the nucleus by DNA to carry the genetic information to the ribosomes for the construction of a protein.

mutation A change in the DNA base sequence that alters the formation of a protein in the cell.

nitrogen-containing base Nitrogen-containing compounds found in DNA and RNA: adenine (A), thymine (T), cytosine (C), guanine (G), and uracil (U).

nucleic acids Large molecules composed of nucleotides, found as a double helix in DNA and as the single strands of RNA.

nucleoside The combination of a pentose sugar and a nitrogen-containing base.

nucleotides Building blocks of a nucleic acid consisting of a nitrogen-containing base, a pentose sugar (ribose or deoxyribose), and a phosphate group.

phosphodiester bond The phosphate link that joins the 3′-hydroxyl group in one nucleotide to the phosphate group on the 5′-carbon atom in the next nucleotide.

primary structure The sequences of nucleotides in nucleic acids.

replication The process of duplicating DNA by pairing the bases on each parent strand with their complementary base.

retrovirus A virus that contains RNA as its genetic material and that synthesizes a complementary DNA strand inside a cell.

RNA Ribonucleic acid, a type of nucleic acid that is a single strand of nucleotides containing adenine, cytosine, guanine, and uracil.

rRNA Ribosomal RNA; the most prevalent type of RNA; a major component of the ribosomes.

substitution A mutation that replaces one base in a DNA with a different base.

transcription The transfer of genetic information from DNA by the formation of mRNA.

translation The interpretation of the codons in mRNA as amino acids in a peptide.

translocation The shift of a ribosome along mRNA from one codon (three bases) to the next codon during translation.

tRNA Transfer RNA; an RNA that places a specific amino acid into a peptide chain at the ribosome. There is one or more tRNA for each of the 20 different amino acids.

virus Small particles containing DNA or RNA in a protein coat that require a host cell for replication.

ADDITIONAL QUESTIONS AND PROBLEMS

17.55 Identify each of the following nitrogen bases as a pyrimidine or a purine:
  a. cytosine   **b.** adenine   **c.** uracil
  d. thymine   **e.** guanine

17.56 Indicate if each of the nitrogen bases in problem 17.55 are found in DNA only, RNA only, or both DNA and RNA.

17.57 Identify the nitrogen base and sugar in each of the following nucleosides:
  a. deoxythymidine
  b. adenosine
  c. cytidine
  d. deoxyguanosine

17.58 Identify the nitrogen base and sugar in each of the following nucleotides:
  a. CMP
  b. dAMP
  c. dGMP
  d. UMP

17.59 How do the bases of thymine and uracil differ?

17.60 How do the bases of cytosine and uracil differ?

17.61 What is similar about the primary structure of RNA and DNA?

17.62 What is different about the primary structure of RNA and DNA?

17.63 Write the complementary base sequence for each of the following DNA segments:
  a. —G—A—C—T—T—A—G—G—C—
  b. —T—G—C—A—A—A—C—T—A—G—C—T—
  c. —A—T—C—G—A—T—C—G—A—T—C—G—

17.64 Write the complementary base sequence for each of the following DNA segments:
  a. —T—T—A—C—G—G—A—C—C—G—C—
  b. —A—T—A—G—C—C—C—T—
            T—A—C—T—G—G—
  c. —G—G—C—C—T—A—C—C—T—T—A—
            A—C—G—A—C—G—

17.65 Match the following statements with rRNA, mRNA, or tRNA.
  a. is the smallest type of RNA
  b. makes up the highest percent of RNA in the cell
  c. carries genetic information from the nucleus to the ribosomes

17.66 Match the following statements with rRNA, mRNA, or tRNA.
  a. have two subunits
  b. brings amino acids to the ribosomes for protein synthesis
  c. acts as a template for protein synthesis

17.67 What are the possible codons for each of the following amino acids?
  a. threonine
  b. serine
  c. cysteine

17.68 What are the possible codons for each of the following amino acids?
  a. valine
  b. proline
  c. histidine

17.69 What is the amino acid for each of the following codons?
  a. AAG
  b. AUU
  c. CGG

17.70 What is the amino acid for each of the following codons?
  a. CAA
  b. GGC
  c. AAC

17.71 Endorphins are polypeptides that reduce pain. What is the amino acid order for the following mRNA that codes for a pentapeptide that is an endorphin called leucine enkephalin?
  —AUG—UAC—GGU—GGA—UUU—CUA—UAA—

17.72 Endorphins are polypeptides that reduce pain. What is the amino acid order for the following mRNA that codes for a pentapeptide that is an endorphin called methionine enkephalin?
  —AUG—UAC—GGU—GGA—UUU—AUG—UAA—

17.73 What is the anticodon on tRNA for each of the following codons in an mRNA?
  a. AGC   **b.** UAU   **c.** CCA

17.74 What is the anticodon on tRNA for each of the following codons in an mRNA?
  a. GUG   **b.** CCC   **c.** GAA

UNDERSTANDING THE CONCEPTS

17.75 Answer the following questions for the given section of DNA.
a. Complete the bases in the parent and template strands.

Parent strand

Template strand

b. Using the template strand, write the mRNA sequence

c. Write the 3-letter symbols of the amino acids that would go into the peptide from the mRNA you wrote in part b.

17.76 Suppose a mutation occurs in the DNA section in problem 17.75 and the first base in the parent chain, adenine, is replaced by guanine.
a. What type of mutation has occurred?
b. Using the template strand that results from this mutation, write the order of bases in the altered mRNA.

c. Write the 3-letter symbols of the amino acids that would go into the peptide from the mRNA you wrote in part b.

d. What effect, if any, might this mutation have on the structure and/or function of the resulting protein?

QUESTIONS FOR TEAMWORK

17.77 Oxytocin is a nonapeptide with nine amino acids. How many nucleotides would be found in the mRNA for this protein?

17.78 A protein contains 35 amino acids. How many nucleotides would be found in the mRNA for this protein?

17.79 a. If the DNA double helix in salmon contains 28% adenine, what is the percent of thymine, guanine, and cytosine?

b. If the DNA double helix in humans contains 20% cytosine, what is the percent of guanine, adenine, and thymine?

17.80 Why are there no base pairs in DNA between adenine and guanine, or thymine and cytosine?

17.81 What is the difference between a DNA virus and a retrovirus?

ANSWERS

Answers to Study Checks

17.1 a. Guanine is found in both RNA and DNA.
b. Uracil is found only in RNA.

17.2 deoxycytidine 5′-monophosphate (dCMP)

17.3

Free 5′ phosphate

Cytidine

Cytidine

Free 3′ OH ⟶ OH OH

17.4 —C—C—A—A—T—T—G—G—

17.5 —C—G—T—T—A—G—

17.6 Each type of tRNA brings specific amino acids to the ribosome for protein synthesis.

17.7 —C—C—C—A—A—A—T—T—T—

17.8 UGA is a stop codon that signals the termination of translation.

17.9 at UAG

17.10 If the substitution of an amino acid in the polypeptide affects an interaction essential to functional structure on the binding of a substrate, the resulting protein could be less effective or nonfunctional.

17.11 A protease inhibitor modifies the three-dimensional structure of protease, which prevents the enzymes from synthesizing the proteins needed to produce more viruses.

Answers to Selected Questions and Problems

17.1 a. pyrimidine   **b.** pyrimidine

17.3 a. DNA   **b.** both DNA and RNA

17.5 deoxyadenosine 5′-monophosphate (dAMP), deoxythymidine 5′-monophosphate (dTMP), deoxycytidine 5′-monophosphate (dCMP), and deoxyguanosine 5′-monophosphate (dGMP)

17.7 a. nucleoside   **b.** nucleoside
     c. nucleoside   **d.** nucleotide

17.9

17.11 The nucleotides in nucleic acids are held together by phosphodiester bonds between the 3′ OH of a sugar (ribose or deoxyribose) and a phosphate group on the 5′ carbon of another sugar.

17.13

17.15 The two DNA strands are held together by hydrogen bonds between the bases in each strand.

17.17 a. —T—T—T—T—T—T—
     b. —C—C—C—C—C—C—
     c. —T—C—A—G—G—T—C—C—A—
     d. —G—A—C—A—T—A—T—G—
                             C—A—A—T—

17.19 Ribosomal RNA, messenger RNA, and transfer RNA

17.21 In transcription, the sequence of nucleotides on a DNA template (one strand) is used to produce the base sequences of a messenger RNA.

17.23 —G—G—C—U—U—C—C—A—A—G—U—G—

17.25 A three-base sequence in mRNA that codes for a specific amino acid in a protein.

17.27 a. leucine   **b.** serine
     c. glycine   **d.** arginine

17.29 When AUG is the first codon, it signals the start of protein synthesis. Thereafter, AUG codes for methionine.

17.31 A codon is a base triplet in the mRNA. An anticodon is the complementary triplet on a tRNA for a specific amino acid.

17.33 a. —Lys—Lys—Lys—
     b. —Phe—Pro—Phe—Pro—
     c. —Tyr—Gly—Arg—Cys—

17.35 The new amino acid is joined by a peptide bond to the peptide chain. The ribosome moves to the next codon, which attaches to a tRNA carrying the next amino acid.

17.37 a. —CGA—AAA—GUU—UUU—
     b. GCU, UUU, CAA, AAA
     c. Using codons in mRNA: —Arg—Lys—Val—Phe—

17.39 A base in DNA is replaced by a different base.

17.41 If the resulting codon still codes for the same amino acid, there is no effect. If the new codon codes for a different amino acid, there is a change in the order of amino acids in the polypeptide.

17.43 The normal triplet TTT forms a codon AAA, which codes for lysine. The mutation TTC forms a codon AAG, which also codes for lysine. There is no effect on the amino acid sequence.

17.45 a. —Thr—Ser—Arg—Val—
     b. —Thr—Thr—Arg—Val—
     c. —Thr—Ser—Gly—Val—
     d. —Thr—STOP   Protein synthesis would terminate early. If this occurs early in the formation of the polypeptide, the resulting protein will probably be nonfunctional.
     e. The new protein will contain the sequence —Asp—Ile—Thr—Gly—.
     f. The new protein will contain the sequence —His—His—Gly—.

17.47 a. GCC and GCA both code for alanine.
     b. A vital ionic cross-link in the tertiary structure of hemoglobin cannot be formed when the polar glutamine is replaced by valine, which is nonpolar. The resulting hemoglobin is malformed and less capable of carrying oxygen.

17.49 DNA or RNA, but not both.

17.51 a. Viral RNA is used to synthesize viral DNA, which produces the proteins for the protein coat that allows the virus to replicate and leave the cell.
     b. retrovirus

17.53 Nucleoside analogs such as AZT and ddI are similar to the nucleosides required to make viral DNA in reverse transcription. However, they interfere with the ability of the DNA to form and thereby disrupt the life cycle of the HIV-1 virus.

17.55 a. pyrimidine   **b.** purine   **c.** pyrimidine
     d. pyrimidine   **e.** purine

17.57 a. thymine and deoxyribose
     b. adenine and ribose
     c. cytosine and ribose
     d. guanine and deoxyribose

17.59 They are both pyrimidines, but thymine has a methyl group.

17.61 They are both polymers of nucleotides connected through phosphodiester bonds between alternating sugar and phosphate groups with bases extending out from each sugar.

17.63 a. —C—T—G—A—A—T—C—C—G—
b. —A—C—G—T—T—T—G—A—T—C—G—A—
c. —T—A—G—C—T—A—G—C—T—A—G—C—

17.65 a. tRNA   **b.** rRNA   **c.** mRNA

17.67 a. ACU, ACC, ACA, and ACG
b. UCU, UCC, UCA, UCG, AGU, and AGC
c. UGU and UGC

17.69 a. lysine   **b.** isoleucine   **c.** arginine

17.71 START—Tyr—Gly—Gly—Phe—Leu—STOP

17.73 a. UCG   **b.** AUA   **c.** GGU

17.75
a.

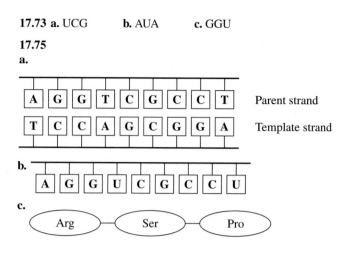

b.

c.

18 Metabolic Pathways and Energy Production

"I am trained in basic life support. I work with the ER staff to assist in patient care," says Mandy Dornell, emergency medical technician at Seaton Medical Center. "In the ER, I take vital signs, do patient assessment, and do CPR. If someone has a motor vehicle accident, I may suspect a neck or back injury. Then I may use a backboard or a cervical collar, which prevents the patient from moving and causing further damage. When people have difficulty breathing, I insert an airway—nasal or oral—to assist ventilation. I also set up and monitor IVs, and I am trained in childbirth."

When someone is critically ill or injured, the quick reactions of emergency medical technicians (EMTs) and paramedics provide immediate medical care and transport to an ER or trauma center.

the Chemistry place

Visit **www.aw-bc.com/chemplace** or use the CD-ROM in the book for extra quizzes, interactive tutorials, career resources, PowerPoint slides, math help, and case studies.

All the chemical reactions that take place in living cells to break down or build molecules are known as *metabolism*. In a metabolic pathway, reactions are linked together in a series, each catalyzed by a specific enzyme. In this chapter, we will look at these pathways and the way they produce energy and cellular compounds.

When we eat food such as a tuna fish sandwich, the polysaccharides, lipids, and proteins are digested to smaller molecules that are absorbed into the cells of our body. As these molecules of glucose, fatty acids, and amino acids are broken down further, energy is released. This energy is used in the cells to synthesize high-energy adenosine triphosphate (ATP). Our cells utilize ATP energy when they do work such as contracting muscles, synthesizing large molecules, sending nerve impulses, and moving substances across cell membranes.

In the *citric acid cycle*, a series of metabolic reactions in the mitochondria oxidize acetyl CoA to carbon dioxide, which releases energy to produce NADH and $FADH_2$. These reduced coenzymes enter the *electron transport chain* or *respiratory chain* where they provide hydrogen ions and electrons that combine with oxygen (O_2) to form H_2O. The energy released during electron transport is used to synthesize ATP from ADP.

Although glucose is the primary fuel for the synthesis of ATP, lipids and proteins also play an important role in metabolism and energy production. Almost all our energy is stored in the form of triacylglycerols in fat cells. When our caloric intake exceeds the nutritional and metabolic needs of the body, excess carbohydrates and fatty acids are converted to triacylglycerols and added to our fat cells.

The degradation of proteins provides amino acids, which are needed to synthesize nitrogen-containing compounds in our cells. Although amino acids are not considered a primary source of fuel, energy can be extracted from amino acids if carbohydrate and fat reserves have been depleted.

LEARNING GOAL

Describe three stages of metabolism and the role of ATP.

18.1 Metabolism and ATP Energy

The term **metabolism** refers to all the chemical reactions that provide energy and the substances required for continued cell growth. There are two types of metabolic reactions: catabolic and anabolic. In **catabolic reactions**, complex molecules are broken down to simpler ones with an accompanying release of energy. **Anabolic reactions** utilize energy available in the cell to build large molecules from simple ones.

We can think of the catabolic processes in metabolism as consisting of three stages. (See Figure 18.1.) Consider a tuna fish sandwich for our example. In stage 1 of metabolism, the processes of digestion break down the large macromolecules into small monomer units. The polysaccharides in bread break down to monosac-

Stages of Metabolism

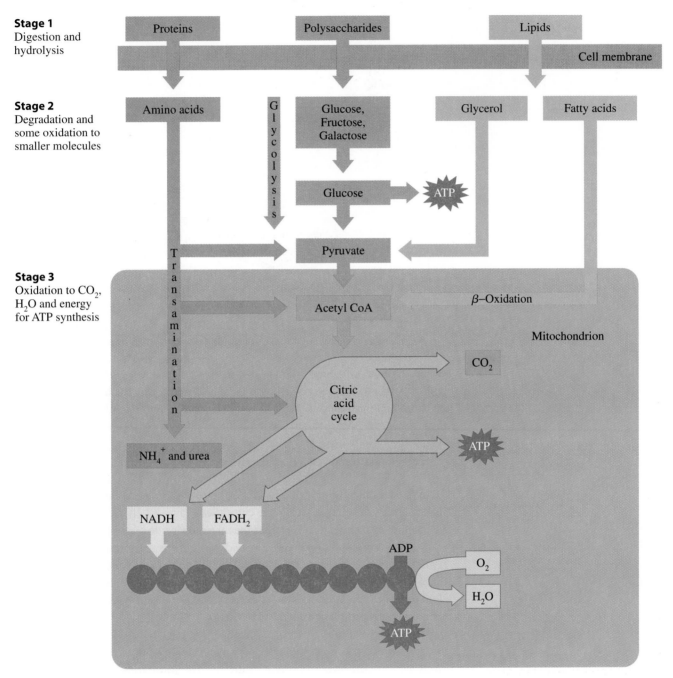

Stage 1
Digestion and hydrolysis

Stage 2
Degradation and some oxidation to smaller molecules

Stage 3
Oxidation to CO_2, H_2O and energy for ATP synthesis

FIGURE 18.1

In the three stages of catabolic metabolism, foods are digested and degraded into smaller molecules, which are oxidized to produce energy.

Q Where is most of the ATP energy produced in the cells?

charides, the lipids in the mayonnaise break down to glycerol and fatty acids, and the proteins from the tuna yield amino acids. These digestion products diffuse into the bloodstream for transport to cells. In stage 2, they are broken down to two- and three-carbon compounds such as pyruvate and acetyl CoA. Stage 3 begins with the

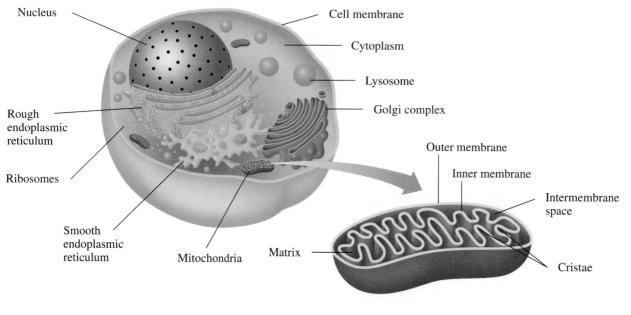

FIGURE 18.2

The diagram illustrates the major components of a typical animal cell.

Q What is the cytoplasm in a cell?

oxidation of the two-carbon acetyl CoA in the citric acid cycle, which produces several reduced coenzymes. As long as the cells have oxygen, the hydrogen ions and electrons are transferred from the coenzymes to the electron transport chain, where most of the energy for the cell is produced.

Cell Structure for Metabolism

To understand the relationship of metabolic reactions, we need to look at where metabolic reactions take place in the cells. (See Figure 18.2.)

In animals, a *membrane* separates the materials inside the cell from the aqueous environment surrounding the cell. The *nucleus* contains the genes that control DNA replication and protein synthesis of the cell. The **cytoplasm** consists of all the materials between the nucleus and the cell membrane. The **cytosol**, which is the fluid part of the cytoplasm, is an aqueous solution of electrolytes and enzymes that catalyze many of the cell's chemical reactions.

Within the cytoplasm are specialized structures called *organelles* that carry out specific functions in the cell. We have already seen (Chapter 17) that the *ribosomes* are the sites of protein synthesis. The **mitochondria** are the energy-producing factories of the cells. A mitochondrion consists of an outer membrane and an inner membrane with an intermembrane space between them. The fluid section surrounded by the inner membrane is called the *matrix*. Enzymes located in the matrix and along the inner membrane catalyze the oxidation of carbohydrates, fats, and amino acids. All of these oxidation pathways lead to CO_2, H_2O, and energy, which is used to form energy-rich compounds. Table 18.1 summarizes some of the functions of the cellular components in animal cells.

TABLE 18.1

Locations and Functions of Components in Animal Cells

Component	Description and Function
Cell membrane	Separates the contents of a cell from the external environment and contains structures that communicate with other cells
Cytoplasm	Consists of all of the cellular contents between the cell membrane and nucleus
Cytosol	Is the fluid part of the cytoplasm that contains enzymes for many of the cell's chemical reactions
Endoplasmic reticulum	Rough type processes proteins for secretion and synthesizes phospholipids; smooth type synthesizes fats and steroids
Golgi complex	Modifies and secretes proteins from the endoplasmic reticulum and synthesizes glycoproteins and cell membranes
Lysosomes	Contain hydrolytic enzymes that digest and recycle old cell structures
Mitochondria	Contain the structures for the synthesis of ATP from energy-producing reactions
Nucleus	Contains genetic information for the replication of DNA and the synthesis of protein
Ribosomes	Are the sites of protein synthesis using mRNA templates

SAMPLE PROBLEM 18.1

■ Metabolism and Cell Structure

Identify the following as catabolic or anabolic reactions.

a. Digestion of polysaccharides
b. Synthesis of proteins
c. Oxidation of glucose to CO_2 and H_2O

SOLUTION

a. The breakdown of large molecules is a catabolic reaction.
b. The synthesis of large molecules requires energy and involves anabolic reactions.
c. The breakdown of monomers such as glucose involves catabolic reactions.

STUDY CHECK

What is the difference between the cytoplasm and cytosol?

■ ATP and Energy

In our cells, the energy released from the oxidation of the food we eat is used to form a "high energy" compound called *adenosine triphosphate* (**ATP**). The ATP molecule is composed of the nitrogen base adenine, a ribose sugar, and three phosphate groups. When ATP undergoes hydrolysis, the cleavage of one phosphate group releases energy of 7.3 kcal per mole of ATP. The products are adenosine diphosphate (**ADP**) and hydrogen phosphate ion (HPO_4^{2-}) abbreviated as P_i (inorganic phosphate).

the
Chemistry
place

WEB TUTORIAL
ATP

Adenosine

Adenine

Ribose

—————————————————— Adenosine triphosphate (ATP) ——————————

ATP (Adenosine triphosphate)

Adenosine—O—P—O—P—O⁻ + HO—P—O⁻ + H⁺ + 7.3 kcal/mole

ADP (Adenosine diphosphate)   Inorganic phosphate P_i

$CO_2 + H_2O$
$+ NH_4^+$

Catabolic reactions
(energy producing)

Oxidation of
carbohydrates,
fats, and proteins

ADP + P_i ATP
(energy used)   (energy stored)

Anabolic reactions
(energy requiring)

Muscle contraction,
transport, and synthesis
of cellular compounds

We can write this reaction in its simplified form as

$$ATP + H_2O \longrightarrow ADP + P_i + 7.3 \text{ kcal/mole}$$

Every time we contract muscles, move substances across cellular membranes, send nerve signals, or synthesize an enzyme, we use energy from ATP hydrolysis. In a cell that is doing work (anabolic processes), 1–2 million ATP molecules may be hydrolyzed in one second. The amount of ATP hydrolyzed in one day can be as much as our body mass, even though only about 1 gram of ATP is present in all our cells at any given time.

When we take in food, the resulting catabolic reactions provide energy to regenerate ATP in our cells. Then 7.3 kcal/mole is used to make ATP from ADP and P_i.

$$ADP + P_i + 7.3 \text{ kcal/mole} \longrightarrow ATP$$

SAMPLE PROBLEM 18.2

■ **Hydrolysis of ATP**

Write an equation for the hydrolysis of ATP.

SOLUTION

The hydrolysis of ATP produces ADP, P_i, and energy.

$$ATP + H_2O \longrightarrow ADP + P_i + \text{energy}$$

STUDY CHECK

What are the components of the ATP molecule?

■ ATP Energy and Ca²⁺ Needed to Contract Muscles

Our muscles consist of thousands of parallel fibers. Within these muscle fibers are fibrils composed of two kinds of proteins called filaments. Arranged in alternating rows, the thick filaments of myosin overlap the thin filaments containing actin. During a muscle contraction, the thin filaments slide inward over the thick filaments causing a shortening of the muscle fibers.

Calcium ion, Ca^{2+}, and ATP play an important role in muscle contraction. An increase in the Ca^{2+} concentration in the muscle fibers causes the filaments to slide, while a decrease stops the process. In a relaxed muscle, the Ca^{2+} concentration is low. When a nerve impulse reaches the muscle, calcium channels in the membrane open and Ca^{2+} flows into the fluid surrounding the filaments. Myosin binds to actin and starts to pull the actin filaments inward. The energy used for the contraction is provided by splitting ATP to $ADP + P_i$ and energy.

Muscle contraction continues as long as both ATP and Ca^{2+} levels are high around the filaments. Relaxation of the muscle occurs as the nerve impulse ends and the Ca^{2+} channels close. The Ca^{2+} concentration is lowered as ATP energy is used to pump the remaining calcium ions out of the area, which causes the thin filaments to return to their relaxed state. In rigor mortis, Ca^{2+} concentration remains high within the muscle fibers causing a continued state of rigidity for approximately 24 hours. After that, cellular deterioration causes a decrease in calcium ions and muscles relax.

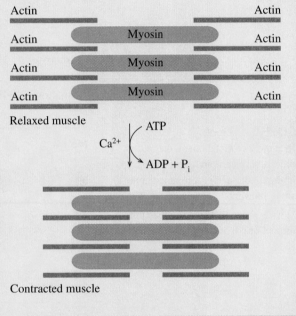

QUESTIONS AND PROBLEMS

■ Metabolism and ATP Energy

18.1 What stage of metabolism involves the digestion of polysaccharides?

18.2 What stage of metabolism involves the conversion of small molecules to CO_2, H_2O, and energy?

18.3 What is meant by a catabolic reaction in metabolism?

18.4 What is meant by an anabolic reaction in metabolism?

18.5 Why is ATP considered an energy-rich compound?

18.6 How much energy is obtained from the hydrolysis of ATP?

■ 18.2 Digestion: Stage 1

In the first stage of catabolism, foods undergo **digestion**, a process that converts large molecules to smaller ones that can be absorbed by the body.

LEARNING GOAL

Give the sites and products of digestion for carbohydrates, triacylglycerols, and proteins.

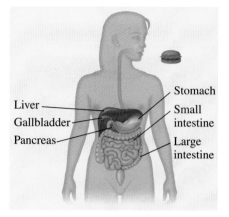

Liver
Gallbladder
Pancreas

Stomach
Small intestine
Large intestine

Digestion of Carbohydrates

We begin the digestion of carbohydrates as soon as we chew food. (See Figure 18.3.) An enzyme produced in the salivary glands called **amylase** hydrolyzes some of the α-glycosidic bonds in amylose and amylopectin, producing smaller polysaccharides called dextrins, which contain three to eight glucose units; maltose; and some glucose. After swallowing, the partially digested starches enter the acid environment of the stomach, where the low pH soon stops further carbohydrate digestion.

In the small intestine, which has a pH of about 8, an α-amylase produced in the pancreas hydrolyzes the remaining polysaccharides to maltose and glucose. Then enzymes produced in the mucosal cells that line the small intestine hydrolyze maltose as well as lactose and sucrose. The hydrolysis reactions for the three common dietary disaccharides are written as follows.

$$\text{lactose} + \text{H}_2\text{O} \xrightarrow{\text{lactase}} \text{galactose} + \text{glucose}$$

$$\text{sucrose} + \text{H}_2\text{O} \xrightarrow{\text{sucrase}} \text{fructose} + \text{glucose}$$

$$\text{maltose} + \text{H}_2\text{O} \xrightarrow{\text{maltase}} \text{glucose} + \text{glucose}$$

FIGURE 18.3

In stage 1 of metabolism, the digestion of carbohydrates begins in the mouth and is completed in the small intestine.

Q Why is there little or no digestion of carbohydrates in the stomach?

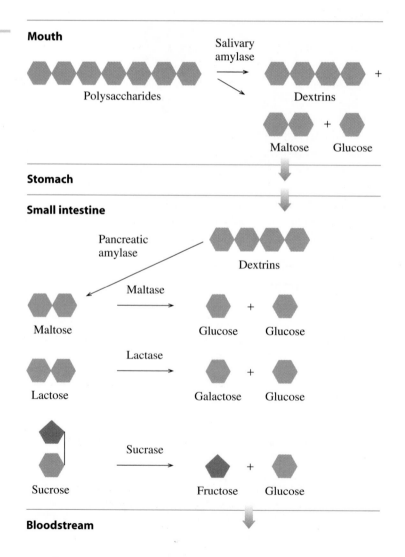

The monosaccharides are absorbed through the intestinal wall into the bloodstream, which carries them to the liver where fructose and galactose are converted to glucose.

SAMPLE PROBLEM 18.3

▪ Digestion of Carbohydrates

Indicate the carbohydrate that undergoes digestion in each of the following sites:

a. mouth   **b.** stomach   **c.** small intestine

SOLUTION

a. Starches amylose and amylopectin.
b. Essentially no digestion occurs in the stomach.
c. Dextrins, maltose, sucrose, and lactose.

STUDY CHECK

Describe the digestion of amylose, a polymer of glucose molecules joined by α-glycosidic bonds.

▪ Carbohydrate Digestion

1. Obtain a cracker or small piece of bread and chew it for 4–5 minutes. During that time observe any change in the taste.
2. Some milk products contain Lactaid, which is the lactase that digests lactose. Look for the brands of milk and ice cream that contain Lactaid or lactase enzyme.

QUESTIONS

1a. How does the taste of the cracker or bread change after you have chewed it for 4–5 minutes? What could be an explanation?
1b. What part of carbohydrate digestion occurs in the mouth?
2a. Write an equation for the digestion of lactose.
2b. Where does lactose undergo digestion?

▪ Digestion of Fats

The digestion of dietary fats begins in the small intestine, when the hydrophobic fat globules mix with bile salts released from the gallbladder. In a process called *emulsification*, the bile salts break the fat globules into smaller droplets called micelles. Then *pancreatic lipases* released from the pancreas hydrolyze the triacylglycerols in the micelles to yield monoacylglycerols and free fatty acids. These digestion products are absorbed into the intestinal lining where they recombine to form triacylglycerols, which are coated with proteins to form lipoproteins called **chylomicrons**. The chylomicrons transport the triacylglycerols through the lymph system and into the bloodstream to be carried to cells of the heart, muscle, and adipose tissues. (See Figure 18.4.) In the cells, enzymes hydrolyze the triacylglycerols to yield glycerol and free fatty acids, which can be used for energy production. We can write the overall equation for the digestion of triacylglycerols as follows:

$$\text{Triacylglycerols} + 3H_2O \xrightarrow{\text{lipase}} \text{Glycerol} + 3 \text{ Fatty acids}$$

The products of fat digestion, glycerol and fatty acids, diffuse into the bloodstream and bind with plasma proteins (albumin) to be transported to the tissues. Most of the glycerol goes into the liver where it is converted to glucose.

SAMPLE PROBLEM 18.4

▪ Fats and Digestion

What are the sites, enzymes, and products for the digestion of triacylglycerols?

SOLUTION

The digestion of triacylglycerols takes place in the small intestine where pancreatic lipase catalyzes their hydrolysis to monoacylglycerols and fatty acids.

The triacylglycerols that re-form in the intestinal wall bind to proteins for transport through the lymphatic system and bloodstream to the cells.

Q **What kinds of enzymes are secreted from the pancreas into the small intestine to hydrolyze triacylglycerols?**

Small intestine

$$
\begin{array}{c}
H_2C - \text{Fatty acid} \\
| \\
HC - \text{Fatty acid} \quad \xrightarrow{\text{lipase}} \\
| \qquad\qquad + 2H_2O \\
H_2C - \text{Fatty acid}
\end{array}
\qquad
\begin{array}{c}
H_2C - OH \\
| \\
HC - \text{Fatty acid} \\
| \qquad + 2\ \text{Fatty acids} \\
H_2C - OH
\end{array}
$$

Triacylglycerol   **Monoacylglycerol**

Intestinal wall (epithelial cells)

Monoacylglycerols + 2 Fatty acids ⟶ Triacylglycerols

Protein

Lipoproteins

Chylomicrons

Lymphatic system

Bloodstream

Cells   **Glycerol + Fatty acids**

STUDY CHECK

What happens to the products from the digestion of triacylglycerols in the membrane of the small intestine?

Stomach

$$\text{Pepsinogen} \xrightarrow{\text{HCl}} \text{Pepsin}$$

$$\text{Proteins} \longrightarrow \text{Polypeptides}$$

Small intestine
  Trypsin
  Chymotrypsin

          Amino acids

Intestinal wall

Bloodstream

Proteins are hydrolyzed to polypeptides in the stomach and to amino acids in the small intestine.

Q **What enzyme, secreted into the small intestine, hydrolyzes peptides?**

Digestion of Proteins

In stage 1, the digestion of proteins begins in the stomach, where hydrochloric acid (HCl) at pH 2 denatures proteins and activates enzymes such as pepsin that begin to hydrolyze peptide bonds. Polypeptides from the stomach move into the small intestine where trypsin and chymotrypsin complete the hydrolysis of the peptides to amino acids. The amino acids are absorbed through the intestinal walls into the bloodstream for transport to the cells. (See Figure 18.5.)

■ Digestion of Proteins

What are the sites and end products for the digestion of proteins?

SOLUTION

Proteins begin digestion in the stomach and complete digestion in the small intestine to yield amino acids.

STUDY CHECK

What is the function of HCl in the stomach?

QUESTIONS AND PROBLEMS

■ Digestion: Stage 1

18.7 What is the general type of reaction that occurs during the digestion of carbohydrates?

18.8 Why is α-amylase produced in the salivary glands and in the pancreas?

18.9 Complete the following equation by filling in the missing words:
   a. _____ + H_2O ⟶ galactose + glucose
   b. Sucrose + H_2O ⟶ _____ + _____
   c. Maltose + H_2O ⟶ _____ + _____

18.10 Give the site and the enzyme for each of the reactions in problem 18.9.

18.11 What is the role of bile salts in lipid digestion?

18.12 How are insoluble triacylglycerols transported to the tissues?

18.13 Where do dietary proteins undergo digestion in the body?

18.14 What is the purpose of digestion in stage 1?

■ 18.3 Important Coenzymes in Metabolic Pathways

LEARNING GOAL

Describe the components and functions of the coenzymes FAD, NAD^+, and coenzyme A.

Before we look at the metabolic reactions that extract energy from our food, we need to review some ideas about oxidation and reduction reactions (Chapters 5 and 12). An **oxidation** involves the loss of hydrogen and electrons by a substance or an increase in oxygen. **Reduction** is the gain of hydrogen and electrons or a decrease in oxygen.

When an enzyme catalyzes an oxidation, hydrogen atoms are removed from a substrate as hydrogen ions, $2H^+$, and electrons, $2\,e^-$.

$$2 \text{ H atoms (removed in oxidation)} \longrightarrow 2H^+ + 2\,e^-$$

Those hydrogen ions and electrons are picked up by a coenzyme, which is reduced. As we saw in Chapter 16, the structures of many coenzymes include the water-soluble B vitamins we obtain from the foods in our diets. Now we can look at the structure of several important coenzymes.

■ NAD^+

NAD^+ (nicotinamide adenine dinucleotide) is an important coenzyme in which the vitamin *niacin* provides the *nicotinamide* group, which is bonded to adenosine diphosphate (ADP). (See Figure 18.6.) The NAD^+ coenzyme participates in reactions that produce a carbon–oxygen (C=O) double bond such as the oxidation of alcohols to aldehydes and ketones. The NAD^+ is reduced when the carbon in nicotinamide accepts a hydrogen ion and two electrons leaving one H^+. Let's look at the reactions that take place when ethanol is oxidized in the liver to acetaldehyde using NAD^+.

Overall oxidation–reduction reaction

$$CH_3-CH_2-OH + NAD^+ \xrightarrow[\text{dehydrogenase}]{\text{alcohol}} CH_3-\overset{\displaystyle O}{\overset{\|}{C}}-H + NADH + H^+$$

ethanol   acetaldehyde

FIGURE 18.6

The coenzyme NAD$^+$ (nicotinamide adenine dinucleotide), which consists of a nicotinamide portion from the vitamin niacin, ribose, and adenine diphosphate, is reduced to NADH + H$^+$.

Q Why is the conversion of NAD$^+$ to NADH and H$^+$ called a reduction?

FAD

FAD (flavin adenine dinucleotide) is a coenzyme derived from adenosine diphosphate (ADP) and riboflavin. Riboflavin, also known as vitamin B$_2$, consists of ribitol, a sugar alcohol, and flavin. As a coenzyme, two nitrogen atoms in the flavin part of the FAD coenzyme accept the hydrogen, which reduces the FAD to FADH$_2$. (See Figure 18.7.) FAD typically participates in oxidation reactions that produce a carbon–carbon (C=C) double bond.

Coenzyme A

Coenzyme A (CoA) is made up of several components: pantothenic acid (vitamin B$_5$), adenosine diphosphate (ADP), and aminoethanethiol. (See Figure 18.8.) The main function of coenzyme A is to activate acyl groups (indicated by the letter A in CoA), particularly the acetyl group.

$$FAD^+ + 2H^+ + 2\,e^- \longrightarrow FADH_2$$

FAD (flavin adenine dinucleotide)

FIGURE 18.7

The coenzyme FAD (flavin adenine dinucleotide) made from riboflavin (vitamin B_2) and adenine diphosphate is reduced to $FADH_2$.

Q **What is the type of reaction in which FAD accepts hydrogen?**

SAMPLE PROBLEM 18.6

■ **Coenzymes in Metabolic Pathways**

What vitamin is part of the coenzyme FAD?

SOLUTION

FAD, flavin adenine dinucleotide, is made from the vitamin riboflavin.

STUDY CHECK

What is the abbreviation of the reduced form of FAD?

Aminoethanethiol   Pantothenic acid   Phosphorylated ADP

Coenzyme A

FIGURE 18.8

Coenzyme A is derived from a phosphorylated adenine diphosphate (ADP) and pantothenic acid bonded by an amide linked to aminoethanethiol, which contains the —SH reactive part of the molecule.

Q **What part of coenzyme A reacts with a two-carbon acetyl group?**

Lactose Intolerance

The disaccharide in milk is lactose, which is broken down by *lactase* in the intestinal tract to monosaccharides that are a source of energy. Infants and small children produce lactase to break down the lactose in milk. It is rare for an infant to lack the ability to produce lactase. As they mature, many people experience a decrease in the production of the lactase enzyme. By the time they are adults, many people have little or no lactase production, causing lactose intolerance. This condition may affect 25% of the people in the United States. A deficiency of lactase occurs in adults throughout the world, but in the United States it is more prevalent among the African-American, Hispanic, and Asian populations.

When lactose is not broken down into glucose and galactose, it cannot be absorbed through the intestinal wall and remains in the intestinal tract. In the intestines, the lactose undergoes fermentation to products that include lactic acid and gases such as methane (CH_4) and CO_2. Symptoms of lactose intolerance, which appear approximately $\frac{1}{2}$ to 1 hour after ingesting milk or milk products, include nausea, abdominal cramps, and diarrhea. The severity of the symptoms depends on how much lactose is present in the food and how much lactase a person produces.

TREATMENT OF LACTOSE INTOLERANCE

One way to reduce the reaction to lactose is to avoid products that contain lactose. However, it is important to consume foods that provide the body with calcium. Many people with lactose intoler-

ance seem to tolerate yogurt, which is a good source of calcium. Although there is lactose in yogurt, the bacteria in yogurt may produce some lactase, which helps to digest the lactose. A person who is lactose intolerant should also know that some foods that may not seem to be dairy products contain lactose. For example, baked goods, cereals, breakfast drinks, salad dressings, and even lunchmeat can contain lactose in their ingredients. Labels must be read carefully to see if the ingredients include "milk" or "lactose."

The enzyme lactase is now available in many forms such as tablets that are taken with meals, drops that are added to milk, or as additives in many dairy products such as milk. When lactase is added to milk that is left in the refrigerator for 24 hours, the lactose level is reduced by 70–90%. Lactase pills or chewable tablets are taken when a person begins to eat a meal that contains dairy foods. If taken too far ahead of the meal, too much of the lactase will be degraded by stomach acid. If taken following a meal, the lactose will have entered the lower intestine.

QUESTIONS AND PROBLEMS

Important Coenzymes in Metabolic Pathways

18.15 Give the abbreviation for the following:
  a. the reduced form of NAD^+
  b. the oxidized form of $FADH_2$

18.16 Give the abbreviation for the following:
  a. the reduced form of FAD
  b. the oxidized form of NADH

18.17 What coenzyme picks up hydrogen when a carbon–carbon double bond is formed?

18.18 What coenzyme picks up hydrogen when a carbon–oxygen double bond is formed?

LEARNING GOAL

Describe the conversion of glucose to pyruvate in glycolysis and the subsequent conversion of pyruvate to acetyl CoA or lactate.

18.4 Glycolysis: Stage 2

Our major source of energy is the glucose produced when we digest the carbohydrates in our food, or from glycogen, a polysaccharide stored in the liver and skeletal muscle. Glucose in the bloodstream enters our cells for further degradation in a pathway called glycolysis. Early organisms used this pathway to produce energy from simple nutrients long before there was any oxygen in Earth's atmosphere. Glycolysis is an **anaerobic** process; no oxygen is required.

In **glycolysis**, which is one of the metabolic pathways in stage 2, a six-carbon glucose molecule is broken down to yield two three-carbon pyruvate molecules. In the first five reactions (1–5), the energy of 2 ATPs is invested to add phosphate

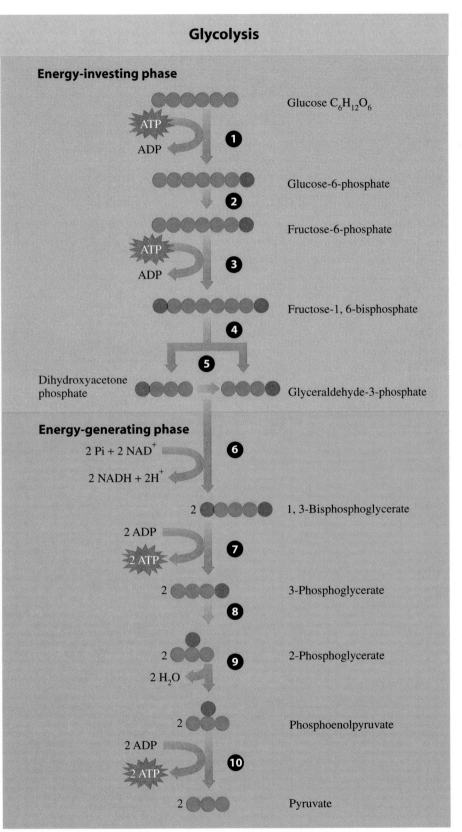

Glycolysis

Energy-investing phase

Glucose $C_6H_{12}O_6$

① ATP → ADP

Glucose-6-phosphate

②

Fructose-6-phosphate

③ ATP → ADP

Fructose-1, 6-bisphosphate

④

⑤

Dihydroxyacetone phosphate

Glyceraldehyde-3-phosphate

Energy-generating phase

$2 Pi + 2 NAD^+$

⑥

$2 NADH + 2H^+$

2 1, 3-Bisphosphoglycerate

2 ADP

⑦

2 ATP

2 3-Phosphoglycerate

⑧

2 2-Phosphoglycerate

⑨

$2 H_2O$

2 Phosphoenolpyruvate

2 ADP

⑩

2 ATP

2 Pyruvate

FIGURE 18.9

In glycolysis, the six-carbon glucose molecule is degraded to yield two three-carbon pyruvate molecules. A net of two ATPs are produced along with two NADH.

Q **Where in the glycolysis pathway is glucose cleaved to yield two three-carbon compounds?**

groups to form sugar phosphates. Then the six-carbon sugar phosphate is cleaved to yield two three-carbon sugar phosphate molecules. In the last five reactions (6–10), the phosphate groups in these energy-rich trioses are hydrolyzed, which generates energy in the form of 4 ATPs. (See Figure 18.9.)

the **C**hemistry place

WEB TUTORIAL
Glycolysis

■ Energy-Investing Reactions: Steps 1–5

Reaction 1   Phosphorylation: first ATP invested ——

Glucose is converted to glucose-6-phosphate by a reaction with ATP catalyzed by *hexokinase*.

Reaction 2   Isomerization————————————

The enzyme *phosphoglucoisomerase* converts glucose-6-phosphate, an aldose, to fructose-6-phosphate, a ketose.

Reaction 3   Phosphorylation: second ATP invested ——

A second ATP reacts with fructose-6-phosphate to give fructose-1,6-bisphosphate. The word *bisphosphate* is used to show that the phosphates are on different carbons in fructose and not connected to each other.

Reaction 4   Cleavage: two trioses form ————————

Fructose-1,6-bisphosphate is "split" into two triose phosphates—dihydroxyacetone phosphate and glyceraldehyde-3-phosphate—catalyzed by *aldolase*.

Reaction 5   Isomerization of a triose ————————

In reaction 5, *triosephosphate isomerase* converts one of the triose products, dihydroxyacetone phosphate, to the other, glyceraldehyde-3-phosphate. Now all 6 carbon atoms from glucose are in two identical triose phosphates.

Energy-Generating Reactions: Steps 6–10

Reaction 6   First energy-rich compound

The aldehyde group of glyceraldehyde-3-phosphate is oxidized and phosphorylated by *glyceraldehyde-3-phosphate dehydrogenase*. The coenzyme NAD^+ is reduced to NADH and H^+. The 1,3-bisphosphoglycerate is an extremely high-energy compound that releases 12 kcal per mole when the phosphate bond is broken.

Reaction 7   Formation of first ATP

The energy-rich 1,3-bisphosphoglycerate now drives the formation of ATP when *phosphoglycerate kinase* transfers a phosphate group to ADP. This process is called a substrate-level phosphorylation. At this point, glycolysis pays back the debt of two ATP invested in the early reactions.

Reaction 8   Formation of 2-phosphoglycerate

A *phosphoglycerate mutase* transfers the phosphate group from carbon 3 to carbon 2 to yield 2-phosphoglycerate.

Reaction 9   Second energy-rich compound

An *enolase* catalyzes the removal of water to yield phosphoenolpyruvate, a high-energy compound that releases 15 kcal per mole when the phosphate bond is hydrolyzed.

Reaction 10   Formation of second ATP

This last reaction, catalyzed by *pyruvate kinase,* transfers the phosphate to ADP in a second direct substrate phosphorylation to yield pyruvate and ATP.

Summary of Glycolysis

In the glycolysis pathway, glucose is converted to two pyruvates. Initially, two ATP molecules were required to form sugar phosphate. Later, four ATP were generated, which gives a net gain of two ATP. Overall, glycolysis yields two ATP and two NADH for each glucose that is converted to two pyruvates.

$$C_6H_{12}O_6 + 2\,NAD^+ \xrightarrow[\quad]{2\,ADP\ +\ 2\,P_i \quad\quad 2\,ATP} 2\ CH_3-\overset{\displaystyle O}{\overset{\|}{C}}-COO^- + 2\,NADH + 4H^+$$

glucose   pyruvate

SAMPLE PROBLEM 18.7

■ Glycolysis

What are the steps in glycolysis that generate ATP?

SOLUTION

ATP is produced when phosphate groups are transferred directly to ADP from 1,3-bisphosphoglycerate (step 7) and from phosphoenolpyruvate (step 10).

STUDY CHECK

If four ATP molecules are produced in glycolysis, why is there a net yield of two ATP?

Pathways for Pyruvate

The pyruvate produced from glucose can now enter pathways that continue to extract energy. The available pathway depends on whether there is sufficient oxygen in the cell. During **aerobic** conditions, oxygen is available to convert pyruvate to acetyl coenzyme A (CoA). When oxygen levels are low, pyruvate is reduced to lactate.

Aerobic Conditions

In glycolysis, two ATP molecules were generated when glucose was converted to pyruvate. However, much more energy is still available. The greatest amount of the energy is obtained from glucose when oxygen levels are high in the cells. Under aerobic conditions, pyruvate is oxidized and a carbon atom is removed from pyruvate as CO_2. The coenzyme NAD^+ is required for the oxidation. The resulting two-carbon acetyl compound is attached to CoA, producing **acetyl CoA**, an important intermediate in many metabolic pathways. (See Figure 18.10.)

$$CH_3-\overset{\displaystyle O}{\overset{\|}{C}}-\overset{\displaystyle O}{\overset{\|}{C}}-O^- + HS-CoA + NAD^+ \xrightarrow[\text{dehydrogenase}]{\text{pyruvate}} CH_3-\overset{\displaystyle O}{\overset{\|}{C}}-S-CoA + CO_2 + NADH$$

pyruvate   acetyl CoA

Anaerobic Conditions

When we engage in strenuous exercise, the oxygen stored in our muscle cells is quickly depleted. Under anaerobic conditions, pyruvate is reduced to lactate.

FIGURE 18.10

Pyruvate is converted to acetyl CoA under aerobic conditions and to lactate under anaerobic conditions.

Q **During vigorous exercise, why does lactate accumulate in the muscles?**

NAD^+ is produced and used to oxidize more glyceraldehyde-3-phosphate in the glycolysis pathway, which produces a small but needed amount of ATP.

The accumulation of lactate causes the muscles to tire rapidly and become sore. After exercise, a person continues to breathe rapidly to repay the oxygen debt incurred during exercise. Most of the lactate is transported to the liver where it is converted back into pyruvate. Under anaerobic conditions, the only ATP production in glycolysis occurs during the steps that phosphorylate ADP directly, giving a net gain of only two ATP molecules:

$$C_6H_{12}O_6 + 2\,ADP + 2\,P_i \longrightarrow 2\,CH_3\!\!-\!\!\overset{\displaystyle OH}{\underset{}{CH}}\!\!-\!\!COO^- + 2\,ATP$$
glucose   lactate

Bacteria also convert pyruvate to lactate under anaerobic conditions. In the preparation of kimchee and sauerkraut, cabbage is covered with a salt brine. The glucose from the starches is converted to lactate. This acid environment acts as a preservative that prevents the growth of other bacteria. The pickling of olives and cucumbers gives similar products. When cultures of bacteria that produce lactate are added to milk, the acid denatures the milk proteins to give sour cream and yogurt.

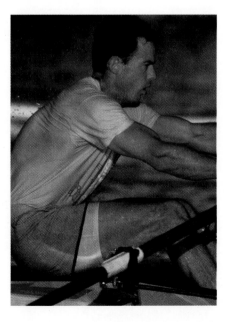

Glycogen

We have just eaten a large meal that has supplied us with all the glucose we need to produce pyruvate and ATP by glycolysis. Then we use excess glucose to replenish our energy reserves by synthesizing glycogen that is stored in limited amounts in our skeletal muscle and liver. When glycogen stores are full, any remaining glucose is converted to triacylglycerols and stored as body fat. When our diet does not supply sufficient glucose, or we have utilized our blood glucose, we degrade the stored glycogen and release glucose.

SAMPLE PROBLEM 18.8

■ Fates of Pyruvate

Is each of the following products from pyruvate produced under anaerobic or aerobic conditions?

a. acetyl CoA   **b.** lactate

SOLUTION

a. aerobic conditions   **b.** anaerobic conditions

STUDY CHECK

After strenuous exercise, some lactate is oxidized back to pyruvate by lactate dehydrogenase using NAD^+. Write an equation to show this reaction.

QUESTIONS AND PROBLEMS

■ Glycolysis: Stage 2

18.19 What is the starting compound of glycolysis?

18.20 What is the end product of glycolysis?

18.21 How is ATP used in the initial steps of glycolysis?

18.22 How many ATP molecules are used in the initial steps of glycolysis?

18.23 How does direct phosphorylation account for the production of ATP in glycolysis?

18.24 Why are there two ATP molecules formed for one molecule of glucose?

18.25 How many ATP or NADH are produced (or required) in each of the following steps in glycolysis?
   a. glucose to glucose-6-phosphate
   b. glyceraldehyde-3-phosphate to 1,3-bisphosphoglycerate
   c. glucose to pyruvate

18.26 How many ATP or NADH are produced (or required) in each of the following steps in glycolysis?
   a. 1,3-bisphosphoglycerate to 3-phosphoglycerate
   b. fructose-6-phosphate to fructose-1,6-bisphosphate
   c. phosphoenolpyruvate to pyruvate

18.27 What condition is needed in the cell to convert pyruvate to acetyl CoA?

18.28 What coenzymes are needed for the oxidation of pyruvate to acetyl CoA?

18.29 Write the overall equation for the conversion of pyruvate to acetyl CoA.

18.30 What is the product of pyruvate under anaerobic conditions?

18.31 How does the formation of lactate permit glycolysis to continue under anaerobic conditions?

18.32 After running a marathon, a runner has muscle pain and cramping. What might have occurred in the muscle cells to cause this?

◼ 18.5 The Citric Acid Cycle: Stage 3

LEARNING GOAL

◼ Describe the oxidation of acetyl CoA in the citric acid cycle.

As a central pathway in metabolism, the citric acid cycle uses acetyl CoA from the degradation of carbohydrates, lipids, and proteins. The **citric acid cycle** is a series of reactions that degrades acetyl CoA to yield CO_2, NADH + H$^+$, and $FADH_2$.

There are a total of 8 reactions in the citric acid cycle, which we can separate into two parts. In part 1, a two-carbon acetyl group bonds with four-carbon oxaloacetate to yield citrate. Then decarboxylation reactions remove two carbon atoms as CO_2, which produces a four-carbon compound. Then the reactions in part 2 convert the four-carbon compound back to oxaloacetate, which bonds with another acetyl CoA and goes through the cycle again. (See Figure 18.11.) In one turn of the citric acid cycle, four oxidation reactions provide hydrogen ions and electrons, which are used to reduce FAD and NAD$^+$ coenzymes.

Reaction 1   Formation of Citrate

In the first reaction, the two-carbon acetyl group in acetyl CoA bonds with the four-carbon oxaloacetate to yield citrate and CoA. (See Figure 18.12.)

Reaction 2   Isomerization to Isocitrate

In order to continue oxidation, citrate undergoes isomerization to yield isocitrate. This is necessary because the secondary hydroxyl group (Chapter 12) in isocitrate can be oxidized in the next reaction, while the tertiary hydroxyl group in citrate cannot be oxidized.

Reaction 3   First Oxidative Decarboxylation (CO_2)

This is the first time in the citric acid cycle that both an oxidation and a decarboxylation occur together. The oxidation converts the hydroxyl group to a ketone and the **decarboxylation** removes a carbon as a CO_2 molecule. The loss of CO_2 shortens the carbon chain to yield a five-carbon α-ketoglutarate. The energy from the

WEB TUTORIAL
Krebs Cycle

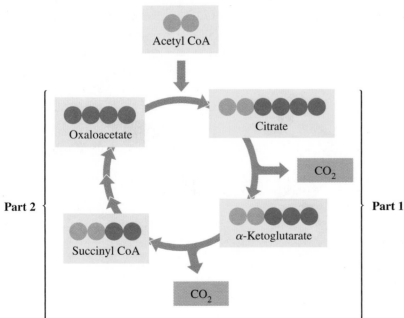

FIGURE 18.11

In part 1 of the citric acid cycle, two carbon atoms are removed as CO_2 from six-carbon citrate to give four-carbon succinyl CoA, which is converted in part 2 to four-carbon oxaloacetate.

Q **What is the difference in part 1 and part 2 of the citric acid cycle?**

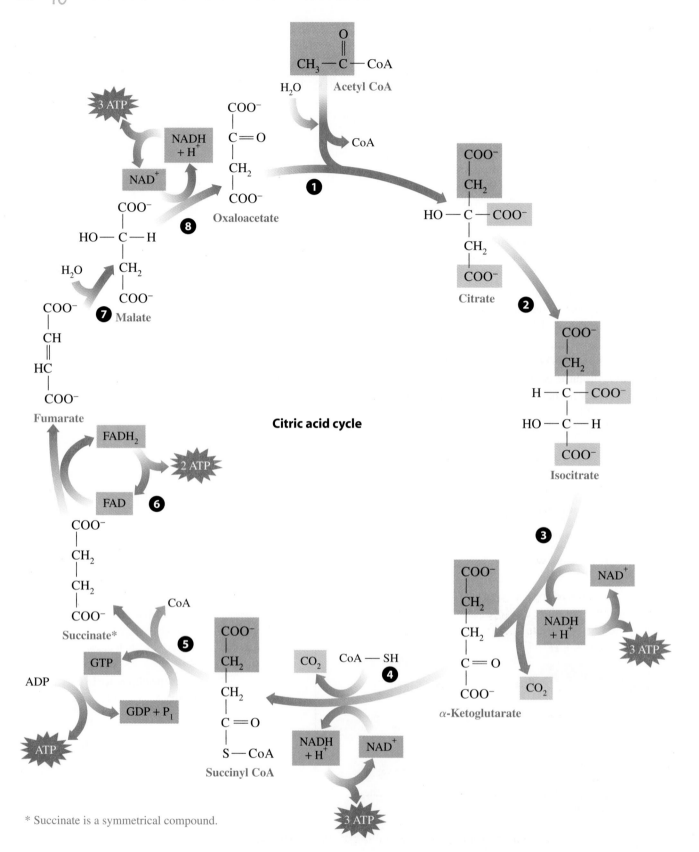

FIGURE 18.12

In the citric acid cycle, oxidation reactions produce two CO_2, reduced coenzymes NADH and $FADH_2$, and regenerate oxaloacetate.

Q **How many reactions in the citric acid cycle produce a reduced coenzyme?**

oxidation is used to transfer hydrogen ions and electrons to NAD^+. We can summarize the reactions as follows:

1. The hydroxyl group (—OH) is oxidized to a ketone (C=O).
2. NAD^+ is reduced to yield NADH.
3. A carboxylate group (COO^-) is removed as CO_2.

Reaction 4   Second Oxidative Decarboxylation (CO_2)

In this reaction, a second CO_2 is removed as α-ketoglutarate undergoes oxidative decarboxylation. The resulting four-carbon group combines with coenzyme A to form succinyl CoA and hydrogen ions and electrons are transferred to NAD^+. The two reactions that occur are as follows:

1. A second carbon is removed as CO_2.
2. NAD^+ is reduced to yield NADH.

Reaction 5   Hydrolysis of Succinyl CoA

The energy released by the hydrolysis of succinyl CoA is used to add a phosphate group (P_i) directly to GDP (guanosine diphosphate). The products are succinate and GTP, which is a high-energy compound similar to ATP.

The hydrolysis of GTP is used to add a phosphate group to ADP, which regenerates GDP for the citric acid cycle. This is the only time in the citric acid cycle that a direct substrate phosphorylation is used to produce ATP.

$$GTP + ADP \longrightarrow GDP + ATP$$

Reaction 6   Dehydrogenation of Succinate

In this oxidation reaction, hydrogen is removed from two carbon atoms in succinate, which produces fumarate, a compound with a trans double bond. This is the only place in the citric acid cycle where FAD is reduced to $FADH_2$.

Reaction 7   Hydration

In a hydration reaction, water adds to the double bond of fumarate to yield malate.

Reaction 8   Dehydrogenation Forms Oxaloacetate

In the last step of the citric acid cycle, the hydroxyl (—OH) group in malate is oxidized to yield oxaloacetate, which has a ketone group. The coenzyme NAD^+ is reduced to $NADH + H^+$.

Summary of Products from the Citric Acid Cycle

A summary of the products from one turn of the citric acid cycle is as follows:

2 CO_2 molecules
3 NADH molecules
1 $FADH_2$ molecule
1 GTP molecule used to form ATP
1 HS—CoA molecule

We can write the overall chemical equation for one turn of the citric acid cycle as follows:

$$\text{Acetyl CoA} + 3\ NAD^+ + FAD + GDP + P_i + 2H_2O \longrightarrow$$
$$2CO_2 + 3\ NADH + 2H^+ + FADH_2 + HS—CoA + GTP$$

■ **Citric Acid Cycle**

When one acetyl CoA completes the citric acid cycle, how many of each of the following are produced?

a. NADH
b. ketone group
c. CO_2

SOLUTION

a. One turn of the citric acid cycle produces three molecules of NADH.
b. Two ketone groups form when the secondary alcohol groups in isocitrate and malate are oxidized by NAD^+.
c. Two molecules of CO_2 are produced by the decarboxylation of isocitrate and α-ketoglutarate.

STUDY CHECK

What substance is a substrate in the first reaction of the citric acid cycle and a product in the last reaction?

■

QUESTIONS AND PROBLEMS

■ **The Citric Acid Cycle: Stage 3**

18.33 What are the products from one turn of the citric acid cycle?

18.34 What compounds are needed to start the citric acid cycle?

18.35 Which reactions of the citric acid cycle involve oxidative decarboxylation?

18.36 Which reactions of the citric acid cycle involve a hydration reaction?

18.37 Which reactions of the citric acid cycle reduce NAD^+?

18.38 Which reactions of the citric acid cycle reduce FAD?

18.39 Where does a direct substrate phosphorylation occur?

18.40 What is the total NADH and total $FADH_2$ produced in one turn of the citric acid cycle?

18.41 Refer to the diagram of the citric acid cycle to answer each of the following:
    a. What are the six-carbon compounds?
    b. How is the number of carbon atoms decreased?
    c. What are the five-carbon compounds?
    d. Which reactions are oxidation reactions?
    e. In which reactions are secondary alcohols oxidized?

18.42 Refer to the diagram of the citric acid cycle to answer each of the following:
    a. What is the yield of CO_2 molecules?
    b. What are the four-carbon compounds?
    c. What is the yield of GTP molecules?
    d. What are the decarboxylation reactions?
    e. Where does a hydration occur?

LEARNING GOAL

Describe the electron carriers involved in electron transport.

■ 18.6 Electron Transport

In **electron transport**, hydrogen ions and electrons from NADH and $FADH_2$ are passed from one electron carrier to the next until they combine with oxygen to form H_2O.

$$\text{FMN}^+ + 2\text{H}^+ + 2\,e^- \longrightarrow \text{FMNH}_2$$

FMN (flavin in mononucletotide)   **FMNH₂**

FIGURE 18.13

The electron carrier FMN consists of a flavin ring system containing the reactive center, ribitol, and a phosphate group.

Q What part of the FMN molecule is reduced when hydrogen ions and electrons are accepted?

Electron Carriers

There are four types of electron carriers that make up the electron transport system.

1. **FMN (flavin mononucleotide)** is a coenzyme derived from riboflavin (vitamin B₂). In riboflavin, the ring system is attached to ribitol, the sugar alcohol of ribose. (See Figure 18.13.) The reduced product is FMNH₂.

2. **Fe–S clusters** is the name given to a group of iron–sulfur proteins that contain iron–sulfur clusters embedded in the proteins of the electron transport chain. The clusters contain iron ions, inorganic sulfides, and several cysteine groups. The iron in the clusters is reduced to Fe^{2+} and oxidized to Fe^{3+} as electrons are accepted and lost. (See Figure 18.14.)

3. **Coenzyme Q (Q or CoQ)** is derived from quinone, which is a six-carbon cyclic compound with two double bonds and two keto groups attached to a long carbon chain. (See Figure 18.15.) Coenzyme Q is reduced when the keto groups of quinone accept hydrogen ions and electrons.

the **Chemistry place**

WEB TUTORIAL
Electron Transport

Typical Fe-S cluster

FIGURE 18.14

In a typical iron–sulfur cluster, an iron ion bonds to sulfur atoms in the thiol (—SH) groups of four cysteines in proteins.

Q In an iron–sulfur cluster, what are the ionic charges of the oxidized and reduced iron ions?

$$Q + 2H^+ + 2\,e^- \rightleftharpoons QH_2$$

Oxidized coenzyme Q (Q)   Reduced coenzyme Q (QH₂)

FIGURE 18.15

The electron carrier coenzyme Q accepts electrons from $FADH_2$ and $FMNH_2$ and passes them to the cytochromes.

Q How does reduced coenzyme Q compare to the oxidized form?

4. **Cytochromes (cyt)** are proteins that contain an iron ion in a heme group. The different cytochromes are indicated by the letters following the abbreviation for cytochrome (cyt): cyt b, cyt c_1, cyt c, cyt a, and cyt a_3. In each cytochrome, the Fe^{3+} accepts a single electron to form Fe^{2+}, which is oxidized back to Fe^{3+} when the electron is passed to the next cytochrome. (See Figure 18.16.)

$$Fe^{3+} + 1\,e^- \rightleftharpoons Fe^{2+}$$

Simplified structure of cytochromes c and c_1

FIGURE 18.16

The iron-containing proteins known as cytochromes are identified as b, c, c_1, a, and a_3.

Q What are the reduced and oxidized forms of the cytochromes?

SAMPLE PROBLEM 18.10

■ Oxidation and Reduction

Identify the following steps in the electron transport chain as oxidation or reduction:

a. $FMN + 2H^+ + 2e^- \longrightarrow FMNH_2$
b. Cyt c $(Fe^{2+}) \longrightarrow$ cyt c $(Fe^{3+}) + e^-$

SOLUTION

a. The gain of hydrogen is reduction.
b. The loss of electrons is oxidation.

STUDY CHECK

Identify each of the following as oxidation or reduction:

a. $QH_2 \longrightarrow Q$ **b.** Cyt b $(Fe^{3+}) \longrightarrow$ cyt b (Fe^{2+})

■ Electron Transfer

A mitochondrion consists of inner and outer membranes with the matrix located between. Along the highly folded inner membrane are the enzymes and electron carriers required for electron transport. When these membranes are broken up, four distinct protein complexes are obtained.

Two electron carriers, coenzyme Q and cytochrome c, are not firmly attached to the membrane. They function as mobile carriers shuttling electrons between the protein complexes that are tightly bound to the membrane. (See Figure 18.17.)

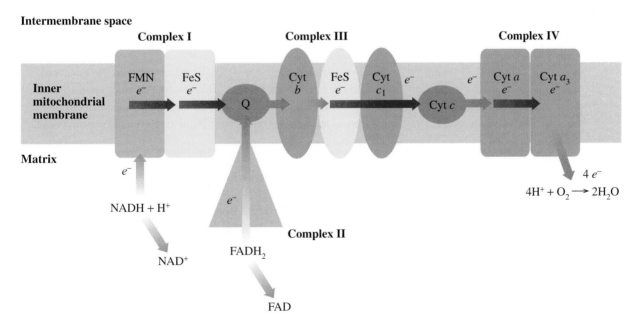

FIGURE 18.17

Most of the electron carriers in the electron transport chain are found in protein complexes bound to the inner membrane of the mitochondria. Two are mobile carriers that carry electrons between the protein complexes.

Q **What is the function of the electron carriers coenzyme Q and cytochrome c?**

Complex I   NADH Dehydrogenase

At complex I, NADH transfers hydrogen ions and electrons to FMN. The reduced $FMNH_2$ forms, while NADH is reoxidized to NAD^+, which returns to oxidative pathways such as the citric acid cycle to oxidize more substrates.

$$NADH + H^+ + FMN \longrightarrow NAD^+ + FMNH_2$$

Within complex I, the NADH electrons are transferred to iron–sulfur (Fe–S) clusters and then to coenzyme (Q).

$$FMNH_2 + Q \longrightarrow QH_2 + FMN$$

The overall reaction sequence in complex I can be written as follows:

$$NADH + H^+ + Q \longrightarrow QH_2 + NAD^+$$

Complex II   Succinate Dehydrogenase

Complex II is specifically used when $FADH_2$ is generated by the conversion of succinate to fumarate in the citric acid cycle. The electrons from $FADH_2$ are transferred to coenzyme Q to yield QH_2. Because complex II is at a lower energy level than complex I, the electrons from $FADH_2$ enter electron transport at a lower energy level than those from NADH.

$$FADH_2 + Q \longrightarrow FAD + QH_2$$

Complex III   Coenzyme Q-Cytochrome *c* Reductase

The mobile carrier QH_2 transfers the electrons it has collected from NADH and $FADH_2$ to an iron–sulfur (Fe–S) cluster and then to cytochrome *b*, the first cytochrome in complex III.

$$QH_2 + 2 \text{ cyt } b \text{ (Fe}^{3+}) \longrightarrow Q + 2 \text{ cyt } b \text{ (Fe}^{2+}) + 2H^+$$

From cyt *b*, the electron is transferred to an Fe–S cluster and then to cytochrome c_1 and then to cytochrome *c*. Each time an Fe^{3+} ion accepts an electron, it is reduced to Fe^{2+}, and then oxidized back to Fe^{3+} as the electron is passed on along the chain. Cytochrome *c* is another mobile carrier; it moves the electron from complex III to complex IV.

Complex IV   Cytochrome *c* Oxidase

At complex IV, electrons are transferred from cytochrome *c* to cytochrome *a*, and then to cytochrome a_3, the last cytochrome. In the final step of electron transport, electrons and hydrogen ions combine with oxygen (O_2) to form water.

$$4H^+ + 4\,e^- + O_2 \longrightarrow 2H_2O$$

SAMPLE PROBLEM 18.11

■ Electron Transport

Identify which of the following electron carriers are mobile carriers.

a. Cyt c
b. FMN
c. Fe–S clusters
d. Q

SOLUTION

a. and **d.** Cyt c and Q are mobile carriers.

STUDY CHECK

What is the final substance that accepts electrons in the electron transport chain?

■

QUESTIONS AND PROBLEMS

■ Electron Transport

18.43 Is cyt b (Fe^{3+}) the abbreviation for the oxidized or reduced form of cytochrome b?

18.44 Is $FMNH_2$ the abbreviation for the oxidized or reduced form of flavin mononucleotide?

18.45 Identify the following as oxidation or reduction:
   a. $FMNH_2 \longrightarrow FMN + 2H^+ + 2\,e^-$
   b. $Q + 2H^+ + 2\,e^- \longrightarrow QH_2$

18.46 Identify the following as oxidation or reduction:
   a. Cyt c (Fe^{3+}) $+ e^- \longrightarrow$ cyt c (Fe^{2+})
   b. Fe^{2+}–S cluster $\longrightarrow Fe^{3+}$–S cluster $+ e^-$

18.47 What reduced coenzymes provide the electrons for electron transport?

18.48 What happens to the energy level as electrons are passed along in electron transport?

18.49 Arrange the following in the order they appear in electron transport: cytochrome c, cytochrome b, FAD, and coenzyme Q.

18.50 Arrange the following in the order they appear in electron transport: O_2, NAD^+, cytochrome a_3, and FMN.

18.51 How are electrons carried from complex I to complex III?

18.52 How are electrons carried from complex III to complex IV?

18.53 How is NADH oxidized in electron transport?

18.54 How is $FADH_2$ oxidized in electron transport?

18.55 Complete the following reactions in electron transport:
   a. $NADH + H^+ +$ _____ $\longrightarrow$ _____ $+ FMNH_2$
   b. $QH_2 + 2$ cyt b (Fe^{3+}) $\longrightarrow$ _____ $+$ _____ $+ 2H^+$

18.56 Complete the following reactions in electron transport:
   a. $Q +$ _____ $\longrightarrow$ _____ $+ FAD$
   b. 2 cyt a (Fe^{2+}) $+ 2$ cyt a_3 (Fe^{3+}) $\longrightarrow$ _____ $+$ _____

HEALTH NOTE

■ Toxins: Inhibitors of Electron Transport

There are several substances that inhibit the electron carriers in the electron transport chain. Rotenone, a product from a plant root used in South America to poison fish, blocks electron transport between the NADH dehydrogenase (complex I) and coenzyme Q. The barbiturates amytal and demerol also inhibit the NADH dehydrogenase complex. Another inhibitor is the antibiotic antimycin A, which blocks the flow of electrons between cytochrome b and cytochrome c_1 (complex III). Another group of compounds including cyanide (CN^-) and carbon monoxide inhibit cytochrome c oxidase (complex IV). The toxic nature of these compounds makes it clear that an organism relies heavily on the process of electron transport.

rotenone

amytal

antimycin A

When an inhibitor blocks a step in the electron transport chain, the carriers preceding that step are unable to transfer electrons and remain in their reduced forms. All the carriers after the blocked step remain oxidized without a source of electrons. Thus, any of these inhibitors shut down the flow of electrons through the electron transport chain.

LEARNING GOAL

Describe the process of oxidative phosphorylation in ATP synthesis; calculate the ATP from complete oxidation of glucose.

■ 18.7 Oxidative Phosphorylation and ATP

We have seen that energy is generated when electrons from the oxidation of substrates flow through electron transport. Now we will look at how that energy is used in the production of ATP for the cell, which is a process called **oxidative phosphorylation**.

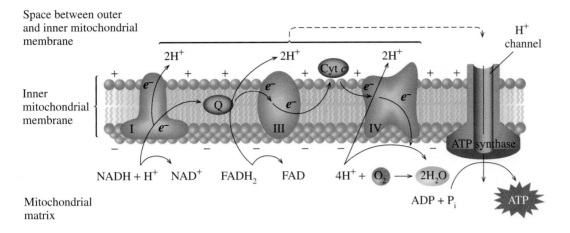

Space between outer and inner mitochondrial membrane

Inner mitochondrial membrane

Mitochondrial matrix

FIGURE 18.18

In electron transport, protein complexes oxidize and reduce coenzymes to provide electrons and protons that move into the intermembrane space where they create a proton gradient that drives ATP synthesis.

Q What is the major source of NADH for the electron transport chain?

Chemiosmotic Model

In 1978, Peter Mitchell received the Nobel Prize for his theory called the **chemiosmotic model**, which links the energy from electron transport to a proton gradient that drives the synthesis of ATP. In this model, three of the complexes (I, III, and IV) extend through the inner membrane with one end of each complex in the matrix and the other end in the intermembrane space. In the chemiosmotic model, each of these complexes act as a **proton pump** by pushing protons (H^+) out of the matrix and into the intermembrane space. This increase in protons in the intermembrane space lowers the pH and creates a proton gradient. Because protons are positively charged, both the lower pH and the electrical charge of the proton gradient make it an electrochemical gradient. (See Figure 18.18.)

To equalize the pH between the intermembrane space and the matrix, there is a tendency by the protons to return the matrix. However, protons cannot diffuse through the inner membrane. The only way protons can return to the matrix is to pass through a protein complex called **ATP synthase**. We might think of the flow of protons as a stream or river that turns a water wheel. As the protons flow through ATP synthase, energy generated from the proton gradient is used to drive the ATP synthesis. Thus the process of oxidative phosphorylation couples the energy from electron transport to the synthesis of ATP from ADP and P_i.

$$ADP + P_i + \text{energy} \xrightarrow{\text{ATP synthase}} ATP$$

Electron Transport and ATP Synthesis

We have seen that oxidative phosphorylation couples the energy from electron transport with the synthesis of ATP. Because NADH enters the electron transport chain at complex I, energy is released from the oxidation of NADH to synthesize three ATP. However, $FADH_2$, which enters the chain at a lower energy

High energy

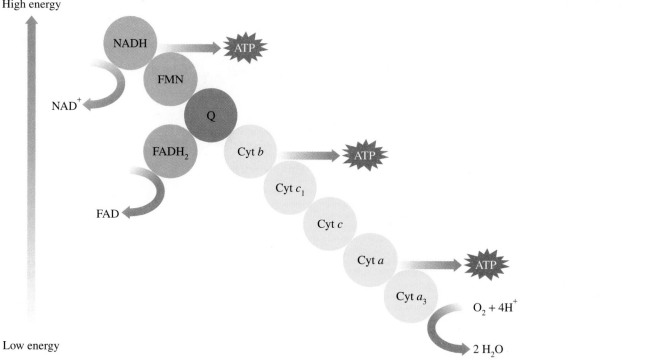

Low energy

FIGURE 18.19

As the energy levels decrease in the flow of electrons along the major electron carriers, electron transfers release sufficient energy to drive ATP synthesis.

Q Why does electron transfer from FADH$_2$ provide less energy than from NADH and H$^+$?

level at complex II, provides energy to drive the synthesis of only two ATP. (See Figure 18.19.) The overall equation for the oxidation of NADH and FADH$_2$ can be written as follows:

$$\boxed{\textbf{NADH} + \textbf{H}^+} + \tfrac{1}{2}O_2 + 3\text{ ADP} + 3\text{ P}_i \longrightarrow NAD^+ + H_2O + \boxed{\textbf{3 ATP}}$$

$$\boxed{\textbf{FADH}_2} + \tfrac{1}{2}O_2 + 2\text{ ADP} + 2\text{ P}_i \longrightarrow FAD + H_2O + \boxed{\textbf{2 ATP}}$$

SAMPLE PROBLEM 18.12

■ **ATP Synthesis**

Why does the oxidation of NADH provide energy for the formation of three ATP whereas FADH$_2$ produces two ATP?

SOLUTION

Electrons from the oxidation of NADH enter the electron chain at a higher energy level than FADH$_2$, providing energy to pump three pairs of protons to the intermembrane. However, FADH$_2$ transfers electrons to Q that go through two sites that pump protons and therefore provide energy for two ATP.

STUDY CHECK

Which complexes act as proton pumps?

ATP from Glycolysis

In glycolysis, the oxidation of glucose stores energy in two NADH molecules as well as two ATP molecules from direct substrate phosphorylation. However, glycolysis occurs in the cytoplasm, and the NADH produced cannot pass through the mitochondrial membrane to the electron transport chain.

Therefore the hydrogen ions and electrons from NADH in the cytoplasm are transferred to compounds that can enter the mitochondria. In this shuttle system, $FADH_2$ is produced. The overall reaction for the shuttle is

$$\underset{\text{cytoplasm}}{NADH + H^+ + FAD} \longrightarrow \underset{\text{mitochondria}}{NAD^+ + FADH_2}$$

Therefore the transfer of electrons from NADH in the cytoplasm to $FADH_2$ produces only two ATP, rather than three. In glycolysis, four ATP from two NADH and two ATP from direct phosphorylation add up to six ATP from one glucose molecule.

$$Glucose \longrightarrow 2 \text{ pyruvate} + 2 \text{ ATP} + 2 \text{ NADH } (\longrightarrow 2 \text{ FADH}_2)$$
$$Glucose \longrightarrow 2 \text{ pyruvate} + 6 \text{ ATP}$$

ATP from the Oxidation of Two Pyruvate

Under aerobic conditions, pyruvate enters the mitochondria, where it is oxidized to give acetyl CoA, CO_2, and NADH. Because each glucose molecule yields two

---

ATP Synthase and Heating the Body

Some types of compounds called *uncouplers* separate the electron transport system from the ATP synthase. They do this by disrupting the proton gradient needed for the synthesis of ATP. The electrons are transported to O_2 in electron transport, but ATP is not formed by ATP synthase.

Some uncouplers transport the protons through the inner membrane, which is normally impermeable to protons; others block the channel in the ATP synthase. Compounds such as dicumarol and 2,4-dinitrophenol (DNP) are hydrophobic and bind with protons to carry them across the inner membrane. An antibiotic, oligomycin, blocks the channel, which does not allow any protons to return to the matrix. By removing protons or blocking the channel, there is no proton flow to generate energy for ATP synthesis.

dicumarol

2,4-dinitrophenol (DNP)

When there is no mechanism for ATP synthesis, the energy of electron transport is released as heat. Certain animals that are adapted to cold climates have developed their own uncoupling system, which allows them to use electron transport energy for heat production. These animals have large amounts of a tissue called brown fat, which contains a high concentration of mitochondria. This tissue is brown because of the color of iron in the cytochromes of the mitochondria. The proton pumps still operate in brown tissue but a protein embedded in the cell wall allows the protons to bypass ATP synthase. The energy that would be used to synthesize ATP is released as heat. In newborn babies brown fat is used to generate heat because newborns have not stored much fat. The brown fat deposits are located near major blood vessels, which carry the warmed blood to the body. Infants have a small mass but a large surface area and need to produce more heat than the adult. Most adults have little or no brown fat, although someone who works outdoors in a cold climate will develop some brown fat deposits.

Plants also use uncouplers. Some plants use uncoupling agents to volatize fragrant compounds that attract insects to pollinate the plants. Skunk cabbage uses this system. In early spring, heat is used to warm early shoots of plants under the snow, which helps them melt the snow around the plants.

pyruvate, two NADH enter electron transport. The oxidation of two pyruvate molecules leads to the production of six ATP molecules.

$$2 \text{ Pyruvate} \longrightarrow 2 \text{ acetyl CoA} + 6 \text{ ATP}$$

ATP from the Citric Acid Cycle

One turn of the citric acid cycle produces two CO_2, three NADH, one $FADH_2$, and one ATP by direct substrate phosphorylation.

$$
\begin{array}{l}
3 \text{ NADH} \times 3 \text{ ATP} = \quad 9 \text{ ATP} \\
1 \text{ FADH}_2 \times 2 \text{ ATP} = \quad 2 \text{ ATP} \\
\underline{1 \text{ GTP} \times 1 \text{ ATP} \quad = \quad 1 \text{ ATP}} \\
\qquad \text{Total (one turn)} = 12 \text{ ATP}
\end{array}
$$

Because two acetyl CoA molecules are produced from each glucose, two turns of the citric acid cycle produce 24 ATP.

$$\text{Acetyl CoA} \longrightarrow 2CO_2 + 12 \text{ ATP (one turn of citric acid cycle)}$$

$$2 \text{ Acetyl CoA} \longrightarrow 4CO_2 + 24 \text{ ATP (two turns of citric acid cycle)}$$

ATP from the Complete Oxidation of Glucose

We can now determine the total ATP for the complete oxidation of glucose by combining the equations from glycolysis, oxidation of pyruvate, and citric acid cycle. (See Figure 18.20.) The details of the reaction pathways are given in Table 18.2.

TABLE 18.2

ATP Produced by the Complete Oxidation of Glucose	
Reaction Pathway	**ATP for One Glucose**
ATP from Glycolysis	
Activation of glucose	-2 ATP
Oxidation of glyceraldehyde-3-phosphate (2 NADH)	6 ATP
Transport of 2 NADH across membrane	-2 ATP
Direct ADP phosphorylation (two triose phosphate)	4 ATP
Summary: $C_6H_{12}O_6 \longrightarrow 2$ pyruvate $+ 2H_2O$ glucose	6 ATP
ATP from Pyruvate	
2 Pyruvate $\longrightarrow$ 2 acetyl CoA (2 NADH)	6 ATP
ATP from Two Turns of the Citric Acid Cycle	
Oxidation of 2 isocitrate (2 NADH)	6 ATP
Oxidation of 2 α-ketoglutarate (2 NADH)	6 ATP
2 Direct substrate phosphorylations (2 GTP)	2 ATP
Oxidation of 2 succinate (2 $FADH_2$)	4 ATP
Oxidation of 2 malate (2 NADH)	6 ATP
Summary: 2 Acetyl CoA $\longrightarrow 4CO_2 + 2H_2O$	24 ATP
Overall ATP Production for One Glucose	
$C_6H_{12}O_6 + 6O_2 + 36 \text{ ADP} + 36 \text{ P}_i \longrightarrow 6CO_2 + 6H_2O + 36 \text{ ATP}$ glucose	

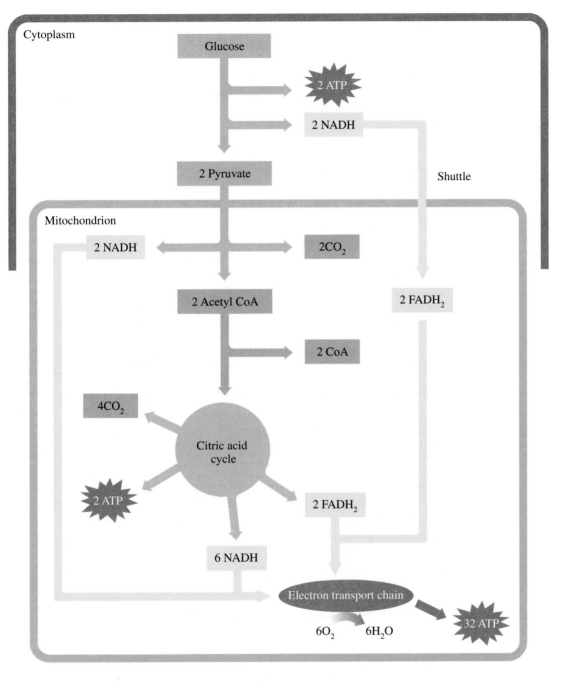

FIGURE 18.20

The complete oxidation of glucose to CO_2 and H_2O yields a total of 36 ATP.

Q What metabolic pathway produces most of the ATP from the oxidation of glucose?

SAMPLE PROBLEM 18.13

■ **ATP Production**

Indicate the amount of ATP produced by each of the following oxidation reactions:

a. pyruvate to acetyl CoA
b. glucose to acetyl CoA

SOLUTION

a. The oxidation of pyruvate to acetyl CoA produces one NADH, which yields three ATP.

b. Six ATP are produced from the oxidation of glucose to two pyruvate molecules. Another six ATP result from the oxidation of two pyruvate molecules to two acetyl CoA molecules.

STUDY CHECK

What are the sources of ATP in the citric acid cycle?

When glucose is not immediately used by the cells for energy, it is stored as glycogen in the liver and muscles. When the levels of glucose in the brain or blood become low, the glycogen reserves are hydrolyzed and glucose is released into the blood. If glycogen stores are depleted, some glucose can be synthesized from noncarbohydrate sources. It is the balance of all these reactions that maintains the necessary blood glucose level available to our cells and provides the necessary amount of ATP for our energy needs.

QUESTIONS AND PROBLEMS

■ Oxidative Phosphorylation and ATP

18.57 What is meant by oxidative phosphorylation?

18.58 How is the proton gradient established?

18.59 According to the chemiosmotic theory, how does the proton gradient provide energy to synthesize ATP?

18.60 How does the phosphorylation of ADP occur?

18.61 How are glycolysis and the citric acid cycle linked to the production of ATP by the electron transport chain?

18.62 Why does $FADH_2$ have a yield of two ATP via the electron chain, but NADH yields three ATP?

18.63 What is the energy yield in ATP molecules associated with each of the following?
   a. $NADH \longrightarrow NAD^+$
   b. glucose $\longrightarrow$ 2 pyruvate
   c. 2 pyruvate $\longrightarrow$ 2 acetyl CoA + $2CO_2$
   d. acetyl CoA $\longrightarrow$ $2CO_2$

18.64 What is the energy yield in ATP molecules associated with each of the following?
   a. $FADH_2 \longrightarrow FAD$
   b. glucose + $6O_2 \longrightarrow 6CO_2 + 6H_2O$
   c. glucose $\longrightarrow$ 2 lactate
   d. pyruvate $\longrightarrow$ lactate

■ 18.8 Oxidation of Fatty Acids

LEARNING GOAL

■

Describe the metabolic pathway of β oxidation; calculate the ATP from the complete oxidation of a fatty acid.

A large amount of energy is obtained when fatty acids undergo oxidation in the mitochondria to yield acetyl CoA. In stage 2 of fat metabolism, fatty acids undergo **beta oxidation (β oxidation)**, which removes two-carbon segments, one at a time, from a fatty acid.

β Oxidation occurs here

$$CH_3-(CH_2)_{14}-CH_2-CH_2-\overset{\overset{O}{\|}}{C}-OH$$

$\qquad\qquad\qquad\quad \beta \qquad \alpha$

stearic acid

Each cycle in β oxidation produces acetyl CoA and a fatty acid that is shorter by two carbons. The cycle repeats until the original fatty acid is completely degraded to two-carbon acetyl CoA units. Each acetyl CoA can then enter the citric acid cycle in the same way as the acetyl CoA units derived from glucose.

■ Fatty Acid Activation

Before a fatty acid can enter the mitochondria, it undergoes activation in the cytosol. The activation process combines a fatty acid with coenzyme A to yield fatty acyl CoA. The energy released by the hydrolysis of two phosphate groups from ATP is used to drive the reaction. The products are AMP and pyrophosphate (PP_i), which hydrolyzes to yield two inorganic phosphates ($2\ P_i$).

$$R-CH_2-\overset{\overset{O}{\|}}{C}-O^- + ATP + HS-CoA \xrightarrow{\text{acyl CoA} \atop \text{synthetase}} R-CH_2-\overset{\overset{O}{\|}}{C}-S-CoA + AMP + 2\ P_i$$

fatty acid $\qquad\qquad\qquad\qquad\qquad\qquad\qquad\qquad\qquad$ fatty acyl CoA

■ Reactions of β-Oxidation Cycle

In the matrix, fatty acyl CoA molecules undergo β oxidation, which is a cycle of four reactions that convert the $-CH_2-$ of the β carbon to a β-keto group. Once the β-keto group is formed, a two-carbon acetyl group can be split from the chain, which shortens the fatty acyl group.

■ β-Oxidation Pathway

Reaction 1 Oxidation (Dehydrogenation)
In the first reaction of β oxidation, the FAD coenzyme removes hydrogen atoms from the α and β carbons of the activated fatty acid to form a trans carbon–carbon double bond and $FADH_2$.

Reaction 2 Hydration
A water molecule now adds across the trans double bond, which places a hydroxyl group ($-OH$) on the β carbon.

Reaction 3   Oxidation (Dehydrogenation)

The hydroxy group on the β carbon is oxidized to yield a ketone. The hydrogen atoms removed in the dehydrogenation reduce coenzyme NAD^+ to $NADH + H^+$. At this point, the β carbon has been oxidized to a keto group.

Reaction 4   Cleavage of Acetyl CoA

In the final step of β oxidation, the C_α—C_β bond splits to yield free acetyl CoA and a fatty acyl CoA molecule that is shorter by two carbon atoms. This shorter fatty acyl CoA is ready to go through the β-oxidation cycle again.

■ Fat Storage and Blubber

Obtain two medium-sized plastic baggies and a can of Crisco or other type of shortening used for cooking. You will also need a bucket or large container with water and ice cubes. Place several tablespoons of the shortening in one of the baggies. Place the second baggie inside and tape the outside edges of the bag. With your hand inside the inner baggie, move the shortening around to cover your hand. With one hand inside the double baggie, submerge both your hands in the container of ice water. Measure the time it takes for one hand to feel uncomfortably cold. Experiment with different amounts of shortening.

QUESTIONS

1. How effective is the bag with "blubber" in protecting your hand from the cold?
2. How does "blubber" help an animal survive starvation?
3. How would twice the amount of shortening affect your results?
4. Why would animals in warm climates, such as camels and migratory birds, need to store fat?
5. If you placed 300 g of shortening in the baggie, how many moles of ATP could it provide if used for energy production (assume it produces the same ATP as myristic acid)?

The reaction for one cycle of β oxidation is written as follows:

$$R-CH_2-CH_2-\overset{\overset{\displaystyle O}{\|}}{C}-S-CoA + NAD^+ + FAD + H_2O + HS-CoA \longrightarrow$$

fatty acyl CoA

$$R-\overset{\overset{\displaystyle O}{\|}}{C}-S-CoA + CH_3-\overset{\overset{\displaystyle O}{\|}}{C}-S-CoA + NADH + H^+ + FADH_2$$

new fatty acyl (–2 C) CoA   acetyl CoA

SAMPLE PROBLEM 18.14

■ β Oxidation

Match each of the following with reactions in the β-oxidation cycle:
(1) first oxidation   (2) hydration   (3) second oxidation   (4) cleavage

a. water is added to a trans double bond
b. an acetyl CoA is removed
c. FAD is reduced to $FADH_2$

SOLUTION

a. (2) hydration   **b.** (4) cleavage   **c.** (1) first oxidation

STUDY CHECK

Which coenzyme is needed in reaction 3 when a β-hydroxyl group is converted to a β-keto group?

Fatty Acid Length Determines Cycle Repeats

The number of carbon atoms in a fatty acid determines the number of times the cycle repeats and the number of acetyl CoA units it produces. For example, the complete β oxidation of myristic acid (C_{14}) produces seven acetyl CoA groups, which is equal to one-half the number of carbon atoms in the chain. Because the final turn of the cycle produces two acetyl CoA groups, the total number of times the cycle repeats is one less than the total number of acetyl groups it produces. Therefore, the C_{14} fatty acid goes through the cycle six times.

Fatty acid	Number of Acetyl CoA	β-Oxidation cycles
myristic acid C_{14}	7 acetyl CoA	6
palmitic acid C_{16}	8 acetyl CoA	7
stearic acid C_{18}	9 acetyl CoA	8

We can write an overall equation for the complete oxidation of myristyl-CoA as follows:

$$\text{Myristyl-CoA} + 6\,\text{CoA} + 6\,\text{FAD} + 6\,\text{NAD}^+ + 6\text{H}_2\text{O} \longrightarrow$$
$$7\,\text{acetyl CoA} + 6\,\text{FADH}_2 + 6\,\text{NADH} + 6\text{H}^+$$

ATP from Fatty Acid Oxidation

We can now determine the total energy yield from the oxidation of a particular fatty acid. In each β-oxidation cycle, one NADH, one $FADH_2$, and one acetyl CoA are produced. Each NADH generates sufficient energy to synthesize three ATP, whereas $FADH_2$ leads to the synthesis of two ATP. However, the greatest amount of energy produced from a fatty acid is generated by the production of the acetyl CoA units that enter the citric acid cycle. Earlier we saw that one acetyl CoA leads to the synthesis of a total of 12 ATP.

So far we know that the C_{14} acid produces seven acetyl CoA units and goes through six turns of the cycle. We also need to remember that activation of the myristic acid requires the equivalent of two ATP. We can set up the calculation as follows.

ATP Production for Myristic Acid

Activation	-2 ATP
7 acetyl CoA	
7 acetyl CoA $\times$ 12 ATP/acetyl CoA	84 ATP
6 β-oxidation cycles	
6 $FADH_2 \times$ 2 ATP/$FADH_2$	12 ATP
6 NADH $\times$ 3 ATP/NADH	18 ATP
Total	112 ATP

SAMPLE PROBLEM 18.15

■ ATP Production from β Oxidation

How much ATP will be produced from the β oxidation of palmitic acid, a C_{16} saturated fatty acid?

SOLUTION

A 16-carbon fatty acid will produce 8 acetyl CoA units and go through 7 β-oxidation cycles. Each acetyl CoA can produce 12 ATP by way of the citric acid cycle. In the electron transport chain, each $FADH_2$ produces 2 ATP, and each NADH produces 3 ATP.

HEALTH NOTE

Stored Fat and Obesity

The storage of fat is an important survival feature in the lives of many animals. In hibernating animals, large amounts of stored fat provide the energy for the entire hibernation period, which could be several months. In camels, such as dromedary camels, large amounts of food are stored in the camel's hump, which is actually a huge fat deposit. When food resources are low, the camel can survive months without food or water by utilizing the fat reserves in the hump. Migratory birds preparing to fly long distances also store large amounts of fat. Whales are kept warm by a layer of body fat called "blubber" under their skin, which can be as thick as 2 feet. Blubber also provides energy when whales must survive long periods of starvation. Penguins also have blubber, which protects them from the cold and provides energy when they are sitting on a nest of eggs.

Humans also have the capability to store large amounts of fat, although they do not hibernate or usually have to survive for long periods of time without food. When humans survived on sparse diets that were mostly vegetarian, the fat content was about 20%. Today, a typical diet includes more dairy products and foods with high fat levels, which increases the daily fat intake to as much as 60% of the diet. The U.S. Public Health Service now estimates that in the United States, more than one-third of adults are obese. Obesity is defined as a body weight that is more than 20 percent over an ideal weight. Obesity is a major factor in health problems such as diabetes, heart disease, high blood pressure, stroke, and gallstones as well as some cancers and forms of arthritis.

At one time we thought that obesity was simply a problem of eating too much. However, research now indicates that certain pathways in lipid and carbohydrate metabolism may cause excessive weight gain in some people. In 1995, scientists discovered that a hormone called *leptin* is produced in fat cells. When fat cells are full, high levels of leptin signal the brain to limit the intake of food. When fat stores are low, leptin production decreases, which signals the brain to increase food intake. Some obese persons have high levels of leptin, which means that leptin did not cause them to decrease how much they ate.

Research on obesity has become a major research field. Scientists are studying differences in the rate of leptin production, degrees of resistance to leptin, and possible combinations of these factors. After a person has dieted and lost weight, the leptin level drops. This decrease in leptin may cause an increase in hunger, slow metabolism, and increased food intake, which starts the weight-gain cycle all over again. Currently, studies are being made to assess the safety of leptin therapy following a weight loss.

ATP Production for Palmitic Acid ($C_{16}H_{32}O_2$)

Activation of palmitic acid to palmitoyl-CoA	-2 ATP
8 acetyl CoA $\times$ 12 ATP (citric acid cycle)	96 ATP
7 $FADH_2$ $\times$ 2 ATP (electron transport chain)	14 ATP
7 NADH $\times$ 3 ATP (electron transport chain)	21 ATP
Total	129 ATP

STUDY CHECK

In β oxidation, why is more ATP produced from acetyl CoA than from reduced coenzymes?

Ketone Bodies

When carbohydrates are not available to meet energy needs, the body breaks down body fat. However, the oxidation of large amounts of fatty acids can cause acetyl

FIGURE 18.21

In ketogenesis, acetyl CoA molecules combine to produce ketone bodies: acetoacetate, β-hydroxybutyrate, and acetone.

Q What condition in the body leads to the formation of ketone bodies?

CoA molecules to accumulate in the liver. Then acetyl CoA molecules combine to form keto compounds called **ketone bodies**. (See Figure 18.21.)

Ketone bodies are produced mostly in the liver and transported to cells in the heart, brain, and skeletal muscle, where small amounts of energy can be obtained by converting acetoacetate or β-hydroxybutyrate back to acetyl CoA.

$$\beta\text{-Hydroxybutyrate} \longrightarrow \text{acetoacetate} + 2\,\text{CoA} \longrightarrow 2\,\text{acetyl CoA}$$

Ketosis

When ketone bodies accumulate, they may not be completely metabolized by the body. This may lead to a condition called **ketosis**, which is found in severe diabetes, diets high in fat and low in carbohydrates, and starvation. Because two of the ketone bodies are acids, they can lower the blood pH below 7.4, which is **acidosis**, a condition that often accompanies ketosis. A drop in blood pH can interfere with the ability of the blood to carry oxygen and cause breathing difficulties.

Fatty Acid Synthesis

When the body has met all its energy needs and the glycogen stores are full, acetyl CoA from the breakdown of carbohydrates and fatty acids is used to form new fatty acids. Two-carbon acetyl units are linked together to give a 16-carbon fatty acid, palmitic acid. Several of the reactions are the reverse of the reactions we discussed in fatty acid oxidation. However, fatty acid oxidation occurs in the mitochondria and uses FAD and NAD^+, whereas fatty acid synthesis occurs in the cytosol and uses the reduced coenzyme NADPH. NADPH is similar to NADH, except it has a phosphate group. The new fatty acids are attached to glycerol to make triacylglycerols, which are stored as body fat.

QUESTIONS AND PROBLEMS

■ Oxidation of Fatty Acids

18.65 Where in the cell is a fatty acid activated?

18.66 What coenzymes are required for β oxidation?

18.67 Capric acid, CH_3—$(CH_2)_8$—COOH, is a C_{10} fatty acid.
  a. Write the formula of the activated form of capric acid.
  b. Indicate the α and β carbon atoms in the fatty acid.
  c. Write the overall equation for the first cycle of β oxidation for capric acid.
  d. Write the overall equation for the complete β oxidation of capric acid.

18.68 Arachidic acid, CH_3—$(CH_2)_{18}$—COOH, is a C_{20} fatty acid.
  a. Write the formula of the activated form of arachidic acid.
  b. Indicate the α and β carbon atoms in the fatty acid.
  c. Write the overall equation for the first cycle of β oxidation for arachidic acid.
  d. Write the overall equation for the complete β oxidation of arachidic acid.

18.69 Why is the energy of fatty acid activation from ATP to AMP considered the same as the hydrolysis of 2 ATP $\longrightarrow$ 2ADP?

18.70 What is the number of ATP obtained from one acetyl CoA in the citric acid cycle?

18.71 Consider the complete oxidation of capric acid CH_3—$(CH_2)_8$—COOH, a C_{10} fatty acid.
  a. How many acetyl CoA units are produced?
  b. How many cycles of β oxidation are needed?
  c. How many ATPs are generated from the oxidation of capric acid?

18.72 Consider the complete oxidation of arachidic acid, CH_3—$(CH_2)_{18}$—COOH, a C_{20} fatty acid.
  a. How many acetyl CoA units are produced?
  b. How many cycles of β oxidation are needed?
  c. How many ATPs are generated from the oxidation of arachidic acid?

18.73 What are ketone bodies?

18.74 If a person is fasting, why would he or she have high levels of acetyl CoA?

18.75 What is ketosis?

18.76 Why do diabetics produce high levels of ketone bodies?

LEARNING GOAL

■ Describe the reactions of transamination, oxidative deamination, and the formation of urea from ammonium ion.

■ 18.9 Degradation of Amino Acids

When dietary protein exceeds the nitrogen needed by the body, the excess amino acids are degraded. The α-amino group is removed to yield a keto acid, which can be converted to an intermediate of other metabolic pathways. The carbon atoms from amino acids are used in the citric acid cycle as well as the synthesis of fatty acids, ketone bodies, and glucose. Most of the amino groups are converted to urea.

■ Transamination

The degradation of amino acids occurs primarily in the liver. In a **transamination** reaction, an α-amino group is transferred from an amino acid to an α-keto acid, usually α-ketoglutarate. A new amino acid and a new α-keto acid are produced.

▪ Ketone Bodies and Diabetes

The blood glucose is elevated within 30 minutes following a meal containing carbohydrates. The elevated level of glucose stimulates the secretion of the hormone insulin from the pancreas, which increases the flow of glucose into muscle and adipose tissue for the synthesis of glycogen. As blood glucose levels drop, the secretion of insulin decreases. When blood glucose is low, another hormone, glucagon, is secreted by the pancreas, which stimulates the breakdown of glycogen in the liver to yield glucose.

In *diabetes mellitus*, glucose cannot be utilized or stored as glycogen because insulin is not secreted or does not function properly. In Type I, *insulin-dependent diabetes*, which often occurs in childhood, the pancreas produces inadequate levels of insulin. This type of diabetes can result from damage to the pancreas by viral infections or from genetic mutations. In Type II, *insulin-resistant diabetes*, which usually occurs in adults, insulin is produced, but insulin receptors are not responsive. Thus a person with Type II diabetes does not respond to insulin therapy. *Gestational diabetes* can occur during pregnancy,

but blood glucose levels usually return to normal after the baby is born. Mothers with diabetes tend to gain weight and have large babies.

In all types of diabetes, insufficient amounts of glucose are available in the muscle, liver, and adipose tissue. As a result liver cells synthesize glucose from noncarbohydrate sources (gluconeogenesis) and break down fat, which elevates the acetyl CoA level. Excess acetyl CoA undergoes ketogenesis and ketone bodies accumulate in the blood. As the level of acetone increases, its odor can be detected on the breath of a person with uncontrolled diabetes who is in ketosis.

In uncontrolled diabetes, the concentration of blood glucose exceeds the ability of the kidney to reabsorb glucose, and glucose appears in the urine. High levels of glucose increase the osmotic pressure in the blood, which leads to an increase in urine output. Symptoms of diabetes include frequent urination and excessive thirst. Treatment for diabetes includes a change to a diet to limit carbohydrate intake and may require medication such as a daily injection of insulin.

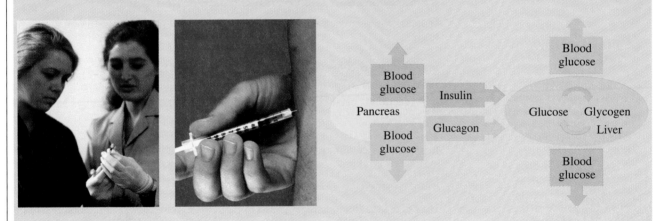

We can write an equation to show the transfer of the amino group from alanine to α-ketoglutarate to yield glutamate, the new amino acid, and the α-keto acid pyruvate.

CASE STUDY
Diabetes and Blood Glucose

Oxidative Deamination

In a process called **oxidative deamination**, the amino group in glutamate is removed as an ammonium ion NH_4^+. This reaction is catalyzed by *glutamate dehydrogenase*, which uses either NAD^+ or $NADP^+$ as a coenzyme.

$$\overset{\overset{+}{N}H_3}{\underset{\text{glutamate}}{^-OOC-CH-CH_2-CH_2-COO^-}} + H_2O + NAD^+ \text{ (or } NADP^+) \xrightarrow{\text{glutamate dehydrogenase}}$$

$$\overset{O}{\underset{\alpha\text{-ketoglutarate}}{^-OOC-\overset{\|}{C}-CH_2-CH_2-COO^-}} + NH_4^+ + NADH \text{ (or NADPH)} + H^+$$

Therefore the amino group from any amino acid can be used to form glutamate, which undergoes oxidative deamination converting the amino group to an ammonium ion. Then the ammonium ion is converted to urea.

SAMPLE PROBLEM 18.16

■ Transamination and Oxidative Deamination

Indicate whether each of the following represents a transamination or an oxidative deamination:

a. Glutamate is converted to α-ketoglutarate and NH_4^+.
b. Alanine and α-ketoglutarate react to form pyruvate and glutamate.
c. A reaction is catalyzed by glutamate dehydrogenase, which requires NAD^+.

SOLUTION

a. oxidative deamination   **b.** transamination   **c.** oxidative deamination

STUDY CHECK

How is α-ketoglutarate regenerated to participate in more transamination reactions?

Urea Cycle

Ammonium ion is highly toxic. In a series of reactions called the **urea cycle**, NH_4^+ combines with CO_2 to yield urea, which is excreted in the urine. There are several steps of the urea cycle with several intermediates, but the overall reaction converts ammonium ions into urea.

$$2NH_4^+ + CO_2 \longrightarrow \underset{\text{urea}}{H_2N-\overset{\overset{O}{\|}}{C}-NH_2} + 2H^+ + H_2O$$

In one day, a typical adult may excrete about 25–30 g of urea in the urine. This amount increases when a diet is high in protein. If urea is not properly excreted, it builds up quickly to a toxic level. To detect renal disease, the blood urea nitrogen (BUN) level is measured. If the BUN is high, protein intake must be reduced, and hemodialysis may be needed to remove toxic nitrogen waste from the blood.

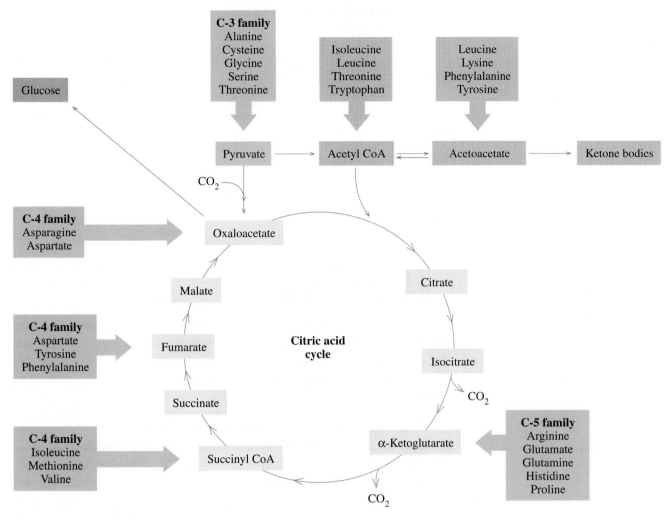

FIGURE 18.22

Carbon atoms from degraded amino acids are converted to the intermediates of the citric acid cycle or other pathways.

Q **What compound in the citric acid cycle is obtained from the carbon atoms of alanine and glycine?**

Fates of the Carbon Atoms from Amino Acids

The carbon skeletons from the transamination of amino acids are used as intermediates of the citric acid cycle or other metabolic pathways. We can classify the amino acids according to the number of carbon atoms in those intermediates. (See Figure 18.22.) The amino acids with three carbons, or those that are converted into carbon skeletons with three carbons, are converted to pyruvate. The four-carbon group consists of amino acids that are converted to oxaloacetate, and the five-carbon group provides α-ketoglutarate. Some amino acids are listed twice because they can enter different pathways to form citric acid cycle intermediates.

Energy from Amino Acids

Normally, only a small amount (about 10%) of our energy needs is supplied by amino acids. However, more energy is extracted from amino acids in conditions such as fasting or starvation, when carbohydrate and fat stores are exhausted. If amino acids remain the only source of energy for a long period of time, the breakdown of body proteins eventually leads to a destruction of essential body tissues.

SAMPLE PROBLEM 18.17

■ Carbon Atoms from Amino Acids

What degradation products of amino acids are citric acid intermediates?

SOLUTION

Carbon atoms from amino acids enter the citric acid cycle as acetyl CoA, α-ketoglutarate, succinyl CoA, fumarate, or oxaloacetate.

STUDY CHECK

Which amino acids provide carbon atoms that enter the citric acid as α-ketoglutarate?

■ Overview of Metabolism

In this chapter, we have seen that catabolic pathways degrade large molecules to small molecules that are used for energy production via the citric acid cycle and the electron transport chain. We have also indicated that anabolic pathways lead to the synthesis of larger molecules in the cell. In the overall view of metabolism, there are several branch points from which compounds may be degraded for energy or used to synthesize larger molecules. For example, glucose can be degraded to acetyl CoA for the citric acid cycle to produce energy or converted to glycogen for storage. When glycogen stores are depleted, fatty acids are degraded for energy. Amino acids normally used to synthesize nitrogen-containing compounds in the cells can also be used for energy after they are degraded to intermediates of the citric acid cycle. In the synthesis of certain amino acids, α-keto acids of the citric acid cycle enter a variety of reactions that convert them to amino acids through transamination by glutamate. (See Figure 18.23.)

QUESTIONS AND PROBLEMS

■ Degradation of Amino Acids

18.77 What are the reactants and products in transamination reactions?

18.78 What types of enzymes catalyze transamination reactions?

18.79 Write the structure of the α-keto acid produced from each of the following in transamination:

a. $\overset{\overset{+}{N}H_3}{\underset{|}{H-CH-COO^-}}$ glycine   b. $\overset{\overset{+}{N}H_3}{\underset{|}{CH_3-CH-COO^-}}$ alanine

18.80 Write the structure of the α-keto acid produced from each of the following in transamination:

a. $\overset{\overset{+}{N}H_3}{\underset{|}{^-OOC-CH_2-CH-COO^-}}$ aspartate

b. $\overset{CH_3\ \ \overset{+}{N}H_3}{\underset{|\ \ \ \ \ \ |}{CH_3-CH_2-CH-CH-COO^-}}$ isoleucine

18.81 Why does the body convert NH_4^+ to urea?

18.82 What is the structure of urea?

18.83 What metabolic substrate(s) can be produced from the carbon atoms of each of the following amino acids?

    a. alanine   **b.** aspartate   **c.** valine   **d.** glutamine

18.84 What metabolic substrate(s) can be produced from the carbon atoms of each of the following amino acids?

    a. leucine   **b.** asparagine   **c.** cysteine   **d.** arginine

18.85 What do we call the amino acids that humans can synthesize?

18.86 How do humans obtain the amino acids that cannot be synthesized in the body?

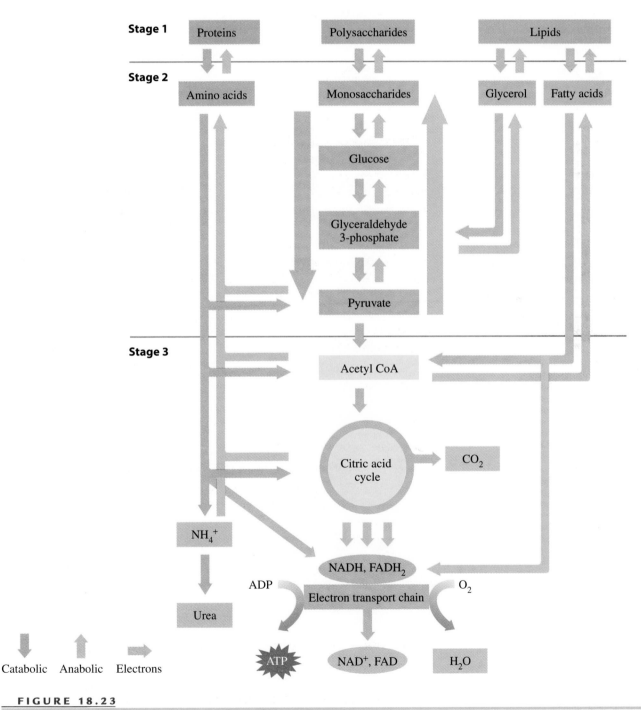

FIGURE 18.23

Catabolic and anabolic pathways in the cells provide the energy and necessary compounds for the cells.

Q **Under what conditions in the cell are amino acids degraded for energy?**

CONCEPT MAP

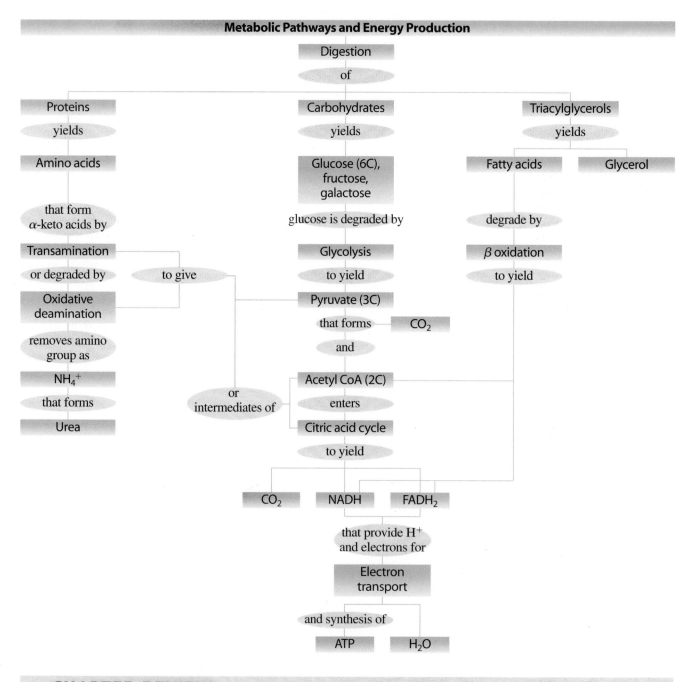

Metabolic Pathways and Energy Production

Digestion

of

Proteins — Carbohydrates — Triacylglycerols

yields — yields — yields

Amino acids — Glucose (6C), fructose, galactose — Fatty acids — Glycerol

that form α-keto acids by — glucose is degraded by — degrade by

Transamination — Glycolysis — β oxidation

or degraded by — to give — to yield — to yield

Oxidative deamination — Pyruvate (3C)

removes amino group as — that forms — CO_2

NH_4^+ — and

that forms — or intermediates of — Acetyl CoA (2C)

Urea — enters

Citric acid cycle

to yield

CO_2 — NADH — $FADH_2$

that provide H^+ and electrons for

Electron transport

and synthesis of

ATP — H_2O

CHAPTER REVIEW

18.1 Metabolism and ATP Energy

Metabolism includes all the catabolic and anabolic reactions that occur in the cells. Catabolic reactions degrade large molecules into smaller ones with an accompanying release of energy. Anabolic reactions require energy to synthesize larger molecules from smaller ones. The 3 stages of metabolism are digestion of food, degradation of monomers such as glucose to pyruvate, and the extraction of energy from the two- and three-carbon compounds. Energy obtained from catabolic reactions is stored in adenosine triphosphate (ATP), a high-energy compound that is hydrolyzed when energy is required by anabolic reactions.

18.2 Digestion: Stage 1

The digestion of carbohydrates is a series of reactions that breaks down polysaccharides into glucose, galactose, and fructose. These monomers can be absorbed through the intestinal wall into the bloodstream to be carried to cells where they provide energy and carbon atoms for synthesis of new molecules. Triacylglycerols are hydrolyzed in the small intestine to monoacylglycerol and fatty acids, which enter the intestinal wall and form new triacylglycerols. They bind with proteins to form chylomicrons, which transport them through the lymphatic system and bloodstream to the tissues. The digestion of proteins, which begins in the stomach and continues in the small intestine, involves the hydrolysis of peptide bonds to yield amino acids that are absorbed through the intestinal wall and transported to the cells.

18.3 Important Coenzymes in Metabolic Pathways

FAD and NAD^+ are the oxidized forms of coenzymes that participate in oxidation–reduction reactions. When they pick up hydrogen ions and electrons, they are reduced to $FADH_2$ and $NADH + H^+$. Coenzyme A contains a thiol group that bonds with a two-carbon acetyl group (acetyl CoA) or a longer chain acyl group (acyl CoA).

18.4 Glycolysis: Stage 2

Glycolysis, which occurs in the cytosol, consists of 10 reactions that degrade glucose (six carbons) to two pyruvate molecules (three carbons each). The overall series of reactions yields two molecules of the reduced coenzyme NADH and two ATP. Under aerobic conditions, pyruvate is oxidized in the mitochondria to acetyl CoA. In the absence of oxygen, pyruvate is reduced to lactate and NAD^+ is regenerated for the continuation of glycolysis.

18.5 The Citric Acid Cycle: Stage 3

In a sequence of reactions called the citric acid cycle, an acetyl group is combined with oxaloacetate to yield citrate. Citrate undergoes oxidation and decarboxylation to yield two CO_2, GTP, three NADH, and $FADH_2$ with the regeneration of oxaloacetate. The direct phosphorylation of ADP by GTP yields ATP.

18.6 Electron Transport

Electron carriers that transfer hydrogen ions and electrons include FMN, iron–sulfur proteins, coenzyme Q, and several cytochromes.

Both iron–sulfur proteins and cytochromes contain iron ions that are reduced to Fe^{2+} and reoxidized to Fe^{3+} as electrons are accepted and then passed to the next electron carrier. The reduced coenzymes NADH and $FADH_2$ from various metabolic pathways are oxidized to NAD^+ and FAD when their protons and electrons are transferred to the electron transport chain. The final acceptor, O_2, combines with protons and electrons to yield H_2O. At three points in electron transport, the energy decrease provides the necessary energy for ATP synthesis.

18.7 Oxidative Phosphorylation and ATP

The protein complexes in electron transport act as proton pumps to move protons into the inner membrane space, which produces a proton gradient. As the protons return to the matrix by way of ATP synthase, energy is generated. This energy is used to drive the synthesis of ATP in a process known as oxidative phosphorylation. The oxidation of NADH yields three ATP molecules, and $FADH_2$ yields two ATP. The energy from the NADH produced in the cytoplasm is used to form $FADH_2$. Under aerobic conditions, the complete oxidation of glucose yields a total of 36 ATP.

18.8 Oxidation of Fatty Acids

When needed as an energy source, fatty acids are linked to coenzyme A and transported into the mitochondria where they undergo β oxidation. The fatty acyl chain is oxidized to yield a shorter fatty acid, acetyl CoA, and the reduced coenzymes NADH and $FADH_2$. Although the energy from a particular fatty acid depends on its length, each oxidation cycle yields 5 ATP with another 12 ATP from the acetyl CoA that enter the citric acid cycle. When high levels of acetyl CoA are present in the cell, they enter the ketogenesis pathway forming ketone bodies such as acetoacetate, which cause ketosis and acidosis.

18.9 Degradation of Amino Acids

When the amount of amino acids in the cells exceeds that needed for synthesis of nitrogen compounds, the process of transamination converts them to α-keto acids and glutamate. Oxidative deamination of glutamate produces ammonium ions and α-ketoglutarate. Ammonium ions from oxidative deamination are converted to urea. The carbon atoms from the degradation of amino acids can enter the citric acid cycle or other metabolic pathways. Certain amino acids are synthesized when amino groups from glutamate are transferred to an α-keto acid obtained from glycolysis or the citric acid cycle.

SUMMARY OF KEY REACTIONS

HYDROLYSIS OF ATP

$$ATP + H_2O \longrightarrow ADP + P_i + 7.3 \text{ kcal/mole}$$

FORMATION OF ATP

$$ADP + P_i + 7.3 \text{ kcal/mole} \longrightarrow ATP$$

REDUCTION OF FAD AND NAD^+

$$FAD + 2H^+ + 2\,e^- \longrightarrow FADH_2$$
$$NAD^+ + 2H^+ + 2\,e^- \longrightarrow NADH + H^+$$

HYDROLYSIS OF DISACCHARIDES

$$\text{Lactose} + H_2O \xrightarrow{\text{lactase}} \text{galactose} + \text{glucose}$$

$$\text{Sucrose} + H_2O \xrightarrow{\text{sucrase}} \text{fructose} + \text{glucose}$$

$$\text{Maltose} + H_2O \xrightarrow{\text{maltase}} \text{glucose} + \text{glucose}$$

DIGESTION OF TRIACYLGLYCEROLS

$$\text{Triacylglycerols} + 3H_2O \xrightarrow{\text{lipases}} \text{glycerol} + 3 \text{ fatty acids}$$

DIGESTION OF PROTEINS

$$\text{Protein} + H_2O \xrightarrow{H^+} \text{amino acids}$$

GLYCOLYSIS

$$C_6H_{12}O_6 + 2\,ADP + 2\,P_i + 2\,NAD^+ \rightarrow 2\,CH_3-\overset{\overset{\displaystyle O}{\|}}{C}-COO^- + 2\,ATP + 2\,NADH + 4H^+$$

glucose   pyruvate

OXIDATION OF PYRUVATE TO ACETYL-CoA

$$CH_3-\overset{\overset{\displaystyle O}{\|}}{C}-COO^- + NAD^+ + HS-CoA \longrightarrow CH_3-\overset{\overset{\displaystyle O}{\|}}{C}-S-CoA + NADH + CO_2$$

pyruvate   acetyl CoA

CITRIC ACID CYCLE

$$\text{Acetyl CoA} + 3\,NAD^+ + FAD + GDP + P_i + 2H_2O \longrightarrow 2CO_2 + 3\,NADH + 2H^+ + FADH_2 + HS-CoA + GTP$$

ELECTRON TRANSPORT CHAIN

$$NADH + H^+ + 3\,ADP + 3\,P_i + \tfrac{1}{2}O_2 \longrightarrow NAD^+ + 3\,ATP + H_2O$$

$$FADH_2 + 2\,ADP + 2\,P_i + \tfrac{1}{2}O_2 \longrightarrow FAD + 2\,ATP + H_2O$$

COMPLETE OXIDATION OF GLUCOSE

$$C_6H_{12}O_6 + 6O_2 + 36\,ADP + 36\,P_i \longrightarrow 6\,CO_2 + 6\,H_2O + 36\,ATP$$

REDUCTION OF PYRUVATE TO LACTATE

$$CH_3-\overset{\overset{\displaystyle O}{\|}}{C}-COO^- + NADH + H^+ \longrightarrow CH_3-\overset{\overset{\displaystyle OH}{|}}{CH}-COO^- + NAD^+$$

pyruvate   lactate

β OXIDATION OF FATTY ACID

$$\text{Myristyl-CoA} + 6\,CoA + 6\,FAD + 6\,NAD^+ + 6\,H_2O \longrightarrow 7\ \text{acetyl CoA} + 6\,FADH_2 + 6\,NADH + 6H^+$$

TRANSAMINATION

$$CH_3-\overset{\overset{\displaystyle \overset{+}{N}H_3}{|}}{CH}-COO^- + {}^-OOC-\overset{\overset{\displaystyle O}{\|}}{C}-CH_2-CH_2-COO^- \xrightarrow{\substack{\text{alanine}\\\text{aminotransferase}}}$$

alanine   α-ketoglutarate

$$CH_3-\overset{\overset{\displaystyle O}{\|}}{C}-COO^- + {}^-OOC-\overset{\overset{\displaystyle \overset{+}{N}H_3}{|}}{CH}-CH_2-CH_2-COO^-$$

pyruvate   glutamate

OXIDATIVE DEAMINATION

$$\overset{\overset{+}{\underset{|}{NH_3}}}{^-OOC-CH-CH_2-CH_2-COO^-} + H_2O + NAD^+ (NADP) \xrightarrow{\text{glutamate dehydrogenase}}$$

glutamate

$$\overset{\overset{O}{\underset{||}{}}}{^-OOC-C-CH_2-CH_2-COO^-} + NH_4^+ + NADH \ (NADPH) + H^+$$

α-ketoglutarate

UREA CYCLE

$$2NH_4^+ + CO_2 \longrightarrow H_2N-\overset{\overset{O}{||}}{C}-NH_2 + 2H^+ + H_2O$$

KEY TERMS

acetyl CoA A two-carbon acetyl unit from oxidation of pyruvate that bonds to coenzyme A.

acidosis Low blood pH resulting from the formation of acidic ketone bodies.

ADP Adenosine diphosphate, a compound of adenine, a ribose sugar, and two phosphate groups. It is formed by the hydrolysis of ATP.

aerobic An oxygen-containing environment in the cells.

amylase An enzyme that hydrolyzes the glycosidic bonds in polysaccharides during digestion.

anabolic reaction A metabolic reaction that requires energy.

anaerobic A condition in cells when there is no oxygen.

ATP Adenosine triphosphate, a high-energy compound that stores energy in the cells and that consists of adenine, a ribose sugar, and three phosphate groups.

ATP synthase An enzyme complex that links the energy released by protons returning to the matrix with the synthesis of ATP from ADP and P_i.

beta (β) oxidation The degradation of fatty acids that removes two-carbon segments from an oxidized β carbon in the chain.

catabolic reaction A metabolic reaction that produces energy for the cell by the degradation and oxidation of glucose and other molecules.

chemiosmotic model The conservation of energy from transfer of electrons in the electron transport chain by pumping protons into the intermembrane space to produce a proton gradient that provides the energy to synthesize ATP.

chylomicron Lipoproteins formed by coating triacylglycerols with proteins for transport in the lymph and bloodstream.

citric acid cycle A series of oxidation reactions in the mitochondria that convert acetyl CoA to CO_2 and yield NADH and $FADH_2$. It is also called the tricarboxylic acid cycle and the Krebs cycle.

coenzyme A (CoA) A coenzyme that transports acyl and acetyl groups.

coenzyme Q (CoQ, Q) A mobile carrier that transfers electrons from NADH and $FADH_2$ to cytochrome *b* in complex III.

cytochromes (cyt) Iron-containing proteins that transfer electrons from QH_2 to oxygen.

cytoplasm The material in eukaryotic cells between the nucleus and the plasma membrane.

cytosol The fluid of the cytoplasm, which is an aqueous solution of electrolytes and enzymes.

decarboxylation The loss of a carbon atom in the form of CO_2.

digestion The processes in the gastrointestinal tract that break down large food molecules to smaller ones that pass through the intestinal membrane into the bloodstream.

electron transport A series of reactions in the mitochondria that transfer electrons from NADH and $FADH_2$ to electron carriers, which are arranged from higher to lower energy levels, and finally to O_2, which produces H_2O. Energy changes during three of these transfers provide energy for ATP synthesis.

FAD A coenzyme (flavin adenine dinucleotide) for dehydrogenase enzymes that form carbon–carbon double bonds.

Fe–S (iron–sulfur) clusters Proteins containing iron and sulfur in which the iron ions accept/release electrons during electron transport.

FMN (flavin mononucleotide) An electron carrier derived from riboflavin (vitamin B_2) that transfers hydrogen ions and electrons from NADH entering in the electron transport chain.

glycolysis The ten oxidation reactions of glucose that yield two pyruvate molecules.

ketone bodies The products of ketogenesis: acetoacetate, β-hydroxybutyrate, and acetone.

ketosis A condition in which high levels of ketone bodies cannot be metabolized, leading to lower blood pH.

metabolism All the chemical reactions in living cells that carry out molecular and energy transformations.

mitochondria The organelles of the cells where energy-producing reactions take place.

NAD$^+$ The hydrogen acceptor used in oxidation reactions that form carbon–oxygen double bonds.

oxidation The loss of hydrogen as hydrogen ions and electrons or the gain of oxygen by a substrate that is degraded to smaller molecules or a coenzyme.

oxidative deamination The loss of ammonium ion when glutamate is degraded to α-ketoglutarate.

oxidative phosphorylation The synthesis of ATP from ADP and P_i using energy generated by the oxidation reactions in the electron transport chain.

proton pumps The enzyme complexes I, III, and IV that move protons from the matrix into the intermembrane space, creating a proton gradient.

reduction The gain of hydrogen ions and electrons or the loss of oxygen by a substrate or a coenzyme.

transamination The transfer of an amino group from an amino acid to an α-keto acid.

urea cycle The process in which ammonium ions from the degradation of amino acids are converted to urea.

ADDITIONAL QUESTIONS AND PROBLEMS

18.87 Write an equation for the hydrolysis of ATP to ADP.

18.88 At the gym, you expend 300 kcal riding the stationary bicycle for 1 hr. How many moles of ATP will this require?

18.89 How and where does lactose undergo digestion in the body? What are the products?

18.90 How and where does sucrose undergo digestion in the body? What are the products?

18.91 What are the reactant and product of glycolysis?

18.92 What is the general type of reaction that takes place in the digestion of carbohydrates?

18.93 When is pyruvate converted to lactate in the body?

18.94 When pyruvate is used to form acetyl CoA, the product has only two carbon atoms. What happened to the third carbon?

18.95 What is the main function of the citric acid cycle in energy production?

18.96 Most metabolic pathways are not considered cycles. Why is the citric acid cycle considered to be a metabolic cycle?

18.97 If there are no reactions in the citric acid cycle that used oxygen, O_2, why does the cycle operate only in aerobic conditions?

18.98 What products of the citric acid cycle are needed for electron transport?

18.99 In the chemiosmotic model, how is energy provided to synthesize ATP?

18.100 What is the effect of proton accumulation in the intermembrane space?

18.101 How many ATP are produced when glucose is oxidized to pyruvate compared to when glucose is oxidized to CO_2 and H_2O?

18.102 What metabolic substrate(s) can be produced from the carbon atoms of each of the following amino acids?
a. leucine
b. isoleucine
c. cysteine
d. phenylalanine

UNDERSTANDING THE CONCEPTS

18.103 Lauric acid, CH_3—$(CH_2)_{10}$—$COOH$, which is found in coconut oil, is a saturated fatty acid.

a. Write the formula of the activated form of lauric acid.
b. Indicate the α and β carbon atoms in the fatty acyl molecule.
c. Write the overall equation for the complete β oxidation for lauric acid.
d. How many acetyl CoA units are produced?
e. How many cycles of β oxidation are needed?
f. Account for the total ATP yield from β oxidation of lauric acid (C_{12} fatty acid) by completing the following calculation:

activation	$\longrightarrow$	-2ATP
____acetyl CoA	$\longrightarrow$	____ATP
____FADH$_2$	$\longrightarrow$	____ATP
____NADH	$\longrightarrow$	____ATP
Total		____ATP

18.104 A hiker expends 450 kcal. How many moles of ATP does this require?

18.105 Identify the type of food as carbohydrate, fat, or protein that gives each of the following digestion products:

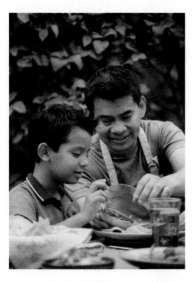

a. glucose
b. fatty acid
c. maltose
d. glycerol
e. amino acids
f. dextrins

18.106 Identify each of the following as a 6-carbon or a 3-carbon compound and arrange them in the order in which they occur in glycolysis.
a. 3-phosphoglycerate
b. pyruvate
c. glucose-6-phosphate
d. glucose
e. fructose-1,6-bisphosphate

QUESTIONS FOR TEAMWORK

18.107 One cell at work may break down 2 million (2 000 000) ATP molecules in one second. Some researchers estimate that the human body has about 10^{13} cells.
a. How much energy in kcal could be produced by the cells in the body in one day?
b. If ATP has a molar mass of 507 g/mole, how many grams of ATP are hydrolyzed?

18.108 State if each of the following processes produce or consume ATP.
a. Citric acid cycle
b. Glucose forms two pyruvates
c. Pyuvate froms acetyl CoA
d. Glucose forms glucose-6-phosphate
e. Oxidation of α-ketoglutarate
f. Transport of NADH across mitochondrial membrane
g. Activation of a fatty acid

18.109 Match the following ATP yields to reactions a–g.
2 ATP   3 ATP   6 ATP   12 ATP
18 ATP   36 ATP   44 ATP
a. Glucose forms two pyruvates
b. Pyruvate forms acetyl CoA
c. Glucose forms two acetyl CoAs
d. Acetyl CoA goes through one turn of the citric acid cycle
e. Caproic acid (C_6) is completely oxidized
f. NADH + H^+ is oxidized to NAD^+
g. $FADH_2$ is oxidized to FAD

18.110 Identify each of the following reactions in the β oxidation of palmitic acid (C_{14}), a fatty acid, as
(1) activation   (2) first dehydrogenation (oxidation)
(3) hydration   (4) second dehydrogenation
(5) cleavage of acetyl CoA
a. palmityl CoA and FAD form α, β-unsaturated palmityl CoA and $FADH_2$
b. β-keto palmityl CoA forms myristyl CoA and acetyl CoA
c. palmitic acid and ATP form palmityl CoA
d. α, β-unsaturated palmityl CoA and H_2O form β-hydroxy palmityl CoA
e. β-hydroxy palmityl CoA and NAD^+ form β-keto palmityl CoA and NADH + H^+

18.111 Which of the following molecules will produce the most ATP per mole?
a. glucose or maltose
b. myristic acid, CH_3—$(CH_2)_{12}$—COOH or stearic acid, CH_3—$(CH_2)_{16}$—COOH
c. glucose or two acetyl CoA
d. glucose or caprylic acid (C_8)
e. citrate or succinate in one turn of the citric acid cycle

18.112 If acetyl CoA has a molar mass of 809 g/mole, how many moles ATP are produced when 1.0 μg acetyl CoA completes the citric acid cycle?

ANSWERS

Answers to Study Checks

18.1 The cytoplasm is all the cellular material between the cell membrane and the nucleus. The cytosol, the aqueous part of the cytoplasm, is a solution of electrolytes and enzymes.

18.2 Adenine, ribose, and three phosphate groups.

18.3 The digestion of amylose begins in the mouth when salivary amylase hydrolyzes some of the glycosidic bonds. In the small intestine, pancreatic amylase hydrolyzes more glycosidic bonds, and finally maltose is hydrolyzed by maltase to yield glucose.

18.4 In the membrane of the small intestine, monoacylglycerols and fatty acids recombine to form new triacylglycerols that bind with proteins to form chylomicrons for transport to the lymphatic system and the bloodstream.

18.5 HCl denatures proteins and activates enzymes such as pepsin.

18.6 $FADH_2$

18.7 In the initial reactions of glycolysis, energy in the form of two ATP is invested to convert glucose to fructose-1,6-bisphosphate.

18.8

$$CH_3-\underset{\underset{OH}{|}}{CH}-\underset{\underset{O}{||}}{C}-O^- + NAD^+ \xrightarrow{\text{lactate dehydrogenase}}$$

$$CH_3-\underset{\underset{O}{||}}{C}-\underset{\underset{O}{||}}{C}-O^- + NADH + H^+$$

18.9 oxaloacetate

18.10 a. oxidation   **b.** reduction

18.11 Oxygen (O_2) is the last substance that accepts electrons.

18.12 The protein complexes I, III, and IV pump protons from the matrix to the intermembrane space.

18.13 Three NADH provide nine ATP, one $FADH_2$ provides two ATP, and one direct phosphorylation provides one ATP.

18.14 NAD^+

18.15 In β oxidation, each acetyl CoA gives 12 ATP, while the reduced coenzymes give 5 ATP.

18.16 The process of oxidative deamination regenerates α-ketoglutarate from glutamate.

18.17 Arginine, glutamate, glutamine, histidine, and proline provide carbon atoms for α-ketoglutarate.

Answers to Selected Questions and Problems

18.1 The digestion of polysaccharides takes place in stage 1.

18.3 In metabolism, a catabolic reaction breaks apart large molecules, releasing energy.

18.5 The hydrolysis of the phosphoric anhydride bond (P—O—P) in ATP release energy that is sufficient for energy-requiring processes in the cell.

18.7 Hydrolysis is the main reaction involved in the digestion of carbohydrates.

18.9 a. lactose   **b.** glucose and fructose   **c.** glucose

18.11 The bile salts emulsify fat to give small fat globules for lipase hydrolysis.

18.13 The digestion of proteins begins in the stomach and is completed in the small intestine.

18.15 a. NADH   **b.** FAD

18.17 FAD

18.19 glucose

18.21 ATP is required in phosphorylation reactions.

18.23 ATP is produced in glycolysis by transferring a phosphate from 1,3-bisphosphoglycerate and from phosphoenolpyruvate directly to ADP.

18.25 a. 1 ATP required
b. 1 NADH is produced for each triose.
c. 2 ATP and 2 NADH

18.27 Aerobic (oxygen) conditions are needed.

18.29 The oxidation of pyruvate converts NAD^+ to NADH and produces acetyl CoA and CO_2.

Pyruvate + NAD^+ + CoA $\longrightarrow$
Acetyl CoA + CO_2 + NADH + H^+

18.31 When pyruvate is reduced to lactate, NAD^+ is produced, which can be used for the oxidation of more glyceraldehyde-3-phosphate in glycolysis. This recycles NADH.

18.33 $2CO_2$, 3 NADH, $FADH_2$, GTP (ATP), and HS—CoA.

18.35 Two reactions, 3 and 4, involve oxidative decarboxylation.

18.37 NAD^+ is reduced in reactions 3, 4, and 8 of the citric acid cycle.

18.39 In reaction 5, GDP undergoes a direct substrate phosphorylation.

18.41 a. citrate and isocitrate
b. In decarboxylation, a carbon atom is lost as CO_2.
c. α-ketoglutarate
d. isocitrate $\longrightarrow$ α-ketoglutrate; α-ketoglutarate $\longrightarrow$ succinyl CoA; succinate $\longrightarrow$ fumarate; malate $\longrightarrow$ oxaloacetate
e. steps 3, 8

18.43 oxidized

18.45 a. oxidation   **b.** reduction

18.47 NADH and $FADH_2$

18.49 FAD, coenzyme Q, cytochrome b, cytochrome c

18.51 The mobile carrier Q transfers electrons from complex I to III.

18.53 NADH transfers electrons to FMN in complex I to give NAD^+.

18.55 a. NADH + H^+ + FMN $\longrightarrow$ NAD^+ + $FMNH_2$
b. QH_2 + 2 cyt b (Fe^{3+}) $\longrightarrow$ Q + 2 cyt b (Fe^{2+}) + $2H^+$

18.57 In oxidation phosphorylation, the energy from the oxidation reactions in the electron transport chain is used to drive ATP synthesis.

18.59 As protons return to the lower energy environment in the matrix, they pass through ATP synthase where they release energy to drive the synthesis of ATP.

18.61 The oxidation of the reduced coenzymes NADH and $FADH_2$, provided by glycolysis and the citric acid cycle, in the electron transport chain generates energy to drive the synthesis of ATP.

18.63 a. 3 ATP   **b.** 2 ATP   **c.** 6 ATP   **d.** 12 ATP

18.65 In the cytosol at the outer mitochondrial membrane

18.67 a. and **b.** $CH_3-(CH_2)_6-\underset{\beta}{CH_2}-\underset{\alpha}{CH_2}-\overset{\displaystyle O}{\overset{\|}{C}}-S-CoA$

c. $CH_3-(CH_2)_8-\overset{\displaystyle O}{\overset{\|}{C}}-S-CoA + NAD^+ + FAD + H_2O + HS-CoA \longrightarrow$

$CH_3-(CH_2)_6-\overset{\displaystyle O}{\overset{\|}{C}}-S-CoA + CH_3-\overset{\displaystyle O}{\overset{\|}{C}}-S-CoA + NADH + H^+ + FADH_2$

d. $CH_3-(CH_2)_8-COOH + 4CoA + 4FAD + 4NAD^+ + 4H_2O \longrightarrow$

$5 Acetyl\ CoA + 4FADH_2 + 4NADH + 4H^+$

18.69 The hydrolysis of ATP to AMP provides the same amount of energy as the hydrolysis of 2 ATP to 2 ADP.

18.71 a. 5 acetyl CoA units
b. 4 cycles of β oxidation
c. 60 ATP from 5 acetyl CoA (citric acid cycle) + 12 ATP from 4 NADH + 8 ATP from 4 FADH$_2$ −2 ATP (activation) = 80 −2 = 78 ATP

18.73 Ketone bodies are formed from excess acetyl CoA when the body breaks down large amounts of fat for energy.

18.75 High levels of ketone bodies lead to ketosis, a condition characterized by acidosis (a drop in blood pH values), excessive urination, and strong thirst.

18.77 The reactants are an amino acid and an α-keto acid, and the products are a new amino acid and a new α-keto acid.

18.79 a. $H-\overset{\displaystyle O}{\overset{\|}{C}}-COO^-$ **b.** $CH_3-\overset{\displaystyle O}{\overset{\|}{C}}-COO^-$

18.81 NH_4^+ is toxic if allowed to accumulate in the liver.

18.83 a. pyruvate   **b.** oxaloacetate, fumarate
c. succinyl CoA   **d.** α-ketoglutarate

18.85 nonessential amino acids

18.87 $ATP + H_2O \longrightarrow ADP + P_i + 7.3$ kcal/mole

18.89 Lactose undergoes digestion in the mucosal cells of the small intestine to yield galactose and glucose.

18.91 Glucose is the reactant and pyruvate is the product of glycolysis.

18.93 Pyruvate is converted to lactate when oxygen is not present in the cell (anaerobic) to regenerate NAD^+ for glycolysis.

18.95 The oxidation reactions of the citric acid cycle produce a source of reduced coenzymes for the electron transport chain and ATP synthesis.

18.97 The oxidized coenzymes NAD^+ and FAD needed for the citric acid cycle are regenerated by the electron transport chain.

18.99 Energy released as protons flow through ATP synthase back to the matrix is utilized for the synthesis ATP.

18.101 The oxidation of glucose to pyruvate produces 6 ATP whereas the oxidation of glucose to CO_2 and H_2O produces 36 ATP.

18.103 a. and **b.** $CH_3-(CH_2)_8-\underset{\beta}{CH_2}-\underset{\alpha}{CH_2}-\overset{\displaystyle O}{\overset{\|}{C}}-CoA$

c. Lauryl-CoA + 5 CoA + 5 FAD + 5 NAD$^+$ + 5 H$_2$O $\longrightarrow$ 6 acetyl CoA + 5 FADH$_2$ + 5 NADH + 5H$^+$
d. Six acetyl CoA units are produced
e. Five cycles of β oxidation are needed
f.

activation	$\longrightarrow$	−2 ATP
6 acetyl CoA × 12	$\longrightarrow$	72 ATP
5 FADH$_2$ × 2	$\longrightarrow$	10 ATP
5 NADH × 3	$\longrightarrow$	15 ATP
	Total	95 ATP

18.105 a. carbohydrate   **b.** fat   **c.** carbohydrate
d. fat   **e.** protein   **f.** carbohydrate

Credits

Glossary/Index

Heat The energy associated with the motion of particles in a substance. 41–44, 66

 denaturation of proteins with, 549
 specific. *See* Specific heat
 transfer of, 47

Heat equation, 53
Heat extractor, 58

Heating curve A diagram that shows the temperature changes and changes of state of a substance as it is heated. 62–66

Heating fuels, 347
Heating oil, 349
Heat of condensation, 60–62

Heat of fusion The energy required to melt exactly 1 g of a substance at its melting point. For water, 80. cal are needed to melt 1 g of ice; 80. cal are released when 1 g of water freezes. 57–58, 66

Heat of reaction, 184

Heat of vaporization The energy required to vaporize 1 g of a substance at its boiling point. For water, 540 calories are needed to vaporize exactly 1 g of liquid; 1 g of steam gives off 540 cal when it condenses. 60–62, 66

Heatstroke, 51
Heavy metals
 denaturation of proteins by, 549
 enzyme inhibition by, 559
Helicase, 582
Helium, 76, 79, 81, 94–95, 99–101, 111, 215
 uses for, 113
Helium balloon, 113, 210–11
Heme group, 546–47
Hemlock, 447

Hemodialysis A mechanical cleansing of the blood by an artificial kidney using the principle of dialysis. 258, 261

Hemoglobin, 217, 533–34
 binding of carbon monoxide to, 348
 sickle-cell, 550, 596
 structure of, 547

Hemolysis A swelling and bursting of red blood cells in a hypotonic solution due to an increase in fluid volume. 256–57, 261

Hemophilia, 596

Henry's law The solubility of a gas in a liquid is directly related to the pressure of that gas above the liquid. 237–38, 261

Hepatitis, 556
Hepatitis A, 597
Hepatitis B, 597, 600
Hepatitis C, 573, 597

Heptane, 336, 347
Heroin, 447, 450
Herpes simplex virus, 597, 600

Heterogeneous mixture A mixture of two or more substances that are not mixed uniformly. 72, 74, 104, 252

Hexane, 336, 347
Hexanoic acid, 432
1-Hexanol, 394
Hexene, cis–trans isomers of, 367
Hexokinase, 622
Hexosaminidase A, 596
3-Hexyne, 364
Hibernation, 505, 646
High-density lipoproteins (HDL), 510, 518–19
High-protein diet, 404
Histamine, 354, 427, 444
Histidine, 536, 546
Histologist, 58
HIV, 598–99
Hodgkin's disease, 600

Homogeneous mixture A mixture of two or more substances that are mixed uniformly. 72, 74, 104, 252

Honey, 470
Hormones, 520, 533–34
Horns (animal), 533
Hot-air balloon, 204
Hot pack, 39, 184
Household cleaners, 354
Human immunodeficiency virus (HIV), 598–99
Human T-cell lymphotropic virus, 600
Hunger center, 47
Huntington's disease, 596

Hydration The process of surrounding dissolved ions by water molecules. 227–28, 230, 261, 371, 380
 of alkenes, 371, 379
Hydrobromic acid, 267–68, 271–72

Hydrocarbons Organic compounds consisting of only carbon and hydrogen. 334, 356
 saturated, 331–53
 unsaturated, 362–78, 380
Hydrochloric acid, 244, 249–50, 266–67, 269, 272–73, 280, 286, 288–89
Hydrocyanic acid, 272
Hydrofluoric acid, 270, 272
Hydrogen, 76, 79, 81–82, 88–89, 94–95, 99–101, 130
 in body, 79
 diatomic molecule, 128–29, 137, 140
 nuclear fusion, 324
 transfer in biological oxidation-reduction reactions, 164–65

Hydrogenation The addition of hydrogen (H_2) to the double bond of alkenes or alkynes to yield alkanes. 369–70, 380, 508–9, 526
 of alkenes, 369–70, 379
 of alkynes, 369–70, 379
 complete, 508–9
 partial, 508–10
 of unsaturated fats, 186, 370, 508–9

Metric and SI Units and Some Useful Conversion Factors

Length SI unit meter (m)	Volume SI unit cubic meter (m^3)	Mass SI unit kilogram (kg)

Length   SI unit meter (m)

1 meter (m) = 100 centimeters (cm)

1 meter (m) = 1000 millimeters (mm)

1 cm = 10 mm

1 kilometer (km) = 0.6214 mile (mi)

1 inch (in.) = 2.54 cm (exact)

Volume   SI unit cubic meter (m^3)

1 liter (L) = 1000 milliliters (mL)

1 mL = 1 cm^3

1 L = 1.06 quart (qt)

1 qt = 946 mL

Mass   SI unit kilogram (kg)

1 kilogram (kg) = 1000 grams (g)

1 g = 1000 milligrams (mg)

1 kg = 2.20 lb

1 lb = 454 g

1 mole = 6.02×10^{23} particles

Water

density = 1.00 g/mL

Temperature   SI unit kelvin (K)

$°F = 1.8(°C) + 32$

$°C = \dfrac{(°F - 32)}{1.8}$

$K = °C + 273$

Pressure   SI unit pascal (Pa)

1 atm = 760 mm Hg

1 atm = 760 torr

1 mole (STP) = 22.4 L

Energy   SI unit joule (J)

1 calorie (cal) = 4.184 J

1 kcal = 1000 cal

Water

Heat of fusion = 80. cal/g

Heat of vaporization = 540 cal/g

SH = 4.184 J/g°C

Prefixes for Metric (SI) Units

Prefix	Symbol	Power of Ten
Values greater than 1		
giga	G	10^9
mega	M	10^6
kilo	k	10^3
Values less than 1		
deci	d	10^{-1}
centi	c	10^{-2}
milli	m	10^{-3}
micro	μ	10^{-6}
nano	n	10^{-9}